FRESH KILLS

MARTIN V. MELOSI

FRESH KILLS

A HISTORY OF CONSUMING AND DISCARDING IN NEW YORK CITY

Columbia University Press

New York

Columbia University Press
Publishers Since 1893
New York   Chichester, West Sussex
cup.columbia.edu

Copyright © 2020 Columbia University Press

All rights reserved

Library of Congress Cataloging-in-Publication Data
Names: Melosi, Martin V., 1947- author.
Title: Fresh Kills : a history of consuming and discarding in New York City /
Martin V. Melosi.
Description: New York : Columbia University Press, 2020. | Includes bibliographical
references and index.
Identifiers: LCCN 2019023877 (print) | LCCN 2019023878 (ebook) | ISBN
9780231189484 (cloth) | ISBN 9780231189491 (paperback) | ISBN 9780231548359
(ebook)
Subjects: LCSH: Sanitary landfill closures—New York (State)—New York. | Fresh Kills
Landfill (New York, N.Y.) | Freshkills Park (New York, N.Y.)
Classification: LCC TD788.4.N72 M45 2020 (print) | LCC TD788.4.N72 (ebook) |
DDC 628.4/45640974726—dc23
LC record available at https://lccn.loc.gov/2019023877
LC ebook record available at https://lccn.loc.gov/2019023878

Cover image: Digital composite. © Diane Cook and Fotosearch/Getty Images
and Len Jenshel/Getty Images

Cover design: Lisa Hamm

For Gianna and Angelina

CONTENTS

Preface ix
Acknowledgments xiii

Introduction:
The Dilemma of Consuming 1

PART I: THE BACKDROP

1. Island City 15
2. Wasting Away 38

PART II: STATEN ISLAND: BOROUGH OF LAST RESORT

3. The Quarantine 69
4. The Garbage War 87

PART III: SEEKING A DISPOSAL SINK

5. The Go-Away Society 123
6. One Best Way 145
7. Futile Protests 169

PART IV: LIVING WITH AND SURVIVING THE LANDFILL

8. The Burning Question   199

9. The End of Isolation   232

10. An Environmental Turn   252

11. Fiscal Crisis and Disposal Dilemma   275

12. Fresh Kills at Midlife   303

13. Barge to Nowhere   328

14. A New Plan   350

PART V: THE ROAD TO CLOSURE

15. Secession   373

16. Closure   395

17. Now What?   422

PART VI: THE POST-CLOSURE ERA

18. 9/11   463

19. Regeneration   491

20. Crossroads   520

Conclusion   547

Notes   557
Index   759

PREFACE

The idea for this book began a more than ten years ago as a narrative history of the largest landfill (or one of the largest) that the world had ever seen—Fresh Kills on Staten Island, New York. A 1991 op-ed piece in the local newspaper noted that Staten Islanders "live side by side with a landfill that's larger than some cities."[1] My awareness of the site, however, goes back to the late 1970s. At that time I was doing research in New York City for a book project that ultimately became *Garbage in the Cities: Refuse, Reform, and the Environment, 1880–1920* (College Station: Texas A&M University Press, 1981). On the research trip I became acquainted with the work of the Department of Sanitation, which was responsible for the collection and disposal of refuse for the city. I mined what documents were available, and I met a few people associated with the department, including a former commissioner. I did not venture to Staten Island, nor did I visit the wastescape at the time.

In *Garbage in the Cities* I briefly discussed Fresh Kills, mostly as a "gee whiz" example of a massive dumping ground for Gotham's never-ending stream of solid waste. By the 1980s, landfills were becoming too expensive to build or facing increasing criticism as environmental failures. Yet Fresh Kills was going strong and remained a marvel of large-scale engineering for many more years, much to the frustration of most Staten Islanders. As the book made the academic and public rounds, I got to know many engineers, city officials, lawyers, and business leaders associated with the waste management field. Colleagues dubbed me the "garbage historian"—a moniker I was not very happy with. I soon came to respect and appreciate the variety of people in the solid waste field for the work they did, and I reveled in the small but growing field of urban environmental history (which included my good friend and mentor Joel Tarr) interested in city services and urban technology. I also had the opportunity to meet Mierle Laderman Ukeles, an imaginative artist, who celebrated the efforts of New York sanitation workers and provided trenchant commentary on the implications of consumption and waste in our society. On a personal note, I learned that my dad's first cousin had been a sanitation worker for Sunset Scavenger Company in San Francisco, and in the 1990s my brother began working for Browning-Ferris, Inc., in Montana and Wisconsin, and later for Allied Waste in Utah, as a safety inspector.

Although my scholarship on the urban environment and solid waste was equally fascinating and quirky to me, I soon came to realize the daunting challenge of confronting the waste issue as a vital health and environmental imperative. On a trip to Newark in 1983 to give a talk at the New Jersey Institute of Technology, a colleague took me to the Meadowlands, where I got my first glimpse of a vast landfill site—not as big as Fresh Kills but nevertheless gigantic. In 1988, I chaired a small session at Hunter College in New York City, which pitted an incinerator executive against the famous environmentalist Barry Commoner. The session centered on reintroducing incineration in the Big Apple, and I marveled (and was stunned) at the verbal fireworks that ensued, especially from the unwavering Commoner. I was able to observe firsthand the intensity and divisiveness that differences over waste disposal could provoke.

For the next ten years, I devoted most of my time to topics other than waste management, but I returned to it as part of my research for *The Sanitary City: Urban Infrastructure in America from Colonial Times to the Present* (Baltimore, MD: Johns Hopkins University Press, 2001). The book dealt with solid waste, water, and wastewater services, and since I knew less about the latter two, most of my original research focused there.

What brought me back to Fresh Kills was a phone call from the archeologist/garbologist Bill Rathje in late 2001 or early 2002. Bill was engaged in work related to the September 11 disaster, especially the decision to transfer debris and human remains from the wreckage of the Twin Towers to Staten Island. He wanted to pick my brain about approaching the research, especially since I had recently conducted a major project for the National Park Service on the historical significance of the Fresno, California, Landfill (an amazing story for another day).[2] After our conversation, I filed away what Bill had told me about Fresh Kills and its ties to September 11, briefly reflecting on this curious turn in the landfill's history.

Probably two or three years later, around the time that I was trying to decide on a future book project, I recalled my conversation with Bill Rathje. The idea of developing a book on Fresh Kills grew beyond a story of the building and closing of the site and its "gee whiz" image. I envisioned work along two parallel lines: First, my long-standing fascination with the massive landfill, its role in New York's refuse-management history (and solid waste history in general), and its place in the changing physical landscape of Staten Island. Second, and more recent to my thinking, was the desire to explore Fresh Kill broadly in terms of the causes and effects not only as a site but as a symbol. This meant, among other things, trying to understand how rampant consumption of goods and services, especially since World War II, created entirely novel waste disposal problems. Taking some hints from the new field of critical discard studies,[3] I also wanted to examine a little more deeply the social construction of waste and wasting in the United States, especially the value (or lack of value) that people place on material things. Such an exploration might help explain what kinds of disposal were acceptable and how people responded to such a massive facility as Fresh Kills. Of course, such exposure also would be shaped by perceptions based on the site's nuisance characteristics and its potential health and environmental risk factors.

A gem of an essay, "The Town Dump," written by the novelist and environmentalist Wallace Stegner, captures some of the imagery of landfills as social structures. Albeit somewhat romantically, Stegner provides a vivid portrayal of the importance of the town dump in Whitemud, Saskatchewan: "The place fascinated us, as it should have. For this was the kitchen midden of all the civilization we knew; it gave us the most tantalizing glimpses into our lives as well as into those of the neighbors. It gave us an aesthetic distance from which to know ourselves."[4] My friend and colleague from Beijing, the philosophy professor Song Tian, told me that a landfill "is a space (land) that lost its dignity." This is quite a contrast to Stegner, but such polarities make the study of Fresh Kills quite enticing.

I also have gained unusually valuable insights from historical geography and landscape studies. There is no denying that "place" is very important. In the case of Fresh Kills Landfill, it is not simply that the facility was built but where it was built. Gulping up the Staten Island salt marsh to construct the landfill was itself an important aspect of rampant consumption and a change agent for the borough and greater New York City. Timing of construction is equally important, as every historian knows. I initially envisioned a timeline for the book extending from the opening of the landfill in 1948 through its closure in 2001, including its place in the story of September 11. I soon began to see that I needed to build a backstory about significant steps leading to Fresh Kills Landfill before 1948, emphasizing how disposal was treated before then and describing what Staten Island and its marshes were like in the late nineteenth century, before the borough was part of New York City's consolidation. I also needed to extend my narrative past September 11 to include the drive to convert the landfill space into parkland, which gained momentum by late 2001. What I sought, therefore, was a way to incorporate the transience of time and space into my story, which ventured from landscape to wastescape to ecoscape (or parkscape).

A narrative approach seemed the best way to track this long evolutionary path and to identify key markers along the way. The expanded timeline, the role of Fresh Kills as site and symbol, and New York's disposal dilemma are in many ways uniquely local but also implicitly and broadly global.

ACKNOWLEDGMENTS

Typing the acknowledgments (although I only can type with two fingers) is one of my favorite parts of writing a book. It signals the project coming to an end and also allows me to thank all of the people who made the project possible. My wife, Carolyn, comes first. She tolerated my frustrations and enthusiasms daily as I delved into my research and slogged away at writing. That is true love!

I cannot thank Phil Papas (and his wife, Lori Weintrob) enough for their hospitality in guiding me through the experience of Staten Island and for Phil's (and in some cases Lori's) tireless efforts to read every page of this study with an eye to correcting my most egregious errors and typos in telling the story of the island, its people, and its notorious dump. Jonathan Soffer also read the whole manuscript with an excellent critical eye and alerted me to shortcomings and mischaracterizations about New York City politics, among other things. My dear friend Joel Tarr read several chapters, always exhorting me to edit carefully and be economical with my verbiage. Although I usually write way more than he ever thinks is necessary, I'm pleased he is at my shoulder trying to elevate the quality of what I try to accomplish.

Ben Miller was a constant source of information and perspective on various topics, providing insights that only a participant in the story and a well-schooled author has at first hand. Cheryl Bontales gave me some wonderful material about her dad, Bill Criaris, who had been general superintendent at Fresh Kills. Other colleagues offered a range of services, including Ted Steinberg, who graciously allowed me to read a final draft of *Gotham Unbound*, and Steve Corey, who has unlimited knowledge of the New York refuse experience and provided me with pictures and documents. I also want to thank Richard Flanagan, Michael Gerrard, Joe Pratt, Julie Cohn, Daniel Gallacher, Adam Zalma, Melissa Zavala, Michael Miscione, Mark Simpson-Vos, Brandon Proia, Chaz Miller, Michael Rawson, Themis Chronopoulos, Carl Zimring, Jordan Howell, Robin Nagle, and Samantha McBride for their many kindnesses and critiques.

Several interviewees were generous with their time and delivered some very useful leads. Included here are Mierle Laderman Ukeles, Norman Steisel, Brendan Sexton, Paul Casowitz, Philip Gleason, Robin Geller, Kurt Reike, George Pataki, Carrie Grassi, Eloise Hirsh, and Mariel Vallere. Ellen Schuble completed outstanding transcripts of many of the interviews. Several additional interviews were available through the Columbia Center for Oral History, the College of Staten Island Oral History Collection, and the Oral History Project of the DSNY/Freshkills Park. I only wish I had time to gather even more interviews.

I was blessed with several first-rate graduate assistants who helped build my bibliography and carry out tireless—and often mind-numbing—research. I could not have completed this book without them. Lisa Ng and Faith D'Alessandro in the New York City area were indispensable, talented, and trustworthy. I could not be in more than one place at a time, so their efforts really maximized my research capabilities. Kevin Tang also contributed to my research in New York in his brief time with me, as did my amazing daughter Gina Melosi. In Houston, I received strong support—especially in constructing bibliographies—from Mao Da, Deanne Ashton, Alex LaRotta, and Lindsay Scovil. Mario Lopez was responsible for most of the outstanding maps in the book.

No scholar can function without the help of archivists and librarians. I was lucky to get excellent support from the staffs of numerous depositories. I simply would not have been able to gather many of the necessary works I needed without the assistance of the outstanding Interlibrary Loan Department at the University of Houston. On Staten Island I depended heavily on Faith D'Alessandro at several locations and also on Cara DeLatte, Pat Salmon, and Gabriella Leone at the Staten Island Museum; Jeff Coogan and others at the Archives and Special Collections of the College of Staten Island, CUNY; and staff at the Staten Island Historical Society and the New York Public Library's St. George Library Center. (Phil Papas and Lori Weintrob also acquired material for me.) In Manhattan I used several collections, often with the help of Lisa Ng, foremost, materials at the Municipal Archives and New York City Hall Library of the New York City Department of Records and Information Services, with thanks to the late Leonara A. Gidlund and to Christine Bruzzese. Douglas Dicarlo was very helpful with material from the LaGuardia and Wagner Archives at LaGuardia Community College. I also spent valuable time at Archives and Manuscripts of the New York Public Library and at the Columbia

Center for Oral History at the Columbia University Libraries. Special thanks to Reference Services at the New York State Archives in Albany. I also visited the Library of Congress in Washington, DC, early in the project.

My professorship, the Hugh Roy and Lillie Cranz Cullen Professorship in History, and the College of Liberal Arts and Social Sciences at UH provided necessary funding for my research on this book. And I certainly would not have been able to deal with myriad logistics for the project without the steady hand of Wes Jackson at the Center for Public History.

My editor at Columbia University Press, Bridget Flannery-McCoy, was the best editor I have worked with during my career (and that's saying something, given the excellent editors I have encountered). Her detailed comments and superb critical sense made the manuscript far superior to my initial drafts. Since she left for a new position at Princeton University Press, I have been in the capable hands of Stephen Wesley and others on the Columbia University Press staff. The three outside reviewers that the press engaged to read the original manuscript also provided encouragement and very useful criticism and insight.

I dedicate this book to my wonderful granddaughters, Gianna and Angelina, respectively eight years old and four years old at the time of its completion. They preserved my sanity during much of the research and all of the writing process. Their unconditional love sustains me every day.

FRESH KILLS

INTRODUCTION

THE DILEMMA OF CONSUMING

F*resh Kills: A History of Consuming and Discarding in New York City* centers on the mundane yet profound problem of solid waste disposal in the modern metropolis.[1] Waste from mass consumption was a serious dilemma for cities in the twentieth century—and into the twenty-first. This is true whether we are talking about fossil-fuel emissions and climate change or accumulating tons of refuse from discarded goods and not knowing what to do with them. Societies in both the developed and developing worlds have faced the challenge of pursuing economic growth and, at the same time, coping with the unwanted residue of material accumulation. The rise of Greater New York City and its decision to build Fresh Kills Landfill offer a supreme example of the dilemma of consuming in recent years. Fresh Kills was a consequence of mass consumption and the embodiment of massive waste.

New York City's story about disposal is told through the history of a monumental structure, the 2,200-acre Fresh Kills Landfill on Staten Island, which is one of the largest human-engineered formations in the world. It is located along the Fresh Kills Estuary in the northwestern part of Staten Island on what had been extensive salt marsh. The landfill was opened in 1948, but its story begins several decades before the consolidation of New York City (1898). Tracking the transformation of the site and the disposal practices in Gotham in the mid-to-late nineteenth century explains how Staten Island's land uses changed well before the landfill was laid out and what other disposal technologies fell out of favor.

The teeming boroughs making up the city first emptied their vast amounts of accumulated wastes into nearby watercourses, including the Atlantic Ocean, and also onto land throughout the city. The municipal government also began experimenting with incinerators in 1885 and reduction plants in 1895. Eventually the city delivered its refuse to Staten Island (only a short scow trip from Manhattan, Brooklyn, Queens, and the Bronx), which was neither as populous nor as economically productive as most of its neighbors. Refuse traveled from where many people lived and worked to where they did not (at least initially), making the landfill at Fresh Kills a justifiable destination for things no longer of value to the people living in the swelling metropolis. The landfill served as a primary disposal facility for Greater New York's solid waste from 1948 to 2001. For Staten Islanders, the endless citywide creation and delivery of thousands of tons of garbage and trash each day was a curse, leaving its citizens to wonder why "the Dump" abutting their homes and neighborhoods had been designated a sacrifice zone.

Fresh Kills reopened briefly in late 2001 through June 2002 to provide a receiving point for human remains and building rubble from the destroyed Twin Towers after the September 11 attack (and it opened up again after Hurricane Sandy in 2012). What had been a notorious disposal site of unwanted trash now also became a cemetery for the remains of many people who had lost their lives in the nation's worst terrorist tragedy. Today the landfill is the heart of a mammoth reclamation project to construct an expansive parkland known as Freshkills Park, three times the size of Central Park.

Using the history of Fresh Kills Landfill as a focal point, *Fresh Kills* tells the intersecting histories of New York City, Staten Island, and the challenges of waste management. The long timeline and attention to the landscape is crucial. Observers most often visualize Fresh Kills Landfill from the back end of its history, that is, from the gigantic mounds of smelly waste spreading all along Staten Island's North Shore, or, more positively, as the raw material for a grand park. Such a vantage point gives a sense of inevitability to the state of the place. The pre- and posthistory of the landfill, however, suggests otherwise. The land upon which it was built was a salt marsh used for many generations by Lenape American Indians and European immigrants for growing salt hay and for hunting and fishing. About one hundred years before the landfill began, the city placed a quarantine station and a supporting medical complex on Staten Island to distance infected sailors and others from New York's main population. During World War I, the city sited an unwanted waste disposal plant on the very spot where the landfill would be built several decades later.

This backstory particularly highlights Staten Island's long period of alienation and the sense of its citizenry that their borough was regularly being exploited by the core city and that their homes were valued differently than those on Manhattan and elsewhere in Greater New York. Over the years, much of New York City's marshland had been transformed into "useful" or taxable land or into various dumps and landfills. On Staten Island, marshland destruction led to the creation of an enormous waste site. For those who despised it, Fresh Kills Landfill would always be "the Dump." For those who viewed it as a necessary evil, it was a sanitary landfill. There was nothing ordained about Fresh Kills as a disposal facility, other than the city's will to build it. Why it became so and how it evolved into something else is at the heart of this book.

New York City's hunt for an ultimate disposal option is the consequence of more than rational decision making. Timing, circumstance, options, impulse, and accident are the stuff of history. Sanitation has always been and will always be political, and politics as much as anything else led to Fresh Kills. The landfill also is a supreme example of the serious consequences of massive environmental transformation; it is an altered landscape and socialscape that affects not only Staten Island but the whole city—and beyond.

Fresh Kills is also as much symbol as site. The space that the landfill occupies was transformed from an apparent marginal landscape of salt marsh into a wastescape and then into an ecoscape or reimagined park space. The landfill is a tangible reminder of how human habits and societal behaviors are caught between material wants and valueless remnants. Stepping back from its remarkable history, the artist Mierle Laderman Ukeles looked upon Fresh Kills as "a social sculpture," a reflection of our material culture, our consumer practices, and our sense of value and worthlessness.

Why should we care what a landfill—even a gigantic one like Fresh Kills—tells us about our history? Most obviously, consumption and waste are inextricably connected, but that relationship is much more complex and broadly significant than simple cause and effect. Consumption of goods rarely if ever results in nothing left over. Material goods generate residue, which must be reused in some form or disposed of in some way. Even in an era of greater environmental consciousness, the lion's share of waste is discarded. In modern cities in particular, the individuals who discard feel no further responsibility for their waste, are usually not the ones who decide how and where to dispose of it, and ultimately relinquish all ownership of it. Historically, the dilemma of consumption (or consuming) needs to be understood as a triangular relationship among the consumption of goods, the creation of waste, and disposal (often meaning place). Recently, efforts to alter this relationship call for reducing consumption, putting responsibility on manufacturers to create products that do not have one-time use only, encouraging reuse and recycling, and more carefully monitoring disposal. So far, the aspirations for Zero Waste remain just that—aspirations.

Putting Fresh Kills and New York's disposal history at the heart of this book offers an important case study beyond the "gee whiz" value of the site. Production of massive quantities of waste, certainly from the nineteenth century onward, was linked to finding a workable disposal option. The meaning of "workable" constantly changed. Sometimes it simply implied "out of sight, out of mind." Often it entailed inconveniencing or harming those unable to muster sufficient resistance against it. More recently it required (or was impelled) to consider potential environmental risks. For much of New York City's history, major portions of the waste stream were merely stuff no longer wanted. "Clean fill" and construction rubble were used to create new land, some organic material might be reused as animal feed, and recycling efforts came and went—and came back. The immutable connections between consumption and waste took time to recognize, let alone address, especially the squandering of natural resources and the recognition that production processes affect what and how materials can either be reused or discarded.

Fresh Kills is a dramatic example of consumption gone wild. But the landfill also sits at the third point in the triangular relationship among consumption, waste, and disposal (place). As a location for disposal, it created a political storm between the city proper and Staten Island; it raised the specter of social injustice for its border neighborhoods (despite the fact that they primarily were white); and it exposed environmental risks from methane gas, leachate, and other substances.

If this book only told the story of Fresh Kills as a disposal site and as part of the chain of consumption, waste, and disposal, it would be useful but incomplete. The events that led to the creation of the landfill certainly inform our understanding about the massive problem of solid waste, the decision-making processes influencing disposal practices, and the environmental and social implications of creating a refuse sacrifice zone. But history is a temporal discipline, and it requires attention to temporal matters. The construction and use of Fresh Kills (1948–2001) is only a part of the story of those 2,200 acres of Staten Island. To appreciate the landfill's place in the history of Staten Island and New York is, of course, to understand it in context. But each phase in the history of the site—from salt marsh to park—evokes powerful memories of what the site had been before the landfill, memories of the landfill itself (a nagging reality of vast discards left behind), and aspirations for a changing role for the space once it became a park. Fresh Kill's history is cumulative, not episodic.

Fresh Kills as a transient space and bundle of memories adds layers of significance to the relationship among consumption, waste, and disposal (place). In any city or country, the generation of waste is a vast physical, social, and political occurrence with long-term implications. The book uses Fresh Kills to impress upon us the broad consequences of wasting and wastefulness but also to make clear that the landfill was not a static land use. Fresh Kills contains multitudes: narratives about seeking a waste sink, political rivalries, NIMBY-ism and social justice, pollution and toxic threats, land use and planning, terrorism and remembering, and restoration ecology and park building. The connection among consumption, waste, and place is essential to explain how materiality and its effects can help broaden our understanding of urban culture. A "waste history" of this kind exposes a different way to explore a city's past, in this case New York City.

Since the relationship among consumption, waste, and disposal (place) is the centerpiece of *Fresh Kills*, it might be useful to explore these concepts a little further—for the sake of clarity, not abstraction.

CONSUMPTION

As a 1999 study on consuming cultures rightly stated, "We consume to live; yet we also consume to do much more than just live." It also argued that "consumption is one of the basic ways in which society is structured and organized, usually unequally, sometimes incredibly so."[2] Any generalization about consumption needs to take into account important disparities separating the rich and the poor throughout the history of the United States and elsewhere. The Industrial Revolution did a great deal to restructure our consuming culture and to promote urban growth. Until about the 1870s most Americans were not yet wide-ranging consumers; instead, they spent most of what they earned on food, clothing, and housing. By the late nineteenth century, however, a new consumer society was emerging in which the average person could aspire to—if not always achieve—upward mobility. Among the growing middle class, especially, the new consumerism was most obvious.[3]

Rising wages or farm income opened up the possibility for a larger number of Americans to acquire more material goods. By 1900, 65 percent the country's population lived and worked in and around the industrial centers in the Northeast and Upper Midwest—the so-called Rust Belt—where factory and office jobs were most plentiful. The oft-repeated observation that a new middle class was rising at the time was borne out by the changing character of work. In the 1870s, 52 percent of Americans were employed in agriculture, 21 percent in manufacturing and mining, and 27 percent in other fields. By the 1890s, 41 percent worked in agriculture, 22 percent in production, and 37 percent in services. This pattern continued until the eve of World War II, when only 21 percent were farmers and service employees reached 54 percent.[4]

Industrial capitalism ramped up the mass production of goods and expedited mass consumption. Mass production through factory assembly lines increased worker productivity, made consumer goods affordable to many workers, and led to a consumer revolution. Into the twentieth century, increased consumerism was spurred by the new business of mass advertising. Economic growth in the late nineteenth and early twentieth centuries, however, was not a steady climb, and boom-and-bust cycles took their toll.[5] Marxists and other critics of capitalism argued that consumption was "the stalking horse of capitalism," which focused on profits rather than the social good and led to a variety of societal and environmental ills beyond the undulations of the marketplace.[6]

While such an assessment might give too big a role to consumerism in the overall impact of the capitalist system, consumption practices and patterns were related directly to a growing waste problem in the United States beginning in the mid-to-late nineteenth century. Like Europe, the United States in its industrial era was confronted with huge amounts of refuse, with two distinct dimensions. One was tethered to the physical distress caused by overcrowding, poor sanitation, and primitive methods of collection and disposal; the other was linked to the rising affluence of the middle class, an abundance of resources, and consumerism, which continued into America's postindustrial era.[7]

In a period of mass production, structural changes in the consumer market encouraged acquisition of all kinds of goods. New factories produced massive amounts of primary materials, such as pig iron, steel, railroad freight cars, and cottonseed oil. But consumer goods (such as toothpaste, chewing gum, safety razors, and breakfast cereal) also became readily available to many people who until recently had had no access to them, and the practice of wearing homemade clothes was being challenged by ready-mades. The cheap cotton-clothing industry, for example, was a leader in promoting industrial capitalism, beginning in the early nineteenth century. New York City was the center of this industry. In 1860, it had 1,286 companies making or retailing clothing. Other cities, such as New Orleans (982 firms), Philadelphia (560), Cincinnati (345), and San Francisco (271), were clothing centers as well. Mass food production, such as meat from Chicago and flour from Minneapolis, changed eating habits. For those living at a distance from the new department stores and retail shops, the Montgomery Ward and Sears catalogues offered mail-order access to the new world of manufactured goods.[8] The historian Susan Strasser notes, "Americans everywhere and of all classes began to eat, drink, clean with, wear, and sit on products made in factories."[9]

Over time, the goods produced in American factories and imported from abroad would become more diverse and technically more intricate (such as automobiles, televisions, computers, and innumerable plastic products) and the discards more plentiful and complex. Consumer demand was propped up by frequent model and style changes, not only in cars and clothes but eventually in cell phones and other electronic products. "Planned obsolescence" of washers and driers, refrigerators, and dishwashers kept consumers coming back for more and newer products. As the United States moved from an industrial to a service economy in the mid-twentieth century, changes in consumer trends were not easy to keep up with. And at times, critics and reformers pushed back against buying sprees to question materialism, to assert the dangers of a "throwaway society," or to assert "Small Is Beautiful."[10]

The home was not the only "consumption space." New goods and services were becoming available at an array of major sites, such as restaurants, parks, festivals, sports arenas, art galleries, cultural centers, and, later, cinema complexes, shopping malls, and theme parks. Aside from offering social interaction, arts and culture, and just plain fun, such consumption spaces became important mechanisms for accessing more and different commodities.[11] While the discards of the nineteenth century in the various consumption spaces were largely food wastes, wood and coal ash, rubbish, and horse manure, the more recent waste stream included a mix of hard-to-replace and recyclable materials—as well as a variety of toxic substances. Of all the current discards, paper, plastics, and aluminum have increased most rapidly. In addition, the collection of these wastes has been made more difficult not only by the volume and composition of discarded materials but also because of the large populations to be served and the greater territory that sanitation workers were required to cover.

By World War I if not before, many Americans had come to view some goods as necessities, not luxuries. Aspiring for greater upward economic mobility and a desire to enhance one's standard of living brought with it a demand for conveniences like indoor plumbing. At least by the end of World War II, an automobile and a modest house were the status symbols of the rising middle class. In 1920, only 26 percent of households owned a vehicle; by 2000, that number was 89 percent. Having electricity, running water, and multiple appliances was expected.[12] American affluence was widely envied throughout the world, even though not everyone shared in the bounty, and poverty was far from ended in the United States. With affluence came the distinction of discarding more. In the early twenty-first century, the United States consumes about one-third of the world's resources, producing 50 percent more solid waste per person than any other Western economy. Americans discard more than 1,600 pounds of refuse per person per year, or 4.5 pounds each day.[13]

WASTE

Turning the discussion on its head: examining the waste stream is a good way to understand systems of production and consumption.[14] After all, wasting is an inevitable part of life and an integral part of consuming. The urban planner Kevin Lynch stated, "Wasting pervades the living system. Organisms appropriate substance and energy, use what they need, and then expire or dispose of what they cannot use."[15] For humans, the process is a little more problematic, as the journalist Heather Rogers observed: "Consumption lies at the heart of American life and economic health, and intrinsic to consumption is garbage. Such high levels of waste are the product not of any natural law or strange primordial impulse but of history, of social forces."[16] Waste, also, is subject to moral judgment. The sociologist Gay Hawkins argued that "Waste makes us feel bad, its presence disgusts and horrifies us, it wrecks everything." There is a tendency, therefore, to place blame on the disposed object and not on the one doing the disposing.[17]

Scholars and philosophers of all stripes have expended much energy on trying to explain exactly what waste is and if it has value—materially or symbolically. Garbage, trash, rubbish, offal, refuse, junk, debris, clutter, litter, rejectamenta: solid waste goes by many names. In the natural world, waste is simply part of the life cycle; it is a change of substance returned to the physical environment in a different form. In human civilization, waste has an entirely different role and connotation. While humans are not alone in waste making, they are the only species that passes judgment on it. Probably the best known of these thinkers is the British social scientist Michael Thompson, through his book *Rubbish Theory*, first published in 1979 (revised in 2017). Thompson believes that "we need a theory of people *and* stuff—particularly now that we are faced with seemingly intractable discard-generated problems such as climate change—and that . . . is precisely what rubbish theory gives us."[18]

Thompson's basic premise is that there are two cultural categories that are "socially imposed" on objects—the durable and the transient. The former (for example, antique urns) have "increasing value and infinite expected lifespans"; the latter (for example, milk cartons) have "decreasing value and finite expected lifespans." The transient, however, has the possibility of rebounding into the durable category (an old postcard perhaps?). Included in "rubbish" is the item that falls neither into the durable or transient category but, rather than reaching zero value and zero expected life, usually "continues to exist in a timeless and valueless limbo where at some later date (if it has not by that time turned, or been made, into dust) it has the chance of being discovered . . . and successfully transferred to durability." Thompson concludes that "in order to study the social control of value, we have to study rubbish."[19] For the purposes of *Fresh Kills*, Thompson's conclusion that "the boundary between rubbish and non-rubbish is not fixed but moves in response to social pressures" highlights the difficulty of what disposal actually accomplishes.[20]

In its typical context, "to consume" gives the impression of finality, that is, completing the process of acquiring some material thing and using it up. In many cases, the lifespan of waste is longer than its useful phase. The sociologist John Scanlan is right when he states that "what we consume never really disappears."[21] Some societies, past and present, used many goods until they were worn out, but that is much less the case in industrial and postindustrial America, where worn-out items are discarded along with no-longer-desirable goods. Even food items rarely get wholly consumed.[22] Society is faced with several, albeit not wholly satisfactory options in dealing with waste: ignore it, store it, sell it, give it away, break it down, combine it with something else, or reuse as much of it as possible. In Thompson's view, and that of others, society assigns a value to waste (which sometimes changes), and that value dictates its disposal or reuse.[23]

For New York City, as with most other cities, the chief aim of refuse disposal for much of its history was unburdening the streets, alleyways, and vacant lands of piles of rubbish for the sake of mobility, safety, and health. The advent of the modern environmental movement shifted attention to resource recovery and recycling but did not appreciably change disposal practices for what might not be reused. Today New York City ships most of its refuse out of state to other people's landfills and to other people's incinerators, although efforts to recycle and reuse materials have increased. The debate over what we throw away and what we waste, nevertheless, goes on.

DISPOSAL/PLACE

The third part of the consumption-waste-disposal (place) system as developed in this book speaks explicitly to the central role of landscape in understanding Fresh Kills. First, it suggests the importance of Fresh Kills Landfill as a disposal choice for the City of New York, and, second, it discusses how the 2,200-acre space shaped and shapes the history of Staten Island in particular (the evolution of landscape into wastescape into ecoscape). In broad terms, as the sociologist Kevin Hetherington points out, "Disposal . . . is not primarily about waste but about placing. It is as much a spatial as a temporal category."[24]

Sites like Fresh Kills Landfill are spaces within spaces, trash footprints, and areas of waste displacement. Even as a landfill it faced constant change given the ceaseless need to deal not only with great volumes of refuse but also with a waste stream that frequently altered in composition and environmental impact. Landfills operate as long-term storage facilities but are also "consumption junctions" where the remnants of consuming merely reach a way station, not a final resting place. They are part of a provisional process because they frequently do not remain a landfill forever.[25] The landfill is characterized as an infrastructure having a mediating role between nature and city in dealing with waste, but its permanence or impermanence as a land use is decided by a variety of objectives, desires, and aspirations for the space.[26] Landfills like Fresh Kills also are archives of material and memories—collectively the accumulation of real and imagined pieces of, in this case, every borough.[27]

Some have treated landfills as "antilandscapes," that is, material places that cannot sustain life, or places that have become "culturally sterile or metaphorically invisible." But this may be true only if landfills are examined in a limited time span. Their place in the consumer-waste-disposal cycle is rarely permanent or unchanged, but their long-term significance can be linked to an altogether different identity.[28] As the geographer Anna Storm states, "There are wounded landscapes that are difficult to heal. However, the physical scars are also part of the stories that are the fabric of heritage, and these landscapes of waste may reemerge as new landscapes of memory and community identity that are directed toward the future."[29] The site's past also is a defining feature. In this case, Staten Island's vast marshland existing before the landfill is central to understanding the borough's environmental and human history and how that was changed by Fresh Kills Landfill.

Landfills as disposal sites make it clear that place matters. Heather Rogers hits on a critical point: "There's a reason landfills are tucked away on the edge of town, in otherwise un-traveled terrain, camouflaged by hydroseeded, neatly tiered slopes. If people saw what happened to their waste, lived with the stench, witnessed the scale of destruction, they might start asking difficult questions."[30] On Staten Island people asked difficult questions. Ultimately, the eyesore could no longer be overlooked by others in New York City. Its scale was impossible to ignore, which was much more dramatic than the dozens of other dumps and landfills that had populated Gotham over the years.

But Rogers's interpretation of a landfill's location is only part of a bigger issue about landfills in the landscape. As a highly engineered space, it bears the mark of a building and design tradition that more than hints at the lengths that city leaders go to to dispose of the city's waste.[31] Landfills, incinerators, and reduction plants are elaborate and intricate capital-intensive technologies created to hide, destroy, or remake wastes, with little concern about how that waste came to be. They are marvelous inventions, but they were not developed to stem the tide of consumption; they are only reminders that the residuals from consumption need to be tolerated. In many ways, recovering value from landfills in particular required turning the wastescape into something else. As the landscape architect Mira Engler asserts, in turning it into a park and golf course, "the dump was redeemed, the repugnance covered, and the bad memory erased.... At times, taking a utilitarian approach, the land is assigned some recreation use to recover its value and add a sense of communal worth."[32]

In the case of Fresh Kills, at least, the recovery value of constructing a park neither eliminated the memory of the landfill nor restored the marshland's ecological integrity. The specters of 9/11 continued to haunt it as well. Fresh Kills as a landfill was one particular iteration for that particular piece of Staten Island real estate, but it hardly seemed so at the time. The dump would always be the Dump, but the site's meaning and value would not so much change as be modified by the powerful forces of history represented by the Twin Tower disaster and the building of Freshkills Park. The landfill was never really an isolated space, hiding the discards of a great city. Its recognition grew out of its relationship not simply to Staten Island but to Greater New York and the surrounding area—and also to the world of waste and wastefulness. Staten Island paid the price for its location—as did, for example, neighborhoods near transfer stations—but eventually the forgotten borough could reap the benefit of the promised Freshkills Park.

The story of Fresh Kills is about the dilemma of consuming, to be sure, but is its future linked more to the resurrection and preservation of a natural setting or to a human artifact encasing memories of generations of New Yorkers?

* * *

The backstory of Staten Island, the building and closing of Fresh Kills Landfill, and the future of Freshkills Park are parts of a sweeping history that deals with the life of one dump, albeit a very large and important one, and its broader implications.

Chapters 1 and 2 discuss New York City as an island city that was land poor but economically and demographically dynamic, where outlets for waste from mass consumption faced severe limits and the mounting piles of refuse created the city's first major "garbage problem." Chapters 3 through 6 address the backstory of New Yorkers, Staten Island, and Fresh Kills before the landfill. Chapter 3 provides a glimpse of early Staten Island and also the notorious "Quarantine War" in the 1850s, when locals burned down the quarantine station and adjacent buildings located on the island for fear of future epidemics of yellow fever. The event was a precursor to other mainland incursions on the "Forgotten Borough." Chapter 4 relates the "Garbage War" before World War I, in which the building of a refuse plant—on the site where Fresh Kills Landfill would be constructed several decades later—set off a major controversy. Chapters 5 and 6 return to New York City's struggles with disposal, especially the experiments with incineration and the turn toward landfills.

Chapter 7 discusses the emergence of Fresh Kills Landfill, especially through the machinations of the "master builder" Robert Moses. Chapters 8 through 14 treat the history of Fresh Kills Landfill and New York City's chronic battles over disposal. Chapter 15 discusses Staten Island's rebellion through secession to protest its years as a sacrifice zone. Chapters 16 and 17 take on the decision to close the landfill and the debates over what to do next. Chapter 18 introduces the tragedy of September 11 and how it intertwined with the history of Fresh Kills. And chapters 19 and 20 transition to Freshkills Park and the Department of Sanitation's efforts to change its disposal practices in a post–Fresh Kills world. As a whole, the book is more than a study of Fresh Kills, but the book's story could not be told without it.

PART I
THE BACKDROP

CHAPTER 1

ISLAND CITY

Situated on an island which I think it will one day cover, it rises like Venice from the sea, and like that fairest of cities in the days of her glory, receives into its lap tribute of all the riches of the earth.

So wrote Frances "Fanny" Trollope about Manhattan in 1827. While not typically laudatory of Americans (her 1832 *Domestic Manners of the Americans* offered a disparaging portrayal of the fledgling United States, and she said of Americans aside from her friends and her "friends friends," "I do not like them. I do not like their principles, I do not like their manners, I do not like their opinions"), the snobbish author praised New York City for its promising economic future and its beauty, elevating it to the grandeur of an Old World icon.[1]

The comparison of Manhattan with Venice was apt; both cities' physical settings offered opportunities but also limitations. There is no question that New York City was one of the world's greatest ports in modern times, as Venice had been during its heyday. For decades, New York City was faced with a predicament shared with other great cities, in its case magnified because of its geography, scale, and the density and size of its population: the disposal of massive waste. The city's extraordinary economic productivity and its voracious consumption of goods and services made Gotham the envy of the world. The same forces hindered it logistically, politically, and environmentally because of the quandary of what to do with its enormous amounts of discards.

As economically dynamic as New York City became, it was faced with the immutable fact that it was land poor—an island city situated on an estuary. As the historian Ted Steinberg observed, "The city tried to grow in every way it could. But it found that there were limits to island life. New York was trapped by water."[2] Early in its history, however, waterways not only provided effective transportation but an obvious refuse basin. The downside of that option quickly became visible: waste washed ashore on beaches, and benthic life, such as oysters, disappeared from New York Bay. Reclaiming materials in the waste stream or burning garbage were tried in the late nineteenth century and later, but these options presented many obstacles too. Dumping on land was simple but not effective—and certainly not popular. After numerous attempts to deal with ever-mounting piles of garbage and rubbish, in the late 1940s the city built Fresh Kills Landfill on Staten Island, which became New York City's main disposal site for more than half a century. However, by 2001 the controversy that the decision inspired and the landfill's environmental implications doomed this disposal solution as well.

After Fresh Kills closed, New York City resorted to sending its solid waste outside the city limits (for others to deal with), eliminating the need to use precious local real estate as wasteland. To appreciate the magnitude and intensity of New York City's long-standing dilemma over waste disposal—and its commitment to Fresh Kills—is to grasp more deeply the role of the city's environment,

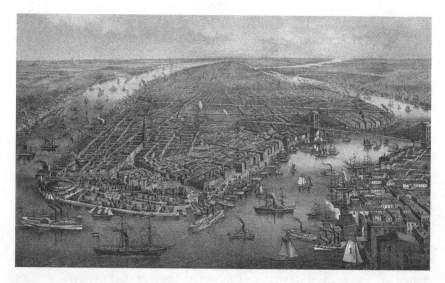

FIGURE 1.1 Manhattan cityscape, 1873.

Source: Library of Congress, Prints & Photographs Online Catalog.

its penchant for growth, and its geography and topography in shaping its sanitation history.

NEW YORK CITY'S ENVIRONMENT

Located on the Atlantic Coastal Plain in southeastern New York State, the city is positioned at the conjunction of the Hudson and East Rivers, with a naturally sheltered deepwater harbor leading to the Atlantic Ocean. The Hudson River flows out of the Hudson Valley to the north of the city and into New York Bay, where it becomes a tidal estuary separating the Bronx and Manhattan from northern New Jersey. The Harlem River is also a tidal strait located between the Hudson and East Rivers, dividing Manhattan from the Bronx. Both the Hudson and East Rivers rise and fall with the tides. The city also contains numerous bays, such as Jamaica Bay; a variety of smaller islands; and Long Island Sound. The larger metropolitan area today includes the eastern portion of Long Island, northern New Jersey, and southwestern Connecticut.[3] The Atlantic Ocean clearly affects the city's humid, continental climate, giving it cold winters and hot, moist summers.[4]

The importance of New York City's location in the estuary of the Hudson River cannot be overstated. "Estuaries are very special environments and, from an ecological perspective, highly productive ones. They are located at the point where freshwater and salt water join together, and play a role not only as habitat for birds and other wildlife but also in the health of the oceans, by filtering water and acting a nursery grounds for fish. They tend to be crammed with life." Yet such an estuary "with a high natural density" ultimately was replaced with "an astonishingly high unnatural (for a lack of a better word) density."[5] New York City's natural and human history, therefore, are linked to its distinctive waterside setting.

THE BOROUGHS

The five boroughs of New York are physically, politically, and economically interdependent. They are connected by common watercourses but remain somewhat diverse topographically. The elevation is less than fifty feet above sea level for much of Manhattan, Brooklyn, and Queens but nearly three hundred feet in northern Manhattan and the Bronx. Todt Hill on Staten Island is the highest point in the city, at 412 feet above sea level.[6] Politically, the boroughs are linked by charter (in 1898), yet today they still retain somewhat individual identities and personalities. When we speak of the "city" of New York as a

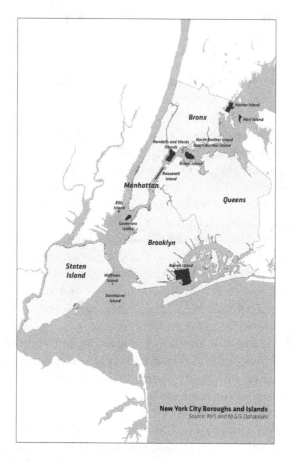

FIGURE 1.2 New York City boroughs and islands.

Source: Courtesy of Jose Mario Lopez, Hines School of Architecture, University of Houston.

political unit before the late nineteenth century, we mean the developed areas of Manhattan; the other boroughs remained rural or dotted with a few towns.[7]

Manhattan, the smallest in area (22.6 square miles, or 7.1 percent of the city), has become the most densely populated of the boroughs. It is by far the heart of business activity and offers many cultural attractions. It is bordered by the Hudson River to the west, by the Harlem River to the north, and by the East River and New York Bay on the south. Within its boundaries one also finds Governors, Randalls, Wards, Roosevelt, and U Thant Islands, among others.[8] At just 13.4 miles long and 2.3 miles wide, Manhattan ascends ever upward with a profusion of tall buildings. Kurt Vonnegut referred to it as "Skyscraper National Park." Because of its amazing drawing power for businesses and commercial development, Manhattan's upward thrust became a necessity. Lower and northern Manhattan are essentially protrusions of schist, an incredibly durable bedrock. Schist runs from the Battery north to Henry Hudson Bridge. The southern tip and the center of the island are solid schist, with abrupt dips and rises of several hundred feet along the way. In all, Greater New York has five bedrock layers, aside from the Manhattan schist, which include Fordham gneiss primarily in the Bronx; the Hartland Formation in central Manhattan, the Bronx, Brooklyn, and Queens; Staten Island serpentinite; and Inwood marble in Manhattan and below the rivers that surround it. On Staten Island a ridge of serpentine follows a northeastern line, erupting to the surface in the middle of the island at Todt Hill.[9]

Above Manhattan to the north, the Bronx is the only borough on the North American mainland. At forty-two square miles, it is the fourth-largest borough in area and today ranks fourth in population among the boroughs and third in population density. The western half of the Bronx includes undulating hills and valleys, and to the east of the Bronx River the land slopes toward Long Island Sound.[10]

Brooklyn, located on the southwestern tip of Long Island, is situated on New York Harbor across the East River from Manhattan. At seventy-one square miles in area, it boasts sixty-five miles of natural shoreline. Before it was settled in the late seventeenth century, Brooklyn was largely marshland. It has become Greater New York's most populous borough in recent years; it was an independent city until consolidation in 1898, after a very close vote.

To Brooklyn's north, also on Long Island, is Queens. At 120 square miles, it is the largest of the boroughs geographically—almost as large as Manhattan, the Bronx, and Staten Island combined. It is bounded on the north by the East River and Long Island Sound, to the east by Nassau County, to the west by the East River, to the southwest by Brooklyn, and to the south by the Atlantic Ocean. Ethnically diverse and a haven for immigrants, Queens had long been a grouping of small towns and villages. As the New York area modernized, Queens became a focal point for industrialization.[11]

The fifth borough is Staten Island. Before consolidation, it was called "a little piece of the country in the city." For many years it functioned as a retreat for the wealthy and as an essentially rural community, and then as a suburb. In the early twentieth century, industrial growth on Staten Island became more pronounced, especially along the shores of Arthur Kill and Kill Van Kull on the north side, across from New Jersey. Richmond Borough, as it was called at the time of consolidation, has always been the outlier in Greater New York. The southernmost borough, Staten Island is third largest in area, at sixty square miles, but the least populous and the least densely inhabited. The western shore is marshy and is bisected by the tidal entrance of the Fresh Kills, which drains the borough's northwestern hills and central Greenbelt. A ridge of hills extends from the present-day ferry landing at St. George to the center of the island.

Staten Island is linked economically and culturally to the northern coast of New Jersey to a greater extent than it is to the other boroughs. Within its boundaries lie several smaller islands, including Prall's, Hoffman, Swinburne, Shooters, and the Isle of Meadows. It is the most geographically separate of the boroughs, economically different, and politically conservative. Lying five miles (at its closet point) southwest of Manhattan at the convergence of Upper and Lower New York Bay and just west of Brooklyn, it has been the least accessible historically to the other boroughs. Until 1713 there was no public ferry; people had to go back and forth by private boat. The Bayonne Bridge, connecting Staten Island and New Jersey, did not open until 1931. And the Verrazzano-Narrows Bridge,[12] connecting Staten Island to Brooklyn—and thus to the rest of Greater New York—was not opened until 1964.[13] For those who do not live there, Staten Island seems to migrate into the category of the "other" islands more often than it is treated as a borough.

ECONOMIC ENGINE

New York City's physical environment played an important role in its development as an economic-growth machine. Its setting and the human-made changes to it over time also imposed limits; land was lacking for a variety of potential uses. This raises several questions about New York City's dynamic economy and its enormous population: What accounted for New York City's growth? To what degree was it sustainable? As dizzying economic production led to a new level of mass consumption, what problems did New York City's environment pose for large-scale waste disposal? (The last question will be more forthrightly addressed in subsequent chapters.)

The economist Edward L. Glaeser made a persuasive case that the rise of New York City in the early nineteenth century was the result of

> technological changes that moved ocean shipping from a point-to-point system to a hub and spoke system; New York's geography made it the natural hub of this system. Manufacturing then centered in New York because the hub of a transport system is, in many cases, the ideal place to transform raw materials into finished goods. This initial dominance was entrenched by New York's role as the hub for immigration.[14]

These factors formed the foundation of the city's economy, an economy that remained vigorous even after transitioning in more recent years from being a manufacturing center to a financial, business services, and corporate-management center.

The primary focus of New York City's success was its premier port, connecting it to trans-Atlantic trade, and its waterborne connection to the Great Lakes.

New York's great manufacturing industries in the nineteenth century—sugar refining, publishing, and the garment trade—benefited from the city's place at the center of a major transportation hub. As an entryway for immigrants, the city drew in hundreds of thousands of people. Producing goods in a central location, and on such a grand scale, was attractive to workers. Immigrants stayed in New York City because its industries (especially the garment trade) "were able to increase in scale to accommodate extra labor without a huge drop in wages."[15]

The birth of modern New York City in the late nineteenth century was marked by furious economic expansion, generated great wealth, and inundated Manhattan, Queens, the Bronx, and Brooklyn with scores of people, buildings, and business activity. At the time the city transitioned from a mercantile economy to a modern industrial juggernaut.[16] The lawyer and New York society leader Ward McAllister proclaimed, "Here, all men are more or less in business. We hardly have a class who are not."[17]

Parts of each borough escaped these rapid inflows for a time, none more so than Staten Island. It remained bucolic and underdeveloped throughout the nineteenth century and into the twentieth, especially in comparison with Manhattan. The overall dynamic growth of Greater New York could be seen in attempts to occupy as much usable space as possible with little thought about how the city might contend with the deleterious effects of its "progress," particularly how to deal with mounds of discards. What became the physical margins of this bustling society—along its land and water boundaries and onto outlying islands—were at first largely unplanned, unnoticed, or obscured locations to be transformed, among other things, into waste zones. The city's extraordinary productivity would be turned upside down: its vast absorption of goods and materials created a genuine dilemma of consuming.

THE PORT OF NEW YORK

At the very center of the city's economic development was New York Harbor. In Ernest Poole's 1915 novel *The Harbor*, Billy the narrator called it "the threshold of adventures."[18] Beginning in the late eighteenth century, New York City had a dominant role in the sugar industry because of extensive trade with the West Indies, and it soon engaged in sugar refining.[19] As early as 1815 the city already was showing signs of operating as a national as opposed to a regional economic center. The completion of the Erie Canal in 1825 stimulated unprecedented expansion of the port by linking New York City to the Great Lakes and beyond. Heavy investment in railroad construction also tightened the city's grip on commercial activity in the country's interior. Important were the first regularized shipping lines from New York to Europe, which gave the port a substantial advantage in imports over its competitors such as Boston and

Philadelphia, which did not possess the same access to European ports. Before the Civil War, the Port of New York already had acquired a large segment of the American export market, including cotton.[20] As early as 1830 the port handled 37 percent (by value) of U.S. foreign trade; by 1870 that had risen to about 58 percent. In the 1890s, its share of the foreign trade dropped below 50 percent (although the total value continued to increase), in large measure because of a shift in national population westward and the emergence of cities such as Chicago and St. Louis.[21] Nevertheless, the role of the Port of New York in the local and national economy remained strong by any measure.

By the 1870s or so, New York's port had become the world's busiest, attracting commodities such as coffee, tea, wine, clothing, and jewelry. In 1874, almost 61 percent of all American exports flowed through the Port of New York, including grain, lumber, oil, and coal; until 1910 50 percent of all U.S. imports arrived first in New York Harbor.[22] In 1870 the city created the Department of Docks, charged with the responsibility to devise a port master plan and to supervise and regularize improvements, including the remaking of waterfront structures. While the master plan was never completely realized, the department's vision of great numbers of piers jutting into the harbor, welcoming untold goods from around the world, was achieved.[23]

At the time of the Department of Docks' founding, the *New York Times* carped that the port facilities were "mean, rotten and dilapidated wooden wharves."[24] Under the leadership of its engineer-in-chief George B. McClellan, former general-in-chief of the Union armies in the Civil War, the department adopted a plan in 1871 to construct a masonry wall with docks and piers at regular intervals for more than twenty-seven miles around Manhattan. Work began in 1872 but was not completed until World War I.[25] The author and publisher Moses King paints a more flattering picture of the harbor in 1893 than the *New York Times* in the 1870s:

> It is as beautiful . . . as it is interesting. . . . On all sides the assembled cities encircle the waters with their masses of buildings, the forests of masts by the waterside, the immense warehouses and factories along the pier-heads, and the spires, domes and towers of the beautiful residence-quarters beyond. At night, the harbor is girded about by myriads of yellow and colored lights and white electric stars, and dotted with the lanterns of vessels in motion or at anchor.[26]

Such a portrayal masked waterfront crime, congestion, and derelict structures housing the poor and the homeless. Critics of several stripes and social reformers decried the underside of life along the waterfront. The *Real Estate Record and Guide* in 1891 referred to it as "our abominable waterfront."[27] Others were deeply concerned about claims in the 1890s that the Department of Docks was too cozy with Tammany Hall, the political machine of Boss Richard Croker.[28] Yet the economic vibrancy of the port overshadowed and concealed a host of ills.

MONEYLENDER, TRADER, AND MANUFACTURER TO THE WORLD

Wealth was to be made in the Empire City in almost every economic sector, and New Yorkers placed a high value on commercial achievement. Wealth equaled vast consumption by plutocrats and workers alike. The success of the port stimulated the upsurge of the financial district now located on and around Wall Street in Lower Manhattan. All types of business transactions were conducted there. By 1883 the number of private and state-chartered commercial banks stood at 506. One reason that the banking houses grew so rapidly and so large was that other American banks kept a good deal of their own reserves in this convenient financial nerve center. In addition, related services such as credit-verification offices established themselves there, as did numerous insurance enterprises. Established in 1817 as the New York Stock Exchange and Board, it soon surpassed competing exchanges because of its enormous cash flow, its volume of transactions, and its aggressive speculation. In the 1850s the railways proved important to the exchange's success. By the end of the century, a variety of industrial activities (including mining and oil drilling) could be added to the list.[29]

In 1886 more than one million shares were traded in one day. The triumphs of the banking houses in New York City were reflected in their dominance of Wall Street and their vast influence on the nation's economy. In 1895 Greater New York had almost three hundred mercantile and manufacturing firms worth in excess of $1 million; Chicago had only eighty-two; Philadelphia, sixty-nine; and Boston, sixty-four. By 1900, New York boasted the headquarters of sixty-nine of the leading hundred largest corporations. Its status as an economic powerhouse helped the city and state survive the panics of 1873 and 1893 and also gave financial institutions great influence in society generally.[30]

Yet neither finance nor commerce employed as many workers as manufacturing. While New York's economic image was identified with finance and commerce, industrial enterprise grew very rapidly by mid-century largely because of the boost from commerce, cheap transportation, and good local markets. In 1890 New York had 25,399 factories producing 299 different industrial products. Many of the larger mechanized industries and noxious trades—textile mills, sugar and kerosene refineries, ironworks, foundries, brewing, tanneries, and slaughterhouses—moved away from Manhattan to Brooklyn and neighboring New Jersey as the decades passed. Manhattan, nevertheless, retained much of its printing, publishing, garment making, tobacco, and furniture businesses.[31] According to net value of product in 1890, New York City's six leading industries were men's clothing, newspapers and periodicals, women's clothing, tobacco, malt liquors, and printing and publishing.[32] Garments, as a manufacturing category, constituted 398 establishments in 1860 and 8,266 in 1900.[33]

A CONSUMING POPULATION

The expanding population in New York City made up a consumer base unrivaled anywhere else in the country and filled much of the available residential space. Between 1850 and 1900, New York City's population increased from 696,115 to 3,437,202. By 1950 it reached almost 8 million. In 1860 nearly half of the city's population was foreign-born, with the Irish being the largest contingent. Soon after the turn of the century 41 percent of the city was foreign-born, representing a wide array of ethnicities and races.[34]

The thriving economy produced great wealth for a small upper tier of society but also broadened the middle class and increased opportunities among the working classes. Emblematic of the emerging consumer culture in the late nineteenth century, the 1876 Centennial Exposition, held in Philadelphia, highlighted industrial production and a new wave of consumption. Fair exhibits included the newest furniture, porcelain, ready-made clothes, sewing machines, home appliances, and more. The main building—solely dedicated to consumer goods—stretched over twenty acres of space.[35]

In the late nineteenth century, an "upper tenth" had emerged in the United States, made up of wealthy capitalists, manufacturers and merchants, a variety of landowners, and professionals. The historian Michael McGerr stated that this tiny minority of about 1 or 2 percent of the population "were people who owned the majority of the nation's resources and expected to make the majority of its key decisions." They were concentrated in the Northeast, especially in New York State and Manhattan.[36] By 1902, most million-dollar fortunes in Greater New York were being made in banking and brokerage, wholesaling and importing, real estate, manufacturing, and railroads. New York's modern industrial aristocracy was not drawn primarily from those of inherited wealth but in large measure from the sons and grandsons (not daughters or granddaughters) of older merchant enterprises.[37] As early as 1863, the top 1 percent of income earners in the city reaped 61 percent of the city's wealth. Not all of that money was reinvested. They shopped for luxury items and partook of everything that the retail market had to offer, they built mansions, and they hired servants.[38]

A growing middle class in New York City pursued professional, managerial, clerical, and entrepreneurial positions as white-collar (nonmanual) jobs continued to rise. Many commuted to Manhattan from the suburbs of Brooklyn, New Jersey, and elsewhere. Not surprisingly, Manhattan had more office jobs than the other boroughs. The historian David Hammack has suggested that "the economic resources of middle-income New Yorkers varied widely, but even the most prosperous had to organize to participate effectively in local affairs."[39] Ethnic background made a difference in the occupations opened to those in this group, but nonmanual workers organized along lines of economic interest. One such organization was the Retail Dealers Protective Association; others included charitable, mutual-protection, and self-help organizations.[40] As would be expected, middle-income families consumed goods and services in line with their economic status.

Manual workers had nowhere near the financial resources of the wealthy upper tier of white-collar workers. The standard of living varied widely among them, with skilled workers at the top. In the late 1890s about 60 percent belonged to trade unions. Semiskilled and unskilled workers earned much less and were rarely organized at that time.[41] What New York City's largest class lacked in income, they made up for in numbers. The city's population increasingly engaged in manufacturing. In 1820, New York had only 9,523 workers in manufacturing and 3,412 in commerce. By 1850, 43,340 people worked in manufacturing and 11,360 in commerce. The garment industry clearly led the way in manufacturing employment in New York until 1970. In the mid-twentieth century, about 27 percent of workers were employed in garments.[42]

As consumers, working-class (and most white-collar) New Yorkers did not run to Fifth Avenue to buy jewelry and fur coats, but their numbers contributed mightily to the economy—and to the waste stream. Although seafood in particular was abundant in the nineteenth century, food in general was not necessarily cheap. Day laborers, whose work was seasonal or subject to the whims of prevailing economic conditions, had to survive on what they could. Pork by far was the cheapest type of meat available.[43] When times were good and the economy steadily grew, waste production from mass consumption was particularly high. In the 1890s, for instance, New York City residents consumed approximately 750,000 watermelons during the warmer months of each year. Waste also came from ash residue from burning wood and coal and from ever-present horse manure in the streets.[44]

LAND-USE PATTERNS

With the transition to industrial capitalism, land-use patterns favored business development over residential communities in several parts of the city, especially in Manhattan. As such, land available for other uses (such as disposal sites) became scarcer and encouraged extensive destruction of marshland and the use of watercourses, smaller islands, and outer boroughs for a variety of activities unwelcome in the urban core. One estimate claimed that in the early twentieth century there were about three hundred square miles of wetlands within a twenty-five-mile radius of downtown Manhattan. The original marshland in the city had declined by almost 80 percent.[45] Such changes in land use (not only in marshes) had occurred to some extent before that transition, but the pace had now considerably stepped up.

In 1811, the New York State legislature approved a sweeping plan for Manhattan as a basis for the systematic sale and development of land between Fourteenth Street and Washington Heights. The plan called for laying out a rectangular grid of streets and property lines, with no regard for topography. While the plan was frequently altered, it was praised by some as orderly and reviled by others as arbitrary. It nevertheless assured a form of high-density living and working space that characterized Manhattan's development for years. At its best, the grid meant easy proximity to a variety of business activities and "closely built" living conditions; at its worst, it meant crowding and congestion.[46] Among other things, a philosophy of high-density development in such a water-bound environment as New York City might help sanitation workers collect the trash, but it made it all the more difficult to find a place to put it down.

The separation of business and residence became more pronounced with the new capitalist economy, and in several cases New Yorkers who could afford it (and some who could not) were forced to move away from commercial and industrial districts because of high rents or the conversion of residential property to business use.[47] Villages and towns sprung up along railways for miles around the city. Jersey City and Hoboken, New Jersey, on the west bank of the Hudson, sported industrial and residential suburbs. Brooklyn was regularly absorbing great numbers of the city's population. Charles Dickens called it "New York's dormitory."[48]

Property values soared in the city despite severe national economic dislocations in the 1870s and early 1890s.[49] Members of the new Real Estate Exchange and Auction Room (owners, developers, auctioneers, builders, and brokers), formed in 1885, were thrilled with the spectacular rise in land and building values since the 1860s. New York's property market directed investment and land-use practices to shape the development of specialized districts so typical of the city's growth in the late nineteenth century. This was not urban planning in any conventional sense, since intermingled among the districts was plenty of spatial disorder, including a high concentration of tenements, transportation gridlock, and chronic sanitation problems.[50] Yet property was a commodity, and thus in areas where commerce and other forms of business took hold, the activities that offered the greatest economic returns gained priority by demonstrating the ability to pay rent for the space. Land values increased accordingly.[51]

CHANGING DEMOGRAPHIC PATTERNS IN THE CENTRAL CITY

Escalating real estate prices, while a boon to some, were a curse to many. The crush of humanity made the search for living space in this urban, capitalist paradise a serious challenge. In 1900, New York City was far and away the most populous city in the United States, at 3.4 million. Chicago was a distant second at almost 1.7 million.[52] The rate of growth was even more astounding. In 1830 Manhattan and Brooklyn accounted for 220,000 inhabitants; in 1860, one million; and in 1890 2.5 million.[53] Interestingly, the rate of growth among the boroughs between 1850 and 1900 placed Brooklyn first, Queens second, Staten Island third, and Manhattan tied for fourth (even though total numbers favored Manhattan by a wide margin). In 1900, Manhattan's population (in county figures) exceeded 1,850,000, while Kings County (in which Brooklyn is located) had approximately 1,166,000; the Bronx, 201,000; Queens, 153,000; and Richmond (Staten Island), 67,000.[54] What may account for these figures was the availability and affordability of residences away from the specialized districts for those who were financially able to move.

The scale of immigration has been a central story of New York's population growth for generations and a key dynamic in its contest for occupying the city's available space. Between 1892 and 1924, 16 million immigrants were processed on Ellis Island alone.[55] Through much of the nineteenth century, especially from 1850 to 1875, 80 percent of the immigrants in New York came from the German states and Ireland (70 percent in Brooklyn). Others came from England, Wales, Scotland, France, and Canada. Beginning in the 1870s there was a major geographic tilt to southern and eastern Europe: Italy, Russia, Bohemia, Poland, and elsewhere. What had been a German-Irish city by mid-century was much more diverse by the early 1900s. Places like Little Italy became icons of the changing ethnic makeup of various neighborhoods. In 1880 New York City had 80,000 Jews, primarily from German towns and cities. With increasing persecutions in Russia in the early 1880s, the numbers of Jews emigrating rose dramatically. Between 1881 and 1910 more than 1.5 million came to the United States, with a majority remaining in New York and large numbers flocking to the Lower East Side. The proportion of African Americans, which was one in ten New Yorkers in 1820, was less than 2 percent in 1890. A steady migration from the South in the following decades would reverse that decline.[56]

Early on Americans welcomed the influx of new workers to fill the factories and provide a variety of services. Things sometimes got ugly (as they had for earlier immigrants and migrants) when the waves of newcomers began exceeding the labor demand. The Swiss journalist Max Frisch once notably remarked about his country's guest-worker program that "We asked for workers. We got people instead."[57] Such an observation was many times lost by those in the host country, who increasingly viewed the southern and eastern Europeans as competitors for their jobs or as simply "the other." New York was at once a human cauldron of opportunity, gamble, and risk for the new immigrants.

While there were clear spatial gaps between middle- and upper-class residential and commercial/business districts, those unable to move were too often bound to crowded and substandard living within close proximity to their jobs. In working-class districts, housing could be set among the workshops, warehouses, slaughterhouses, and factories. In the mid-nineteenth century, the notorious tenement houses on the Lower East Side packed thousands of people into cramped quarters. The tenements were supposed to help solve New York's housing crisis but seemed destined to make it worse. The late 1870s saw the building of teeming "dumbbell tenements" (narrowed in the middle, thus looking like a dumbbell), a reform effort that failed. The buildings were constructed on twenty-five-foot-wide spaces to limit the proportion of lots to be occupied; this made each living quarter minuscule. Sanitation was a serious problem, especially when people dumped garbage into the spaces between the buildings. Such construction also made the buildings major fire traps.[58] More than twenty thousand dumbbell tenements were built between 1879 and 1901. Each building housed about four hundred people, approximately four thousand per block. In 1900 Manhattan had 42,700 tenement houses with a combined population of 1,585,000.[59] The concentration of tenements spoke to the competition for space. This concrete jungle also exposed the vast differences among the classes and races when it came to dominating that space and to how real estate would be utilized.

The times and circumstances did not favor housing reform. Through the 1840s, the most populous wards were in Lower Manhattan, with a shift to Midtown in the 1850s. The densest wards consistently remained in the Lower East and West Sides through the turn of the century. Wards 1 through 6 are all located in the "downtown" district, which was then the financial center of the city. Businesses in those wards crowded out dwellings during the period, accounting for Wards 1, 4, and 6 falling out of the high-density category. Ward 12 was located above Eighty-Sixth Street, where the migration north was slowly heading.[60]

Elite status in New York as elsewhere was increasingly tied to economic opportunity and success—along with being born into an acceptable race, ethnic group, and religion. New York City saw the greatest concentration of wealth in the United States in these years.[61] Distinctions among rich, middle class, working class, and poor were reflected by physical segregation as well—tenement living, for example. Some European observers regarded the New York slums worse than those found in London or other European cities.[62]

The mercantile class could not be happy to see swarms of underclass workers seeking living space in close proximity to them. As early as the 1840s New York's mercantile elite had become a relatively unified group well aware of its high status. The members attended the same churches, sent their children to the same schools, joined the same clubs, and shared common business interests. The move to the suburbs was one option for remaining separate. For those wanting to stay in Manhattan, addresses above Fourteenth Street were most "fashionable," none more so than Millionaires' Row on upper Fifth Avenue. By 1860, 344,000 people (mainly of the middle and upper classes) lived north of Fourteenth Street, bringing with them their religious, social, and educational institutions, such as Grace Church, the Union Club, and Columbia College.[63]

These gentry also could easily isolate themselves by moving to the northernmost precincts of Harlem. Intracity rail transportation did not become more accessible to the lower classes until the late 1870s, which complicated residence-to-work patterns.[64] Although wealth alone did not always determine the dividing line between rich and poor dwellings and neighborhoods, class distinctions were graphically displayed along property lines. The political economist Henry George was a severe critic of the powerful and wealthy, as his widely read *Progress and Poverty* (1879) made clear. George was adamant about the destructive influence of private property: "Poverty deepens as wealth increases, and wages are forced down while productive power grows, because land, which is the source of all wealth and the field of all labor, is monopolized."[65] Controlling real estate and its uses was particularly crucial in a place like New York City, where land was so dear and so limited.

RESHAPING THE CENTRAL CITY'S LANDSCAPE

Dynamic economic growth demanded more space and new construction for warehouses and counting houses, law offices and banks, clothing plants and printing shops; it also called for more sophisticated infrastructure such as transit and power lines. By 1900 essentially all vacant land south of Central Park had been developed.[66] And the expansion was rapid. In 1825 urbanized areas in Manhattan extended well short of First Street at Canal Street. By 1860 the city was closing in on Fiftieth Street. On the East Side urban growth reached Ninetieth Street, and by 1890 135th Street.[67]

As early as the mid-nineteenth century, Manhattan's Wall Street, Broadway, and Fifth Avenue were rounding into their modern form.[68] The city's only remaining heavy industry, shipbuilding, and its other foremost industries, such as garment making, printing, and publishing, were located in Lower Manhattan near the financial district. In several cases, there was geographic concentration of specific businesses (districts formed exclusively for one type of activity). Mercantile businesses could be found along the riverfronts of the Hudson or East Rivers. Warehouses congregated along Pearl Street. The financial district was centered on Wall Street. By the 1880s new shopping districts formed in various areas.[69] Also, for example, in 1900 wood plants were highly concentrated in Lower Manhattan, along the Hudson to Central Park, and along the East River as far north as the Bronx. Textile and printing plants, clothing businesses, chemical factories, and wholesale groceries were also located primarily in Lower Manhattan. Printing plants could be found in Brooklyn, and food-processing plants also clustered there. Economic segregation in Lower Manhattan was matched by an expansion of business activity to the north. By the 1920s, the pattern changed significantly for several industries, and business moved north toward the Bronx or into Brooklyn, and some began to decline altogether.[70] As early as the 1870s, commercial districts had begun to move several miles to the north and to the east along the East River from the original business center in the Battery.[71] Competition for available land never ended.

Complementing horizontal growth was vertical growth, extending the possibilities of building density, especially on Manhattan's precious land. Manhattan has long been known for its skyscrapers. Yet as late as 1876 the island was still primarily an agglomeration of low-lying buildings no higher than five or six stories. The only spikes protruding from the horizontal city were the spires of Trinity Church on Broadway and Wall Street and Saint Paul's Church on Broadway and Fulton.[72] The trend to go vertical began in the late 1860s. The first skyscraper towers were built because of the high demand for commercial space, the lack of building-height regulations, and the development of the grid. New technology such as elevators and steel girders also made such construction possible. The city's many relatively small lots, as the architecture professor Carol Willis has suggested, "created an economic logic" for the first towers. Large sites were difficult and expensive to amass.[73]

Buildings steadily grew taller because of the demand for space by various businesses, and such elevations also were made possible by advances in engineering that made the structures more efficient and safer. "Ego and advertising also played a role," Willis added, with some highly competitive builders and companies seeking prestige through elevation.[74] The Western Union Telegraph Building, constructed in the early 1870s, rose one hundred feet from street level. The clock tower perched on the Tribune Building (1875) was 285 feet above ground. More such buildings were to follow. By 1900, Manhattan had six structures taller than three hundred feet; the Park Row Business Building (1898) had twenty-nine stories and stood 392 feet above the ground. Most tall buildings were clustered at City Hall Park, Wall Street, and Lower Broadway, with an average height of ten stories. By 1891 this area was a legitimate high-rise district. The new skyline, dotted with tall buildings, soon expanded beyond Lower Manhattan to the west of Central Park and to the north of Fifty-Ninth Street. By 1897 there were ninety-six tall buildings in Manhattan. In the early years of the new century, skyscraper development reached its peak in the city, with 366 such buildings by 1908.[75]

In a 1933 interview, the Spanish poet and dramatist Federico Garcia Lorca (1898–1936) said, "The only thing that the United States has given to the world are skyscrapers, jazz, and cocktails. That is all. And in Cuba, in our America, they make much better cocktails."[76] Lorca obviously had more on his mind than skyscrapers, but New York's contribution to one of his "big three" was central to creating a high-density environment like few others. The reactions to the vertical thrust into the New York skyline inspired awe but also unease and misgiving. *King's Handbook* referred to the area around City Hall Park as "the grandest architectural square in America."[77] Others were unhappy that these imposing structures were blotting out the sky. A disgruntled British observer described the Tribune Building as "a sort of brick and mortar giraffe." Various writers regarded them as "outrageous," "savage anarchy," "towers of babel," and "human beehives."[78] Probably most apt was the moniker "cathedrals of commerce,"[79] for the skyscrapers represented the drive to provide more space for economic productivity every bit as much as establishing various business districts did. Together, they also squeezed out residents and effectively changed the cityscape to reconfirm the major commitment to material progress, couched as modernity. Building density was not equally distributed among the boroughs, but physical changes in Manhattan presaged a variety of intensified infrastructural changes throughout Greater New York City.

LAND AND WATER, WATER AND LAND

With all of its economic success, it was difficult to comprehend what a major drawback land scarcity would be for New York City. Perched on a coastline open to the sea (like Venice and other port cities), Gotham's links to fresh and salt water defined it, but so did its land masses. There is little doubt that environmental factors (particularly its aquatic heritage) shaped New York City's past, present, and future. Its unique geography, however, was also a leading reason for its waste disposal woes once the area became heavily populated.[80] As a conglomeration of islands, a virtual archipelago, New York City benefited from plentiful riverine and oceanic transport for moving people and goods. It had a large, deep, protected port central to its commercial and industrial prowess.[81] It also attracted vast numbers of migrants from the hinterland and immigrants from across the Atlantic. Collectively the islands served as essential spaces for a variety of activities, but the same waterways that afforded such important transport of people and goods also separated the city into a number of unique parcels. The need to overcome water barriers to utilize every strip of earth was indispensable for the city's growth and development.

Its situation on delimited land masses meant that New York was increasingly hard-pressed to accommodate adequately its growing population and its emerging commercial and industrial activity. Hardened political boundaries, especially along the New Jersey border, prohibited opportunities to expand outward, especially to New York City's west. This reality required building upward, consolidating its control around a valuable harbor, adding land to existing islands, and procuring or constructing some smaller islands. The city regularly created new land wherever it could through flattening rugged space, draining marshes, or a process of fill and reclamation extending into the bay. New York City refused to view its topography as static, despite its obvious limitations.

Of the five boroughs that encompass the city today, four are situated on three islands. Manhattan and Staten Island are islands; Brooklyn and Queens are part of western Long Island. Only the Bronx is attached to the mainland.[82] The complex of urban entities comprising Greater New York is modest in area (given its huge, dense population) at about 305 square miles. It barely makes the top twenty-five American cities in land area.[83]

THE "OTHER" ISLANDS

Glacial erosion more than twelve thousand years ago created dozens of islands, sandbars, and drowned rocks in the area of New York City. Some, like Liberty Island, Ellis Island, and Coney Island, are well known. Others, such as Castle Clinton, Wards Island, and Hunter Island are more obscure. Some of the islands were privately owned, some New York City simply claimed as its own, a few were human-made, and others were purchased by the city and readily absorbed into the metropolis. Between 1828 and 1892, the municipal government acquired Blackwell's Island, Randalls Island, Wards Island, Hart Island, North Brother Island, Rikers Island, and Ellis Island.

Much of the land in the boroughs proper was reserved for commercial and industrial activity and for housing workers and others connected to the economic life of the city. Some small islands were used for a variety of specialized wants and needs, if at all. A few originally were locations for family farms and military forts; others hosted government institutions, recreational centers and parks, and wildlife preserves. Some islands were connected with fill to the mainland. A few islands were human-made with material from construction sites, ashes, and other discards.

In almost every instance, the small patches of real estate in the New York archipelago augmented the limited acreage of the five boroughs. Specific islands were valuable in this land-poor city as sites for people and activities regarded as unsavory or unwanted—intentionally isolating them from existing neighborhoods, shops, and factories located in the urban centers. These included quarantining the sick, confining criminals, detaining immigrants, and dumping waste. For a good portion of its history, Staten Island—more than the other outer boroughs—found itself considered as one of the "other" islands.

The uses of the offshore islands changed over time and in some cases caused controversies. In 1800 the federal government gained control of Governors, Bedloe's, and Ellis Islands, intent on having them serve as harbor defense lines. While Governors Island remained one of America's continuously active military posts, Bedloe's Island became Liberty Island and the site of the Statue of Liberty (dedicated in 1886). Ellis Island is best known as a gateway for southern and eastern European immigrants.[84]

While Ellis Island symbolized refuge and opportunity in a new land, a more typical use of several of the New York islands was to cordon off the sick, the poor, and the criminal. As we will see later, there is an eerie similarity between using the islands as a dumping ground for troubled and undesirable people and using some of the islands as disposal sites for waste. This is not to suggest that rapacious action always drove mainlander New Yorkers to treat humans as so much flotsam and jetsam but to note that some islands functioned as societal safety valves or sacrifice zones for the unwanted in the name of protecting the general population.

Blackwell's Island (now Roosevelt Island), a two-mile-long protrusion located in the East River between Queens and Manhattan, was the site for New York City's prison (1832) and a workhouse (1850). Blackwell's was not only the site for the prison but also for the city's "lunatic" asylum (1841), almshouse (1847), and New York City's first hospital (1849). Penitentiary Hospital (later Island Hospital) was a charity operation serving both citizens and convicts, mostly for treating infectious diseases. Charles Dickens visited some of the island's facilities in 1842, noting that "the terrible crowd with which these halls and galleries were filled so shocked me, that I abridged my stay within the shortest limits, and declined to see that portion of the building in which the refractory and violent were under closer restraint."[85] In 1858 Charity Hospital (later City Hospital) replaced the dilapidated Island Hospital, which had been destroyed by fire. Charity was then the largest facility of its kind in the country; workhouse prisoners served as its nurses and servants. Other hospitals and related institutions followed, focusing on the treatment of smallpox, epilepsy, and "incurable" diseases.[86]

Rikers Island, located in the East River directly north of Queens, is actually part of the Bronx. New York's Department of Public Charities and Correction bought it in 1884 for the purpose of building a new prison (completed in 1935). Promising to be an important break with the past in terms of prison reform, Rikers devolved into an overcrowded cesspool for many years. Throughout the nineteenth century, Randalls and Wards Islands and North Brother Island (all in the East River) also were sites for a variety of government institutions used for isolating the sick, the poor, and others unwelcome in the city proper. Hart Island, in Bronx waters, became a potter's field beginning in 1869.[87]

MAKING MORE LAND

The islands in the New York archipelago, which served a variety of purposes for a city strapped for space, were insufficient to meet all of the city's needs. In some cases, expanding those islands with fill or constructing new ones created several potential uses. Extending the shoreline along the edges of the cityscape and turning marshland into "usable" land were other ways to provide New York City with needed space. There was a difference, of course, between utilizing islands for specific purposes and adding land to the main islands to make room for more people, businesses, and the needs of the port. Taken as a whole, enlarging the city's physical environment became an important way to offset limitations to growth. As we will see later, using construction rubble, ashes, and other discarded materials as fill was (among other things) one of the earliest forms of waste disposal for the city, as was the conversion or destruction of marshland as dumpsites.[88]

FIGURE 1.3 Manhattan land reclamation.

Source: Courtesy of Jose Mario Lopez, Hines School of Architecture, University of Houston.

Using fill to increase land was a widely practiced enterprise for cities across the country and around the world. The precise objectives may have been different, but the basic practice was the same, that is, to create wanted space out of unwanted space (or water). At one time, filling marshes and swamps to make "usable" land was viewed as a way to convert a noisome or useless site into solid ground to build upon. Extending fill land into oceans and rivers replaced what appeared to be an inexhaustible aquatic space with a more serviceable terrestrial one. There was little regard (or understanding) of the possible ecological impacts of creating new land, let alone of declaring certain landscapes like marshes as worthless. (The conversion of marshes and wetlands in New York City also will be discussed more fully in later chapters.) But short-term goals trumped long-term consequences. Large parts of the Netherlands and the city of Rio de Janeiro, for example, are fill lands. Parts of Dublin, Saint Petersburg, Helsinki, Beirut, Mumbai, Shenzhen, Manila, Singapore, Montevideo, and Mexico City were constructed on human-made acreage. In North America, Toronto, Montreal, New Orleans, San Francisco, Chicago, and Boston relied on the creation of fill land.

Since Dutch New York, Manhattan has seen major changes in its shoreline through excavation-and-fill projects. Peter Stuyvesant, the Dutch director-general, prompted the digging of a canal along present-day Broad Street in New Amsterdam for the purposes of waste disposal and transportation. The Dutch also filled in swamps for building sites and constructed piers into the East River.[89] The original shoreline of Greater New York was dotted with small bays, tidal estuaries, and salt marshes. In 1686 the charter of Governor Thomas Dongan gave the city all of the state-owned land that was under water between low and high tides.[90]

Through the late nineteenth century, New York Harbor and its nearest environs were the prime focus of turning water into land. Landfill operations (intentionally or unintentionally, refuse might sometimes be mixed with fill material such as ashes) made Manhattan south of City Hall one-third larger than it had been at the time of its Dutch founding.[91] The motivation was quite clear: the determination of the merchant elite, a variety of entrepreneurs, and political leaders to exploit the vast economic growth that commerce (and later industry) provided through the magnificent port.[92] The harbor as the center of a waterway system comprises 1,500 square miles, with more than 770 miles of waterfront.[93] New Amsterdam began to develop its waterfront in the 1650s. To combat shoreline erosion the city built a seawall extending along the East River shore, which linked to the defenses near modern-day Wall Street by the 1660s.[94]

The Dongan Charter (1686) paved the way for the major alteration of Lower Manhattan. The English provincial governor Thomas Dongan granted to the corporation of the city "all the waste, vacant, unpatented, and unappropriated lands within the city and island, extending to low-water mark in all parts," which established the limits of the city to land exposed at low tide.[95] It also allowed vacant land on Manhattan to be developed by private users. This resulted in the city providing "water lot" grants on parcels lying between marks of high and low tide (essentially land under water half of the time). The grants were used to extend land along the shallow shoreline to provide space for warehouses and as mooring places for larger vessels. This was called "wharfing out" by the English colonists. Soon after the Dongan grant reached its physical limits (in the 1720s), the Montgomerie Charter (John Montgomerie was the colonial governor of New York from 1728–1731) broadened the definition of exploitable underwater land. This allowed for the expansion of Manhattan further out into the bay.[96]

The Port of New York faced chronic problems, and the harbor waters suffered as a result. Well before an elaborate port project was completed during World War I, the harbor had undergone environmental changes. These were caused by fill and other activities in the Hudson Estuary and Long Island Strait, where some plant and animal life had disappeared and the thriving oyster industry had collapsed. In 1888, given the continual formation of land along the East and Hudson Rivers, a federal law set limits on authorized expansion.[97] In 1899, Section 13 of the Rivers and Harbors Act (commonly called the Refuse Act) forbade the discharge of most wastes into navigable waters without a permit. Directed particularly at New York Harbor, the law superseded an 1890 act that prohibited only dumping that would "impede or obstruct navigation." While the new law did not end the vast fouling of the harbor with sewage, industrial wastes, and a variety of refuse, it did set a precedent for future action concerning water pollution.[98] Yet lively trade, commerce, and industry remained the primary measure of the harbor's success.

Making new land, filling marshes and bogs, leveling hills, and aggressive real estate speculation assured a massive physical transformation of New York City early in its history. Establishing the city on a grid in 1811 added a pathway for further expansion by allowing straight and aligned streets, among other things, to connect to the waterfront more easily and to encourage further real estate development.[99]

New York's smaller islands were also subject to earth-fill practices to expand their area and increase their usefulness. In many respects, the utility of islands like Castle Clinton reflect their time and the shifting needs and wants of the city. Castle Clinton was human-made, initially connected to the mainland by a wooden bridge. Completed in 1811, just before the War of 1812, the Southwest Battery, as it was then called, was a fortress site off the tip of Manhattan. One of four forts in the harbor, it helped keep the British at arm's length during the war. It was renamed Castle Clinton in 1817 after the mayor of New York City and later governor of New York Dewitt Clinton. The island was an entertainment center between 1823 and 1854 and then became an immigrant landing depot or gateway between 1855 and 1890. Castle Clinton even served as the site for the city aquarium in 1896 until it was moved to Coney Island. In 1975 it opened as Castle Clinton National Monument; this is where the ticket office for the Statue of Liberty is located.[100]

Terraforming through the use of fill was a notable and environmentally significant way to remake and reshape New York City for the use of its citizens and business activity. Such actions, in a broad sense, reveal how dissatisfied humans can be with taking nature on its own terms. More practically, the comprehensive process of absorbing the islands into the city proper and turning water lots into usable land demonstrated the relentless effort to expand urban spatial resources. Such resources were necessary for the kind of population and economic growth that leaders envisioned or desired, despite potential political, social, or environmental costs.

The rise of modern New York City was shaped by impressive economic expansion, explosive and dense population growth, and incredible mass consumption of resources. A waste disposal dilemma of considerable proportions was a price paid for such dynamism, made worse by a city landscape short on land and surrounded by water. While geographic determinism cannot entirely explain the city's remarkable ascendancy, land-poor New York was chronically vulnerable to accumulating refuse problems that defied easy solutions.

CHAPTER 2

WASTING AWAY

In no branch of the municipal service has so little progress been made in the United States as in the disposal of garbage.
—League of Municipalities, *Proceedings*, August 21–24, 1901

New York City's surging economy and swelling population in the mid-to-late nineteenth century had a significant downside. Mounds of solid waste were graphic reminders of the disagreeable consequences of unrelenting growth and consumption. The amount and variety of wastes made disposal challenging, especially given the city's geographic setting. The harbor and the ocean became the primary sinks for the boroughs through the early twentieth century. This easy but objectionable solution was a constant drawback and irritant, obstructing water traffic, strewing rejected items on the beaches, and destroying oyster beds.

By the early 1890s, after generations of struggles with its refuse, New Yorkers had become exasperated by the "garbage problem." The appointment of Colonel George E. Waring Jr., the "Apostle of Cleanliness," as commissioner of street cleaning in 1895 finally brought about real change. He has long been regarded as the man who turned New York City's sanitation fortunes around. Although Waring made great strides in refuse collection and also in recycling, even he was unable to find an effective long-term disposal solution. Waring buoyed up hopes, but only temporarily.

THE GARBAGE PLIGHT

New York City rarely had a day in its history without a waste problem, and the issues intensified (at least in scale) during those years when rapid population and economic growth defined the city. Before the consolidation of Greater New York City in 1898, much of what was understood to be "the garbage problem" focused primarily on Manhattan and the Bronx (and sometimes Brooklyn). Before the emergence of the massive New York metropolis, personal responsibility for disposing of refuse was the practice, leading to, at best, tolerable sanitary conditions—and a very inexact definition of "tolerable."

Preindustrial American towns fared slightly better than the much larger European urban areas, but this probably had more to do with their less crowded conditions than with any enlightened outlook about sanitation. The "healthiness" of American towns was a matter of degree. Tolerance for nuisances and the almost unexpected occurrence of epidemic disease played major roles in determining the sanitary quality of the communities. Pigs roamed freely, slaughterhouses spewed noxious fumes and effluent, and vermin infested dwellings. Few towns and cities were free of nuisances, and they showed little resolve to move against the "noxious trades" (soap makers, tanners, slaughterhouses, butchers, and blubber boilers), especially if they were located in the poorer areas. Animals resident in urban communities were a part of preindustrial life. Horses for transportation; cattle, hogs, and chickens for food; and dogs and cats as pets and strays roamed freely through vacant lots, streets, and alleys. Pigs and turkeys were widely accepted as useful scavengers. Manure and dead animals were simply annoyances balanced against the value of sharing space with contributors to the town's welfare.[1]

Like many cities, New York struggled with the need to promote its commercial growth yet somehow contend with its problems of filth. In 1657 authorities prohibited citizens from emptying "tubs of odour and nastiness" into the streets, but where such waste was to go was unclear. The city did not award its first street-cleaning contract until thirty-nine years later.[2] In 1702, a local ordinance required householders to gather dirt in mounds in front of their dwellings on Friday morning and have them removed before Saturday night. Failure to do so meant a six-shilling fine. Private cartmen charged three cents per load to collect the discards, but their dumping practices were not explicitly regulated. Some scavengers worked independently of the cartmen in seeking business to remove garbage and rubbish.[3]

The historian Catherine McNeur paints a graphic picture of Manhattan's sanitary state in the early nineteenth century:

> As the city's population grew, so too did its garbage heaps. Householders dumped their trash onto the streets in anticipation of its collection by

scavengers. Rotten food such as corn cobs, watermelon rinds, oyster shells and fish heads joined with dead cats, dogs, rats, and pigs, as well as enormous piles of manure, to create a stench particularly offensive in the heat of the summer. Between the piles of odoriferous waste, deep mud puddles, and sauntering hogs, New Yorkers had much to complain about.[4]

The cartmen proved extremely important to the functioning of the city in the nineteenth century; they also were keen on protecting the right of free labor and their role in moving goods into the city and moving dirt and waste materials out.[5]

Until the late nineteenth century, the city alternated between running its own street cleaning and refuse collection (also dealing with disposal) and leaving that responsibility to private cartmen and scavengers.[6] The city inspector initially had jurisdiction over sanitation matters in early-nineteenth-century New York. This was a controversial post, since several inspectors operated through nepotism and corruption. Street cleaning was an important form of municipal patronage. Contracts that the city awarded were lucrative, since manure and other waste products could be sold at a good price.[7] The Metropolitan Health Act of 1866, passed in March, transferred authority for cleaning and waste collection from the city inspector to the Metropolitan Board of Health. In 1872, the Metropolitan Board of Health shifted responsibility for cleaning and collection to the Board of Police. The assumption was that the police were best positioned to monitor street cleaning.[8] The change in authority was not enough to execute the job efficiently. Private contractors did not perform well, even when supplemented by unskilled laborers hired by the police. The city gave street cleaning a low priority and provided insufficient funds to see that the streets were cleaned properly.[9]

Nationwide, municipal control of street cleaning made its greatest strides in the 1880s. Many cities with a population larger than 30,000 placed contract street cleaning and collection under municipal authority or assumed direct responsibility for the services. On May 26, 1881, New York City created the Department of Street Cleaning for oversight and/or direct responsibility for most residential and some commercial street cleaning and collection operations.[10] But whether under private or public control, Queens and Staten Island did not get the same level of service that was afforded Manhattan or the Bronx, since they had smaller populations and less political clout.[11] According to the sanitary engineer Dr. George A. Soper, the organization of the Department of Street Cleaning as a separate branch of the city's administration "was the result of much agitation on the part of citizens, physicians, and political reform associations."[12] In fact, the Ladies' Health Protective Association, organized in 1844, was possibly the most influential group of its kind in the country. It undertook school and slaughterhouse sanitation as well as street-cleaning reform and improvements in waste disposal methods.[13]

Street cleaning was not an easy job. Not only did the new department face the daily problems associated with dust and debris of all kinds, but in winter it was also responsible for snow removal. In 1890 New York City had 575 miles of streets and alleys, of which 358 miles were paved. (Brooklyn, an independent city at the time, had 653 miles total, with 375 miles paved.) The new department did hire private contractors, who used their own equipment to clean Manhattan streets below Fourteenth Street.[14]

Collection was another matter. Even in this era, the waste stream was complex. In addition to organic matter (garbage, manure, human excrement, and dead animals), there were also tons of coal and wood ashes, street sweepings, wastepaper, cans, old shoes, and other assorted rejectamenta. Some waste, such as kitchen and restaurant garbage, obviously had to be collected more frequently than rubbish. With improvements in carts and the addition of night collection service in 1887, the tasks became more efficient. At the turn of the century in Manhattan, scavengers were collecting between 612 to 1,100 tons of garbage daily, depending on the season. Beginning in 1900, each citizen in Manhattan, Brooklyn, and the Bronx annually produced about 160 pounds of garbage, 1,231 pounds of ashes, and ninety-seven pounds of rubbish.[15]

REFUSE DISPOSAL AND HARBOR FILL

While street cleaning and collection were monumental tasks in New York City, disposal posed its own entirely different set of problems. A sizable portion of the total amounts of waste generated in the preindustrial years went for fill or, in some cases, were recycled as secondhand goods or as commercial products. "Scow trimmers"—mostly Italian immigrants—rummaged through the heaps of waste on the wooden dumping scows (barges for disposing of refuse and dredging material moved in the water by tugboats) searching for rags, shoes, carpets, paper, or anything else with a resale value. Until 1878, the city had paid the trimmers for their services, allowing them to keep what they salvaged. From 1878 to 1882, the subsidies were curtailed, and the trimmers were allowed to take what they wanted. Beginning in 1882, the city charged a flat rate for the privilege of trimming, realizing that the business had heated competition among rival scavenging crews. Scrap-metal dealers and resale businesses of various kinds also operated in the city and accepted a wide array of used materials.[16]

The harbor was the obvious terminus for almost everything in the waste stream that could not be recycled or reused in some fashion. It was the focus of economic activity for the city but also the major destination for fill and a cheap spot for tossing out discards.[17] Initially water-lot grants stipulated that only "dock mudd" (mud from the water adjacent to the docks) could be used along the shoreline to make new land, but the quality of fill varied greatly there and elsewhere in the New York City area. While some fill in the harbor projects came from "clean" sources, others were silt dredged up in the slips, rocks, ship ballast, wood and debris from demolished buildings, and rubble from fires. In various locations some type of solid waste (organic and inorganic) was utilized. The city often deployed cartmen to collect fill-quality material separated from garbage and rubbish, and city ordinances in the early nineteenth century demanded that clean sand be used for fill. Such laws often were broken.[18]

There were some inland dumps that could be used for refuse, but they served most frequently as depositories for fill or fertilizer. In 1860, of the more than 798,000 loads of garbage and refuse removed from Manhattan's streets, most headed for the harbor for final disposition.[19] Cartmen sold some salvageable materials but also were responsible (along with scavengers hired by the city) for hauling household and trade wastes and manure to disposal sites on the waterfront. These sites were referred to as dumping boards (or often just dumps) and were located at berths where scows were moored. At street-level platforms refuse was pushed into the troughs of the scows, which were used for dumping wastes in open waters.[20]

Filling in areas along the harbor with refuse and other materials proved problematic. Signs of trouble began to appear as early as the 1850s as Manhattan built out farther into the East River and Hudson River, encroaching on shipping lanes and severely polluting the waterfront and adjacent beaches. In 1855, the state legislature appointed a commission to investigate the concerns. It established state bulkhead lines to limit further landfilling operations into the Hudson River. Authorities levied fines for those intentionally dumping waste materials into the docks or slips. Cartmen also faced possible fines for negligence.

The scale of the operation was enormous: approximately 15 percent of all piers and bulkheads along the East and Hudson Rivers served as dumping boards. The scows transported manure and other salvageable wastes to outlying farms and other sites and dumped the remainder into the harbor or out to sea. Some refuse never made it to the scows but spilled from the dumping boards into the slips or spaces between the wharves. By the late 1870s the slips saw increased siltation, averaging three to four feet per year, affecting their depth. Solid waste also posed a nuisance to incoming and outgoing vessels and to businesses along the waterfront. Steam-powered ships tipped their cinders, ash, and other discards into the harbor, and ballast from sailing ships found its way into port waters.[21]

Despite its potential environmental drawbacks, fill was one of the most important mechanisms for dealing with a significant portion of waste materials in preindustrial and early industrial New York City, especially if it was largely composed of ash and construction materials. New land was created sporadically as fill became available. The simplest way to hold the landfill in place was to use wooden-plank bulkheads. A better solution was a cobb wharf, a large wooden wharf made of log frames filled with stones or cobbles (thus "cobb") sunk into place. In other cases the hulls of abandoned ships sunk with sand and granite offered support for holding fill. It nevertheless took a good deal of time for the fill to settle and allow for structures to be built on the new land. By the nineteenth century, water lots and marsh filling added 137 acres of land to Lower Manhattan. Modifications to Brooklyn's waterfront proceeded more slowly than Manhattan's but experienced extensive modification.[22]

Over time, fill was deposited in bands around the southern tip of Manhattan Island. In the early nineteenth century, land was added in the East River to make room for port facilities. Waterfront streets such as Water Street on the East River and Greenwich Street on the Hudson River are now five hundred feet from the water. Five Points and Washington Square in Lower Manhattan are well-known neighborhoods that were built on landfill.[23] Soundview Park in the Bronx, Marine Park in Brooklyn, and Great Kills Park on Staten Island are former marshlands. Market spoils and debris from demolished buildings and fires were utilized to build several blocks between Greenwich and West Streets and between Vesey and Liberty Streets. Most maritime-related fill was completed by the Civil War. Later fill was used to create municipal facilities, highways, parks, and industrial sites. In the late nineteenth century, the Brooklyn Ash Removal Company extended northern Queens seaward using Manhattan coal ash in Flushing Bay. Fill also moved Riverside Park westward into the Hudson River. Landfilling operations were conducted in every borough and along much of the collective shoreline.[24]

DIVERTING WASTE FROM THE HARBOR

City leaders made attempts to divert refuse from the harbor as a means of limiting obstructions and pollution. In 1857, the Board of Pilot Commissioners chose Oyster Island (now Ellis and Liberty Islands) as the city's first official dumping ground. This spot saw continuous use for fifteen years. As the population of the city expanded and migration northward in Manhattan quickened, available inland disposal sites shrank even more. Despite various efforts, the refuse problem reached crisis levels by the 1870s. Municipal officials had accomplished little to protect maritime operations or to improve collection and disposal practices. Dumping at sea was more active than ever. In the 1870s, a U.S. Coastal Survey showed that water depth in the harbor had changed significantly between 1855 and 1871. The legislature attempted to curtail the tossing of wastes into the East and Hudson Rivers, Upper New York Bay, and parts of Raritan Bay. (Refuse could be dumped into Lower New York Bay south of the Narrows, affecting the Staten Island shoreline communities and beaches, if a party obtained a permit.) In July 1872 legal dumping grounds were moved from Oyster Island to an area southeast of Staten Island. The shore inspector was given more power to enforce the new laws.[25]

Another site, privately run, had been established in 1849 on South Brother Island, between the Bronx and Queens and west of Rikers Island. It was located near North Brother Island, which was the home to hospitals serving patients with serious contagious diseases such as tuberculosis, typhoid fever, and smallpox. For more than twenty years beginning in 1916, Typhoid Mary (Mary Mallon) was quarantined there.[26] The druggist and city inspector Alfred W. White wanted to remove garbage from the city because of a cholera epidemic. He set up a company (as a secret, silent partner because the venture was in direct conflict with his city job) to take animal feces, bones, carcasses, and offal for resale to the island. There the castoffs were sorted, some resold, others rendered (converting waste into saleable byproducts) or fed to hogs. The company, Baxter, Brady, Lent & Company, got the materials from several sources. It bought bones, offal, and carcasses (now banned from the city) from the noxious trades (for example, slaughterhouses). It purchased horse and cow carcasses at the city inspector's pier, contracted for hotel kitchen waste, and bought other items where it could. After collecting these rejects, scows delivered them within several hours to the island.

This kind of scavenging quickly became a profitable business. Disputes among the partners and constant complaints from neighbors about the stench and floating debris in the river undermined its success. The island was a considerable distance from the harbor and the core of Manhattan's population, but it was only half a mile from the Bronx's south shore, where a few country estates were scattered about; it also was close to villages in northern Queens. The Queens-shore Board of Health issued a resolution claiming that the practice was a health hazard. As an offshoot of the furor, the company shut down its operations on South Brother Island. A new group with some of the original partners from Baxter, Brady, Lent & Company (including White) began a new venture away from the East River. They found a location along the southern shore of Brooklyn, in the mouth of Jamaica Bay. The new site at Barren Island, five miles from Canarsie and one mile from Coney Island, had an uploading dock, a factory building, access to workers, and scores of hogs. In 1876 the *Brooklyn Union* referred to it as "a place of swamps, some hills, fertilizing factories, and bad odors."[27] White's new operation opened soon after work ceased at the previous site. The facility, isolated from the Barren Island population, had the largest concentration of offal businesses in the world and became a significant disposal business (of a sort) in New York City.[28]

TROUBLES OVER OCEAN DUMPING

In the late nineteenth century New York City was dealing with its refuse disposal in a piecemeal fashion and continued to rely on practices that mostly were carried out in the private sector or under public contract: fill operations, rendering plants, hog feeding, some recycling and resale, and, most harmfully, ocean dumping. Illegal dumping added to the navigational problems of refuse in the harbor and other watercourses; waste drifting onto beaches was a serious nuisance that frustrated homeowners, businessmen, and bathers. From 1876 to 1880, the first so-called Garbage Wars took place, set off by ocean-dumping fiascos. In the summer of 1876, several employees of the city's street-cleaning bureau were arrested because scows in the vicinity of a disposal site near Sandy Hook were dumping waste on the incoming tide. This action caused garbage and trash to wash up on nearby beaches on Staten Island and in New Jersey. The relatively minor incident touched off a major controversy over disposal. Not only were Manhattan scows tipping at proscribed dumping grounds, but other boroughs such as Brooklyn were discarding refuse at sea with few hindrances.

Without any coordination or common plan, the various municipalities squabbled over one another's dumping practices, leading to a flurry of injunctions and restraining orders. While New York City considered alternative

FIGURE 2.1 Scow being unloaded by hand off Sandy Hook (1898).

Source: Department of Sanitation, Office of Engineering, "Refuse Disposal in New York City," February 3, 1954.

practices, the street-cleaning bureau (then under the New York City Board of Police) decided to empty their scows in Long Island Sound beyond the jurisdiction of the shore inspector. The legislature put an end to the practice not only because of complaints from the shoreline residents but also because of what the dumping was doing to oyster and clam beds. In the summer of 1877 the shore inspector began issuing permits for scows moving through the Narrows as long as they agreed to dump during ebb tide. From time to time the dumping grounds were moved further from the cities. But refuse habitually washed up on the beaches each summer, and cries for change arose each time. Governor Alonzo B. Cornell intervened, ordering the scows to dump further out to sea. This was hardly a permanent solution, but the action temporarily quelled the immediate tensions.[29]

In the mid- to late 1880s the city owned twenty-three deck scows and contracted an additional ten "self-dumpers," which were loaded at the dump boards and towed out to sea. At the time, there were several dumping grounds, the most important of which was about six miles south-southeast of Coney Island Point.[30] While collection and disposal statistics were not always reliable at the time, quantities collected and disposed by the city (which excluded amounts from contract work) climbed steadily in the late nineteenth century by almost any measure. An 1892 report noted the general composition of the refuse collected as about 62 percent ashes, another 12 percent garbage, 24 percent street sweepings, and 2 percent ice and snow.[31]

Fairly good comparative statistics are available for the 1880s. In 1882, the city collected 818,957 cartloads of ashes and garbage (or 1,228,436 cubic yards) and 269,922 cartloads of street sweepings (or 404,833 cubic yards). Of that number, about 67.6 percent was dumped at sea; about 29 percent was used as fill behind bulkheads or in lots. In 1887, the city collected 1,044,895 cartloads of ashes and garbage (1,561,342 cubic yards) and 311,169 cartloads (466,754 cubic yards) of street sweepings (statistics for snow were kept separately). Of that amount almost 75 percent was dumped at sea, and only 18.6 percent was used as fill behind bulkheads or in lots. These percentages oscillated through the late nineteenth century, depending on the ratio of dumping to filling and the amount of materials recycled.[32]

In 1885, the mayor of Kansas City queried about Gotham's solid waste collection and disposal "method[s] in vogue." The *Sanitary Engineer* reported, "The city has not as yet attempted to institute any system of burning garbage or of disposal, other than conveyance out to sea, and dumping in deep water, as the law requires.... By law [the scows] cannot dump their contents in any part of the harbor of New York, but must go out to the open sea—a provision which, it is common rumor, is often evaded in stormy and foggy weather."[33] Yet, according to Commissioner of Street Cleaning J. S. Coleman, speaking about the city's disposal practice in 1889, there is, probably, no city in Europe or America that possesses such admirable facilities, in an economical sense, for disposing of ashes, house waste, street dirt, etc., as New York. Twenty-five miles of towage brings us to a point so far seaward that when the refuse is dumped, as it is now done, twice a day, from three and a half to six miles from the nearest of land and with from three to five hours outward flow of the tide, the offensive drift is carried to the northeast beyond the influence of the next return tide.[34]

Notwithstanding Coleman's optimism about his disposal practices, city leaders and others looked to the state, and especially to the federal government, to deal with the predicament of harbor pollution and the cost of ocean dumping. These were problems affecting economics as well as navigation (primarily a federal responsibility) and sanitation (primarily local). Between 1878 and 1888, the cost of disposing refuse had doubled for the city. The creation of a new federal office of the supervisor of the harbor seemed promising, but it never received the authority to do anything significant about the dumping problem. The city had to face the prospect of finding its own solution.[35]

By the late nineteenth century, New York City had tried many available disposal options, including dumping waste on land and in water, using it as fill or as animal feed, and selling waste as fertilizer to farmers. All had come up wanting in some fashion or did not suit the needs of the island city's vast population.[36] Incineration, one of the newer (and potentially most promising) disposal technologies, began to receive attention in the United States in the 1880s. Industrialized Great Britain led the way because of its limited open space for land dumping and constraints on sea dumping given English maritime interests and potential conflicts with neighbors. Also, Great Britain's population was sufficiently dense to make a centralized system of disposal potentially economical. Between the late 1860s and 1910s, the so-called British destructor (a furnace or incinerator used for burning refuse) went through three stages of development, starting with low-temperature, slow-combustion units; then moving to destructors that operated at higher temperatures, capable of producing steam for various purposes; and finally destructors providing power for generating electricity or for pumping liquids.

First-generation incinerators were designed simply to burn waste. In about 1870 the earliest municipal furnace was erected in a suburb of London, but it produced disagreeable smoke and operated poorly. In 1874–1875, a series of destructors were built in Manchester. The poor operation of this first-generation technology undermined confidence in cremation. By 1885 efforts to reduce the smoke led to technical adjustments, but the cost of retrofitting destructors remained high. Newer destructors in the late 1880s and 1890s generated higher temperatures, reduced smoke, and produced steam for usable power. The transfer of incineration technology to the United States raised hopes but also caused problems. Beginning in 1885, some communities tried low-temperature furnaces. After the turn of the century, companies built high-temperature destructors with adjacent boilers and power-generating capability.

Both failed to catch on: they suffered from the same weaknesses the British had experienced and from a lack of understanding about the different physical, economic, and social environment into which the new technology was being introduced. Many American-produced furnaces were inferior in construction; the United States had more cheap land for dumping than in Europe; it had abundant fuels, which made long hauls of waste less expensive; and American refuse normally had a higher water content than refuse in Europe, which meant that the destructors ran less efficiently.[37] The first American commercial-grade incinerator was built on Governors Island in 1885, and the first municipal furnace was built in 1886 in Allegheny City, Pennsylvania. By the 1890s, cremators were popping up all across the country. But for the reasons just mentioned, plus the generation of deleterious smoke, incomplete combustion, and excessive use of external fuels, the equipment never became popular in the United States. Of the 180 furnaces erected between 1885 and 1908, 102 were abandoned or dismantled by 1909.[38]

Incineration was not the answer to turn-of-the-century disposal woes in New York City, but it would prove to be the focus of many public battles in Gotham for years to come. Manhattan Refuse Cremating Company built a small blast furnace on Front Street in the late 1870s (predating the incinerator on Governors Island), but this experimental cremator was too small to compete with ocean dumping.[39] Commissioner Coleman considered cremation in the late 1880s but remained convinced that ocean dumping was more economical and sanitary.[40] Bay and ocean waters were too vast to lose out to this fledgling technology, even before the limitations of cremators were demonstrated elsewhere.

Another possibility, one that seemed to offer a real alternative to ocean dumping, was the plan to enlarge Rikers Island through the use of city garbage. In February 1893, Mayor Thomas Gilroy announced a plan to expand Rikers Island from eighty-eight to four hundred acres by this means. The New York Department of Public Charities and Correction bought the island in 1884 for the purposes of eventually building a new prison there. Residents in Queens, the Bronx, and Manhattan, who were less than one mile away, wanted nothing to do with the scheme to enlarge the island with dumped refuse. They feared the heightened stench from the large-scale operation would wreak havoc. Nevertheless, operations began later that year. The Woolf Electric Disinfecting Company operated a plant there to yield sodium hypochlorite via electrolysis (substances broken down into simpler substances as an electric current is passed through them).[41] The solution produced was then sprayed over the refuse to deodorize it. This attempt to mitigate the odors was not completely successful. Complaints mounted, and charges of unsafe health conditions because of the new disposal practice multiplied. A temporary restraining order briefly stopped operations, but the court vacated the injunction because the city promised to seek an alternative dumping strategy. Ultimately, the protests and the cost of the operations resulted in only ash and rubbish being dumped on Rikers Island and the ocean still used for swallowing garbage.[42] No new disposal panacea was appearing on the horizon.

ENOUGH IS ENOUGH

In 1894, the physician, lawyer, and author Henry Smith Williams attempted to explain why New York City (meaning Manhattan) "is not the cleanest city in the world," although he believed it ought to be. He argued that its location on a narrow strip of land "with tide-swept waters on either side"; its "natural watershed" running almost the whole length of the island; "its highest portions are northerly, from whence the water supply must come"; and the municipality's "enormous sum" devoted to sanitation were the features that should have made the city "what it is not." Reasons for its unsanitary lapses, he added, were a result of citizen neglect, companies tearing up streets and giving little attention to cleanliness, municipal authorities that built "uncleanable" road beds (and neglected other infrastructure), and Tammany Hall's (the local political machine) dominance of the Department of Street Cleaning.[43] In several respects, Williams's broadside aptly presaged the appointment of Colonel Waring.

Keeping the streets clean was an understandable obsession for a city so dependent on unobstructed transit routes and traffic lanes. Disposal of waste (or final disposition, as it was often called) was more confounding and far more significant in the big picture of things than other urban shortcomings. Sanitarians certainly recognized waste disposal as a universal problem as industrialization accelerated New York City's economic and population growth. "The removal of dirt and filth from a great city," one observer affirmed, "is just as essential as the cleansing of the pores of a man's skin."[44] Williams stated that "it is quite in order that New York should be grappling with the garbage problem at this time, for almost every other large city in the civilized world is in a similar predicament."[45] Cleanliness meant order and health and even social and cultural advancement.

Where Williams insisted that "New York ought to be the cleanest city in the world" because of its greatness,[46] others saw the city's size, its complexity, its exploding population, and its seamy politics as hindrances to cleanliness. Sanitarians such as Dr. Thomas H. Manley and Dr. Douglas H. Stewart were disgusted and frustrated by the city's failed attempts at improving the disposal of household garbage. Despite "considerable discussion" of the waste problem over the years, they saw "nothing further than some unimportant, abortive changes" having been tried, the results being that "the vast, unwieldy nuisance continues; its underlying, fundamental defects still untouched; our city's scavenging yet remaining far behind that of all the principal cities of Europe, and, indeed, it may be said, with all truth, it in no sense is equal to that of many of our rival cities of the West." They concluded that "the conditions of things demand a radical revolution of the whole system."[47]

Easy solutions were hard to come by. With some resignation, Manley and Stewart concluded, "Remember, this garbage has been the Waterloo of hardworking, conscientious men with good intentions; who have come, have seen, and have been utterly conquered by it. The trouble must certainly lie somewhere in the system, not the men."[48] The 1893 report of the Department of Street Cleaning was quick to note that the lack of adequate facilities, not the poor performance of its workers, hampered their success. "Final disposition of the city refuse is the most difficult problem to be solved. No adequate provision for it has ever existed."[49]

Yet there was plenty of additional blame to go around: the lack of public enthusiasm, the venality of machine politics that sold favors for votes, and the problem of a large population consuming (some wastefully) vast amounts of food and available goods. Williams was quick to blame the garbage problem on "the extravagance and wastefulness of servants," failing to make a distinction among the classes as to what were necessities for some and overindulgences for others.[50] For their part, Manley and Stewart made a sweeping and overgeneralized indictment that Americans were acquiring "habits of reckless extravagance" over the recent years of growth and prosperity. "From extravagance we have drifted into indolent habits, and through sheer carelessness and indifference have permitted the waste to go on unrestrained."[51] Although Williams, Manley, and Stewart made no exceptions for the needy, the poor, or the working classes in their reproaches, they did recognize a connection between consumption and waste. What they recognized as overconsumption, however, was in fact an insoluble dilemma of what to do with the discards of a materially productive society.

No little reason for the mounting disposal problem was the increasing scale of waste and its changing composition—issues clearly related to consumption patterns and population. As long as the bulk of discards was ash and other materials essential to "clean fill," the creation of new land absorbed much of the refuse. In the opening years of the 1890s, ashes still dominated New York's discards. One study stated that 62 percent of the material deposited at local dumps (not all waste was collected, collected efficiently, or placed in dumps) was ashes. Garbage made up 12 percent, street sweepings another 24 percent, and snow and ice 2 percent. In those same years, estimates suggest that New Yorkers were burning about 700,000 tons of coal in the city annually, resulting in residue of about 17 percent of the total weight (1,190,000 tons).[52]

The times were changing as increased electrical-power generation moved the burning of coal from most inner cities to the urban periphery and as gas and kerosene competed with coal for home and business heating. Residue persisted, to be sure, but was largely replaced by smoke and air pollution. However, as ash and clinker from coal and other heat sources declined, the volume of organic material was on the rise, especially in the spring and summer. Foodstuffs, fruits, and vegetables entered the urban market on a scale never seen before.[53] Restaurant discards added greatly to those produced in the home.

Some major types of organic waste, such as horse manure, declined because of the phasing in of mechanical and electric sources of motive power and the phasing out of animal power.[54] During their heyday in the nineteenth and early twentieth centuries, 120,000 horses resided in New York City. Since the average, healthy horse produced between fifteen and thirty-five pounds of manure and one quart of urine daily, the cleanup difficulties were profound, with horse carcasses only adding to the problem. In 1880 New York City scavengers removed 15,000 dead horses from the streets.[55] While street cleaners applauded the demise of the urban horse as a source of waste, other types of refuse did not go away, and new materials entered the waste stream in large volumes. Human excrement, more commonly called "night soil," had to be dealt with constantly. Night-soil collection commenced in the city in 1887.[56]

As industrialization accelerated, the United States was becoming a new kind of consumer culture. Handcrafts and homespun clothes made way for factory goods, from garments to cleaning products, from toothpaste to boxed cereal. While the population doubled between 1880 and 1910, industry produced nine times as much paper. Among the poor and working classes, time to make handcrafts was limited and land for private gardens scarce. Accompanying mass-production techniques was mass distribution, making access to a wide range of products and packaging easier for urbanites. Frequenting the more and different shops and stores was matched by scouring the pages of Sears and Montgomery Ward catalogues.[57]

Extra stuff combined with limited places to store it meant the need to discard what was not essential to keep. Those with few resources still tried to, or were forced to, be thrifty and reuse what they could. Circumstances were making it harder to do so. For those on the rise financially in the middle and upper classes, access to diverse merchandise provided opportunities for acquiring belongings difficult to resist. Critics of new immigrants were quick to criticize their "uncivilized" practices and lack of cleanliness—and, by implication, their wastefulness. For example, an article in an 1891 issue of *Popular Science Monthly* asserted, "In those cities and parts of cities where the people of the laboring class and the poor are crowded in tenement-houses and where a considerable part of the population is foreign-born and from countries where personal and public cleanliness have not been enforced by proper police regulations, it is no trifling task to secure cleanliness of the streets."[58]

It was easy to see the difference in living conditions between countless immigrants stuffed into grimy and insalubrious tenements and the relatively clean and orderly neighborhoods of the "respectable classes." It also was easy to blame the poor for the waste problem because of where they lived.[59] In truth, uptown restaurants and hotels serving the well-to-do added copiously to the waste stream without so much as a second thought. More production begat more consumption, which begat more waste.[60] Some items in the waste stream were new, others more plentiful than in earlier eras: wastepaper, discarded clothing, old shoes and other leather, broken pottery, rags, glass, and more.[61] Items normally recycled, such as animal and vegetable fats, now ended up in the dustbin, to be replaced with newer products.[62]

Having to contend with new types of refuse (new glass and metal containers, petroleum products, and so forth) raised many questions about disposal strategies, especially regarding what should be dumped and what should be salvaged. Such questions brought to bear debates over methods, economic issues, and sheer convenience. The generation of refuse was relentless and cumulative. Ignoring the daily requirements of collection and disposal compounded a chronic problem made increasingly difficult to ignore. Estimates varied on the amounts of waste produced by New Yorkers in any given day near the end of the nineteenth century. The Department of Street Cleaning estimated that in 1892 it collected on average 6,137 loads of ashes, garbage, and sweepings daily. "This year [1893] the quantity will be greater," the report noted. The material was collected along 592 miles of streets, paved and unpaved, then hauled to dumps a distance averaging one mile for each load. The refuse then was carried by scow (as many as one hundred) and ditched at sea or on other dumping grounds.[63] To miss just a few days of collection and/or disposal was to court bedlam.

Soper noted that "for the first ten years [since 1881], the department of street cleaning did little to justify its existence, and by 1891 the operation of the department had become entirely unsatisfactory." In January 1891, Mayor Hugh J. Grant appointed a five-man committee headed by the businessman Morris K. Jesup to study the operations of the Department of Street Cleaning. The Jesup Committee found the department inefficiently managed and called for its reorganization, recommending an increase in appropriations. The report was the foundation upon which changes were to be made later in the 1890s.[64] The committee's suggestion to continue ocean dumping as "the only practicable method" of solid waste disposal was not surprising. The justification for such a conclusion was based purely on cost, and the committee was skeptical about the ability to recoup expenses by selling ashes, street sweepings, and garbage. Landfilling was acceptable, but space for dumps was dear. Incineration had yet to prove successful.[65] Could New York ever hope to move beyond the resignation of the Jesup Committee?

THE APOSTLE OF CLEANLINESS

Growing public resentment of Tammany Hall (fueled in part by the Lexow Committee report exposing corruption and brutality among the city police) propelled Mayor William L. Strong, a reformist Fusion Party candidate, into office in 1894.[66] Tammany was notorious for using government bodies like the Department of Street Cleaning for patronage and payoffs. For example, the cost per capita for street cleaning varied widely across the country, with little or no correlation with the cleanliness of streets. One survey indicated that in the 1880s, citizens of Buffalo, New York, paid an average of 5 cents per year for street cleaning; New York City residents paid 71 cents per year. Translated into the cost of street cleaning per mile, people in Buffalo paid $34; New York City residents paid a staggering $1,870.[67] Graft and corruption played a large part in this disparity. In an 1895 article in *Harper's Weekly*, F. W. Hewes echoed other criticisms of Tammany Hall's mishandling of street-cleaning funds. In Hewes's survey, other cities spent less than one-fifth what New York did, and the streets of New York were still deplorable.[68]

The new mayor wasted little time in addressing the problems in the police force by appointing the rising Republican star Theodore Roosevelt to the Board of Police Commissioners, a position that later propelled him into the governorship of the state.[69] But also on Strong's mind was fixing the street-cleaning department. As Roosevelt later stated in his autobiography, "Mayor Strong had already offered me the Street-Cleaning Department.... For this work I did not feel that I had any especial fitness. I resolutely refused to accept the position, and the Mayor ultimately got a far better man for his purpose in Colonel George E. Waring [Jr.]."[70] Roosevelt's sentiments were repeated by other reform-minded leaders, who saw Waring's appointment as a first step to solving the city's monumental solid waste problem.[71]

With the endorsement of the Street-Cleaning Society, Strong offered Waring the post in January 1895. At the time, the new appointee was serving as assistant engineer for New Orleans.[72] Born in Pound Ridge, New York, in 1833, Waring became a student of the renowned agricultural scientist James J. Mapes. He spent his early career experimenting with farming techniques and developed skills as a drainage engineer. In 1855 he managed Horace Greeley's farm near Chappaqua, New York, and then accepted a similar post on the noted landscape architect Frederick Law Olmsted's farm on Staten Island in 1857. Waring's connection with Olmsted secured him a position that same year on the Central Park project as drainage engineer.

His moniker "colonel" came from his army service during the Civil War, receiving the title with the Fourth Missouri Cavalry. In 1867, he became the manager of Ogden Farm near Newport, Rhode Island. In the late 1860s and early 1870s Waring turned his attention to drainage and sewerage engineering.

He developed the "separate system," or "Waring system," which diverted raw sewage and household water into small pipes and then combined these sources with runoff in larger pipes or surface drainage ditches. Although it was a source of controversy in battles over the best available sewerage approaches, he achieved great notoriety with a system he built for Memphis, Tennessee, in 1880. The colonel promoted his separate system and other collection and disposal plans and devices nationally, gaining a reputation as an innovator but also as a self-promoter. He was appointed to a special commission of the short-lived National Board of Health (1879) and compiled social statistics of cities for the 10th U.S. Census (1880).

Throughout his career Waring was a major proponent of the "filth theory," or "miasmatic theory," which purported that diseases were noncontagious but emanated from smells (sewer gas) or decaying matter. There is some question as to whether he ultimately converted to (or began to appreciate) the germ theory (or contagionist theory) that became more popular in the 1880s and 1890s. Nonetheless, he was a great believer in effective sanitary practices as the key to public cleanliness—an idea that drove his thinking and practices in his new position in New York City.[73]

FIGURE 2.2 George E. Waring Jr., "Apostle of Cleanliness" (1897).

Source: Library of Congress, Prints & Photographs Online Catalog.

Waring recalled in 1896, "I met a friend of Mayor Strong at lunch in the autumn of 1894, and sent word to him that if he was looking for a commissioner of street cleaning I was at his service. He called me to New York and offered me the position. I told him that I would accept it on precisely the same conditions that I had indicated before, that I should have my own way."[74] This was the colonel to a tee. He could be flamboyant, always projecting a military bearing, and unflaggingly self-confident and stubborn. He was leery of partisan politics, believing "we are generally led by the nose" by political parties. Playing off Abraham Lincoln's famous quote about fooling the people, he stated, "The sad truth that underlies this shrewd statement is that the majority of the people can be fooled nearly all the time."[75] Waring also shared with Strong disdain for Tammany and all it represented. "I made it a condition of my acceptance," he later noted, "that I should be entirely exempt from interference and free of all political obligations."[76] Waring vowed to put "a man instead of a voter" behind each broom and to expel political cronies who did not accept his way of doing business.[77]

Despite regularly provoking Tammany's ire, in practice he retained approximately half of the current employees he inherited in the department. In one instance, Waring summoned the holdover superintendent in charge of snow removal into his office, stating that the employee had been accused of being a "rank Tammany man." "Whenever you want my resignation, it is at your service," the man replied. "Don't be quite so fast," the commissioner responded. "Let me hear your version of the case." The superintendent spoke up, "Do you know what a Tammany man is? It is a man who votes for his job. I have been a Tammany man, and a faithful one. I have worked for the organization; I have paid regular contributions to it. But I am a Waring man now." Detecting a smile on the colonel's face, he added, "Don't misunderstand me. If Tammany comes into power again, I shall be a Tammany man again." Impressed by the man's candor, Waring kept him on.[78] The new commissioner was sometimes exasperated by criticism, however. "I found that my suggestion that no political appointments would be made and that no political removals would be made was received as so much utter nonsense. I think it was then that they began to say that I was from Rhode Island and did not know what New York really was."[79]

As a major figure in the reformist government of the city, Waring was a regular target of reproach by political enemies. He sometimes was party to disagreements and controversies within the engineering profession because of his unorthodox methods. In April 1895 local critics charged him with conflict of interest, alleging that his engineering firm, Waring, Chapman, and Farquar, had connections with the Sanitary Security Company, which inspected and certified houses for sanitation compliance. (He was cleared of the charge.) Some politicos connected with Tammany accused him of increasing the cost of administering the department without increased benefits to the community. It did not help when Waring referred to them in the 1898 election as "a thievish and ruffianly gang of good fellows" and "a very bad lot."[80] In 1897 he fought back against Tammany forces beyond exchanging insults, bringing charges of malicious libel against Boss Richard Croker.[81] He nevertheless often helped make the case for his opponents and cause problems for the mayor, on one occasion referring to the Grand Army of the Republic as an aggregation of "pension bummers."[82]

Often photographed astride a well-groomed horse, dressed in riding gear and pith helmet, and sporting a waxed handlebar mustache, Waring gave the appearance of a parody of the military officer he actually had been during the war. When he assumed the commissionership, the press mocked his appointment as much for his image and bearing as for his actions. How Waring handled the street-cleaning corps offered much fodder for the newspapers. To impress upon the citizenry the sanitary service that his "soldiers of the public" provided, Waring dressed street cleaners in white as a symbol of health and cleanliness. His "White Wings" not only carried out their duties in this garb (although some resented the inconvenience) but participated in citywide parades to gain public attention.[83] The first employee parades in May 1896 comprised 1,400 sweepers, six hundred drivers, and twenty-three bands. And when, within his first year, the streets became much cleaner and collection of wastes improved in many neighborhoods, the derision turned to lavish praise.[84] An editorial in *Garden and Forest* proclaimed, "[The parade] was an inspiring sight when it was recalled through what a storm of distrust and abuse Colonel Waring had to make his way when he proposed to turn politics out of his department, and especially when he put his men in uniforms—and white uniforms at that."[85]

A master at harnessing human resources, Waring often enlisted the public in his efforts at sanitary reform. After a year and a half in office, he noted,

> I believe that if the people were once interested in [the refuse problem], and it were known that it was possible to disregard the ideas of the politicians, they would show such a desire for reform in this particular, that political influence would have no weight against them. This was the reason I did so many things that were considered "injudicious," "dramatic," and perhaps, even foolish. My plan was to force the department on to the attention of the people in every possible way, and I knew that the easiest way to do this was to introduce the personal element, and to make myself as Commissioner as conspicuous as I could.[86]

His most arresting effort at civic involvement was the formation of the Juvenile Street Cleaning League. Initially, more than five hundred youngsters participated in this program to disseminate information about proper sanitation and to inspire community participation in keeping the streets clean. Waring hoped that the children, especially from "the ignorant populations in some East Side districts," would set an example for their "less enlightened" parents. Despite the dubious class and ethnic connotations of the plan, Waring's ultimate goal was to spread knowledge about proper sanitation and foster a sense of personal commitment to city cleanliness. The leagues eventually spread to other cities across the country, with varying degrees of success.[87]

Public relations was one thing, but Waring brought real change to the way New York City dealt with its wastes after years of poor management and ineffective collection and disposal. He reorganized the Department of Street Cleaning by bringing in graduates from technical schools or men with military background for supervisory positions. "Knowing that organizations of men are good or bad according to the way in which they are handled," he stated, "that 'a good colonel makes a good regiment,' I paid attention first to those at the top—to the colonels."[88] He viewed the street sweepers and other workers as links between the public and the department. The white uniforms were the outward display of this sentiment. Waring favored a top-down approach and demanded strict discipline as a mark of efficiency.[89] Initially he supported pay cuts to economize the work of the department but encountered serious resistance among the employees. The potential political fallout of an austerity effort put on the backs of the workers moved him toward more accommodating labor relations. He instituted an arbitration system, established an eight-hour work day, and set salaries to sixty dollars per month (almost double what unskilled labor then received). Ever the promoter, he turned the possible liability of a labor fight into a positive by regularly touting the virtues of his workforce, who were instead "fighting daily battles with dirt."[90]

Change came quickly in street cleaning and collection of refuse. The street-sweeping program was regularized along 433 miles of paved streets in the city, relying on 60 percent of his work force (1,450 sweepers) and absorbing 40 percent of his budget. He also revised the snow-clearing program. In the area of collection he consolidated some activities and innovated in others. His goal was to collect household waste quickly and efficiently, to recover whatever economic value the discarded material might have, and to dispose of the rest. His most notable achievement was "primary separation" (or, more generically, "source separation"), an activity advocated for years but never attempted. Each householder and business was required to keep ashes, rubbish, and garbage (organic waste) in separate containers for pickup. Ideally, the three basic types of waste were to be treated differently: ashes could still be employed as fill, rubbish sorted for resalable or reusable items, and garbage utilized as swine feed or other potential products. Whatever remained could be discarded. This ambitious system had some successes, especially in reducing the amount of solid waste that ultimately had to be disposed.[91]

FIGURE 2.3 Waring's sanitation workers.

Source: NYC, Department of Sanitation, New York City Recycling-Context, August 2001.

WARING AND FINAL DISPOSITION

Above all Waring advocated for the strict public control of collection and disposal of solid waste, an outlook that flew in the face of decades of past practices. In an 1895 article he stated stridently, "The city should, in short, assert its right to an absolute monopoly of the garbage business, for all garbage is a nuisance unless brought under proper control. Such control cannot be exercised by the city unless it takes possession of the entire field."[92] Such a perspective framed his view that dividing responsibility for collection and disposal between private and public entities ensured failure, bred inefficiency, and reduced the possibility for a city to recoup some of the cost of its services. He argued that "public authorities might with advantage take control of the whole business of the collection of rubbish. This would probably be necessary to the securing of the great pecuniary return." He viewed pushcart men as carrying on "a more or less illicit traffic with domestic servants" and sought to take over the carting trade itself.[93] Under Waring, the city took charge of scow trimming, undercutting the private system in place, and in January 1898 the department established the first American rubbish-sorting plant.[94]

Waring's disposal plan depended on his ability to remove as much as he could from the waste stream before final disposition. This was a strategy built less on health concerns and more on the potential economic benefit to the city and the efficiency of his overall operation. He experimented with several methods, dismissing some, and he made real (if temporary) progress in reducing the amounts of waste New York City dumped into the sea. In January 1896 (possibly earlier), however, it was clear that Mayor Strong was growing impatient with Waring's progress on granting private garbage contracts where needed and his seeming caution in devising a new plan for final disposition of wastes. Strong may have believed that Waring was too focused on public relations and not engaged enough in the practical problems that collection and disposal posed.[95] In his 1896 report to the mayor, Waring stated that his department had conducted "an active investigation and consideration" of questions related to final disposition of waste. "There are no two opinions," he added, "as to the barbarism of the practice of depositing these matters in the sea—a practice that is both wasteful and pernicious. Thus far, however, no better means has been found for getting rid of them."[96] (On another occasion he noted the "very serious menace" of the fouling of the beaches of New York, Brooklyn, and Jersey City because of sea dumping.)[97] One change was to use newer scows less likely to strew their cargo across the waters, but this was only a modest adjustment.[98]

Incineration was an option that others had embraced. But for Colonel Waring, this method offered little benefit, most especially in the case of garbage disposal. At the time, cremating waste materials took place in only a very few locations in New York City, and for specific purposes such as dealing with light rubbish.[99] He objected to the practice because it was costly and wasteful; that is, it provided no opportunity to reuse or recycle the discards. He conceded that newer models "were far less objectionable than they formerly were," in the sense that they polluted the air less. In his inimical elitist fashion, he added that "the older furnaces were usually located in the poorer parts of the town, and among the population not especially squeamish about foul odors," while the newer, improved ones were sometimes built in the neighborhood of "the better class" and proved less objectionable.[100] Beyond this questionable explanation, he was concerned that some investigations showed that even after cremation, the resulting ash included organic residue or other unburned materials. He raised additional concerns, especially about the functioning of the labor force at the units, but his final assessment came down to "cremation means destruction and loss of matter which may be converted into a source of revenue." He concluded: "With a view to economy, our attention in New York was early given to the rival process of utilization."[101]

In line with the commitment to utilization, Waring decided to continue the fill program at Rikers Island. Complaints about the foul smells coming from the island had not subsided. The stench of rotting garbage disgusted everyone close enough to experience the prevailing winds. Under the colonel's direction, a bulkhead was constructed around a shoal at the island in the East River.

Dumping scows regularly dropped loads of ashes and street sweepings there. But after the source-separation program was in place, only inorganic material made its way into the bulkhead, and the odors decreased.[102] The city also provided fill material at no cost to private owners of shore flats and participated in experiments to turn ashes and organic materials into fireproofing blocks.[103]

The work at Rikers Island did not address Waring's wish to find an effective utilization process to handle garbage and other organic material. The answer was reduction. The technology grew out of traditional agricultural reuse practices, but it was a method best suited to large cities that could afford it.[104] Originating in Europe (especially England and Germany), reduction was introduced into the United States in 1886 by a company in Buffalo, New York. It employed the so-called Vienna or Merz process for extracting oils from city garbage to manufacture fertilizer and other byproducts such as fragrance base. Organic waste was "cooked" with the water content vaporized, and the remaining material (about 35 percent of the original bulk) was placed in closed steel tanks flooded with naphtha. The resulting mass was pressed to extract grease and oils, and the residue, called tankage, dried for fertilizer. Inventors fashioned various forms of the process, each with slightly different approaches to extraction. The reduction method was intended to provide cities with salable materials, which would offset part of the cost of disposition.[105]

Appearing in the United States about the same time as incineration, reduction underwent a similar early development: impulsive implementation, criticism, and reevaluation. But several engineers and some city officials saw promise in the method as a way to return revenue to the city. After a period of operation, undesirable side effects (especially escaping foul odors) led to rising censure and protests. At the meeting of the League of American Municipalities in 1898, the New Orleans councilman Quitman Kohnke complained, "We have been seduced by the glowing promises of rich rewards which the reduction process has failed to give us."[106] Plants built on the Merz design in the 1880s in Milwaukee, Saint Paul, Chicago, Denver, and elsewhere were proving to be failures. Newer plants built in the 1890s were faring little better. By 1914 only twenty-one of the forty-five reduction plants in the country were still in use.[107]

Despite the spotty results, and before the extent of the failures were known, Waring was convinced that reduction was what New York City needed. The department began tests on reduction technologies in August 1895, and in June 1896 the Board of Estimate awarded a five-year contract to the New-York Sanitary Utilization Company (NYSUC, based in Philadelphia) for reduction facilities to be located on Barren Island.[108] The NYSUC built the first works in 1896, adding new buildings as more garbage arrived from Manhattan, Brooklyn, and the Bronx. Employing some five hundred men (Irish, Italians, Poles, and African Americans), it became the largest plant of its kind in the world, with a capacity of three thousand tons per day.[109] "With a view to economy," Waring later stated, "our attention in New York was early given to the rival processes of utilization."[110] For him, "The rubbish had hardly enough value to pay its separation, and the water has no value at all." But to eliminate these waste products and to have saleable items remaining made reduction worthwhile in his mind. In saying this, he all but dismissed (or ignored) the impact of the escaping fumes and the "dark-colored caramel refuse" discharged into Jamaica Bay.[111] The eight hundred tons of garbage made into fertilizer and oil daily by 1898 was Waring's major interest.[112]

Reforms to the disposal program in the Department of Street Cleaning focused on utilization and attempted to wean the city off of ocean dumping. As Soper concluded, Waring "accomplished, at least, a beginning."[113] Statistics in 1899 show that Waring achieved only modest changes in the area of final disposal. The total number of cartloads of garbage, ashes, and street sweepings collected by the department had steadily increased. In that year the total was 1,325,107, with private carters collecting another 424,119. Of the total collected (and this is a little skewed because it includes cartloads from the private carters), the city still was dumping 58.8 percent. Barren Island took in only about 10 percent, with the remainder dumped in lots or at sites in other parts of the city. Some of the refuse collected by private cartmen ended up in New Jersey.[114]

CONSOLIDATION

The Strong administration survived only one term, and with its demise a turn back to Tammany was in the offing. The mayor had become increasingly isolated. He faced constant pressure from machine politicians, had limited patronage to dole out, and alienated reform Democrats by supporting anti-Catholic school proposals and insisting on the enforcement of Sunday blue laws. He chose not to run for reelection in 1897.[115] Boss Croker then "nominated" as the Democratic standard-bearer Robert A. Van Wyck, an obscure Tammany judge, who became the first mayor of Greater New York City.[116] Waring may have had aspirations as mayor, but the Citizens Union (which ran the reformer Seth Low against the Democrat Van Wyck) passed on that opportunity because of the colonel's negative standing among labor groups.[117]

Consolidation of New York City's boroughs in 1898 brought with it the beginning of a new political configuration and administration that would influence service delivery of all kinds, including refuse collection and disposal. On a terribly rainy day, January 1, 1898, unifying an array of cities and neighborhoods in the Greater New York area became a reality. Some forty separate local governments in Manhattan, Brooklyn, Queens, the Bronx, and Staten Island (ninety-six governmental units in all) were combined into a single urban giant comprising 360 square miles, three thousand miles of streets, and 3.2 million people. Gotham was now the second-largest city in the world, behind only London.[118] Many regard the feat as the most notable municipal consolidation ever undertaken in North American history.[119]

Thoughts of consolidating New York City and Brooklyn began as early as the 1820s.[120] Actual movement toward merger started modestly in 1857, when New York State combined the police, fire, and health departments of Manhattan and Brooklyn.[121] Economics, always tinged with politics, drove much of the consolidation initiative in the late nineteenth century. At the core, mercantile and political elites in Manhattan saw the opportunity to enhance the commercial advantage of the city by improving and maintaining the port through unifying its control and its activities.[122] In general terms, those who favored consolidation believed that unification of the various entities would end needless competition among them, provide new political clout on the state level against the rural power base in the legislature, and boost leverage with municipal rivals outside of the state. A unified city also would open up vast economic opportunity, enhancing growth through an almost limitless borrowing capacity.[123]

Not everyone so easily fell in line with the proconsolidation vision. Some cities, especially large and influential Brooklyn, would lose autonomy, and schisms within various communities complicated the picture further. Brooklyn became "the hotbed of resistance" in the merger story.[124] Consolidation evolved into a clash between Manhattan and Brooklyn; other communities marked for merger were largely relegated to the fringes of the dispute.

The lawyer, planner, and civic leader Andrew Haswell Green is recognized as the "forgotten visionary" who shaped much of New York City in the late nineteenth century; he was the inspiration and spearhead for consolidation. Green was first and foremost a trailblazing proponent of urban (and regional) planning who favored the orderly development of the large metropolitan area through central control. He and others strongly believed that the city needed a planning strategy to accommodate the massive expansion of the population.[125] In 1890, the reform Democrat stated the destiny of New York City in lofty terms: "As the commercial capital of a great nation we have a great trust in charge. The temptations which might lure a community from the path of healthful progress were never so lavishly spread as upon our path. More than any other community we have had greatness thrust upon us."[126] This meant ending the ceaseless political conflicts among the various municipal entities strung along the harbor.[127]

After a number of setbacks, proconsolidation forces received a solid majority of 57.22 percent favoring merger in the November 1894 election.[128] They scored impressive victories in New York County (61.78 percent), Queens County (61.93 percent), Richmond County (Staten Island, 78.6 percent), Westchester (64.48 percent), and the Town of Pelham (62.13 percent). Eking out a razor-thin victory in Kings County (Brooklyn, 50.1 percent) was the biggest relief for those favoring the merger. In general, the wealthy and middle classes supported merger, as did places needing help with public works, such as Staten Island. The poorer classes were more skeptical and cast largely "against" votes. On Manhattan, Tammany strongholds voted against consolidation but were outflanked by upper- and middle-class districts and native-born groups. In Queens, the largest number of votes came from urban areas close to Manhattan. Since the referendum was worded so vaguely, it is difficult to know exactly why many voters favored or opposed the measure. The flush of victory, however, was short-lived, since turning the vote into a practical plan for a united city faced numerous hurdles.[129]

AFTER THE COLONEL

Colonel Waring's abrupt and shocking death in 1898 made him a martyr for the cause of sanitation and also ended his future participation in reforming New York in the new era of consolidation. Soon after Waring's retirement as commissioner in 1898, President William McKinley appointed him as special commissioner to investigate health conditions in Havana, Cuba, which had been occupied by the United States during the Spanish-American War. A treaty ending the war was signed in December. Before then, in the fall of 1898, McKinley feared that an epidemic of yellow fever could sweep the island unless something was done immediately. While in Havana on his mission, Waring contracted yellow fever and died on October 29, soon after returning to New York City.[130]

A memorial service was held at Cooper Union on November 22, attended by many prominent political and civic leaders, possibly as many as five thousand in all.[131] Albert Shaw, the editor of *Review of Reviews*, remembered Waring as "peculiarly the apostle of cleanliness, the scourge of dirt."[132] Others noted that he "awoke popular appreciation of the importance of sanitary science."[133] He also was described as "a born organizer"[134] and "frank, open, and outspoken" with "the courage of his opinions."[135] Most poignant of all was an observation in the *New York Times*: "What more cruel stroke could there be of the irony of fate than the death of George Waring from a filth-disease?"[136] The real irony was that Colonel Waring had been struck down by a *viral* disease carried by mosquitoes, a form of contagion that was obscured in his long-time endorsement of filth diseases as the primary effect of poor sanitation.

With Waring's death and the reemergence of Tammany rule, New York fell upon old habits very quickly. Van Wyck initially appointed James McCartney as commissioner of street cleaning, and he attempted carry out some of Waring's practices in the new and different political environment of postconsolidation New York. But McCartney died in February 1900 and was replaced by a machine hack, Percival E. Nagle, a real estate developer and contractor.[137] The total percentage increase for the street-cleaning department for the three years of restored Tammany rule was 24 percent greater than in the previous administration. A swing toward reform occurred again in 1902 with the election of another Fusion reform candidate, Seth Low. He appointed the military sanitarian Major John McGaw Woodbury to replace Nagle, and the new commissioner sought in earnest to restore much of the Waring program.[138]

The seesaw politics brought little stability to the Department of Street Cleaning, let alone solving disposal problems left unresolved under Waring. As one observer noted in *Municipal Affairs* in 1900, "A most vital part of the work of city cleaning, and particularly in a town so densely populated and at the same time surrounded by water, as is New York . . . is the removal and disposal of the wastes collected from the inhabitants."[139] An experimental rubbish-disposal plant on East Eighteenth Street was discontinued. Troubles existed on Rikers Island between the Department of Corrections, which occupied the island, and the War Department, which had control of harbor lines, over the proper enclosing of the cribwork where ashes and other material were being dumped. Criticism arose over the source-separation program in Manhattan and the Bronx. And the on-again, off-again decisions related to sea dumping persisted. The Barren Island works—purported to be the biggest of its kind in the world—continued to be dogged by protests over its noxious smells, resulting in the formation of the Anti–Barren Island League.[140]

A story in the *New York Times* in November 1898 described the Barren Island problem graphically: "On the island there is not only a stench, but a heated compound of gases, vapors, volatized oils, organic particles, and germs."[141] A bill presented to the state legislature in the 1898–1899 session called for the termination of operations. Governor Theodore Roosevelt vetoed the bill, but in the next session he signed a similar one. The courts overturned the law as unconstitutional because it broke the existing contract.[142] Having the support of the courts and substantial political leverage, the New-York Sanitary Utilization Company was able to secure contract renewals until 1917.[143]

Complaints about the stench from Barren Island's reduction plant had started during its construction and then had never stopped. New York City Health Commissioner Dr. Ernst J. Ederle declared in 1902 that the odors on the island "must go." The *Brooklyn Eagle* defended the waste facility in a slightly backhanded way by arguing that despite the complaints, "As the processes for reducing garbage to fertilizer have been improved, the odors have grown less and less perceptible, until now the area of complaint, which used to include the whole city of Brooklyn, has been reduced to the summer resorts along the Rockaway Coast." The paper did not deny that "when the wind blows their way these summer residents do become unpleasantly aware of the existence of Barren Island." "But the fact remains," it added, "that city garbage must be reduced somewhere." Dumping at sea was regarded by many as a much more serious nuisance.[144]

In January 1902 Commissioner Woodbury had boarded the U.S. government tugboat *Vigilant* for an inspection of Barren Island. He was unwilling to state if he favored the city purchasing the plant from the NYSUC, as had been proposed the summer before. "I shall go over the plant thoroughly," he stated, "but I cannot say what I am going to do any more than I am going to observe the methods pursued in disposing of the city's garbage."[145]

Incineration companies and their advocates tried to take advantage of the attacks on the rival reduction plant but failed. Even after the eastern end of the island (constructed with fill in 1905) broke away and fires ravaged the island (in 1906), the unpopular facilities continued to operate. The reduction plant was rebuilt after the fires, even larger than before.[146] A letter to the editor in the *New York Times* in May 1907 mused, "Two destructive fires and the sinking of a slice of Barren Island into the depths of Jamaica Bay, within a period of only a few weeks, give rise to the thought that perhaps Providence has intervened to do for us what the city and State authorities have been unable to accomplish—do away with this group of noisome factories at the gateway of New York Harbor."[147] That would not happen anytime soon. One observer colorfully referred to the island as "land of inverted perfumes and of the altar of roses gone wrong" where wild hogs were running free by the hundreds.[148]

Consolidation of Greater New York City had opened up additional problems in unifying its collection and disposal programs.[149] This would be no easy problem to confront. The contract for collection and disposal in Brooklyn was set to expire in January 1902. A new contract called for ashes and rubbish to be moved by trolley lines to landfills on salt marshes located on the southern side of the borough.[150] On Staten Island (Richmond Borough), a furnace for disposal purposes was being used in New Brighton, and four were operating in Queens. In February 1900 the City of New York bought those in Queens supposedly because the work by the contractor had been "unsatisfactorily handled."[151] Other wastes on Staten Island and in Queens were dumped away from the shore (by contract labor). Sea dumping persisted for both boroughs intermittently.[152] None of the boroughs much beyond Manhattan (and maybe Brooklyn) had received the same level of collection and disposal service before consolidation.

The disjointed approach to disposal at the turn of the century was aggravated by the ballooning amounts of waste. In 1906, the total refuse and ashes collected topped 3 million tons. Expressed in cubic yards, this was 8,359,648, compared with 2,204,560 in 1888. And the volume increased in subsequent years.[153] The composition of the waste stream, however, remained remarkably static in the first century of the twentieth century—although that would change, especially after World War I. A 1906 survey stated that almost 80 percent of residential refuse by weight was ash, with food representing slightly more than 13 percent (total organic content was 19.6 percent). Paper was a distant third at 5 percent.[154]

New York's garbage plight in the 1890s had been building for years. There was hope of resolution with the appointment of Colonel Waring as commissioner of street cleaning. And he proved to be successful in setting some new paths for the city in its collection and disposal programs. For all its faults, the Barren Island reduction plant had shown that ocean dumping was not the only disposal option available. But as quickly as the colonel sought to transform or upgrade the sanitary services, his end of term and untimely death unraveled many of his plans.

* * *

The account of the city's disposal woes, starting in the eighteenth century, frames this examination of what led to the Fresh Kills Landfill (and beyond) in the years lying ahead. But also central to this story was the recurring exploitation of Staten Island as an emerging sacrifice zone before and after Waring and several decades before the Fresh Kills Landfill was constructed. Especially noteworthy, and to be explored in the next two chapters, was the siting, first, in the mid-nineteenth century, of a notorious quarantine station there and, next, of a smoke-belching reduction plant in the years leading up to World War I.

PART II
STATEN ISLAND

BOROUGH OF LAST RESORT

CHAPTER 3

THE QUARANTINE

To many, Staten Island is the consummate "Forgotten Borough" or merely the adjective in "Staten Island Ferry." These reductions obscure the island's history: its rural beginnings, its eventual industrialization, and its suburbanization. The creation of Fresh Kills Landfill in 1948, however, defined Staten Island for many years, much to the consternation of its citizens. Yet the overwhelming sense that the island had been unfairly pegged as a central location for New York City's refuse (and unwanted people) is older than that. The "Quarantine War" in the 1850s (set off because of a feared epidemic of yellow fever) and the "Garbage War" (the result of a protest over the siting of a reduction plant in the salt marshes) in 1916 to 1918 were the prologues to the unwelcome placement of "the Dump" on Staten Island. This chapter tells the story of the Quarantine War.

THE PHYSICAL SETTING AND EARLY HISTORY

Staten Island, five miles southwest of Manhattan, is about fifty-nine square miles in land area and is surrounded by estuarine waters.[1] Valleys and marshland, inlets and bays give Staten Island a diverse topography. A ridge of hills runs across the length of the island. The coastline totals thirty-five miles of waterfront: skirting the Narrows is the east shore; from St. George to Arthur Kill, the north shore; along Arthur Kill is the west shore; and the beach between Fort Wadsworth and the southern end of the island is the south shore. The western shore is fairly level and quite marshy in many places, and toward the

70 ■ STATEN ISLAND: BOROUGH OF LAST RESORT

FIGURE 3.1 Staten Island's topography.

Source: Courtesy of Jose Mario Lopez, Hines School of Architecture, University of Houston.

middle this shore is deeply cut by a tidewater inlet (stream and freshwater estuary) known as Fresh Kills. The name comes from the Middle Dutch word *kille*, which means riverbed or water channel.[2]

Much of the surface soil on the island was deposited during the most recent glaciation period. The deposits range in thickness from ten to eighty-four feet and can be seen at numerous clay pits, including areas in the western and southwestern portion of the island.[3] As late as the 1930s, Staten Island still enjoyed a high level of biodiversity: more than 1,300 plant species and over five thousand acres of wetlands. In 2000, wetlands accounted for 3,407 acres, or 9 percent of Richmond County. At one time, there was a vast wetland in New York Harbor made up of three great marshes: the coast of Staten Island, Jamaica Bay, and New Jersey's Meadowlands.[4]

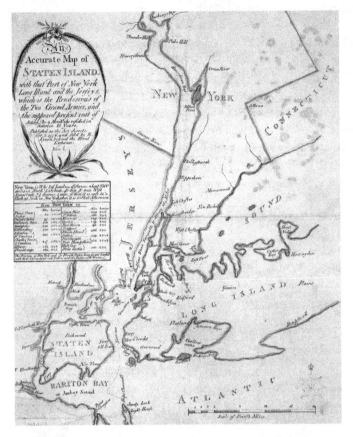

FIGURE 3.2 An "Accurate Map" of Staten Island, October 3, 1776.

Source: Wikimedia Commons.

Archeological evidence of the first permanent Native American settlements and agriculture date back about five thousand years. In the sixteenth century, Staten Island was part of an area called Lenapehoking, which was inhabited by the Lenape (Lenni-Lenape), an Algonquin people also known as the Delaware. The Lenape did not live in permanent encampments but moved seasonally, employing slash-and-burn agriculture and relying on shellfish as a staple. The Italian explorer Giovanni da Verrazzano made the first recorded European contact with the island in 1524. In 1609, Henry Hudson established a Dutch trade area there and named it "Staaten Eylandt," after the Dutch parliament, "Staten-Generaal." The first permanent Dutch settlement was established at Oude Dorp (Dutch for "Old Village"), south of the Narrows. At the end of the Second Anglo-Dutch War (1667) the island was ceded to England. In 1670, Native

Americans ceded all claims to Staten Island to the English. The colony of New York (of which Staten Island was a part) was divided into ten colonies in 1683. Staten Island along with several smaller islands became Richmond County.[5]

At some point in the seventeenth century, New Jersey contended that it had the rightful claim to Staten Island. According to legend, the dispute lasted until the Duke of York decreed that any land that could be circumnavigated in under twenty-four hours belonged to New York. When Captain Christopher Billopp did just that, the debate was over.[6] Some would contend that by proximity, economy, and disposition, Staten Island is more like New Jersey than New York to this day.

AN UNDERDEVELOPED ECONOMY

The commercial and industrial dynamism that overtook Manhattan and Brooklyn did not quite reach the southern edge of New York Harbor at Staten Island. In one sense, this was surprising, given that the island had valuable clay deposits, a long waterfront, potential ship channels, an inland waterway, and considerable open land. On the other hand, Staten Island's relative isolation worked against its economic growth, particularly before World War II.[7]

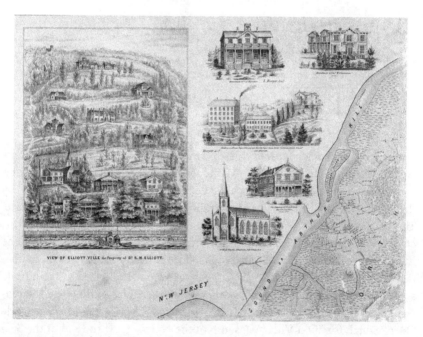

FIGURE 3.3 Staten Island, Richmond County, 1853.

Source: New York Public Library, Digital Collections.

Between the late eighteenth century and 1810, the island was predominantly rural. Most Staten Islanders were engaged in farming, fishing, oystering, making handicrafts, milling, and trading around the harbor. Staten Island's commercial ties were strong with New Jersey, western Long Island, and Manhattan. Agriculture, fisheries, and shipping remained relatively important into the twentieth century. In fact, as late as 1880 more than 44 percent of the land was being used for agriculture, consisting of 342 individual farms. Several mills (powered by the movement of fresh water from streams and ponds or the flow of the tides) operated on the island throughout the nineteenth century.[8]

Farming was never easy, given the restrictive size of the island, its rocky and sandy soil, and the many years of overtaxing the fields. Early in the nineteenth century, farm plots were small, averaging about eighty acres in area, and the farmers grew a variety of crops primarily convenient for the local urban markets. Production of wheat and other cereal grains declined as truck farming increased for berries and other fruits and for dairying. Hay was the principal crop, either "fresh" (cultivated) hay or "salt hay" grown in the extensive marshy areas.[9] As long as horses were necessary for transportation locally and in the other boroughs, hay cultivation was quite profitable. As early as 1850, agriculture on the island underwent changes stemming from the increased population and the development of resorts for the wealthy, a variety of institutions, and suburban residential property, all of which drove up land prices and restricted areas for farming on the north and east shores.[10]

FIGURE 3.4 Old Place Mill in the Staten Island salt marshes (1890).

Source: *Harper's Weekly*, September 6, 1890.

In 1810 industries were modest, consisting of two textile-carding machines, two tanneries, three distilleries, and fifty-nine looms. Manufacturing on a large scale began in 1819 with the opening of New York Dyeing and Printing for textiles in what became known as Factoryville (West New Brighton). In 1835 the dye works was the largest (of seven) operating in New York State. Beginning in the 1880s, the original factory began losing out to the newer Cherry Lane factory, both on the island. Other burgeoning industries included breweries in Stapleton, Clifton, and Castleton Corners and clay mines and brickworks in Kreischerville (Charleston). Established in 1842, the Jewett White Lead Company was located at Port Richmond. Bottling works, cabinet making, carriage manufacturing, and wallpaper manufacturing also were on the island.[11]

In the late 1860s and early 1870s the first linoleum factory in the United States was established in Travis on the island's western shore. Travis had been called Long Neck at the time but then changed its name to the aptly descriptive Linoleumville. Throughout the nineteenth century, a significant portion of mills and rural factories served the local economy, including a huge tidal gristmill at Fresh Kills near Richmond.[12] By mid-century, the addition of steam-ferry service and the development of suburban and industrial villages on the northern and eastern shores were slowly remaking the agrarian and small-town character of Staten Island. In 1880 there were about one hundred principal manufacturing establishments there, but they employed under 1,600 people.[13]

As late as the 1920s many still bemoaned Staten Island's untapped potential for economic expansion—despite its strategic location along New York Harbor, more than 15,000 people employed in the factory workforce, eighteen shipyards and repair plants, and the opening of Procter & Gamble's eastern plant in 1904.[14] The Industrial Bureau of the Merchants' Association of New York believed that "Staten Island, or the Borough of Richmond, is one of New York City's greatest undeveloped industrial assets." With its "navigable waterfront, extensive piers, direct rail connections with the great trunk lines, its proximity to the unequaled markets of the Metropolitan District and its wide area of relatively inexpensive land for industrial purposes," Staten Island had advantages "which can hardly be duplicated anywhere else on the Atlantic seaboard."[15]

Staten Island's location appeared to give it "unusual natural advantages" for industry and commerce. It was not as if the business community had bypassed industrial development altogether, but it was clearly confined to a few stretches on the island. In 1921 there were 120 industrial plants in Richmond Borough with five or more workers. Twenty-nine of them usually employed one hundred or more people. About half of the industrial activity (less before this time) was located on the north shore, especially along Kill Van Kull in close proximity to the New Jersey cities of Elizabeth and Bayonne.[16]

By contrast, the eastern shore, with a substantial number of piers, was better situated for commercial activity because of its proximity to Lower New York

Bay. The southern shore, despite directly facing the Atlantic Ocean, lacked infrastructural improvements such as a dredged ship channel, a breakwater, or rail facilities. And the western shore along Arthur Kill, with its limited railroad facilities, kept it from competing effectively with its industrial New Jersey neighbors. The Industrial Bureau concluded that the fundamental reason for Staten Island's "retarded growth undoubtedly has been its comparative isolation."[17] What was an advantage for wealthy summer-home owners was an economic setback for Richmond Borough in general.

AN ISLAND NOT SO FAR

The bureau recognized that Staten Island's isolation "has been more imaginary than real," but it nevertheless was "a powerful influence in limiting the island's industrial development."[18] The relatively short distance from Manhattan to Staten Island, however, seemed hardly insurmountable given the island's economic potential. What prompted the perception of isolation was something else. One might speculate that (fairly or unfairly) some people considered Staten Island to be outside the mainstream of Greater New York's heady development because of the overwhelming success of Manhattan and to a lesser extent Brooklyn. Alternatively, some contemporary observers celebrated Staten Island's physical separation from Manhattan and its so-called isolation as signs of its uniqueness and its ability to stay aloof from runaway changes.

Yet the reality seemed to lie somewhere in between. Charles W. Leng, the director of the Staten Island Institute of Arts and Sciences, and William T. Davis, president of the Staten Island Historical Society, noted in their 1930 history of Staten Island that "the extensive waterfront thus possessed by Staten Island has from the earliest period in its history been of importance." Its "wealth of seafood" attracted Indians and aided white settlers. Its tides operated saw and grist mills. Close to Manhattan, it became a summer resort and also a location for commercial development. "On the other hand," they concluded, "the isolation of the Island and the need for ferries to reach it from the mainland have operated powerfully to retard its growth in population."[19]

The lack of transportation options, of course, was crucial during the period when Leng and Davis were writing their history. Before 1931, three bridges connected Staten Island to New Jersey (the Goethals Bridge, Outerbridge Crossing, and Bayonne Bridge). For all of the economic value that these bridges added to the west of the island, however, no bridge connected Staten Island to Greater New York City (Brooklyn being closest) until 1964, with the construction of the Verrazzano-Narrows Bridge. Since access directly to Manhattan could only be accomplished by ferry or other boat and ship traffic, Staten Island was more separated from the rest of the city of which it was politically bound than it was to the neighboring state of New Jersey.

Leng and Davis also understood that the hierarchy of the boroughs was well established before consolidation in 1898. Manhattan and Brooklyn, in particular, were the heart of the "mainland"; Staten Island was more or less an appendage. Especially in providing services, Staten Island (and sometimes Queens and the Bronx) could be overlooked. But it also was not uncommon in the mid-nineteenth to early twentieth centuries for observers to celebrate Staten Island's separation from the rest of New York City. Henry David Thoreau proclaimed, "The whole Island is like a garden, and affords fine scenery."[20] An 1893 history referred to Staten Island as "pleasantly situated five miles southwest of the city of New York." "The island," it added, "is sought by many New-York business men as a place of residence; it is dotted over with homes of all varieties, from the modest cottage to the pretentious country-seat."[21] One poet referred to the island as "an untapped suburban empire."[22] Distance could be a curse but also a hedge against disrupting Staten Island's unique identity.[23]

REAL ESTATE VS. NATURE

Cornelius Geertruyus Kolff (1860–1950) was an important figure caught between the desire to maintain Staten Island's splendid isolation and willingness to take advantage of its many assets. Kolff was a prominent civic leader in his time, deeply involved in a variety of Staten Island institutions, ranging from the Chamber of Commerce to the Staten Island Historical Society. He liked to be called "Staten Island's Most Obedient Servant." He also was a realtor who boosted the image of the place at every opportunity. He had the foresight to propose a bridge across the narrows between Staten Island and Brooklyn in 1926 as a way of enhancing the island's commercial development. Such a bridge proved to be several decades off. Late in his life Kolff was torn between promoting new businesses (thus opening Staten Island to growth not unlike Manhattan) and wanting to preserve the pristine beauty of the place. In fact, boasting of being a disciple of Ralph Waldo Emerson, he built a log cabin on Emerson Hill called "Philosopher's Retreat," close to his own farm and to the property of Judge William Emerson, the philosopher's brother.[24]

In the conclusion of his book *Early History of Staten Island* (1918) Kolff proclaimed, "Where in the American continent is there a place more favored by nature than Staten Island? Its beautiful hills, commanding views of the ocean, its picturesque valleys, its magnificent forests, its pretty lakes, its sea shores, its climate, tempered by the proximity of the ocean, all combine to make it an ideal place for human habitation." He added that "those who live here love it" and that those who left often "turn with an affectionate memory" to it.[25] Such was the declaration of a proud citizen, a realtor, and a resident caught in ambivalent ambitions for his home. With large tracts of land available, Staten Island had the potential to be a realtor's paradise, and others like Kolff extolled the virtues of "this most Beautiful Isle of the Sea." In the mid-nineteenth century, however, recurring problems with malaria, competition from other boroughs, and inaccessibility limited the response to the boosterism and sales pitches.[26]

The Industrial Bureau's 1921 report had focused as much on future hopes as past accomplishments in nineteenth- and early-twentieth-century Staten Island. Comparatively or in absolute terms, Staten Island did not demonstrate the economic dynamism of Manhattan and Brooklyn. But it was hardly an untouched Eden, despite what promoters were saying. Staten Island was a canvas yet to be filled in, with space available for all kinds of uses. Change came incrementally,

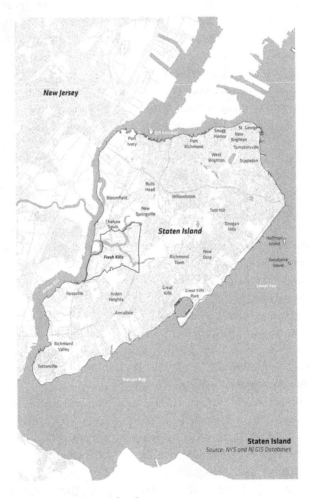

FIGURE 3.5 Major towns on Staten Island.

Source: Courtesy of Jose Mario Lopez, Hines School of Architecture, University of Houston.

and even into the mid-twentieth century it still possessed (at least from an economic viewpoint) great amounts of undeveloped, underutilized, or inhospitable land.

THE FEAR OF YELLOW FEVER

The lack of economic development on Staten Island and the sense that its growth was out of time with the rest of New York does not fully explain Staten Islanders' contention that they lived on a "Forgotten Borough" or on one of the "other" islands. Nor does it provide a singular link to the eventual placement of Fresh Kills Landfill on the island in the mid-twentieth century. The Quarantine War in the 1850s, however, is an excellent starting point in both respects.

In the mid-nineteenth century, health officials in many towns and cities close to maritime traffic established quarantine stations to prevent the spread of yellow fever and other communicable diseases. Epidemics had been rampant in many port cities at the time. Well before the germ theory (bacteriological theory) was accepted as a way of understanding contagion, many people (including Colonel George E. Waring Jr.) believed that disease was contracted through miasmas (smells from decaying matter), by the wind, or through specific carriers. Thus some form of quarantine was necessary, it was believed, to cordon off the infected from the general population.[27]

As early as 1716 ships from the West Indies were required to set anchor off Staten Island because of fears of disease; those found to be infected were moved to quarantine facilities. In 1758 the colonial legislature in New York designated Bedloe's (or Bedlow's) Island as the first quarantine station (or lazaretto, as it was sometimes called) in the state. The facility was moved to Governors Island (Nutten Island) in 1794 (or 1796, some suggest). Citizens there protested the facility, especially after attacks of yellow fever broke out. The state decided to move it again, to a location farther from Manhattan. The legislature then passed an act appointing commissioners to find a new site on Staten Island, appraise the land to be appropriated, and begin constructing hospitals there. Island representatives attempted to defeat the bill, and many refused to sell their land. They wanted nothing to do with the facility.[28]

A group of merchants and ship owners, however, concurred that the Staten Island hospital complex possessed "peculiar advantages of situation, and furnishes all the facilities required for Quarantine purposes, both as respects the paramount necessities of the public health, and the shipping arriving in our port."[29] The siting of the New York Marine Hospital Grounds in their backyard made Staten Islanders nervous; they feared that proximity to the facility might pose a health hazard to the community. To property owners in Tompkinsville, it threatened real estate values and curbed opportunities for other economic growth.[30]

FIGURE 3.6 View of the Quarantine, 1833.

Source: Library of Congress, Prints & Photographs Online Catalog.

The New York Marine Hospital, known as the Quarantine (or the Tompkinsville Quarantine), opened in 1799. It was built on the northeastern section of Staten Island in the village of Tompkinsville, which was encircled by the town of Castleton and close to the modern-day disembarkation dock for the Staten Island Ferry. Set on a rustic thirty-acre tract called Duxbury Glebe from land that belonged to the Church of St. Andrew, the nation's largest quarantine hospital had several buildings (including St. Nicholas Hospital, the Smallpox Hospital, and the Female Hospital) surrounded by a wall and four piers for unloading patients and equipment. The land adjacent to the compound was dotted with farms.[31]

The Quarantine was managed jointly by the state and city.[32] Since the facility dealt with infectious diseases but prevailing medical practice knew of very few ways other than quarantine to deal with them, the compound was set at a careful distance from New York City's congested center. Staten Island, while not altogether secluded, was large and had a sparse population (about 20,000) and rural surroundings, making it an attractive choice for the Quarantine. Patients sent to the Marine Hospital came from vessels in the harbor blighted with disease or were individuals expelled from New York City because they showed signs of infection. As immigrant numbers swelled, so did the patient load at the compound, which could house as many as 1,500 people. The fear that patients and hospital workers were transmitting disease into Staten Island communities was not an idle debate point. In 1821, twenty-nine cases of yellow fever struck the island, which were blamed on a large number of infected vessels driven to shore by a storm. Smaller outbreaks of both yellow fever and cholera hit the island in subsequent years.[33]

THE QUARANTINE WAR

In the late 1840s to mid-1850s, the so-called Quarantine War or Staten Island Rebellion pitted locals wanting to rid themselves of the Quarantine against the city and state governments who had foisted it on them.[34] The Quarantine War was the first of several clashes over the years between Staten Island and the City of New York stemming from the islanders' sense that they were pawns rather than partners in city affairs.

A very serious outbreak of yellow fever occurred in 1848 on Staten Island, estimated at 180 cases, with several fatalities among the townspeople. At the close of the Mexican War several vessels arrived in New York Harbor from Vera Cruz, and they were cited as the cause. While locals speculated that the wind blowing offshore carried the pestilence from ships directly to Staten Island, they also were suspicious that staff at the Quarantine and stevedores working the docks were disease carriers. More generally, bias and suspicion arose over newcomers, who were regarded as an unsavory element intruding on the relatively secluded island.[35]

The 1848 outbreak prompted the first major effort to purge the island of the Quarantine. Staten Island residents sent a petition to Albany exhorting the legislature to remove the compound. The deliberative body appointed a committee headed by Wessel S. Smith to study the problem, leading to the examination of the health officer, physicians, shipping merchants, sea captains, and various citizens. A New York legislative committee report the following year supported some of the claims by asserting that the Tomkinsville Quarantine Station did little to protect the people of New York or Brooklyn and that the complex was a menace to Staten Island and an "unjustifiable burden" on Richmond County. A retort that would come to be repeated often on Staten Island was that the imposed facility was "a prejudice against the whole Island."[36] The committee "unhesitatingly" recommended the Quarantine's removal from Staten Island. But an effort to relocate the Marine Hospital to Sandy Hook, New Jersey, failed because of resistance from New York shipping interests and the state of New Jersey. Thus nothing came of the legislature's response.[37]

Another epidemic of yellow fever broke out on Long Island, Governors Island, and Staten Island in the summer and fall of 1856. Dr. Elisha Harris, physician-in-chief of the Marine Hospital, reported 538 cases of yellow fever in the vicinity of the Port of New York, with ninety-three on Staten Island. More than a third of all the sick died of "black vomit." Among its population at the time, the Marine Hospital on Staten Island was quarantining 177 yellow fever patients, 325 smallpox patients, 269 patients with typhus fever, and another eight with cholera.[38]

Again, the legislature passed an act (March 6, 1857) calling for "the removal of the Quarantine station" from Staten Island. The health officer, Board of Underwriters of New York, the Commissioners of Emigration, the New York Chamber of Commerce, and many merchants opposed the decision. As a report of the Executive Committee of Staten Island stated, "The citizens of Richmond congratulated themselves that they had at length got rid of a deadly nuisance." But they were "doomed to a speedy disappointment."[39]

The report reflected the cynicism and frustration with the quarantine process:

> Coasting vessels coming from south of Cape Henlopen, as well as vessels from foreign ports, sometimes numbering between one and two hundred a day, were subjected to visitation at enormous expense. The exclusive privilege of lightering was given to a monopoly. Ship loads of passengers from foreign and domestic ports were unnecessarily quarantined for days, and compelled to pay large sums for board and the most trifling accommodations. In many instances, it is alleged, persons and vessels so quarantined were, for a pecuniary consideration, permitted to leave their anchorage, and proceed to the city of New York. The Quarantine grounds were frequently overcrowded, the sick and well were placed in dangerous proximity to each other, and promiscuous intercourse was permitted between those inside and outside the walls. Stevedores and other employees were suffered to mingle indiscriminately with other citizens, and to pass at pleasure on the ferries to and from their families in the cities of New York and Brooklyn. All the time there were published in the daily papers, and the reports of the Health officer to the Board of Health of New York, the most exaggerated statements of the rigid enforcement of the Quarantine regulations.[40]

The idea of moving the Quarantine to Sandy Hook—"the most eligible site"—came up again. A report of the commissioners restated the bad news: "New Jersey, through her legislature, has refused to entertain the proposal, and has distinctly intimated, at the same time, that she would never consent to the establishment, by New-York, of a Quarantine station on the barren sand Spit at Sandy Hook." It added, in a self-serving way: "New-York has no such site within her boundaries, if she had, she would not ask of a sister State, her nearest neighbor, whose people are daily mingled with her own, in the pursuits of commerce and business, to aid her in an undertaking which would bring safety to New-York without danger to New-Jersey."[41]

The commissioners thus devised a plan to set up a temporary quarantine facility on the south side of Staten Island, near the town of Westfield at Seguine's Point (Prince's Bay). The intention was to use the site until a permanent one could be established. Residents of Westfield, however, burned down the new hospital structures and during rebuilding in 1857 burned down the partially completed hospitals. The decision to move the Quarantine to Seguine's Point on a permanent basis was deferred. Stopgap measures were implemented in 1857 and 1858. They included creating a harbor police force to constrain the movement of stevedores, building a high fence around the Quarantine, and passing an injunction to keep residents from blocking the Quarantine's actions. None of these efforts reduced the growing tensions and in fact exacerbated them.[42]

Inspired by the actions of the citizens at Seguine's Point, mounting threats to burn down the Quarantine reached their peak in September 1858. On the afternoon of September 1, local citizens and property holders in the Castleton area issued a series of resolutions posted throughout the island and also on the walls of the Quarantine itself. Among them: "Resolved. That [The Quarantine] is a nuisance too intolerable to be borne by the citizens of these towns any longer." Under the leadership of Ray Tomkins, a wealthy landowner and chief of the firehouse in Tomkinsville,[43] and John Thompson, the owner of a general store, plans were made for a raid on the Quarantine. Given the long-standing grievances over the facility, the event was hardly spontaneous. Late that evening about thirty men set fire to the buildings in the complex, including the St. Nicholas. Stevedores and others got patients out of the burning buildings. The scene was chaotic, but only two people died during the melee: a stevedore shot by a fellow employee settling an old score and a patient with yellow fever. By daylight, it appeared that the worst was over, but not because of the efforts of local or state authorities. The general superintendent of city police had decided not to jeopardize his forces by sending them into the disease-riddled compound. The state commissioner of emigration did little more than refer the situation to a committee, and the federal government sent in sixty marines, but only to defend federal inspection buildings.[44]

On the following evening, more than two hundred people attended a meeting of Richmond County citizens to celebrate the victory and make additional plans. Later that night a mob appeared at the Quarantine grounds, ready to complete what had begun the night before. The facility staff was better prepared this time, including having moved all of the patients to safety, but the remaining structures were razed. After the two days of bedlam, the wall encircling the grounds and a number of buildings were hurriedly reconstructed. The Quarantine continued to operate for a time, with guards posted around the property to head off any additional attacks.[45]

The *New York Times* interpreted the Quarantine War in hostile terms. It labeled the Staten Island protesters "mobocrats" and called for a complete

FIGURE 3.7 The burning of the Quarantine, 1858.

Source: Library of Congress, Prints & Photographs Online Catalog.

rebuilding of the facility on the original site. Its opinion was in line with the "official" view stated by Governor John A. King:

> Hence great excitement on Staten Island, so that at last, on two successive nights of the 1st and 2d September last [1858], an armed mob broke forcibly into the enclosure of the Quarantine grounds, fired the hospitals, houses and outhouses, in some of which were patients in various stages of mortal disease, and dragging these sufferers from their beds, and placing them unsheltered on the ground, exposed to the midnight air and to the scorching heat of the burning edifices in close proximity, utterly demolished the whole establishment, and then departed, unrestrained and unquestioned by any of the law officers or magistrates of the county.[46]

The Commissioners of Emigration concurred with the governor, referring to the melee as "lawless violence," and exaggerated the crowd that stormed the compound as "a thousand strong." According to their report, the superintendent, physicians, and other officers on site were threatened with death and that "the sick were dragged from the hospitals, men, women and children, some afflicted with typhus, and others with yellow fever, were thrown indiscriminately upon the earth. Not a few were placed between two of the burning buildings not widely distant from each other."[47]

The *New York Times* printed a letter from one Edward Cunningham, who recounted the death of his wife attributed to the Quarantine War. He stated that on September 2 his house was burned by the mob and that his sick wife "was obliged to flee before the flames." Soon after the fire, he added, her symptoms of "confirmed consumption" became worse, and she died. "The crime of committing arson," Cunningham concluded, "and thereby turning a sick and disabled woman into the streets at night, is a sin and an outrage of sufficient magnitude, without adding slander thereto to tarnish her memory."[48]

Retelling the story of the burning of the Quarantine was subject to intensely partisan and biased sentiments. Below the surface of the controversy but possibly contributing to it were demographic factors and the bitter nativism of the time. A large number of the protesters were Irish, either transplants who had been residents in the United States for more than ten years or members of prominent Staten Island families. They had felt the sting of prejudice in their time. For example, cholera sometimes went by the colloquialism "the Irish disease."[49] "Coffin ships" had delivered poor and ill Irish immigrants (some with smallpox, yellow fever, and dysentery) to the Quarantine regularly. Those that died were buried in one of two Quarantine graveyards. Rumors that "un-Christian burials" took place at the cemetery particularly infuriated Christian immigrants.[50] Locals and others linked the outbreaks of yellow fever associated with the Quarantine in the late 1850s to the large numbers of non-Europeans from the West Indies and Asia now housed in the facility. The Quarantine represented an unwanted institution for more reasons than its intrusion on Staten Island real estate.[51]

Some observers, especially those living on the island, had a more tolerant view of the arsonists than the *New York Times* had displayed. Leng and Davis described one scene on the first night of the unrest as follows: "The hospital for men, with a wooden statue of a sailor on top, was the next [to burn], after it had been cleared of every living thing, even to a cat and a canary bird. There were only three yellow fever patients who were carefully carried out and placed on beds under an open shed; it was a very warm night and they laid [sic] enjoying the scene and ultimately got well."[52]

The *New York Herald* also sided with the Staten Islanders and recommended that a new facility be placed elsewhere, preferably in New Jersey. The *Herald*'s report on September 3 clearly indicated the paper's view: "There appears to be no doubt in the way the fire originated. As was to be expected, considering the intense hostility which the residents of the place feel toward the hospitals, in consequence of the infection they breed, and the corrupt practices which are said to be prevalent within their precincts, the incident has been ascribed to incendiarism."[53]

New York City and New York State took belated steps on the mob action, sending one hundred metropolitan police on September 3 and then the Eighth Regiment of the state militia on September 11.[54] The troops remained until January 4, 1859. Several of the arsonists were taken to jail, but only on the minor charge of breaking Quarantine regulations, the only action over which the city had jurisdiction. The industrialist Cornelius Vanderbilt, a native of Staten Island, bailed out many of the defendants, and the cases simply faded away.[55]

The state, however, began a trial of the ringleaders of the arsonist group on September 15 in Richmond County.[56] The defense of the arsonists' actions essentially was an indictment of the Quarantine itself: the buildings had been located on Staten Island despite the protests of its citizens; land was taken by force of law to build the facilities; contagious and infectious diseases had spread over the island on numerous occasions; the Quarantine had been badly managed for years, thus endangering locals; a grand jury had twice presented the facility as a nuisance; and two legislatures had decided that it should be removed.[57]

On November 11, the sympathetic judge H. B. Metcalfe (a Staten Islander who had earlier favored removal of the Quarantine) found the defendants not guilty: "The arson is charged to have been committed on buildings. But it could not be arson in the *first* degree, because, though some of them were inhabited dwelling-houses, and although they were burned in the night-time, yet the proof fails to show that there was in them at the time, any human being." He elaborated on the additional unlikely possibility of second- and third-degree arson and made several additional observations, including whether the burning of the buildings per se suggested guilt: "And this is whenever the act is *wilfully* done. I have always said it was done *intentionally*; but *wilfulness* implies malice, and malice is inferred from the intention, unless it be disproved."[58] He added:

> Undoubtedly the city of New York is entitled to all the protection in the matter that the State can give, consistently with the health of others; she has no right to more. Her great advantages are attended by corresponding inconveniences; her great public works by great expenditures; her great foreign commerce by the infection it brings. But the Legislature can no more apportion upon the surrounding communities her dangers, than her expenses; no more compel them to do her dying, than to pay her taxes; neither can be done.[59]

The idea that the Quarantine posed a menace to Staten Island won out over the prosecution's claim of the wanton destruction of the property.[60]

A BATTLE WON?

The station was moved from Staten Island in 1859, but the state, paradoxically, charged to the County of Richmond the expenses for transferring the hospitals.[61] Floating hospitals (the *Illinois* and the *Florence Nightingale*), anchored offshore in New York Harbor, took victims of yellow fever. Those contracting smallpox were sent to Blackwell's Island, typhus sufferers to Wards Island. Seven years later, quarantine stations were set up on Swinburne (originally Dix) and Hoffman Islands. These were two artificial islands made from rubble and dredged sand in the lower bay, one mile southeast of South Beach and west of the main channel leading to the harbor. In the 1920s the Marine Hospital moved to Ellis Island, the consummate symbol of immigration to the United States.[62]

The burning of the Quarantine was not a spontaneous act by a group of unruly citizens. There were actual threats to the health of locals because of the decision to locate the New York Marine Hospital on Staten Island, and citizens had been upset with the decision since its inception at the end of the eighteenth century. The so-called rebellion also was the result of an ongoing struggle on Staten Island to control its own affairs and to determine the use of its property. Occurring well before the 1898 consolidation, the Quarantine War raised serious questions about the value of place to longtime residents versus the use of underdeveloped land for other purposes. For the city and state, Staten Island offered a fortuitous choice for potential exploitation as it was distant enough from major centers of population and only modestly developed at the time. Staten Islanders, by contrast, felt that Gotham placed little value on their home beyond the city's own needs, which they viewed as a "prejudice against the whole island." This difference in perspective would create tension between Staten Island and Greater New York City for generations.

CHAPTER 4

THE GARBAGE WAR

Staten Islanders remained bitter over the Quarantine episode for many years. The consolidation of Greater New York in 1898, however, brought optimism about the new Richmond Borough's place in the gigantic metropolitan creation and the chance for island residents to gain support for, among other things, crucial infrastructure and city services that had been neglected or ignored. Hopes for better treatment at the hands of political officials in Manhattan and Albany were soon dashed by the Garbage War of 1916–1918.

CONSOLIDATION

Manhattan and Brooklyn dominated the consolidation negotiations, with other entities marked for merger relegated to the fringes. Staten Island might have been an afterthought had it not been situated along New York Harbor.[1] The amalgamation of the communities there and cohesive economic control over the port were pivotal to the new metropolis. The isolation and estrangement of the forgotten borough made some of its citizens wary of joining the collective. Others wanted to take advantage of new economic opportunities and to connect Staten Island securely to the rest of the city by bridges, roads, tunnels, and subways.

As early as 1868, Andrew Haswell Green, the father of consolidation, saw how infrastructure needs could draw partners into his idea of an "imperial city": "Westchester is demanding ways to transmit her population to the city;

Richmond County by her ferries and railways, is exerting herself in the same direction."[2] Green gave several public statements stressing the absolute necessity of including the outer boroughs in the merger. To its advantage, he noted in 1890, "Staten Island has already felt the touch of modern enterprise. It will require but a few years to develop some special advantages to such a degree as to render her an indispensable auxiliary to the growing business of the port."[3] But he also believed that "the names of outlying districts are so obscure that their significance can only be revealed by expressing their relation to the great metropolis."[4] Of the three "outlying districts" (the Bronx, Queens, and Richmond), the gap between expectation and reward was widest for Richmond. It was the most remote and least urban of the new boroughs.[5]

Also in 1890, the humorist and Staten Island native Bill Nye, with tongue in cheek, questioned the inclusion of Richmond County in the consolidation: "As a resident and taxpayer of Staten Island I know that I but voice the sentiments of many of my neighbors when I say that we do not wish to have New York annexed to us." Anxious to get out of the City of New York "and its temptations," Nye understood that change was coming but that Manhattan should look elsewhere for expansion: "It is true that New York has outgrown her garments, and that she is oozing gradually through the pores of her Jersey, but why not extend back up the Hudson to Albany, where they are used to political corruption and mussed up virtue?"[6]

In 1894 a spirited public debate on Staten Island took place in the spacious University Temple in the North Shore neighborhood of Westerleigh.[7] Many landholders favored consolidation.[8] Those with less of a stake in the island's potential economic boon, such as the mayor of Middleton, Dr. John Feeney, feared that in the merger Staten Island would be "a drop in the bucket of New York." The English-born (although he claimed to be born in Ireland and involved in radical activities) dime-novel author and local tax collector John De Morgan declared that he would "shudder to think of the time when our lovely hills and beautiful valleys shall be made as unsightly as the dirty streets of New York." He added that a time would come when "the neat little cottages with their tasteful gardens shall be swept away, and row after row of brick or stone houses, factories and gin mills occupy their place."[9]

After some failed attempts, proconsolidation forces prevailed in the November 1894 election. They scored impressive victories in New York County, Queens County, Westchester, and Pelham. Richmond County, however, provided the largest margin of victory, with 5,531 for and 1,505 against. Staten Island gave consolidation its most lopsided vote, reflecting its eagerness to benefit from integration into a metropolis. The flush of victory was short-lived: turning the vote into a practical plan for a united city (and one that would benefit Staten Island) faced numerous hurdles.[10]

The first bill to create a commission to draft a charter failed to pass the legislature in 1895, especially because of Brooklyn's lukewarm support for consolidation. In January 1896, the proconsolidation *New York Times* reported that "Greater New-York has again become a political football."[11] But in May 1897 Governor Levi P. Morton signed a complex and cumbersome charter bill into law, with the first election of officers set for November 2.[12]

Under the new charter, the Municipal Assembly would be made up of a Council and a Board of Aldermen, replacing the Common Council. The new Council would have twenty-nine members, chosen from ten districts. Five of the districts were to be in New York County (then including Manhattan and the Bronx), three in Brooklyn, and only one each in Richmond and Queens. The position of mayor was strengthened considerably. The Board of Estimate (the Board of Estimate and Apportionment until 1938) was retained from the pre-1898 system but now was clearly dominated by the mayor and wielded the most influence over the budget. The boroughs had no representation of their own on the Board of Estimate. Instead, the office of borough president would be created—which would only be responsible for local administration and public works.[13]

Staten Island's overwhelming support for consolidation might suggest unqualified commitment to a collective approach to governing the metropolitan region, but its needs were more practical than theoretical. Staten Islanders needed sewers and reduced costs for maintaining fire and police departments. Possibly the biggest argument that swayed votes was the expectation of better transportation connections to Manhattan and the other boroughs—meaning bridges, ferries, rail, and subways.[14] But the local citizens of Richmond Borough were certainly aware of the "six miles of water between it and the battery in New York City." Or as the *Staten Island Advance* stated, "The Island was tied to 'the city' only by ferryboat, not by law."[15]

After years of energetic debate over consolidation, time arrived for the transition to a single government.[16] Republican George Cromwell was elected as the first president of Richmond Borough. Dr. John Feeney, who had opposed the merger, was Cromwell's opponent, in what became a donnybrook of an election. It was so close (an especially curious result given the island's support for consolidation) that Cromwell had to wait six months after the voting in November 1897 to be declared the winner.[17] Soon after Cromwell became borough president, the island's population stood at 65,000, from migration from other boroughs and immigration. As a result, demand for new services was mounting. By 1913, at the end of his three terms, the population was more than 86,000.[18]

During Cromwell's tenure, Staten Island saw several physical changes, including the first municipal ferry terminal in St. George, a new civic center, the construction of Borough Hall (1906), a courthouse on Stuyvesant Place, a high school (1904), its first New York City fire department building (1905), and the opening of St. George Library (1907).[19] Cromwell also had responsibility for maintaining island highways, a practice that remained intact until 1963, when a new city charter took highway authority away from the borough presidents. In many parts of the borough, citizens remained on septic systems rather than on a citywide sewerage system, and many roads built by developers were merely a thin layer of asphalt over dirt. The five towns now were designated as wards; home rule for those communities was replaced by borough administration and a centralized government based in Manhattan.[20] All in all, the initial impact of consolidation on Staten Island was mixed.

The most significant obstacle to keeping Richmond Borough painfully isolated from the rest of Greater New York was the absence of a regular flow of people and goods over high-capacity bridges or tunnels. This issue was not addressed by consolidation. Ferry service improved, and forty-five miles of roadways enhanced internal travel on the island. Yet despite the bridges connecting to New Jersey, none linked the island to any of New York's other boroughs until the 1960s. There was no tunnel, subway connection, or rail link to Manhattan.

In 1923 construction began to connect the subway in Brooklyn to Staten Island, with Brooklyn-Manhattan Transit (BMT) digging a tunnel from Bay Ridge (in southwestern Brooklyn) to the island, to meet with what was to become the Staten Island Railway. The Democratic mayor John Hylan (a former Brooklyn subway motorman) supported the tunnel project. His administration had been instrumental in modernizing Staten Island's cargo capacity in World War I, but when it was completed, the war had ended, and demand for the docks (at least temporarily) diminished. Hylan now was stuck with an unprofitable venture, one (along with local ambivalence) that undermined the idea of a connecting tunnel. He was ousted in the Democratic primary in 1925, essentially dooming the project. Revival of the idea never gained political traction.[21]

Under charter revisions in 1901, the leverage of the mayor was reduced, shifting more power to the borough governments. The Board of Estimate was redesigned to include equal representation for the boroughs regardless of population. This shift placated at least four of the five boroughs. The Board of Estimate, moreover, would continue to increase in influence over the years, especially in financial matters. This new governmental arrangement was a balancing act between a modicum of borough autonomy and centralized control.[22]

For Richmond Borough the strong vote of support for merger masked uneasiness for many who questioned the tradeoffs between autonomy and inclusion. (Autonomy no more meant self-sufficiency because inclusion meant equal treatment.) Staten Island's isolation made its people crave attention from the central city, but its citizens did not realize that a modicum of attention did not necessarily lead to benefits. One recent author believed the merger "turned out to be something less than the miracle it was expected to be."[23] Exactly how Staten Island would be treated—or used—became clearer during the Garbage War.

FINDING NEW SPACE FOR REDUCTION

In the 1910s, disposal of New York City's wastes was finding several outlets. Ashes, rubbish, and street sweepings were discarded mostly in land dumps, incinerators, or at sea. Between 1910 and 1914, land dumping was the most significant but was declining as the Great War approached (from 88 percent in 1910 to 72 percent in 1914), incineration was limited and disappearing (from less than 1 percent in 1910 to 1912 and not registered in 1913 and 1914), and dumping at sea was again on the rise (from less than 1 percent in 1910 to 17 percent in 1914). Reduction of garbage on Barren Island represented disposal of 10.5 percent of the city's total waste in 1910 and a little over 11 percent in 1914.[24]

Beginning in 1916, an effort was underway to build a large waste reduction plant on Staten Island to replace the one on Barren Island. The site selected for

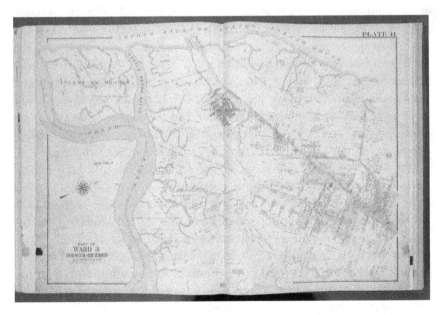

FIGURE 4.1 Map of Ward 3 of Richmond Borough with the Island (Isle) of Meadows, 1917, in the top-left-hand corner.

Source: New York Public Library, Digital Gallery.

this facility, which was meant to serve Manhattan, Brooklyn, and the Bronx, was Lake's Island, along the Fresh Kills Estuary. Located near the northwestern part of the island, it sat directly opposite the Isle of Meadows (a hundred-acre island located where the Fresh Kills Estuary and the Arthur Kill Waterway met) and was one mile south of Linoleumville (today the neighborhood of Travis). Its siting set off the notorious Garbage War. Thirty-two years later, a massive landfill blanketed 2,200 acres of salt marsh in the same location.

Some contemporaries had a long memory about the Garbage War, as a 1946 newspaper story attests. Mrs. Charles E. Simonson of Dongan Hills (in her seventies at the time of the interview) headed the Women's Anti-Garbage League, which had opposed the plant in 1916–1917. "People don't get out today and fight like we did," she stated.[25] The league broke up before the plant was closed by court order in October 1918, but its efforts ultimately were rewarded. Simonson and other Staten Island citizens put home first in opposing the reduction plant. At the time of the earliest public protests against the plant in 1916, one leading citizen declared that the Quarantine was the first attempt to defile Staten Island, and "God knows what they may do next."[26] Locals do not forget.

BARREN ISLAND'S WOES

Before (and even a few years after) the new plant was built, animal, vegetable, and other organic waste matter from Manhattan, Brooklyn, and the Bronx made their way on scows to Barren Island and then were reduced to grease and fertilizer by the New-York Sanitary Utilization Company (NYSUC).[27] The contracts for the site, held from 1896 to 1917, were fraught with political intrigue and corruption. Under Tammany, the new NYSUC agreement kept much of the city's waste disposal in private hands while filling the coffers of Street Cleaning Commissioner Percy Nagle (appointed in 1898). Major John McGaw Woodbury, appointed commissioner in 1902, faced similar allegations of corruption.[28]

The Barren Island facility survived contract battles, unstable fill land, and devastating fires. Additional waste materials not sent there were sent to some incinerators (such as the Delancey Street incinerator under the Williamsburg Bridge), and approximately 60 percent of the ashes, rubbish, and street sweepings of the three boroughs continued to be used as fill on Rikers Island.[29] Brooklyn Ash Removal Company, organized in the late nineteenth century, was one of several companies contracted by New York City to haul cinders and ash from coal-fired furnaces used to heat homes and businesses. It received the lucrative ash-removal contract in 1904 and used ash to fill in the marshlands and tributaries of Barren Island. In 1909 it directed dumping activities to the former salt marshes at Corona Ash Dump (eventually labeled Mount Corona) and eventually at Flushing Meadows, both in Queens.[30]

FIGURE 4.2 Presses used by the New York Sanitary Utilization Company facility at Barren Island in 1898.

Source: Google Books, from *The Cosmopolitan* (1898).

FIGURE 4.3 Unloading conveyors at the New York Sanitary Utilization Company facility at Barren Island in 1898.

Source: Google Books, from *The Cosmopolitan* (1898).

FIGURE 4.4 Unloading garbage from scows at Barren Island.

Source: *Scribner's Magazine*, October 1903.

Flushing Meadows was transformed into an enormous, smoldering ash wasteland. In *The Great Gatsby* (1925), F. Scott Fitzgerald described the place: "This is a valley of ashes—a fantastic farm where ashes grow like wheat into ridges and hills and grotesque gardens; where ashes take the forms of houses and chimneys and rising smoke and, finally, with a transcendent effort, of men who move dimly and already crumbling through the powdery air."[31]

Available space for NYSUC's work on Barren Island was becoming scarce as early as 1909. In April 1910 an exploding digester blew a ten-foot hole in the roof of the plant, killing one man and severely wounding five more.[32] In early 1913 the NYSUC faced a strike of its workers. During the summer of 1913 the Merchants' Association formally complained about the odors on Barren Island to the health commissioner, and in 1915 the Health Department threatened a lawsuit against the contractor.[33] But none of this was enough to stop the dumping and reducing. The protests were a constant reminder of the odiferous place, and debate over what to do with the facility seemed endless.

From 1859 to 1933 Barren Island served in some capacity as a producer of fertilizer and fish oil and as a garbage dump site. It was connected to the mainland in 1926 by filling marshland and became the location for the private Barren Island Airport. The city took that opportunity to add eight hundred acres of landfill and then renamed the area Floyd Bennett Field, which became New York City's first municipal airport (1930). The site also was part of Robert Moses's plan to expand Marine Park in Brooklyn. After a better and more convenient airport opened in 1939 (North Beach, then renamed LaGuardia Airport), Floyd Bennett Field was sold to the U.S. Navy. Today it is part of Jamaica Bay's Gateway National Recreation Area.[34]

REFUSE DISPOSAL: WHERE TO TURN?

The city realized well before the Moses era that the existing waste facilities and practices were inadequate. In 1912 the Board of Estimate began an inquiry about changing how and where waste disposal should be conducted in New York City. A bill proposed "the disposal of all classes of refuse under one general arrangement," which, among other things, intended that the city share profits from the sale of byproducts rather than incur more debt from constructing disposal facilities. The legislature passed the bill, but the governor vetoed it. Since the NYSUC contract was due to end by 1917, and given the ceaseless complaints about Barren Island, Mayor John Purroy Mitchel announced in early 1916 that a new reduction facility would be constructed on Staten Island.[35]

Mitchel's announcement opened a new and dramatic chapter in New York City's waste disposal story. It comes as little surprise, in hindsight, that Staten Island would be targeted as an optimum location for a waste facility. Distance from the city's centers of population and relatively easy access by water made it attractive. So too was the availability of open space in the form of marshland, which was considered otherwise useless. In a 1913 Department of Docks and Ferries report, Commissioner of Docks Calvin Tomkins noted that in the New York district of the port, "the most extensive opportunities for riparian public reclamation are to be found in the marshes surrounding Jamaica Bay." He also suggested that operations for reclamation might be available at other locations, including the Bronx shoreline, Queens, and New Jersey. "At the Fresh Kill district and Arthur Kill district of Staten Island, also at Great Kills and back of South Beach and Midland Beach," he added, "extensive reclamation opportunities may be found."[36] Yet sometimes fill operations to reclaim land and disposal of refuse became entangled.

In the body of the Docks and Ferries report, Chief Engineer Charles W. Staniford recommended a further extension of Rikers Island but asserted:

> In the meantime many thousands of acres of waste lands exist in all the Boroughs of Greater New York, as well as in the adjoining sections of New Jersey,

which are susceptible of being filled in by the City waste, and are now being filled in as rapidly as the necessary arrangements can be made by the street cleaning bureaus of the different Boroughs, as practically none of it is going to sea.

He recommended to the commissioner that the development of new areas of filling needed to be undertaken "on account of the large areas now available in Jamaica Bay district, in Flushing Bay, and in Staten Island itself, and what is already being used in Newark Bay."[37] With the Jamaica Bay and Flushing Bay operations winding down or under public attack, Staten Island must have looked very promising. Data in the report listed the area of land under water and marshland near the bulkhead lines in the Metropolitan District (New York City and New Jersey) presumably available for fill and/or waste disposal. Table 4.1 shows the square footage of marshland in the rear of bulkhead lines and shore lines in the various boroughs and New Jersey.

In the City of New York, Jamaica Bay held the largest amount of marshland. The ongoing battles over Barren Island, however, made siting a new facility there politically untenable. The Bronx and Queens had less available marshland and were relatively close to population centers. Manhattan and Brooklyn were not even in the running. New Jersey, although tempting because of the vastness of its available marshland, posed its own problems of jurisdiction. But Richmond Borough (Staten Island) had undeveloped land, was within the city limits, could be accessed by barges and scows, and was sufficiently distant from the metropolitan core.

It also is important to remember that ocean dumping was wildly unpopular. Incineration had its advocates (including Mayor Mitchel), but few were

TABLE 4.1 Marshland in the Municipal District, 1913

LOCATION	MARSHLAND IN SQUARE FEET
The Bronx	57,056,572
Jamaica Bay (Brooklyn and Queens)	243,570,000
Brooklyn (excluding Jamaica Bay)	—
Queens (excluding Jamaica Bay)	45,482,497
Richmond	237,459,843
New Jersey	1,402,328,309

Source: New York City, Department of Docks and Ferries, *Report on the Disposal of City Wastes* (1913), 19.

pushing it as a universal solution to the garbage problem. Constructing a new reduction plant was the obvious answer—especially if it was not placed in your community. Mayor Mitchel had stated in a January 26, 1916, affidavit that "the interest of the City demand that a garbage disposal plant should not be located within the confines of Jamaica Bay" because of the lengthy journey of scows to the current site, which, if there were delays, led to putrefied garbage causing offensive odors.[38] (Mitchel most likely assumed that a viable location required less political and environmental resistance.)

Jamaica Bay was the target of opportunity for some enterprising individuals. The Brooklyn native William H. "Billy" Reynolds, a noted and notorious "mover and shaker," made a career as a real estate developer, politician, and entrepreneur. He was the builder of the Dreamland amusement park on Coney Island. Along with his partner, the banker Frank Bailey; his business associate William M. Greve; Alderman James Gaffney; and the president of Brooklyn Ash, Billy Reynolds schemed to acquire the next reduction-plant contract out from under NYSUC.[39] Given their real estate interests in the Jamaica Bay area, such a plan also would fatten their wallets: land values there would rise once the Barren Island plant closed. But when a spot in Jamaica Bay became untenable, they looked elsewhere.[40]

TARGET: STATEN ISLAND

Rumors of a Staten Island plant began in January 1916. The newly elected borough president, Calvin D. Van Name, had heard rumors about a huge disposal facility planned for the island that would accept refuse from the other boroughs. Commissioner John T. Fetherston disingenuously denied knowledge of any such plan on January 8. Yet on January 18, the Board of Aldermen announced a request for proposals from private contractors to operate the plant. Bill Franz in the *Staten Island Register* many years later charged, "Not only had the city early on decided to make Staten Island its garbage pit, but it lied about it from the beginning as well."[41]

Fetherston signed the new contract with the firm of Gaffney, Gahagan & Van Etten, who assigned the building and operation of the plant to the newly created Metropolitan By-Products Company. James Gaffney gave a bond of $250,000 to guarantee that the plant would be operated without nuisance. In addition, the Gaffney firm agreed to pay the city $900,000 during the term of the contract for the right to utilize the refuse. The plant (a Cobwell reduction plant to be built by C. O. Bartlett and Snow Company) would be constructed on privately owned land and was to begin operation on January 1, 1917.[42] The city had the right to buy the plant after the five-year contract. The next step to seek a location for it on relatively isolated Staten Island.[43]

Mayor Mitchel had a say in the building of a new reduction plant on Staten Island. An ambitious and competitive anti-Tammany Republican (whom some regarded as an elitist), he was elected mayor in 1913 at the age of thirty-four. Originally supported by the yellow journalist William Randolph Hearst, Mitchel lost the endorsement of the newspaper magnate over a squabble and because of his close friendship with Billy Reynolds. At one gathering the new mayor told reporters, "Those who do not like my friends can go to hell."[44] Giving up his plan for a web of incinerators, Mitchel made sure that the Reynolds bid was not sidetracked. Van Name cast a negative vote against offering the contract to Metropolitan By-Products Company at the hearings in April, but he was outflanked.[45]

Staten Islanders fought a spirited battle against the reduction plant, one they would relentlessly pursue even after the decision had been formalized. On March 8, Van Name and Henry P. Morrison, the borough commissioner of public works, warned locals that the probable location for the plant was Staten Island. At a meeting of the New Brighton Board of Trade on the same day, "a prevailing rumor" about the placement of the reduction plant was discussed, and by unanimous vote the board members advised the mayor that "the property owners, taxpayers and residents of Richmond Borough will most strongly protest against and vigorously oppose such action."[46]

On March 17, William Wirt Mills, vice president of the Civic League, longtime resident of Staten Island, and former commissioner of the Department of Plant and Structures for the City of New York, called a meeting attended by representatives of twenty-one civic groups. A vigilance committee was formed with Francis F. Leman, president of the West New Brighton Board of Trade, as chair.[47] A letter to the editor in the *Staten Islander* stated that those present listened to a statement from Commissioner Fetherston concerning the merits of the proposed plant. The writer asked why the locals had to take care of their own waste in incinerators in New Brighton and Rosebank and also "endure" the large plant and its "enormous quantity of offensive material" from boroughs "many miles away." "A man may endure a boil on his own face—he cannot prevent it—he has to endure it; but by what reason should he be obliged to endure the placing or transfer of boils from the faces of three of his friends or enemies?"[48]

Six days after the meeting, Mills requested that Van Name urge the mayor to "grant a public hearing" before any action was taken to establish a reduction plant on the island.[49] William G. Willcox, president of both the Civic League and the Board of Education, wrote the mayor, stating "in thirty years residence I have never known the people of Staten Island more aroused and united about anything than they are about this." Willcox expected "a square deal" from Mitchel, realizing that the city faced serious problems with waste disposal. But, he added, "in the face of widespread and well-nigh universal opinion" that a public nuisance will be created with "great and lasting injury [to Richmond Borough], the effort to foist this plant upon Staten Island, without even giving us an opportunity to be heard, has outraged the sense of justice of our best citizenship."[50] Citizens wanted "fair play" for what some leaders called the "weakest Borough."[51] This was David versus Goliath.

Given the mayor's entrenched position on this issue, citizens employed a number of tactics in an attempt to reverse the tide. The Vigilance Committee tried to meet with Mitchel, to no avail. Injunctions to stop the plant secured over several weeks in March and April also failed.[52] The mere mention of the mayor's name triggered a hostile reaction. Mitchel not only defended the refuse contract but questioned the opposition's motives. He was hissed at a protest meeting of several hundred Staten Islanders at Stapleton's German Club Rooms on April 14. "I have heard it said of the Mayor," the first speaker proclaimed, "that he was either a fool or a knave. I have looked up the dictionary definitions of these two terms and I have concluded that the Mayor is not a fool. A fool has no brains."[53] Leman called Mitchel "Mayor of Manhattan, Brooklyn and the Bronx, but not of us." He was trying to make "our fair island the dumping ground" for those three boroughs. Leman added that the district attorney of Richmond County should bring criminal charges to "prevent any invasion of our rights." If the "outrage is to be perpetuated on us," it was time to "come to the parting of the ways."[54]

The Vigilance Committee, including representatives of the Richmond County Medical Society, the Bankers' Committee, clergy, and boards of trade, issued a "Warning to the Mayor" on April 20. It was proposed by Dr. Louis A. Dreyfus, a chemist deeply involved in local activities and eventual president of the Civic League.[55] In it, the undersigned stated that they "give you fair warning that the people of Richmond will NOT have the refuse of other boroughs brought here for any purpose, and will NOT allow the erection and operation of a garbage reduction plant on Staten Island." They threatened to appeal in the courts, secure indictments against responsible parties, and petition the federal government and even considered separating Staten Island from Greater New York. They added that proximate cities in New Jersey were equally aroused. "We are fighting for our homes and our wives and children and their right to live in clean and healthy surroundings, and we will fight to the last ditch. You cannot under cover of official might pollute our air and destroy our homes. If you do not stop this outrage you will create a feeling that Richmond Borough should withdraw from the City."[56] The good and hopeful feelings in Richmond Borough over the 1898 consolidation were dead.

A short time before "The Warning," State Senator George Cromwell (a Republican who served from 1915 to 1918) introduced a bill in the legislature to prevent any borough from disposing of refuse in another borough without the consent of the majority of the local boards of improvement. The bill passed and reached the mayor on April 22. On May 2, about two thousand Staten Islanders gathered at City Hall (cordoned off by police toting nightsticks) to urge the mayor to sign it. Many of the protesters wore badges declaring "Fair Play for Staten Island. No Garbage."[57] Somewhat ingenuously, Mitchel stated that his administration was willing to take the plant from Staten Island if an alternative was available. A suggestion to utilize Swinburne Island was rejected because it was too small, and another proposal to move the plant further down the kills also was not feasible. A third idea, from a subcommittee of the Vigilance Committee, was to construct an artificial island for the plant on Romer Shoal (in Lower New York Bay, about two miles north of Sandy Hook and a good distance east of Staten Island). At a session of the Board of Estimate on May 5, contractors rejected the Romer Shoal plan.[58] By that time, Mitchel had exercised his veto power on the Cromwell bill (May 3). Not to be deterred, a Staten Island Separation Committee was organized on May 11, and the Vigilance Committee pressed the Board of Estimate the following day to abrogate the garbage contract on the grounds that it violated public health law. The battle raged on.[59]

BATTLES OVER PRALL'S AND LAKE'S ISLANDS

While the Vigilance Committee intensified its protests over the siting of the waste plant, other dramatic events unfolded on Prall's Island, the location where contractors initially proposed to build the reduction plant.[60] The contractors had agreed to buy the Prall's Island property from the Manhattan real estate developer J. Sterling Drake, who believed he had the right to sell the land.[61] But this claim came into dispute in 1916, when Edward P. Doyle asserted ownership of the island. Doyle was a realtor, Democratic politician, and bureaucrat; he also was a Mariners Harbor native and Port Richmond resident. He strongly identified with those resisting the siting of the plant on Staten Island and became one of the leaders of the anti-garbage group and also a member of those promoting secession.[62]

Prall's Island included a salt meadow where some cattle grazed. In 1910, it became a depository for dredge spills from Arthur Kill.[63] Edward Doyle maintained that his title (now in the possession of his son Roger) came from a quitclaim deed obtained from L. D. Rourke of Greenville, Connecticut. Allegedly it was based on a 1708 land grant from Queen Anne to Lancaster Symes. In a letter to the *Staten Island Advance*, Edward Doyle stated, "Queen Anne granted all the land not previously granted under water around Staten Island to Major Symes, a friend of hers. The rent was to be a bushel of wheat paid every year at Windsor." On this rather thin pretext, the Doyles laid claim to Howlands Hook, Prall's Island, and Lake's Island on the North Shore of Staten Island.[64] This was no mere real estate dispute. In an April 1916 edition of the *Richmond County*

Advance, Doyle proclaimed, "We do not belong in Greater New York. We are a home-loving and home-owning community. The Greater City is neither."[65]

To hold his claim to the proposed site of the reduction plant on Prall's Island, Doyle constructed a shack there in late April and surrounded it by barbed wire. He then placed the seventy-one-year-old twins James and William Farmer as watchmen in the shack. No sooner had they settled in when three men were arrested on April 30 for cutting the fences on the proposed plant site. Late on the evening of May 18, launches carrying fifteen men with shotguns landed on the island and kidnapped the elderly Farmer brothers. According to the watchmen, they were taken to Brooklyn and let loose but were threatened with death or injury if they returned to the property. At about 4 a.m. on that day, police discovered fifteen "raiders" in the island shack "shivering around the stove." They were rounded up and promptly taken to jail on a trespass complaint by Roger Doyle. The men were released on bail and the charges dismissed. (The Richmond County grand jury later indicted Drake in connection with the alleged kidnapping.)

George Stewart, the ringleader of the group, told authorities that Drake had hired them to take possession of the property. Another raider stated that he went along because he was told he could make a few dollars pulling down a fence on the island. He also had helped carry a food order consisting of two hundred sandwiches and forty quarts of coffee, which seemed suspiciously large, to Pier 6 on the East River. On arrival he saw Drake and a group of men preparing for the launch. "I saw that Drake had a roll of money that would choke an elephant." He then retold the story of the night's events, bemoaning the fact that he never got paid. James Gaffney denied any involvement in the raid or in the shanghaiing of the Farmers. The reaction of Staten Islanders was more clamorous. Around the same time, about two hundred wives and mothers formed the Anti-Garbage League for Women, to back the fight against the plant. Mrs. Charles Simonson was named chair.[66]

While Drake still claimed ownership of Prall's Island, the Doyles were not easily moveable—placing nine men on guard there. The contracting firm, resigned to the fact that only a long court battle might win them Prall's Island, were frustrated that Drake was unable to sell them the property. In mid-May, they took possession of nearby Lake's Island (Simonson Island, which also was claimed by the Doyles), close to Rossville, as the new location for the reduction plant. They placed fifteen men from the Schmittberger Detective Agency and the police department (of which Max F. Schmittberger was chief inspector!) on Lake's Island to guard it. The Gaffney firm began surveying the island, which included about seventy-five acres of marshland. At the time, they had neither a building permit nor permission from the Board of Estimate to change the plant site. The men occupying the island burned dead grass and salt hay but did not build structures. According to the *Staten Islander*, District Attorney

Albert C. Fach, who had visited the site, was told the men "were employed by a moving picture company and that they were on the island to make a thrilling episode in a feature film." While the silent-picture industry often used Staten Island locations, in this case no photographic equipment was in evidence.[67]

On May 26, the Board of Estimate again agreed with the contactors' plan, this time granting their request to change the site of the proposed garbage plant to Lake's Island. The *Staten Islander* proclaimed: "From the moment that the request of the contractors was taken up until the vote was called for by the mayor it was most apparent that the desires of the garbage plant people would be granted, and that once more Staten Island would have to play the part of the city's 'stepchild.'" The Vigilance Committee called for mass meetings to be held immediately at Rossville and Tottenville.[68]

At the Board of Estimate meeting, Edward Doyle protested to the mayor about the new occupation: "Well, that Island is the property of my son, Roger. These men are squatting on it. I desire to build a bungalow on that Island. Will the police protect me if I start work there, and see that myself and my men are not injured[?]" Mitchel responded that "they will" and added that the contractors' men "are merely there to preserve the peace." That statement, the *Staten Islander* noted, "caused the only laughter from the Staten Island men and women present, who seeing the way things were going, did not feel any too happy." President Van Name reminded the mayor that his fellow citizens would not allow anything forced on them without serious consequences, dredging up the burning of the Quarantine fifty years earlier. After Van Name spoke, the board nevertheless voted in favor of the contractors.[69]

With the political system continuing to thwart the goals of the anti-garbage protesters, Edward Doyle again took matters into his own hands, urged on by his friends and supporters. Although the story (like many of the incidents in the Garbage War) varies somewhat in the telling, the gist of the most recent Doyle adventure is as follows: On the evening of May 27, Doyle ran a barge with seven men on it (including Roger Doyle) to Lake's Island. In what appeared to be a rather thinly veiled act of subterfuge, the vessel bore a sign—"Doyle's Bathing Pavilion"—with ten small bath houses on deck. The barge actually contained a load of lumber to build Doyle's "bungalow" (he had a building permit, to boot). However, the occupying party of private detectives and police stationed on the island tried to keep Doyle from landing his supplies, and so the tug pulling the barge set it aground, which impeded water traffic.[70]

Subsequently, Doyle and his seven men were arrested on a misdemeanor for obstructing navigation in the narrow channel around the island.[71] Doyle was ordered to remove the barge. At home later in the evening of May 27, he declared victory over the contractors, adding (and continuing his charade) that he would put together a moving-picture barge, a café barge, a merry-go-round and carousel barge, and turn Lake's Island into what the *New York Times* called "a wonderful Summer resort on barges." Doyle firmly believed that the garbage contractors would never build the plant there and that if they did locals would tear it down.[72] The barge affair made Doyle even more popular among Staten Islanders.

The comic opera played on in June. On the 8th, Edward Doyle and fifty or more men set out from Mariners Harbor into Staten Island Sound on a barge rigged with a thousand-ton derrick and boom. Reaching Prall's Island (now occupied by Drake's men, who had reclaimed it from Doyle's guards), the boom was swung out over the hut and a hawser (thick cable) secured around and under it. Some in Doyle's party approached the hut and insisted that Drake's men inside leave. A second warning followed: "You are intruders on this property and we demand that you leave. Are you going to do it?" In response they heard, "We belong here and here we stick." At that point the hut was slowly lifted and swung over the water with the men in it. The *New York Times* reported, "A frantic commotion within the hut gave evidence that the prisoners were not especially comfortable during their aerial voyage, the derrick not working as well as it might, but they were finally deposited without mishap on the deck of the lighter, and the lighter steamed slowly again out into the stream."[73] A police launch in pursuit overtook the barge. Policemen along with Drake boarded it. Drake barked, "You will please release my men and at once." Doyle complied, and the men were let go.[74]

Thirty of Doyle's armed men then occupied a new cabin on the island and did not allow Drake or his men to return. Of the latest encounter, Doyle stated, "It was too easy for words." He then proclaimed that he would use the same methods to retake Lake's Island. The *New York Tribune* painted a vivid picture of the latest Doyle-Drake clash, calling it "the most dramatic day in the history of Staten Island's war against the proposed garbage reduction plant."[75] Ill will between the anti-garbage protesters and the contractors rose to a new high.

APPEAL TO THE GOVERNOR

The anti-garbage protesters on Staten Island viewed Doyle's actions as courageous and flocked to support him. On June 3, Lake's Island Realty Company had been formed, which took control of Doyle's deeds to Lake's Island and the Isle of Meadows.[76] Enraged Vigilance Committee members had hurriedly organized the company, hoping to give more leverage to the Doyle land claims.[77] Not coincidentally, Edward Doyle was first vice president of the new company, which also included William Willcox, president of the Civic League; Charles Simonson, the League's finance chair; William Mills, league vice president; and other key protesters.[78]

On July 1, however, police barred them from the Lake's Island property, and on June 5 Justice James C. Cropsey denied writs on their behalf. In his opinion, he noted, with some irony:

> It is not surprising that the residents of the Borough of Richmond are opposed to the erection of such a structure. And it is still less surprising that they make strong opposition when, as here, the contemplated structure is to dispose only of the garbage from the boroughs of Manhattan, The Bronx and Brooklyn and NOT of the garbage from the borough in which the other garbage is to be destroyed.[79]

But the Staten Islanders got no satisfaction that day in his decision, despite his observations. Doyle responded, "The fight will go on.... We are far from being defeated."[80] The contractors, for their part, moved forward with plans to build the facility, despite lacking a permit. And they sought damages from the Vigilance Committee for obstruction of their alleged right to build.[81]

The Vigilance Committee had tried everything. In a strongly worded letter on July 7 to the governor, mayor, surgeon general of the U.S. Public Health Service, and the Commissioners of Health for both the city and state, it queried: "What profiteth it a City if it makes a few dollars out of garbage and slays hundreds of babies?" It claimed, as others had before them, that flies bred and feeding on waste matter caused infantile paralysis (poliomyelitis). The solution was "quicker and more efficient" collection of garbage and "prompt provision for the destruction of all garbage by incineration in small plants so located as to necessitate only short hauls in covered wagons." This plan reinforced the cremation process already in place in Richmond and Queens Boroughs.[82] The letter sent a message that the proposed "uncivilized" reduction plant would only contribute to the crisis. It was the city's duty, it concluded, "to abandon the idea of sucking a penny out of filth at the cost of babies' lives."[83]

After Doyle's failed attempts to gain control of possible sites for the plant and the Vigilance Committee's unsuccessful efforts to get a sympathetic ear from the mayor, anti-garbage forces appealed to Governor Charles S. Whitman. About two hundred protesters traveled to Albany in July to do just that. George Scofield, chair of the Richmond Borough Republican Committee, arranged for local "garbage warriors" to make their case to Whitman. A wealthy shipbuilder from West Brighton chartered a special train to take the group to Albany.[84] Thirteen of the protesters and a Gaffney attorney made arguments. A biased story in the *Staten Island Civic League Bulletin* reported that Staten Islanders "performed his or her allotted share of the work with brevity and skill" while the Gaffney lawyer offered "specious pleas of the contractors" that were "shattered so completely that their advocate was reduced to a state of speechless helplessness."[85] After a three-hour hearing, at which time each side made its case,

Whitman declared that he would order an investigation on the grounds of public health concerns.[86]

Two weeks before the hearing in Albany, a Richmond grand jury had recommended against the Staten Island plant siting, stating, "The information in our possession would indicate that there was a ruthless disregard of the sentiments of this community with a consequent irreparable damage and injury to the people of this County, in that through the acts of the aforesaid officials, a garbage reduction plant is to be located in said county."[87] The Vigilance Committee submitted a brief to the governor in support of the grand jury's recommendation, making clear that the committee was formed to "resist the invasion" of the contractors, which "discriminates against" Richmond Borough, resulting in "irreparable injustice and injury" to the citizens and property thereof. It also argued that Richmond County as a "county" was not included in the 1898 consolidation and thus fell under state not city jurisdiction.[88]

A Republican and an attorney by profession, Whitman had been propelled into statewide notoriety in 1912 by his campaign against police corruption in New York City. A relatively modest reformer, he was seeking reelection in 1916 and was not prepared to ignore some of his constituents.[89] In a formal letter to State Commissioner of Health Hermann M. Biggs, Whitman acknowledged the results of the grand jury and the hearing, authorizing the commissioner "to make a complete and thorough examination of the entire subject and to report to me as speedily as possible your findings and conclusions upon this matter."[90] Ironically, at the same time as the governor's charge, the Metropolitan By-Products Company secured approval of the Borough of Richmond's building superintendent to erect two structures on Lake's Island.[91]

TWELVE DAYS OF HEARINGS: AUGUST–NOVEMBER, 1916

Dr. Linsly R. Williams, deputy state commissioner of health, was assigned to conduct the hearings. Williams was well known in the field of social and preventive medicine and the allied field of public health. He was the first director of the New York Academy of Medicine.[92] The hearing board also included Theodore Horton, director of the Division of Sanitary Engineering for the State Department of Health; Wilber W. Chambers, deputy attorney general; and Professor George C. Whipple, special consulting engineer. The governor, Williams, and Horton agreed that including a sanitary expert who was not employed by the Department of Health was prudent.[93] Whipple was a protégé of William T. Segwick, a key figure in shaping public health practice in the United States. The Gordon McKay Professor of Sanitary Engineering at Harvard University, Whipple also was a partner in the celebrated engineering firm of Hazen,

Whipple and Fuller in New York City. His influential publications focused especially on water supplies. Waste disposal, however, was not his specialty.[94]

The hearings were held in Surrogate's Court at Borough Hall in the St. George section of Staten Island. Metropolitan By-Products Company was quick to protect its interests by offering to assist the health commissioner in the investigation. In a statement issued in mid-July, L. Fletcher Snapp, a representative for the contractors, seemed to be arguing two sides of the issue:

> Even if Staten Island were a thickly built-up community residents there would have nothing to fear from our garbage disposal plant. But the island, while it comprises one-fourth the area of the city, has but 1 ½ per cent of the population, and the entire Fifth Ward, in which the garbage plant is to be built, has on an average but one person to the acre. Such isolation of the plant or any isolation, for that matter, is wholly unnecessary.[95]

Most, if not all, of the population of Staten Island likely disagreed with this assessment.

The hearings were conducted between August 28 and November 23, 1916. Albert Fach represented the people of Richmond County, with Vigilance Committee attorneys also in attendance. For the defendants, R. P. Chittenden and John F. Collins were corporate counsel for the City of New York, and Liston L. Lewis, of Lewis and Kelsey, represented the Metropolitan By-Products Company.[96] During the extensive and exhausting proceedings, counsels for the complainants and defendants called twenty-one witnesses apiece. Among them were chemists and engineers, sanitary experts, government officials, business people, local residents, a photographer, a game warden, and an oyster dealer.[97]

The prominent sanitary engineer Colonel William F. Morse testified for the complainants. Morse had gained a reputation as a prolific writer on the subject of refuse collection and disposal and as a designer of modern incineration systems. Those systems had been in direct competition with garbage-reduction plants, which made him a useful witness for those opposed to the Metropolitan By-Products Company plant. Rudolph Hering, the "father of American sanitary engineering," was a witness for the defense. His investigations of water supplies alone had resulted in reports for more than 150 cities. A winner of awards too numerous to mention, he had been elected president of the American Public Health Association in 1913.[98]

The presence of such powerhouse witnesses and the detailed probing of issues on both sides (resulting in 1,694 typed pages of testimony and a large number of exhibits) clearly indicated the seriousness of the inquiry. Whipple noted in his final report that the attorneys had "more than ordinary latitude" in introducing evidence because the proceedings complemented a general investigation of the subject under debate by the commissioner of health.[99] The hearings would be no whitewash or a pro forma event. Neither were they a referendum on reduction. They focused instead on the merits of the Cobwell system to be employed on Staten Island and the degree to which it might pose a nuisance or health risk.

An extraordinary amount of time during the proceedings was devoted to technical and scientific details about possible odors and other nuisances emanating from the proposed plant and the scows delivering the waste. Lewis exhorted the hearing's board members "to put yourselves in our [the Metropolitan By-Products Company] shoes and consider our practical position." "It does not take a far-sighted man," he added, "to know that we are compelled to transport that garbage, operate that plant, in such a way as not to constitute a nuisance."[100] The strategy of the defense was to keep issues as narrowly focused as possible on the quality of the plant. For this reason, the Cobwell system itself (the latest reduction technology for processing waste) received more scrutiny than the reduction process in general. Both sides gave substantial attention to the three places currently employing the new technology (New Bedford, Massachusetts; Los Angeles, California; and the Panama Exposition in San Francisco).

For the Vigilance Committee, it was not the state of the technology but its placement that mattered: "Why Staten Island?" This issue received only limited attention. Before the opening of the hearings, William Wirt Mills of the Civic League reminded the board that the people of Richmond County "have been under great stress" over the matter at hand "and feel very keenly the way they have been made victims of a series of lawless acts upon the part of City Officials" and the actions of the Corporation Counsel.[101]

Mills's outpourings did little to influence the momentum of the hearings. In an early exchange between the sides, John F. Collins, attorney for the City of New York, argued that the location of the plant had been determined by the Board of Estimate and other duly constituted officials, and thus siting was "beyond the scope of this inquiry." Health Commissioner Williams chimed in that "it does not seem to me to make any difference if there were fourteen places where it could be put." But Fach disagreed: "It does make a great deal of difference to the health of the people of this community, and that is the fact you have to determine." Williams was steadfast: "I do not think so."[102] Fach was frustrated by this exchange and its implications. In questioning Henry P. Morrison, Richmond's commissioner of public works, Collins stated, "I simply want to show that the President of the Borough of Richmond voted to put this same Cobwell system on Rykers Island." Fach responded, "Exactly, where it belonged. Let them take care of their own swill; we are taking care of ours."[103]

From the outset, the spokesperson for the Metropolitan By-Products Company was not going to give an inch on the quality of their proposed operation. Fach questioned Charles R. Van Etten, speaking for the Gaffney firm, about the notorious Barren Island site. Van Etten agreed with Fach that "it has actually been a nuisance." But, Van Etten added, "I believe a legal nuisance has never been established against the plant."[104] Day after day, the main issues remained: the odors emanating from existing plants and the proposed plant, nuisance caused by shipping waste on open scows, and the potential health risks of the disposal option to be carried out by the Metropolitan By-Products Company.

Dr. John R. Hicks, a bacteriologist and medical officer testifying for the complainants, stated, "There is not an authority in the world that will state positively that there are no odors from any reduction plant ever built. If there is, we have several of the most eminent ones that we can quote easily." Whipple did not challenge the remark.[105] Fach repeatedly tried to link the limitations of the reduction process to its possible impacts on Staten Island. During the examination of the consulting engineer Olin H. Landreth, Fach asked pointedly if the proposed plant would constitute a public nuisance. Landreth answered, "In my opinion it will." "Will it constitute a menace to public health?" Fach asked. "In my opinion it will," Landreth answered.[106] Later in his testimony Landreth told Whipple that the odor from the new plant "would be 50 times as great in volume" than at Barren Island, given the greater amount of garbage to be processed.[107]

Seeking comparisons with the proposed plant for Lake's Island, the hearing frequently turned to experiences with reduction past and present. Barren Island was a consistent focal point. The defense attorneys played down, when possible, the health risk of the island's notorious past. In questioning Hicks, Lewis asked, "Do you know of any worse—I might say 'stink-pot', if you will pardon the expression—than Barren Island immediately close to the Barren Island plant?" He answered, "No." Lewis continued, "Do you know of any cases of infantile paralysis on Barren Island?" He answered, "Not that I have heard of." Fach had painted a picture of Barren Island as a worst-case example of a disposal facility and inferred that conditions were right for polio to exist there. Lewis pounced on this argument: "If I state to you, which I think is a fact, that Barren Island has a population of 450 boys and girls and children, many under ten years of age, and not a single case of infantile paralysis developed on Barren Island, the very worst spot according to the contention of Mr. Fach, which could be found; there is not a case of infantile paralysis, how would you account for it?" Hicks shot back, "Mr. Lewis, that counts nothing on earth; that does not show one thing, not one thing."[108]

Without the plant on Lake's Island yet constructed, defense turned to showing how effective the new reduction system worked in New Bedford and Los Angeles. The complainants used Paul Hansen, a sanitary engineer from Illinois with experience on odor nuisances, as an expert witness. He submitted a report on the New Bedford plant and reduction but did not testify at the hearing.[109] Hansen's statement, read into the record, however, was not a strong piece of evidence for the complainants. He had visited the New Bedford reduction works on November 15, and the cold weather, he stated, inhibited his ability to judge the intensity of odors. While he speculated that privies near the garbage works could attract flies and spread some disease, he found it "rather difficult to get a measure of the danger of fly infection."[110]

Infection from rats did concern him. In a rather big speculative leap, he stated:

> Should Bubonic plague get a foothold in the port of New York, and should there be a great congregation of rats on Staten Island, due to the presence of the reduction works, it is easy to perceive that Staten Island might be hard hit by an epidemic of the dread disease. This line of testimony might be utilized to show why the reduction works should be placed on a relatively uninhabited island like Rikers Island, and not in a populous residence district like Staten Island.[111]

Rather than making his statements in a direct manner, Hansen continually editorialized, offering tactics and strategy for the complainant counsel. Hansen asserted that "I do not believe you will gain much" by questioning the Cobwell process's ability to sterilize garbage and that "time and effort would be better spent" focusing on "nuisance conditions and property damage."[112] Further undercutting his own side, he stated that while he disagreed with "the great superiority" of the Cobwell process, "it must be acknowledged that [it] has a number of points distinctly superior to other methods."[113]

Other complainant witnesses were more predictable. Olin Landreth skirted the defense attorney's question about the suitability of the Cobwell process by arguing that "it is impossible for me to draw a comparison because there are no other plants of just the same size of the New Bedford, and capacity has quite as much to do as thoroughness of method, in the production of odors."[114] Louis L. Tribus, consulting engineer, was asked point blank if he knew of any better system than the Cobwell. He answered, "I do not."[115] The complainant's star expert witness, Colonel Morse, suggested that odors from the reduction process at New Bedford were "inevitable." "No odor can be prevented from arising from it."[116] As to whether the new plant on Lake's Island would be offensive, he said yes for two reasons: odors emanating from the plant and pollution along Fresh Kills. The odor, he stated, "would be objectionable, offensive, and a detriment to health to every person so far as it was carried."[117]

The defense pressed Morse to clarify why he believed the odors were detrimental to health and received an ambiguous response. They got him to admit that there was no better reduction method than the Cobwell, although as a backer of incineration his admittance had an implied qualifier.[118] August E. Hansen, a hydraulic and sanitary engineer, was more emphatic than Morse.

Asked if he knew of a better reduction system than the Cobwell, he stated, "I think all of the reduction systems of garbage are undesirable." But he did not want to venture an opinion on the Cobwell because he considered it still in the experimental stage.[119]

Defendants mustered their own witnesses to demonstrate the viability of the Cobwell system. W. J. Springborn, general manager of the New Bedford Extractor Company, testified to what the attorney called "the practical impossibility of creating odors under the Cobwell process as one of the distinguishing features from the other processes in use."[120] The eminent Rudolph Hering spoke about a Cobwell plant operating at the Panama Exposition in San Francisco. "Personally, I did not at any time observe an odor, either at the plant or away

FIGURE 4.5 Blackwell's Island convicts unloading scows of ashes at Riker's Island.

Source: *Scribner's Magazine*, October 1903.

from it."[121] There was a "decided odor" of the tankage, "but not at all offensive."[122] Raymond Wells, chemist and inventor of the Cobwell system, was one of the strongest witnesses for the defendants. Fach did not challenge Wells's design of the reduction plant.[123] The questioning and debate over the Cobwell system and its potential risk factors ultimately resulted in a draw.

Siting the plant on Staten Island was, at best, a minor factor in the hearings. Fach plugged away at showing how specific risks at the proposed site were sure to come. But sometimes defense witnesses got in the way. Joseph A. Shears, a sanitary expert for the Health Department of the City of New York, denied that a menace to health could occur, "providing the housekeeping and the operation of the plant is carefully done."[124] Dr. Haven Emerson, commissioner of health for the city, categorically denied that "a plant similarly operated at the site of the location on Staten Island [as at New Bedford] would not cause any nuisance to the residents of Staten Island in any of the places you have mentioned."[125] He did think, however, that Rikers Island was "a more logical, convenient place for the location of the plant."[126] Despite the length of the hearings, the likely outcome was not going to appease the anti-garbage protesters. There would be neither a referendum on reduction nor any path toward an alternative disposal method. Proponents of placing a waste plant on the island would not heed the witness John R. Eustis, sanitary inspector for the New York State Department of Health: "a garbage plant in my opinion is a nuisance, and always will be a nuisance."[127]

THE WHIPPLE REPORT AND THE GOVERNOR'S DECISION

Professor George Whipple's report (December 9, 1916) gave Biggs his money's worth. After consulting with the other commissioners, Whipple began writing.[128] He had evaluated the pertinent laws; examined the disposal contract; inspected garbage dumps in Manhattan; visited Fresh Kills and nearby water routes, Barren Island, and the Cobwell works at New Bedford; and attended the hearings. In the report he discussed the present methods of disposal in New York City and the quantities of waste. He offered a thorough assessment of the contractor's obligations, especially with respect to preventing emissions from the plant and efficiently handling the garbage. He provided detail on the scows, the Cobwell process, and the New Bedford and Los Angeles plants. Whipple spoke about complaints concerning the Panama Exposition plant. And he suggested what legally constituted a nuisance.[129]

The learned engineer's conclusions reflected the scope of the testimony over the twelve days of examination and cross-examination. Whipple did not avoid addressing the one issue that the defendants wanted played down—the suitability of the Fresh Kills site as compared to elsewhere in Greater New York City.

He stated emphatically that the transport of garbage by scows and the disposal by the Cobwell process "will not, in any appreciable or material way, affect the security of life or endanger the health of the people of Staten Island."[130]

He admitted that the disposal works will, "by reason of odors emitted at certain times," be a nuisance in the Fresh Kills marshes, around Linoleumville, and along highways but "will not extend generally over Staten Island." On their trek to the site, the scows will sometimes produce odors and litter. With respect to the Cobwell process, Whipple gave it a ringing endorsement: "the best method now known and available for the disposal of the garbage of the city of New York." But he sided with the complainants on one key point, asserting that from a standpoint of sanitation there were other sites "more suitable and satisfactory" for the location of the plant, especially continuing to use Rikers Island or Barren Island.[131]

On balance, the report had something for both sides, with Whipple's conclusions reflecting the strengths of each case. Complainants had to be pleased that Whipple had addressed the issue of "Why Staten Island?" and acknowledged the likely nuisance of both the scows and the facility. Defendants had to take heart in Whipple's endorsement of the Cobwell process as the best possible technology and his argument that a "health menace" could not be proved (although whether Whipple was qualified to have expert opinions on health matters is not so clear). The report left room for interpretation by the commissioner and the governor; how much room remained to be seen.

Curiously, the cover letter to the report that Williams submitted to Biggs listed only three of Whipple's conclusions: (1) transport of the garbage and the operation of the plant would not constitute a health risk, (2) transport by scows will at times prove offensive to Staten Islanders, and (3) the plant may at times disseminate objectionable odors. There was no mention of the Cobwell system or of preference for other disposal sites. It seemed clear that the decision to make revolved around whether the new facility under contract would pose a nuisance—and nothing more.[132]

Commissioner Biggs submitted the report to Governor Whitman on December 27, 1916. Leaks of portions of the report before the governor's decision inspired false hope on Staten Island. A headline in the December 29 *New York Tribune* proclaimed: "Richmond Wins Garbage Fight: Staten Islanders Hear Report to Whitman Favors Them." The story went on to suggest that "so widespread was the rumor and apparently so well authenticated, that plans were set afoot immediately to have a tremendous celebration, to which Governor Whitman should be invited." The story not only misread the governor's response but referred to the new plant as an "incinerator."[133] In an interview with the *Staten Islander*, the Civic League's Louis Dreyfus stated that published excerpts of the report in some of the Manhattan dailies before the governor's decision "would to my mind indicate that Staten Islanders had scored a clean-cut victory in this investigation." He added with no little tinge of bitterness that they must await the governor's action, but "we must also expect a belittlement of our victory by the officials of the city of New York, the contractors and their lawyers. All that they can say, however, will detract not one iota from the findings of facts of the commissioners conducting that investigation."[134]

Rumors of victory for Staten Island were dashed on January 25. The governor's decision left the anti-garbage crowd disheartened. It also brought into question why the hearings had been held at all. Governor Whitman first requested an opinion from the state's attorney general as to whether the governor had authority "to direct the cessation of the construction of the garbage plant on Staten Island on the grounds that its use and operation would become a nuisance within the meaning of the law."[135] The attorney general's opinion was that the existing statute law only allowed the governor to declare the plant a public nuisance after it had been certified as such by the commissioner of health. Since no plant yet existed, no ban was possible. All that Whitman could do was to direct the sheriff and district attorney of Richmond County to inform him "when the operation of the proposed garbage disposal work is commenced and garbage is transported" so that he may take such action that "shall seem just and proper at that time."[136]

Thus no injunction was forthcoming, and only after the operation of the plant had begun could the sheriff summon a posse comitatus (men conscripted to assist him) if necessary to enforce the law against the plant or shut it down if a nuisance occurred.[137] One has to wonder why information on the statutory limits was not sought before the hearings. Putting the best face possible on the governor's decision, the *Staten Islander* ran the headline "Garbage Report: Governor Whitman Approves Health Commissioner's Findings."[138]

Despite Whitman's decision, Staten Islanders continued efforts to rid themselves of the reduction plant. State Senator Cromwell again tried a legislative route. On January 30 he introduced two bills to amend public health statutes, which in effect would check the building of the plant on Staten Island (or any plant without local permission) and provide compensation for the Metropolitan By-Products Company if the contract was rescinded. Since the bills were statewide measures, Mayor Mitchel could not veto them if passed.[139] A hearing on the bills was set for February 21 in the Cities Committee of the Senate. A large delegation from Richmond County traveled to the state capital to attend it. Noted civic leaders such as Willcox, Dreyfus, Fach, and Mrs. Simonson planned to address the committee.[140]

The tone and substance of the anti-garbage protesters at the meeting was encapsulated in Mrs. Simonson's assertion: "God help us if Staten Island becomes known as the dumping ground of the great city of New York."[141] As might be expected, the latest confrontation over the plant produced some heated retorts. Greve, speaking for the contractors, believed that the passage of the legislation would cost the city millions in damages. "This bill," he stated angrily, "and all this talk is utter foolishness." He added, with a warning, "If they want to put us out of Staten Island all right, but they will have to pay us. We are operating the Barren Island plant only under sufferance, and if the Staten Island project is killed we'll shut down the Barren Island plant, and the city will have to pay damages as well as find some other means of disposal of the garbage of Manhattan, The Bronx and Brooklyn."[142]

Despite such threats, the *Staten Islander* was certain that the bills were "sure to pass."[143] But as the debate dragged on into April, the protesters faced rocky times in the Senate. For his part, Cromwell stated that "I feel very, very much encouraged."[144] On April 13, the so-called garbage bill advanced to a third reading by a sizeable margin. But Cromwell faced strong opposition from Brooklyn Republicans, who saw the defeat of the reduction plant in Staten Island as a necessary prolongation of the Barren Island facility. In response to the Brooklyn state senator Charles F. Murphy's attack, Cromwell replied with a Great War reference: "The situation in a nut shell is this. We have no mayor in New York. We have a Kaiser and he is trying to make a Belgium of Staten Island."[145] (The Germans had invaded and occupied Belgium in 1914.) The Brooklyn delegation, however, could not prevent passage of the anti-garbage bill, and the Senate approved it on April 24 by a vote of 28 to 19.[146]

The story in the Assembly was different. Offered as the Cromwell-Seesselberg bill, the legislation to prevent transport of garbage to Staten Island failed. The Richmond County Democratic assemblyman Henry A. Seesselberg believed Cromwell had been "double-crossed" by Assembly Speaker Thaddeus C. Sweet, an upstate Republican.[147] The *Staten Island Civic League Bulletin* claimed that the speaker usurped "the constitutional prerogatives of the 150 members of the Assembly" and "tucked the bill in a pigeon hole in his office." Seesselberg's efforts to take the bill out of the Committee on Rules failed when Sweet called a Republican caucus, defeating Seesselberg's motion. This was clearly a party measure.[148] On May 9, the bill was dead.

THE PLANT UNWANTED OPENS

The contractors were of no mind to wait for an end to the political wrangling in Albany before building and operating their plant, and they had the work ongoing before Whitman rendered his decision on the Health Department hearings. Initially, the garbage from the three boroughs (Manhattan, Brooklyn, and Queens) was to be delivered by scow to the new reduction plant beginning January 1, 1917. Delays pushed the opening date back to April, with the Staten Island facility receiving all the refuse by December.[149] (From January until April, the New-York Sanitary Utilization Company would dispose of the city's garbage at Barren Island.) Ashes, rubbish, and street sweepings from the boroughs were disposed of by contractors at sea, at Rikers Island, at Hunts Point (on the Hudson River), at Staten Island, and at landfills within the various boroughs. Incinerators were in operation along almost all of the waterfront and at railroad stations in Brooklyn.[150]

Months before the first loads of garbage reached Staten Island, preparations for the plant were underway. Two buildings were already in place (constructed without a permit). During the legislative battle over the Cromwell bills, Greve boasted that he would have the plant in operation before they were passed. In March, the Richmond Light & Railroad Company had already completed an application forwarded to the War Department seeking permission to lay two power cables across Fresh Kills to connect with the plant.[151] But labor problems arose in early March, with trade-union workers disputing agreements with the contractors.[152] Neither labor problems nor community protests, however, kept the plant from opening. Two hundred reducer units were to be employed to handle the hefty garbage load.[153]

On April 20, 1917, the first scow loaded with garbage arrived at Lake's Island, and the next day the reduction plant was put into partial operation, with four reducers. The whole process was complicated. The scows (uncovered, as many had feared) traveled mainly at night through the Kill Van Kull and Arthur Kill. At the plant, they pumped out swill liquor (liquid organic waste), which ran into the Kill, and then workers picked over and removed items such as bottles, cans, and rags (which were washed and boiled before sale). The green garbage was lifted by cranes onto a hopper, then moved by conveyor to the top of the digester building. (The effluent was simply discharged into Fresh Kills.) The green garbage was exposed to petroleum naphtha or kerosene in steam-jacketed containers (digesters), boiled, and finally dehydrated for several hours. The naphtha grease was drawn off with the garbage now washed with solvent. The grease was separated from the naphtha in stills. In the third stage live steam was introduced to eliminate the naphtha from the garbage, which was then dried. The solid portion of the garbage was removed from the containers as tankage. This was sifted and ground. The tankage (for fertilizer and feedstuffs) was shipped in bulk, the grease and oil (for lubricants and fragrance base) placed in barrels, and the pickings collected. These three types of material constituted the saleable products of the reduction process.[154]

Emotions of all kinds swirled around Staten Island after the garbage battle was lost. The Vigilance Committee vowed not to end its opposition.[155] The *Staten Islander* announced, "We have fought the good fight. We have met the enemy and we are theirs." Then it added, "We must wait and not injure Staten Island's fair name by talking about it as a dumping ground when such talk can be no good."[156] A letter in the newspaper on May 12 made sport of Mayor Mitchel, who had become a constant target of criticism. The writer suggested that Mitchel escort the French marshal Joseph Joffre, victor at the First Battle of the Marne (1914), to "Garbageland" on Staten Island. "History will forever and forever laud Joffre as the hero of the Marne, so let us all now hail the hero of Garbage Island, the Honorable John Purr Mitchel, the man who carried out his desire to make us a dumping ground."[157] At this time, signs of a proposed Staten Island secession from Greater New York were in the air, via a legislative bill drawn up by Mills.[158]

THE ROAD TO CLOSURE

In late July, the *New York Times*, tinged with wishful thinking, announced, "the bitter opposition to the new plant, so manifest some months ago, has almost entirely abated, while many prominent people are strong supporters of the enterprise." Taking a stance in line with the Metropolitan By-Products Company's own defense, the paper went on to state that "misinformation" and "the persistent influence of opposing interests" had inspired antagonism toward the plant. Staten Islanders would not experience a nuisance, "but, on the contrary, [the plant] will comprise an institution of great value to the island."[159] In early August, the *Staten Islander* reported that the facility was "the Mecca early this week for big garbage plant owners from all over the country" who wanted to see it in operation. Van Etten, vice president of the Metropolitan By-Products Company, entertained dozens of guests. The approximately $2.5 million plant, with a capacity of two thousand tons a day, was purportedly the largest in the world.[160] Later in the month, city officials, civic leaders, and others toured the new plant as well as Barren Island. The *New York Tribune* reported that "as a matter of fact, the plant is so nearly odorless that Staten Island did not generally know it was running until yesterday."[161] More skeptical, Dreyfus called the gathering "the dress parade at the garbage plant."[162]

The anti-garbage group did not share the optimism expressed in the Manhattan newspapers. They awaited a "legally evidenced nuisance." Less public were the financial pressures affecting the contractors as a result of the war in Europe. Labor costs skyrocketed as more American soldiers went off to battle. With the overall rise in the cost of doing business, the Metropolitan By-Products Company was finding it difficult to live up to the city contract.[163] At the end of December 1917 and again on January 1, 1918, fires broke out at the plant. The latter was caused by an explosion of unknown origin, which damaged the docks and destroyed three scows. City garbage had to be temporarily dumped at sea or burned.[164] The contractors also faced relentless complaints from locals.

In June, 1918, a Richmond County grand jury began investigating complaints by villagers about fumes and odors originating from the plant. The Civic League and other organizations renewed their protests.[165] On June 21 the grand jury handed a statement to Supreme Court Justice Charles H. Kelby in Brooklyn condemning the plant as a nuisance but providing an extension of sixty days to install abatement equipment.[166] When the contractors did not comply, in July the grand jury demanded the revocation of the garbage contract. In a statement to the press, E. Steward Taxter, a foreman, declared that if the contract was not revoked the grand jury had the right to indict the Board of Health of the City of New York and the commissioner of street cleaning. "We also reserve the right" he added, "to prevent any more garbage scows being brought to the plant. We will, if necessary, resort to forcible resistance on the ground that such scows constitute a menace to health."[167]

THE PLANT FIZZLES

Before the newest round of protests intensified, a change in the mayor's office a few months after the United States entered the war in 1917 seemed to signal a fresh start in Staten Island's relationship with the city government. The Republican mayor Mitchel had been the focus of the protesters' ire, especially his being in cahoots with the Metropolitan By-Products Company.[168] His Democratic challenger, the Kings County judge John Francis Hylan, was supported by the Hearst publications and Tammany. Hylan was no mental giant, but his timing was right for election, and he had backers who could make a difference. Mitchel's campaign stressed "Hearst, Hylan, and the Hohenzollerns," trying to link Hylan to Hearst's anti-interventionist, pro-German sympathies. Despite the support of presidents Taft, Roosevelt, and Wilson and New York's governor Charles Evans Hughes, Mitchel was swept out of office in the November 6 election by a crushing margin (313,956 for Hylan, 155,497 for Mitchel). The Socialist candidate Morris Hillquit, who supported Wilson's preparedness policy, received an impressive 145,332 votes and severely undercut Mitchel's middle-class base.[169]

John ("Red Mike") Hylan, the former locomotive engineer and then lawyer, was as expressionless as they came but utterly loyal to his party.[170] To Staten Islanders, he was a vast improvement over the hated Mitchel.[171] In a letter to all of the members of the Fusion Committee of 250, the Women's Anti-Garbage League chastised the group for supporting Mitchel. "As mothers, homemakers and taxpayers," the letter stated, "we view with alarm the possibility of another four years of such autocratic rule. Mr. Mitchel has been our '*Kaiser*,' and Staten Island a second '*Belgium*.'"[172] In meetings in the borough in late October, Hylan assailed the so-called garbage ring and promised to appoint a Corporation Counsel "to attack the validity of that iniquitous garbage contract." He added, "The health of the people of Staten Island will be given prior consideration to the profits of Reynolds, Greve, and Bailey. The next Health Commissioner of this city will deal with the garbage situation with an iron hand."[173] In none of the meetings in different parts of the borough did he promise to shut down the plant.

True to his word, Hylan began an investigation of the Metropolitan By-Products Company plant, threatened to rescind the contract if the company did not resolve the nuisance and health issues, and considered having the city take over operations to produce glycerin needed for the war effort.[174] The mayor even diverted fifteen scows from the plant, dumping the garbage in the ocean instead.[175] He temporarily turned back to Barren Island and the New-York Sanitary Utilization Company in 1918, which was terribly unpopular, and the following year surprisingly authorized ocean dumping as the city's official disposal practice.[176] Also in 1918 (and especially significant) Metropolitan By-Products Company abandoned its Lake's Island reduction plant, stating that it was not profitable. (In part of July and August, the plant was forced to close down, for example, because the company could not afford to purchase coal and solvent.) The company had fought so hard to acquire, build, and operate the plant only to quickly walk away. On July 19, the Richmond County grand jury indicted Frederick Jeffries, superintendent of the plant, on the grounds of operating it in an "improper and inefficient manner, causing it to become a public nuisance [because of the odors]." At a Board of Estimate meeting the idea was floated to have the federal government take over the plant because of the need for glycerin and grease in wartime—and because the city had no funds to run the plant. But nothing came of this idea.[177]

The receiver in the bankruptcy proceedings briefly attempted to operate the facility, but the Board of Health shut it down on October 2, 1918. State Commissioner of Health Dr. Royal S. Copeland explained that a department investigation "demonstrated that the odors from the Lake Island plant carried ten miles—in some case as far as St. George."[178] In an even more sweeping decision, Copeland added that no boat carrying garbage could be moored in any dock in the city for more than twenty-four hours. The *New York Tribune* reported that "Staten Island for the first time in more than a year is breathing easier."[179]

A PYRRHIC VICTORY

Vigilance Committee members had no way of knowing that their triumph over the reduction plant on Lake's Island—a battle that lasted so long, but a plant that failed so quickly—was only provisional. Dumping of New York City's refuse on Staten Island would take another form in less than thirty years (and on a scale never imagined), and Fresh Kills would become the site of the world's largest landfill. The Garbage War demonstrated just how bitterly Staten Islanders were willing to fight to resist becoming the disposal heap for Greater New York, just as they had fought the Quarantine in the 1850s. Staten Island's resistance, however, did not only arise out of the arbitrary use of its land or fear of public health consequences but over image and identity. Statement after statement, harangue after harangue spoke to how the people in Richmond Borough firmly believed that they were being singled out to solve a citywide problem. They felt invaded. Measured against the controversies surrounding Barren Island, the locals had ample reasons for concern and apprehension.

Yet the Garbage War also exposed in graphic fashion how difficult New York City officials had found realizing a solution to a refuse-disposal problem bulging out of control. Dumping at sea was unsound and unpopular but too often became a fallback option. Incineration was a partial answer at best. Barren Island had become an albatross. Building a new reduction plant (still regarded by many as a workable solution) and seeking a location away from Jamaica Bay appeared to offer multiple benefits. But what appeared to be a viable technical decision, and one assumed to offer the least political resistance, had blown up in the officials' faces. They miscalculated the viability of the plant and underestimated the political pushback. As New York City continued to grow, its waste problems grew with it. The island city had no definitive disposal plan when the Cobwell plant shut down. It turned again to sea dumping and waited for a solution not yet in sight.

Also developing, especially after the end of sea dumping, was a constant round-robin of disposal options—reuse or reduction, burning, and landfilling—which would be tried and abandoned and tried again for generations.

PART III
SEEKING A DISPOSAL SINK

CHAPTER 5

THE GO-AWAY SOCIETY

Staten Islanders could breathe easier after the waste plant on Lake's Island was shut down. But for citizens of the other boroughs (and coastal New Jersey), the hesitant return to ocean dumping offered little to rejoice about. While ocean dumping would continue until the early 1930s, city leaders sought a better disposal plan—a better "disposal sink"—that everyone (meaning every borough) could live with.[1] The Waring-era approach to utilizing various forms of reuse and waste reduction had faded by the 1910s, and incineration emerged as the new panacea for New York City in the 1920s. The city seemed to be leaping from one impulsive choice to another, offering little chance for cooperation among the boroughs.

SEEKING A NEW REFUSE PLAN

In the wake of the reduction-plant fiasco on Staten Island, Mayor John Hylan in March 1919 proposed building a series of incinerators throughout New York City. Incinerators had been employed on a small scale before then but had not challenged either ocean dumping or reduction as a citywide disposal method. Hylan was impressed with the independently operated plants in the boroughs of Queens and Richmond and felt that burning refuse was superior to the maligned reduction plants and increasingly unpopular ocean dumping. Commissioner of Street Cleaning Arnold B. MacStay, who was appointed from within the department by Hylan in 1918, argued that he was hindered by outdated methods and equipment. He announced plans to erect six incinerator plants in Brooklyn.[2]

These plans were soon thwarted by the same local pressures that had bedeviled the reduction plant on Lake's Island. Brooklyn residents and others feared that under the proposed disposal policy they would wake up to find waste burners in their neighborhoods and vehemently protested the decision. The plan was dropped. MacStay then put out a bid request for a new reduction facility, but he found no takers.[3] Another scheme in the Bronx to fill swampland with waste was blocked in 1920 when Thomas F. Cavanaugh, president of the Bronx Taxpayers' Alliance, obtained a temporary injunction to restrain contractors from filling in swampland and "making a dumping ground out of the Bronx."[4] The disposal situation seemed untenable. Inner-city incinerators or reduction plants incurred the wrath of locals, and sea dumping threatened nearby beachfronts.

In a 1920 report directed at the Board of Estimate (the Board of Estimate and Apportionment until 1938), the Brooklyn Chamber of Commerce's Health and Sanitation Committee urged action on a new disposal plan for New York City, especially in light of the unsuccessful practices of the recent past and the abrupt dropping of Hylan's incineration strategy. It maintained that the board had not "considered any concrete proposition for disposing of the waste of the city" for two years, then "wisely discarded as impracticable and uneconomical" an attempt to build incinerators throughout the city. (Public resistance to such a proposal certainly did not help promote it.) The report added that conditions surrounding the collection and disposal of waste had grown "steadily worse"; contracts expired in one borough, while vendors surrendered a contract in another; temporary agreements with existing companies had led to "exorbitant rates"; "experiments with discarded and untried systems" and "piecemeal plans" had not worked, "while the city drifts aimlessly from bad to worse, seemingly unable to help itself, dependent on private contractors who own his equipment and control the cars, barges and piers."[5]

The Health and Sanitation Committee was critical of the charter, which kept collection and disposal functions "scattered and decentralized." The Bronx, Manhattan, and Brooklyn services were under the commissioner of street cleaning, appointed by the mayor, but the Queens and Richmond services were under superintendents appointed by the borough presidents. "This decentralized administration and this unorganized, unstandardized and unscientific system of collecting and disposing of the waste of this city of over 5,000,000 people," the report declared, "can result only in inefficiency and abnormally high cost for the service."[6]

The report went on to suggest that New York City had one of the most expensive and inefficient systems in the country because "the city does not avail itself of its natural advantages." This referred to a compact population, the concentration of large quantities of waste, a nearby waterfront, potential for short hauls, and water transportation. In this context, a return to the Waring approach was needed, that is, a system based upon reuse and reduction. The committee called for source separation of garbage, rubbish, and ashes; separate collection of the waste products; rubbish sorting for salvage and sale; and reduction of garbage and dead animals by either a city-owned or private facility.[7]

Not only did the report call for resuming the Waring system; it also recommended the resurrection of the Staten Island plant.

> No better location for a reduction plant for this city can be found than the shores of the Kill van Kull which separates Staten Island from New Jersey. It is within the quiet waters of the port; transportation would never be delayed by rough weather; it is farthest removed from residential districts; and it is already bordered with odor producing plants, soap factories, glue works, oil refineries, copper mills, and linoleum factories.[8]

The turn back to Staten Island seemed a little disingenuous, especially coming from a committee in Brooklyn. The report had declared that reduction technology could provide valuable byproducts and save money (its most important attribute) while still being odorless. The Lake's Island facility, it asserted, had not lived up to expectations, but the stench usually could be traced to "wrong construction, careless operation, or overloading." This assessment flew in the face of how the facility had been viewed during the Garbage War. Nevertheless, the report argued that the Staten Island equipment was still in good condition: "In the opinion of this committee, it will be an inexcusable oversight and wasteful negligence to permit this modern and well constructed plant to be dismantled and moved, when it can so readily be utilized to the advantage of the city."[9] How fortunate to have such a perfect location for disposal, at such a distance from Brooklyn. How easy to rewrite the history of a place and time, in such variance with the experience of Staten Islanders.

In making its case for reduction, the committee criticized every other disposal option, particularly incineration, because of its high costs, its inevitable distribution throughout the city, and the "complete destruction of the byproducts."[10] Such arguments were reinforced by a strong dose of Not in My Back Yard (NIMBY). "While the citizens of a crowded municipality can justly be called upon to surrender individual convenience and even comforts for the common good," the report declared, "it is hardly fair for the city to ask them to accept and tolerate any nuisance which threatens to affect property values especially when there is some other practical and economical method of performing this particular public work" (especially on Staten Island).[11] The Brooklyn plan was significant not specifically for its recommendations but for its perspective, that is, its emphasis on shifting disposal responsibility away from its own borough to somewhere else. For decades to come, that "somewhere else" was Staten Island.

Mayor Hylan and the Board of Estimate neither accepted nor embraced the Brooklyn plan. It stands to reason that they were leery of the political tempest that reopening the reduction plant on Staten Island would stir up. Hylan and the board did explore and consider several different options (public and private), some requiring major changes and others largely relying on what was in place. The choices included combinations of incineration, landfills, reclamation, reduction, and feeding garbage to hogs, but no sea dumping. The board, however, could not reach an agreement.[12]

Any hint of a return to reduction made Staten Islanders nervous. Representatives of Richmond Borough in the legislature presented two bills in 1921 to keep city officials from reopening the plant at Lake's Island. Commissioner of Health Dr. Royal S. Copeland and others opposed legislation, fearing that such measures would hamstring city officials in future siting decisions.[13] Late in December 1921, the *Staten Island Advance* reported (with little hard evidence) that the reduction plant was being "tuned up" for use again.[14] In fact, the Cobwell facility was neither reinstated nor torn down but sat rusting in place for many years.

Throughout the early 1920s, with an absence of a better plan, ocean dumping continued. The mayor's office submitted a positive but self-serving report to the Board of Aldermen in March 1925. It gave every indication that collection and disposal of refuse was moving along well in New York City, leading to the end of sea dumping. Five new dumping boards for scows along the waterways in Manhattan, Brooklyn, and the Bronx had replaced "old worn-out" structures. New disposal plants (small cremators for burning dry and wet refuse) were being constructed in Brooklyn and Manhattan (with more anticipated in the future), and the city had purchased the plant at Rikers Island for unloading scows. There was an ample number of seagoing scows for the city's needs (although the report conceded that disposing garbage at sea was "archaic"). Motor trucks were replacing horse-drawn carts. Snow was being promptly removed. All in all, the operation was presented as smooth and effective (but curiously moving toward more disposal plants, despite the previous rejection of Hylan's citywide incineration proposal).[15] The report's favorable spin on refuse services camouflaged the actual drift and uncertainty in the Department of Street Cleaning (NYDSC). The prevailing disposal program was essentially piecemeal and lacked an overall strategy. A solution to the problem in the early 1920s was no closer than it was during the Garbage War.

THE CHRONIC DISPOSAL CHALLENGE

Under Hylan's commissioners of street cleaning MacStay and Alfred A. Taylor, a decentralized, uncoordinated plan for refuse disposal persisted through the 1920s. Although there were modest variations in those years, the system in 1923 was typical. The NYDSC's Bureau of Final Disposition supervised the discarding of all waste material collected in Manhattan, Brooklyn, and the Bronx. The practice was largely divided among trucking refuse to designated unloading plants (such as Rikers Island; Port Ivory, Staten Island; Bayonne, New Jersey; and Flushing, Long Island), directly dumping material at landfills and in marshlands, or barging waste out to sea.[16] This work was performed by "semimunicipal" operations, with the city providing municipally owned or chartered scows and tugboats and contractors unloading the vessels.

In the outlying areas of the Bronx, material collected was deposited on landfills if requested by property owners or utilized as fill on city-owned property or roads. In Brooklyn all material collected was delivered to waterfront dumps, railroad stations, and landfills controlled by the Brooklyn Ash Removal Company to be reclaimed or incinerated. Garbage deposited at sea was brought to a location fifteen miles southeast of Scotland Light (about seventeen miles off the Long Island and New Jersey shores) under the direction of the U.S. supervisor of the harbor. Contracted scow trimmers picked through the rejectamenta for recyclables before the remainder found its way to waterfront sites in Manhattan and the Bronx.[17] Plans for a new incinerator plant at Fifty-Sixth Street and Twelfth Avenue in Manhattan were underway to cut down on long hauls of refuse from the area.[18]

Meanwhile, borough authorities managed refuse disposal in Richmond and Queens strictly through the use of incineration. On Staten Island ashes were mixed with garbage and rubbish and then shoveled into city-owned incinerators. The residue was used for road building or other types of fill. In Queens the rubbish also was burned with garbage in city-owned incinerators.[19]

The return to ocean dumping after the closing of the reduction plant on Staten Island came under particularly harsh criticism in the public arena. New Jersey and Long Island resorts were particularly affected by the practice.[20] On July 24, 1923, Commissioner Taylor stated that there had been no change in the NYDSC's methods of disposal, since garbage was carried forty miles out to sea (Taylor used other distance estimates in a later statement) and then dumped. "I am certain," he stated, "it is not our garbage that is being thrown up on the beaches. It may come from steamships, but I do not know just where it does come from. We are going to make a thorough investigation and as soon as we find the source the nuisance will be stopped at once."[21]

Governor George S. Silzer of New Jersey appealed to the federal government for action later that month after he accused the NYDSC of dumping refuse only five miles off the state's coast.[22] Taylor responded that "Governor Silzer must have been misinformed. I'm sure the Governor's investigators saw no city scows dumping garbage within five miles of the New Jersey coast. They probably saw mud and ashes being dumped."[23]

By 1925 about 25 percent of New York City's refuse was being discarded at sea.[24] In July 1925 Acting Secretary of War Dwight F. Davis told the Hylan administration that this practice must stop, especially because of what it was doing to the harbor and environs. Although the federal government had jurisdiction over the harbor, it was trying to give the city time to arrive at a solution to its disposal problem. Supervisor of the Harbor Captain John C. Fremont estimated that the city dumped 2,000 to 2,500 tons of waste at sea each day, but the NYDSC refuted those estimates, claiming that waste dumped only amounted to between 450 to 1,000 tons daily. Fremont's assessment, further investigation, and complaints over beach pollution led Davis to demand an immediate city response.[25]

Governor Silzer concurred with Davis's and Fremont's assessments of ocean dumping rather than Taylor's. "If the people of Rockaway and Long Beach and Long Branch carted their garbage on scows to New York City and dumped it on Forty-second Street," he declared, "I could imagine what the people of New York City would say, and what a protest they would make." Fremont added, "The sea cannot possibly serve as a dumping place for New York's immense accumulation of refuse." He favored incineration as the best alternative to ocean dumping, although he realized that there was local resistance to it. Commissioner Taylor had to know that he was fighting a losing battle by defending his department's sea-dumping practices. Acknowledging that there was "public prejudice" against disposal plants, he made clear that supporting incineration as opposed to ocean dumping or other methods was taking hold within the Department of Street Cleaning.[26]

While the city's disposal system lacked careful coordination, the problem was not simply the absence of an agreed-upon plan. The burdensome workload of the department extended well beyond disposal. Inadequate equipment, labor shortages, and strikes were debilitating. Incessant contract disputes and renegotiations with private parties were disruptive.[27] The market for recyclables was unreliable or absent, leading to contractor defaults on agreements with the city.[28] Not only sea dumping but land dumping raised public ire, hampering possible alternative disposal options.[29] The weather frequently was a curse, especially if an unpredictably large snowfall hit the city in winter, as it did in 1920. More streets meant a greater demand for snow removal and refuse collection. Between 1928 and 1929 alone, the total length of streets in Manhattan, Brooklyn, and the Bronx increased by sixty-one linear miles.[30]

THE CITY GROWS—AND CHANGES

Complicating the refuse problem was New York City's relentless population growth after 1920, the intricacy of the metropolitan region, and the mounting quantities of refuse. U.S. Census figures for New York City in 1900 ranked it first in the nation, with a population of 3,437,202. The numbers rose to 5,620,048 in 1920 and to 7,454,995 in 1940.[31] The growth between those years represented an increase of more than 1.8 million consumers and producers, not to mention regular commuters who moved in and out of the city each day. The impressive surge in population was not the whole story (see table 5.1). The manner in which the population grew, and where it grew, altered the city's demographic patterns and affected the spatial configuration of the metropolis. In 1920 Brooklyn was quickly catching up to Manhattan as the most populous borough and by 1930 surpassed it. The Bronx and Queens vied for third, with Queens moving ahead by 1950. Richmond was never a contender. In terms of population density, no other borough came close to Manhattan.[32] Uneven population growth and varying densities in the boroughs posed difficulties for any refuse-collection and -disposal program, especially one that was often in flux.

The outward push of the New York metropolis created new living and working patterns, and these had political, economic, social, and environmental repercussions. (The very definition of what actually constituted the region was cause for some confusion.) Before 1920, the city center and the counties in the region had grown at about the same rate, but the former held about 75 percent of the total population. As the years passed, the metropolitan nucleus could not continue to absorb major additions to its numbers. Living space shifted from single-family structures to multiunit dwellings, and people moved outward. Nassau County (adjacent to Queens), for example, doubled its population between 1920 and 1930. After 1920 or so, boroughs contiguous to Manhattan became more crowded, as people sought to live close to where the bulk of the jobs were found. Mass transit was essential, as were bridges. In the late 1920s the first highways were built in the region. For Queens, especially, the

TABLE 5.1 Population and Area of New York City Boroughs, 1920–1950

BOROUGH (SQ. MI.)	POPULATION			
	1920	1930	1940	1950
The Bronx (41.4)	732,016	1,265,258	1,394,711	1,451,277
Brooklyn (79.8)	2,018,356	2,560,401	2,698,285	2,738,175
Manhattan (22.3)	2,284,103	1,867,312	1,889,924	1,960,101
Queens (118.6)	469,042	1,079,129	1,297,634	1,550,849
Richmond (57.0)	116,531	158,346	174,441	191,555

Source: Wallace S. Sayre and Herbert Kaufman, *Governing New York City: Politics in the Metropolis* (New York: Russell Sage Foundation, 1960), 18–19.

development of a subway offered a crucial link; Staten Island, by contrast, remained unconnected other than by ferry.[33] The Great Depression and World War II temporarily retarded movement to the suburbs until the postwar building boom reinvigorated home construction.[34]

Changes in the ethnic, racial, and social patterns of growth also reshaped the region. Foreign immigration dropped off precipitously in the 1920s, after the passage of tough national anti-immigration laws. The Immigration Act of 1924 set severe quotas on immigrants from Southern and Eastern Europe and excluded all Asians. At the time, 75 percent of New York City's population was foreign-born or of foreign stock, with 40 percent of the total number either Jewish or Italian. Many new in-migrants to the core now came from the American South (in the Great Migration) and Puerto Rico, with growing numbers from Latin America and fewer from Northern Europe. Black migration from the American South and the West Indies was steadier after 1920, though African Americans and West Indians still constituted only a small portion of the overall core population. By contrast, the communities of the outer area had fewer minorities and were generally more affluent by the 1940s.

While New York's demographic shifts were not entirely unique from a national perspective, the local impact of outward growth and population diversity could not help but influence how the city intended to provide and distribute city services, including refuse disposal. The particular needs of the boroughs, for example, would be reflected in their different population densities and makeup and the changing economic base. Demands for services in the urban core also were challenged by suburban expansion.[35] It was difficult enough devising a citywide disposal policy, but it was further muddled by an evolving metropolis and consumer base.

Rapid city growth meant more consumers, more buildings, more paved streets, and a greater area to cover, exacerbated by large accumulations of refuse to collect. From at least 1900 forward, national trends in consumption of all kinds of goods was on the rise. Between 1900 and 1950, consumer expenditures in the United States increased six times.[36] New technologies enabled industry to produce goods of greater variety. By the end of the 1920s, the consumption of mass-produced products reached an all-time high in the United States, with automobiles, household appliances, ready-made clothes, and processed foods leading the way.[37]

The changing patterns of consumption were graphically represented in the nature of the waste. There was less ash and horse manure and more paper, packaging, and disposable items, which required variations or outright changes in collection and disposal practices.[38] (One does not pick up old cardboard boxes in the same way one collects horse manure.)

And there was more waste altogether. In the decade following World War I the amount of solid waste produced in New York City rose by 70 percent; by

1929 discards reached 15.5 million cubic yards annually.[39] In 1921 the Street Cleaning Department collected 1.5 million more cubic yards of refuse than in 1920 and about 570,000 yards more in 1922 than in 1921.[40] In 1924, to make collection easier, the department began "a determined drive" to get citizens to properly separate ashes, garbage, and rubbish at the source.[41] Its efforts met with only modest success, if at all.

THE NYDSC'S REVOLVING DOOR

A revolving door in the commissioner's office in the early 1920s added to departmental instability. MacStay submitted his resignation to Mayor Hylan on January 3, 1921, about one year after accepting the appointment, citing health problems (an explanation belied by his immediate appointment as deputy commissioner of public welfare).[42] The mayor went outside the NYDSC with his appointment of John P. Leo on January 5, 1921. Leo was chairman of the Board of Standards and Appeals but had spent most of his working life as a builder and architect. He was not a stranger to the waste field, having served on a special mayoral committee to study rubbish and garbage disposal options, and was a member of the Snow Removal Committee.[43]

Leo did not last a year. He resigned on November 18, stating in his letter that the decision was triggered by disputes with contractors furnishing dumping scows. The quarrels stemmed from Leo's decision to cut rental prices in half for scows provided by O'Brien Brothers, friends of Hylan. The mayor subsequently—and predictably—withdrew his support for the new commissioner, weakening Leo's standing. His short tenure was further marred by an ongoing dispute with Commissioner of Accounts David Hirshfield, who questioned Leo's honesty as part of an investigation of the NYDSC. While these issues were the proximate cause of the commissioner's resignation, Leo also had faced ongoing problems with other politically connected contractors, organized labor, and the Brooklyn Democratic establishment.[44]

Political infighting plagued many municipal governments in New York throughout the years, and untangling accusations and counteraccusations was never easy. After he left his post, in 1921 and 1922 Leo testified before a legislative committee investigating the affairs of New York City that the department should have political protection from the mayor and his supporters (a reflection on his own troubles); he also strongly supported centralizing the NYDSC to "cover all boroughs."[45] Leo became another voice for change in the department's operations—a view slowly gathering momentum over the decade.

Leo's experience exemplified the disruptions that political infighting brought to the department, and as long as the commissioner position was treated as a spoils job, such turmoil was to be expected. Recurring accusations drew attention to the mayor's actions. In December 1920 Leonard Wallstein, chair of the Citizens Union Committee on City Government, resurrected the uproar over the Staten Island reduction plant. He charged that after the closure of the plant, Hylan "revived the Barren Island nuisance, continued it as long as he dared, and finally reverted to the primitive method of towing to sea and dumping." This resulted, Wallstein concluded, in a loss of revenue to be derived from salvage of the waste materials.[46]

In October 1921 the Coalition candidate for mayor, Henry H. Curran, accused Hylan of breaking his promise to provide municipally owned garbage reduction and waste distribution plants. He blasted the practice of ocean dumping: "Those of you who live in Brooklyn, near some of the city beaches, probably remember how you have been awakened morning after morning, during the last three or four years, to find that, like chickens come home to roost, this garbage, which you have finally shipped out to sea, has returned on the waves to plague you." While he exaggerated the timeline by a year or more, Curran made his point and added that he advocated an "up-to-date" disposal plan.[47] Hylan, however, survived the onslaught of criticism with reelection in 1921 and remained in office until 1925.

The appointment of Alfred Taylor as NYDSC commissioner late in 1921 brought a degree of stability, if not radical change, to collection and disposal of refuse for the remainder of the decade. Taylor, then general superintendent, had started in the Street Cleaning Department in 1900 as a mechanic and was the first to rise from the ranks to become commissioner.[48] Although he stated his intention to bring improvements to the NYDSC, he was a loyalist, not a reformer. As such, he was unlikely to ruffle many political feathers.

Accusations of inefficiency, corruption, and cronyism persisted. For example, in the spring and summer of 1928, Commissioner of Accounts James A. Higgins leveled allegations of graft against nineteen department employees in Manhattan with respect to snow-removal practices. Four others in the Bronx faced charges of "payroll padding." These instances led to a citywide investigation of the department, relentless exposure in the newspapers, and new evidence of wrongdoing uncovered in Brooklyn. The incidents were branded collectively as "a street cleaning scandal" by the *New York Times*. Arrests were made and officials suspended. But Taylor was never a target of criticism.[49]

AN INCINERATION STRATEGY, NOT A POLICY

While a new official disposal plan was slow to evolve in the 1920s and early 1930s, the steady construction of incinerators (disposal plants) in Manhattan, Brooklyn, and the Bronx and continued use of them in the boroughs of Queens and Richmond produced a de facto strategy, one that reinforced Hylan's previous goal of developing a citywide incineration network.[50]

The timing of the decision to favor the burning of waste coincided with the design of a new generation of incinerators that purportedly was an improvement over previous types. Earlier models had been unreliable and noxious. The incinerator used on Staten Island, for example, was shut down by the court in 1898 largely because of the odors; it was followed by a second similar plant, which also failed. These and other cremators built before 1908 were limited to burning moist organic materials at low temperatures with the aid of fuel such as coal. Incomplete combustion was common, as were offensive odors. The lack of standardized designs, poor management, and the absence of good data exacerbated the problems. As stated in chapter 2, of the 180 furnaces erected in the United States between 1885 and 1908, 102 were abandoned or dismantled by 1909.[51]

In 1906, North American engineers made the first successful adoption of an English-style destructor in Westmount, Quebec, followed by similar projects in Vancouver, Seattle, Milwaukee, and West New Brighton on Staten Island. By 1910, many engineers were claiming that a new generation of burning waste had arrived. By World War I approximately three hundred plants were in operation in the United States and Canada, eighty-eight of them built between 1908 and 1914. Before 1920, however, incineration failed to maintain a prominent place among disposal options. The poor performance of the first generation of burners and the prohibitive cost of the second-generation plants weakened their broad-scale appeal. In the early 1920s, incinerators were most likely to replace open dumps in suburbs and smaller urban communities where suitable plant locations were available.[52]

As a major metropolis, New York City was an exception to the rule. It moved to incineration gradually, targeting select locations instead of blanketing the city with the plants all at once. The facility located on Governors Island in 1885, although pioneering, did not inspire further construction. In 1903 a small experimental rubbish burner in Manhattan led to the construction of two more plants there and four more in Queens in 1906. A major goal for these early incinerators was to convert the waste heat into usable power, but difficulties in operation undermined this idea. Siting furnaces introduced an additional tier of difficulty and often led to pitched battles between the department and targeted neighborhoods.

A good example was friction between the NYDSC and residents of University Heights in the Bronx in 1925, with respect to an incinerator scheduled for placement on the Harlem River. The proposed siting became an interborough issue because the plant was to be located between 205th and 206th Streets on the Manhattan side of the river, thus affecting both boroughs. New York University had a campus in the Bronx, and it feared that billowing smoke would affect it. Business leaders and others in the Bronx sided with NYU. Manhattan groups soon joined the protest. Mayor Hylan bristled at the opposition but saw the need for compromise. A long and protracted process finally resulted in the decision to locate the plant at 215th Street and the Harlem River, but it was not completed until 1934.[53]

While delays were not always so lengthy, such difficulties characterized the development of an incinerator network. The NYDSC was constantly engaged in legal squabbles with contractors, taxpayers, and others. Despite their initial enthusiasm for burning wastes, Hylan and his Board of Estimate and Apportionment did not always lend their full support to the department's efforts to build new incinerators; only two incinerators were constructed during the eight years of the Hylan administration. In some cases, it was difficult to obtain funding given other priorities in the budget. Once the Great Depression hit the city, such capital projects had to compete with more pressing needs. This certainly was the case with the 215th Street destructor.[54] Local resistance to siting because of odors and smoke proved frequent in these years despite the fact that municipal engineers viewed incinerators as acceptable from a sanitary standpoint. If properly installed, they argued, these furnaces did not pose an odor or smoke nuisance and destroyed germ-infected materials.[55] Apparently, they were not always properly installed.

In 1924 the NYDSC announced the operation of a new incinerator on West Fifty-Sixth Street and Twelfth Avenue, which would burn about 350 to 450 tons of garbage and rubbish a day collected primarily from north of Fourteenth Street to 110th Street west of Sixth Avenue in Manhattan. Commissioner Taylor announced that with the modern incinerators, the department "has inaugurated the most improved and sanitary method of garbage disposal."[56] Another incinerator of somewhat smaller capacity (300 to 320 tons daily) was to be installed in 1925 at Fifth Avenue and 139th Street for waste collected above 106th Street and parts of the Bronx.[57]

The NYDSC was thrilled with the availability of the two new incinerators not only because they lowered the cost of towing and scow charter but because they reduced the risk of fire hazards near the city dumps and unloading plants. The 1926 *Annual Report* declared that "disposal of garbage by incineration will eventually end all complaints regarding the pollution of the water and beaches."[58] By 1927, the two incinerators handled approximately 20 percent of the rubbish and 25 percent of the garbage collected in the three boroughs. A third (320-ton capacity) was near completion between Seventy-Third and Seventy-Fourth Streets on Exterior Street.[59]

In 1928, the department optimistically announced that while disposal of wastes had always been a problem, "we are nearing a complete solution." Incineration was that solution: The new incinerator at East 73rd Street and

Incinerators 1929 and 1930

	Type	No. of Furnaces	24 Hrs. Rat. Cap. Tons.	Description of Plant
....	Decarie	6	300	Two turbo steam generators, steam auxiliaries for fans, blowers, etc., four electric cranes, electric ash hoist and hopper, scales.
....	Sterling	4	320	Motor-driven compressors, electric fans and blowers, four ash hoppers, scales, oil-burning heating boilers.
....	Sterling	4	320	Motor-driven compressors, electric fans and blowers, electric ash hoist, scales.
....	Decarie	2	100	Steam-driven auxiliaries (pumps, fans, blowers), one Universal crane.
....	Decarie	3	270	Steam-driven auxiliaries (pumps, fans, blowers), electric ash hoist, traveling crane.
es............	Decarie	2	100	Steam-driven auxiliaries (pumps, fans, blowers), one Universal crane.
....	Decarie	3	225	Steam-driven auxiliaries (pumps, fans, blowers), electric ash hoist, one tractor loader.
hing.........	Decarie	2	70	Steam-driven auxiliaries (pumps, fans, blowers), one Universal crane.
maica........	Decarie	2	100	Steam-driven auxiliaries (pumps, fans, blowers), one Universal crane.
....	Decarie	3	150	Steam-driven auxiliaries (pumps, fans, blowers), one Universal crane, electric ash hoist.
ach..........	Decarie	3	150	Steam-driven auxiliaries (pumps, fans, blowers), electric ash hoist, one tractor loader.
Beach........	Decarie	2	70	Steam-driven auxiliaries (pumps, fans, blowers).
at Brighton..	Nye	2	150	Motor-driven forced draft fans, electric ash hoist, scales, one tractor plow.
	Heenan	1	60	Steam-driven auxiliaries.
....	Nye	4	90	Motor-driven forced draft fans, electric ash hoist, scales.
	Heenan	2	90	Steam-driven auxiliaries.
....	Nye	4	150	Motor-driven forced draft fans, electric ash hoist, scales, one tractor plow.
....	Sterling	2	100	Leased to Brooklyn Ash Removal Co., operated by them—motor-driven compressors, fans and blowers.
....	Heenan-Froude	2	100	Leased to Brooklyn Ash Removal Co., operated by them—motor-driven compressors, fans and blowers.
....	Hiler	4	500	Owned and operated by the Brooklyn Ash Removal Co., to be recaptured in 1933.
....	Hiler	4	500	Owned and operated by the Brooklyn Ash Removal Co., to be recaptured in 1933.
....	Hiler	4	500	Owned and operated by the Brooklyn Ash Removal Co., to be recaptured in 1933.

FIGURE 5.1 Incinerators in New York City in 1929–1930.

Source: NYC Department of Sanitation, *First Annual Report* (1930).

Exterior had been completed, and sites had been secured for two more in Manhattan and two in the Bronx (one was the plant planned for the Harlem River that was subject to such intense interborough debate). Three incinerators (five-hundred-ton capacity each) were under construction in Brooklyn. About 55 percent of all rubbish and 32 percent of all garbage was being disposed of in the city's five operating incinerators at that time. The NYDSC confidently concluded: "When the construction of all the incinerators now planned has been completed all garbage and rubbish collected in the three boroughs under the jurisdiction of this Department will be consumed in these plants."[60] In addition to the large destructors, contractors operated small incinerators at several receiving points.[61]

By the end of the decade, 70 percent of the garbage from Manhattan, Brooklyn, and the Bronx was incinerated.[62] Despite the enthusiasm from within the department, Commissioner Taylor faced ongoing protests from communities chosen as sites for the plants. In response to criticisms, he stated in the summer of 1927: "Long delays are caused by the populace becoming alarmed at the locations selected for incinerators, and using all kinds of schemes to prevent progress, through indignation meetings, and finally resorting to court injunctions to prevent the department from going ahead with the program."[63] Taylor and commissioners to follow would discover that citizen pushback to incineration was not going away.

REVAMPING REFUSE MANAGEMENT

Despite the department's proclamations of success, calls for radical organizational change had only grown louder over the course of the decade. Many municipal departments were being consolidated and reorganized, and a special committee that included the Manhattan borough president, the city comptroller, and the chief engineer of the Board of Estimate was appointed to consider what could be done to improve sanitation practices.

The committee drafted a plan for a new sanitation department that would fold in sewerage service and end the independent refuse programs of Queens and Richmond by bringing them under city control. "This great house-cleaning department," as a writer for the *New York Times* called it, would be divided into three bodies: a division of administration, a division of street cleaning and waste disposal, and a division of intercepting sewers and sewage disposal. The mayor would appoint a Sanitary Commission of three members (one of which had to be a physician and another an engineer, or people equally qualified in those fields). The Sanitary Commission would oversee the whole collection and disposal operations with each commissioner (or someone designated) responsible for one division apiece.[64]

On November 5, 1929, the plan was put before the voters of New York City. The decision to centralize had the backing of many sanitary engineers and civic organizations.[65] The long-talked-about revamping of the city's collection and disposal services was at hand. On December 1, 1929, the Sanitary Commission became a reality. The Tammany-backed Democratic mayor James J. (Jimmy) Walker had made the NYDSC a campaign issue in his successful 1925 bid for mayor against the incumbent Hylan, and he wanted to improve incinerator capacity (although few new incinerators were built during his administration).[66] Walker swore in Dr. William Schroeder Jr. as the first chairman at the mayor's apartment in the Ritz-Carlton Hotel, "just before he left to attend the Army–Notre Dame football game."[67] Schroeder had practiced medicine and surgery in Brooklyn and had served for a year as commissioner of the newly created Department of Hospitals (or Hospital Commission). The new sanitary commissioner was Walker's close friend, his personal physician, and a Tammany Democrat. The second member of the commission was Leonard C. L. Smith, who had been engineer-in-charge of water supply for Queens and an expert in sewer treatment. The third, Charles S. Hand, a journalist and Walker's press secretary, had a very good understanding of street cleaning and citizens' concerns about it.[68] Taylor ceded his power to Schroeder and the Sanitary Commission, retiring in July 1933.[69]

The first annual report (1930) of the newly created Department of Sanitation (DSNY) clearly indicated the determination to coordinate street cleaning, collection, and disposal activities for all the boroughs. That task would not be easy.

FIGURE 5.2 LaFrance truck (1929) dumping refuse.

Source: NYC Department of Sanitation.

In Richmond Borough, for instance, consolidation meant that trucks needing major repairs had to be sent to Manhattan, rather than being fixed in local shops. This caused a constant shortage of vehicles. Also, a "floating seasonal population" of about three hundred thousand visitors along the beachfront on Staten Island made keeping the beaches clean quite difficult. Adding sewerage and sewage disposal as responsibilities of the department created another layer of complexity.[70]

The 1930s brought an even stronger commitment to incinerators (built by the city but largely without municipal regulation) as the primary disposal option for New York City.[71] Comptroller Charles W. Berry frankly stated in late 1928 that "proper disposal" of refuse was the most important issue facing the current administration.[72] Mayor Walker defended incineration even in lieu of neighborhood resistance because "we cannot kick the garbage around the city."[73] A writer for the *Brooklyn Daily Eagle* reported:

> The policy adopted by the Walker Board of Estimate was that incinerators must be built and the site must be within the city limits. And now it is up to the Sanitation Commission to build them—enough of them to destroy the last pound of the city's refuse in the only sanitary way it can be done. It can be said that these incinerators, when properly located, are not a public nuisance. Every precaution has been taken to make them as inoffensive as possible.[74]

Of course, what determined "when properly located" was up for debate.

In February 1930, Schroeder announced a comprehensive three-year budget to build sewage plants, garages for the vehicle fleet—and incinerators.[75] The report indicated that it was necessary to continue sea dumping even though it caused "endless controversies." While steamships and communities in adjoining states had contributed to the beach littering, "New York City, because of its size, has not been able to escape much of the blame for these conditions. In order that the City may be held blameless, this Department has under way a program for the building of incinerators sufficient in size and number to make it unnecessary to send any garbage or rubbish to sea."[76]

The decision to do away with ocean dumping was soon made mandatory. After years of constant criticism, New Jersey filed a bill of complaint against New York City on May 20, 1929, and by October the case reached the U.S. Supreme Court. Retired federal judge Edward Campbell was appointed as a special master to investigate the claims of polluting the Jersey shoreline. Campbell ruled in favor of New Jersey, and on December 7, 1931, the court ordered New York to cease ocean dumping operations by July 1, 1934, or face a daily $5,000 fine (although only municipal, not commercial or industrial, wastes were prohibited). He also decreed that the city needed to increase its incinerator inventory beyond the twenty already in use. The final municipal scow went out to sea on June 28, 1934.[77]

The task ahead was not an easy one. Incinerators were located in every borough save the Bronx, but many had limited capacity or were antiquated. The DSNY sought bids to build new units and also looked to upgrade older ones.[78] It overstated (some might say misstated)[79] the value of incinerators, saying they were "universally recognized" as the "correct solution" to disposal. The department carped on delays caused by voluble opposition from citizens and neighborhood protest groups. "When the maximum of good that incineration offers is weighed against the minimum of inconvenience involved," the report added, "it seems to this Department that enlightened public opinion should support whole-heartedly our efforts in this work."[80]

It also drew a line between reduction plants, which it conceded as "objectionable," and incinerators, which it maintained "most vigorously" were "neither a nuisance nor an injury to a neighborhood but on the contrary are a great benefit."[81] In some ways, this sounded like the defense of the Staten Island reduction plant just a few years before, but now with incineration the panacea. Kenneth Allen, a sanitary engineer for the Board of Estimate and a proponent of incineration, might have been correct when he stated in 1930 that "New York may claim to be the pioneer in [incineration] use in the country," going back to the Governors Island plant (1885).[82]

With plans for a citywide incineration program in place by 1930,[83] New York City signed its last private contracts for refuse disposal in 1933 and did not sign another for sixty-four years. Onsite incinerators in apartment houses, schools, and hospitals continued to multiply. By 1933 there were more than two thousand private burners in operation in the city.[84] The eminent sanitarian George A. Soper looked forward to serious improvement in the city's disposal methods despite the trend toward a citywide incineration system. "The results of this patchwork method of garbage disposal," he stated, "have long been subject to criticism." He pointed to Rikers Island in particular, where all kinds of refuse were being dumped in the early 1930s, as "perhaps the worst and most extensive nuisance of its kind anywhere to be found. It is a mountain of ill-smelling waste."[85]

The task ahead was formidable. The department needed to complete two incinerators already on line, reconstruct equipment on Rikers Island for handling ashes from Manhattan, develop a system of small landfills to accommodate ashes from the Bronx and Brooklyn, seek lowland fill sites, and design new incinerators that could withstand the higher temperatures caused by burning a large amount of paper (70 percent of the total volume).[86] In 1934, a cancelled contract with Brooklyn Ash Removal Company meant that railroad dumps, landfills, and waterfront dumps previously used in Brooklyn were no longer available.[87] In 1935, there were in effect forty-four permits issued by the DSNY for private dumps throughout New York City, but no new permits had been issued since January 1.[88] The city still operated eighty-nine small municipal landfills, but some experts hoped that the city would turn to new methods of refuse salvage beyond using ash for fill.[89] Wishing for a new disposal system had yet to make one appear.

The DSNY also faced its usual range of problems in the 1930s, and these diverted attention from necessary changes in its disposal practices. They included ongoing political machinations. The dapper Walker, known as "Beau James," had a talent for his position but never really treated it as his full-time job. In his first two and a half years in office Walker took 149 vacations, rarely acted as a hands-on chief executive, and allowed Tammany partisans to run amok. After winning reelection in 1929, he got caught up in a scandal that led to his resignation in 1932.[90] The Tammany-supported Democrat John P. O'Brien won the election in 1932 and, having no loyalty to Walker, abolished the Sanitary Commission in 1933 on the grounds that some departments required reorganization. Control of the Department of Sanitation returned to a single individual rather than to the three-man body. Constant squabbles within and without the commission mired it in partisan politics.[91] The *New York Times* reported on March 21, 1933: "As soon as Mayor Walker resigned, the general feeling among organization Democrats was that Dr. Schroeder's days in city service were numbered."[92]

Continued shuffling of leadership was far from the only problem. There were frequent complaints that the department was undermanned—especially in

Queens and the Bronx—even though the commission obtained six hundred additional drivers and sweepers over the budget allowance for 1930. The economic pressures of the times were intensifying, however, which made layoffs a constant worry.[93] Traffic, razing of old buildings, abandoned automobiles, and subway construction were constant impediments to street cleaning. Consistent with the bigotry of the day, the DSNY was not averse to blaming the foreign-born and certain "racial groups" (especially in Manhattan) who were "manifestly accustomed to disposing of household waste by throwing it into the roadway," resulting in "constant supervision" required to keep the streets in decent condition. In 1931, the department issued more than 55,000 warnings and about seven thousand summonses for throwing refuse into the streets and improperly preparing it for collection.[94]

The Great Depression played a pivotal role in DSNY activities, undermining plans to float the long-term bonds needed for plant construction. Although the stock-market crash had occurred during Walker's term, the full impact of the Depression hit New York City while O'Brien was mayor. He was forced to sign an agreement under which clearinghouse banks funded part of the city's operations budget through 1937. Although Gotham avoided bankruptcy, it carried an unbalanced budget and a severely lowered credit rating. The staggering rise in unemployment hit the citizenry hard, with 16 percent living on relief. The physical city was showing serious signs of wear (hospitals, schools, and parks were outmoded, and many needed repair or replacement to their structures or equipment). Only with the election of Fiorello La Guardia in 1933 and the advent of the New Deal did the city's fortunes take a turn for the better.[95]

La Guardia was born in Lower Manhattan but raised in the American West. A Republican at odds with Tammany Hall, he was the first Italian American to be elected to the U.S. Congress, serving a predominantly working-class district in the 1920s. After losing his seat in the Democratic landslide of 1932, he turned to municipal politics and was elected mayor on a Fusion ticket in 1933. He became known for modernizing and centralizing the local government, consolidating departments, and reducing the power of borough and county offices. For example, the Department of Public Works absorbed a variety of responsibilities, including the planning and constructing of sewage-disposal plants. La Guardia also devoted considerable effort to garnering funds from Washington for local projects. Honest and charismatic, La Guardia confronted New York City's crushing debt by moving beyond the older laissez-faire doctrine.[96]

In the tradition of the several reform governments that came before it, the La Guardia administration focused on a variety of urban social and physical needs, including relief efforts and infrastructure improvements. And it functioned within a setting where the power of the political machine, for a time at least, was on the downturn.[97] La Guardia was by no means a perfect leader.

THE GO-AWAY SOCIETY ■ 141

FIGURE 5.3 New York Mayor Fiorello La Guardia confers with the new Works Progress administrator, Col. F. C. Harrington, in Washington, DC, December 29, 1938.

Source: Wikimedia Commons.

While he was forceful and decisive, he could not delegate authority effectively and was thin-skinned in facing criticism. Often a visionary, he was disinterested in routine and detail, thus allowing others to take charge of mundane but important concerns.[98] This proved too often the case with the work of the DSNY and the important issues it would confront going into the 1940s—most notably, a sudden departure from incineration in 1936.

THE (MIS)FORTUNES OF THE DSNY

Changes in DSNY leadership during the 1920s through the mid-1930s had created some political diversions and disruptions, but they did not deter plans for an incineration network. Attracting federal funds from New Deal programs—the Public Works Administration (PWA) and the Works Progress Administration (WPA)—the DSNY was able to surmount budget shortfalls and other preoccupations to build more incinerators and carry out other construction projects. The department also was developing a good reputation as a quality government body.[99]

This was all the more impressive considering the continued turnover in leadership. Ernest P. Goodrich (who served as commissioner in 1933–1934) aimed, among other things, to build on the incineration mandate.[100] He had an excellent reputation as a consulting engineer and planner, having designed harbors and zoning plans throughout the world, including for Nanking, China, and Bogota, Colombia. He also was the first president of the Institute of Traffic Engineers of the American Society for Civil Engineers. But in 1934, as a holdover from the previous administration, Goodrich was soon replaced.[101]

The newly elected Mayor La Guardia wanted his own people serving within the administration and believed that the DSNY was riddled with corruption.[102] Goodrich took offense at La Guardia's characterization of the department, but the mayor asserted, "The whole department has to be reorganized. It is thoroughly demoralized and in terrible condition. It is rotten with politics and I ask the patience and indulgence of the public until we have time to clean it out."[103] In Goodrich's place, La Guardia appointed Colonel Thomas W. Hammond, a retired West Pointer who was currently deputy commissioner of sanitation. An "old personal friend" of the mayor going back to their childhood, Hammond had been an engineer, served on the general staff of the American Expeditionary Force in France during World War I, commanded the Twenty-Eighth Infantry at the front, and been stationed at Governors Island for two years in command of the Sixteenth Infantry.[104] La Guardia made clear that he liked Hammond's experience as a military man and engineer (à la Col. George Waring, it might be presumed) and whose "aptitude and ability to handle men" he knew based on their long association.[105]

The mayor charged Hammond with cleaning up the department.[106] Hammond noted that the mayor said "the Department was demoralized by graft and politics and that my first job would be to develop an organization and a system of control at the top that would not only insure keeping the City clean, but that would get rid of the graft and politics and build up the morale of the personnel."[107] Hammond immediately demoted three high officials and later ousted Gordon C. Poole, the last of the Tammany deputy commissioners. He also forced three hundred employees holding "soft" desk jobs to assume positions as sweepers and drivers.[108] Hammond broke up "the racket on the waterfront dumps"—the practice of collecting fees from the private cartmen who had permits to dump for free—but then had the department set its own fee.[109] He pledged to clean the streets more effectively but exhorted citizens to adopt "habits of cleanliness" as well.[110] "No doubt the streets can be improved," he said. "Any one can see that. The people must stop littering them."[111] Shifting the blame and responsibility to locals was never a workable plan for sanitation people, and Hammond's department received its share of criticism when streets got cluttered and dirty and the snow began to fall.[112] Yet Hammond lived up to several of his promises by bringing more efficiency to the department, controlling costs, improving collection, and trying to buck up morale.[113]

The year 1934 was auspicious because Hammond's department had to confront the required shift from ocean dumping to an alternative refuse-disposal system. He freely admitted that the New Jersey lawsuit was "absolutely justified," asserting that "rivers and seas are a constant temptation to carelessness in treatment of waste."[114] For final disposition of refuse Hammond favored "improving" city land through fill projects and continuing the use of destructors. La Guardia wanted "scientific" disposal methods and more efficient incineration, but other than a brief and modest return to resource-recovery schemes, the existing methods prevailed. In essence, Hammond's disposal plans varied little from his predecessors.[115] Between 1934 and 1936, incineration did increase from 40 to 46 percent as a method of disposal, but Rikers Island and all other landfills remained steady at 54 percent.[116]

In May 1936 Hammond was transferred from DSNY to the Department of Water Supply, Gas, and Electricity. He died suddenly of a heart attack in September. The *New York Times* reported that "winter weather proved a more difficult problem than either inefficiency or dishonesty" for Hammond. Earlier in March 1936, Mayor La Guardia had called for yet another reorganization of the department and made clear that he wanted a businessman to run it.[117]

That businessman would be William F. Carey, who was appointed commissioner on May 27, 1936.[118] Carey was brawny and hard-living. He loved boxing and Dixieland jazz and was a promoter of some notoriety. A *Saturday Evening Post* article stated:

> Meeting him for the first time, you might consider him to be a human bullroarer, a year-round Scrooge with a single talent—that of howling unmercifully at the hired help. Such a conception would be altogether false. He does roar a lot—roars like a buffalo with an impacted wisdom tooth. But back of his roaring exists a personality in which pure sentiment and vigorous humor are the main ingredients.[119]

Carey was living the American Dream, rising from a poor boy in Hoosick Falls in upstate New York to become an international millionaire industrialist, civil engineer, and contractor. He had worked on the Panama Canal and built railroads, canals, and bridges in China and South America. Seims-Carey Railway and Canal Company, operating in Asia, was his first company. In 1924 his construction company built the Andes Mountain Railway in Bolivia. W. F. Carey & Company erected the world's highest earth dam in Vermont. During the Depression he engaged in strip mining in Pennsylvania. In New York City, he had been responsible for constructing the North Beach airport. His most famous endeavor was building the new Madison Square Garden with the sports promoter Tex Rickard. Carey became vice president and treasurer of the Madison Square Garden Corporation and then president after Rickard died. His construction company also built Boston Garden, of which he became director.[120]

In appointing Carey, La Guardia stated, "Mr. Carey is an experienced builder and contractor and has undertaken projects of enormous magnitude all over the world. At my request he has made a survey and study of street cleaning and sanitation methods so that in assuming his new duties he will have the advantage of specialized preparation as well as his splendid experience and engineering background."[121] La Guardia disagreed with Carey over his disposal strategy (the mayor initially favoring incineration and waste-to-energy facilities, while the new commissioner saw them as too costly), but he wanted Carey's star quality as a businessman more than he wanted consensus.[122] A story in the *New York Times* about Carey's appointment observed that the Department of Sanitation "enters upon another phase of its development."[123] It was yet unclear what this "phase" might represent. Unanticipated circumstances, departmental and mayoral leadership, the economy, the role of the public, the weather, and politics had all played their part in determining the success or failure of this giant enterprise. What Carey's role would be remained to be seen.

A NEW DISPOSAL PATHWAY?

The end of New York City's ocean dumping of municipal solid waste proved to be just one reason for establishing incineration as the DSNY's disposal choice. Until the late 1930s all signs pointed in that direction, but under William Carey's leadership things would change drastically in the department. The shift from burning waste to dumping in sanitary landfills occurred so rapidly that few could have predicted such a sea change. As in the Garbage War and as wished for in the 1920 report of the Brooklyn Chamber of Commerce's Health and Sanitation Committee, Richmond Borough eventually would emerge as the newest destination for the discards of the metropolis. There would be less and less sharing of the disposal burden in years to come (something the boroughs would come to accept for many years) and little need for neighborhood protests in Manhattan, Brooklyn, Queens, or the Bronx. The first steps toward the behemoth Fresh Kills Landfill and the designation of Staten Island as the city's singular destination for refuse would take place in 1937.

CHAPTER 6

ONE BEST WAY

Three years after ocean dumping ended in 1934, incineration of New York City's refuse appeared to be the obvious disposal solution. Problem solved.

Little did New Yorkers realize that its huge, consumption-fed disposal dilemma was far from ending. In little over a decade, incinerators would be out and Fresh Kills Landfill would be in. But burying the city's refuse on Staten Island was to confound the city's waste problem in unanticipated ways, by creating a behemoth too big, too hard to ignore, and too indispensable to abandon.

FROM INCINERATION TO LANDFILLING

The number of municipal incinerators in the city peaked in 1937 at twenty-two; some were among the largest in the world. Twenty-one of the plants burned more than 2 million metric tons of refuse in that year. Yet, late in the decade, what appeared to be a full-scale commitment to incineration abruptly faltered. Of the fifteen new incinerators planned for construction in 1930, only four were built. Combustion rates in New York City plunged 65 percent between 1937 and 1944.[1] In 1937 and 1938 alone, the amount of New York City's refuse burned in cremators dropped by 10 percent.[2] Older manually operated incinerators soon were phased out.[3]

Several forces converged to push incineration out of favor. Local protests about siting the plants were chronic. Economic conditions during the Great

Depression were a major blow, since incinerators were expensive to repair, upgrade, or build. A change in refuse composition created technical problems for combustion. Most American incinerators in the early twentieth century burned organic material such as food waste (not so much in New York City) and nonorganic discards such as paper and rags. As paper and nonorganic materials increased (especially those laced with chemicals and other toxic substances), air-pollution conditions worsened.[4] During World War II reusing and recycling took many materials out of the waste stream that otherwise might have been burned. In addition, the federal government rescinded Public Works Administration (PWA) grants in 1941–1942, and shortages in machine parts and manpower because of the war meant that many of the operating incinerators had to be closed.[5]

Although incineration crested in 1937, its fate was foretold the year before when New York City Department of Sanitation (DSNY) Commissioner William F. Carey decided that burning refuse was just too expensive. Landfills as the primary form of disposal (which would offer an inexpensive place to dump refuse and also create usable land) would become the new normal in New York City, pushed hard in 1937 and 1938 and sanctioned by the department as the single best method by 1940. Tensions rose between Carey (whose objective was to utilize existing refuse landfills and to construct new ones) and Parks Commissioner Robert Moses (who preferred to use only "clean fill"—dirt and ashes but little or no organic material—for reclaiming land for parks and other purposes). The clash intensified at Staten Island's Marine Park at Great Kills, which proved to be a prelude to building Fresh Kills Landfill several years later. In 1937 the DSNY was still managing disposal to compensate for the termination of ocean dumping. Aside from incinerating refuse, nonburnable materials and residue from destructors were deposited on Rikers Island or in various landfills throughout the boroughs.[6] Waste in eleven waterfront dumps was transported by scows to Rikers. Landfills in "low swampy areas within the City limits" also were being filled, although several sites had reached capacity. The DSNY annual report of that year assured municipal officials that the city-owned landfills were "sanitary in every respect."[7] Few neighborhood groups agreed.

Carey's decision to move away from incineration was essentially economic, but the circumstances that led him to landfilling as the primary disposal alternative were more complex. The cost of building and maintaining burners during the Great Depression strongly influenced Carey to deemphasize (or possibly abandon) incineration in favor of landfilling. He began closing the smallest and the oldest plants, allowed others to rust, suspended future waste-to-energy facilities, and moved toward transforming land dumps (about eighty-seven currently in use) into "sanitary landfills" or at least augmenting existing land dumps with new sanitary landfills.[8] Not only did Carey regard landfills as cheaper than incinerators, but he justified them on the grounds that they would be operated using a "scientific method." Facing a civil suit for nuisances caused by the existing dumps, the new practice was preemptive, if not essential. Yet Carey could never convince many New Yorkers that the landfills were less of a nuisance than incinerators.[9]

The 1938 DSNY *Annual Report* stated that the process of developing upgraded landfills consisted of selecting "mosquito-breeding marsh lands at locations within economic hauling distance of the regular collection, preference given to city-owned lands." After dumping the refuse, compacting it, and leveling it off with bulldozers, the top of the dumpsite would be covered with "clean earth" largely acquired from the area being filled or hauled in from nearby sites. The DSNY justified the plan as both sanitary and economical. "By this method of disposal," the report stated, "great value in reclaimed waste mosquito-breeding lands are being created and the disposal cost is only a fraction of what it is by other methods."[10] By the summer of 1938, the department was using seven tracts in low-lying areas for demonstration purposes.[11]

Landfilling was carried out in close cooperation with the New York City Department of Parks, which became the beneficiary of ample reclaimed land. In addition, some hoped that making use of available land would at least delay the return to incineration and that the new "usable" land would add to the city's tax base.[12] Approximately two hundred acres per year of "submarginal land" such as swamps, tidal flats, and marshes were converted in this way, which meant an immense loss of wetlands areas within the city.[13] The practice also contributed to the dwindling availability of land disposal sites in the most heavily populated parts of New York City; this required scouting out new sites in the outlying boroughs.[14]

These were among the concerns voiced by the consulting engineer Harrison P. Eddy Jr., who stated that the landfill method in Gotham "is undoubtedly a big improvement over the uncontrolled dumping of mixed refuse. It seems desirable, however, to sound a note of caution before many cities adopt this method of disposal, apparently so attractive because of its cheapness."[15] Eddy warned about depending on mixed-refuse fill as a building foundation (such land became notoriously unstable), the speed of decomposition of the waste, the seepage and leaching from landfills, vermin control, the possibility of fire, and (most prophetic for New York) the scarcity of potential dumping areas. "If the typical American city were to adopt the sanitary-fill method of disposal today," he asserted, "it would not be many years before all the convenient areas would be filled up." In that event, the city would either have to "use dumps in its outskirts," with a commensurate increase in the cost of hauling, or turn to other methods of disposal. "The sanitary fill method thus uses up a valuable present, natural resource at the expense of posterity."[16]

Carey's decision to move from incineration to land filling, while suited to local exigencies, coincided with the rise in prominence of the sanitary landfill in the United States at the time.[17] The "sanitary landfill" was the breakthrough, beginning in the 1930s, that ultimately made filling the primary disposal option in the United States until the late 1970s or 1980s.[18] Seattle, New Orleans, and Davenport, Iowa, made early attempts at sanitary filling beginning in the 1910s, but these were little more than land-based dumps. The modern practice originated in Great Britain in the 1920s under the name "controlled tipping." Yet London and cities in the vicinity simply were dumping wastes between houses and covering the piles with street sweepings. The American equivalent to the British practice appeared in the 1930s in New York City, San Francisco, and Fresno, California. In New York, refuse was placed in deep holes covered with dirt primarily located in marshes. In San Francisco, layers of refuse were deposited in tidelands to produce additional land. Fresno's method relied on trenches in which alternating layers of refuse and dirt were dumped, followed by compaction of the layers to keep out moisture and vermin. This trench system became the model for modern landfill construction.[19]

Sanitary landfilling was an encouraging land disposal technology that combined features of both filling and dumping. To some authorities this method was initially regarded as nothing more than glorified open dumping, and they criticized it as labor intensive, unsafe, and unsanitary. The idea of a "sanitary" fill was intriguing to other experts because of its promise to deal with a wide array of refuse and the potential economy of the disposal method.[20]

The person most responsible for developing, implementing, and disseminating the sanitary landfill in the United States was Jean Vincenz (1894–1989), who served as commissioner of public works, city engineer, and manager of utilities in Fresno, California, from 1931 to 1941.[21] He came to believe that a true sanitary landfill required different elements than those utilized elsewhere, especially the systematic construction of refuse cells, a deeper cover of dirt between layers of refuse, and compaction of both the earth cover and the waste. The initial sanitary landfill in Fresno, an experimental or demonstration fill, began on October 15, 1934; the permanent site opened in 1937 and closed in 1987.[22]

A commitment to sanitary fills in New York began in 1936 (the year of Carey's appointment). The design was different from the Fresno enterprise and was intended for land-reclamation projects. In the New York fills, a primary goal was to create usable land, not to devise—at first at least—a new disposal option per se. Additionally, the front of the New York sanitary fills were not sealed like the Fresno fill, which produced an airtight closure around the whole area. Pleased with the sanitary-fill project first attempted at Rikers Island, Carey authorized other sites, hoping to reclaim substantial additional land. The debate that broke out over whether the sites were indeed "sanitary" did not undermine the practice in the city, which lasted for years.[23]

Sanitary landfilling redefined how New York City disposed of its refuse. Since 1937, more than 50 million cubic yards of material had gone to landfills

FIGURE 6.1 Distributing waste materials with tractors and Athey wagons.

Source: NYC Department of Sanitation, *Annual Report, 1939.*

in New York City, which the sanitary engineer Rolf Eliassen calculated was about half the yardage of the world's largest dam, located in Fort Peck, Montana.[24] In 1940 the DSNY was disposing nine million cubic yards per year through the use of landfills. The department completed five landfill sites, recovering 325 acres of land. They also were working in ten additional areas meant to reclaim one thousand more acres.[25] In 1941 there were thirteen landfills in operation, accepting a total of 11,481,992 cubic yards of waste.

Scale of disposal was only one measure of the impact of landfills on New York City. Landfills were distributed unevenly throughout the metropolis until the Fresh Kills Landfill dominated the city's disposal in the 1970s. By county, the largest area devoted to fills was in Queens County. Following Queens in descending order were Kings, Bronx, Richmond, and New York County, with a significant drop-off after Kings. The effects of eliminating wetlands and of leaching pollution and methane release were little appreciated at the time.[26]

RIKERS, CAREY, AND MOSES

Carey was convinced that New York City's commitment to sanitary landfills addressed the limitations of incineration as a disposal option. Sanitary landfills would take all refuse not reused for other purposes. (Presumably, "clean fill" would not go to landfills.) Upon completion the sanitary landfills would be suitable for reclamation because the layers of waste and dirt would pose no sanitation risk, only produce usable land.

It proved difficult for the DSNY to find new landfill sites near the major population centers of the city. The search for outlying locations soon began and quickly threatened Carey's program. Before new sites could be identified, what to do with Rikers Island was paramount. In the mid-1930s, it was the city's chief disposal facility, incorporating twelve dumps of about four thousand feet apiece. Each dump was from forty to 120 feet high. Complaints from the boroughs of Queens, Manhattan, and the Bronx about noxious odors were neverending.[27] The opening of the prison there in 1935 prompted the NYS Commission of Correction to call for the end of dumping, but nothing happened immediately.[28]

Its location, as much as the prison, proved to be the beginning of the end for Rikers Island as a principal disposal destination. Rikers Island is in the East River, south of Hunts Point in the Bronx and north of East Elmhurst in Queens. More significant for this story was the island's proximity to Flushing Bay and Flushing Meadows and to the planned route for the Grand Central Parkway, which ran through Flushing Meadow Corona Park—the site of the 1939 World's Fair. The parkway and World's Fair, both orchestrated by Parks Commissioner Robert Moses, elevated Rikers Island above eyesore status to an impediment.[29]

Moses feared that fetid refuse mounds on Rikers Island would be visible from the proposed fairgrounds site, and he attempted in 1936 to get all disposal activity to cease. For Carey, the island's dumps were necessary to help realize his landfill plans. The differences between Carey and Moses over Rikers Island set off a larger dispute between the former's disposal policy and the latter's urban-planning strategy. It also set in motion the deterioration of relations between Carey and Moses (who had been seen as kindred spirits) and also put the DSNY at odds with the Parks Department over several other land-use issues.[30]

Robert Moses's ambitious plans for Flushing Meadows and the surrounding area grew out of his grandiose aspirations for New York City in general. Although it is unclear how expansive a vision Moses held for the metropolis in the 1930s, there is little doubt about his commitment to a broad-scale remaking of its infrastructure. He began to participate in the inner workings of city affairs in 1913 at the Municipal Research Bureau, which was involved in restructuring the civil service system, and, in the 1920s, as president of the Long Island State Park Commission and member of the State Council of Parks. He got a taste of politics serving under Governor Alfred E. Smith (his friend and benefactor) as secretary of state beginning in 1923 and through his failed run for governor in 1934. Over his career, he was never elected to public office.[31] Beverly Smith, writing for *American Magazine* in 1934, said to Al Smith, "Strange that a man of Moses' curious talents ever went into politics." The governor responded, "But Moses has never been in politics."[32] This may have been the truest statement ever uttered about Moses, who built his own almost sovereign empire through the piling of one appointive position upon another. At one time he held as many as ten or twelve positions simultaneously.[33]

FIGURE 6.2 Robert Moses, pictured here as sponsor of Battery Bridge (1939).

Source: Library of Congress, Prints & Photographs Online Catalog.

From 1934, when Mayor La Guardia appointed him the first citywide parks commissioner, to 1968, when his power slipped away, Moses was first and foremost a builder.[34] During his career, he was responsible for constructing thirteen bridges, 416 miles of parkways, 658 playgrounds, two hydroelectric dams, vast amounts of housing, Lincoln Center for the Performing Arts, Shea Stadium, Jones Beach State Park, and more, while preserving more than 2.5 million acres of parks in New York State.[35]

With the deepening of the Depression, New York City faced public works needs on a grand scale. The 1929 *Regional Plan of New York and Its Environs*, undertaken by prominent business and civic leaders, offered a blueprint for remaking the city.[36] How to develop New York City was probably the most widely discussed issue in national planning circles at the time, with a heavy emphasis on transportation—initially focusing on rail. For Moses, execution of the plan would be achieved in his own fashion and through his own priorities, with a reliance on parkways and highways.[37]

La Guardia was in no position to spearhead the variety of projects necessary to rebuild the city, assuming that Moses's expertise and drive (and the largesse of FDR's New Deal) offered the best approach.[38] Appointing Moses, however, had its risks. While the new commissioner had remarkable talent for the job, he possessed a personality and style that clashed jarringly with the mayor, others in the administration, and the New Deal leadership in Washington.[39]

While Moses and La Guardia developed a certain respect for each other, the mayor once confided to a colleague, "I can't get along with him, and I can't get along without him."[40] Moses expressed a love/hate attitude as well when talking about the mayor. Remarking on La Guardia's political acumen, he stated in his autobiography: "It must be admitted that in exploiting racial and religious prejudices LaGuardia could run circles around the bosses he despised and derided. When it came to raking ashes of old World hates, warming ancient grudges, waving the bloody shirt, tuning the ear to ancestral voices, he could easily outdemagogue the demagogues."[41]

In essence, Moses had a panorama of traits and skills that garnered him deep admiration in some corners and animus in others. Born into an affluent family in Connecticut (1888), he graduated from Yale College, attended Oxford University, and received a PhD from Columbia University.[42] He also demonstrated a strong legal aptitude, which served him well given the cornucopia of positions he held during his career.[43] Moses coveted power over wealth, pragmatism over theory, and tangible accomplishments over façade. No professional planner, he preferred to carry out projects on a grand scale rather than paying particular attention to the needs of specific neighborhoods. He was an advocate of the "public realm" but did not oppose public-private partnerships. He harbored many of the racial biases of his day and had contempt for the poor, but he did not view his projects as elitist mandates. He favored self-help and evolutionary change rather than social equality.

Moses was a product of the automobile age, seeking to enhance mobility through the parkway rather than by retaining or improving mass transit. Bulldozers were the best answer to slums. Money was no object. For Moses, getting things done with high-quality, durable infrastructure was worth the expense. All told, his public works projects cost an estimated $27 billion (in 1968 dollars). He liked to say, "Once you sink your first stake, they'll never make you pull it up."[44]

Moses's personality and how he got things done often rankled people. He was arrogant, often bullying, and sometimes spiteful.[45] He was dismissive of opposition and occasionally rancorous. He rationalized, "I have not lost friends but made enemies. But the enemies I have made do not harbor their anger. Go to any of the parks and beaches we laid out on Long Island and you'll see the people who objected to our turning the places into public resorts rubbing elbows with the folks they wanted to keep out." He also stated, "I know the old saying about catching more flies with sugar than with vinegar, but I am not in the fly-catching business. In order to build parks and highways some blasting is necessary."[46]

Moses operated in a world (especially in the 1930s and the 1940s) desperate for jobs, hoping for an economic turnaround, and witnessing technical and social changes of enormous scale. He approached those issues in his own way, remaking the very look of the New York metropolitan region, altering its natural environment, and redistributing its people.[47] No single person, with the possible exception of Andrew Haswell Green, has had such a major impact on New York City. But as the shopworn cliché has it, "The devil is in the details."

MOSES AND THE WORLD'S FAIR

The World's Fair was an exciting prospect for New Yorkers, and Moses was not prepared to see anything stand in the way of its success—not Rikers Island, not Carey, and not the Corona dump. The location selected for the parkway linking to the World's Fair site extended south from Flushing Bay along the banks of the Flushing River. The west bank of the Flushing River was then overtaken by a vast, notorious dump in Corona, Queens.[48] The "valley of ashes" in what was Flushing Meadows, famously depicted in F. Scott Fitzgerald's *The Great Gatsby*, was an agglomeration of furnace ash, cinder, and household waste. In the novel, Fitzgerald referred to it as "a fantastic farm where ashes grow like wheat into ridges and hills and grotesque gardens."[49] Reaching a height of at least ninety feet above the old marsh level, it was derisively called "Mount Corona."[50] The site was even more grotesque than the operations on Rikers Island. In line with Mayor La Guardia's policy of ending privately owned dumping grounds in the city and based on Commissioner Moses's recommendation that it be condemned, the Corona dump was acquired and designated as parkland.[51]

Moses had been anxious to acquire the land because in addition to developing additional parkland, he wanted to build the Grand Central Parkway Extension (1932) on the property.[52] Flushing Meadows was to be Moses's great park—a second grand public space for New York City.[53] The commissioner regarded the extension as a crucial project, since the parkway linked Manhattan, the Bronx, and Westchester, via the Triborough Bridge (1936) crossing the East River, to the Long Island Parkway system.[54] The parkway extension would cut a six-hundred-foot-wide band through the dumpsite.[55]

The decision to sponsor the World's Fair fit nicely into plans to develop Flushing Meadows and was made easier by the extension of the Grand Central Parkway. As Martin Charles in *Review of Reviews* noted in 1936,

> New York will have its World's Fair—an act of God alone forbidding. And Moses will have much to do with its creation. There will be rumpled tresses

and injured feelings, if precedent is to be followed, but New York is destined to enjoy the Fair and the $100,000,000 that is to be spent on and at the exposition. And when it is all over and the curtain is drawn Bob Moses will be ready, with a little preparation, to turn over another park to the City of New York. The prophet will be seeking more wilderness to conquer.[56]

Moses described how the project of the World's Fair came to him, both in accounts at the time and later in life. In a 1966 retelling, the story began when the Park Department tried to obtain right-of-way for the Grand Central Parkway. Moses's wanted the whole meadow for the project, but this seemed impossible. "Then the miracle happened—the idea of a World's Fair," he declared.[57] In 1935 a nonprofit corporation was formed by business and civic leaders, with the fair to be financed through federal, state, municipal, and private funds and in collaboration with foreign governments. On May 1, 1936, the Board of Estimate authorized Moses to move forward with the plan.[58] The World's Fair would open in 1939, marking the 150th anniversary of the inauguration of George Washington in New York City. In all, Moses developed 1,216 acres (half the size of Central Park) for the fair.[59]

Moses now had his right-of-way, was able to subdue the Corona dump, and began the process of developing a park in Flushing Meadows. From a financial standpoint, building the fair had certain advantages. Money from the fair would pay for leveling the dump site and other landscaping projects. It was possible that the funds could pay for various park costs as well.[60] The land-poor status of the island city played a major role in the location of the World's Fair. As the fair's president, Grover A. Whalen, suggested, the location was "surprisingly close" to the Manhattan skyline, and "when we came to look for a place to build a great exposition—and vacant land was the scarcest commodity in New York, particularly if it had to be near the center of the city, near public transit—we could find no better spot. It must be made to serve the purpose."[61]

In 1938, a year before the fair opened, Moses and Carey reached a compromise over Rikers Island, diverting waste material from there to the eastern Bronx as fill for a new park in Sound View;[62] tons of ash also were utilized as fill for North Beach (later LaGuardia) Airport.[63]

But in June 1939, Moses reacted with great hostility to a city plan to extend Rikers Island (by fill) closer to Queens and further blocking Flushing Bay by adding a six-hundred-acre dump. The parks commissioner bristled at the thought of "another lousy garbage dump." "I am opposed to any further extension of Rikers Island and especially through garbage dumping," he erupted, "which has caused such a public nuisance in the past." North Shore property owners and civic groups were poised to join the fight. Carey responded by stating that extending the island would be conducted under "the same methods employed for the past 40 years."[64]

ONE BEST WAY   ■ 155

Until the dumping operation ended and the piles of refuse were removed, Rikers Island was camouflaged by building screening berms (ridges) around the mounds and planting groves of trees on and around them. By 1940, a fifty-acre nursery on the island included seven thousand trees.[65] Carey also eliminated the burning masses of waste to keep ashes from wafting over the fairgrounds.[66] In 1943, landfills on Rikers Island finally were closed and the bargefill operation moved to a small site at Great Kills on Staten Island, which opened up a new front in the clash between Carey and Moses.[67]

THE BATTLE FOR JAMAICA BAY

Before Great Kills became another point of tension between the two commissioners, Carey poked the proverbial hornets' nest by first attempting to develop additional landfills, beyond those already active in the area, near residential areas in Jamaica Bay. In April 1938 the DSNY announced that dumping would begin on marshes in Jamaica Bay once the department received new steel barges. The proposed site was part of a plan to convert more than 18,000 acres of the marshland into a container port.[68] Moses, however, envisioned a gigantic

FIGURE 6.3 New steel barges placed in service in 1939.

Source: NYC Department of Sanitation, *Annual Report, 1939*.

recreational area along the shore, with bathing beaches, playgrounds, and a wildlife refuge. Jamaica Bay was to Moses what Central Park had been to its builders, Frederick Law Olmsted and Calvert Vaux.[69]

The parks commissioner inquired of the DSNY what area they had in mind for the project, how the work was to be accomplished, and what would be done to prevent refuse from washing up on shore. "There is just no sense in cleaning up a mess in one place on our boundary waters," Moses wrote, "and then creating a worse one in an equally important recreational area somewhere else." He concluded that making Jamaica Bay entirely into an industrial section "is just plain bunk."[70] In July the Parks Department published *The Future of Jamaica Bay*, in which Moses repeated his objections to the DSNY's plans. In the cover letter to Mayor La Guardia, he stated, "This is not a question to leave to land speculators and crackpot industrial theorists." He pointed out that there was adequate port space beyond Jamaica Bay, "particularly along the shores of Staten Island" (giving away the difference he drew between the development of Jamaica Bay and the Richmond borough).[71] Ultimately, Moses got the state legislature to transfer Jamaica Bay to the jurisdiction of the Parks Department.[72]

In making his case for Jamaica Bay, Moses also criticized Carey's disposal strategy:

> It seems incredible that after the recent experience of this Administration in reclaiming Rikers Island and Flushing Meadow, an attempt should be made by a member of the same Administration to create a worse situation in an equally important recreational area. Having spent millions to remove two great dumps at Flushing Bay before the World's Fair, are we so stupid as to fill up the middle of Jamaica Bay with garbage, rubbish and ashes.[73]

The Future of Jamaica Bay also included contrasting images of Jamaica Bay: "unlimited possibilities" for leisure activities, on the one hand, with an artist's conception of a huge refuse dump—a "civic nightmare"—spanning the bay, accompanied by pictures of the Rikers Island and Flushing Meadows dumps, on the other.[74]

Local residents also sided with Moses against Carey and the DSNY. They believed that property values would be jeopardized and feared the health consequences of further disposal operations along Jamaica Bay. About two hundred people met in protest in Howard Beach, Queens, on May 5, 1938. Other groups also congregated in other parts of Queens and in Brooklyn. The Community Councils of New York (representing one hundred organizations), which had spoken out against disposal issues before, vigorously challenged Carey.[75]

On a July visit to an existing garbage-disposal site at Edgemere on Jamaica Bay's marshy shoreline, the commissioner was met by "forty irate residents" of the Rockaways, who arrived "on foot and by rowboat" to protest odors and truck noise from the disposal site, which was scaring away summer tourists. One man grumbled that fumes from the dump had stripped the new paint from two of his cottages. Carey was disturbed and irritated by the challenge but conceded nothing. Andrew J. Kenny, president of the Chamber of Commerce of the Rockaways, said he would seek an injunction to prevent the DSNY from continuing to use the marsh as a dumpsite, and he certainly did not favor any new plans along Jamaica Bay.[76]

Protests extended to the Board of Estimate and the mayor, followed by legal action. But no immediate relief came from the courts about closing the dump. The protests escalated into civil disobedience when fifty housewives, several with children in tow, blocked the entrance to the Lefferts Boulevard landfill in Queens. They left peacefully, and the landfill remained open. Rancor was directed explicitly at Carey. First the local groups exhorted Mayor La Guardia to fire the commissioner, then they turned to publicly scorning his actions, doubting his leadership, and questioning his credibility.[77]

In September, Carey bent under pressure and, according to the *New York Times*, "postponed indefinitely" the plan to dump waste in the contested marshes of Jamaica Bay. He came to an agreement with Moses to get the Board of Estimate's approval to acquire sixty acres of land for a dumpsite near Sound View Park. The site Moses wanted at the mouth of the Bronx River would be reclaimed to create more parkland. Moses purportedly had secret plans to participate in home building in the area (later announced publicly), which accounted for the specific location of the dumpsite in the Bronx.[78]

With the advent of the Sound View Park dumpsite, the DSNY no longer had any disposal options in the Jamaica Bay area. Carey, however, declined to disclose another location (on Staten Island) where he proposed to dump refuse "on a large scale." This decision was soon to stir up a new controversy.[79] In the aftermath of the disposal fracas over Jamaica Bay, Carey's and Moses's relationship deteriorated further. The sanitation commissioner resented that the *New York Times* and others had sided with Moses and that his resolve to focus on landfills instead of incinerators had met with criticism.

The two commissioners engaged in a "sharp-tongued exchange" over Jamaica Bay dumping at an executive session of the Board of Estimate in early October.[80] Sensitive to the constant questioning of his actions, Carey turned more aggressively on Moses at the meeting, claiming that the parks commissioner surpassed the others in the administration by being "a much more artistic manipulator of noise and fanfare." He stated that "Bob is a ruthless spender" and called him "a propagandist." For his part, Moses abhorred that Carey had focused so much attention on landfilling north of Jamaica Bay, since building his parkway could not be sidetracked.[81] Privately, Moses thought that the decision to discontinue using incinerators and to create new landfills was "stupid and barbaric."[82]

ANOTHER STATEN ISLAND GAMBIT

The war of words between Carey and Moses at the Board of Estimate meeting and after signaled a break that seemed impossible to mend. Undergirding Moses's reaction to the DSNY disposal plans was the parks commissioner's grandiose scheme—not always publicly acknowledged—to crisscross the boroughs with roads, parks, and parkways and to use "clean fill" to produce taxable land on which to do so.[83] Carey was less opaque, bent on finding some practical solution to the disposal quandary while working through the political and public maze. He continued to insist that using refuse for landfilling was the best bargain for the city, was carried out in "useless marsh sections," and was the product of scientific methods.[84] Despite the acrimony, deception, politics, and hidden agendas, new deals lay ahead for the battling commissioners.

Opportunities for finding an adequate landfill site (or sites) were scant in late 1938. Carey's desire to continue at Rikers Island was overridden by the World's Fair. Mount Corona was shrinking to a molehill. Carey, Moses, and local residents contested Jamaica Bay. A formal agreement over Sound View was delayed until 1939. Carey, in particular, did not share Moses's enthusiasm for using the site in the eastern Bronx. The authorization was directly connected to the dumping at Rikers Island but also to the WPA project to remove ash from the island to serve as fill for North Beach Airport.[85] The DSNY desperately needed some new, acceptable disposal locales.

Vacant land was at a premium all over the region. Manhattan was out, the populations of Queens and Brooklyn were growing in population, and the Bronx had few available sites. New Jersey had never been happy to be a dumping ground for New York. What still remained were large chunks of Richmond Borough.

Whether drawn to Staten Island because of its previous history, its vast expanse of salt marshes, its relatively isolated location, Moses's defense of Jamaica Bay, or a combination of these, Carey tipped his hand about a landfill site there at the October 11 Board of Estimate meeting. If options were no longer tenable in the Jamaica Bay area, the DSNY would begin operations in the Fresh Kills section of the island in April 1939. Moses quickly responded that "As far as Staten Island is concerned, I am going to oppose anything that I think is going to be harmful to the park system." He reminded the group that the proposed site was ringed by the New Springville, Willowbrook, and La Tourette parks.[86]

In late 1938, however, the DSNY already had begun dumping refuse at Oakwood on the southern shore of Staten Island, near what is now Great Kills Park.[87] Moses had designs on filled-in Fresh Kills marshes as a possible location for a bridge foundation but was unwilling to take a public stand in favor of Carey's plans for dumping refuse in Richmond Borough.[88] At the board meeting, Moses claimed that the project for Fresh Kills "was entirely new to me and to the engineers who have been working with me." But evidence supports that the original idea for landfilling in Fresh Kills actually had come from Moses in 1937.[89]

Moses gave the strong impression, whether accurate or not, that "dumping without any notice to anybody concerned on an area arbitrarily picked out by the Sanitation Department" was unacceptable.[90] "It is my opinion that nothing will ever meet with his [the parks commissioner's] approval unless he is first permitted to suggest it," Carey observed at the board meeting.[91]

The proposal to utilize Fresh Kills intensified the animus between the two commissioners. That decision, along with the landfilling at Oakwood, set off local protests not unlike what was happening in the Jamaica Bay area and what had happened on Staten Island two decades earlier. Oakwood had been the site of an incinerator, but after it was shut down, clean fill was dumped there. Once it was discovered that not only clean fill but garbage was being disposed of at the site, citizens became irate.[92]

On the day after the now infamous October Board of Estimate meeting, the *Staten Island Advance* included an editorial stating, "If a private property owner dumps garbage and offal on a vacant lot, he is hauled to court for erecting a public nuisance and unsanitary condition. Yet that is precisely what Sanitary Commissioner Carey proposes to do at Fresh Kills next spring." It also declared Fresh Kills and its salt meadows as unspoiled "remnants of the old Staten Island" and "historic ground" where the first English settlers founded the village of Richmond (omitting the earlier contributions of the Lenape Indians). Such a dump would also destroy scenic views and landscapes and require dredging the waterway. Beyond the environmental and historical repercussions, "It won't be Staten Island garbage," which was reminiscent of the observation sternly made during the Garbage War. And in a last bit of irony, the editorial excoriated Carey while praising Robert Moses as "our one staunch friend" in the La Guardia administration.[93]

Two days later, the Democratic Club of Oakwood Heights passed a resolution, which was sent on to Carey, condemning the landfilling activity in the area. Nearly everyone at the meeting stood up to speak. "The city is going back on us again," one citizen asserted. "First they railroad an insane asylum on us and now they're going to dump on us, as if the asylum wasn't enough."[94] Another decried the depreciation of real estate values. The Columbian Rotarians also forwarded a resolution to Carey and planned to circulate a petition among several local communities.[95] The 1938–1939 *Annual Report* of President of the Borough Joseph A. Palma reminded everyone that on several occasions in the past, Staten Island "had been forced to oppose the dumping of garbage within the Borough." Public opinion now opposed "the rat infested garbage dumps" that should not be placed near developed residential areas. Palma claimed that there was no place on Staten Island that could serve as a garbage dump "without destroying residential or industrial property, and the value of adjoining properties."[96]

At separate locations on the island, protesters grew in number within a week of the Oakwood action. The Staten Island Vigilantes threatened to "prevent by force" (if there were no alternatives left) the dumping of refuse in Richmond Borough. They selected the former district attorney Albert C. Fach, a central player in the hearings of the Metropolitan By-Products Company reduction plant, as the leader of their campaign. Groups as diverse as the Staten Island Society of Architects joined the protest.[97] A Staten Island Citizens Committee of 10,000 was formed by the local teacher and author Dr. Vernon B. Hampton. The committee's slogan was "Remember the Old Garbage Plant," referring to the Cobwell reduction facility of the late 1910s.[98] Its petition proclaimed that the island was primarily residential and that dumping raw garbage from other boroughs was "dangerous to the health and comfort of the community, destructive of scenic and property valuations, and unsanitary and unwholesome."[99] Hampton passionately stated that Carey's action was a "horrible example for other cities" and that the commissioner was "the health menace of twentieth century America."[100]

On January 10, 1939, the stakes got higher when a Richmond County grand jury, in an unusual step, indicted Carey, three of his aides, and two officials of the Health Department, charging that they were contributing to the public nuisance by dumping garbage at Oakwood.[101] (Only one of the officials, Borough Sanitation Superintendent Joseph A. McDonald, lived on Staten Island.) In November, members of the grand jury and District Attorney Frank H. Innes inspected the site. After their investigation, the grand jury handed a sealed indictment to Supreme Court Justice Alfred V. Norton in St. George. Bench warrants were issued for Carey and the others.

Mayor La Guardia branded the action as strictly political, prompted by the Staten Island Democratic leader William T. Fetherston. "I think that the city will pretty soon get an inkling of what I mean by opposition to the normal and regular administration of city affairs," the mayor stated. "Mandamuses by the wholesale, injunctions by the scores, and now starting to arrest officials for doing their duty." He added that Carey had been "taken from a $100,000 position and brought into city service, and he is doing an excellent job, rendering great service. If he had been taken from a green-covered table in the back room of a political club, I suppose he would have been given a medal or a monument."

Fetherston, of course, regarded the accusations made against him as ridiculous. Carey was distressed by the action (although the charge was merely a misdemeanor), but when reporters congregated at his door, he said: "What's this—you're not ashamed to mingle with a convict?" All but one of the officials (one was ill) were booked and fingerprinted, then released. Asked afterward about the proceedings, Carey said, "Well, I don't want to get facetious about it, because the people of Staten Island are serious about it, or I'm sure they wouldn't have done it. It is an interesting experience, though."[102] He added in his defense, "The fact is that ever since there has been a Richmond, refuse has been dumped promiscuously and we are now doing it scientifically."[103]

Carey not only faced enraged Staten Island protesters but also a festering controversy in Queens in early 1939. On February 1, the *New York Times* reported: "Irate Queens citizens tried to take City Hall by storm yesterday, with garbage pails—one of them full—as their weapons, in a new protest against the depositing of garbage and rubbish in open dumps at various points in South Queens."[104] Rene Robitaille, chair of the anti-dumping committee of the South Side Allied Associations, carried a full bucket of refuse "draped in black crepe." The protesters, however, were denied a meeting until a future date. The story was the same in Queens as in Oakwood. Clean fill had been initially used at the dump sites, soon followed by unburned garbage and rubbish viewed as health hazards.[105]

Tensions continued to escalate on other fronts, further demonstrating the DSNY's ongoing tribulations over obtaining disposal sites. La Guardia ousted the Community Councils from the offices they occupied in the Manhattan Municipal Building, stating that the city could not afford to keep tenants who paid no rent in the overcrowded building. The Community Councils, founded in 1917, originally assisted soldiers, sailors, and marines but most recently had led the anti-garbage protests in Richmond Borough. They saw the action simply as retaliation and vowed to continue their work. In the meantime, Carey was called before a Queens County grand jury to testify on dumping in that borough.[106] In March, the appellate court in Brooklyn ordered a reinstatement and trial of a taxpayers' suit to restrain dumping refuse in Jamaica Bay.[107] This had come on the heels of the latest agreement between Carey and Moses to shift refuse from Rikers Island to Sound View, which was in question again.[108]

Even the City Council pushed back against Carey's decision to emphasize landfilling over incineration. In an overwhelming vote, they supported a bill to restrict intercounty dumping of waste. The bill died in committee. Similar bills were sent to Mayor La Guardia, who defended his DSNY commissioner and vetoed them all.[109] In May 1941, Governor Herbert H. Lehman vetoed a bill on the issue of intercounty disposal, deferring to La Guardia's decision on the subject.[110]

As the NIMBY ring around the city got tighter, Carey faced yet another legal setback. This time, on March 28 he and Commissioner of Health John L. Rice were held for trial in Queens County Court on April 17 on charges of maintaining a nuisance at five different dumps.[111] Both men also had been named in an indictment by a Queens grand jury. They were arrested and booked in Long Island City; they pleaded not guilty.[112] As the *Brooklyn Daily Eagle* succinctly reminded its readers about Carey's dilemma: "His fix: Can't find a place to dump the city's garbage. Trouble has been brewing for some time."[113]

In June, Carey and Rice got some good news when the indictment against them in Queens was dropped. The decision came after an agreement to have Surgeon General Dr. Thomas Parren Jr. appoint a commission to arbitrate the Queens County dumping dispute and suggest possible improvements. Speaking for the mayor, Assistant Corporation Counsel Frederick P. Bryan declared that "if the board finds that these conditions constitute a danger to public health, the Mayor is eager to take steps necessary to remedy this condition."[114] Whether La Guardia was willing to continue to support Carey's approach to disposal in the wake of all of the disputes was unclear.

THE PARREN COMMISSION LIBERATES LANDFILLS

Commissioner Carey had faced the brunt of the criticism in Queens and Staten Island for what was considered arbitrary choices over landfill sitings. The findings of the Parren Commission, however, gave Carey's landfill policy a big boost. In June 1939, after dropping the charges against Carey and Rice in the Queens case, New York Supreme Court Justice Isidor Wasservogel turned over the question of landfill safety to a special panel for review. Dr. Parren selected several nationally known health experts to serve on the panel.[115] The panel studied numerous technical reports on refuse disposal and visited six landfill sites in Jamaica Bay between July 1939 and March 1940. The Parren Commission issued its report in March.[116]

While the experts believed that no disposal method was flawless, they concluded that landfill practices did not create a hazard to public health or safety. The experts, however, did make a clear distinction between regulated landfills and ordinary town dumps.[117] With respect to the five landfills under review, the report was just as certain; they found none of them creating a public health or safety danger. "On the contrary," it stated, "the committee is convinced that certain potential public health hazards obtaining previously at the sites of these fills have been diminished, for filling is one of the best methods of controlling rats and mosquitoes in marshes and swamps."[118] In addition to its findings, the commission made several technical recommendations concerning the construction of the landfills.[119]

The Parren report gave the green light and stamp of approval to the current fills in Queens and to future locations that would employ sanitary landfilling methods. Carey had a clearer mandate than ever before. In practice, returning to incineration was not economically possible, given the need to add substantially to the current inventory at great cost. In collecting refuse throughout the city, he shifted the practice away from source separation to refuse mixing, which favored the landfill option. With the legal challenges behind him as well, he could seek new disposal sites with less concern about public reprisals.[120]

THE MARINE PARK AT GREAT KILLS

Unlike Carey, Commissioner Moses usually received the benefit of the doubt from citizens for his park-building efforts and his less-than-enthusiastic stance on refuse fills. (When he utilized refuse for fill, he liked to make it clear that offensive organic and other materials were not included or at least dealt with in a sanitary fashion.) Moses's plans for developing Marine Park at Great Kills on Staten Island's South Shore (which required an extensive landfill operation) led to a major confrontation with Carey. That Staten Islanders appeared to trust Moses more than Carey substantially blunted the public outcries when the project was announced. The goodwill that the parks commissioner engendered at Great Kills, however, did not help him in promoting the eventual new landfill at Fresh Kills in the late 1940s after Marine Park was underway. Part of the reason was growing complaints about odor problems surfacing at Great Kills.

Moses had been interested for some time in building "marine parks" in Brooklyn and Staten Island, places where parkland was scarce. (Both areas had undeveloped marshland.) In 1930 the Jimmy Walker administration had agreed to issue $30 million in bonds to buy new parkland, and the city had acquired 2,530 acres in Staten Island for proposed sites at Great Kills and Willowbrook, plus at other locations in Queens and Brooklyn. But in 1931, pressure to spend city resources for more immediate needs stalled the financing.[121] In 1934 work began on constructing a bulkhead, filling in Crookes Point (a peninsula at Great Kills), and then attaching it to the mainland. New Deal relief forces conducted the work, but Moses believed that the project was too big to be completed in this way; financing had been and continued to be a problem.

In 1936 the city acquired title to an adjacent meadow "to round out the park area." The plan called for filling the entire expanse, including the salt meadow from Oakwood Heights to the harbor and the property where an abandoned incinerator was located. He believed that there was "a considerable amount of fill available" to be brought in by scows after filling Sound View Park. Finishing the bulkhead and other facilities had to be in place before the filling operation. He told La Guardia that when the work was completed "no serious inconvenience to the neighborhood will be experienced."[122]

As he wrote La Guardia in May 1940: "From time to time the Park Department has called attention to the importance of the development of the Marine Park at Great Kills, Staten Island. This is one of the great future shorefront recreation areas of the city."[123]

In the summer of 1940, Moses worked to get dredging going for Great Kills as an important step in improving the larger site. The project had been approved by Congress, but the War Department (in charge of the dredging operations) still awaited word on funding. Moses was informed that local cost sharing with the federal government would be needed for the project. In addition, because it was not listed among projects for the ensuing fiscal year, the city would need to advance all the funds to get the marine park underway.[124]

Moses wrote to La Guardia on June 4 arguing that the position of the War Department was "enough assurance for the City of New York to justify us in going ahead. May I ask you to let me know if you agree?"[125] Thus began a drawn-out effort to build the marine park, which Moses hoped to complete by 1944. In addition, Moses was lining up more local endorsements to help make his case for the project, such as the Staten Island Real Estate Board and the Staten Island Chamber of Commerce (the groups he originally thought were the primary opposition).[126]

The parks commissioner's efforts in remaking the marshland on Staten Island's South Shore were viewed quite positively by the public, especially compared to the wary reception that the DSNY had received at Oakwood. The *Staten Island Advance* stated on June 7, "When Parks Commissioner Moses announced his plans for 'controlled' dumping of garbage and other refuse at the site of the Great Kills Marine Park our immediate reaction was one of opposition." It added,

> Mindful of the recent attempt by Sanitation Commissioner Carey to dump the city's offal in the Fresh Kills meadows, we suspected that Mr. Moses's proposal might contain some of the elements of a "Trojan Horse"; that it might be utilized by the city administration as an entering wedge to convert this borough into a depository for all the filth collected from Riverdale in the Bronx to Far Rockaway in Queens.

Moses had invited "anyone interested" in the park to inspect similar projects elsewhere in the city (particularly Sound View) as a way of assuaging their fears. The *Advance* was convinced that Moses was "an uncompromising critic of the dumping methods advocated by Commissioner Carey" and thus could be trusted to do the right thing.[127]

The *Advance* also characterized Moses as having shown himself "sympathetic toward Staten Island, keenly interested in its future," and opposed to anything that might detract from its "scenic and natural advantages."[128] A reporter and a photographer were sent to the Bronx to check out Sound View. The reporter's feedback convinced the local paper that "the Island could without reservation rely on Mr. Moses to fill the South Shore waterfront area with minimum of nuisance or inconvenience to the residential communities nearby." One of the Queens civic leaders interviewed by the reporter stated, "We'll take Moses's dumping anywhere, any time."[129]

From July and into September 1940 the bureaucratic wheels began to turn, and the Great Kills project now seemed to be moving forward.[130] Some outstanding issues remained, especially whether the city would advance the necessary funds for dredging.[131] On October 11 the *Advance* reported that the City Planning Commission "has thrown a monkey wrench into the gear wheels of Parks Commissioner Robert Moses's program." The commission threatened to destroy the park project by rejecting the request to complete the bulkhead at the park site. The structure was necessary to provide a retaining wall in order to fill in low-lying areas along the waterfront. "The bulkhead means the difference between a Moses landfill and a Carey garbage dump." Staten Islanders had accepted the dumping program for the marine park "on the express understanding that it would be done the Moses way." The newspaper concluded: "Staten Island doesn't want the Marine Park if its development involves the creation of a vast Carey garbage dump at Great Kills." Without the bulkhead, the project would break down.[132]

Moses wrote to La Guardia, "We had considerable difficulty in getting the people of Staten Island to a point where they would not object to Commissioner Carey's dumping at Marine Park and I don't think we should give them any opportunity to say that we are changing the program as planned."[133] There is little reason to believe, therefore, that Moses made much of a distinction between Carey's dumping program and his own. What mattered to the parks commissioner was getting the project underway without resistance and with full public confidence in his actions. Just before Christmas, the Board of Estimate authorized filling and grading at Marine Park, with an advance in city funds.[134]

The Great Kills project slowly moved forward into the early 1940s. Massive amounts of sand from New York Harbor and DSNY fill from Manhattan were to be employed to complete the work.[135] In a letter to Frederick J. Spender with the *Staten Islander*, Moses explained in detail the "controlled landfill" process being employed at Great Kills, which he believed would take two to three years to complete. The use of DSNY fill, he argued, "is the only possible way" to prepare the park for future development, especially given the size of the project. He likened the process to a "similar operation" in Sound View Park, which had taken place over the last three years "without nuisance and without objection on the part of the adjacent property owners, except for one complaint which was corrected." He added, "I am certain that you will not find the operation in Great Kills Park objectionable."[136]

Support for the new park from Staten Islanders was based on a big assumption that Moses gave priority to "clean fill," which mitigated against their worst fears of yet another unsightly and unhealthy dump tainting their shores. They underestimated Moses's top priority: building his park at all cost—and doing so during a war that drew the country's primary attention and garnered its major resources.

The project bogged down during World War II, and efforts to ensure that the landfilling was accomplished with the least amount of inconvenience to locals faltered. Moses wrote to Dan Harper, the editor of the *Staten Island Advance*, in November 1945: "I agree that the method of operation [at Great Kills Park] is not satisfactory at the present time, and we are taking active steps to correct the situation. There was a good excuse during the war, but now that the personnel situation is improving and new equipment can be purchased, we all recognize that the present condition can and will be corrected."[137] Carey also agreed with Moses that at the time "conditions at Great Kills are not as good as we'd like to see them" and that "you certainly put your finger on the causes when you said this could be attributed to a lack of proper equipment."[138] At least that was one reason.

Murmurings, nonetheless, had been getting louder since 1944 as more refuse was being diverted to Great Kills from Rikers Island and elsewhere. In 1945 protests got even more intense over the pests and odors at Great Kills, despite Moses having assured the residents that the project would be carried out with the greatest care. Speaking at a meeting of campaign workers, John A. Lynch, a former borough president, severely criticized the city officials who "authorized the use of garbage for landfill." The work at Great Kills Park, he stated, "is a disgrace to the City of New York." He did not spare Moses, saying: "All the efforts of Park Commissioner Bob Moses and other city officials to justify the use of garbage are refuted by the smell that fills the air of our South Shore." The protests continued as the project lumbered on.[139]

Moses explained the obstacles he faced in completing the marine park in a June 22, 1946, letter to the editor of the *Staten Island Advance*:

> We have used every kind of material, money, personnel and aid we could lay our hands on legitimately.... The immense fill required in this area could only have been provided by transferring to Great Kills the sanitation skow [sic] filling operations which were used so successfully in the development of LaGuardia Airport, and the Orchard Beach and Sound View Parks in the Bronx. It is true that the sanitation operation during the war period was nothing to be proud of, but it was carried on under the most trying and unusual conditions. In any event, the operation is under control now and better equipped.[140]

Great Kills Park finally would open in 1949.

For his part, Carey's stock had dropped to a low point. In 1943, the bargefill operations at Rikers were moved formally to the 460-acre site at Great Kills. The relatively small landfill on Staten Island was absorbing about 35 percent of the city's total solid waste and was only a temporary solution at best.[141] But as long as it had the DSNY's—and Carey's—name associated with it, locals were not happy.

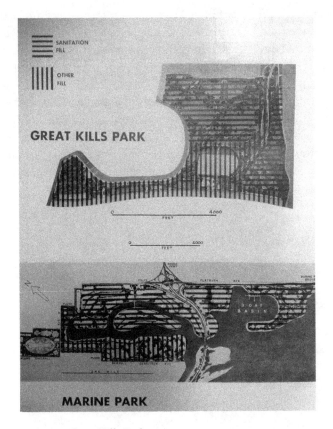

FIGURE 6.4 Great Kills Park, 1950.

Source: Department of Parks, *Reclamation*, 1950.

FROM GREAT KILLS TO FRESH KILLS

The journey that led from Great Kills to Fresh Kills was not determined by one person, nor was it the result of an inevitable chain of events. As disposal options for New York City continued to narrow (or change) in the 1930s, a variety of circumstances (including those concerning Rikers Island, Corona, the World's Fair, and Jamaica Bay) led to Staten Island. William Carey's and the DSNY's commitment to landfills over incinerators was validated by the conclusions of the Parren Committee. Once landfilling became a priority, the phasing out and then closing of historic dump sites like Rikers Island and Mount Corona set off a search for new disposal venues.

The determination to dispose of refuse in an environmentally sound way and also to convert pestilential marshland into "tax-generating" property or parks offered justification for landfilling. Within this context, powerful leaders such as Carey, Moses, and La Guardia sought opportunities for their own projects and purposes. And finally, Staten Island's history as "the other" and its political impotence limited what it could do to reverse the policies and proposals originating in Manhattan. As William O'Dwyer, who served as mayor of New York City from 1946 to 1950, noted in his autobiography, "The voting strength of the various localities, or lack of it, too often was the final determining factor as to where garbage was to be dumped, and Staten Island, sparsely populated and with little voting strength, became one of the likely spots."[142] Its salt marshes became a tantalizing landing place for a major landfill and, perhaps, terraforming for future projects.

CHAPTER 7

FUTILE PROTESTS

Staten Island was extremely attractive as a major disposal site for the City of New York in the mid-twentieth century because it was close enough to the other boroughs for hauling waste by barge but far enough from their backyards. Richmond Borough also possessed an extraordinary amount of salt marsh, at the time regarded as useless land, which could provide the necessary space for a landfill. These particulars were terribly enticing for a city seeking a solution to its chronic waste problems. The opening of Fresh Kills Landfill in 1948—much to the locals' dismay—would connect the boroughs in new and more intimate ways. Staten Island now was an island not so far.

CONNECTING WITH AND RELATING TO NEW YORK CITY

The lack of railroad, subway, or automobile connections to the rest of New York City (until the 1960s) said much about Staten Island's politics, space, and accessibility. Those people living, working, and sightseeing in Manhattan, Brooklyn, Queens, and the Bronx had several options when moving from one borough to the next. But if you wanted to go to Staten Island, you needed the ferry.

An article in a 1919 issue of *Scientific American* had made a case for connecting Staten Island with Manhattan by subway. "Topographically Staten Island belongs to New Jersey. . . . In every way Staten Island is disconnected from the other boroughs of Greater New York." "And yet," the article emphasized, "it is a part of the big city and should, therefore, be entitled to some of the rapid transit facilities which have been developed to such an extensive degree in all the rest of the boroughs.... It was clearly evident," it added, "that something must be done, and done soon, to connect Staten Island more closely with the city."[1]

Calvin Van Name, the Richmond borough president, echoed these sentiments. The borough, he argued, was "an integral part" of New York City and "entitled to the same favorable treatment that has been bestowed upon the other boroughs." "The city government has been prodigal in providing millions upon millions for the development of transportation facilities in other boroughs," he observed, "and has almost ignored all applications for aid in our efforts to develop Staten Island. The City of New York must use the shores of Staten Island in its competition with the advantageous shores of New Jersey," implying that a tunnel for a subway connection would quickly prove beneficial for the city as a whole.[2]

The state legislature concurred that a connection was necessary, although the one it envisioned was not a direct route to Manhattan. In 1921 it passed a bill requiring the city to build a tunnel connecting Brooklyn to Staten Island for subway trains and Baltimore and Ohio Railroad (B&O) freight trains.[3] The construction was to be part of the proposed Brooklyn Fourth Avenue subway, which would run across the lower end of New York Bay to Staten Island. In 1923, the tunnel project began with great enthusiasm, at least on the Staten Island side.[4] In anticipation of the tunnel, all three Staten Island passenger railroad lines (there was no subway on the island) were electrified, and construction on tunnel headings started in Staten Island and Brooklyn by 1925.[5]

Citywide politics and economic rivalries terminated the project before completion.[6] The Port Authority of New York and New Jersey's (PANYNJ) Comprehensive Plan in the 1920s envisioned rail routes to Staten Island as connectors to a metropolitan freight system, but it was never implemented.[7] The idea for the project arose again under Mayor Fiorello La Guardia, who pledged to revive it during his 1933 campaign. Without federal funds and because of competing priorities, he did not deliver on his promise and, in fact, fought efforts by Staten Islanders to circumvent the city by trying to secure legislative support. Robert Moses never warmed to a rail route or rail tunnels, preferring bridges to carry automobiles and trucks.[8]

The PANYNJ's Comprehensive Plan did lead to the building of three bridges (vehicular traffic only) crossing the Kill Van Kull and the Arthur Kill from Staten Island into New Jersey: the Goethals Bridge (1928), from Old Place to Elizabeth; the Outerbridge Crossing (1928), from Tottenville to Perth Amboy; and the Bayonne Bridge (1931), from Port Richmond to Bayonne. That Staten Island's isolation was reduced by these bridges—with respect to New Jersey at least—suggests the vitality of the relationship along what was becoming a significant industrial corridor. Since the Port Authority was a two-state body that

FIGURE 7.1 Old Goethals Bridge, from the Staten Island side looking southwest, spanning Arthur Kill from New Jersey to Staten Island, 1991.

Source: Wikimedia Commons.

FIGURE 7.2 The Outerbridge Crossing, looking east toward Staten Island, spanning Arthur Kill from New Jersey to Staten Island, 1991.

Source: Wikimedia Commons.

constructed, operated, and maintained the bridges, this made political and border issues essentially disappear.[9]

The overall lack of rail and vehicular traffic from Greater New York City to Staten Island in some strange way reinforced the value of the place as a major disposal site. Roads and bridges would not need to be clogged with garbage trucks. Instead, barges and scows could do the work, a practice New York City had long employed in its days of ocean dumping.

Staten Island was a curious combination of farmland, a slowly developing industrial zone, and a suburban enclave. Richmond Borough's population grew steadily but modestly from 116,531 in 1920 to 191,555 by 1950. In 1900 its residents represented only 1.9 percent of the total population of New York City, and by 1957 only 2.7 percent.[10] In 1929, only 31 percent of the area was developed, much less than the other boroughs, because of its relative isolation.[11]

Economic activity varied across the island. Agriculture remained small but flourishing in some sections. Because of the ongoing activity of the Port of New York and New Jersey, industrialization was expanding along the shorelines of the Arthur Kill and the Kill Van Kull waterways. By the end of 1927 there were two large factories in operation on Staten Island, Richmond Airways, Inc., and Bellanca Aeroplane Corporation. The chemical industry was well represented by businesses related to shipping, oil storage and refining, brick making, and manufacturing. Procter & Gamble Manufacturing Company at Port Ivory employed about 1,200 people.[12] Commercial access to the island increased by the early 1930s thanks to the opening of the three bridges to New Jersey. The Great Depression, however, slowed economic growth, with several industrial plants compelled to shut down. A temporary spike in the local economy during World War II raised hopes for improvement after the war.[13]

The industrialization of parts of Staten Island and surrounding areas provided jobs but also had environmental repercussions. Air pollution from stationary and motive power sources was bad in several parts of the island. In the early 1930s truck farms (farms that produced vegetables for the market) in New Springville, Bulls Head, and Greenridge noted crop damage from airborne pollution. In response, farmers organized the Staten Island Growers' Association as a means of both protest and promotion. Some farmers abandoned raising vegetables altogether and turned to greenhouse flowers. Industry on New Jersey's shore opposite Staten Island and the B&O freight yards polluted the air and water, making the Richmond Terrace Corridor and other nearby neighborhoods less suitable for homes, hotels, and retail establishments.

The waters around much of the island had been contaminated especially because of the intense activity in New York Harbor. Only about 40 percent of the sewage dumped into its surrounding waters in 1935 was treated, and no sewage-disposal plant was erected on Staten Island until 1950. Clamming, oystering, and commercial fishing suffered from the tainted water. Oyster beds were condemned or temporarily shut down in Jamaica Bay, New York Harbor, and the Arthur Kill from a fear of typhoid fever.[14] A colonial-era letter that proclaimed "Oysters, I think, would serve all England" no longer applied to the New York City area.[15] In the minds of some residents, Staten Island remained pristine in the face of its neighbor's rush to modernity, but the reality was somewhat different.

After World War I, about two-thirds of the island's citizens lived in the northern and eastern areas. The hills on Staten Island were being developed by real estate groups, who actively subdivided parcels on Emerson Hill, Lighthouse Hill, and areas around New Dorp and Midland Beach. Staten Island was increasingly becoming a community of home owners.[16] The North Shore was the first section of the island to become more densely populated. The East Shore was primarily farmland until many small, single-family homes multiplied after World War II. The South Shore lies along the Lower New York Bay and includes Great Kills but was essentially undeveloped.[17] The central western shoreline (West Shore) continued to be rural and became the site for the Fresh Kills Landfill.[18] Yet viable communities existed there for many generations. South of Fresh Kills the town of Kreischerville (now Charleston) manufactured fire bricks, gas retorts, drainpipes, and other products made of clay. North of Fresh Kills was the company town of Linoleumville, home of the American Linoleum Company.[19] As one resident put it, "Imagine going into some of them big Manhattan department stores to buy and giving your home address as Linoleumville!"[20]

In 1931, Linoleumville changed its name to Travis (for local hero Colonel Jacob Travis). Despite its identity with the commodities it produced, the change emphasized that it was not just an industrial zone but a town of living, breathing people, whatever the name. Regrettably, Travis soon would be associated with a site much more ignoble than the American Linoleum Company: the Fresh Kills Landfill.

THE SALT MARSHES

What made Staten Island attractive as a dumpsite, aside from its relative isolation, small population, and political vulnerability, was its abundance of salt marshes.

Today marshlands are recognized as important environmental resources. Native wetlands are flooded or saturated areas where water drenches the soil recurrently. They are a collection of wet environments that include salt marshes, inland freshwater marshes, wet meadows, bogs, swamps, and seasonally inundated floodplains. Ponds and the shallow-water zones of lakes also can be considered wetlands.[21] These spaces provide numerous environmental, economic, and cultural benefits to proximate communities. Probably most significant is their role as water filters, helping maintain surface-water quality in rivers, streams, and reservoirs—and even improving degraded water. Wetlands do this by removing or retaining key nutrients, processing organic and chemical wastes, and reducing sediment to receiving waterways. Wetlands also are important in flood control by storing water temporarily from overflowing riverbanks or by collecting water in depressions. They are among the most productive and essential natural ecosystems in biomass production, which serves as food for small invertebrates, fish, other aquatic animals, and humans. Wetlands are key habitats, providing cover, breeding areas, and nurseries for all kinds of animals. As areas of biodiversity for plants and animals they have few rivals.[22]

Marshlands make up 11 percent of the lower forty-eight states. But the modern understanding of their importance took a long time to develop. Until late in the twentieth century many people regarded marshes and swamps as worthless, bothersome, mosquito infested, and unhealthy. Neither water nor land, they were seen as having value only when drained, covered over, and turned into "usable" real estate. And Americans were successful in doing so. Between 1780 and the 1980s, 53 percent of U.S. wetlands were destroyed—or in many people's minds, reclaimed. Rising sea levels also have had detrimental effects on wetlands.[23]

New York City's salt marshes, or tidal marshes, exist as a boundary between solid ground and the bodies of water surrounding the city. They are regularly flooded and drained by the tides. Their salt-tolerant grasses grow vertically through layers of sediment. They shelter birds, snakes, toads, turtles, and fish. Slow in developing, they have existed along the edges of New York Harbor for only four thousand years, which is a mere instant in geologic time—and at their peak they made up the vast wetlands of Jamaica Bay, the Meadowlands in New Jersey, and coastal Staten Island.[24]

Staten Island contains both tidal and nontidal wetlands. The tidal wetlands (salt marshes and tidal mudflats) are low-lying lands along the coast and tidal rivers flooded with brackish water at high tide. The nontidal wetlands (freshwater marshes, swamps, vernal pools) occur beyond the ocean tides along streams, ponds, and lakes and in isolated depressions. Estuarine wetlands (an estuary is a water passage where the tide meets a river current), where Fresh Kills Landfill would be constructed, are tidal wetlands affected by the daily tides or less frequent floods. Most of these wetlands are vegetated and are referred to as salt marshes.[25]

The most extensive salt marshes spread along Great Kills and northward from Rossville beside the Arthur Kill. Most of the island's inland wetlands are deciduous forested wetlands; that is, they contain trees that lose their leaves in the fall. Currently remaining on Staten Island (as measured in the 1990s) are approximately 3,407 acres of wetlands, or 9 percent of Richmond County.[26] Even before people began to recognize their environmental importance, Staten Island's marshes were never regarded as useless. Three tribes of the Lenape nation lived on or near Staten Island. (The Lenape were among the first people in North America, dating back 10,000 years.) The Lenape grew corn, squash, and beans in the salt marshes, hunted and fished there, and harvested oysters from the creeks.[27] After the Lenape, early white farmers grew corn, wheat, potatoes, turnips, and flax and grazed sheep and cattle in the salt marshes; hunters tracked a variety of waterfowl and other quarry.[28]

The wetlands also were home to *Spartina patens* (salt hay or salt meadow hay), which grows on the peat at the bottom of the marshes. The short, tangled grass was used by Dutch and English colonists in Staten Island, Brooklyn, Queens, and elsewhere as fodder for cattle. The settlers also used it as bedding for horses and livestock, thatch for barn roofs, packing material, insulation for ice houses, and traction on roads. In northern New Jersey's Hackensack Meadowlands, salt hay was harvested and shipped to southern New Jersey glass manufacturers (presumably for packing), in what was a lucrative trade until the early twentieth century.[29] New York City was a large market for Staten Island salt hay to feed the thousands of horses pulling carts, wagons, and streetcars.[30] "Marshland, a rolling plain of mud and wild grass stretching from Travis to New Springville, has for years been nothing but the fly in the real estate developer's ointment but today it has justified its existence," went a story in the January 30, 1932, edition of the *Staten Island Advance*. The article then profiled the brothers Charles and Burton Cortelyou, who gathered salt grass along Fresh Kills and sold it to contractors as bedding or for use on freshly laid concrete roads to protect them from the sun.[31]

The closer that marshlands were in proximity to urban settlement, the greater the chance they would be drained and filled. As the New York City area industrialized, the demand for feed for working horses diminished, but the need for human food products (and places to grow them) increased. Acreage that had been diked for harvesting salt hay often made way for truck farms. The outward thrust of the metropolis also created demand for more housing and other buildings, which called not only for filling marshes but for digging drainage ditches, constructing tidal gates, and undertaking vast mosquito-elimination operations.[32] These objectives changed the priorities for the use of the salt marshes and reinterpreted their value. As E. J. Cleary, the assistant editor of *Engineering News-Record*, put it, landfilling was converting "hitherto unusable swampy and low-lying areas" into "first-class properties."[33]

Momentum for increasing marshland conversion on Staten Island accelerated in the 1940s. Between 1936 and 1942, the total acreage of fresh- and saltwater marshes in New York City dropped from 29,000 to 15,000 because of vigorous development. Almost half of the city's remaining marshland was located on Staten Island.[34] Such opportunity for producing usable land could not slip the notice of Moses, Carey, and others.

AN END AND A BEGINNING

As William Carey's tenure as sanitation commissioner was coming to a close (he resigned in 1946), sanitary landfilling as New York City's primary disposal option was well entrenched, bolstered by the Parren report and court action over the Carey indictments. Waste reclamation and recycling in support of the war effort temporarily modified the Department of Sanitation's (DSNY) overall disposal plans from 1941 to 1945 but did not transform them.[35] The DSNY still depended heavily on wetlands filling, which was integrated into department policy. In 1940 more than two hundred acres of marsh were filled with refuse, increasing to 414 acres in 1947 and 690 acres in 1949. By the mid-1940s, however, major new disposal sites had not been identified, were off limits, or were otherwise not available.[36]

Carey had been safe in his city position as DSNY commissioner during much of his tenure, thanks in large part to Fiorello La Guardia's loyalty to him. They did not always agree, and the squabbles with Robert Moses sometimes put the mayor in a difficult spot, but Carey lasted until the end of the La Guardia era. He had hinted at leaving his post on a couple of occasions, but nothing came of it.[37] Some political maneuverings in and around La Guardia's waning time in office, however, drew Carey into some embarrassing circumstances. For example, he was charged with comingling public and private funds in developing a Long Island resort (Sanita Hills) for use by DSNY employees. The commissioner declared that he wanted to build good relations with his men by spearheading the resort. He admitted that the accusation was true, that he had likely violated the City Charter, but that he intended for city funds invested in the project to be repaid from the proceeds of the department's annual baseball game. "Sanita Hills," Carey stated, "won't cost the city a red cent."[38] When it was learned that Carey and his wife had loaned more than $100,000 to continue the work at Sanita, public sympathy shifted to the commissioner, and he was exonerated.[39] Lost in the end-of-term food fight were concerns about where (but not how) New York City was likely to dispose of its refuse after World War II ended.

It was not inevitable that Staten Island would be the chosen site for the world's largest landfill; only hindsight makes it appear so. Too many signs, however, were pointing to Fresh Kills as the next depository for city waste. How much of that waste, and for how long it would be dumped there, no one was prepared or able to say. Immediate needs took priority over long-term possibilities. The change in the city's administration in 1945, among other things, seemed to produce a repudiation of the Carey landfills-only policy. While the new mayor,

William O'Dwyer, and the new sanitation commissioner, William J. Powell, set a different course for the DSNY, it was Moses who tipped the balance toward Fresh Kills.

Ireland-born O'Dwyer shared working-class roots with La Guardia. Immigrating to the United States at age twenty (1910), he held a number of jobs before attending Fordham Law School. His legal training took him into the public sphere as a magistrate in Brooklyn, a county judge, and, in 1939, as district attorney in Kings County. There he won a national reputation for prosecuting members of the Murder Incorporated crime syndicate. He parlayed his success into a run for mayor against La Guardia in 1941 but was defeated. He then joined the U.S. Army during World War II as chief of the Allied Control Commission in Italy, attaining the rank of brigadier general. It was not until the "Little Flower" stepped down in 1945 that the Tammany-backed Democrat won the mayor's office, in which he served for five years. A scandal in the police department and alleged links with the underworld led to his resignation in 1950. In 1952 President Harry S. Truman appointed him ambassador to Mexico, but he was dogged until his death about the growth of organized crime in New York City.[40]

The postwar years would witness unprecedented growth and prosperity in New York City but also changes, especially large-scale suburbanization and eventually the loss of central-city manufacturing. The population in 1950 sat at almost 7.9 million. Mayor O'Dwyer faced a variety of challenges, including the unenviable task of dealing with an unbalanced budget, overcrowded schools, understaffed city departments, crumbling subways and other infrastructure, traffic congestion, and a variety of pollution problems—all of which had been neglected during the war years. The desperate need for veterans' jobs and living quarters led O'Dwyer to focus attention on housing, especially rental apartments and rent control. He knew the city well and appreciated the human side of what was a difficult transition from war to peace.[41]

The new mayor offered some continuity with his predecessor by keeping on some of La Guardia's appointees. Of singular importance was Robert Moses. The parks commissioner was powerful enough at the time without O'Dwyer's help, but the two men entered into a strong partnership—at least initially. A key to that apparent bond was the mayor's decision to establish the Office of the City Construction Coordinator, which gave Moses one more avenue to shape future building practices and to enhance the influence and independence embedded in his string of positions. The new post put him at the center of every building project in the city, including where fill might be needed for new projects.[42]

For sanitation commissioner, O'Dwyer appointed Powell, who had solid qualifications but was hardly a match for Moses. In 1936 Mayor La Guardia had appointed Powell deputy commissioner under Carey, referring to him as a "typical career man."[43] Born in Brooklyn in 1906, Powell joined the DSNY as a garbage-truck driver, then rose through the ranks all the way to deputy commissioner. He was known as a model employee, and as the *New York Times* stated in his obituary, "few Commissioners have worked at a disagreeable and discouraging job with greater personal devotion, or so uncomplainingly."[44] Eager to resurrect incineration as the city's primary disposal option, Powell (with O'Dwyer's backing) was bent on dismantling Carey's reliance on landfilling. Powell, like several others in the DSNY, believed that seeking new landfill sites was a short-term objective given the paucity of potential locations. This new direction did not necessarily coincide with Moses's longer-range building goals. A complex triangular ruckus among the DSNY, Moses, and the people of Staten Island over the latest disposal crisis began percolating in 1945 and burst into the open the following year.

THE DECISION

Nothing had come of developing Fresh Kills as a disposal or fill site in the late 1930s, despite the interest expressed by both Carey and Moses. Fresh Kills, however, had not been forgotten. Great Kills Park, which would include 467 acres of marsh and underwater land, was moving slowly toward completion in 1944–1945, and Moses was interested in further development on the island, where plenty of additional salt marsh was available. (In the late 1940s, Staten Island possessed about 65 percent of New York City's remaining marshland and also the last major wetlands complex at Fresh Kills.) In 1945 Moses successfully got the state to transfer to the city the title of some property in Fresh Kills. At Moses's request, State Senator Robert Bainbridge (Staten Island) had sponsored a bill in March 1945 transferring the property; he did not state what use was in store for the area. Bainbridge later claimed that Moses told him that the bill would be used for the purposes of acquiring more parkland for the borough.[45]

In confidential correspondence with Governor Thomas E. Dewey, urging him to sign the Bainbridge legislation, Moses revealed a rationale for the bill, emphasizing the means and ends of its purpose:

> As you know, the disposal of garbage, trash and other waste materials is a serious and expensive problem in the city of New York. The Department of Sanitation has found that *the most economical method of disposal is the landfill on valueless marshlands and swamps on the outskirts of the city.* This practice has been carried out successfully for a number of years and has resulted in the elimination of many low-lying, mosquito-breeding areas, the creation of new residential areas, and a decided improvement in living conditions in the surrounding neighborhood.

FIGURE 7.3 Tug and barge headed to the Staten Island marine loading station at Great Kills.

Source: NYC Department of Sanitation, *Annual Report, 1947*.

> The land to be transferred is a low-lying area lying generally between Arthur Kill and Richmond Ave. which has remained unimproved and unused since colonial times. It will be used by the city for refuse disposal under the present system of landfill operations of the Department of Sanitation. Filling operations will eliminate the unsanitary mosquito breeding swamp and will provide additions to La Tourette and New Springville . . . Parks *and at the same time will solve the city's disposal problem for a number of years to come.*[46]

Perhaps Moses was trying to sell his idea to the governor by offering a sanitation benefit for New York City. But it seems more probable that the commissioner's view reflected the necessity to use refuse along with clean fill, as he did at Great Kills, to produce—under prevailing conditions—the developed land he desired.[47] In a letter to the editor of the *Staten Island Advance* (mentioned in chapter 6), Moses argued that he knew "of no other way of reclaiming this area for municipal and industrial use than to use Sanitation fill."[48] The secret nature of his correspondence with Dewey suggested that Moses feared his motives in developing the Fresh Kills site might be questioned (or misunderstood) in an open forum.[49]

In many respects, Moses had reaped substantial capital with the citizens of Staten Island by publicly opposing an attempt to open a dumpsite at Fresh Kills in the late 1930s and by distinguishing his plans for Great Kills from those of Carey and the DSNY—especially with respect to utilizing "clean" fill as opposed to refuse there in 1940. He knew that the project at Great Kills would be linked to whatever he wanted to achieve at Fresh Kills. The public outcries over unsanitary conditions at Great Kills during the war were warnings to tread carefully in his public statements about his latest venture.

In correspondence with Deputy Mayor George E. Spargo on January 7, 1946, Moses was very clear about his goals and strategy. He wanted to "revive the Fresh Kills plan . . . just as soon as Commissioner Powell has demonstrated that he can do an unobjectionable job at Great Kills." That plan constituted extensive dumping of fill to turn substantial acreage of Staten Island's salt marsh on its western shore into useable land for parks, parkways, and other city building projects. Moses trusted his own practice of gaining public approval for his reclamation projects even in environments "without great enthusiasm and sometimes in the face of considerable local opposition, simply because we were able to persuade them that parks would result in the end which they could not otherwise expect."

He strongly criticized the DSNY for its "pretty rotten job" (its poor management and execution of the filling process) at Great Kills and for "second and third rate jobs elsewhere," wanting to make sure that Powell got on board with the proper strategy for cultivating the public. "If Commissioner Powell is not prepared to recognize these facts," he warned, "I would suggest that he trundle his waste somewhere else and see where he gets along without the park idea to support him." Great Kills was the linchpin: "It was because of the complete breakdown at Great Kills that so much opposition arose against the Fresh Kills project." Moses wanted a plan in place for Fresh Kills in 1946 and recommended staying away from starting any new dumping area "with all the hullabaloo and opposition this would stir up."[50] For Moses, justifying the creation of a new landfill without the promise of a park (or other beneficial land use to follow) was untenable.

Powell struggled to improve conditions at Great Kills after the rise in public protests in the mid-1940s and was willing to cooperate with the engineers from the Park Department working there. Yet he faced what seemed to be overwhelming problems in developing an adequate overall disposal program for the city.[51] He also was pushed by Moses to carry out filling operations "properly," bringing into question to whom Powell was expected to answer in the O'Dwyer administration.[52]

Moses, too, was in an uncomfortable spot. Public disgust over the unsanitary conditions at the Great Kills site was matched with disillusionment over Moses's fading promises for a model project there. The talk about a new Fresh

Kills location for dumping would undermine what remaining goodwill the parks commissioner had previously secured. The first inkling of a renewed Fresh Kills project was disclosed in a July 7, 1945, story run in the *Staten Island Advance*: "The Fresh Kills marshland . . . will be reclaimed by the city through the landfill operations of the Sanitation Department—refuse covered by clay method—after the work at Great Kills is completed[,] it has been revealed by the Park Department."[53]

At a public hearing held in the fall of 1945, attended by "scores of indignant islanders," the City Planning Commission approved a DSNY request to develop preliminary plans for acquiring land for a dump at Fresh Kills. The local protesters from Staten Island reacted so strongly that the Board of Estimate eliminated the item from the city's capital outlay budget for 1946. Borough President Cornelius A. ("Neil") Hall fought against the planning commission's request, as did O'Dwyer, who was campaigning for mayor at the time and certainly did not want to upset potential voters on the island.[54] Public opposition to the Fresh Kills plan on Staten Island spread rapidly. In November 1945, protesters chanting "Don't make Staten Island a dumping ground" distributed flyers at the St. George Ferry terminal and then marched on City Hall.[55]

In June 1946 the Board of Estimate reversed its stance on a Fresh Kills plan. The planning commission announced that it was seeking an amendment to restore its previous request in the already approved budget specifically regarding a $650,000 marine unloading plant for Fresh Kills.[56] Moses advocated restoring the funds for an unloading platform "because I knew of no other way of meeting a problem which the city will have to meet immediately, and because I honestly believed that it was a constructive long-range, intelligent way of facing the issue [of disposal]." Borough President Hall also supported moving forward with the facility at Fresh Kills.[57]

Dissent over Fresh Kills turned to fury. On June 5, 1946, the *Advance* listed Moses and Hall as "two of the outstanding advocates of the Fresh Kills garbage dump" and noted that these were reversals of position for both.[58] Hall fancied himself an independent Democrat, even though he had succeeded Joseph A. Palma as borough president in 1946 as the candidate of the Republican, American Labor, and Fusion parties. He won the election over his good friend and former borough president Jack Lynch, a Democrat. Lynch had made Hall assistant commissioner of public works for Richmond in 1930, and he also held that position in the Palma administration. On the Board of Estimate, Hall sided with Mayor O'Dwyer in most cases. However, this loyalty (and his various associations) in and of itself does not explain his stance on Fresh Kills.[59]

Hall had reversed his opposition to a fill site at Fresh Kills but wanted "the operations limited to a period not to exceed three years." He viewed the project, like Moses, as reclaiming marshland, which would permit the building of the West Shore Expressway and open opportunities for more industrial development. Hall also envisioned a site for airports and parks to serve all the citizens of the island. The *Advance* concluded, "And the residents of Staten Island—Who cares about them?"[60]

It appears that Hall's support for a fill project (not a permanent refuse dumping site but an area of about 750 acres at the more than two-thousand-acre site) rested upon the assumption that the ultimate improvements in Staten Island's infrastructure would vindicate his unpopular decision. Many others viewed Hall as favoring the interests of New York City over those of the locals in Richmond County, whether for altruistic or self-seeking reasons.[61] On June 15, the *Advance* came down particularly hard on Moses. It stated that he was a great engineer, "but he is not a great engineer in human relations, and not someone who could be counted on as a friend of Staten Island ALL THE TIME."[62]

The *Advance* jumped on the decision on Fresh Kills, immediately linking it to Great Kills: "Doesn't this mean the same dumping of garbage, etc., that has been partially successful in filling up Great Kills Marine Park and virtually 100 per cent successful in annoying residents of that section of the Island?" While there was uncertainty about the exact nature of the plans for Fresh Kills (reclamation? refuse dumping?), it appeared that the original proposal for preliminary plans submitted to the planning commission called for ten years of dumping on the West Shore of Staten Island, not the three years that Hall had wanted. The thought of such an extended period of dumping became a major issue—and a confusing one—in the local protests against the proposed site. So did Hall's change of heart (and change of vote), the sense that Staten Island had been targeted because its population was small and its political leverage weak, and a growing mistrust of Moses's objectives.[63]

Staten Island residents hardly needed any prompting to rise in dissent against the latest incursion from the city government. The *Advance* sought to rally the troops. In a June 10 editorial, "Let's Fight!" it reminded citizens, "We cannot forget that the Great Kills dumping was to be closely supervised according to promises from the city administration then in power. The Great Kills landfill is the Island's laboratory for this kind of dumping. What happened there is proof that this kind of thing is not wanted in the Borough of Richmond."[64] Aside from urging the city to dump elsewhere, the debate over switching from landfill sites to incinerators fulminated anew, as some feared that available land disposal sites were becoming scarcer.[65]

A chorus of protests broke out across the island—not just around the Fresh Kills area—over the new disposal site and exactly how it was to be used. Citizens of Oakwood Heights, Great Kills, and Bay Terrace continued to decry odiferous conditions at the marine park area. Even the sighting of a rare snowy owl at the dumpsite provided an opportunity to criticize the landfill operations. Howard Cleaves of the Staten Island Bird and Nature Club noted, "It is almost certain the bird stationed himself in such uncouth surroundings for the very good and practical purpose of capturing rats attracted in numbers to the unprecedented banquet being provided for them at Great Kills."[66] The Richmond chapter of the New York State Society of Professional Engineers condemned the methods used at Great Kills and the potential nuisance of a landfill proposed for a ten-year stint at Fresh Kills Meadow; they also were shocked at Hall's retreat on the subject.[67] The American Legion, the Veterans of Foreign Wars, the Midland Beach Democratic Association, and the PTA also got involved.[68]

Councilman Fred Schick and Dr. Natale Colosi, of the Interstate Sanitation Commission and professor of bacteriology at Wagner College, became key spokespersons in the growing dispute over the new disposal site. Former borough president Joseph A. Palma and Democratic County Chair Jeremiah A. Sullivan lent their support to the protest. "Staten Island is essentially a community of small home owners," Sullivan stated, "and any plan which would make a garbage dump of the island simply could not be tolerated." Republican County Chair Edward A. Ruppell was the only public official to stand by Hall in supporting the dumpsite.[69] Mayor O'Dwyer, who in a pre-election speech had stated that disposal by landfilling "must cease," declined to comment on the Fresh Kills dispute in mid-June.[70]

With the increasing criticism and accusations, Moses felt compelled to defend his position. Interviewed by the *Advance* on June 11, the parks commissioner stated that the landfill proposed for Fresh Kills would be a "long-term project" but different from others in that "papers, refuse—clean fill" would be dumped there, along similar lines as used at Rikers Island, LaGuardia Airport, Sound View Park, and Great Kills. (This seemed quite speculative and contradictory, given the persistent complaints at locations like Rikers Island and Great Kills that material other than clean fill was being dumped at the sites.) As to how long the project would last, he made no specific promises, implying that it would be akin to other multiyear ventures like Great Kills.

Moses added that incineration was too expensive and that the Arthur Kill (Staten Island Sound), the tidal strait situated between New Jersey and Staten Island's West Shore, offered a practical water route by scow from Manhattan to Fresh Kills. He explained—with a new twist on the subject—that a marshland dumping site had been considered in Jamaica Bay, but because it was not free of ice in winter like the Arthur Kill it posed a transport difficulty. In a common refrain, he saw Fresh Kills as having the advantage of reclaiming worthless land and making way for a marine arterial highway system and an airfield.[71] It was probably not a coincidence that Moses sounded very much like Borough President Hall, but some of the justifications (such as the Arthur Kill passageway) seemed pulled out of thin air.

MORE PROTESTS

A City Planning Commission hearing on the funding of the Fresh Kills Landfill scheduled for mid-June became the focal point of Staten Island resistance. More than one hundred protesters marched on City Hall for the event. The vote was due on June 21, but observers gave little chance of a decision favoring the opponents.[72] They were correct, and the approved program now moved to the Board of Estimate. Dr. Colosi pointed out that even if the city accelerated plans for more incinerators in the wake of dwindling available land, most

FIGURE 7.4 Robert Moses, 1956.

Source: New York Public Library, Digital Collections.

refuse would still end up in landfills, which did not speak well for the chances of abandoning Fresh Kills. Other observers remarked that the ten-year time span for dumping at the proposed site hardly was firm, given that so much of Staten Island was marshland. Such dumping could go on for twenty or thirty years.[73]

On the day after the planning commission voted, Moses sent a lengthy letter to the *Advance*. "May I ask you to print the attached letter?" he queried in a churlish manner. "It is rather long, but certainly no longer than the space you have devoted to the other side."[74] Though he feigned indifference, he was clearly upset that the newspaper had compared him so unfavorably with Newbold Morris, a City Planning Commission member and former president of the City Council. The *Advance* depicted Morris as a friend of Staten Island who opposed the landfill. Moses asserted, "What I do object to in your editorial is the complete absence of a direct, frank and realistic approach to the City's problem, disposal and reclamation, and your failure to evaluate properly the indispensable part which disposal of waste has played in the reclamation of such areas as Great Kills." He then outlined how he had turned around the sanitary operations there, which had been "nothing to be proud of" during the war. He added that Morris had not been around then, but when the improvements were completed, "he will be able to pat the little children, examine the bathing beauties, and refer to the undeviating support which he gave to this enterprise during the early days of trial and tribulation." He also acknowledged the "fair-weather friends" who will "be on hand when they play the Star Spangled Banner and dedicate the park."

Turning to Fresh Kills, he stated again, as he had done in the previous interview with the *Advance*, that he knew of "no other solution" to the "City's problem of disposal," and neither did his critics. He could not ignore a problem "that would have come back to haunt the present administration a year from now without any preparation or means of meeting it." In anticipation of what appeared to be a reversal of his public stance not so many years ago about Fresh Kills, he played to the crowd: "Staten Island is in many ways the most attractive part of the whole city. The last thing I would do is anything which would retard its progress or spoil its natural beauties." But since "no one is living anywhere near the operation we propose" and the immense stretch of land in the area is "valueless," there is an opportunity to substantially increase industrial development "opposite the heavy industrial area of New Jersey."

He added a rather gratuitous statement about how he persuaded New York State to turn over control of property in the Fresh Kills area to the city. This "gift" to the city actually cloaked the means by which he had established a foothold in the marshland to fulfill some of his own goals. The Fresh Kills plan for landfilling, he went on, was "precisely the same kind of construction reclamation which has been successful elsewhere in the City"—that is, like in Great Kills.[75]

On the surface, the tone of Moses's frontal assault on his opponents was reminiscent of other declarations of his vision for the city. It had the air of being all things to all people. Yet he had stood the issue on its head by looking to solve "the problem" of refuse disposal first and foremost by using the tools available to him. This was not parks and parkways first. In what was becoming a true war of words, the *Advance* responded to the letter from "the city's most ardent official advocate of landfill." In a mocking tone, it stated: "He's an honest man, this Robert Moses. A forthright man. That quality helps make him a good public official. And his defense of landfill is not a new pose. In fact, it is not a pose. He means what he says. Richmond should feel indebted to him for telling us exactly what to expect—'The offscourings and leavings of millions.' "[76]

A cartoon appeared in the newspaper that day showing a funnel attached to Fresh Kills Meadow with a garbage truck pouring refuse onto Staten Island.[77] How far things had come since praise for Moses over Great Kills Marine Park. Another *Advance* editorial on June 26 reinforced what had been said two days before, in even more colorful prose. It stated that Moses made his argument "with the skill of an Eisenhower pounding the coast of Normandy." It also asked why the dumping issue was not in the hands of the Department of Sanitation instead of the parks commissioner (reversing the scorn given to Carey and the praise to Moses over Great Kills) and why Moses characterized the dumping issue as "your problem" and not the city's problem It also suggested that the rationale to turn the marshes into industrial lands was dubious. The bottom line: "Must Staten Island be 'the goat'?"[78]

On June 27, the Board of Estimate voted to move the Fresh Kills project forward. The newly organized Staten Island Anti-Garbage Dump Association (at one point four thousand members strong), Councilman Schick, and other protesters (including Great Kills campaigners) had little influence on the outcome but continued to renounce the decision. The Garbage War dispute of the late 1910s was dredged up to plead the case for Fresh Kills, as was the advocacy for incineration. Republican County Committee members (aside from Hall and Ruppell) wanted their representatives in Albany to take the fight to the floor of the legislature.[79]

In Travis, the town closest to the proposed dump site, the reaction on the street had been uniformly negative: "Hell no! Why should we have a dumping ground for the rest of the city? . . . The odor . . . would ruin our town. Travis is the garden spot of Staten Island." "We think it's a terrible idea." "We had enough experience with the Mitchel administration's garbage plant [the reduction plant built during World War I] in Fresh Kills." "I'm new to Travis, and I like the town the way it is. . . . Travis is liable to become a deserted village."[80] The Fresh Kills feud was clearly a NIMBY issue, but it also had become more than that.

Staten Islanders were stubbornly unwilling to let the Board of Estimate have the final word on something they had chosen not to tolerate. New tactics sprung up constantly to fight the extensive dumping project. The Republican congressman Ellsworth B. Buck, representing Staten Island, moved to have the federal government intervene because the facility, he asserted, would affect the migration of wild birds and severely damage one of the last wildlife areas in the region.[81] The locals were handed another blow when Mayor O'Dwyer reversed himself and supported the Board of Estimate's ruling. Sounding like a combination of Moses and Hall, O'Dwyer stated that there was no other way to dispose of the waste from Brooklyn and Lower Manhattan than the Fresh Kills Landfill. But he added that this disposal process would be as limited as possible and would be used only for odorless, nonflammable materials. Hopefully the facility would be closed in two years if the incineration program could be revived quickly enough, he conjectured. He ordered Powell to "Open up the incinerators you can use and do it right away." There were no provisions in the current budget for incinerator construction.[82]

Councilman Schick requested a public hearing before the finance committee of the City Council to plead the anti-dump case, and in the meantime he invited the mayor and others to tour the Great Kills facility for purposes of comparison with the proposed Fresh Kills plan. He did not want to disclose the date of the visit for fear that the DSNY would make sure that the site was well covered and odorless. "This, apparently, is what the Board of Estimate members want to find," he observed, "when they come down here: the borough president and the mayor are making sure that the foul conditions which exist at this dump at present do not exist when they visit there. Then they can go back and say that the people are wrong, that the dump is not a nuisance—a complete whitewash."[83] On another front, District Attorney Farrell M. Kane announced on July 1 that a Richmond County grand jury would investigate the city proposal to continue dumping refuse at Fresh Kills. The announcement disclosed, to many people's amazement, that dumping activity already was underway.[84] There was little follow-up to determine if this was an accurate assertion.

Another hearing, another protest. Joseph W. McCallum, vice chair of the Staten Island Anti-Garbage Dump Association, began organizing for the council meeting and also circulating petitions. This had become a well-oiled process in late 1945 and into 1946, and efforts to make the opposition island-wide moved well beyond a local NIMBY event.[85] Anti-dump rhetoric ramped up. On July 8, 1946, the *Advance* sounded the charge of betrayal that had ended in the Board of Estimate's decision and Hall's and O'Dwyer's reversals: "This is a considerable price to pay for stench. It is a tremendous price to pay for an invasion of rats that will follow the city's garbage scows to this new picnic ground for itinerant vermin. It is a tremendous price to pay for a Nagasaki of broken promises so glibly aired in the hungry days of electioneering last fall." And it did not forget Moses: "The proposed Fresh Kills dump was spawned a year ago in the same air of deception." Nor was it willing to believe that the landfill would be temporary.[86]

Other critics, including Schick, also pilloried Moses, underscored problems at Great Kills, raised the almost forgotten issue of the 1938–1939 fight to keep a dump away from Fresh Kills, and acknowledged additional environmental threats like water pollution, which would accompany the airborne odors.[87] The *Advance* reluctantly noted that "Staten Island is grievously handicapped in this battle" by a borough president on the other side of the issue and the island's "relative smallness politically."[88] Despite Schick's efforts to secure nine other councilmen to his side, the vote to keep Fresh Kills free of the dump failed by one vote. The desire to avoid such a blight in their own neighborhoods carried the day for the other boroughs, which now could look to Staten Island to bear a growing responsibility for accepting the city's refuse.[89] NIMBYism was alive and well in all the boroughs.

In an interview, Mayor O'Dwyer said that operations would begin at Fresh Kills officially sometime after January 1. None of the city leaders, however, knew when the dumping would end. Even Moses hinted that the deadline was uncertain, now having fallen in line with the mayor and others who looked to incinerators to assume more of the disposal responsibility in the future. (This was hypothetical, or even disingenuous, because funds had yet to be made available for a rigorous incineration plan.) O'Dwyer asked Moses, "Is it a safe time target to say that in not later than three years, there will be no garbage dumping in any part of the city?" The man of many hats replied, "Yes, assuming there is no serious delay in building incinerators." He added the caveat that if there were no delay in the plans or "no controversies like those that have arisen here," the timetable was reasonable. This was a speculative rather than a deceptive answer, and it put Moses on the record—rather inaccurately—of promising that Fresh Kills would close in three years. O'Dwyer was referring to garbage, not necessarily all types of nonburnable refuse, and so was Moses.[90]

Staten Islanders tried another tactic in trying to stop the landfill. District Attorney Farrell M. Kane announced that on July 10 a Richmond County grand jury would investigate the city's proposal to continue refuse dumping in the Fresh Kills area. The inquiry was established under the state public health law to determine whether the landfill would constitute a health menace or a public nuisance. "If either condition exists," he stated, "those responsible will be prosecuted."[91] On July 11, Powell was scheduled to appear; Moses and Hall were to follow.[92] On July 16, the grand jury announced that it was steadfastly opposed to the dumping of garbage at Fresh Kills, in line with citizen opposition. It presented the Great Kills experience as a prime example of why the city should avoid using the West Shore site.[93]

The grand jury concluded that the landfill would cause a public nuisance and that future grand juries should stay abreast of that possibility. In the finding of facts, it asserted that if the available 2,100 acres were all used for the landfill, dumping would continue for about ten years; if the city used only 750 acres, as Borough President Hall had proposed, the operation would last about three years. Even if the city chose Hall's alternative, it added, the dumpsite could always be extended when the 750 acres were filled—thus recognizing that the only real limit to dumping was the size of the site. The grand jury concluded that it did not discover any definite plan for how the reclaimed land ultimately would be used once the dumping ceased.[94] The findings made it clear that if the Fresh Kills operation was completed (whatever its size), Richmond would be the refuse depository for all of the boroughs, protests would continue, and the landfill would infringe on the rights of citizens "to live in a clean, airy, wholesome, healthful community free of rats, flies, vermin and nauseating stench."[95]

While strongly opposing the project, the presentment revealed the latitude that the city was likely to possess on beginning and ending this project, which went well beyond any best case Moses, Powell, or Hall might render. Hall continued to defend his proposal but insisted repeatedly that it was Palma, in the previous administration, who really began the process leading to the Fresh Kills dump. Hall produced a letter (August 12, 1943) to Palma, which he claimed showed that the Palma administration was exploring ways in which the area around Fresh Kills Creek could be reclaimed by the fill method. Hall added, "I recognized that the map that Mr. Nelson [topographical engineer Harold Nelson] presented with his report as being identical with the map now being used by the City Planning Commission."[96]

Palma asserted that there was "never any scheme during my tenure of office to create any garbage fill, any place on Staten Island."[97] Palma also was on record as voting against the proposal for a Fresh Kills Landfill at the 1945 Board of Estimate meeting.[98] Hall's finger pointing was a way to put the borough president in a position of reluctant advocate of the landfill, even though his own aspirations for Fresh Kills amounted to trading island development for a dump.[99]

The *Advance*'s assessment of the grand jury findings was poignant: "Why were not details of this project disclosed long ago? The answer, of course, is that those public officials who have nursed this scheme along so far were afraid of public opinion. They were afraid because they knew their plot was unfair in outvoted Staten Island."[100] The comments suggest a somewhat overstated conspiracy, but the gap was wide between the public knowledge of the details of the project and private intentions concerning future land uses at Fresh Kills.

To calm down the outraged citizens of Staten Island, Mayor O'Dwyer, key commissioners, and members of the Board of Estimate traveled to Marine Park at Great Kills on July 18. They were met with unpleasant smells. Declining an offer to spend the night, they attempted to make the best of a bad situation. Moses unhelpfully asserted, "It was much worse when we were working at

Orchard Beach and Rikers Island." Acting Borough President of Queens Maurice A. Fitzgerald chimed in, "It seemed a lot worse years ago when they were dumping garbage in my borough." O'Dwyer, more sensitive to the political moment, told Commissioner Powell, "It's foolish to say we can't get covers for these scows [delivering refuse]. It will cost some money, but we can do it. Do it at once." A goal of buying new incinerators was mentioned, but no one was able to rebut Dr. Weinstein's claim that "great, green waste" at Fresh Kills was a health hazard. So much for a publicity opportunity. Staten Islanders remained rightly suspicious of what was to come at Fresh Kills.[101]

CAMPAIGN TO NOWHERE

Staten Island protesters were a persistent lot, and even with all of the roadblocks, the most ardent among them carried on the fight late into the summer of 1946 and beyond. Schick planned to take the dumping issue to the governor.[102] The Staten Island Anti-Garbage Dump Association also kept up the pressure on the city and state.[103] Additional concerns arose over the value and potential uses of fill land, questioning the long time it took for filled land to settle before building could be undertaken.[104]

The biggest upshot in broadening the resistance to the new landfill was urging New Jersey forces to join in the protest. Congressman Buck appealed to three New Jersey colleagues to enroll in the fight against the Fresh Kills project in late July, arguing that their constituencies would face odors from the dump and would have to deal with the scow traffic as well.[105] On August 2, the *Carteret Press* reported that Mayor Steven Skiba of the borough of Carteret, New Jersey, stated that his citizens "don't want to be annoyed by more odors than they already have."[106] In early August, the Carteret Board of Health asked the help of the New Jersey State Board of Health in blocking the dump. Interestingly, the community was currently being pressured by the Interstate Sanitation Commission to solve sewage problems affecting the common waterway with Staten Island. John Turk, president of the Carteret Borough Council, presented a resolution pledging his community's support to block the Fresh Kills dump, and other New Jersey communities added to the numbers of protesting groups.[107] The *Advance* announced, "City Hall occupants who weeks ago brushed off Staten Island's opposition to the Fresh Kills garbage dump with the wisecrack, 'They're against everything down there,' will find it less easy to belittle the New Jersey residents who have joined the fight."[108]

Middlesex County, Perth Amboy, South Amboy, Woodbridge, Bayonne, Elizabeth, Linden, and Rahway all saw the project as a health threat or a nuisance. With the announcement that the DSNY soon would begin preliminary work at Fresh Kills, federal action along with local protest was contemplated on the grounds that the facility could affect New Jersey municipalities.[109] Although New York and New Jersey communities rallied against the Fresh Kills project into the fall of 1946, New York City authorities retained the upper hand.[110] On August 22, the Board of Estimate endorsed a request from Borough President Hall for a $50,000 appropriation to prepare drainage and grading plans for the Fresh Kills dump. Earlier in the month, Commissioner Powell stated that $200,000 was being used to foreclose on properties within the 2,100-acre Fresh Kills site on which it had tax liens. Additional funds would be used for dredging the creeks, rebuilding an unloader, and constructing bulkheads.[111]

In 1946, New York City had eight landfills open and eleven operable incinerators. Of the 22,699,100 cubic yards of refuse disposed of by the DSNY in that year, landfills accounted for 76.5 percent (17,354,669 cubic yards); incinerators burned 22.8 percent (5,164,147 cubic yards) of garbage and other combustible waste.[112] A July 18, 1946, report, "Waste Disposal in New York City," from the Department of Parks files treated Fresh Kills Landfill as a foregone conclusion. It outlined the DSNY's current landfill program, noting that about 50 percent of Manhattan's waste went to landfills, 65 percent for the Bronx, 60 percent for Brooklyn, 75 percent from Queens, and 75 percent from Richmond. It added that about 55 percent of the refuse collection was unburnable materials ill-suited for incinerators, which "must be disposed of by land fill operations under any program of disposal." Even with complete incineration of combustible refuse, approximately 44,000 cubic yards of material daily "cannot be burned and which must be disposed of by land fill in areas such as Fresh Kills, Staten Island, along the water from Jamaica Bay . . . and along the East River and Long Island Sound in the Bronx."[113]

Such a conclusion was a clear indication that no matter what city officials told the public, some form of land disposal was here to stay—regardless of the number of incinerators in operation. The report went on to discuss the existing incineration program: five new units since the beginning of the year, with five more needed. Estimates suggested that it would take about three additional years to design and construct the new incinerators to handle the burnable waste. Great Kills was heading toward capacity, with a scheduled completion date of August 1947. It would take about one year to ready the Fresh Kills site, which had to be available when Great Kills closed. In a most telling remark, the report stated that "because of the substantial sums involved in the preparation and acquisition of the site, the City must dispose of refuse at this location *for a number of years.*" It recommended that the site be used for mixed refuse for three years, and after the incineration program was in place it could handle "unburnable waste and incinerator residue."[114]

Such an assessment made clear that Moses (and the DSNY) did not view Fresh Kills as a temporary or makeshift disposal site but as one that would evolve into a long-term facility accepting (as noted in another part of the report) "unobjectionable, unburnable material" as fill for park improvement, airport expansion, and industrial development, as he had already stated publicly.[115]

For some reason, many people had assumed that Moses promised three years of dumping at Fresh Kills and no more. They either were not listening, or he was obscuring the facts.

On the one hand, the idea that Fresh Kills would be a three-year temporary facility was specious. On the other, while garbage hopefully would not be deposited there after three years, dumping of some types of wastes would continue in perpetuity. Rationalized or not, this strategy was more than splitting hairs. The "benefits" of the disposal plan were pretty much what Moses had told Governor Dewey in 1945, albeit sometimes intimated, cloaked, or concealed in public utterances: elimination of swampland, completion of the grading for the first section of the Shore Expressway, startup for an airport, more industrial development, new parkland, and fill for Richmond Parkway. Solving the "problem" of the city's waste disposal, in Moses's eyes among others, was not going to happen without incineration for burnable waste and attention to long-term land reclamation for what nonburnable refuse remained.[116] Building new incinerators apparently was a way to avoid dumping all kinds of residential and commercial waste at Fresh Kills. But plans were scanty in developing a full-fledged incinerator program. Moses had to know that.

Fresh Kills would be included in the inventory of Moses's projects from Orchard Beach to Bayswater Park, from Flushing Meadow Park to Sound View Park. "All of these add to the resources of the City and result in a general increase in the valuation of properties for park and tax purposes," the waste disposal report concluded.[117] In a letter to Mayor O'Dwyer on August 20, Moses stated that the Fresh Kills dumping program "seems well under way in the manner best calculated to cause the least objections from the residents of Staten Island." He added that the incineration program, "which will eliminate the dumping of garbage in Staten Island," was being accelerated.[118] There was little doubt that Moses was calling the shots in this latest project, with Commissioner Powell and others falling into line or following orders, but little real action at the DSNY vis-à-vis incinerators had yet taken place.[119]

Largely in response to the Staten Island protests, Mayor O'Dwyer in June 1946 had ordered the reopening of every possible refuse incinerator in the city and the construction of new ones.[120] The DSNY's 1946 *Annual Report* confirmed the degree to which the city was at once trying to move away from Carey's and La Guardia's landfills-only policy, wanting to build more incinerators, but also embracing Fresh Kills as an immediate solution to disposal needs. The report touted its efforts to recondition three idle incineration plants, "in compliance with the policy of the new Administration to cease the provocative practices of dumping refuse containing garbage into landfills."[121] Among the issues driving the decision to return to incineration was the declining landfill space.[122]

The stage was set for the debut of Fresh Kills Landfill. It remained to be seen whether its construction was simply a stopgap measure, the first step in a vast land-reclamation project, or something not anticipated. But the seesawing over incinerators versus landfills—in theory at least—continued unabated.

SECESSION

Some Staten Islanders viewed secession from New York City as a last-ditch response to the dumping conflict. This was reminiscent of a similar reaction during the Garbage War of 1916–1918 (and later in the 1980s and 1990s). Was it simply a way of getting attention? Was it the final recourse to a group with anemic political clout? Or was it an assertive response to the imposed will of the city? The Republican assemblyman from Stapleton, Edward P. Radigan, announced in early January 1947 that he would introduce a secession bill in Albany if he obtained broad approval from Staten Island citizens and civic groups. He already had the support of the Staten Island Anti-Garbage Dump Association. Chair Joseph McCallum stated, "I and the members of my association are for anything that would prevent the City of New York from dumping its filthy and dirty garbage on Staten Island." Adding to the unrest on the island were many complaints over ferry service, taxes, and the need for a variety of civic and infrastructural improvements.[123] Radigan believed that his bill would end the plan to place a citywide garbage dump on the island, eliminate payment of the city sales tax, and attract new industries.[124]

A story on secession in the *New York Times* began: "Staten Island—New York City's Borough of Richmond—is almost as little known to the inhabitants of the other four boroughs as is northern Bukovina." Such a lead-in seemed to make the locals' case about their sense of being neglected. A week earlier, it asserted, "the wrath of Staten Islanders found expression" in the call for secession.[125]

Legislators had mixed feelings about the secession tactic. The Republican assemblyman Arthur T. Berge declared that he would oppose any such bill, calling it "ridiculous."[126] In response, Radigan replied, "It ill becomes my colleague, my own party colleague, to oppose me so ferociously in a matter of so much concern to the people of Staten Island before his constituents have had an opportunity to pass judgment."[127] But Radigan faced an uphill battle with other political confederates as well. Councilman Schick and Borough President Hall criticized the plan as "impractical and unsound." So did the Republican senator Robert E. Johnson. GOP County Chair Ruppell branded it "unworkable and unsound."[128]

Johnson and Berge proposed legislation to prohibit the dumping of garbage within New York City limits, and they called for the use of incinerators. In an

April memorandum to Governor Dewey, Johnson urged him to sign the legislation so that New York City "may be given lawful protection against the evil which threatens their homes." Johnson characterized the Great Kills project as "a public scandal" that had threatened the health of locals.[129]

The dumping debate writ large obviously was not going away. Four Queens legislators introduced bills to prohibit the use of garbage landfills on city or private property and to ban building on fill land for twenty-five years.[130] Such action may have inspired Radigan's persistence. In late February the assemblyman stated that since his announcement of the proposed bill on January 3, he had received one hundred letters showing "live interest" in the proposal and in the prospect that locals could develop Staten Island themselves.

While Radigan was making his case, the Board of Estimate awarded a contract to the Giardino-Lenart Corporation, a Manhattan firm, to build a marine unloading plant at Fresh Kills.[131] Part of the contract was dredging the waterways at the site, building a bulkhead, and constructing docking facilities and walkways. The board appropriated funds for acquiring tracts of land for dumping. In a related matter, it also rejected a plan to acquire a possible dumpsite in the Blanchester section of the Bronx.[132] The buzz about the Radigan bill was that it did not stand a chance of passing, and if it did and the governor signed it, the measure would face a tough referendum. Radigan's path was tortuous. His bill died in committee on March 18, 1947.[133]

DESPERATELY SEEKING DUMPING SITES

For many reasons, there was little Staten Island could do to derail the Fresh Kills plan. However, a bill pushed by Johnson and Berge (similar to the Radigan bill) forbidding disposal of refuse in New York City unless it was first incinerated passed the New York State Legislature. The governor's veto ensured that landfilling would not be a thing of the past. O'Dwyer had lobbied Governor Dewey to veto the bill, fearing anything that would enhance borough control of dumping. Despite his support for incineration, O'Dwyer stood up for the landfill approach at this juncture, viewing it as risk free and buying time for the DSNY to build more incinerators.[134]

The DSNY continued to push for more incinerators, but in the short term at least the demand for landfill sites was high in New York City. In early 1947, it asked for assistance from the Health Department in obtaining more of them. A current survey ranked Staten Island first in the city, with 65.2 percent of the available marshland (Jamaica Bay was not included in the survey). And a 1947 report stated, "Since Richmond has 65% of the acreage, it is apparent that a substantial portion of the city's refuse may be disposed of there."[135]

FUTILE PROTESTS   ■ 195

Stories had circulated in late January that the City of New York was studying the possibility of establishing a new dumpsite in an undisclosed borough to divert large amounts of waste from Fresh Kills. Mayor O'Dwyer "with dramatic suddenness" asked the Board of Estimate to put off the funding of a new marine unloading plant at Fresh Kills until the new site could be evaluated. The possible location, in the Bronx, as it turned out, was deemed too small and too costly to pursue further, and thus nothing came of it.[136]

No compelling arguments or actions now blocked the opening of Fresh Kills. On April 16, 1948, the first scow, loaded with refuse, would arrive at the site. But complicating Fresh Kills' opening was a lack of clarity as to what building the landfill actually was meant to accomplish. The duration of its use (and for what ends) was a point of contention. From the start, there were confusing and conflicting views about what purpose Fresh Kills would serve: Would it be primarily a reclamation project to produce a vast amount of usable land? Would

FIGURE 7.5 Fresh Kills topography and ecology, 1947.

Source: Courtesy of Jose Mario Lopez, Hines School of Architecture, University of Houston.

it be a space reserved to handle the city's long-term needs for disposing nonburnable materials and incinerator ash in order to complement a still notional program of extensive incineration? Or would it simply become a dump, concentrating New York City's solid wastes in a single place, a single borough? At various times in its history, Fresh Kills was any one of these approaches, and possibly all of them, depending on whom you asked.

But at its inception, it was debatable that Fresh Kills was part of a comprehensive plan for disposal in New York City. The metropolis had other dumpsites and several incinerators. And although this was not immediately apparent, the determination to develop a giant landfill (and phasing out smaller ones) but also to construct more incinerators was not feasible. Instead, it pitted a one-borough disposal approach against an incompatible citywide approach. The presence of Fresh Kills Landfill was just the beginning of this incongruity. The city's unrelenting increase in dumping at the Staten Island site would result in a one-borough approach for many years to come. Fresh Kills became the default plan.

PART IV
LIVING WITH AND SURVIVING THE LANDFILL

CHAPTER 8

THE BURNING QUESTION

On January 26, 1948, Mayor William O'Dwyer, flanked by Hollywood celebrities, spoke on the Mutual Broadcasting Network about New York City's forthcoming fiftieth-anniversary celebration. In 1898 the five boroughs had been consolidated into "the greatest city in the greatest country in the whole wide world." He welcomed everyone to join in on the festivities planned for the next several months.[1]

A Fifth Avenue parade on June 12 marked the official opening of the Golden Jubilee celebration.[2] On August 22 the Golden Anniversary Exposition opened at Grand Central Palace "with a burst of atomic power and a prayer" provided the night before. An "old-fashioned torchlight procession" had advanced down Lexington Avenue to launch the ceremony. The *New York Times* observed:

> In front of the palace they stood by a block-long ribbon down the center of the avenue. Atop the Empire State Building and in a Navy "truculent turtle" bomber overhead telescopes were trained on the star Alioth, just fifty light-years away.
>
> At 8:30 o'clock light which left the star in the year that New York's five boroughs were united, was admitted to the scopes.
>
> It activated photoelectric cells and sent radio impulses to the Exposition where it triggered an atomic pile, split a uranium atom and ignited, electrically, a mass of magnesium in the ribbon.
>
> With a flash and loud crack the ribbon flew apart and the invited guests, 6,000 strong, poured into the building to see the four floors of exhibits put on by the city departments and atomic energy agencies. Another 50,000 watched from the sidewalks.[3]

What an extraordinary, if potentially dangerous, opening ceremony.

Officials hoped that those partaking in the exhibits and shows would appreciate the contrast between the New York of 1898 and 1948.[4] For many Staten Islanders the celebration was bittersweet. No borough had embraced the 1898 consolidation more enthusiastically than Richmond. No part of the city had higher expectations for what that consolidation might bring them. In 1948 their hopes remained unfulfilled. Richmond by far was still the smallest borough in population, with fewer than 200,000 inhabitants.[5] It was still a borough "without a subway." What Staten Islanders got instead was empty promises about a land link to the rest of the city, many protests, talk of secession (and possibly joining New Jersey), and Fresh Kills Landfill. An observation in the *New York Times* from two years earlier rang hollow: "More than any other of the five boroughs of New York City, Richmond's greatness lies in its future."[6] Now that the landfill officially was due to open, a major question remained unanswered: What role was this depository meant to play? A temporary waystation for refuse until new incinerators came on line? A new filled marsh to undergird a Robert Moses infrastructure project? Or a permanent blight on the island?

THE LANDFILL OPENS

From the landfill's opening day on April 16, 1948 (fifty years after consolidation), Staten Islanders were confronted with another Quarantine controversy, another reduction-plant debacle, but on a larger scale and for a much longer duration. In the first years of its opening, Fresh Kills' place in the disposal plans of the city was not being defined primarily as a dumpsite but largely in terms of its role as a reclamation project and complement to incineration. As long as the Department of Sanitation (DSNY) believed that the city could dispose of its burnable garbage and refuse through incineration, Fresh Kills would not be the heart of the disposal program.[7] But when the incineration system began to falter in the late 1950s, Staten Island would offer a solution to the DSNY's quandary. The largely unwanted Richmond landfill would emerge as the dominant disposal option for New York City for years to come.

Initially, refuse from the other boroughs was dumped at the Fresh Kills site in an area bordering the nearby creek, about one mile from Richmond Avenue and Arthur Kill Road and near the old Metropolitan By-Products Company buildings. A spokesperson for the DSNY announced that the unloading operations would be stepped up a week after the opening and that crews would be able to handle seven or eight bargeloads of waste each day. (Activities at Great Kills were set to end in May.) The department hoped to dump about 7 million yards of waste per year at the new site, as compared with the peak of 5 million a year at Great Kills.[8] In the first eight months of operations, Fresh Kills accepted

15 percent of the city's waste volume. This increased to 28 percent in two years and 33 percent in three.[9]

In August, the Board of Estimate approved in principle a $44 million program for refuse disposal submitted by Construction Coordinator Robert Moses (one of his many roles), which called for discontinuing dumping of garbage and combustible material on private property, building new incinerators and increasing the efficiency of existing ones, constructing new parks, and restating his opposition to dumping garbage and burnable waste in landfills. (O'Dwyer had ordered the practice to cease two years earlier.)[10] The board was keen on establishing an updated refuse-disposal policy and endorsed Moses's ideas. At a public hearing before the City Planning Commission, the DSNY announced that its capital-budget request would be modified in line with the board's action. There were differences—again—over the time required to carry out the program. Moses said three years; an official with the Budget Bureau said five.[11] In the cover letter to Mayor O'Dwyer with the attached report, Moses had stated, "By adopting this program the mistakes of the past will be largely corrected, and at the same time constructive results in the form of reclaimed recreation and other areas will be obtained."[12]

The DSNY now was committed to limiting landfills to city-owned areas it could directly control, monitor, manage, and have "under Park Dept. surveillance." The city worked to end private dumping of refuse (most often controlled by organized crime) and also projects that the department could neither manage nor use for city reclamation. For the first time the DSNY's land disposal activities were officially linked to city reclamation—a policy carried out informally in previous years and demonstrating Moses's influence on land-use issues.[13]

The determination to build more incinerators continued, as well as a decision to undertake a major departmental reorganization. One goal was to rethink a comprehensive long-range disposal program, taking into account the respective value of alternative methods, such as sea dumping, landfilling, hog feeding, reduction, composting, garbage grinding, and incineration.[14] Much of the urgency behind these decisions came because the department believed that at the current rate of disposal, total space in existing landfills (pre–Fresh Kills) was dwindling quickly.[15]

There was little doubt that Fresh Kills was a central component of the DSNY's disposal plans in the late 1940s. Its size and location were important, reflecting the need for large spaces away from major population centers to dispose of the city's mounting volume of waste. The DSNY accelerated its incineration program, among other things, to conserve space in sites like Fresh Kills for the future disposal of ashes, nonburnables, and incineration residue.[16] Exactly what future role Fresh Kills would play in balancing disposing garbage and unburnables versus accepting ash and incineration residue was unclear. The length of service of the site overall was mere guesswork in 1948.

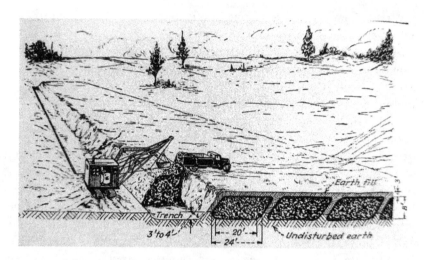

FIGURE 8.1 Landfills in New York City in 1954.

Source: NYC Department of Sanitation, Office of Engineering, *Sanitary Landfills in New York City*, May 12, 1954.

Given what officials were saying publicly, including Moses, Sanitation Commissioner William J. Powell, Mayor William O'Dwyer, and Staten Island Borough President Cornelius A. Hall, the use of Fresh Kills for disposing refuse over the long haul seemed highly unlikely, but private utterances, conjecture, deception, and ambiguity did not rule out a long life for New York City's newest and biggest landfill. Also unclear was whether Fresh Kills could rightly be described as a "dump" or a "sanitary landfill," given the connotations of both. Would the site meet current disposal needs in a nonthreatening manner and provide a future bonus of reclaimed land, or would it be the eyesore, nuisance, and health threat many citizens feared? Interestingly, Fresh Kills emerged as a result of the momentum of Carey's landfills-only policy but opened its gates as incineration regained its status and following in New York City. And now Fresh Kills was regarded as absolutely necessary by city officials, whether disposal practices shifted to incineration or not.

There is little doubt that Robert Moses played a key role in the opening of Fresh Kills, if not in its future operations. He frequently repeated his narrative about the short-term use of Fresh Kills for mixed refuse and as a long-term location for fill material. Such proclamations, however, were misconstrued by Staten Islanders to mean that Fresh Kills as a dump would have a very short (three-year) lifespan. As time passed, they came to believe that the parks commissioner had broken his promise or lied to them.

Cynics viewed Moses as a manipulator who birthed the Staten Island disposal site for his own ends. But Moses was less a manipulator and more an opportunist who took advantage of a chance to fulfill some of his construction plans for the city. At the celebration of the seventy-fifth anniversary of the Staten Island Institute of Arts and Sciences in 1957, Moses stated, "I should like to talk informally about the future of this fortunate borough, the only one which can still be planned intelligently."[17] He viewed Staten Island as a place with many vacant spaces for him to shape. Fresh Kills became a means to an end, not an end in itself. Moses—while shifting the justification for the site on several occasions—did not come out of the controversy over the landfill unscathed. In time, he would be nudged out of his power base, lose his authority, and be held responsible for Staten Island's biggest eyesore.

In the largest sense, what was at stake were issues of both immediate and long-term importance: for some, placing a landfill where it was not wanted; for others, finding a way to move refuse away from its source. Few could have imagined that neither three nor ten years would define the Fresh Kills Landfill's life span or its significance.

A LANDFILL DESIGNED TO LAST

The landfill was built to be immense: more than three times larger than the disposal area at Great Kills. About one year before the first scows dumped their loads at the new site, the Board of Estimate authorized the Department of Public Works to seek bids for dredging at Fresh Kills in order to clear a path for the scows and for erecting bulkheads, wharfage facilities, and foundations for a marine unloading plant. The contracts called for rebuilding old structures left on site from the World War I–era Cobwell reduction plant and for electrical, plumbing, heating, and ventilation work.[18]

Bids let in 1947 stated that the landfill was to be developed in two sections, together representing about six hundred acres of land. The second would be mapped out at a later date and would require an additional marine unloader, new buildings, a service road, and more equipment. One of two unloaders at Great Kills would be disassembled and transported to Fresh Kills. The DSNY also planned for two or three movable unloaders to be barged to new locations as the dumping work progressed.[19]

Two additional sections for dumping were designated as Section 3 (about 270 acres) and Section 9 (about 330 acres). The former adjoined the Brown Oil and Chemical plant in Travis; the latter extended from Muldoon Avenue south toward Rossville. Thirteen sections in all (totaling approximately 2,600 acres of marshland in the Fresh Kills area) had been mapped. Only sections 1 to 4 were immediately to be put into use.[20]

The gigantic area was bounded by a line just south of Victory Boulevard in Travis, west of Travis Avenue, Richmond Avenue, and west of Arthur Kill Road as far as Rossville. The area also included marshland east of Richmond Avenue, between LaTourette Park and Arthur Kill Road, as far as St. Andrew's Church in Richmond. The Arthur Kill near the dumpsites was the most industrialized section of the entire port area, accounting for approximately 40 percent of the commerce (in tonnage) of the Port of New York and New Jersey and already responsible for a good deal of its pollution.[21]

According to the *Advance*, "Some of these areas [of the Fresh Kills site] . . . are not slated to be used for dumping for some years to come until long after the city is scheduled to complete its incinerator-construction program."[22] This suggests the DSNY's open-ended plans for the use of Fresh Kills at the time, with no closing date in mind. A 1949 training-course document stated, "The probable life of the ultimate Fresh Kills sites has been estimated . . . at eighteen (18) years," and with proper control, sanitary landfill operations "are neither a health menace nor a nuisance."[23] The department would reclaim marshland, but the task at hand was to dump as much city refuse—not clean fill—as possible.[24]

By early 1949 the DSNY was disposing eight scowloads of refuse each day at Fresh Kills, more than 30 percent more than at peak times at Great Kills. (That number jumped to twelve barges a day by late 1952.) The waste came primarily from Manhattan (about 60 percent), with the remainder from northwestern Brooklyn and Staten Island itself. Since operations had begun in April 1948 through early 1949, forty-five acres of marshland had been reclaimed.[25] The land reclamation and related work required extensive support from the Division of Marine Operations of the DSNY. Collection trucks delivered refuse to eight waterfront stations in Manhattan and Brooklyn (with more to come, including in Queens), where it was loaded onto barges. Three diesel-driven tugs, forty-two

FIGURE 8.2 Marine unloading plant, Fresh Kills, 1948.

Source: NYC Department of Sanitation, *Annual Report, 1948*.

FIGURE 8.3 Marine unloading station, Fresh Kills, 1954.

Source: Department of Sanitation, Office of Engineering, "Refuse Disposal in New York City," February 3, 1954.

steel-well barges, and auxiliary equipment were used to transport the refuse to the marine unloading plants at Fresh Kills. The tugs traveled about 75,000 nautical miles annually in carrying about 26 percent of the city's refuse to Staten Island.

Workers conducted unloading operations at Marine Unloading Plant #1, the site of an old factory on the south side of the junction of Great Fresh Kills and Little Fresh Kills, and eventually at Marine Unloading Plant #2 (in 1950), upstream on the north side of Fresh Kills near Richmond Creek. Beginning in May 1950, Plant #1 was being operated in three shifts, twenty-four hours per day. The Brookfield Avenue site (opened in 1966), north of Arthur Kill Road and Brookfield Avenue, was a truck-fill facility.

At Plant #1 and at the adjacent landfill site, between 225 and 250 employees (nearly three hundred by late 1952) transferred the refuse to its final destination. Five diesel or steam-powered diggers (bucket cranes) removed the waste from the scows and loaded it onto Athey Wagons waiting on the shore.[26] Tractors hauled the wagons to the dumpsites. Trucks working on Staten Island collection routes deposited their refuse at Fresh Kills and Brookfield.[27]

The waste material in the wagons was dumped at the working face of the fill and pushed down the bank by bulldozers. Travel over the fill area by the heavy wagons and bulldozers compacted the fill. A completed area would be from fifteen to thirty-five feet in depth, with refuse compacted to about 40 percent of its volume. Insecticide and deodorant often were sprayed over the fill, and then a layer of clean sand and gravel would follow. Eventually the completed fill area was covered with two feet of clean dirt. The surface was finished to a two-percent slope for purposes of drainage, and then ditches, culverts, and drains were added.[28]

Fill operations were initially intended to be limited in the southern part of La Tourette Park and extend along a line to the town of Richmond. The area contained important historic sites (old public docks, an old tide mill, and so forth) and buildings and especially narrow streets at the heart of the old town of Richmond (Richmondtown).[29] As early as 1952, however, Moses, Hall, and the DSNY made efforts to extend the landfill to include portions of the historic village of Richmond as well as other locations.[30] Pushback from the Chamber of Commerce, community groups, and private-property owners became heated but did not stop the expansion.[31] In March 1953 the Board of Estimates authorized the acquisition of about 1,100 acres to add to the landfill. Set aside, however, were twenty-six historic buildings in Richmondtown. The Staten Island Historical Society subsequently proceeded with plans for its Richmondtown restoration project.[32]

Different projections about how long Fresh Kills Landfill would remain open—two, three, even eighteen years—made it clear that no absolute end date was certain. Moses's forecasts were no better than the others. Mayor O'Dwyer's promise that garbage would only be dumped there for two years smacked of political expediency. And the DSNY's longer timeline dealt with the capacity of the space, availability of other dumpsites, and the relationship to incinerators coming on line. The layout of the site plainly reflected long-term use.[33]

In the late 1940s, Fresh Kills was one of thirteen city landfills, then one of ten as Great Kills (Richmond), Lefferts (Queens), and Kissena Park (Queens) were phased out by the end of 1948. Yet, by design and use, Fresh Kills had more staying power and greater size than all of the rest. Excluding Fresh Kills at approximately 1,200 acres at the time—and with the closure of the Great Kills/Oakwood fills—the next largest was Edgemere in Queens with a little more than 322 acres. Others ranged in size from fifty-one to under three hundred acres.[34]

Also in 1949, city refuse managers for the first time increased the elevation of active landfill mounds (ten to fifteen feet above sea level) to expand their capacity.[35] While it was widely believed that all other landfills would close before Fresh Kills, the Richmond site was clearly emerging as the only land destination that would be available for the city's solid waste—either with or without an extensive incineration program.[36]

Encroachment of the landfill into the heart of Staten Island drew immediate protests from locals. As early as 1949, and possibly before then, citizens complained about the odors rising from the decomposing garbage despite DSNY claims that the site posed no health menace or nuisance.[37] In some cases, fill spilled onto the roads from trucks headed to dump sites, and the trucks themselves stirred up massive clouds of dust.[38]

Arleen McNamara Brahn's family moved to Travis in the summer of 1945. "Our house," she said, "was a large farmhouse built over 100 years before, with four acres of property, at 3920 Victory Boulevard. Later it became McNamara's Picnic Grove." In 1947 the city bought a half-acre of the property for the landfill project. The family shared the space with the heaps of waste until they sold their remaining property in 1986. Over the course of many years, her childhood memories were marred by the changing landscape. She recalled, "It is sad the landfill changed our whole way of life. The wildlife is gone, the waterways and marshes are filled in, and of course the crabbing, clamming and fishing are gone with them."[39]

Hansine Bowe was a girl when Fresh Kills opened and lived on land along what is now Richmond Avenue. "Before they put the mounds [of the landfill] kind of built up to Richmond Avenue," she stated, "you could look at it and see it getting bigger and bigger and bigger. And also from Travis you could also see it. Because there were thousands and thousands of tons of garbage coming." "At one time before they started capping it," she added, "[refuse] used to fly on a wind storm, not where we lived, but you would go along Richmond Ave and you would see paper and stuff all over the place."[40]

Periodic fires broke out on scows or on the piles of waste. The *Advance* reported on April 29: "Clouds of thick acrid smoke pouring from garbage afire on eight scows tied up at the Fresh Kills dump yesterday made dozens of Greenridge residents nauseous. The foul-smelling smoke penetrated houses blocks away and even today traces of the smell could be detected in the area."[41] Another scow fire broke out on May 15 and a dump fire on June 12, which covered Travis and Springville with smoky skies.[42] When offensive odors drifted across Arthur Kill from New Jersey, Staten Island residents first looked to the dumpsite as its cause.[43]

INCINERATION: THE TOP PRIORITY

The DSNY was far from solving its disposal problems by the opening of Fresh Kills. And at the time, no matter how large the landfill operations grew there, developing incineration as the primary disposal option for New York City was still the goal. On October 1, 1948, the department underwent another reorganization. The objective (once again) was to minimize the political nature of department leadership and to enhance technical expertise and efficiency in carrying out its various responsibilities. This meant an unabated commitment to incineration.

The changes dealt primarily with operational, staff, and administrative procedures, while borough arrangements remained unchanged. Commissioner Powell was tasked with appointing a director of operations (a professional engineer) to supervise four bureaus and two divisions. Henry Liebman was the first such appointee. The reorganization reinforced the ongoing disposal strategy to move toward incineration and away from landfills.[44]

On August 19, 1948, the Board of Estimate made official what had been an informal policy of the DSNY when it passed a resolution meant "to increase the efficiency of existing incinerators" and to make sure that "the program of repairs, rehabilitation, expansion and new construction of incinerators be accelerated with the end in view of having sufficient incinerators in operation by the end of 1951 to enable the Department of Sanitation to terminate all dumping of garbage and combustible material in landfills."[45] (So much for Bill Carey.) The tone of the resolution suggested urgency, especially given the results of the department's 1946–1947 salt-marsh acreage survey, which estimated that sufficient marshland for dumping and filling would be available for only nine and a half more years.[46] Once again, speculation about a firm date for ending landfilling was adjusting the timeline.

Despite the reaffirmation in the early 1950s of incineration as the DSNY's major disposal objective, the city faced the difficult task of rehabilitating existing incinerators, increasing their capacity, and building new ones.[47] Struggling over whether to place the new incinerators at waterfront or interior locations, the department opted for several interior sites. This suggests that the DSNY had little concern about the possible nuisance value of the burners to the neighborhoods or at least that it deflected such criticism.[48]

Complementing the incinerator plan was research carried out to determine the availability of future fill sites in the boroughs beyond Fresh Kills and the need for marine loading (or transfer) stations.[49] With no workable dump sites in Manhattan, its waste was barged directly to Staten Island. In the Bronx, the available marshland in the eastern part had "decreased substantially," but there were no marine loading stations. There was a lack of marshland also in northern Brooklyn, resulting in the construction of three marine loading stations. In northern Queens available marshland was exhausted, and a new marine station was proposed. The only places for landfill sites were in southeastern Brooklyn and southwestern Queens. Jamaica Bay had enough space to consider landfilling for the near future, but such activity had been blocked by Moses.[50] Although the Staten Island location was not the sole depository for unburnable New York City waste, severe limits on available dumping sites elsewhere were beginning to point in that direction.

Disposal in general was a "controlling factor" in the mayor's capital outlay for 1949, along with waterfront rehabilitation and sewage disposal. Moses, as city construction coordinator, communicated to Mayor O'Dwyer that the garbage-disposal program would cost approximately $44 million over three years. This budget assumed the construction of new incinerators and destructor plants. It also included plans for landfilling shoreline areas around Ferry Point and Pelham Bay landfills in the Bronx by using dikes and wire-mesh screens to keep trash out of the water.[51]

Moses was very pleased with O'Dwyer's commitment to dealing resolutely with the refuse and sewage problems. Asked by the outgoing mayor to take his place in presiding over the dedication of Great Kills Park on July 1, 1949, Moses declared, "I want to say this about your mayor. . . . This park involved many problems. The incinerator program was one of the mayor's headaches. Sewage disposal was the most serious of all. These problems are being tackled by the mayor. . . . The great thing is that the mayor is facing these problems and is doing something about them. I wish he would stay long enough to finish the job."[52]

Changes in the waste stream reconfirmed the value of incineration. Unparalleled growth of the packaging industry—a direct response to flourishing consumerism after World War II—was largely responsible for the creation of countless goods with short useful lives. Packaging took on special importance in the late 1940s because of the rise of self-service merchandising through supermarkets and other consumer outlets, ushering in the "throwaway" society. The use of paper stock rose from 7.3 million tons in 1946 to 10.2 million tons nationally in 1966.[53] Paper and cardboard burned easily in incinerators.

Through the mayor's incinerator construction and modernization program, the DSNY was poised to provide a means to dispose of all burnable garbage and refuse by building five new destructors beginning in 1948 and modernizing the eleven existing plants. The new incinerators were to be located in Queens, Manhattan, Brooklyn, and Richmond.[54] Officials insisted that the incinerator program was not only the city's long-term priority for refuse disposal but was being accelerated to conserve the shrinking acreage of marshland, which was to be set aside for ashes and incinerator residue.[55] This response was echoed by Borough President Hall, who favored the mayor's three-year program to install new incinerators "to end the disgraceful land dumps for all time and create many acres of recreational grounds where dumps are in operation."[56] Where did this leave Fresh Kills?

There remained a major gap between policy and practice in the Department of Sanitation. Landfilling did not diminish in importance; if anything it was becoming more central. From mid-1947 to 1950, landfilling far exceeded incineration (table 1).[57] Although there was a slight increase in material incinerated between 1947 and 1950, the total amount disposed of by this method for each

TABLE 8.1 New York City Refuse Disposal, 1947–1950 (in cubic yards)

DATE	LANDFILLS	MARINE UNLOADING	INCINERATORS	PRIVATE DISPOSAL
July 1, 1947– June 30, 1948	13,089,562 c.y. (53%)	4,973,634 c.y. (20.1%)	6,385,814 c.y. (25.8%)	268,709 c.y. (1.1%)
1948–1949	12,968,975 c.y. (47.8%)	7,131,135 c.y. (26.3%)	6,881,656 c.y. (25.3%)	172,785 c.y. (0.6%)

Source: City of New York, Department of Sanitation, *Annual Report, 1948*, 20; *1949*, 20.

year did not represent much more than one-fourth of the total amount produced. Landfills accepted nearly three-fourths of all disposed materials.[58] Only a very small amount, mostly ashes, was disposed of by private concerns.

AN END TO LANDFILLS? THE INCINERATOR MANDATE

In the early 1950s, the DSNY set out to satisfy the mandate for an accelerated incineration program and for an apparent end to landfills. It was doing so in a turbulent political landscape. Public hostility to a New York Police Department (NYPD) corruption scandal and O'Dwyer's role in it broke in January 1950. No sooner had he won reelection than the mayor fled to Florida, claiming "nervous prostration." O'Dwyer resigned in August. Vincent Impellitteri, president of the City Council, assumed the office until a special election could be held in November. Born in Sicily, Impellitteri grew up in New York and was well known as a tough lawyer and effective prosecutor. He was nominally a Democrat but sought to distance himself from Tammany Hall, although he was not totally free of Italian gang associations. He thus entered the mayor's race as a nominee of the fusion Experience Party. In a three-way race (all Italian Americans) he was viewed as an "independent," which appealed to voters tired of corruption and Tammany pressure. Without a strong party base, Impellitteri was not particularly effective, although he continued to rely on Moses for city-planning decisions. Impellitteri also faced accusations of corruption and links to organized crime. His time as mayor was short-lived, and he lost the Democratic mayoral primary in 1954.[59]

While in office, Impellitteri supported O'Dwyer's incineration program. Nationally, incineration continued to develop as one of the disposal options open to big cities in the 1950s, for the same reasons it did in New York. Either because of concerns over shortages in landfill sites or because of hauling costs, burning waste had a following. In addition, the technology used for incineration was undergoing major changes in these years. Batch-fed furnaces with intermittent discharge of residue were more frequently being replaced by continuously fed, mechanically stoked furnaces with constant ash removal.[60]

In late December 1949, Andrew W. Mulrain replaced William Powell as commissioner of sanitation. The new head was only forty-nine years old and had worked his way up through the ranks. He had served as assistant to Powell before his selection under Mayor O'Dwyer.[61] A *Saturday Evening Post* story declared that "as a lad of twenty, he started cleaning the city's streets with a horse, a cart, a scoop, a broom, a tremendous determination and a winning Irish way."[62] Mulrain came into office under the reorganization, and the incineration program became one of his highest priorities. Moses was on board with the incineration strategy from the beginning, but he did not want the public to lose sight of what reclaiming marshland had meant to the growth of the city. In a 1950 Department of Parks report, he asserted: "For sixteen years the New York City Park Department has been reclaiming waste lands and swamps to create new park areas as part of a long-term program. Already 2,135 acres have been added and additional projects of 1,900 acres are under way."[63]

It was Moses to the rescue then and now, he believed. The current strategy encouraged waste materials only placed on city-owned lands to be developed as "parks or for other public purposes" at sites in Marine Park, Brooklyn; Spring Creek Park, Queens; Ferry Point Park, the Bronx; and Fresh Kills Park. Fresh Kills was thus rationalized as part of a larger plan to produce usable land and also to hold the line until incinerators could be built. The Richmond site was drawn into Moses's master plan, he claimed, for the greater good of the public and little more. For Moses, a central problem that "has plagued the Park Department" during the final development of the filled lands for recreational purposes was "securing at reasonable cost the topsoil required to sustain lawns, shrubs, and trees."[64]

Advocates of incineration increasingly viewed the technology as a disposal end in itself, that is, as an alternative to land dumping. In such a scenario, Fresh Kills had a precarious role—it had come on line as a complement to incineration, not as a replacement for it or for any other disposal method. The Staten Island landfill, like the several other smaller ones in the city, existed in a disposal limbo, where its role was not clearly defined by all parties.

Given current waste disposal goals, the opening of the eight-hundred-ton Betts Avenue incinerator in August 1950 was big news. The DSNY's 1950 *Annual Report* claimed that it "probably is the most modern in the country" and was the world's largest. Bringing the incinerator on line allowed the department to close the outmoded Ravenswood facility to make room for a Long Island City housing development.[65] The Betts Avenue plant was the first of five that were to be built using a continuous feed and continuous-burning technology. Only

FIGURE 8.4 Betts Avenue incinerator.

Source: NYC Department of Sanitation, *Annual Report, 1949*.

three other incinerators in the country met those requirements at the time.[66] Betts Avenue became the symbol of the city's waste disposal policy. Speaking at the formal unveiling of the plant, Public Works Commissioner Frederick H. Zurmuhlen enthusiastically stated that "all open garbage dumping in New York City [presumably including Fresh Kills] will be terminated by 1953 under an accelerated incinerator construction program laid down by Mayor O'Dwyer."[67]

WHAT TO DO WITH LANDFILLS?

In March 1951, Casimir Rogus, director of engineering for the DSNY, added a chapter to the department's December 1950 report that focused specifically on "Landfills—Present and Future." He restated the department's view that it "must perforce continue the operation of the existing five landfill sites [including Fresh Kills] until their completion." They, "through necessity, now accept all types of refuse, burnable and nonburnable." That would change when all the incinerators were completed.[68]

Rogus not only dealt with the transition to incinerators in the department's program but underlined future plans for Fresh Kills in particular. "The Fresh

Kills Landfill . . . now is and will continue to be for the next two decades the disposal point for the previously stated amounts of refuse and incinerator residue, during the incinerator construction period, and solely for ash and residue disposal in the post-construction period."[69] Rogus was unable to calculate the exact fill capacity of the site but estimated that "a reasonably accurate calculation" placed its completion date at 1967. The dates for the active use of Fresh Kills continued to push into the future.

Rogus speculated about opportunities to develop future fill or disposal projects, especially for ash and incinerator residue. He discussed filling in about ten thousand acres of marginal tidal lands in and outside the city. Because they were privately owned and within close proximity to developed areas, there would be "considerable difficulty" in getting required approvals, for example, from many New Jersey municipalities. He explored the idea of dumping nonrefuse waste at sea but concluded that processing and hauling costs added to the "legal difficulties," making that approach unlikely. He referred to parts of Jamaica Bay as a promising site, while noting that Local Law 31 had assigned the land (about 11,560 acres) to the Park Department and prohibited dumping of fill from the DSNY. He wondered, however, if the use of ashes and "well burned incinerator residue" might meet with approval; if so, the site could provide disposal for forty or sixty years.[70] In the end, Rogus made a striking case for the ongoing disposal problems that faced the city even with the implementation of the incineration plan.

Incrementalism ruled the day. In a May 15, 1951, memorandum, Moses recommended to Mayor Impellitteri that the city extend the use of garbage as landfill on city-owned property for an additional four years. This was necessary because the incinerator-construction program "had bogged down completely." Restrictions on building materials because of the Korean War and ever-increasing amounts of waste made further delays likely. Moses counseled the mayor, "You must have some program to cope with the problem [of the slow implementation of incinerators], otherwise it will reach the same state of public turmoil experienced a few years ago." He also warned that the landfill operations had to be carefully planned and monitored. "They can become an intolerable nuisance if there is any skimping on proper site preparation, amount of cover made available for immediate covering of refuse and garbage or constant adequate supervision."[71] Cracks in the incineration plan were starting to widen.

The DSNY gave few hints to the public that its plan to coordinate landfilling and develop incinerators was stalled. In a June 1951 article in *Public Works*, Director of Operations Henry Liebman outlined how 35 percent (or about 1.8 million tons) of refuse from Manhattan, northwestern Brooklyn, and Staten Island was being deposited in Fresh Kills Landfill. In the future, refuse from northern Queens also would be dumped there. Fresh Kills, he stated, "is one of the few remaining areas of this type still available for landfill purposes in the city," but "its location and subsurface conditions presented, and are presenting, many difficult engineering problems both in the preparation of the site and in the daily operations." Liebman was candid in stating that until the incineration program was accomplished (eight or nine years he estimated), refuse that included unburned garbage would need to be dumped at Fresh Kills. He wanted to assure his readers that thanks to new methods of sanitary landfill "we have received practically no complaints but, on the other hand, much praise from civic officials, civic associations and the local newspapers."[72] Such claims did not curb neighborhood murmurings on Staten Island.[73]

A vivid brochure spearheaded by Moses and also signed by Borough President Hall and the DSNY's Mulrain was submitted to Mayor Impellitteri and the Board of Estimate in November 1951. Complete with maps, drawings, and pictures, *Fresh Kills Landfill* put the best spin possible on the use of the Richmond facility. It turned apprehension over the large-scale dumping site into anticipation for economic and recreational opportunities through the reclamation of the existing 2,741 acres. "During the last three hundred years, land-fill operations have played an increasingly important part in the development of the area now compromising Greater New York," the brochure began. Because of "economic pressures caused by this steady growth [of the city]," it continued, "swamps, ponds and streams in the interior of [Manhattan] were filled, so that new streets and building sites could be created and unsightly and unsanitary waste lands made useful and profitable to match rising land values in surrounding districts."[74]

A familiar narrative followed, beginning with the end of sea dumping in 1934 and ending with the "permanent benefits" derived from sanitary landfilling in places like Great Kills Park. The brochure then proclaimed that as important as previous projects had been, "they are overshadowed by one of the world's largest sanitary land-fill operations at Fresh Kills, Staten Island." Anticipating the completion of the city's incinerator program "in about 1960," Fresh Kills provided disposal space not available (or suitable) in Manhattan or Brooklyn and could sustain accepting refuse until about 1968. Parroting others, such as Liebman, the report pointed to "much praise and but few complaints" over the three years of operation at the site.[75]

The focal point of the report was stating what reclaiming the marshes at Fresh Kills would mean for Staten Islanders in particular and for New Yorkers in general. Opportunity was equated with the ultimate completion of the West Shore Highway (begun in the 1950s) running through the center of the landfill site to make way for future development[76] and the eventual extension of the railroad line to industrial property south of Fresh Kills. Fill boundaries would exclude some park lands and private property to be preserved in their present condition.[77]

The current plan would close Fresh Kills to navigation and build a tide gate and dam on which the highway and railroad would cross. Along with these important infrastructural changes came the promise of the development of reclaimed property inside the West Shore Highway, creation of freshwater lakes "doubling the acreage devoted to recreation," and new land for residential and commercial development. It estimated that land area in the park additions would total 575 acres, 160 acres in new lakes. Park belts would offer baseball and softball fields, playgrounds, picnicking areas, boating ramps, and fishing piers, with "tree-shaded paths along the water's edge" and "ample space" for bicycle and bridle paths. Additions to Schmul Park were included "so that residents of Travis will have a connecting corridor to the varied facilities of the new Fresh Kills park system." New Springville Park "with its fine old woodlands" would extend its boundaries, and a protected wildlife refuge would incorporate the existing bird sanctuary. Efforts would be made to preserve "features of historic and scenic interest" and the marsh and area adjacent to Richmond Creek.[78]

In an effort to put the best face on the enormous Fresh Kills Landfill and to offer a carrot to a dubious public, the brochure concluded:

> The Fresh Kills project is not merely a means of disposing of the city's refuse in an efficient, sanitary and unobjectionable manner pending the building of incinerators. We believe that it represents the greatest single opportunity for community planning in this City. The cooperation of the [various governmental parties] will create enough valuable new property in this presently fallow and useless area to pay the cost of the project many times over and to produce a well-rounded and diversified community, practically planned, to meet the future needs of Staten Island.[79]

This was vintage Moses. The West Shore Expressway eventually was completed; much of the rest was not reconsidered until the development of Freshkills Park more than fifty years later.[80]

While Hall signed off on the Moses plan, he had been urged by Moses in June 1950 to commission a study of his own leading to a plan for Fresh Kills' future use.[81] Apparently, Moses wanted to keep the borough president on board for developing the Staten Island site by involving him in the process. In light of the proposed Richmond Parkway and the eventual West Shore Expressway, Hall requested in March 1951 that the Department of City Planning in the City Planning Commission "study [the Fresh Kills area] and propose a suitable plan. It should establish zoning requirements, streets, parks, and public land areas."[82]

The task was given to the Office of Master Planning (OMP), which led to meetings with Hall's employees, the Departments of Parks; Purchase; Water Supply, Gas, and Electricity; and Marine and Aviation, as well as the Baltimore and Ohio Railroad. Hall reviewed the handiwork and "seemed satisfied with the principal recommendations," although he was concerned about the retention of some local recreation facilities. The OMP drafted a preliminary report released to the commissioners on November 21.[83] Unaware of Moses's discussions with

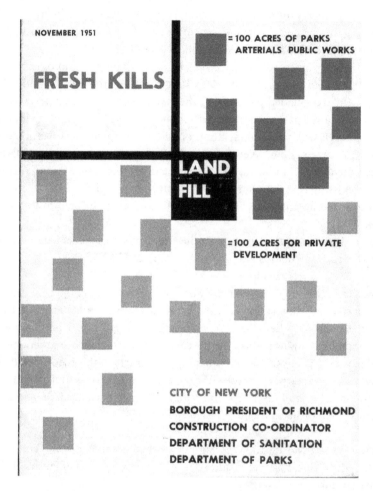

FIGURE 8.5 Robert Moses's brochure promoting Fresh Kills Landfill, 1951.

Source: Courtesy of Freshkills Park and the City of New York.

Hall, Eugene M. Itjen, chief of the Office of Master Planning, stated, "Without our knowledge, a similar project was being carried on," which led to the publication of Moses's 1951 brochure. Itjen added, "While [the brochure] does not answer the Borough President's request [for a report], it covers some of the same territory and is in conflict with some details of our recommendation."[84]

Both the OMP report and the Moses brochure supported heavy-industry development west of the proposed West Shore Expressway. Both agreed that a dike (dam) and tide gate were superior to a bridge crossing the fill area for motor vehicles and the railroad. And both provided for recreational uses of streams east of the proposed dike. Some of the points of disagreement were in the details:

the railroad terminus, the alignment of the expressway, allocations of areas for public purposes, and land-control issues. The "principal conflict" was over the suggested use of lands east of the expressway. The OMP supported medium and light manufacturing; Moses's plan called for residential development. The former pointed to potential problems from land settlement, in which case such occurrences for home owners "often result in mass hysteria with exaggerated complaints and protests to the public agencies."

The Office of Master Planning also raised concerns about the "inevitable subsidence of walks, drives, stoops, porches, stops and other accessory structures" and the possibility of "marsh gas pockets" and "obnoxious fumes" from New Jersey and Fresh Kills industrial areas.[85] Overall, the planning report tilted more toward the "large potential reserve of land available for future heavy industrial expansion," while Moses seemed to stress his vision of more parks and recreational facilities for the people of Staten Island.[86]

Both the report and the brochure implied that Fresh Kills should be a means to an end, not an end in itself. The justification for Fresh Kills Landfill as the best available temporary space in New York City for disposal of its waste (at least until incinerators dominated) made way for visions of remaking and "improving" Staten Island. The reports made it clear that the city would be engaged in the Fresh Kills area for the long haul. The OMP report referenced the DSNY as affirming that the fill operations would take from thirteen to twenty more years.[87]

In the mid-1950s the city released maps of what changes were likely to take place for the Fresh Kills area over time. They showed one thousand acres of new homes, eight hundred acres for parkland, and 875 acres of commercial and industrial property. But, as the *Staten Island Advance* reported years later, "residents living there at the time were more concerned with the trucks spilling the fill dirt on local roads, causing many to make analogies to the infamous Dust Bowl in Oklahoma." Stories circulated about strange items turning up in the landfill. Someone reported in July 1955 that a hand grenade was found there. In July 1949, Bronxite Mitchell Ward recovered some of the thousand dollars that accidentally had been thrown out in a cardboard box by his half-brother. Geiger counters were used to search for eight medical radium capsules apparently buried on a barge in March 1949. And stories periodically circulated about the dumping of dead bodies.[88]

THE CLAY PIT PONDS

While the Moses brochure and OMP report promoted economic growth for Staten Island and the promise of more parkland, preserving existing natural or historical sites in Richmond Borough was not a high priority.[89] In November

1951 the *New York Times* reported, "Staten Island civic groups and wildlife conservationists from the metropolitan area protested yesterday against the city's plan to fill in the Clay Pit Ponds and adjoining marshlands that are the habitat of great blue herons, wood ducks, mallards, bitterns, pied billed grebes, muskrats [still sought by trappers] and other feathered or furred creatures."[90]

Located near the southwestern shore of the island close to Rossville, the seven ponds marked an area shaped by extensive mining for white kaolin clay in the nineteenth century. The abandoned pits filled with rain and spring water and were home to about 180 bird species, including fifty-seven of migratory songbirds.[91] Citizens assumed that Moses would be true to his word when he favored making a rustic park out of the clay-pit ponds and about two hundred acres of an adjacent wooded tract. Title to the land was under the jurisdiction of the DSNY, and not surprisingly its use was contested.[92] Neither Moses nor the DSNY proved advocates of saving the ponds.

Lynn W. McCracken, chair of the Fenways and Brooksides Committee (formed to protect the island's natural areas), insisted that Moses had promised civic leaders to preserve the ponds and adjoining marsh as a sanctuary. Groups including the Fish and Game Protective Association and the Audubon Society of America also opposed the destruction of the ponds.[93] The DSNY's Liebman responded that the two-hundred-acre area under protest would have to be filled in to make room for the Staten Island West Shore Highway. He added that the ponds and marsh were hazards where a drowning and other accidents had taken place.[94] McCracken responded that he could not imagine why garbage would be dumped "into such beautiful ponds," and "our counterproposal is to have garbage dumped in Babylon, L.I. If Mr. Moses likes garbage so much, let him have it at his front door."[95]

In response to protests, spokespersons for Moses and Hall declared that the city had no intention of altering or abandoning its plans for park development in general at Fresh Kills. Stuart Constable, the Park Department's chief designing engineer (who often sat in for Moses at city-planning meetings), stated that detailed park plans soon to be released demonstrated that the disposal project was too large and important to be altered or abandoned, and he was confident that the birds would find a haven elsewhere.[96]

In early January 1952, Moses continued to defend his decision on parkland on Staten Island. Those opposing his position, he stated, "would do the city and themselves a real service if they would forget the small loss and unite in support of our plan, which will give them many more benefits than any park development around the old ponds could possibly produce." He reminded protesters that the clay pits "are, of course, not natural ponds." They "lie between the West Shore expressway and Arthur Kill in an area that cannot logically develop as anything but an industrial neighborhood." Taking away with one hand and giving with another, he added, "The fresh water lakes which will be created behind the tide gate and dam we propose . . . will be many times the size of the old Claypit Ponds and will lie entirely within park areas. We propose to establish wild life sanctuaries . . . which will be far greater in extent than anything that could have been done around the clay pits." But as not to seem indifferent, he acknowledged "the anxiety of various groups" over the loss, promising that the DSNY would "preserve parts of the Claypit Ponds area for as long a time as is practicable."[97]

Hall, again riding the fence with respect to island development versus the community's response, stated that there was a "definite understanding" that the DSNY would not fill in the main Clay Pit Ponds until the construction of the new freshwater lakes and lagoons commenced. Misreading the protest, he added, "What I cannot understand is the attitude of the small group that has been advocating for some time the development of clear, fresh water brooks and fenways, opposing this plan, as it is just what they have been seeking."[98] In an effort to clarify the situation, a DSNY spokesperson stated that the department was filling only a small pond in the west sector of the Clay Pit Ponds area and had no plans to fill the three principal ponds (Long Pond, Round Pond, Shovel Pond) in the near future.[99]

FIGURE 8.6 Fresh Kills topography and ecology, 1955.

Source: Courtesy of Jose Mario Lopez, Hines School of Architecture, University of Houston.

In March 1952, with an apparent arrangement in place for the ponds, top-level engineers and other interested government parties discussed how plans for the West Shore Expressway would be implemented.[100] The protest over the Clay Pit Ponds would not go away, however, and in fact stimulated a broader attack on the landfill. Frank Beals, president of the Community Councils of New York, raised the possibility of disease-infected rats becoming a health hazard at the dumpsite. Beals, McCracken, Assemblyman Edward V. Curry, Myra Barnes (known as "Staten Island's fighting lady"), and a host of representatives from groups like the Fish and Game Protective Association, the Richmond Borough Gun Club, the Staten Island Civic Congress, and the Staten Island Growers Association attended a protest session also in March. The group passed a resolution opposing further dumping in the Clay Pit Ponds area. The chair of the meeting, Harry Cassidy, declared that the politicians who were promising improvements based on the Fresh Kills Landfill project would not be alive when the time came to fulfill those promises. He reminded attendees about the long-forgotten plan to build a subway to Staten Island and declared that "promises made by dead men are not binding on the living."[101]

The closest thing to a solution to the impasse was the DSNY's new promise that the Clay Pit Ponds would not be filled until the freshwater lagoons and ponds had been created and made available to the public.[102] This hardly addressed the deeper apprehensions of the protesters. But Moses simply moved forward, and Hall fell into line.[103]

BUSINESS AS USUAL

Throughout the remainder of the 1950s and into the early 1960s, the DSNY pushed forward with its strategy to make incineration the primary disposal option for New York City and to further develop its reclamation efforts. But progress, although steady, was slower than hoped.[104] What had become a ten-year incineration plan sometime earlier was now extended to a twelve-year plan to construct eleven new incinerators and rehabilitate five others.[105]

Commissioner Mulrain was convinced that the Bureau of Waste Disposal was solving the city's "particularly massive problem" with 1,600 personnel, eleven operating incinerators, six land-reclamation projects, twelve marine loading stations, and a large fleet of tugs and barges. Fresh Kills was absorbing about 28 percent of the refuse generated that year. In fact, in November 1952 the ten thousandth bargeload of refuse was transferred to (according to the DSNY's 1952 *Annual Report*) "the largest and best land-reclamation project of its kind in the world."[106] Not reflected in the report were complaints about local nuisances. The large number of trucks that were carrying dirt to be used as fill and cover material, for example, were kicking up dust and debris across Staten Island, agitating several communities.[107]

DSNY's feel-good mood in 1952 was spoiled the following spring when the novelist and playwright Edna Ferber referred to New York City as the most "disgustingly filthy" city in the world. Returning from a ten-week vacation in Europe in late April, the Michigan transplant (who had lived in New York City since 1913) had nothing good to say about the physical condition of Gotham compared to cities she had just visited: The streets were covered in garbage, the buildings were gray and dirty, parts of the city were rat infested, and Central Park was not fit for "a self-respecting goat." "New York," she asserted, "is a scab on the face of our country." Insisting that she still loved the city, she concluded that it reminded her "of a once exquisitely beautiful woman who had declined into a dirty, degraded, blowzy person."[108]

Ferber's declaration (focusing primarily on street cleaning) received widespread public attention, and she was not alone in this sentiment, as it turned out.[109] A first response from a Sanitation Department spokesperson was that a new cleanup campaign would be underway in May.[110] Both Commissioner Mulrain and Mayor Impellitteri spoke at the Spring Cleanup Campaign, exhorting citizens that "a clean city is a healthy city" and that individuals' actions mattered too.[111] Mulrain also was defensive, perceiving Ferber's comments as an insult to his men. Yet even he had admitted weeks earlier that the streets were "perennially dirty" and that "in Manhattan, the public street is a public dump."[112] But Ferber did not back down from her assertions and declined a role on the citizens' cleanup committee. She conceded that the problem was not simply the fault of the DSNY administration. "It is we, the people of New York, who litter the streets, demolish the parks, tolerate the disorder, permit the blackened air."[113]

Henry W. Taylor, chair of the American Society of Civil Engineers' Refuse Collection and Disposal Committee, directed attention away from Ferber's comments about the streets, stating that "New York yields to no city on the fine job it is doing in increasing municipal incinerator capacity and on the excellence of its garbage land-fill projects."[114] But in the face of Ferber's accusations, the DSNY was confronted with doing more than defending its honor. It now had twelve destructor plants in operation running six days a week around the clock (and one in each borough working on Sundays). The DSNY regarded its incineration program as "probably the most ambitious municipal waste disposal plan ever advanced."[115]

Problems with its overall disposal program continued to mount, however. The department needed to find desirable sites for incineration plants, to build new marine transfer stations and to purchase new barges, and to acquire more city funding. A planned thousand-tons-per-day incinerator in central Brooklyn had to be canceled because of "site-procurement difficulties," and its capacity had to be absorbed by other facilities. Ten older plants were scheduled for abandonment or rehabilitation (which would take them off line for a long period of time). Labor strikes delayed construction of the North Shore marine transfer station and the South Shore incinerator.[116] The good news was that the Gansevoort incinerator opened in May 1953, and with the Betts Avenue incinerator operational, the total inventory stood at twelve.[117]

The year 1954 showed signs of meeting the incineration program's short-term objectives. A report from the DSNY projected that the department would burn more than three times the refuse in 1970 than it did in 1950, increasing from 19.8 percent of the total waste disposed to 49.9 percent.[118] The overall task for the DSNY was not getting any easier, however. The journalist Harrison Salisbury was a sympathetic observer when he stated that Edna Ferber "stirred up a hornet's nest last year by simply putting on public record what many others had felt—that New York is 'the most disgustingly filthy' city in the world." But, he added, it was not difficult to demonstrate that the city was "getting value received" on collection and disposal.[119]

Not everyone was so generous with praise for the DSNY, which regularly faced worker demands for higher pay and better working conditions.[120] One observer noted that refuse collection and disposal in New York City was "an undertaking of staggering proportions. With the job getting bigger and bigger each and every year."[121] Rogus, currently the commissioner's technical adviser, stated that "by any standard of comparison, the New York City Department of Sanitation is a large industry." He added, "This very size presents major problems of management, making heavy demands upon the time of Department executives already busy in administering day to day operations."[122]

The incinerator and landfill programs were well publicized, but the department also contended with selling ash for construction projects and providing garbage for privately operated hog-feeding farms. Even something as routine as hauling residue from incinerators raised problems. This residue was composed of a mixture of ash, clinkers, nonburnable material, and 1 to 3 percent unburned organic waste. The resulting stench was one reason that incinerators were placed near waterfront locations, but this did not by itself alleviate the smell. Seeking the correct hauling method proved chronically difficult and costly.[123] Along the way, Fresh Kills kept getting bigger, covering 669 acres in 1954.[124]

The new mayor, Robert F. Wagner Jr., echoed the overwhelming nature of New York City's refuse problem. In a speech inaugurating "Sanitation Week" in May 1954, he stated:

> But of all the crusades to which the mayor must lend his support, I doubt if there is one more important to more people than that for which we work today: "Sanitation."

> Sanitation is merely another word for cleanliness, and I'm sure all of us subscribe to the biblical admonition that cleanliness is indeed next to godliness.
>
> Cleanliness does not come easy to a city of the size, industry and complexities of New York. It is not illogical that since our city leads the world in so many things, it should also lead in the prosaic matter of garbage and refuse.
>
> Our 8,000,000 population, our immense manufacturing capacity, our great port, and our enormous demand for consumer-goods . . . all make for the biggest housekeeping job in the world.[125]

Elected as a Democrat in 1953, Wagner understood the formidable task of providing adequate city services, having served under Mayor O'Dwyer as tax commissioner (1945–1947) and as commissioner of housing and building (1947–49). He also had been borough president of Manhattan (1950–1953). Wagner had a taste of state politics as an assemblyman (1938 to 1941), also learning a thing or two from his father, Senator Robert F. Wagner Sr., the celebrated New Dealer. Wagner Jr. won his first election as mayor (one of three) with the support of the Manhattan and Bronx political machines. Carmine DeSapio, the head of Tammany, wanted to reform the organization's image in its waning years and sought a spotless candidate to defeat Impellitteri. When "Impy" was unable to collect sufficient signatures to put the Experience Party on the ballot, Wagner was assured victory, taking every borough but Queens.

The new mayor was an effective politician during a lengthy period of prosperity for New York City marked by abundant construction projects, especially in Midtown Manhattan. Committed to liberal social reform, Wagner expanded many services and was best known for launching the largest municipal public-housing program in the nation and for supporting the unionization of city workers. He faced, however, a city with a serious poverty problem amid the prosperity and was exposed to the early stages of the Civil Rights Movement.[126] Going against the practice of putting his people in most key positions, he decided to retain Mulrain as DSNY commissioner because of his experience. He also continued to rely upon Moses.[127]

Concern over street litter—reverberating from Ferber's criticism—was a prevailing issue that the Wagner administration faced almost immediately.[128] There also was a growing interest in tightening municipal regulations on commercial refuse-removal services connected to the private cartmen. In 1955, private companies under city license collected a little over six hundred tons of swill each day, representing 16 percent of the total incinerator capacity of Manhattan.[129] The department wanted to terminate free refuse-collection services to businesses located in residential buildings, in response to complaints about overcharging and monopolistic practices by private cartmen in the trade-waste business.[130]

Slipping by almost unnoticed was the fact that Fresh Kills was now the largest landfill in the world—and that incineration was not moving ahead as planned.[131] The 1955 DSNY *Annual Report* stated that revisions to previously accepted incinerator capacities had to be made in relation to estimated population growth and trends in refuse output.[132] Essentially, earlier optimistic forecasts about incineration had to be revised downward. What was not known at the time was that the total amount of waste being burned was peaking in New York City—and that the amounts of refuse in general was growing.

In 1956, Mulrain's last year as DSNY commissioner, forward momentum in the incineration program appeared to return.[133] The Department of Parks, collaborating with the Department of Sanitation, set contracts for two new landfills to hold incineration ash and nonburnables. The DSNY still was optimistic about reaching its target of sixteen incineration plants despite needing to eliminate "bugs" in the operation of its newest facilities.[134]

INCINERATION FADES

By 1958 the wheels started coming off the wagon, so to speak, of the incineration program. The new sanitation commissioner was now Paul R. Screvane, whom Wagner appointed in February 1957. He was DSNY's youngest commissioner. Screvane began his career in the department as a truck driver at age twenty-two and rose quickly through the ranks after serving with distinction in the military. He was a moving force in the program of sanitary education and antilitter enforcement.[135]

Screvane inherited what appeared to be a solid operation, but by the end of 1958 the context in which the incineration program existed was changing. While the city's capital budget increased 20 percent, no funds were earmarked for new incinerator construction or additional resources for routine maintenance and repairs for existing incinerators.[136] The DSNY acquired a new responsibility (or burden) on January 1, 1958, when it took on the establishment and management of fills to be used for demolition and construction waste. This decision was handed down because Mayor Wagner in mid-June 1957 had revoked the permits for all private dumps (some say sixteen, others eleven) within the city limits and stipulated that disposal on private land be restricted to clean dirt and ashes. All construction, excavation, and demolition waste now would go to public facilities.[137]

For approximately twenty years, the DSNY had issued dumping permits to private individuals with the permission of property owners. Despite regulations requiring sanitary disposal of waste materials, putrescible materials often were found in those private dumps and created local nuisances. The fact that Italian Americans ran a great number of the private carting businesses incited old ethnic prejudices in the city, which reinforced the feeling that corruption dominated the carting operations (which it did).[138]

The construction waste added greatly to the department's disposal responsibilities, resulting in about a half-million cubic yards of additional refuse to be dealt with each year. Three new construction-demolition landfills were created: at Whitestone in Queens, White Plains Road in the Bronx, and Ralph Avenue in Brooklyn. At Fresh Kills a new truck-fill site along Richmond Avenue opened in late 1957. In addition, the closing of three older incinerators in mid-1958 for economic reasons meant shifting approximately sixty thousand tons of refuse to landfills. The DSNY *Annual Report* for 1957–1958 noted that increasing construction costs, difficulties in procuring sites, and the lack of engineering personnel were making it tough to meet the target date of 1961 for the completion of the incineration program.[139]

While the DSNY focused on a wide array of problems, what was happening at Fresh Kills and the surrounding Staten Island neighborhoods sometimes was ignored. In 1958, John J. Marchi, a Republican state senator from Staten Island, introduced a bill to close the landfill. Born in 1921, Marchi was a native Staten Islander. Graduating from Manhattan College in 1942, then receiving a Juris Doctor degree from St. John's University and a doctorate in judicial science from Brooklyn Law School (1953), he practiced general law in Richmond Borough and lectured to Italian jurists through a State Department initiative. A Coast Guard and Navy veteran of World War II, Marchi saw combat in both the Atlantic and Pacific theaters. He was first elected to the New York State Senate in 1956 and went on to serve for fifty years in Albany, focusing on state-government relations with the City of New York. He also ran unsuccessfully for New York City mayor in 1969 and 1973.[140] He was a staunch champion of Staten Island and its fight over Fresh Kills.

Marchi filed "An Act to amend the penal law, in relation to prohibiting garbage disposal in cities having a population of one million or more" in April 1958. The bill would prohibit disposal of garbage, rubbish, or refuse "from any land or land under water" within New York City on or after 1965.[141] The bill passed both houses, but Mayor Wagner, Commissioner Screvane, and City Construction Coordinator Moses opposed it. Moses argued that every safeguard was in place to make the landfill operations "as unobjectionable as possible with sanitation uppermost in the minds of the planners."[142]

Governor Averill Harriman vetoed Marchi's bill, stating that Fresh Kills assuredly would be closed in a few years and would be replaced with incinerators in any event by 1965. Marchi recalled in an interview with the *Advance* in September 2013 that "Harriman . . . vetoed the bill saying he understood how I felt and sympathized, but the problem would be resolved and they would have a technological answer to the whole thing and it would be stopped."[143] In the same interview, Marchi declared, "They dug out fresh dirt and made an enormous hole, like a grand canyon, to accommodate the size of the dump they were planning. We were always concerned about doing something about it."[144]

There was no way for Marchi to know that the incineration program would soon experience major problems. He was acting in the interests of his constituency against a disposal site that Staten Islanders had never wanted. In 1960, refuse disposal accounted for only 16.7 percent of the department's cost of doing business (down from 18 percent in 1958–1959). Of that amount 8.2 percent went to incineration, which also declined from the previous funding cycle (9.3 percent). The cost of landfilling waste was roughly equal to that of burning it, but the city was spending a relatively small amount on any kind of disposal despite its importance.

The DSNY still boasted that incineration maintained "an upward trend" in production in 1959–1960, after a decline the previous year because of the closing of old incinerators. The Division of Incinerator Operations employed almost one thousand workers, making it the largest unit in the Bureau of Waste Disposal. But despite new electronic scales (which were giving the department a more accurate reading of refuse disposed), the amounts incinerated did not match the peak numbers from a few years earlier. Marine operations also faced challenges. A fire in 1959 ravaged the South Bronx marine transfer station, which was nearing completion, and in September 1960 Hurricane Donna damaged the new headquarters of the barge-maintenance unit at the Fifty-Ninth Street Pier. For landfills, the tonnage numbers remained fairly steady. In an effort to conserve disposal space, the department again raised mound elevations. The previous height of about ten to fifteen feet was increased to twenty to forty feet.[145]

Despite the setbacks, the DSNY officials seemed to believe—or at least made such a case publicly—that their ongoing plans for waste disposal were on track. Hindsight proved them wrong. Between 1950 and 1962 the DSNY built only seven new incineration plants and rehabilitated four more.[146] Although New York City had the six largest incinerators in the country, disposal capacity only achieved 70 percent of the original goal. The year 1962 marked the last time the department would complete an incinerator, the new Hamilton Avenue plant. Plans to develop larger facilities never materialized, especially because of the high cost of advanced air-pollution equipment.[147] The 1961–1962 DSNY *Annual Report* noted that the incineration program "now is about 50% completed" in terms of its 1961 target date.[148]

Limited construction funds was not the only problem haunting the incineration program. Technical solutions were not available to limit air emissions substantially. In addition, beginning in 1951 all newly constructed New York City apartment houses taller than four stories had to be equipped with waste combustors. Between 1910 and 1968, approximately 17,000 domestic apartment buildings installed them, and none had adequate pollution-control equipment. The shift to on-site incineration, far from helping solve the waste disposal issues, developed a well-deserved reputation for contributing mightily to the city's serious air-pollution problem. As early as 1952, 30 percent of the city's air pollution was caused by residential incinerators; by the 1960s 35 percent of the city's airborne particulate matter came from incinerators.[149]

In the late 1950s the city imposed limits on open burning, but this did not curb the problem with burning waste. A few power plants and some facilities in the private sector began to employ electrostatic precipitators and other advanced pollution-control equipment in the early 1950s, but municipal incinerators did not utilize such technologies until the early 1970s. Cost was the major rationale. Also, cities faced little if any regulatory pressure from the state or federal government to limit air pollution in this period.[150]

New York's problems with incinerators mirrored national trends. The 1960s witnessed a number of abandonments throughout the country. One estimate suggested that one-third of cities with incinerators (about 175) had abandoned the plants in favor of other methods, primarily sanitary landfills.[151] New York continued to burn waste for several more years, but it never achieved the level of success envisioned by several mayors, the DSNY, or Robert Moses.

WHEN ALL IS SAID AND DONE

New York City's commitment to incineration was not an abject failure, but it neither solved the city's disposal problems nor replaced the need for land disposal. Between 1955 and 1965, the city's waste stream experienced a 78 percent increase, placing acute pressure on incinerator capacity and greater dependence on landfills.[152] As table 8.2 shows, incineration played a significant part in the city's disposal mix from 1950 to 1962. However, it never represented more than 38 percent of the total, even as overall tonnage increased.[153] Combined totals for landfills (truck fills at various sites) and marine loading (the material shipped to Fresh Kills by barge) far exceeded incineration every year, representing as much as 70 percent of all refuse marked for land disposal. The trajectory for tonnage of refuse disposed remained remarkably steady from 1950 to 1962 for all three forms of disposal, despite the DSNY's expectations concerning incineration. Even when the total tonnage spiked in 1962, both marine unloading and incineration made similar leaps.

A NEW FUTURE AWAITS

An October 1961 story in the *Staten Island Advance* ruminated on the continuing use of Fresh Kills Landfill: "The biggest cover-up job in the world is going on scientifically, laboriously and at times it seems, interminably, behind a 10-foot dirt dike along Richmond Ave. in the New Springville-Fresh Kills area."

TABLE 8.2 New York City Refuse Disposal, 1950–1962 (in tons)

DATE	TOTAL TONS	LANDFILLS	MARINE UNLOADING	INCINERATORS	PRIVATE DISPOSAL (ASHES/ TRADE WASTE)	CONSTRUCTION WASTE
1950	6,072,616	1,837,866 (30.3%)	1,808,414 (29.8%)	1,505,130 (24.8%)	921,206 (15.1%) (trade waste)	
1951	4,802,149	1,585,543 (33.0%)	1,779,301 (37.0%)	1,366,482 (28.5%)	70,823 (1.5%) (ashes)	
1952	5,012,909	1,799,223 (35.9%)	1,772,506 (35.4%)	1,391,180 (27.8%)	50,000 (0.9%) (ashes)	
1953	4,835,081	1,769,860 (36.6%)	1,642,159 (34.0%)	1,374,800 (28.4%)	48,262 (1.0%) (ashes)	
1954	4,846,667	1,746,028 (36.0%)	1,510,514 (31.2%)	1,558,339 (32.2%)	31,786 (0.6%) (ashes)	
1955	4,811,475	1,429,413 (29.7%)	1,650,798 (34.3%)	1,698,152 (35.3%)	33,112 (0.7%) (ashes)	

Year	Total				
1955–1956*	4,378,014	1,426,045 (32.6%)	1,259,422 (28.8%)	1,646,604 (37.6%)	45,943 (1%) (ashes)
1957	4,977,177	1,554,972 (31.2%)	1,693,210 (34%)	1,728,995 (34.8%)	
1958	4,816,239**	1,615,330 (33.5%)	1,599,730 (33.2%)	1,601,179 (33.3%)	576,775 cu. yds.
1959	4,681,000	1,498,000 (32%)	1,530,000 (32.7%)	1,653,000 (35.3%)	399,000
1960	4,995,000	1,467,000 (29.4%)	1,848,000 (37%)	1,680,000 (33.6%)	371,000
1961	5,374,000	1,602,000 (29.8%)	1,940,000 (36.1%)	1,832,000 (34.1%)	770,000
1962	5,697,000	1,505,000 (26.4%)	2,080,000 (36.5%)	2,112,000 (37.1%)	548,000

Note: For the years when the city collected private sources of ashes, in this case through 1956, private disposal is calculated as a part of the total disposal tonnage. Construction waste is not calculated as part of the total tonnage disposal because it distorts the relationship among landfills, marine unloading, and incinerators for the purpose of comparison.

Source: City of New York, Department of Sanitation, *Annual Report, 1950*, 20; *1951*, 20; *1952*, 22; *1953*, 17; *1954*, 22, 25; *1955*, 25; *1956*, 86; *1957–1958*, 23; *1959–1960*, 30; *1961–1962*, 21.

*change in reporting process
**excludes construction waste

FIGURE 8.7 Twenty-five million cubic yards of refuse disposed by the DSNY in 1951.

Source: Department of Sanitation, *Annual Report, 1951*.

All claims of turning the dumpsite into parkland, residences, and industrial sites seemed far off into the future. The latest Parks Department estimate placed preparations for new construction at 1980. In a telling remark, the author added, "Despite greater emphasis being placed on the construction of incinerators, landfill operations still are a vital adjunct to the city's garbage disposal operations. Some say they will always be necessary."[154]

In 1961, Fresh Kills covered 1,284 acres.[155] Despite the preference for incineration, disposal by landfilling cost about half as much.[156] The Richmond site was poised to become more than "a vital adjunct" to incinerators in the next few years. It would become the primary disposal option for the city, not simply a space for further economic development on Staten Island. Moses asserted in the 1951 brochure that "the life of the Fresh Kills project will be affected by the progress of the incinerator construction program."[157] This only was partially true. Incinerators were on their way out, and Fresh Kills had become more than a default plan. There was now little doubt that the site would serve an enduring and fundamental purpose for New York City—a purpose not so clear at its opening in 1948.[158] And it would impose a long-lasting blight on Staten Island.

Curiously, some popular literature and promotional material concerning Staten Island at the time failed to mention the landfill, as if it did not exist.

Maybe that was a way to block it from memory.[159] Fresh Kills was sufficiently distant that it also became somewhat invisible to the people of the other boroughs, with the exception of neighborhoods with transfer stations. The average Staten Islander, however, was reminded daily of the existence of the dump.

Eldon P. Koetter, a sanitary-landfill specialist for Caterpillar Tractor Company, wrote an article for *The American City* in 1956 discussing the pros and cons of landfills. "Over 100 cities will adopt sanitary-landfill methods in 1956, but the net gain in the effective type of garbage disposal remains unknown. We have no way of knowing how many sanitary landfills will revert to nothing more than maintained open dumps in that same period of time. However, we do know that this backward step is constantly taking place." The debate over the value and risk of Fresh Kills going into the 1960s would test Koetter's observation.[160]

Changes in Richmond Borough in the 1960s, especially the completion of the Verrazzano-Narrows Bridge, would change the relationship between Staten Island and mainland New York City by connecting them more unequivocally in space and time. Changes on the island also would draw further attention to the importance—and notoriety—of Fresh Kills Landfill.

CHAPTER 9

THE END OF ISOLATION

A press release from the Triborough Bridge and Tunnel Authority on November 21, 1964, announced that the "new $325,000,000 Verrazano-Narrows Bridge with its Staten Island and Brooklyn approaches, the Staten Island Expressway and the Brooklyn-Queens Expressway," was to open to traffic on that day "after ceremonies in which Federal, State and City Officials will participate."[1]

In 1961 Staten Island celebrated its three hundredth anniversary of the first permanent Dutch settlement, but only now was its sense of isolation from mainland New York City mollified. The new bridge was the first of Gotham's sixty-one such structures to connect Staten Island to the rest of the city.[2] The bridge would do much to change Richmond Borough and its relationship with the rest of New York City. The opening helped create a real estate boom, which substantially changed the island's landscape and made the giant Fresh Kills dump even more anomalous. Larger numbers of people living on Staten Island ultimately meant more voters and more people to protest the landfill operations into the 1960s and beyond.

BUILDING A BRIDGE

Demand for a fixed crossing connecting Staten Island to Brooklyn had arisen as early as 1888. In 1923, a subway tunnel was under construction to join Staten Island to Brooklyn but was scrubbed. The plan reemerged in 1929, 1937, and 1945 but each time went nowhere. The last bridge dedication involving a Staten Island

FIGURE 9.1 First land connection to greater New York.

Source: Courtesy of Jose Mario Lopez, Hines School of Architecture, University of Houston.

delegation before the Verrazzano-Narrows was in 1931, when the Bayonne Bridge (connecting Bayonne, New Jersey, with Staten Island) opened.[3]

The Verrazzano-Narrows Bridge was a huge undertaking. The twelve-lane, double-deck suspension bridge was located on the Narrows (a mile-wide channel at the entrance of New York Harbor) and extended from Fort Hamilton in Brooklyn to Fort Wadsworth in Staten Island. It was designed by the celebrated

Swiss-born engineer Othmar H. Ammann, who planned bridges in the New York City area for over thirty-five years. Ammann expected the structure "to last forever."[4]

The bridge would provide fixed access between Staten Island and the other boroughs of New York City and offer a bypass route avoiding congested Manhattan. The Port of New York Authority first announced the proposal for the bridge in 1955 as part of an extensive program of arterial highways favored by Robert Moses. Groundbreaking for right-of-way clearing began August 13, 1959, with construction of the main bridge commencing January 13, 1960. The undertaking employed ten thousand workers. On November 21, 1964, the upper level was opened to traffic. The Verrazzano-Narrows Bridge, the largest and most expensive bridge of its type until 1981, was nearly three miles long, with its center span reaching 4,260 feet (sixty feet longer than the Golden Gate Bridge).[5]

Aside from the joint agreement between the Port Authority and the Triborough Bridge and Tunnel Authority (TBTA) to construct the bridge, a federal permit was necessary to span the Narrows. The Department of the Army approved a plan to relocate and reconstruct military facilities in Fort Wadsworth and Fort Hamilton.[6] Relocation agents also worked to resettle 2,954 families in Brooklyn and Staten Island neighborhoods displaced by the construction. (Many of those families eventually moved to other homes in their original neighborhoods.) A 1957 TBTA brochure distributed to Staten Island residents assured them that "we have proceeded always on the theory that people are human beings, sensitive, and naturally resentful of being uprooted from streets and neighborhoods in which they have spent a good part of their lives. . . . We take no pleasure in disturbing the lives of people, even temporarily."[7] This was little consolation to those who had to uproot.

Powerful groups on Staten Island saw the bridge as a necessary link to Greater New York City.[8] Writing for the *New York Times*, the budding author Gay Talese acknowledged the engineering achievement but added, "Yet there are people who hate the bridge. To them it is not poetry, merely steel—implacable and unnecessary." Most of the displaced lived in the neighborhood of Bay Ridge, Brooklyn, where eight hundred homes were demolished to make room for the expressway connected to the bridge. It took eighteen months to move the approximately seven thousand people in Bay Ridge. Most residents ultimately learned to live with the bridge, like it or not.[9]

MOSES'S LAST HURRAH

Relocation was not on the minds of speakers at the Verrazzano-Narrows Bridge Dedication on the clear, cold morning of November 11, 1964. Moses, chair of the TBTA, presided over the ceremonies, which began with a ribbon cutting on the Brooklyn side. By this time, the master builder's power was waning. To

Moses's surprise, Governor Nelson A. Rockefeller had accepted his resignation from several positions in 1962. In 1966, his only remaining responsibilities were as the city's arterial highway coordinator and, most importantly, chair of the TBTA. Moses was relieved of that job in 1968 and demoted. Several of his projects and his views on growth were criticized or fell out of fashion.[10]

The celebration of the bridge was Moses's last hurrah. Joining him at the ceremony were dignitaries by the dozen, including Governor Rockefeller (New York), Governor Richard J. Hughes (New Jersey), Mayor Robert F. Wagner Jr., Senator Kenneth B. Keating, Senator Robert F. Kennedy, UN Ambassador Adlai E. Stevenson, Francis Cardinal Spellman, Ammann, among many others.[11] A motorcade of fifty-two black limousines proceeded across the bridge to Staten Island for an hour-long program, followed by the motorcade recrossing the bridge, accompanied by Army, Navy, Coast Guard, and New York City Police Department nautical tributes.[12]

In a commemorative report, Moses waxed eloquent: "For more than four years the fabled Statue of Liberty, lifting her Lamp of Welcome beside the Golden Door, has watched with fascination the spinning of cables of this new Colossus which has dwarfed navies and cargoes of commerce and brought a new dimension to the greatest seaport in the world."[13] At the ceremony Mayor Wagner acknowledged the bridge builders (three workmen had died during construction). He then read a congratulatory letter from President Lyndon B. Johnson:

> I am delighted to extend my congratulations on the dedication of the magnificent Verrazano Narrows Bridge. A structure of breath-taking beauty and superb engineering, this is a most fitting monument to Giovanni de Verrazano. Even as the voyage of this renowned Italian navigator, who in 1524 dropped anchor off Staten Island, this bridge, which proudly bears his name, has opened for America another new vista.

The president also acknowledged how the project was "a brilliant example" of government cooperation on several levels.[14]

Giuseppe Lupis, an Italian government official, gave some brief comments in Italian. The naming of the bridge for the Florentine explorer (credited with discovering New York Harbor) was a proud day for Italian Americans, especially those living on Staten Island. The Italian Historical Society had campaigned for the naming honor despite the opposition of the Staten Island Chamber of Commerce. At the time, approximately one-third of Staten Islanders identified themselves as Italian American.[15]

Albert V. Maniscalco, president of the Borough of Richmond, hit all the notes of Staten Island's expectations about the bridge: "Thanksgiving Day will come to Staten Island twice this year, for those of us who live in the Borough of Richmond feel that today is, also, a day which we should be very thankful. All Staten Island is thankful for the presence of this new pathway to progress." He thanked

FIGURE 9.2 Verrazzano-Narrows Bridge, from Brooklyn, looking toward Staten Island, 1991.

Source: Library of Congress, Prints & Photographs Online Catalog.

the locals for their "patience and co-operation," "the imaginative genius of Robert Moses," Ammann, and "the engineers and the brave men who worked in all kinds of weather conditions." Such enthusiasm harkened backed to Staten Island's excitement over the consolidation of New York City in 1898 and to expectations for its future.[16]

Festive parties broke out on both sides of the bridge after the opening ceremonies. Even in Bay Ridge, Brooklynites hung bunting from buildings and waved flags. In the first hour after opening the bridge to traffic, approximately five thousand cars rolled across (mostly from Staten Island).[17] Beyond the enthusiasm for the structure, few people could imagine what its construction would bring to Staten Island—for the better or for the worse.

STATEN ISLAND OR BUST: THE REMAKING OF A BOROUGH

While the impact of the Verrazzano-Narrows Bridge on Staten Island was great, the structure was not built exclusively or primarily for the borough's benefit. It was foremost an important piece in the completion of the citywide and region-wide transportation network Moses had envisioned. On one of numerous occasions, he stated that the bridge was "a link to the future—a future that holds the promise of at least a substantial part of a solution of the metropolitan traffic problem. . . . It is the most important link in the great highway system stretching from Boston to Washington or, if you please, Maine to Florida."[18]

The bridge offered an opportunity to exploit Staten Island for the benefit of New York City as a whole—as a transportation link but also as a source of land for residence and industry. Moses's (and others') previous plans for new parkland and economic and community development on the West Shore (which had Fresh Kills at its core) fit even more firmly in the growth rush after the Verrazzano-Narrows Bridge opened. An October 3, 1963, issue of *Life* proclaimed: "Surging off the ferry boats with the old commuters is a new wave of urban pioneers, buying up acres of open meadow, building new houses, pushing paved city streets down dusty country roads. The feverish activity is going on 25 minutes from Manhattan on New York's last frontier—Staten Island."[19]

Some residents viewed the real estate fever with anticipation, others with trepidation. Some Staten Islanders were concerned that the mostly white borough would face a "foreign invasion" as minorities poured into their community from other parts of the city. While migration to Staten Island escalated after the bridge opened, the gradual transformation in Staten Island's demographics to include more American-born ethnic minorities and Third World immigrants was not linked to it. Beginning in the 1950s, a small number of African Americans from the South had settled in Richmond Borough. The 1965 Immigration and Nationality Act, which increased visas for nationals especially from Asia, also brought in some immigrants. The inflow of minorities (only amounting to about 5 percent of the island's population in 1970 and 11 percent in 1980) was not really felt until the late 1970s.[20]

At the time of the bridge's opening, however, anticipation of growth and development outweighed unease. Several speakers at the bridge dedication celebrated the new-found "opportunity" now open on Staten Island. Mayor Wagner noted, "By this bridge we open Staten Island as though it were a new continent."[21] President Johnson's statement also included a similar observation: "By connecting Brooklyn and Staten Island, the Verrazano bridge will open the Borough of Richmond to new industrial, commercial, and residential development and will give the entire New York–New Jersey Metropolitan Area a much needed transportation link."[22] Reminiscent of sentiments from Staten Islanders following the 1898 New York City consolidation, Borough President Maniscalco added, "This fine span of transportation will be a spark which will bring to fruition the many potentials of which Staten Island is capable, in the fields of commerce, industry, culture and recreation."[23]

A financial bonanza was on the minds of those expecting large-scale real estate speculation. A Triborough Authority "Fact Sheet" about the bridge made this point bluntly:

> With the opening of the Verrazano-Narrows Bridge, most of Staten Island becomes immediately more accessible to the central core of the City. New residential areas, shopping centers, modern industrial plants, warehouses and other distribution centers could be developed on Staten Island to serve the

central area of New York as well as the rapidly developing sections of Long Island and of New Jersey. Nowhere in the New York–New Jersey Metropolitan Region is there as much vacant and undeveloped land available as now exists in Staten Island.[24]

Richmond Borough was poised to make a transition from a rural and agricultural community with a smattering of towns along its northern shore to a burgeoning commuter region with new housing tracts and shopping centers.[25]

The *Wall Street Journal* also acknowledged the connection between the bridge and a building boom. Staten Island real estate men, it stated, understood that the bridge opening would "trigger a new round of land trading and house hunting, possibly far surpassing anything that had occurred so far."[26] A promotional book published by the Triborough Authority concluded that Brooklyn would benefit most from the bridge, but isolated Staten Island "has the chief asset New York needs most—real estate ready for the development of homes, parks, light industry and commerce."[27]

Change for Staten Island was rapid and notable. The borough sustained a growth rate of 33 percent between 1960 and 1970 (from 222,000 to over 295,000), more than double the increase of the 1950s. During the 1960s, Manhattan's population dropped by 9.4 percent, and Queens and Brooklyn also lost population.[28] In 1964, groundbreaking for the Staten Island Mall began on the site of the former Staten Island Airport. The shopping mecca opened in August 1973 (in close proximity to Fresh Kills Landfill), eventually containing two hundred stores and kiosks.[29]

Sometime before and soon after the construction of the Verrazzano-Narrows Bridge, city planners had designs on transforming Richmond Borough. Although new homes were important, the anticipation of new jobs drove most discussions for growth. The Department of City Planning contracted Lockwood Kessler & Bartlett, Inc., to conduct a study about a proposed Staten Island Industrial Park. Disquiet was widespread that industry was fleeing New York City, and there was a need to attract new industry to replace it. The 1962 report stated, "We cannot over-emphasize the importance of making the Industrial Park an intrinsic part of Staten Island, with the purposeful stressing of the advantages of a suburban community in New York City and the attraction this represents to the highest caliber of modern industry." The site selected was in the Mariner's Harbor section on the northwestern side of the island, on vacant, industrially zoned land spanning the proposed West Shore Expressway.[30]

The Lockwood study affirmed that outmigration of industry from New York City to suburban locations such as Nassau and Suffolk counties accounted for the loss of about six thousand manufacturing jobs each year; from January 1959 to February 1962, 119 industrial plants employing 6,900 people had relocated to those two Long Island counties from New York City. Among the reasons for the move was the need for more "horizontal space," something Staten Island could provide. New development on the island, accelerating its growth as a new middle-class residential haven, was just the thing to "reverse the trend of industrial outmigration from the city and to attract new industries."[31] More people meant more employees.[32]

A 1966 Planning Commission report envisioned an even more dramatic effort at remaking the island. It stated emphatically: "On the basis of all the data and trends we have analyzed, we believe that Staten Island's primary function in the next decade or so will be to provide land for new housing and both local and regional recreational and institutional uses." Greater industrial and commercial activity was meant to follow a population rise on the island.[33]

An immediate goal was to develop sound development controls in an environment that had a lengthy land-use history but also special natural features. Like Frederick Law Olmsted's design of Central Park, the commission wanted to "'undo' before we can 'do'—we must re-build even while we build" in Staten Island. (Olmsted had removed most natural features in Central Park before creating his own vision of a "natural" space.) It claimed not wanting to "erase" Staten Island's past and present, but their goals excluded major efforts at preservation.

The commission argued for orderly planning, mapping, schedules, and standards and avowed the need to keep a good portion of city-owned land out of the hands of private developers. But in seeking to amend zoning ordinances, promoting the development of South Richmond, modernizing transit, attracting industry, expanding ports, rehabilitating the Stapleton piers, and encouraging recreational sites, it demonstrated how comprehensive the plan would be in transforming the island. Some of the challenges (beach erosion, hurricane protection, sewage treatment and drainage, water- and air-pollution controls) needed to be addressed, but where the line would be drawn between aggressive growth and environmental protection was unclear, especially when concepts such as "sustainability" did not exist at the time.[34]

In the 1970s Richmond was among the fastest-growing counties in the state. Population increased by 7.5 percent between 1960 and 1970 (from 221,911 to 295,443) and by 8.4 percent from 1970 to 1980 (295,443 to 352,029).[35] Land values that stood at $1,500 per undeveloped acre in 1960 jumped to $20,000 in 1964 and to $80,000 in 1971. Consumer prices also rose, along with housing and service prices.[36]

By the mid-1970s, the major arterial connections across the island were substantially complete. The four-lane West Shore Expressway, between the Outerbridge Crossing to the south and the Staten Island Expressway and Goethals Bridge to the north, ran through the center of the Fresh Kills project site.[37] In 1969, the City Planning Commission had recommended cluster (or planned unit development) projects on Staten Island that would give zoning bonuses to builders for carefully planned subdivisions.

A few days later the mayor announced the start of a community of two thousand townhouses to be built in Arden Heights section of Staten Island, which would be the largest single development ever built within the city limits. In 1970, the Rouse Company of Columbia City, Maryland, the same company that owned the Staten Island Mall, proposed creating a $6.5 billion "new city within a city" on land in the southwestern section of Staten Island. The plan called for twelve new medium-density communities with a combined population of 300,000 to 450,000 within twenty years. This would more than double the existing population of the borough. The South Richmond area was viewed by some as the city's "last land frontier."[38]

Not all of the development projects saw completion. In 1971, State Senator John J. Marchi introduced legislation in Albany for the South Richmond Development Corporation. It called for the city to use its power of eminent domain to buy private property and transfer it to the Rouse Company. The Conservative Party and several property owners opposed the proposal, and the legislation ultimately was defeated. Several locals had voiced concern at 1972 and 1973 hearings about the city's condemnation power threatening their property and about the population density of the twelve projected communities. The Waterfront Watch, a Staten Island environmentalist group, also opposed the South Richmond development. It claimed that the plans to include Great Kills Park and a proposed 3,300-acre landfill off the South Shore would adversely affect wildlife and waterfront activities in the area. In May 1973 the mayor created a task force to devise "a new South Richmond Plan" for "by far the largest undeveloped territory in New York City." A public project languished, and although plans for private development of South Richmond remained under discussion, they lacked a clear design strategy, attention to environmental concerns, or even a master plan.[39]

While marshes had become less available for fill land on Staten Island, a 1966 Planning Commission report considered the "Forgotten Borough" as "the last major land resource within the city."[40] An estimate in 1963 indicated that 14,000 of its 36,000 acres remained to be developed. Staten Island's post-bridge growth, rising costs of housing and real estate, and the large amounts of city-owned land posed threats to the orderly planning process, however, and increased the need for haste in developing a comprehensive plan. Such a plan did not take sufficient account of the needs of locals or of the ways in which other encroachments like Fresh Kills Landfill (amazingly not addressed at all in the 1966 report) influenced future land uses. The final paragraph of the report noted rather insincerely:

> In the last analysis it is the citizen who must set the standards we seek. He is consumer, critic, judge, and jury. If he settles for mediocrity, he will get it. It is the responsibility of this City to remind every citizen that mediocrity is not

part of our heritage; that the City must represent all of the residents of Staten Island as well as the hundreds of thousands of Staten Islanders still to come. The commitment set forth here is not only to satisfy local demands, but to exceed them.[41]

This was the view of an outsider looking in.

Staten Islanders experienced mixed feelings about the remaking of their borough, but migrants continued to pour into Staten Island, which became the newest bedroom community for the city by the late 1960s. Services, such as Richmond Borough's own sanitation needs, had to expand to meet new demand.[42] The land to the east and southeast of Fresh Kills was becoming increasingly suburbanized, especially with the construction of Staten Island Mall in close proximity. Some truck farms remained active, however, and some people still lived and worked in the marshes.[43]

Laurence O'Donnell in the *Wall Street Journal* discussed Staten Island's "dizzy land speculation and growing pains" during the five years leading up to the bridge opening. He also noted that "the small town character of Staten Island is disappearing." Builders subdivided farms and acquired contracts on vacant land to erect small houses "as fast as they can" at inflated prices.[44] In the April 1969 issue of *Natural History*, the Scottish landscape architect Ian McHarg confessed regret that with hindsight Staten Island ("a special place") "would have ranked high among the splendid resources for the city population." Yet, he believed, all was not lost, "even though the Verrazano-Narrows bridge has opened the floodgates to urban development, some splendid residues still remain."[45] Robert Moses and S. Sloan Colt of the Port of New York Authority stated emphatically that Staten Island's isolation, "partly geographical and partly psychological, will disappear the day the Verrazano-Narrows Bridge is opened to traffic." Staten Islanders, they believed, had "a magnificent opportunity to preserve this one remaining unspoiled borough and make it a model of sound and healthy urban development." But economic growth versus preservation was becoming a tenuous balancing act for city planners.[46]

A 1967 report of a mayor's task force charged with evaluating the future design of New York City stated, "Ten years ago an acre sold for $9,000 [on Staten Island]. Today it sells for $40,000, an increase of more than a quarter billion dollars in money value which is waiting to be made real environmentally by the conversion of the land into pleasant places for living, working, and recreation." The report added,

> But at present there are few visible signs that this will happen. . . . Staten Island, even today has too many small neighborhoods of varying character, abutting, contradicting, confusing, beginning and ending so abruptly that they add up to a general air of steaminess. The pleasant places among them point up the

grotesque error—a failure to design the island's land use practically and sensibly beyond the ruthless rules of real estate, well before the new Verrazano Bridge sent land speculation through the ceiling.[47]

Some locals remembered the island nostalgically in those years before the bridge. Flora Reigada grew up on Vanderbilt Avenue (North Shore) in the 1950s. "Richmond Avenue was just a country road with farms along it," she recalled. "My aunt used to go there to get fresh eggs. It was like taking a drive in the country." "There was a security about life back then," she added. "My grandparents would leave their doors unlocked all day while I ran in and out. Moms used to leave their babies sleeping in carriages outside the supermarket while they shopped inside." Frank Scalero simply stated that Staten Island "really was a good place to grow up."[48] Jamie Santoro from Stapleton declared, "Staten Island was 'the forgotten borough.' It really was the best of both worlds. My family was working in the city and I had a tree house and a fort in the woods." In 1974, John E. Rossi and his wife moved from Brooklyn to Travis and purchased a "handyman special" on Victory Boulevard. Across the street he encountered an older man named Rudolf Nagyvathy. "Doc" regaled Rossi about the town as it existed in 1910 and showed him photos of his old house in the woods, which had been his parents' place before then. He said that the house did not get indoor plumbing until 1968. Doc expressed real concern about health, population density, and what was happening to the local environment.[49]

A WEED AMONG THE ROSES

The boundary between the landfill and nearby residential communities was shrinking rather than widening, which may account for the increasing number of complaints arising over the gigantic waste facility. Fresh Kills Landfill was wedged between, on the one hand, concerns about the destruction of the local environment and the degrading of proximate communities and, on the other, the expectation of turning the disposal site into more reclaimed land for economic development. As one local stated, "If [the landfill] keeps growing it will probably block the sun like the pyramids." How could Fresh Kills not be wrapped up in the island's post-bridge future?

Fresh Kills seemed even more incongruous a place after the building of the bridge, especially given the glowing optimism of those looking to future growth prospects on the island. Yet as it became the chief depository of New York City's solid waste, the landfill continued to accept more and more refuse from Manhattan and other boroughs. Operations at Fresh Kills in the 1960s were managed in three locations: Two marine unloading sites, Plant #1 and Plant #2, and the Brookfield Avenue dumpsite known as Section 10/11 (opened in 1966 east of

Richmond Avenue and used exclusively by trucks).[50] Like Fresh Kills, Brookfield had been a salt marsh.[51] Because it was closer to the source of waste, Plant #1 received more barges than Plant #2 and accepted about 10 percent more refuse.[52]

In 1961, the total landfill area spanned 1,284 acres, and in 1966, 1,584 acres.[53] From June 1964 to June 1965, Fresh Kills was accepting about one-third of all of the refuse that the New York City Department of Sanitation (DSNY) collected. Tonnage of refuse grew by 9.4 percent in the mid-1960s to 2.2 million (of 6.2 million total tons).[54] Since Fresh Kills opened, more than 36 million tons of refuse (about 125 million cubic yards) had been dumped on Staten Island.[55]

The landfill site was an increasingly complex human-made landscape incorporating not only refuse mounds but also roads and lighting for ease of access, firefighting equipment, fencing to control windblown waste materials and litter, waterfront dikes to contain floating waste, and grading and ditches to prevent water from stagnating. Athey Wagons continued to dump refuse in planned strips, which were then bulldozed, sprayed with disinfectant, compacted, and covered with earth fill.[56] Techniques for constructing sanitary landfills in the 1940s, when the facility was opened, represented a step forward from previous haphazard land dumps. They lacked, however, the sophistication and environmental soundness of landfills in the 1990s, which utilized liners, methane gas–capture equipment, and other modern technologies. Despite the effort to make Fresh Kills an effective disposal facility, it would not be long before its shortcomings became apparent.[57]

Items that now were being dumped in the landfill—plastics, chemicals, new types of packaging—also created a more diverse refuse stew with which to contend. Fire debris was allowed at all New York City dumpsites for the first time in 1963. Clearwaters, Inc., was granted permission to dump oil from a spill in Arthur Kill into Brookfield in 1970; earlier, such waste normally had been excluded from landfills. To discourage illegal dumping along roadsides and in vacant lots, Borough President Robert T. Connor asked DSNY officials to consider allowing small builders and some retailers in the borough to dump for free. The Sanitation Department clashed with the Highway Department for dumping old pavement (broken stone and asphalt) in landfills instead of using it to reinforce storm dikes or to repair streets.[58]

The Fresh Kills facility was constantly changing.[59] Expectations among developers and planners to use the filled areas as soon as possible ran high, to take advantage of Staten Island's new opportunity to expand its industrial and commercial prowess.[60] Such goals ran up against the desire of the DSNY to make use of every square inch of the site for disposal until it was at capacity. An on-again, off-again plan by the Department of Marine and Aviation to build an airport for small private planes and light cargo aircraft on 120 acres at Fresh Kills was scrapped, among other reasons because of air traffic from nearby

Newark Airport. The airport idea had been part of the effort to make the essentially undeveloped western shore of Staten Island into an industrial park.

The DSNY balked at any nondisposal-related construction in the landfill area. Some officials in the department feared that building the airport would shorten the useful life of the depository by five years, creating a serious disposal problem for New York City. The old Staten Island Airport had been closed some time before the new initiative, and the land was converted into a shopping center. In late January 1970, Robert Gill of the Richmond Pilots Association and the Staten Island Flyers Association conceded that "I hate to say it, but it looks like we're finished." Last-gasp efforts kept the project alive for a time, but the will of the DSNY won out.[61]

At capital-budget hearings in October 1963, Sanitation Commissioner Frank J. Lucia had warned (repeating a common refrain, but one often varying the timeframe) that in ten or twelve years the city might lack the necessary disposal space for the increasing amount of refuse collected each year. "We can't take it to sea," he argued. "We can't truck it out of the city, nor, in the foreseeable future, will we be able to load it into capsules and shoot it into space." He was unwilling to accept the notion that "sanitation can wait," that there were other, more pressing needs of the city. "New York is running out of time," he concluded, "and it is running out of space." The solution, he believed, was, once again, an intensive program of incinerator construction.[62]

Guiding the decisions on how best to use Fresh Kills was the fact that approximately two-thirds of the existing landfill area (1,800 of 2,700 acres) already had been reclaimed by 1965, with estimates of ten to fifteen years before all disposal operations ended there.[63] City officials considered other potential landfill sites but did not develop a concrete plan. Instead, they focused on mounding techniques to elevate the refuse mounds above the twenty-to-forty-foot height established in 1959 or 1960.[64] Aside from saving landfill capacity, vertical filling limited encroachment of the landfill on neighboring property. This approach to saving space resulted in the periodic elevation of mound height, which eventually reached an astounding five hundred feet by the 1980s.[65] The department also explored the use of compaction plants to reduce the volume of the refuse.[66]

Aside from mounding, the DSNY's efforts to improve the site for more efficient disposal included long-developing plans to widen and deepen the Little Fresh Kill to make it navigable for tugboats and barges. The channel from the Arthur Kill unloading plants would be reduced in length by almost half a mile in reconstructing the waterway flowing through Little Fresh Kill, thus shortening overall disposal runs.[67]

By the end of the decade, calls for improving disposal practices (and confronting Fresh Kills' shrinking capacity) had changed little. At a November 1969 City Planning Meeting, a spokesperson for the DSNY asserted, "The Fresh Kills area will be needed for fill until its capacity is reached, and then we will have to look elsewhere for other methods."[68] Before then, other options were being explored. A story in a December 1967 edition of the *Staten Island Advance* declared that discussions were underway between city officials and a private Philadelphia company to ship 10 percent of the city's refuse to abandoned strip mines in east central Pennsylvania. But a contract was never signed.[69]

Also in 1967, the mayor's office authorized a survey to examine new disposal options. One proposal called for incinerator ash to be used to fill an area between the South Beach–Midland Beach shoreline in Staten Island and Hoffman and Swinbourne Islands. As early as the 1950s the city had hoped to acquire title to state-owned underwater land surrounding the islands as a possible dumpsite for clean fill, but the idea never got off the drawing board. State Senator Marchi was wary about the proposal, and Borough President Connor suggested that "it would be an affront to Staten Island to have a garbage disposal area at both its front and back doorsteps."[70] The DSNY also considered reopening the issue of dumping at Jamaica Bay.[71]

The largest-scale proposal for supplementing Fresh Kills Landfill came in 1969. A study by the engineer consultants Foster D. Snell, Inc., paid for by the state but commissioned by the city, called for a proposed new site on vacant meadowland along the West Shore of Staten Island between Travis and the Staten Island Expressway. The firm recommended that the city must have an "additional massive landfill" to replace Fresh Kills upon its closure (about 1975, the DSNY was now saying). It would be larger than the existing landfill. The *Staten Island Advance* reported on November 14, 1969: "The proposed new dumping area for the city's refuse on Staten Island's West Shore would embrace approximately 3,000 acres, including the 750-acre site of the proposed industrial park, Prall's [sometimes referred to as Prall or Pralls] Island in the Arthur Kill and Prall's River between that island and the Chelsea-Bloomfield mainland."[72]

The DSNY saw this area as a "top priority" for future dumping and asked for funds to purchase the land in the next capital budget.[73] While the plan was never enacted (partly because it would interfere with the development of the proposed industrial park), it indicated that the DSNY and city planners were concerned about future disposal prospects in New York City but also willing to keep Staten Island as its focal point.[74] At a December 1969 press conference, Acting Director of City Planning Edward Robin declared, "A crisis is expected in the mid-1970s when planners predict the city's present means of disposing of garbage will no longer be effective."[75]

MAYOR LINDSAY FACES COLLECTION AND DISPOSAL TROUBLES

Under the new mayor John V. Lindsay, coming to grips with an inventive solution to the chronic refuse-disposal problem was only one of the pressing sanitation issues surfacing in the tense political environment of the mid-1960s. The refuse issue, writ large, would dog the new mayor throughout his time in office. Disarray in the DSNY, its funding and labor woes, the declining quality of collection, and the inequity of service delivery had immediate practical and political consequences. As vital as a good disposal program had to be, collection, street cleaning, and snow removal touched most citizens more immediately and more directly than disposal and constituted the lion's share of the DSNY budget (about 80 percent) and time.[76]

Several of the DSNY's problems were chronic (worker shortages, budget shortfalls, inefficient and ineffective management)—or at least recurring—and preceded the 1960s. Some of its difficulties were exacerbated by the turbulence of the times during Lindsay's mayoralty, such as racial unrest, a growing fiscal crisis, and heightened political tensions. Lindsay was clearly a different kind of mayor for New York City. His obituary described him as "the debonair political irregular," "a maverick," "a calm figure of civic dignity," and "an athletic, buoyant man most of his life."[77] The son of a banker, Lindsay was born in the city in 1921 (he was a twin of five children), educated at Yale, trained as a lawyer, and served in the Navy in World War II. By today's standards he was a political anomaly—a liberal Republican who supported civil rights and LBJ's Great Society. Lindsay had been a congressman from Manhattan's East Side "Silk Stocking" district (1959–1965) when he stood for mayor in 1965. Running on a fusion ticket (Republican-Liberal), he defeated the Democrat Abraham Beame (the eventual choice of a party in turmoil) and the Conservative Party's William F. Buckley (the noted author and commentator).

During Lindsay's two terms, New York City was beset with a transit-worker shutdown; plagued with strikes and slowdowns among teachers, police officers, firefighters, and sanitation workers; public housing problems; the near bankruptcy of the city; rising crime rates; and racial strife. Many of his efforts at reform, especially with the city's poor, failed to materialize, and he was constantly criticized for style over substance or for being a "limousine liberal." A press aide to Mayor Wagner chided, "John Lindsay was the best mayor New York ever had before he took office."[78] Supporters countered by emphasizing the importance of Lindsay's stances on racial equality and economic equity and reminded New Yorkers that the problems of the city and the country were too big for any one man to create or to solve.[79]

THE 1968 GARBAGE STRIKE

What became known as the New York City sanitation crisis in the late 1960s and early 1970s further crippled the department, raised the frustration of low-income neighborhoods over poor service, and led to public protests and a notorious sanitation-worker strike. Neighborhood discontent with DSNY service was increasing the pressure on the department and the mayor. Some neighborhoods claimed that they were receiving inferior or less frequent collection than others. In general, the quality of service had long been deteriorating across the city. Collection and street-sweeping schedules were erratic, and the DSNY's administration was chaotic.[80]

In February 1968 members of the Tompkins Square Community Center on the Lower East Side rented two trucks in an attempt to find a place to dump uncollected refuse from their neighborhood. Getting no help from the mayor's office, the DSNY, or the Parks Department, they emptied the material on East Tenth Street, which was followed by small garbage fires and local disturbances. In January 1969, Queens residents and those in other outlying boroughs placed anti-Lindsay signs in their neighborhoods in response to a major snowstorm (the infamous "Lindsay snowstorm"), which buried the city in nineteen inches of snow. Streets were closed for days because of the breakdown or disrepair of snow-removal equipment and the barring of a private contractor because of irregularities in its business. Other incidents in other neighborhoods, not only in Manhattan but also in Queens and Brooklyn, continued into 1969 and 1970 over issues of mounting garbage piles, poor or no snow removal, rat infestations, and chronically dirty streets. Many of these conditions had been extant for years.[81]

The Lindsay administration was derided from all sides for the breakdown in the DSNY's performance, despite the fact that it had set out in earnest to modernize the city's governmental structure. The mayor had promised in his 1965 campaign to restore regular city services in neglected low-income minority neighborhoods and wanted to reorganize and rebuild the Sanitation Department.[82]

In February 1966, soon after Lindsay had entered office, the administration set up a "Crash Clean-Up Campaign" and sent a horde of sanitation, fire, and building inspectors to poor neighborhoods in East New York, Brooklyn. Its effects were fleeting, however, and failed to address the underlying problems of poor sanitation in the area. Likewise, a six-day-a-week collection program in the South Bronx also failed to improve conditions there. Turning inward, the mayor's office ordered the Department of Investigation to evaluate DSNY practices. Private cartmen were found to be using public disposal facilities without paying the appropriate fee; instead they paid off department employees in cash. Superiors were forcing sanitation workers to pay bribes for promotions. These actions led to personnel changes and the mayor's call for his promised reorganization of the department.[83]

While Lindsay administration officials viewed their direct action as in the best interest of the city, sanitation workers were concerned that they would take the fall for discontent in minority neighborhoods such as East New York and failed programs elsewhere. They also resented forceful actions taken against public-employee unions like the Transport Workers Union and the Patrolmen's Benevolent Association.[84] With the old labor contract expiring on June 30, 1967, and the DSNY currently without a commissioner in place, the Uniformed Sanitationmen's Association (USA) recognized this as a propitious time to push for higher pay and better working conditions. They wanted to achieve parity in salaries and benefits with other uniformed municipal employees (mainly police and firefighters).[85]

Some citizens believed that sanitation workers were well compensated, especially compared to other city employees such as teachers.[86] But this response did not take into account the potential hazards associated with trucks and heavy equipment, injuries from carrying heavy loads, and potential health risks associated with exposure to toxic discards. Nor did it account for the lower status afforded sanitary workers compared to firefighters and the police, despite the essential nature of their services.

The growing distrust between the Lindsay government and the sanitation workers was a central cog in the USA strike in 1968. When the city's contract with the union ended on June 30, 1967, mediators recommended a new package that included a higher wage increase than the police or firefighters had received, plus other benefits. John J. DeLury had been the head of the USA for over thirty-seven years. He was brash but compelling and a battler for his constituency. Self-educated, DeLury had left Wall Street during the Great Depression and became a sanitation worker. As he often said, "We may pick up garbage, but we are not garbage."[87]

DeLury built the USA into a strong union by finessing rivals and concentrating on tangible concerns—wages, working conditions, and pensions. "I couldn't tolerate the condition under which honest men in government were being used by politicians for their own ends and at the workers' expense," he later stated.[88] Between 1958 and 1964 he tried to improve the situation for the sanitation workers by pressuring not only the mayor but also the legislature and governor. The election of Lindsay, however, ended the good relationship DeLury had with his predecessor, Robert Wagner Jr.[89]

On January 30 DeLury accepted a contract with a retroactive pay increase of $400 per year for each man, but the shop stewards of the union rejected it, wanting him to bargain more aggressively. DeLury then called for a work slowdown, but it had little effect.[90] On a freezing morning on Friday, February 2, the USA held a rally in the park in front of City Hall. DeLury believed that the threat of a strike, not a strike itself, was the union's best weapon to secure its demands. He was confident that he could cut a deal with the mayor without taking to the streets. Possibly DeLury opposed a strike at that time because summer, not winter, would heighten the stench of heaps of putrefying garbage to much greater effect. And a strike in any season could be a volatile event not only for the city but also for the internal workings of the union. DeLury wanted to avoid members bickering and tussling among themselves over the terms of a new contract.

Failing to get the mediator to raise the city's original offer to $450 in annual wages, as the union demanded, DeLury simply presented the mediator's original offer to the rank and file (a $400 annual wage increase and some other improvements in benefits). He was not prepared for the reaction from his audience of seven thousand disgruntled USA members. Much to his surprise he was booed and shouted down, and someone threw an egg. This action was not simply about money. The workers—from old immigrant stock (many Italian Americans) mixed with a younger generation of sanitation men—were in no mood for what they viewed as outright disrespect. They were upset with city leaders (especially a mayor from the elite class) and with increasingly bad working conditions. They hated being treated as second-class citizens and derided as "garbagemen." And they were offended by what they perceived as a preference shown for minority groups and neighborhoods. DeLury had little choice but to press the union's demands, force the administration's hand, and put his people on strike.[91] Coming out of City Hall, DeLury addressed the workers: "The Mayor has turned us down! No contract—no work! He says 'No, no, no!' So we say—" and the "sanmen" responded, "Go! Go! Go!"[92]

On February 2, the first day of the strike, 11,000 tons of garbage lay on the street uncollected. Then 22,000 tons on the second day, then 33,000 tons on the third. The health commissioner declared a health emergency for the city, and fires started in uncollected trash. On February 4 someone fired a shotgun into the living room of Sanitation Foreman Louis DeStefano, which shattered a window, but no one was hit. When DeStefano arrived at work the next day, his phone rang, and when he answered it a menacing voice said, "You bastard. You're back. We'll put a bomb under your chair."[93] Despite obtaining an injunction, city officials faced a continuation of the strike. On February 6, at the hearings on the injunction, State Supreme Court Justice Saul S. Streit denounced the workers stating, "I say it's not really a strike. It's blackmail; it's extortion. It's an illegal strike to the detriment of the public (based on the Taylor Law passed the previous year)—eight million men, women and children."[94] DeLury was sentenced to fifteen days in jail and levied a $250 fine for defying a court order enjoining the strike. About the prospect of going to jail, he stated, "This is America? My freedom is restrained but my fortitude is not."[95]

Other city workers ignored the mayor's order to replace the sanmen in manning trucks and picking up the trash. Governor Rockefeller balked at Lindsay's request to send in the National Guard to break the strike and to collect the refuse, fearing possible violence and furthering his own agenda against Lindsay, a political rival.[96] The governor, however, arranged for the release of DeLury and helped restart the negotiations, thus gaining USA support.

Ultimately the mayor fought off Rockefeller's attempt to take over the DSNY and held firm on his contract offer. The sides ultimately agreed to binding arbitration, and the USA accepted the new terms, which bumped up the original offer to $425 and promised future arbitration. Public pressure was too great to turn the deal down. As a result, both sides claimed victory. The strike had lasted nine days (February 2–10, 1968); one hundred thousand tons of refuse in all went uncollected.[97]

Acting Sanitation Commissioner Maurice M. Feldman reported on February 16 that operations were close to normal, with five thousand men working days and 1,800 on nights. Lindsay gained a good deal of public approval for his hard-line position. Two weeks after the strike, a *New York Times* poll found that 61 percent of New York City residents supported Lindsay's stance. For their part, sanitation workers came closer to achieving some parity with police and firefighters, despite the latter groups' displeasure over the contract. Some believed that in the long run sanmen had gained increased stature alongside other uniformed workers in the city.[98] But the heaps of uncollected garbage also made the union and the DSNY pariahs among citizens of the city. As an unsympathetic Roger Kahn, editor-at-large for the *Saturday Evening Post*, put it, "As the garbage strike describes a mayor's function, it is to channel forces rather than defy them. We have to work with unions, racists, right-wingers and all the rest, if we want our cities to survive. It has now become a price of American politics occasionally to address a garbageman as 'sir.'"[99] Other workers connected with New York City sanitation were not placated by the USA settlement. A DSNY unloading plant in Travis and eleven city incinerators were shut down in early November because of a strike by stationary firefighters. It was the second strike there in two weeks; it was finally resolved on November 15. In addition, congested and poor areas such as Bedford-Stuyvesant, Harlem, and the Lower East Side still faced severe refuse problems.[100]

Ironically, the USA was the first union to support Lindsay's reelection campaign, not because they liked him but because, as DeLury stated, "they're going to vote where the bread and butter is. In terms of dollars and fringe benefits, Lindsay gave them more between '66 and '68 than anybody before him." DeLury even campaigned for Lindsay in his presidential bid. The USA endorsed Rockefeller for his reelection in 1970, likely because of his efforts to end the 1968 strike peacefully. Strange bedfellows all around.[101]

The 1968 garbage strike was one of the most dramatic and memorable events that the Lindsay administration faced. By this time, Fresh Kills was a permanent fixture in Staten Island, with no end date in sight. In the back of the minds of several city officials (and others) was a gnawing concern about how long dumping space would last at the Richmond landfill and what might come next for Gotham's waste. The hope that incinerators ultimately would prevail as the city's answer to refuse disposal had not died, even though evidence pointed to the contrary. Concerns about disposal lingered while the DSNY coped with a score of other issues—internal organization, economic woes, politics, snow removal, and the unwieldy task of keeping the streets clean. A new role for the federal government, as potential arbiter in departmental practices and regulator of environmental impact, also loomed large.

At the time, the city was generating about 24,000 tons of refuse each day, which was increasing at a rate of 4 percent per year (more than twice the national average). Yet in 1963–1964, the department's budget was declining.[102] Between 1961 and 1976, sanitation ranked last among the city's major functions (police, fire, education, welfare, hospitals, higher education, and sanitation) in increased expenditures and last in tax-levy expenditures. Between 1974 and 1984 reductions in its workforce exceeded 20 percent.[103] But for the moment, how to deal with the variety of refuse challenges in the postincinerator and post-strike city produced sufficient worry.

On Staten Island, the completion of the Verrazzano-Narrows Bridge, in part at least, brought newfound economic vitality to the various communities there. And Fresh Kills Landfill became more visible as a sign of local grievances against the core city—an infringement that islanders could increasingly get behind.

CHAPTER 10

AN ENVIRONMENTAL TURN

The ceaseless production of refuse in New York City and the need to deal with it as quickly as possible made implementing a workable plan for disposal extraordinarily difficult. In 1968 the Department of Sanitation (DSNY) collected 6.8 million tons of solid waste. Of that amount, 2.3 million tons were incinerated, 2.1 million tons found its way to truck landfills in various boroughs, and 2.4 million tons went to Fresh Kills.[1]

In the March 1969 issue of *New York*, the journalist Paul Wilkes stated,

> Every day it comes into the city. . . . The things we eat, the things we use to keep armpits dry, breath fresh, tensions away; the things we sleep on, sit on. It comes into the city fresh and colorful, ready to be merchandised, consumed or used for a short time. Paper bikinis, one-way bottles. Planned obsolescence.
>
> Every day some of the city's waste matter must *leave*. But how the 19,000 tons [this number varied on the telling] of trash that New York generates each day are disposed of, few people know or care about. While goods are brought in through intricate, meticulously paced networks, our refuse is picked up and disposed of through the most primitive and fragile of systems.[2]

Consumption and disposal of goods on a grand scale were part of the city's daily life, and criticism of its Department of Sanitation went with the territory. But in the late 1960s, the disposal woes in New York City were immensely complicated by the upsurge of the environmental movement. Concern and interest in a wide variety of pollution, land-use, and health issues now found a national stage. The influence of modern environmentalism could be felt all the way down to the local level, and certainly in dealing with Fresh Kills.

A TURNING POINT FOR THE DSNY

Throughout its existence, the DSNY and its predecessor agencies had engaged in community activities and been governed by local and state authorities. A new player arrived on the scene in the mid-1960s, the federal government, whose influence over the work of the department grew steadily from that time forward. There always had been the need to seek federal approval for harbor activities related to dumping, which infringed on federal lands or waters. New laws did not establish federal control over local refuse management, but new regulations made clear that many solid waste problems extended beyond the city limits and that a new voice would now be weighing in on how to address them. Counties, regional authorities, other states, special districts, and private agglomerates further complicated the management process.[3]

In a message on conservation and restoration of national beauty in 1965, President Lyndon B. Johnson called for "better solutions to the disposal of solid waste" and recommended federal legislation and research-and-development funds to assist state governments in creating comprehensive disposal programs. Congress then passed the Solid Waste Disposal Act as Title 2 of the 1965 amendments to the Clean Air Act. The law was the first piece of legislation to involve the federal government in local refuse management. Its primary objective was to "initiate and accelerate" a national research-and-development program and to provide technical and financial assistance to state and local governments and interstate agencies in the "planning, development, and conduct" of disposal programs.

The act also was meant to stimulate the promulgation of guidelines for collection, transportation, separation, recovery, and disposal of solid wastes. The focus would be on demonstration projects to develop new methods of collection, storage, processing, disposal, and reduction of unsalvageable wastes. By 1967 the federal government had only awarded a modest $9 million on various projects. While Washington moved in the direction of more involvement on the local level, there was little reliable data on refuse to guide it. As a result the National Survey of Community Solid Waste Practices was undertaken in 1968, the first truly national study of its kind in the twentieth century.[4] The effects of the new act and the survey were not immediately felt in New York City, but federal interest in solid waste management would lead to a new set of environmental regulations for Fresh Kills beginning in the 1970s.

Although new federal regulations were on the horizon for the DSNY, sanitation managers found themselves dealing primarily with daily operations, short-term collection and disposal needs, political sniping, customer complaints—and crisis management. The commissioner's office also faced rapid and recurring turnover beginning in 1966. From March 1961 until the end of 1965, the department enjoyed relative leadership stability with Frank J. Lucia, who had replaced Paul Screvane. Appointed by Mayor Robert Wagner Jr., Lucia

254 ■ LIVING WITH AND SURVIVING THE LANDFILL

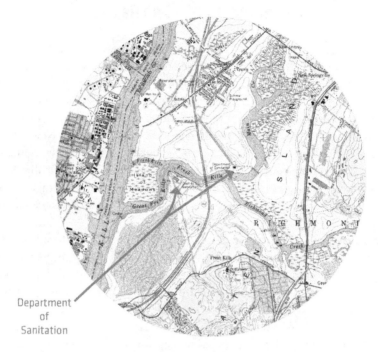

FIGURE 10.1 Fresh Kills topography and ecology, 1966.

Source: Courtesy of Jose Mario Lopez, Hines School of Architecture, University of Houston.

was Screvane's chief of staff. Political observers believed that Lucia's appointment and the elevation of Screvane to deputy mayor were a play for Italian American support, since the mayor was trying to unseat Tammany's Carmine DeSapio as the leader of the Democrats in Manhattan.[5] Lucia was a Brooklyn native, Navy veteran, and career city servant. He joined the DSNY in 1938 and acted in a variety of capacities before assuming his position under Screvane. With the election of Mayor John V. Lindsay in 1965, Lucia retired.[6]

For his first commissioner of sanitation Lindsay named the outgoing Bronx borough president Joseph F. Periconi. He resigned before a rash of payoff scandals rocked the department, although he was not involved in them.[7] Periconi was replaced by Samuel J. Kearing in November 1966. From November of the following year until the end of his mayoralty, Lindsay named four more commissioners or acting commissioners. During his two terms as mayor, nine men in all served as sanitation commissioner.[8] The rapid turnover of commissioners was likely because appointees failed to deal effectively with events such as the "Lindsay snowstorm" in 1969 or other fiascos, which called into question the mayor's leadership. But it also was related to the fact that the Lindsay administration had experimented—not always successfully—with several ways to remake and modernize the local government, including a variety of reorganizations of departments including the DSNY.[9]

One program, which combined elements of departmental reorganization and community involvement, focused on the geographic administrative decentralization of the DSNY. It was in response to citizen demand for improved services, which was one of the issues at the heart of the 1968 strike. The intention was to delegate "command authority to lower organization levels, and [invite] greater community participation in identifying local problems."[10] In 1971, the mayor instituted a pilot project in nine districts, which included programs in citizen involvement on issues related to sanitation.[11] It was a fine idea, but it never developed into a full-fledged citywide program.

AIR POLLUTION, COLLECTION WOES, AND RATS IN THE LANDFILL

The problem of air pollution had blunted the DSNY's lingering interest in incineration well before federal antipollution legislation became more pervasive. Until the 1960s, knowledge of the functioning of refuse furnaces was primarily in the hands of a few pioneer incinerator builders and some engineers. The increasing recognition of environmental problems linked to burning waste put into question what experts had been saying for decades about the safety of the technology. A 1965 study, for example, questioned whether incinerator residue from a public health standpoint was suitable for fill.[12] In the early 1960s New York City probably had the largest network of municipal and residential incinerators anywhere (thirty-six furnaces in eleven incineration plants and about 17,000 apartment-house incinerators).[13]

Lindsay appointed a task force on air pollution in 1966, which issued warnings about the health consequences of airborne toxins and the threat of temperature inversions, which made being outdoors unbearably smoggy on many days. The 1952 "killer fog" in London that produced thousands of casualties was cited as an example of the potential risk from air pollution in New York City. Studies indicated that the smog blanketing Gotham over Thanksgiving weekend in 1966 accounted for 168 deaths and caused 10 percent of the city's residents to suffer from stinging eyes, coughing, or difficulty breathing. In 1969, the U.S. Public Health Service classified New York City as the area having the most severe air-pollution problem in the nation. The city's air quality had worsened over time from the inundation by automobiles, Consolidated Edison's electrical production, and refuse and apartment incinerators.[14]

The first meaningful step in controlling incinerator emissions came when the City Council passed Local Law 14 in May 1966. It prohibited the use of bituminous (soft) coal in the city and required the installation of "scrubber" equipment on all existing incinerators in buildings that were seven stories or taller and on municipal refuse incinerators. Without the upgrade they would have to shut down. The compliance period was one year, and a City Commissioner of Air Pollution Control had the authority to close incinerators if they failed to meet the deadline. As expected, citizens and apartment owners were incensed with having to make the change. The law, however, was largely unenforceable, especially early on, because the technology was not readily available, the cost was high, and emissions standards proved difficult to set.

An amendment to Local law 14 (March 1968) eased the compliance period and gave landlords more options, but only 18 percent of private installations and 20 percent of public facilities managed to meet the requirements in these early years (although this represented a 23 percent cut in particulate emissions between 1966 and 1970). No incinerators were to be built in new residential structures beginning in the mid-1960s.[15] In May 1969, Lindsay announced the closing of the incinerator on College Point Causeway in Queens, which, he stated, "has thrown up into the air of New York 3.5 tons of dirt every working day" since its opening in 1937.[16]

Despite the threat of air pollution, the DSNY still held to the hope that burning waste would be the answer to its disposal problems. Markets Commissioner Samuel J. Kearing Jr., named sanitation commissioner in November 1966, was given a "free hand" to administer his new position under "broad operating instructions" from the mayor. "There's too much litter, too much dirt," asserted the mayor in a time-tested mantra. "It requires better performance. Commissioner Kearing must be tough." A young man of thirty-six, Kearing had been a lawyer and businessman in Binghamton, New York, before turning to city service. From 1961 to 1963, he was counsel to the majority leader of the State Assembly and had worked in Lindsay's last congressional campaign, in 1964–1965.[17]

Kearing immediately began campaigning to improve street cleaning and garbage collection, hoping to get the support and cooperation of the city's ten thousand uniformed sanitation workers, despite their many grievances. He then quickly suspended eighteen department employees who allegedly had accepted payoffs from private carting companies.[18] Kearing insisted that in the Sanitation Department "commissioners come and go. The organization doesn't change. Now the organization must change."[19]

Regarding the disposal question, Kearing saw three possible actions: build more municipal incinerators, delay halting the use of apartment incinerators, and/or search for new landfill sites.[20] At that time, five landfills were still in use: Fresh Kills and Brookfield Avenue on Staten Island, Pelham Bay Park in the Bronx, Fountain Avenue at Beltway Park in Brooklyn, and Edgemere in Queens. Two others in the Bronx and Queens were used exclusively for construction and demolition waste. Most significant, in Kearing's mind, was the imminent closing of Fresh Kills and the other landfill sites. In early 1967 he stated that New York City had only "about eight years left" or less before it filled all of its dumping locations. (On other occasions he said ten years; some officials cited various other numbers.)[21]

All the while, criticisms about the landfills—including Fresh Kills—never eased. Several residents on Staten Island, especially those living near vacant lots and wooded areas, complained about rat infestations. "I'm afraid to let my children go out," a South Beach mother declared. "I just looked outside my window and saw two large rats on the front lawn." The borough sanitarian Sidney Kaufman defended the Staten Island Health Department, stating, "We probably get a lot of complaints but we don't think it [the rat problem] is that terrible. We check every call." But state conservation officials estimated that millions of rodents had migrated to the island on the garbage barges, infesting the landfill and surrounding area. The hordes of rats were sometimes joined by packs of wild dogs.[22]

An Interstate Sanitation Commission (ISC) survey in 1967 noted that a garbage odor at Fresh Kills was "quite strong" and that the DSNY should "be notified to take measures to correct the situation causing the excessive garbage odor from the landfill site."[23] A 1968 ISC report also pointed out that "a considerable amount" of waste intended for the landfill was ending up in Fresh Kills Creek and then in the Arthur Kill.[24] The DSNY kept in contact with the Health Department over practices that might affect the public's health, but differences in evaluating performance sometimes arose. For example, in 1969 the Health Department found some portions of the landfill where the refuse was inadequately covered with dirt, that is, where the covering was "thin."[25]

Kearing believed that the day-to-day job of collection was being conducted under "survival conditions" because of a lack of equipment and other resources. Deputy Commissioner Maurice Feldman described the department's situation in even stronger terms: "We're not experiencing a catastrophe, but a crisis. This means that if we fall behind, we will have a catastrophe. All resources are being used to capacity—there is no reserve."[26] Richmond Borough President Connor was a little more optimistic about the pending disposal limits: "In the next eight years, there will be enormous technological advances which should make it economically feasible to dispose of waste by other than landfill methods." But others tended to agree with Kearing's assessment, some suggesting that the DSNY had no plans at all after Fresh Kills reached capacity.[27] Several people believed the DSNY's problems were wrapped up in corruption; others saw its bureaucratic record to be hopeless. In any event, funds were in perennial short supply, even though from 1957 to 1967 the department's budget increased 47 percent (while the police budget doubled and the welfare budget tripled).[28]

After he was out of office, Kearing made some extremely candid comments about disposal problems. He revealed that behind closed doors specific options for disposal in a post–Fresh Kills era were being considered despite the department's public ambivalence. He quoted a conversation that took place in the fall of 1966 with a career engineer:

> "And after we have filled up Staten Island—then where will we put the garbage?"
> "We will fill Jamaica Bay."
> "But that is parkland and a wildlife refuge."
> "It doesn't matter, Commissioner. If you can't burn it, and you can't export it, then Jamaica Bay is the only place left."[29]

This would not have pleased Robert Moses. At the time of that conversation the city was about to accept a $400,000 federal grant to survey Jamaica Bay to determine how best to preserve it. The hope was that the survey would resolve "a running quarrel" between recreationists and conservationists vying for the use of Jamaica Bay land. Looming over the DSNY was yet another projection that Fresh Kills would fill up by 1974 and that Jamaica Bay was the only "undeveloped place" large enough to absorb the amounts of refuse currently being dumped on Staten Island. Contesting forces over Jamaica Bay were multiplying at a rapid pace, all while the DSNY was planning in secret to start using the site before others got wind of it.

Kearing realized the NIMBY issues at stake but also understood the "politics of garbage." "Almost invariably," he stated, "the area or neighborhood that winds up with the dump in its back yard is politically the weakest or most inept of the potential recipients. Thus, Jamaica Bay in Queens is still there to be saved from becoming a garbage heap because New York City's last great landfill battle was lost by the County of Richmond, our captive rural county, inhabited only by a few Republicans." While "sufficient political strength—and the personal attention of Robert Moses" effectively had defended Jamaica Bay in the past (plus its proximity to Manhattan and Brooklyn), the future was another thing altogether. Yet, Kearing believed, defenders of Jamaica Bay "had not won a victory, merely a temporary reprieve. No site is safe from garbage until a building has been erected on it."[30]

As he had made public, Kearing favored incineration and exploring other disposal options. "I eventually came to understand that municipal incineration was, essentially, an attempt to solve a political problem with technology." But even huge incineration plants, he concluded, "probably cannot save Jamaica Bay." Kearing was in the camp that favored a public solution to disposal, whatever the choice. Private contracting, he argued, allowed for serving many municipalities and was a way to solve the political problem of "moving garbage across political boundaries," but it also was "the basis for control by organized crime of the disposal industry."[31] Disposing waste in public landfills was now the law, but private carting of waste continued as it had for generations.[32]

Jamaica Bay was spared from becoming the new Fresh Kills. Searching for alternative disposal plans beyond current practice was challenging. The DSNY explored various approaches to expanding Fresh Kills or developing other locations on Staten Island and surrounding waters. It considered trying to extend the available disposal space by dumping on sites that were more difficult and costlier to use than other locations, postponing the shutting down of some apartment-house incinerators, and investigating exporting garbage by railcar. Swamps and marshland, which had been desirable dumping places, were now in short supply. Federal regulations protected some of these areas, and conservationists were fighting to save several remaining wildlife refuges. The *Advance* noted, "Once our marshlands are gone, we can't expect people outside the city to let us use theirs as dumping grounds. Sooner or later, we're going to have to do the job the modern way."[33]

Secretary of the Interior Stewart L. Udall wrote to Mayor Lindsay: "I am deeply disturbed over the decision to use the Pelham Bay Park site [the northeastern corner of the Bronx] for the disposal of solid waste." He regarded it as "a living outdoor laboratory that has no counterpart and cannot be replaced." The Park Department offered in response a smaller and less suitable plot adjacent to the landfill as a protected park. Speaking for a Bronx borough group investigating the offer, Dr. Theodore Kazmiroff rejected the park idea and asserted that there was no interest in any new territory being used for landfill. "It would make just as much sense to dump it [in Central Park], and you wouldn't have to haul the garbage so far."[34] A sea change in viewing salt marshes and wetlands not as wasteland but as breeding grounds for fish and wildlife, the shortages of such land for dumping, and the lack of incinerators were now impeding the city's disposal practices. So was the increasingly convoluted perception of what landfills even were supposed to be used for. Were they temporary solutions until incinerators came on line; that is, were incinerators the long-term solution to limitations in landfill space and location? Or, as other disposal options were found unworkable or hazardous, were landfills simply inevitable?

Kearing and others continued to argue that incinerators would save landfill space. In February 1967 New York City Air Pollution Control Commissioner Austin N. Heller proposed that the city turn to waste-to-energy incinerators to produce electricity.[35] This technology was a type of resource-recovery plant that created marketable products in solid, gaseous, or liquid form (and also produced electricity) by burning refuse. Waste-to-energy, he claimed, would burn household refuse rather than coal and thus eliminate sulfur pollution. With effective scrubber technology, smoke and soot could be reduced as well.

A plan to build such a plant at the old Brooklyn Navy Yard was announced on June 25, 1967. While Heller and Kearing agreed on the strategy, it had not been presented to the mayor. Lindsay was furious, especially since he had earmarked the site for an industrial park that would provide many local jobs. The falling out between the two men ultimately led to Kearing's resignation in November.[36]

New national air-pollution legislation further weakened the case for incineration in New York City. The 1970 Federal Clean Air Act inspired New York State to enact limits on particle emissions from incinerators.[37] Between 1966 and 1973, thousands of apartment-building incinerators were shut down, and by about 1980, 90 percent were closed.[38] The number of municipal incinerators declined from eleven to six between 1966 and 1973, with only three remaining after 1981, after the Hamilton Avenue plant closed that year. A federal lawsuit against the city for not meeting Clean Air Act guidelines led to a settlement that resulted in the closure of four municipal plants between 1979 and 1981.[39]

FIGURE 10.2 Betts Avenue incinerator, closed in 1993 (April 10, 2009).

Source: Courtesy of Michael Anton, NYC Department of Sanitation.

FIGURE 10.3 Plaque on Betts Avenue incinerator, closed in 1993 (April 10, 2009).

Source: Courtesy of Michael Anton, NYC Department of Sanitation.

New York City's experience with declining incinerators in the 1960s and 1970s followed national trends.[40] This naturally put more pressure on landfills. It was clear that availability of dumping space was by no means guaranteed once incineration declined, as the DSNY's relentless search demonstrated. At this point, it appeared that New York City was grasping at any disposal option no matter how remote.

REFUSE AS POLLUTION

Concern and interest in a wide variety of pollution, land-use, and health issues were finding a national stage in the 1960s. The air-pollution debate was an early manifestation of the changing character of environmentalism in Gotham, which was to focus on air pollution from industry (including "massive discharges" from heavy industry in New Jersey across from Staten Island), automobile and truck emissions, and incinerator smoke.[41] A 1969 Gallup Poll rated New York City high in a number of categories, including food and nightlife, but it did not even score in cleanliness or health.[42] New York and other cities in the recent past had made progress in connecting good sanitary practices with cleanliness and health, but nationally, the gravity of squandering so many natural resources (the "throwaway society") and the increasing set of risks to public health led solid waste to become designated as the "third pollution" alongside air and water problems.[43]

A March 1973 report of the National League of Cities and the U.S. Conference of Mayors declared, "The disposal of wastes and the conservation of resources are two of the greatest problems to be understood and solved by this nation in the latter third of this century. With almost half of our cities running out of current disposal capacity in from one to five years, America's urban areas face an immediate disposal crisis."[44] Five years earlier, the Regional Plan Association (New York) recognized that refuse was a notable part of the triad of pollution problems facing the region: "Pollution of air and water and the defacement of the landscape with refuse are now major matters of public concern." Its March 1968 report concluded that "a concentration of 19 million people on 2,300 square miles of built-up land has led to increasing concentrations of wastes in water, on land, and in the air."[45]

Modern environmentalism took the solid waste problem well beyond its local roots. Yet at this stage, the Regional Plan Association believed that while pollution issues demanded greater attention, "the potential for improving waste management is so great that higher levels of environmental quality can be achieved even while the Region's population increases substantially."[46] Yet few experts or public officials would so quickly fall in line behind such an optimistic view of dealing with disposal of increasing mountains of refuse.

The "new ecology" and the appearance of the modern environmental movement in the mid-1960s created a setting in which local sanitation issues were understood as being more consequential than nuisances or minor health risks. The basic concept of ecology revolved around "the relationship between the environment and living organisms," particularly the reciprocal relationship between the two.[47] By the 1960s ecology had changed from a scientific concept to a popular one as the questioning of traditional notions of progress and economic growth became more intense.[48] Rachel Carson's *Silent Spring* (1962), a grim warning of the dangers of pesticides, seemed to capture the new spirit best. Career ecologists were making it clear that respecting the biosphere required standing under the law.[49]

At the forefront of the new environmental movement were citizen and public-interest groups as well as a variety of experts. Between 1901 and 1960, an average of three new public-interest conservation groups appeared annually; from 1961, eighteen per year appeared. While the 1970s witnessed the most dramatic rise in the modern environmental movement, the 1960s helped build momentum through older preservationist groups such as the Sierra Club (1892) and the National Audubon Society (1905) and newer organizations such as the Conservation Foundation (1948), Resources for the Future (1952), and the Environmental Defense Fund (1967).[50]

Although environmental groups had yet to identify a common agenda, the tone and spirit of environmentalism were changing. Quality-of-life issues, pollution control, wariness about nuclear power, a critique of consumerism, and an insistence on preservation of natural places indicated a giant step away from "wise use" of resources and a challenge to traditional faith in economic growth and progress.[51] Garbage collection and disposal simply could not be governed by an "out of sight, out of mind" mentality but had to be treated as integral parts of preserving a pollution-free city.

What made the modern environmental movement so remarkable was the speed with which it gained national attention in the 1960s. Nothing epitomized that appeal better than Earth Day. The idea began as a "teach-in" on the model of an anti–Vietnam War tactic. The staff of Environmental Action declared, "On April 22, [1970,] a generation dedicated itself to reclaiming the planet. A new kind of movement was born—a bizarre alliance that spans the ideological spectrum from campus militants to Middle Americans. Its aim: to reverse our rush toward extinction." Across the country, on two thousand college campuses, in ten thousand high schools, and in parks and various open areas, as many as 20 million people celebrated in what was ostensibly "the largest, cleanest, most peaceful demonstration in America's history." As a symbol of the new enthusiasm for environmental matters and as a public recognition of a trend already well underway, Earth Day served its purpose.[52]

President Richard M. Nixon gave his blessing to Earth Day, and on January 1, 1970, he signed the National Environmental Policy Act of 1969 (NEPA). While opposing the bill until it cleared the congressional conferees, the Nixon administration ultimately embraced NEPA as its own. Going on record against "clean air, clean water, and open spaces" served no political purpose. By identifying with environmentalism, Nixon was able to address the issue on his preferred terms, most especially by focusing on solving pollution problems through technical means.

NEPA was far from "the Magna Carta of environmental protection" that some people claimed it was, but it did call for a new national responsibility for the environment. NEPA was not simply a restatement of resource management; it also promoted efforts to preserve and enhance the environment. It particularly emphasized the application of science and technology in the decision-making process. The provision mandating action required federal agencies to prepare environmental-impact statements (EISs) assessing the environmental effects of proposed projects and legislation. NEPA provided substantial opportunity for citizen participation, especially through access to information in agency files. It established the Council on Environmental Quality (CEQ) to review government activities pertaining to the environment, to develop impact-statement guidelines, and to advise the president on environmental matters. This new direction in dealing with environmental issues quickly found its way into many state laws.[53]

After a plan to codify environmental programs into one department failed, the Nixon administration announced that pollution-control programs and the evaluation of impact statements would be the responsibility of a new body, the Environmental Protection Agency. The EPA began operations in December 1970 under the direction of William Ruckelshaus. Initially, it included divisions of water pollution, air pollution, pesticides, solid waste, and radiation. But the EPA did not have overall statutory authority for environmental protection; it simply administered a series of specific statutes directed at particular environmental problems. The agency, nevertheless, was inundated by regulatory responsibility and pressured by industries to skirt those regulations. By the mid-1970s, environmentalism was a well-established national issue. Mainstream environmental groups took the initiative in helping draft new legislation, pressed for implementation of existing legislation, focused on the environmental-impact review process, and monitored government agencies. The courts became a battleground as more litigation tested key regulatory provisions.[54]

LINDSAY, ENVIRONMENTALISM, AND NYEPA

The end of the 1968 garbage strike not only produced a new agreement for New York's sanmen but flowered into a major institutional change for the DSNY. This reflected, in part at least, the ongoing effort by the Lindsay administration to restructure local government and rising concerns stimulated by the environmental movement. The administration had made strides in improving the DSNY.[55] Under a new sanitation commissioner (June 1968), the former Navy engineer Captain Griswold L. Moeller, the DSNY sported an 8.8 percent increase in the 1968–1969 fiscal budget. The funding decision indicated an end to the tight budgets to which the department had become accustomed and also suggested a rising status for sanitation among the basic municipal services. In late 1972, Lindsay also ended a freeze on hiring additional police, firefighters, and sanitation workers, resulting in 780 new sanmen hires.[56]

Moeller would have no honeymoon assignment to inaugurate his tour of office. He managed 14,500 men, three hundred facilities, and 4,400 pieces of equipment. There was a tugboat strike, memories of "the Lindsay snowstorm" from 1969, and of course disposal woes. "By 1975–1976 we're out of business as far as landfill goes," he lamented. At the time, the city was disposing of about 24,000 tons of refuse each day; that would rise to about 28,000 tons a day by 1976. If no changes took place, the Edgemere landfill and Fresh Kills would likely be the only two facility of their kind open after 1976. The imminent closing of the Pelham Bay landfill in the Bronx had the department thinking about how to extend its life. Moeller fell in line in regarding incinerators as an answer to the landfill shortages. "Right now we don't recycle [which was getting national attention]," he added, "because, simply, there's no demand for it. It's not economical."[57]

Moeller and the DSNY were also caught up in a major reorganization, one very different from those before it. The Lindsay administration sought to develop "superagencies" by clustering a variety of broadly related programs and departments into single coordinating units. The DSNY would become part of the New York Environmental Protection Administration (NYEPA), which had preceded similar state bodies and even the federal EPA.[58] In March 1968, the DSNY merged with the Department of Water Resources and the Department of Air Resources into NYEPA, denoting an effort to create an all-inclusive environmental agency in line with current thinking.[59] The commissioner of sanitation was now also the deputy administrator of environmental protection, and the department was divided into six bureaus (including Waste Disposal). The new agency added another layer of bureaucracy to sanitation services and threatened to remove decision making about collection and disposal from its traditional source.[60]

The new superagency requested one-fourth of the city's total capital budget in 1968. The first head of the NYEPA was Merril Eisenbud, an environmental scientist and the first health and safety chief of the U.S. Atomic Energy Commission.[61] He recognized "the need for prompt action in dealing with environmental problems that threaten to overwhelm us with pollution of land, sea, and air." Eisenbud also believed that the city had done a good job moving toward low-sulfur fuels, in promoting nuclear power as a means to reduce air pollution, and continuously monitoring air quality. He noted, "Among our less enviable world records is our extraordinary capacity to generate garbage." Per capita production in Europe and Latin America was one-fifth to one-half of New York City's daily output of 4.5 pounds. "This phenomenon is a by-product of our affluence and our apparent philosophical acceptance of waste as a desirable and necessary economic sink."[62]

Eisenbud advocated finding "new solutions" to disposal via better incineration techniques, providing fill to create more land, and possibly "mining" waste residue for reusable materials.[63] (Although it is debatable whether all of these were "new" solutions.) He told the City Planning Commission that the amount he wanted from the capital budget was "awesome" and the "magnitude of the programs staggering" but that each item was "crucially needed."[64] Such a large allocation, however, did not materialize, given that the city was in the midst of a stinging fiscal crisis.

Lindsay had strong political reasons to reform the sanitation programs. He lost to the conservative State Senator John J. Marchi from Staten Island in the Republican mayoral primary but remained in the race as the nominee of the Liberal-Fusion Party. Marchi and the Democratic candidate, Mario A. Procaccino (also a conservative), had made the street-cleaning problem an important issue in the campaign. In fact, in March 1969, the *New York Daily News* ran a five-day series of articles on the city's sanitation woes. Lindsay now was trying to find a way to blunt that criticism.[65] The mayor understood the basic issue of sanitation quite well, as he stated in his memoirs:

> Every time you shut down an incinerator, you increase the amount of garbage on the city streets. Every time you do that, you either have to put more street-cleaning forces out or face the fact that you are going to have a dirtier city. If incinerators can't burn garbage, it's going to be put out on the street to be collected—or not collected. And that means that we are going to have to find the money to collect that garbage and to dispose of it rapidly, assuming we can find a place to bury it, before it builds up to the point where it engulfs us completely. So you cannot solve the pollution problem without solving the sanitation problem, and you can't solve either problem without money.[66]

The split in the conservative vote between Marchi and Procaccino gave Lindsay the election.[67] For the incumbent mayor, it was only one of many political hurdles in 1968. Given a teachers strike, police slowdowns, threats of "job actions" by firefighters and sanitation workers, and tensions between racial and religious groups, "The last six months of 1968 had been the worst of my public life," the mayor reflected.[68]

Eisenbud's forward-looking views about disposal options in 1968 quickly came crashing down, adding to the Lindsay administration's troubles. By late 1969, NYEPA proposed, as previous DSNY administrations had, simply enlarging the area of Fresh Kills beyond a projected capacity date of 1975. "Our landfill situation is desperate," Eisenbud told the City Planning Commission. "If we don't conserve the existing landfill space and begin immediately to construct the four new incinerators we have asked for, we will literally have no place to put the city's 6½ million tons per year of garbage after 1975." He conceded that "none of the suggested alternatives for waste disposal is available now or even reliably in the future and we have no choice but to proceed with construction [at Fresh Kills]."[69]

Not surprisingly, Staten Islanders were not keen on Eisenbud's temporary fix.[70] They were outright infuriated when the city began to explore a new dumping site, spurred by a report from the engineering consultant firm Foster D.

Snell on vacant meadowland along the West Shore of Staten Island between Travis and the Staten Island Expressway (see chapter 9). There also was discord over how long raising mounding heights could extend the life of Fresh Kills.[71] There was little room for optimism on the island. As the journalist S. S. McSheehy observed, "Staten Island, in its efforts to block the establishment of city-wide refuse disposal facilities here over more than half a century, has a record of zero."[72]

In a curious (daring?) move, the mayor replaced Eisenbud in 1970 with the native New Yorker Jerome Kretchmer as the new commissioner of NYEPA. One environmentalist purportedly stated, "Jerry knows about air pollution because he breathes, he knows about solid waste disposal because he takes care of the garbage, but he doesn't know about the pollution of water because he probably doesn't drink much of the stuff."[73] Kretchmer had no administrative experience and no technical knowledge of environmental problems, but he was politically close to Lindsay and had strong environmentalist leanings.[74] Kretchmer's lack of experience on matters of waste initially raised criticism, but his political skills led him to pay attention to the workers on the street (adding 1,500 to the sanitation force) and to court the Uniformed Sanitationmen's Association's John J. DeLury. The union leader called Kretchmer "the first Sanitation Commissioner with heart" and stated that most commissioners "sit in big offices and push buttons. This guy's a worker."[75]

Kretchmer moved on several fronts to tackle a variety of environmental problems, such as outdoor spraying of asbestos, maintenance of equipment, and even the regularizing of citywide refuse pickups. In 1970, he was named acting sanitation commissioner for a short time in addition to his position with the NYEPA, and he became involved in efforts to begin curbside recycling and implement a resource-reduction system for the city. Kretchmer achieved good levels of productivity in the department and increased service levels—but without controlling spending.[76] Tensions among the mayor, the NYEPA, the sanitation union, and underserved neighborhoods kept street cleaning and collection on center stage into the new decade. Kretchmer rather naïvely stated that "Sanitation was never political before. But garbage has become 1970's police issue."[77] Sanitation had always been and would always be political.

In the early 1970s the DSNY continued to seek ways to expand landfill use as a primary disposal method, but it explored other approaches as well. Negotiations began with Enviro-Chem Systems Inc., a subsidiary of Monsanto Corp., for a possible pyrolysis waste destruction plant to be located at Prall's Island—one of several sites under consideration.[78] Incineration was not completely dead, either. In the late 1960s consideration was given to build four "super incinerators" (one on Staten Island). The current eight municipal incinerators had a combined capacity of five thousand to six thousand tons per day; the proposed super incinerator for the Brooklyn Navy Yard alone had a capacity of six thousand tons per day (with the other three ranging between 3,200 and five thousand tons per day).

The nonpartisan Citizens Budget Commission (CBC), a civic research group, came out against the city building costly super incinerators, the first of which was a 3,200-ton waste heat–recovery plant approved in 1968 at a cost of $72 million. The CBC believed they would add significantly to the city's air pollution and still leave residue to be dumped in landfills. They urged instead using "attractive" resource-recovery technologies (such as pyrolysis, shredding and bailing, pulverizing, and composting). In the interim, its report stated, the city should mound waste at Fresh Kills and develop new landfill sites in the Prall's Island area.[79]

On November 30, 1970, the City Planning Commission formally supported mounding at Fresh Kills—"totally within the confines of the present landfill boundaries"—rather than building super incinerators as the best immediate and short-term response to the disposal problem. The commission recommended raising the mounds to fifty to one hundred feet high, which it believed would prolong the landfill's life until 1986, when presumably new disposal technologies could be implemented. (Yet another projected deadline.) The chair of the commission, Donald H. Elliott, stated, "We've had increasing reservations about the super incinerator program. In our judgment super incinerators are not the answer."[80]

Among other locations, the DSNY continued to have designs on Prall's Island as a disposal site. Denied funds to acquire the property in 1969, it tried again in 1970.[81] Once again, Borough President Connor and other Staten Islanders stood firm in opposing the new landfill. Connor threatened a lawsuit and called for a study on waste disposal methods.[82] "I am not content," he stated, "to have such a large percentage of the total area of the borough indefinitely devoted to the disposal process."[83] Although the 2,200+-acre landfill represented only a fraction of the approximately sixty-one square miles of the island, Connor had made his point. Yet another battle over the Fresh Kills landfill was dragging on.[84]

Kretchmer quit his post as EPA administrator in February 1973 to explore the possibility of a mayoral bid in that year. He left office disagreeing publicly with Lindsay about the extent of the garbage dilemma and when fill space would run out at Fresh Kills. In March 1972 the mayor believed that 1988 was the target year for closure. Kretchmer observed, "I don't know where he's getting the 1988 figure. I think he's adding an extra five years there."[85] But Kretchmer now was leaving the ominous task of dealing with New York City's refuse and its other environmental problems to someone else. He was succeeded by Herbert Elish, the sanitation commissioner who had been appointed in 1971. The talented thirty-something, Harvard-educated lawyer had worked for the Civil Aeronautics Board and Allegheny Airlines and had served on the presidential campaign of Hubert H. Humphrey in 1968. He now assumed the NYEPA administrator's title.[86]

In the 1973 edition of *New York Affairs*, Elish laid out his take on the solid waste problems in New York City. "Solid waste disposal," he asserted, "is the most serious environmental problem facing New York City and it is particularly dangerous because it is a silent crisis that will not grip the imagination of the city until it is too late to find a reasonable solution." New York City's refuse load of thirty thousand tons per day (public and private), he stated, was "more than the combined totals of Tokyo, London, and Paris."[87] The DSNY represented a capital investment of more than $100 million and an annual operating budget of over $30 million. Elish worried about the life of the active dumping sites and agreed with many others that "by the middle 1980s, it will be unreasonable to continue to use the existing landfills."[88]

By now the idea of building super incinerators had been junked and the consensus of moving to incinerators in general put to rest, at least for the present. What else could be done? Ship refuse out of the city? Build new garbage islands in the harbor? Compaction? Shredding? Pyrolysis? Composting? Separation recycling? Such decisions lay ahead. "Environmental problems," Elish concluded, "can indeed be solved. It will take money, planning, and a high degree of political sophistication on the part of the public. Without these ingredients, we are doomed to face an ecological nightmare of New York buried under its own garbage."[89]

THE ALPINE HEIGHTS OF FRESH KILLS

From the mid-1960s into the 1970s, Staten Island sometimes seemed like it was a million miles away from the disputes and quarrels bubbling in Manhattan over a future refuse-collection and -disposal plan and the various labor disputes in the DSNY. It was as if the Verrazzano-Narrows Bridge had never been built or as if Richmond Borough did not share in the city's street-cleaning and collection struggles. But not so very far below the surface of the garbage debate was deep-seated anxiety about what to do with all of the refuse once the streets had been swept and the garbage collected. In this sense, Fresh Kills loomed large. At the core of that concern were projections about how long space would last at Fresh Kills. In 1961 some said the landfill would not be completed for ten to fifteen years. The Parks Department asserted that they would not be able to develop a park there until 1980.[90] In the mid-1960s Commissioner Kearing was guessing eight to ten years—maybe less—before Fresh Kills was filled.[91] Mayor Lindsay wanted to avoid the topic altogether. One thing was certain, even though the predictions on closure were at best speculative: No one was discussing *if* Fresh Kills should be closed, but only *when*. And Staten Islanders just wanted it closed, period.

The various proposals for new off-island dumping made the rounds, and discussions about mounding now took into account ways of contouring the land for future use. By the early 1970s plans to extend the landfill outward into the bay had been abandoned. And the curious idea to construct a ski lift on artificial snow likewise was scrapped. Borough Development Director Holt Meyer and others condemned plans to enlarge landfill operations, most immediately because the DSNY had requested $2 million to acquire additional land to expand Fresh Kills and $300,000 more to prepare the proposed Prall's Island facility. Borough President Connor also found the mounding plan "totally unacceptable." He regarded it as simply a "stalling tactic," not a solution to the disposal problem.[92] Mounding also made evident that the original justification for Fresh Kills, that is, to create more usable land, was becoming more and more specious as garbage was piled higher and higher above the existing terrain. An *Advance* editorial quipped, "The newest proposal of the City Planning Commission concerning city garbage disposal would heap a new indignity—mountains of garbage—on the island scene."[93] In January 1970 one ridge was already fifty-six feet high and several hundred feet long.[94]

What to do after Fresh Kills was one question; how to keep Fresh Kills functioning at the moment was another. For the city at large, the DSNY did not have sufficient numbers of workers to manage the landfill, handle collections effectively, and keep the streets clean. Commissioner Kearing called for a new garage for collection trucks to be located on the southern end of the island, but this was only one in a long list of equipment and logistical needs.[95]

By 1970 Fresh Kills was receiving about half of all of the city's refuse. Landfilling was taking place in four sections of the Fresh Kills Complex, covering about 1,100 of the more than two thousand total acres.[96] Local crews continually dealt with odors emanating from the fill, despite the use of chemical deodorants to mask them. In the early 1970s workers constructed an earthen berm on Richmond Avenue to conceal the waste heaps from the shoppers and other citizens driving by each day.[97] Municipal officials, nevertheless, were exposed to an array of complaints on a regular basis. The town of Woodbridge, New Jersey, for example, sued New York City in federal court because it stated that Fresh Kills was ruining its shoreline.[98]

Within a few years other landfill sites in New York City would close largely because of more stringent regulations. The changes in federal and state landfill laws nationwide had a considerable effect on the practice by the late 1970s. Virtually no regulations governing the construction, operation, and closure of landfills existed before 1970. In 1971, more than 90 percent of the 14,000 communities employing landfills failed to meet new minimum standards. Most landfills were unlined, making them susceptible to leaching toxic materials into nearby land and watercourses, and most produced serious methane-gas emissions. Even if closely monitored, staff could not effectively stop materials from being illegally dumped in the landfills. Soon it would become clear that Fresh Kills suffered from the same shortcomings of other substandard fills—leaching because it lacked a liner, methane emission, and illegal dumping.⁹⁹ Making improvements at the site was becoming more difficult because of the rising costs of construction in the early 1970s.¹⁰⁰

FRESH KILLS AND COMMUNITY

As a long-standing focal point of displeasure for Staten Islanders and as an integral part of the City of New York's disposal troubles, Fresh Kills Landfill was unrivaled. However, it also was an entrenched enterprise in Richmond Borough, which provided a livelihood for hundreds of families. Its workers were a valued (if sometimes underappreciated) part of the community.

The story of William "Bill" Criaris is a good case in point. In 1973, Criaris was named general superintendent in the DSNY's Bureau of Waste Disposal. He oversaw four hundred workers (of which more than ninety were members of the DSNY) in the Marine Unloading Division and landfilling operations at Fresh Kills. Born to Emmanuel and Mathilde Criaris in Manhattan in 1918, he was brought to Staten Island as an infant in 1919. He lived on Staten Island until he died in 1995. Bill Criaris was a resident of Bulls Head, just northeast of the landfill. He was an Army Chief Warrant Officer during World War II, serving as battalion baker for the 94th Quartermaster Battalion in Africa, Sicily, and on the Italian mainland. Much of his adult life was spent in the DSNY after he left the military in 1946 (with seniority retroactive to 1943). He rose to assistant foreman in 1950, foreman in 1954, district superintendent in 1961, senior superintendent and then supervising superintendent in 1967, and finally general superintendent in 1973. Bill retired in 1976, after thirty years of service.¹⁰¹

Criaris was a good company man and a loyal employee. He enjoyed what he did and believed he was rendering an important service to the city. He was an avid gardener, growing tomatoes, corn, squash, pumpkins, and cucumbers on his half-acre plot. He also was an expert at raising evergreens. A committed DSNY employee, he regarded Fresh Kills as "his" two-thousand-plus acres and was overheard saying, "Boy, what a farm I could make out of all this."¹⁰²

Criaris nevertheless appreciated the size of the task at hand during his tenure, realizing that projections for the landfill closure had been too optimistic and that the site would be accepting refuse beyond its expected closing date (1975 at that time), most likely into the mid-1980s. He was committed to having his workers do their jobs carefully, by avoiding spillage into surrounding waters and constructing refuse trenches effectively.¹⁰³ The work was difficult and often dangerous, and his crew, despite its size, seemed continually understaffed to do the job. In 1969 he confided, "We could increase our work force by 25 per cent and still not have enough."¹⁰⁴

FIGURE 10.4 Bill Criaris (second from left), general superintendent of the DSNY's Bureau of Waste Disposal.

Source: Courtesy of Cheryl Criaris-Bontales.

The Staten Islander wanted qualified people, not cronies, around him, thinking particularly about workers in the field such as cranemen and tractor operators. "I hereby propose," he stated in a department memorandum, "that all positions are to be filled by the Bureau of Waste Disposal in the Department of Sanitation, relating to the operations at the various divisions . . . with career Civil Service Employees," hoping that would limit political hangers-on. He questioned the practice of promoting men in the ranks "mainly on their book knowledge." "I feel that some good men," he stated, "who do not have this particular ability relative to that of book learning have other avenues open to them such as operations of various equipment utilized by our department." Testing, therefore, should be heavily weighted to practical and written components and less to seniority.[105]

The landfill was Criaris's office and his ward. It was obvious that he took pride in his work, and he certainly looked forward to the day when the landfill could become productive space again. Criaris willingly gave tours to everyone from students to dignitaries. Michael Belknap, the director of the Mayor's Council on the Environment, thanked the superintendent for a visit in June 1971: "As a result of our visit, the Council's visiting interns came away with a greater understanding of the magnitude of New York's solid waste problems. No doubt, they are now ready to begin their study projects which we hope will bring us all a step closer to solving some of New York's environmental problems."[106]

An American Studies professor from Douglas College brought students from her American City course to the landfill in 1974, which marked the fourth time she had made such a trip. The High Rock Park Conservation Center fairly regularly brought students there on field trips. German visitors were given a tour of the facility as well.[107] All of this, of course, was good publicity for the Fresh Kills operation, while reinforcing its importance to the city. Cooperating with authorities to recover lost items also improved public relations. In 1961 Criaris helped in an investigation of the inadvertent disposal of a large quantity of military documents, and in 1964 he helped the Office of Civil Defense recover federal fallout supplies that had been illegally discarded.[108]

Like others in the DSNY working on Staten Island, Criaris had deep ties to the community. Locally, he was on the board of directors of Holy Trinity Greek Orthodox Church; a member of the DS American Legion Post; and a member of the Greek national organization, the Order of AHEPA.[109] His personal correspondence is rich in letters from grateful islanders for help rendered by "the Marine Unloading Men"—moving a statue for Countess Moore High School, providing bus transportation for children and employees after a fire at St. Michael's Home, or helping Willowbrook Development Center compact and bury some condemned equipment.[110]

The portrayal of Bill Criaris is meant to bring attention to a different side of the Fresh Kills story. It helps to look beyond accounts of questionable working conditions and picketing sanmen and to recall those depending on the landfill for a livelihood despite its unsavory reputation and its function as an infamous symbol of our insatiable consumption. Most importantly, it tells a meaningful story of how the landfill was deeply embedded in Staten Island life.[111]

STATEN ISLAND: FORGOTTEN NO LONGER

As New York City became more dependent on Fresh Kills to absorb its refuse, Staten Island underwent a change in identity. The opening of the Verrazzano-Narrows Bridge had signaled the beginning of the end of the island's isolation. The landfill was no longer banished to a place distant enough and undeveloped enough to justify its initial placement. Available space for new landfills was a citywide issue, but even on Staten Island the seemingly abundant marshland for new dumping sites was becoming sparse or untouchable because of increasing environmental concerns. The combination of efforts to preserve remaining salt marshes and other wetlands and competing uses for the real estate were most obvious beginning in the late 1960s.

The Staten Island Greenbelt–Natural Areas League (SIGNAL), a prominent group of local and off-island conservationists, sought to preserve the island's remaining wetlands. They received support from groups such as the Federated Sportmen's Clubs of Staten Island. A SIGNAL position paper stated:

> The executive committee of SIGNAL deplores the systematic destruction of New York City's vanishing tidal marshes. These irreplaceable natural resources should be protected. Among other things, they are vital to the economy of a multi-million dollar fisheries industry. Yet, using wetlands to dispose of it, all the city's biologically-rich estuarine zone will be gone forever by 1975.[112]

SIGNAL fought to reroute Section I of the Richmond Parkway away from the Greenbelt (a system of contiguous natural areas and public parkland in the central hills of the island). It protested the dredging project at Lemon Creek because it threatened salt marshes there. But most importantly, it sought to preserve the remaining two hundred acres of marsh at Fresh Kills. "These acres are productive of some aquatic life even now," a report stated, "water pollution in the Arthur Kill withstanding."[113]

SIGNAL's chances of success in these battles were considered unlikely, given the booming development of the island. SIGNAL particularly focused on the need to shut down Fresh Kills by the end of 1970. It decried the fact that the landfill complex was once "a healthy, robust marsh" and was deeply concerned that the DSNY was looking at an alternative to a site that might threaten more marshland. Technically advanced waste-to-energy incinerators and particularly recycling, SIGNAL believed, were acceptable alternatives to landfilling marshes.[114]

SIGNAL hoped to convince borough leaders to halt landfill operations by 1971 to save the two-hundred-acre tract, but they were unsuccessful.[115] As one SIGNAL board member stated, "True, some landfills can be useful to the city—they can be turned into parks. We need parks but we also need ecological balance. . . . We need marshes in somewhat the same way we need fresh air. Unfortunately the two resources seem to be mutually incompatible."[116]

What a contradictory and conflicted place Staten Island had become by the mid-1970s. It sported the biggest landfill in the world and also fell to the land-developing avarice of the region. Could such realities coexist? Could the gigantic blemish of a landfill be ignored as plans for creating a suburban sanctuary went forward? The new environmental era that emerged placed Fresh Kills in a new light: not as simply an inconvenience and miscreation but as a health risk and a vast toxic vessel—and as a threat to the remaining open spaces and marshland of the island. As dependence on Fresh Kills grew greater in the late 1970s and 1980s, the landfill itself grew more worrisome and more out of place.

CHAPTER 11

FISCAL CRISIS AND DISPOSAL DILEMMA

Fresh Kills Landfill was an imposing structure. But even its scale and reach were not enough for the city to end the chronic turnover of programs, from dumping to burning to resource recovery and back again, or to satisfy a citizenry that wanted results—just not in its own backyard.

The quest for a favorable disposal outcome was being undertaken in a vacuum. In the mid-1970s, the dominant context setter was the fiscal crisis. The immediate effect of the fiscal crisis on the New York City Department of Sanitation (DSNY) was damaging and in the longer term complicated and uncertain. The department had to handle its daily operations—street cleaning, collection, and disposal—in good economy or bad. The sanitationmen's union, for its part, was preoccupied with keeping workers employed.

The DSNY's regular duties and the shakiness of the city's fiscal state of affairs left little time to rework existing disposal options. Aside from financially strapped city services in the mid-1970s, the city also faced an impending shortage of landfill space that inspired serious discussion of resource-recovery projects built around newer models of waste-to-energy systems. In fact, the mayor would designate the department as the lead agency to implement such a program while it continued to fill the enormous landfill. But for the time being, Fresh Kills remained the mainstay of New York City's disposal efforts despite ceaseless dissent from Staten Island residents.[1]

THE FISCAL CRISIS

In January 1975 the Urban Development Corporation, which had been created by Governor Nelson A. Rockefeller in 1969, defaulted on $100 million in notes. It was the biggest default of any government agency since the Great Depression. The nonpayment dampened prices of New York City bonds, and investors now spurned them. Cut off from access to the credit market, the city faced the worst budget crisis in its history.[2] A *New York Times* analysis on April 11 observed, "New York, America's biggest, richest city, which for many years supported a public budget second in size only to the Federal Government's, last week found itself in the humiliating and dangerous position of being unable to sell a note issue equal to less than half the expected yield from its sales tax." A four-hundred-million-dollar advance of welfare aid from the state allowed the city to postpone loan repayments that month. Despite the timely rescue, the question on people's minds was: "Can New York default on its debts and go bankrupt?"[3]

New York City did not go bankrupt, but its persistent financial woes were unmasked, as was the city's reputation as an unrivaled economic juggernaut. After 1970, numerous American cities (particularly in the East and Midwest) faced an "enduring plight" typified by financial stress, the lack of a sound national urban policy, and intensifying physical and social problems. Federal and state governments mandated that cities perform a wide array of functions for its citizens. The cities also were confronted with subsidizing commuters and suburbanites who took advantage of cultural, recreational, and sports events at or around the core, that is, providing services such as police protection and waste removal that locals—not visitors—had to pay for. Despite intermittent urban-rehabilitation programs and gentrification efforts, many middle-income residents fled the central cities as the influx of lower-income groups increased. This left many cities with dwindling capacity to raise revenue and deliver satisfactory services.[4]

The expense of providing services tended to be higher in central cities than in the surrounding communities. Large numbers of low-income people at the core resulted in higher demand for welfare and health care. The poverty rate was approximately twice as high in central cities as in suburbs. Core cities were substantially older than the periphery, and thus maintaining or replacing existing infrastructure was costly.[5] Sources of general revenue were in decline as white flight continued, and a reluctance to increase taxes led officials to search for other means to balance their budgets or fall into deficits. To meet service obligations several departments borrowed from the city's cash flow, intending to pay off the debt with future revenues. Before long, cities were accumulating uncontrollable internal debt.[6]

New York City's financial collapse in 1975 was the classic case, bringing nationwide attention to the municipal fiscal tailspin. The Big Apple's troubles were at least ten years in the making. In the 1970s it lost 10 percent of its population; between 1950 and 1989, 65 percent of its manufacturing jobs disappeared. Between 1970 and 1980, all the boroughs save Staten Island saw population decreases, and the overall population of greater New York declined from almost 7.9 million in 1970 to under 7.1 million in 1980. The city faced serious racial tensions, infrastructural decay, fires that destroyed more than 100,000 dwellings, and a severe homeless problem.[7] But although New York City had stopped growing, its budget had not. Between 1965 and 1975, there had been at least seven major studies of New York City's fiscal status—virtually all had concluded that the rate of growth in outlays could not be sustained. Expenditures simply were outstripping revenue.[8]

The city government began running budget deficits under Mayor Robert Wagner Jr. (1961–1965), but it had little trouble selling its debt at the time. Bond underwriters simply convinced buyers that the city's problems could be worked out. New York City's debt rating, nevertheless, was downgraded by the mid-1960s. In 1961 the city budget was $2.7 billion; in 1976 it was $13.6 billion. Between those years the city debt tripled.[9]

A major reason for New York City's escalating debt in the early 1960s was a rapid increase of expenditures in many categories and in the funding of some new programs. Important examples were the expansion of public-assistance eligibility and the enactment of Medicaid, which constituted a major leap in the city's mandated costs.[10] In the 1960s spending on public assistance and payments to civil servants accounted for 69 percent of the increases; between 1969 and 1975 82 percent of the increases went to civil servants and to borrowing. Between 1969 and 1975 spending rose 28 percent, while the local economy declined on an average of 1.4 percent annually.[11]

By 1975 the city was running a debt of about $1.5 billion annually (with an operating budget of about $12 billion). It had accumulated its massive debt as a result of borrowing large sums during the economic expansion of the 1960s. The direct cause of the crisis was linked to using expensive short-term financing to fund deficits.[12] Soon after the Richard M. Nixon administration sharply cut federal aid to the cities, the financial balancing act toppled. The city's budget was in disarray, population declined, and five hundred to six hundred thousand jobs were lost.[13]

There was plenty of blame to go around for the fiscal debacle. But fairly or unfairly, Mayor Abraham D. Beame became the focal point of public anger. Many viewed Abe Beame as part of the problem; some regarded him as a rescuer. Interestingly, many city workers perceived him as their defender, despite his draconian cuts.[14] At the very least, he was in over his head, ultimately letting state-created entities run the city.

Beame, born in 1906, grew up under modest circumstances on the Lower East Side. He was the son of Polish Jewish immigrants and became the first Jewish mayor of New York City. After attending City College, he formed an accounting firm with a friend. The Great Depression ruined the business, and he turned to teaching for several years in Queens. He also became active in Democratic politics in Brooklyn. Beame's first government job was as assistant budget director under Mayor William O'Dwyer (1946). He was appointed budget director in 1952, then elected comptroller in 1961 and 1969. The "clubhouse Democrat" and longtime civil servant was no La Guardia when it came to charisma, but he had a passion for politics. Beame was a details man, gentle, cautious, dignified, and honest.[15] In 1973, he mused, "All I ever wanted in life was to be mayor of the greatest city in the world."[16] Beame lost to John Lindsay in the 1965 mayor's race but won in 1973. He served as mayor for only one term.[17]

Beame's campaign slogan had been "He Knows the Buck," yet he had no idea how little that prepared him for what was to come.[18] As early as April 1974 underwriting bankers cautioned Beame about problems in the commercial money market. Believing that the banks contributed to the city's woes, he resisted their advice. In his May 1975 budget address, carried live on television, the mayor stated, "The very institutions which were supposed to be promoting our securities have been out in the street poisoning our wells."[19]

Beame attempted to counter the threat of bankruptcy by cutting the city workforce by 60,000 to 65,000, freezing wages, raising taxes, and restructuring the budget. The city was increasingly unable to pay its police, firefighters, and sanitation workers; to keep open its hospitals; and to meet other obligations.[20] Half of the capital budget in 1975 was used to finance current expenses.[21] This was a tightrope walk, since Beame's austerity plans and budget maneuvers bred public unrest. The actions were insufficient to meet the calamity. He soon felt compelled to turn to the state and federal government for help.

In the governor's office, concern over New York City's plight was hardly altruistic. Governor Hugh L. Carey, a Democrat, worried about the state's own debt rating and that the possibility of abrogating union contracts could result in strikes and public disorder.[22] In mid-April 1975 the state advanced revenue-sharing funds to the city. In June the legislature established the Municipal Assistance Corporation (MAC)—"Big Mac"—an independent corporation authorized to sell bonds to meet the needs of the city. Such actions gave the state increasing control over New York City. The creation of the Emergency Financial Control Board (EFCB), with authority over municipal finances, essentially put the city into receivership. None of this sat well with local officials, but they were powerless to stop it. The EFCB essentially ran the city's finances between 1975 and 1978.[23]

State support proved insufficient to reverse the mounting fiscal crisis. In mid-May Mayor Beame and Governor Carey went to Washington, DC, to meet with President Gerald R. Ford. There seemed to be little concern in the nation's capital about the devastating problems facing New York and other states. Secretary of the Treasury William E. Simon, a hardliner, warned that helping New York City was tantamount to nationalizing municipal debt. The president gave Carey and Beame no immediate assurances of help, writing on his notepad, "24 hours. Must do what's right. Bite bullet." The next day he told Beame and Carey that there was nothing the federal government could do to help. What he really meant was that there was nothing his office *would* do.[24]

On May 14 Ford also wrote to Beame "that the City's basic financial condition is not new but has been a long time in the making without being squarely faced" and that he therefore would deny the request.[25] Ford firmly believed that the city had created its own mess and would have to find its own way out. Furthermore, he did not have the votes in Congress to provide financial support, especially over a decision that had partisan overtones. Vice President Nelson Rockefeller tried to persuade Ford to change his stance, to no avail. Several mayors cautioned the president that New York City's crisis could spread to other major cities.[26]

On October 29, Ford made a speech denying a federal bailout to the city. The next day, the *New York Daily News* ran a headline that would become famous: "FORD TO CITY: DROP DEAD."[27] Ford's actual response was tough but not as scathing as the headline suggested. One month after the speech, the president bent to pressure from a variety of leaders and agreed to extend loans to New York City on the condition that it continue to seek a balanced budget. The funds helped the city stave off bankruptcy but did not provide a permanent fix.[28]

Recovery came slowly. The year 1975 saw the worst slump and the highest unemployment in the United States since the Great Depression.[29] By 1979 the city had entered seasonal financial markets and sold some short-term notes; in 1981 it sold some long-term bonds and finally ran a balanced budget; by 1985 Big Mac was no longer necessary.[30]

The fiscal crisis was as much a political contest of wills as a financial catastrophe. For conservatives, the collapse embodied all of the shortcomings of Lyndon Johnson's Great Society. For liberals, the avaricious business community and its supporters bore much of the blame. Mayor Beame criticized the banks and his predecessors, who had been unwilling to take on the mantle of New York's Herbert Hoover from the Great Depression era.[31] Governor Carey's supporters claimed that his intercession had propped up the city. And Washington leaders argued that federal loan guarantees had saved the day.[32] In the end, Abe Beame proved to be a one-term mayor (although the election in 1977 was surprisingly close), a political casualty of the financial crisis.

FIGURE 11.1 President Gerald Ford's alleged remarks to New York City regarding financial aid to help settle the fiscal crisis in the 1970s.

Source: Wikipedia.

THE DSNY AND THE FISCAL CRISIS

The decline in municipal employment in 1975 and the years following disrupted a variety of services, including sanitation. Reductions in police, fire, transit, education, and sanitation were not necessarily proportional to specific job cuts, but the level or quality of service plainly was affected. City Hall adopted "minimum manning" practices in police, fire, and sanitation to cut costs. In the summer of 1974, the city had reached quick contract settlements with the Uniformed Firefighters Association (UFA) and the Uniformed Sanitationmen's Association (USA) but not with the Patrolmen's Benevolent Association

(PBA). After the PBA rejected a city proposal for contract concessions, Mayor Beame laid off 260 recruits and probationary officers on January 30, 1975; these were the first civil service workers laid off since the Great Depression (180 fire fighters also were to be laid off).[33]

Increasingly urgent was the need to repair or rebuild much of the city's infrastructure (bridges, sewers, sewage pumping stations, paved streets, water-tunnel trunks, and parkland) and to replace many vehicles and other vital equipment, which included subway cars, buses, and sanitation trucks. A 1978 City Planning Commission report cautioned that "renewing the city's capital stock is second in priority only to resolving the fiscal crisis."[34]

Sanitation consistently received the greatest number of complaints among all departments and agencies at this time, until transit overtook it in 1980. Street cleaning and collection drew pointed criticism because results (or lack of results) of that work were basic to most citizens' daily lives and quite visible. The Bureau of Cleaning and Collections (BCC) employed three-fourths of the DSNY manpower, and reductions there were most obvious to the citizenry.[35] In the 1970s, the city lost about five thousand sanitation workers.[36] In 1975 city officials rated 72.5 percent of the streets "acceptably clean"; in 1979 only 52 percent were so classified. This drop was largely the result of layoffs and the austerity budget.[37]

With declining staff, the DSNY was having serious trouble keeping up with the increasing amounts of refuse being produced (about 4 percent more each year). Between 1975 and 1977, the total refuse collection dropped 2 percent. Between 1978 and 1981, amounts collected declined by 10 percent.[38]

While the largest total number of sanitation-employee cuts took place in the Bureau of Cleaning and Collections, the Bureau of Waste Disposal underwent the highest percentage of decreases by far in personal service expenditures. Between 1974 and 1980 the total disposal force dropped from 1,438 to 810. In the mid-1970s the bureau was responsible for Fresh Kills, three other truck-fed landfills, and three incinerators. It also operated nine marine transfer stations located in four boroughs. The cutbacks were possible because of an increased reliance on Fresh Kills, phasing out of other landfill sites, and using BCC staff at the transfer stations.[39]

DISPOSAL BEYOND FRESH KILLS

Aside from labor cutbacks, the flow of refuse to Fresh Kills (10,000 or more of the 25,000 tons collected) taxed not only the site but also the system of marine transfer stations. The number and location of the stations were either insufficient to capture all of the refuse generated in the city; in some cases, they needed repair or updating.[40] For half of 1977 alone, the number of barges unloading at Fresh Kills exceeded 4,770. While facing budget constraints, the DSNY still had to find a way to build a new transfer station and to invest in Fresh Kills to maximize disposal capacity. No new landfills were built in New York City after 1979.[41]

The daily grind of collecting and disposing refuse did not completely overshadow the protracted search for new potential disposal sites. In July 1975 debate resurfaced in Albany over yet another dumping place for New York City—this time the possible construction of an island of refuse to be located in the waterways around the metropolis. Guy V. Molinari, the Republican assemblyman for the Sixtieth District, asserted that the legislation would lead to constructing "garbage islands" off Staten Island or Queens.[42]

Molinari was born in Manhattan in 1928 but grew up on Staten Island. He graduated from Wagner College in 1949, received a law degree from New York Law School in 1951, and practiced real estate law for several years. He served in the U.S. Marine Corps from 1951 to 1953 in the Korean War. Molinari is the only Staten Island official to have been elected at all three levels of government: locally as borough president of Staten Island (1990–2001); on the state level as a New York assemblyman (1975–1980); and in Washington, DC, as a U.S. congressman (1981–1989). A Republican kingpin on Staten Island for many years, he also was influential in state and national politics. State Senator John J. Marchi and Ralph J. Lamberti, the Democratic borough president, were major rivals on the island, and he fell out with others as well. He could be brash and contentious but was wildly popular with his constituency. He would become a key (maybe the key) player in the eventual closing of Fresh Kills Landfill.[43]

The bill concerning the refuse island failed to pass by two votes on July 8. On a second try the bill passed (with opposition from the Staten Island delegation), which would permit offshore landfill operations in the city's waterways. While the garbage-island idea was not popular with any of the nearby boroughs, the measure did allow the city to enter into contracts with private firms to build waste recovery plants (on a much smaller scale than the garbage islands) in existing areas within the city deemed appropriate. This option became more popular over the next few years.[44]

In May 1976 it appeared that the New York State Assembly would pass a bill that guaranteed that no more than one refuse-processing plant could be built on Staten Island. Molinari stated, "I feel that for the first time ever Staten Island can now look forward to the prospect that we will not be bearing an unfair portion of the city's garbage."[45] By the end of June, the legislature included wording in its refuse bill providing an absolute limit to the amount of garbage that could be dumped on Staten Island in future years.[46] Yet the twists and turns in the debate over refuse disposal on Staten Island were far from over.

Also at stake were efforts to seek new dump sites outside of New York and to continue to send a portion of its refuse to New Jersey. In January 1974 the New York Environmental Protection Administration (NYEPA) submitted its 1974–1975 operating budget of $383.2 million (an increase of $23 million over

FIGURE 11.2 Fresh Kills Landfill, March 1973.

Source: Wikimedia Commons.

FIGURE 11.3 Garbage barges from Manhattan unloading at Fresh Kills, March 1973.

Source: Wikimedia Commons.

its current budget). It called for, among other things, $8 million to pay for out-of-state disposal in abandoned strip mines.[47] Administrator Herbert Elish called the budget request "spartan," but he was not prepared for the backlash that the impending fiscal crisis was going to cause.[48]

Compounding the problem, New York found itself in a heated battle with New Jersey over the dumping of some of its refuse in the Hackensack Meadowlands. The site was taking more than 45,000 tons of refuse weekly from 144 local towns and running out of tipping space. New Jersey called for an end to dumping out-of-state refuse there by late 1973 or early 1974, but a New Jersey Superior Court decision temporarily delayed the action.[49]

As much as seven thousand tons of refuse from New York was being privately carted to the Garden State daily, so such a ban would be a serious loss. There were about four hundred private garbage collectors operating in New York City, and disputes over the tipping rates they were charged in New York City often drove them to New Jersey.[50] New York State Commissioner of Environmental Conservation James L. Biggane stated, "There's no use kidding ourselves—it's going to be difficult convincing people to accept someone else's garbage."[51]

Immediate concern, however, squarely focused on crisis management within the DSNY, with respect to layoffs and labor unrest among the city's uniformed employees. In early July 1975, for example, a two-day wildcat strike complicated the delicate negotiations over returning sanitation workers to their jobs. The strike was nothing like the 1968 affair, but it did cause tense moments when a fire in the Travis area of Fresh Kills engulfed a bridge and seven acres of land. At first, nonunion tractor operators refused to cross the picket line, but the union ultimately let some of them through.[52]

The city was in the process of reinstating all of the dismissed sanitation workers at the time of the latest strike but only two thousand of 5,034 laid-off police officers and 750 of 1,650 firefighters. Ken McFeeley, the president of the PBA, was livid because John J. DeLury, the head of the USA, was able to cut a deal with the mayor to return 2,900 laid-off sanitation workers to their jobs after promising that the workers themselves would guarantee their payrolls if the city could not get funds to increase its taxes.[53] McFeeley declared, "DeLury has been sleeping with City Hall for years. Now they've rewarded him by trying to help him after he makes one tremendous irresponsible blunder [the wildcat strike]." Beame's impulsive attempt to quell the controversy by returning police and firefighters to work seemed an obvious ploy to McFeeley: "I don't think that's good news. [Mayor Beame] is throwing back 2,000 as a salve because he's screwed up [through his layoff practices]. Any idiot can see he's been using the police and fire fighters for ransom to get more money from Albany."[54]

The restored sanitation jobs did little to quell public frustration over the delivery of needed services. A *New York Times* editorial stated, "The persisting pile-up of garbage and trash on city streets is a noisome symbol of the city's fiscal crisis and glaringly symptomatic of the shortcomings that helped precipitate it—notably, municipal mismanagement and union greed." Aside from the mayor, the editorial blamed Sanitation Commissioner Robert T. Groh for frittering away "morale and productivity" in the DSNY and DeLury and the USA for performing well below what was expected, given "some of the most generous pay and fringe gains in the city and nation."[55]

Yet the department was laboring over many of the problems it had regularly faced in the past. It took a month just to collect the backlog of garbage accumulated during the two-day wildcat strike. And Groh sounded a common complaint: "I'm going to have an aging fleet, fewer mechanics to repair the trucks, and fewer men to run them." The disposal quagmire continued, with no complement or alternative to Fresh Kills in sight.[56]

Under constant pressure, Groh resigned in September 1975. The engineer Martin Lang, the first deputy administrator of NYEPA and a civil servant for thirty-eight years, replaced the former zoning lawyer and deputy borough president from Queens. Lang wrote to Mayor Beame in October, stating that the department had two "immediate urgent missions": a 3 percent reduction in tax-levy controllable expenditures in the current fiscal year and the need to "ensure maximum effective utilization of present existing resources."[57] Responding to Lang's appointment, DeLury stated, "Unless he's given the tools to do the job, his fate will be the same as Groh's."[58]

KOCH FACES DIRTY STREETS AND POOR COLLECTION

The election of Edward Irving Koch in 1977 proved to be a major turning point in easing and then ending the financial crisis and led to substantive changes in the DSNY. Koch was born in the Bronx in 1924, the second of three children, to Galician Jewish immigrant parents. He was marginally middle class, a decorated infantryman in World War II, and an NYU Law School graduate. Koch entered politics as part of the Village Independent Democrats, defeating Tammany's Carmine DeSapio in a 1963 district race. He then sat on the City Council and served in the House of Representatives for Manhattan's Upper East Side (John Lindsay's old district) from 1969 to 1977. A lifelong bachelor, Koch was tall, a little dumpy, and balding. He was spirited or feisty—depending on your point of view. His common touch and his brutal honesty made him much more than a caricature. "Hizzoner" was a staunch supporter of Israel, a defender of gay rights, and a proponent of civil rights, although he frequently was at odds with African American leaders and others, who sometimes called him a racist. He famously said, "I'm not the type to get ulcers. I give them."[59]

Koch ran against the incumbent Abe Beame and a host of others in the 1977 Democratic primary. Koch was never on very good terms with Beame—believing that he "would make an inadequate mayor"—nor with Governor Carey, with whom he had had his "share of problems."[60] Koch faced and defeated Mario M. Cuomo in the runoff and then won the general election handily against Cuomo (who now was running on the Liberal ticket) and the Republican Roy M. Goodman. The election was a mandate on the fiscal crisis and Mayor Beame, but Koch also courted votes by supporting the death penalty for capital crimes. Although he began his political career as a reform liberal, Koch became a pragmatic conservative (or neoliberal). His trademark "How'm I doin'?" was answered with election to three terms as mayor.[61]

The change in leadership under Koch brought a needed morale boost to New York City, especially since the economy had reached its nadir in 1976, with accompanying social and political strife. He set out to reverse the city's dismal credit rating despite a variety of internal and external limiting factors. For the first year, he continued to make cuts in services on the way to repairing the city's credit. By the end of his first term he was able to balance the budget. Deep into his second term he spearheaded a more than $5 billion locally financed rebuilding program, including in parts of the desperately ramshackle South Bronx. A postindustrial boom resulting in economic and population growth followed. Between 1977 and 1989 total employment in New York City increased by 15 percent (almost five hundred thousand jobs).[62]

The economic boom, especially between 1977 and 1987, raised the city's revenue even as local taxes declined as a proportion of the total municipal economic activity. The infusion of funds thereby allowed Koch and his administration to regain the city's fiscal autonomy, to make easier its ability to govern, and to spend resources on city services. Ultimately, Koch restored services to pre–fiscal crisis levels. His popularity, plus strong support among business leaders in the financial and real estate community, inspired a progrowth environment.[63]

The new mayor was in large degree playing to his constituency—white, ethnic, and middle-class voters—in choosing how to bring the city back. At times, this base was supplemented by more conservative black property owners and poorer and more conservative Latinos.[64] But the economic policies that benefited his constituency in some respects contributed to homelessness, crime, and infrastructure shortfalls, as did the efforts to gentrify the city. Gentrification by its very nature pushed out the poor and the propertyless and had obvious racial overtones. It was not surprising that several black leaders often scorned him.[65]

Despite the uneven record of the Koch years, New York City was in a different place in the late 1970s and into the 1980s than it had been in 1976. This was the work of Mayor Koch, a self-proclaimed "liberal with sanity."[66] Of course, nothing lasts forever. After a successful first two terms, Koch's third was rife with corruption scandals and growing racial divisions. The mayor also was accused of not responding effectively to the growing AIDS crisis in the 1980s. In a somewhat understated manner, Koch concluded after a failed fourth-term run, "People get tired of you. So they decided to throw me out. And so help me God, as the numbers were coming in, I said to myself, 'I'm free at last.'"[67]

In the emerging Koch era, much attention had rightly focused on bringing back a bedraggled economy and navigating through an intense period of social strife. Within the muddled environment of managing the city and providing necessary services, sanitation proved to be a bright spot—at least on the collection side, if not disposal. Koch's initial appointment for commissioner of sanitation, however, was not a good indicator of changes to come. Martin Lang survived as commissioner for only a few months under Mayor Beame, largely because of pressure from the sanitation union. When he was switched to parks commissioner in 1975, Anthony T. Vaccarello got the DSNY job. A career public servant from Brooklyn, Vaccarello understood the enormity of the task before him but came to believe he could do little about it. In 1977 he got national attention for calling New Yorkers "a bunch of slobs." Earlier he said, "I felt all alone out there, particularly when the stark reality sunk home to me that there are over seven million people out there who don't know anything about keeping our city clean—and furthermore don't give a damn."[68]

Vaccarello had written to Mayor Beame in December 1977 about a variety of problems in the department, including the "severe cost exposure" of meeting New York State Department of Environmental Conservation (DEC) regulations to provide dirt covers to the landfills, the limits of landfill space, and the need to explore resource-recovery options. Maximizing the level of service under current budget limits disheartened him. "The important issue which must be confronted," he stated,

> is that most of the [proposed strategies] are functionally unavailable to the Department unless we have full support of the Administration, the workforce and the public. Unless these important ingredients of support are present and we can employ these strategies, I am afraid that you and the public can not truly expect the Department to successfully boost productivity and pursue its missions with the level of staff which will be available to us on July 1, 1978.[69]

The preoccupation of doing more with less tended to mask (or at least delay) any concerted effort to confront the problem of shrinking landfill space. Vaccarello agreed in principle that each borough should dispose of its waste within its own boundaries, but Fresh Kills remained the safety valve for disposal options despite vehement opposition from Staten Island leaders such as Guy Molinari.[70] The commissioner did follow through on his support for resource recovery by having the DSNY accept bids from private industry in September 1978 to construct a plant in Manhattan.[71]

As his first DSNY commissioner, Koch reappointed Vaccarello.[72] But Vaccarello soon left the DSNY for the private sector, prompted by the mayor's unhappiness with his inability to reorganize the department.[73] Koch then moved in a bolder direction and appointed Norman Steisel, as sanitation commissioner (1979–1986).[74] He proved to be the longest serving commissioner since World War II.

Steisel was a different breed of leader for the DSNY. Born in Brooklyn in 1942 (another child of Polish immigrants), Steisel held degrees in chemical engineering from Pratt Institute and Yale University. He studied systems design and administration and had a deep knowledge of public policy analysis and decision making.[75] Before he was appointed commissioner, Steisel was vice president for Griffehagen-Kroeger, Inc., in charge of consulting services and then from 1974 to 1979 served as Koch's first deputy budget director.[76] Self-confident and forceful, Steisel had such a liking for fine food and the good life that his friends referred to him as "Mr. Taste and Mr. Waste." His service to the department, however, was no laughing matter. Upon Steisel's resignation in 1986, Mayor Koch remarked that the commissioner had made the DSNY the epitome of fine public service. The former deputy mayor Nathan Leventhal flatly stated, "He brought the Sanitation Department into the 20th century."[77]

Steisel understood what tools he needed for his new job: "Even during relatively stable times of balanced budgets, labor peace and reliable service delivery, the successful public sector manager must be something of a hybrid showman, businessman, and psychologist."[78] The new commissioner faced a colossal test in his assignment, especially because the fiscal crisis had so damaged the department. By 1978 the quality of service and the perception of the DSNY had dropped to a new low. Steisel also realized that the deterioration of service "threatened to result in serious political repercussions." The overall budget (in constant dollars) was cut 28 percent from its peak in 1974 to 1981. The uniformed manpower levels dropped from 12,231 in 1974 to 9,472 in 1978.[79]

The commissioner's task in the collection and street-cleaning areas encompassed a number of challenges. Major cuts came in street cleaning (essentially 90 percent of the broom operations), which was viewed as the least essential of the department's services.[80] Trash-collection operations were unreliable, and the truck fleet was in bad repair (downtime rates for the trucks could run as high as 50 percent). Collection costs were high and getting higher, and the output of the collection crews was in question. In addition, street cleanliness levels were dropping.

Steisel responded in several ways. Most significantly, he allocated resources where they could be most useful, shifted employees from one function to another for maximum efficiency, and developed new supervisory and disciplinary policies to better manage the workforce.[81] Steisel also set out to improve the truck fleet, establishing the Bureau of Motor Equipment to deal with the long-term question of maintaining reliable vehicles. With the reorganization and new hires of mechanics, downtime rates for the rolling stock declined, and collection reliability improved.

The commissioner also focused on reducing collection unit costs by an unprecedented negotiation with the USA to cut crew size from three to two workers per truck (and introducing performance-based compensation for the workers). The reduction in crew size allowed the city to use more trucks and collect more refuse. This move saved the city approximately $37 million per year. Turning to street cleaning, he increased the number of workers devoted to the task from eight hundred to 1,750 by 1980. The commissioner, sometimes in ways that would have made the flamboyant Colonel George Waring proud, also sought to change the public image of the DSNY. He supported a performance artist–in–residence, Mierle Laderman Ukeles, who ultimately thanked each worker on the force in person for keeping the city clean. (She actually started under Vaccarello.) He integrated female workers into the all-male workforce without incident and found ways to make supervisory positions available to minorities.[82] By 1986 efficiency was up, plans for a $200 million capital-rebuilding program were underway, snow-removal practices had been enhanced, and about three-fourths of the streets were considered clean.[83]

In dealing with chronic problems of low morale, poor discipline, inadequate internal communication, and inordinate sick leave and injury rates, Steisel not only sought to build in better efficiency standards but to do it without being punitive.[84] "What we have to do is manage our resources more effectively," he stated, "and many people have taken that to mean 'work the men harder.' But I think it goes beyond that. What we really have to do is work the men smarter."[85] This is not to say that Steisel never butted heads with the union; he certainly did. Upon retirement he stated that the sanmen did their work "with a great amount of pride. With a modest increase in salary, they are working 50 percent harder than when I came. You have to deal tough and fair with them and you don't demean what they do."[86]

ENOUGH IS ENOUGH: END FRESH KILLS NOW!

Despite improving conditions in the DSNY by the end of the 1970s, environmental regulations, fewer landfills, little success in finding out-of-state venues for waste, and renewed resistance on Staten Island over Fresh Kills heightened the perennial disposal conundrum. The local landfilling plight was the largest instance of what was happening nationwide. Everywhere in the country existing facilities were overflowing, and siting new ones became problematic in many states.[87] Fresh Kills had evolved into New York City's only significant disposal option, all while its existence became more and more abhorrent to Staten Islanders.[88]

Since the 1960s, the area around the landfill had become increasingly suburbanized, and the encroachment of the dumpsites toward residential neighborhoods was of serious concern.[89] In 1976 the city adopted a Special Natural Area District (SNAD) ordinance to preserve aquatic, biologic, geologic, and topographic features considered to have ecological and conservation value. Under SNAD, all new developments and site alterations (primarily on vacant land) were reviewed. In the 1970s landfilling was taking place in four sections (1/9, 2/8, 3/4, and 6/7), across something like 1,100 acres. By 1980 the size of Fresh Kills' area available for fill had decreased to approximately 2,200 acres because of the surrender of former landfill properties east of Richmond Avenue. Such constriction, however, did not preclude discussions on how to extend the landfill site or locate alternative sites. There was some talk in 1977 of the need for a substantial increase in the vertical limits of the mounds, then set at forty feet.[90]

Staten Island community leaders continued to urge the closure of their notorious landfill. In 1978 matters finally started to turn in the island's direction, if ever so slightly. Mayor Koch had originally planned to reopen the Ferry Point landfill in the Bronx (shut down in 1963) to take up the slack from the anticipated closure of the Pelham Bay landfill in January 1979. Under community pressure, he elected instead to temporarily ship part of its waste to the Fountain Avenue site and then ultimately to Fresh Kills.[91] In a common refrain, Deputy Staten Island Borough President Ralph Lamberti stated, "We recognize that we have to take our share of garbage, but we don't think we should be made to take everybody's garbage."[92]

Koch's decision on Ferry Point provoked a strong reaction among many local political leaders. Molinari wrote to Koch on May 20, 1978, that he was "extremely disappointed" that the mayor had decided not to reopen Ferry Point. "Your solution," he stated, "was announced to be an expansion of the Fresh Kills site on Staten Island." Molinari noted that garbage disposal is "an extremely sensitive question, and no community in New York City would sanction the dumping of garbage in its backyard." Staten Islanders "are particularly sensitive since we already have the largest landfill in the entire world and [have now] learn[ed] that our present facility will be expanded to take almost 50% of the garbage produced in New York City." Molinari also explained that Steisel had advised him that a number of resource-recovery plants were ready to be built but were held up from a lack of necessary legislation. But since the legislation passed in 1976, however, "there has been almost no progress made since then."[93]

Koch was getting it from all sides. Councilman Stephen B. Kaufman, a Democrat from the Thirteenth District, in the Bronx, wanted the Pelham Bay landfill closed immediately, not in January 1979 as planned.[94] The mayor urged patience. "You must recognize that as long as the people of the City produce waste, it will be necessary for the City to dispose of it." He again held out the promise of resource recovery as an alternative to current dumping and incineration practices.[95] Even if Koch had satisfied the demands of the Bronx, he was making no friends in Staten Island. It was singling out Staten Island as the city's prime dumping ground that particularly riled the locals, almost more than the long-term siting of the facility in Staten Island's backyard.

CONNELLY AND THE FIRST CONSENT DECREE

In the fall of 1978, Staten Island's Democratic assemblywoman Elizabeth A. "Betty" Connelly (Sixtieth District) filed a lawsuit in New York's Supreme Court contending that Fresh Kills (in her district) had been operating illegally because it did not have a state permit. The lawsuit essentially was intended to force the closure of the landfill, naming the City of New York and the State of New York as defendants. Under a new regulation in 1978,[96] the Department of Environmental Conservation was to oversee statewide compliance requiring municipalities to apply for permits for their landfills to reduce their environmental impacts.[97]

The lawsuit came as a result of new federal and state regulations, the Federal Resource Conservation and Recovery Act (RCRA, 1976) and the New York State Resource Recovery Policy Act (1977). The environmental laws were the most important weapons the public had for fighting the infamous disposal site in what was and continued to be a gigantic NIMBY battle. RCRA was the first comprehensive framework for hazardous waste management, redefining solid waste to include hazardous materials. The law also contained provisions on resource recovery, attempted to close many open dumps, and set minimum standards (including size and location) for waste disposal facilities through rules and regulations promulgated through the states. The New York State law was one such regulatory tool.[98]

With the lawsuit, Connelly moved into a small group of local officials who formed the leadership of the anti–Fresh Kills effort during the next two decades, including Marchi, Molinari, and Assemblyman Eric N. Vitaliano. Born in Brooklyn in 1928 and raised in the Bronx, Connelly moved to Staten Island with her husband in 1954. She worked for Pan American World Airways from 1946 to 1954. Betty Connelly was involved in local schools and community activities on the island, where she was attracted to Democratic politics. In 1973 she won her initial run for the New York Assembly, becoming Staten Island's first female elected official (winning her last election in 1998) and representing most of the North Shore. She also was the first woman Democrat to chair an

Assembly standing committee. Much of Connelly's work centered on social programming, especially support for the mentally disabled. She also got involved in the issue of Fresh Kills in the late 1970s—not surprising for a Staten Island politician.[99]

Connelly's first personal encounter with the landfill was in 1974, when she was invited to the opening of the West Shore Expressway. Driving along the new stretch of road she could see the mounds and smell the garbage. She later reflected, "The smell and the litter coming from [Fresh Kills] just awoke everyone's awareness to what they had not seen before. So people started of course complaining. And we were getting letters, and I ended up asking if I could take a tour out there." Since some landfills still were operating in other boroughs at the time, Connelly added, "all of the legislators were screaming about the landfills and what was happening in the community."[100] Although she had little political clout then, and even though the ongoing fiscal crisis made it clear that "the last thing [Mayor Ed] Koch wanted to spend money on was the landfill, at least the Staten Island landfill," she turned to the lawsuit. "After a lot of frustration," she stated, "we had to sue! I said to my counsel, let's file a suit against the city, I want to close it. Let's challenge them. And, of course he thought at the time I was crazy."[101]

While the lawsuit was being adjudicated, landfilling had become a big national issue in the 1970s. In January 1978 a conference entitled "Defusing the Garbage Time Bomb," sponsored by the Environmental Action Coalition and the National Science Foundation, was held at the College of Staten Island's Richmond Hall. The former City Council president Paul O'Dwyer was the keynote speaker, and he discussed complaints about Staten Island as a dumping ground for the city as a whole and concerns about dumping elsewhere in the city. "The problem goes back to the days of Robert Moses," he stated, "who had a feel for parks, but no affinity for wetlands." He added, "We've come to the point where it is not possible to keep looking for new places to dump where we will be unwelcome." The need for reduction of waste was echoed throughout the meeting.[102]

In the New York State Legislature, the refuse problem had become a regular topic of debate. In late September 1978 Marchi spearheaded a public hearing conducted by the Senate Finance Committee on New York City's solid waste disposal question. Fresh Kills obviously was on the senator's mind.[103] He was prompted because of the pending closure of the Pelham Bay landfill, which would divert about two thousand tons of refuse more per day to Fresh Kills.[104] At the hearing, Connelly railed, "Over the years Staten Island has been bamboozled and preyed upon by state and city officials. Opposition by local residents to the disastrous impact of the landfill going back over many years has been ignored."[105]

FIGURE 11.4 Destruction of wetlands at Staten Island, May 1974.

Source: Wikimedia Commons.

Connelly's efforts to push for a lawsuit were primarily geared toward closing the landfill. But she also wanted to see the development of a park, which she believed had been promised when Fresh Kills first opened. "Debris scatters over the road, and into our waterways. Chemicals leach into our water, rats play in backyards near the dump, and the air smells terrible in the summer. Uncovered city barges continue to haul garbage from all over the city to our shores." Most directly affected were the residents of Travis, Greenbridge, Village Greens, and Annadale, "but all residents of Staten Island are affected in one way or another," she declared.[106]

While the 1979 ruling of State Supreme Court Justice Charles R. Rubin did not restrain the city from dumping at Fresh Kills, it led to a range of new stipulations.[107] The attention his decision raised, along with legal pressure from New Jersey (where several towns across from the Arthur Kill experienced noisome odors from Fresh Kills) and the Interstate Sanitation Commission, helped produce a consent decree (consent order) in 1980 requiring that Fresh Kills comply with current landfill requirements. Under two similar consent decrees, New York City agreed to bring Fresh Kills into compliance with environmental laws by 1984. Rubin's judgment also included a court-ordered "Fresh Kills Landfill Advisory Committee."[108]

While Judge Rubin was attentive to the impact of the landfill on local residents, the decision did not address the important issue of leachate pollution.[109]

As stated earlier, when the landfill was first built—and this was common for all sanitary landfills at the time—it had no liner, no methane gas–retrieval system, and no leachate-recovery plan. The original sanitary landfills in the 1930s and 1940s were thought to be superior to existing dumps because they layered refuse between dirt in a systematic way and then carefully compacted the buried material. Although a step forward in disposal technology at the time, the environmental flaws of the sanitary landfill had become quite obvious by the 1970s. Deep groundwater at the Fresh Kills site was thought to be protected by a thick clay layer below the landfill, but the structure lacked a well-designed liner and other controls meant to shield surface and groundwater from contamination, and thus the purity of the water was degraded. Technically, under current regulations Fresh Kills had become an illegal dumpsite and now was forced to address its issues to meet compliance standards.[110] In practice, such an old landfill would never completely incorporate present-day standards, but a plan for mitigating its worst flaws as best as possible was the goal.

For many Staten Islanders the consent decree was not enough. That Judge Rubin had failed to shut down the landfill immediately was the salient point. In an interview before signing the consent decree, Rubin stated, "None of us can say it's a complete disposition of the matter—it's not."[111] At this point, no definitive deadline for closure was in sight. A report prepared for the DSNY and released earlier in October 1979 predicted that Fresh Kills would need to be open for "several more decades"—a large leap from earlier assessments of a closing date sometime in the mid-1980s. "'Decades' is an unacceptable word for this landfill," Connelly asserted.[112] Molinari added, "The fact that the landfill will continue to be open does not at all surprise me." He now felt compelled to push the city to accelerate its proposal to build waste-to-energy plants, but he sensed that that would not be enough in itself to shut down Fresh Kills.[113]

Responding to an invitation to an escorted tour of Fresh Kills, Connelly wrote to DSNY Commissioner Steisel, "My disdain is not so much that [Fresh Kills] is a dump, and not a landfill anymore, but the enormous expanse has been treated as if it were on a desert island ten thousand miles from civilization, when in reality garbage has been dumped right in our backyards." She found it insulting that the most of the refuse (she stated "all," but Staten Island contributed refuse as well) came from other boroughs—"with Manhattan as a leading 'contributor.'" "It appears," she added, "that the City feels it is dealing with an Island that is uninhabited by allowing the garbage from other boroughs to blow about into our communities."[114] To Staten Islanders, Fresh Kills had been and would always be "the dump"; "sanitary landfill" was merely a euphemism. They were not being "landfilled" in their eyes so much as being "dumped upon" actually, figuratively, and politically. As a March 1979 story in the New York Times declared, "Staten Island was always the one borough every politician could live without."[115]

PANACEA: RESOURCE RECOVERY

Aside from the protests that the Connelly lawsuit helped kindle, Steisel and the DSNY faced day-to-day problems (some routine and some not so routine) in managing the Fresh Kills site. A brief strike of private sanitation workers in December 1978 disrupted some transport of refuse to disposal facilities.[116] Operational problems, such as broken-down or unreliable fixed unloaders, forced Steisel to rent an additional mounted unloader. An out-of-service landfill digger caused a backlog of trash to remain on barges. The equipment malfunctions hampered disposal routines, as did necessary pollution-remediation work. In a kind of domino effect, subsequent restrictions in the use of some marine transfer stations limited access to private cartmen, diverted trucks to other stations, and forced the operation of incinerators on Sundays—a violation of emissions standards.[117]

The most disruptive setback at the time was the eighty-eight-day tugboat strike in the spring of 1979. Tugboats towed city-owned garbage scows to Fresh Kills and to legal offshore dumping sites; without them at least ten thousand tons of solid waste a day could not be hauled to the landfill. Collection service was disrupted, piles of refuse mounted up throughout the city or were being trucked to various sites, and a citywide health emergency was declared. Many workers of the sanitation force were now working overtime, which cost the department an additional $220,000 per week.[118] Desperate to end the strike, the courts put pressure on the union to resume barge service.[119] Because this was a port matter, President Jimmy Carter (urged on by Mayor Koch and Commissioner Steisel) directed the Coast Guard to help haul the refuse in early May. Ferryboat crews also were used on three chartered tugs.[120] In late June the strike finally ended, but the vulnerability of New York City's refuse-disposal system had again been made quite evident.[121]

Several "solutions" to New York City's disposal problems came and went (and came back again) over the years: ocean dumping, reduction, incineration, and landfilling. Across the nation, future prospects for sanitary landfills were fading because of a lack of available land, NIMBY issues in communities unwilling to accept landfills, and environmental concerns. Because of the declining possibilities for landfilling and the unsuccessful building program for conventional incinerators, talk in the DSNY (and among other interested parties) turned to a general interest in resource recovery. In spirit at least, modern resource-recovery technologies and programs were meant to achieve what Colonel Waring and his cohorts wanted to accomplish in the 1890s, that is, the recycling, reuse, or transformation of waste rather than its mere disposal. Ultimately, the options for New York City would narrow to waste-to-energy incineration and recycling.

Seeking to take advantage of the reuse possibilities immediately available, however, the Board of Estimate voted in August 1975 to approve a plan to extract methane gas from the garbage at Fresh Kills.[122] Although landfill sites had always emitted methane gas, it was not until the 1970s that the necessary technical expertise and economic conditions intersected to make extracting the gas commercially promising. The board voted to give NRG New Fuel Company (its successor was Reserve Synthetic Fuel, Inc.) "mining rights" on four hundred acres at the landfill site. In turn NRG New Fuel would sell the methane to Brooklyn Union Gas Company (BUG) and provide the City of New York with a 12.5 percent royalty. The company had a pilot project underway in Los Angeles, but delays plagued the start of drilling at Fresh Kills.[123] In the summer of 1977 the board approved a proposal to permit a joint venture of the NRG Recovery Company and Reserve Synthetic Fuels, Inc., to mine gas from the designated four hundred acres of the landfill. City leaders waited to see if this new project would commence by the predicted start date of April 1978, but in the meantime the project was embroiled in controversy. A competing state-financed project promoted by BUG was allowed to move forward, and some parties questioned if the city was getting enough in return for its concessions.[124]

With the price of natural gas soaring and the volatility of shipping liquefied natural gas (LNG) still an issue, methane appeared to offer a good prospect despite its high cost of extraction. NRG's vice president Richard Manderville stated that "Fresh Kills is like a big digester pot for methane" because so much of the gas was trapped by the smooth clay subsoil.[125] An editorial in the *Advance* endorsed the methane-extraction effort. "Maybe we'll discover that, after all, a silk purse can be made out of a sow's ear."[126]

Testing at the Fresh Kills site for methane was delayed until late 1977, and changes were then made to the initial agreement. In addition, the project developers wanted assurances from the DSNY that a minimum amount of refuse would be deposited annually. This was necessary, they argued, because tests had shown that a steady flow of waste was essential to extend the period for gas recovery. By the end of 1980 the initial contract was revised, and in October 1982 the DSNY signed a joint-venture agreement.[127] On July 28, 1982, the first gas processed at the plant was sold to BUG. The gas-recovery project at Fresh Kills was the biggest in the world at the time and was a model for ventures elsewhere.[128]

The plant operation was not without problems. It shut down temporarily in 1986 because of a spill accident and again in 1987 until a thermal combustor could be acquired.[129] Methane extraction, more than the development of other types of processing facilities, presumed the continuation of landfill usage and thus did not meet, in and of itself, the test of a resource-recovery system meant to be New York City's future primary disposal option.

The proposed increase in waste volume also was contrary to what Staten Islanders had been fighting about all along, namely, closing the landfill. At a

March 1979 meeting of Community Board 3 in Staten Island, Fred Rice from RSF revealed their methane-mining plan. The response from those present was cool at best (some people could see the value of the resource), and in fact they were angered because the project implied keeping open Fresh Kills indefinitely. Richard McGivney, the board's chair, declared, "This is New York City, the greatest city in the world—and we can't handle a landfill?"[130] Similarly, talk about increasing the amount of garbage shipped to Staten Island for processing into refuse-derived fuel (RDF) also was unsettling to locals.[131]

Ongoing misgivings about the clean operation of incinerators led to serious community hostility toward them. Recent disclosures had indicated that many burning facilities across the country had been designed poorly, were operated badly, maintained unsightly locations, and were costly. The number of operating incinerators in the United States was declining: from 265 in 1965 to 241 in 1970.[132] Pollution emissions in particular gave incinerators a bad name with the public, but their unique ability to reduce the volume of waste kept them from disappearing entirely. The capability of newer technologies to raise steam for the production of electricity (plus RDF) caught the attention of interested parties as a consequence of the 1970s energy crisis and the need to develop alternative energy sources to petroleum and natural gas.[133]

For much of the decade, the cost of adding pollution controls to incinerators was prohibitive, and thus burners rarely competed head-on with the cheaper sanitary fills (at least as exclusive disposal options). From an environmental standpoint, incinerators did not appear compatible with the emerging philosophy of resource reuse, despite the promise of recovering metals from residue. Even the more hopeful possibility of producing electricity did not find many popular outlets until the end of the decade, especially because of the difficulty in selling steam byproduct. Those who were most optimistic about incineration believed that the method had a place, particularly where land for sanitary landfills was scarce or unavailable, or in the case of New York City, where the dumping at Fresh Kills had clear limits.

The plants got a modest boost in 1979 with the Public Utility Regulatory Policies Act (PURPA). The law provided guaranteed markets for electricity sales, despite the high cost of generating electricity from garbage. The potential for energy generation made resource-recovery systems of various types intriguing because they held out the possibility of a return on investment. Federal support alone did not give resource-recovery plants a competitive edge in the disposal arena or guarantee a neighborhood's acceptance of such plants.[134]

The waste-to-energy plant was only one option for resource recovery in these years. With the creation of the U.S. Environmental Protection Agency (EPA) in 1970, responsibility for most refuse activities were transferred to it. In that same year, Congress passed the Resource Recovery Act, which shifted emphasis from refuse disposal to recycling, resource recovery of various types (pyrolysis, composting, and so forth), and the conversion of waste to energy. It created the National Commission on Materials Policy to develop a national policy on materials requirements, supply, use, recovery, and disposal. Although federal legislation had limited enforcement power, it caused states to become more involved in updating policies on solid-waste collection and disposal and encouraged alternatives to existing disposal practices.[135]

Recycling, once regarded as a grassroots method of source reduction and an innocuous protest against overconsumption in the 1960s, was at a "takeoff stage" in several American communities by the end of the 1970s and becoming a realistic alternative disposal strategy. Competition from recycling programs raised questions about the economic feasibility of burying resources, the need for preserving virgin materials, and the high cost of landfill construction and maintenance. Concerns about leachate pollution of groundwater and unmonitored methane production also discredited the sanitary landfill.[136]

Many questions arose, however, about a full-throated commitment to recycling: What incentives should be used to foster compliance among householders, businesses, and manufacturers? What about mandatory laws or governmental procurement policy? How much attention should be paid to recycling literacy? And most importantly, can markets be found for the increasing volume of recyclables? In New York as elsewhere these issues would need to be addressed.

In 1978 the City of New York completed its first "Solid Waste Masterplan" for resource recovery from collected refuse. Early in 1974 the NYEPA had recognized the need for such a plan and requested help from the state government.[137] As a condition of this support, the state required the city to develop a Solid Waste Task Force to draw up the master plan and to be responsible to a Citizen Advisory Committee.[138] Also, in December 1974, NYEPA initiated a planning effort concerning markets for refuse-derived energy in the city.[139] Aware of New York City's precarious fiscal situation, it considered the possibility of private financing of resource-recovery technology and realized that it had to act precipitously given the pressure of landfill closings so much on the mind of the DSNY and others. The task force recommended eleven different resource-recovery projects, each with the possibility of being carried out by public and/or private entities. One of the proposed projects was slated to be built on Staten Island (an RDF plant located at Fresh Kills), where preliminary work already had been begun.[140]

Talk about the RDF plant by the task force, a proposed PASNY generating station, and the methane-gas project on Staten Island all pointed to viewing Fresh Kills as a resource generator. The Power Authority was quick to promote the idea that the recovery plant required increasing the flow of refuse to Staten Island, not stemming it. Governor Carey opposed the idea and wanted further study.[141] In October 1979 a group of government officials, energy experts, and environmentalists took a helicopter ride over the landfill. Representative Robert A. Roe, a Democrat from New Jersey, observed, "Here's an enormous resource literally going to waste."[142] In some people's eyes, the shift from land sink to resource quarry added another compelling reason for not abandoning Fresh Kills—much to the consternation of the locals.

Koch and Steisel were not ready to commit to a comprehensive plan for resource recovery in 1978. The mayor proposed a "minimum risk" plan for resource recovery to study one specific site rather than developing a larger overall strategy. Both men contended that the success of resource recovery still remained to be demonstrated and that heavy financial investment was premature.[143] For the DSNY, the move toward a resource-recovery plan based on the mayor's charge began in December 1978 when the Board of Estimate approved consultant contracts, soon to be followed by an identification of promising plant sites.[144]

The DSNY got approval from the Board of Estimate to seek Department of Energy resource-recovery grants for a waste-to-energy facility. Steisel wrote to Koch, "This action has now generally put us on track in this important area."[145] Responsibility for resource recovery and related issues would be assigned to Paul Casowitz, who became deputy commissioner for resource recovery and waste disposal planning.[146] Casowitz had a background in electrical engineering but no familiarity with resource recovery. He gained experience about the city bureaucracy through his work with the fire department, where his mathematical skills helped keep it within its budget and provided other useful analytical tools for his job. Steisel was impressed with this "quant wonk" and sought Casowitz for the new position.[147] "The creation of this new post," Steisel noted, "will provide the Sanitation Department, for the first time, with an organizational structure and mandate necessary to advance resource recovery and also address the often neglected problems of our current waste disposal system."[148] Steisel quickly made public the DSNY's new charge: "We simply do not have the time to waste. We must find a way to dispose of our refuse that will be environmentally sound and economically feasible and we are convinced that resource recovery is the wave of the future."[149]

In 1979 the City of New York retained three consulting firms to advise officials regarding a resource-recovery facility. The initial site-selection process for the first plant was restricted to the boroughs of the Bronx, Brooklyn, and Queens. Staten Island was omitted because its total waste generation would not support a facility and thanks to a policy choice not to increase the volume of refuse that would be sent from other boroughs to Fresh Kills. The project team's recommendation was the Brooklyn Navy Yard in Greenpoint-Williamsburg. During World War II, the Brooklyn Navy Yard (along the East River) was the largest naval construction facility in the United States, with seventy thousand employees. It closed in 1966. In January 1974 Mayor Beame announced that the city had received a federal grant toward the demolition of obsolete structures there and believed that the yard would make available sixteen acres of excellent industrial space for development.[150]

As a justification for using the Brooklyn Navy Yard, Casowitz stressed that the site permitted barge transport of waste instead of relying on truck delivery and that it offered a ready market for steam. He recognized prophetically (given the intense fight ahead over the location, especially with the local Hasidic Jewish population) that the Brooklyn Navy Yard was "not immune from problems. A facility could involve incineration and associated stacks that may provoke local community opposition."[151] In all, the team identified fourteen potential sites, with one in the Bronx, and possibly Hunts Point likely for a second project. Steisel suggested that the mayor issue a statement outlining the city's plans for resource recovery, which would have the strong support of business interests, construction unions, and some key newspapers.[152]

In March 1979 Steisel announced the publication of a white paper on disposal issues faced by the DSNY and the new directions to be taken in resource recovery. The working document, *An Overview of Refuse Disposal and Resource Recovery in New York City: Issues and New Directions*, provided insight on the department's thinking about its future refuse-disposal plans. Steisel candidly stated in the foreword: "Social and legal pressures to change our methods of solid waste disposal are acute, and are justified. Environmental objections to present disposal methods increase continually, and the physical capacity of our existing landfills is limited."[153]

In Steisel's mind, any new project had to protect the city from both financial and technological risk (a plant that would not work or energy that could not be sold). "Resource recovery," he stated, "represents a new field in public administration. These projects, in most cases, are essentially business ventures in which government is in partnership with industry. Because of the lack of experience on both sides, the aggregated risk is substantial."[154] His goal was to offer "a three-way solution" for the city: providing an opportunity to improve disposal practices, mitigate against environmental pollution, and reverse the flight of industry from the city.[155] In a related press release, Steisel made clear that the DSNY would "try hard" to recommend the location of facilities so that each borough "pulls its own weight."[156]

The white paper reviewed the major issues facing the department in implementing the new objectives, including compatibility with current practices, opportunity for economic development, the nature of institutional relationships, and possible legal constraints. With respect to landfills, the biggest issues were taking into account the current capacities (balanced against longer-term disposal needs for resource-recovery residuals and construction waste), the impact of new uses such as methane recovery, and changes in projected termination dates (an issue of supreme importance to Staten Islanders). Incinerators in use also came under scrutiny, as did current collection operations.[157]

The economic viability of resource recovery was a high priority, as was how to manage the current disposal system—three truck-fed landfills (Fountain Avenue, Edgemere, Brookfield), two additional landfills for construction waste (Pennsylvania Avenue and South Avenue), Fresh Kills, nine marine transfer stations, and six incinerators. At the time, approximately 80 percent of the city's refuse was being landfilled, which would increase once the three incinerators were shut down. However, of the $76 million in the DSNY 1979 budget authorized for capital expenditures, $50.5 million was dedicated to waste disposal needs.[158]

The resource-recovery strategy was taking a positive, if modest, step. Out of necessity, this left Fresh Kills as the only major disposal option for the foreseeable future. According to Steisel, "This first resource recovery project will proceed primarily as a waste disposal oriented project, rather than an economic development oriented one." He also was aware of the need to develop a source-separation (recycling) program to reduce the overall volume of disposable materials. This was somewhat of a full-circle return to the Waring years.[159]

A SIMMERING POT READY TO BOIL OVER

Steisel's hopeful vision of a new day for the DSNY in the struggle over disposal would be met with setbacks and roadblocks. But at the very least, the late 1970s was the beginning of a conversation over shifting the city's disposal practices away from Fresh Kills. While the tone of the discourse was not yet frantic, the need for urgent action was setting in. City leaders and the DSNY were becoming aware that their choices were narrowing.

Anticipation about the promise of resource recovery in the late 1970s did not quell concerns and frustration among Staten Islanders about the future state of Fresh Kills.[160] For its part, the DSNY and the Koch administration were using the possibility (warning? threat?) of doubling the current amount of refuse dumped at the landfill by 1985 as justification to support its resource-recovery initiatives. Casowitz was miffed at Molinari for opposing legislation to seek proposals for eight to ten resource-recovery projects. Molinari was balking at the projects because the mayor's office was failing to support mass-transit proposals for Staten Island.

The political logrolling produced fiery language and tangled up a number of unresolved issues. On July 26, 1979, the *Advance* reported: "Last night, before a sympathetic audience, Molinari took the offensive and asked if 'we are going to believe this baloney that the Sanitation Department is handing us that they care about our garbage problem.'" Despite Casowitz's best arguments, Molinari did not believe that building resource-recovery plants should be equated with eliminating landfills.[161] It remained to be seen if the 1980s provided any prospect of moving forward on a workable solution to New York City's disposal debacle.

Fresh Kills had now been around long enough to witness yet another round of interest in replacing landfilling with burning. Waste-to-energy was a newer technology than traditional incinerators, but what seemed to be the same were the persistent insecurities about the existing disposal plan in place. Were city leaders simply looking for a better mousetrap or reacting to a problem—discarding, discarding, and discarding—that never ended?

CHAPTER 12

FRESH KILLS AT MIDLIFE

In the late 1940s, when Fresh Kills Landfill was opened, the dream that incineration would ultimately solve New York City's disposal woes was gaining momentum. City officials waited for the day when combustion would supplant dumping. By the mid-1960s, Fresh Kills continued to grow, and incinerators were shutting down. In the late 1970s and into the 1980s, anticipation that waste-to-energy facilities could eventually replace landfilling brought about a resurgence in the solution being burning refuse, this time in the form of resource-recovery plants. Dumping space was getting scarcer every year, and this new technology was the latest panacea.

New York City was putting its faith in yet another cure-all to solve its refuse-disposal woes. Officials continued to tell themselves and others that Fresh Kills Landfill—in what was now the facility's midlife—was nothing more than a stopgap measure, something that Staten Islanders had been hearing for decades, now.[1] The Department of Sanitation's (DSNY) actions, however, did not treat the facility as if it were makeshift or temporary, but some people were possibly unwilling to accept that fact. The determination that incineration would save the day evoked the tramps of Samuel Beckett's 1953 play *Waiting for Godot*, endlessly anticipating a man who never arrives. Was waiting for new incinerators any less futile?

DUAL TRACKS: FRESH KILLS AND RESOURCE RECOVERY

The operation of Fresh Kills remained a major task for the DSNY in the 1980s. Promoting resource recovery as an alternative solution to landfilling also consumed its time. The Staten Island landfill and the proposed resource-recovery plants therefore ran along parallel tracks in the search for a way to deal successfully with the endless production of waste facing New York City.[2] Maintaining such a dual track was tricky: it required economic, technical, and environmental justification for both disposal options and pitted a one-borough solution against a multiborough approach.

Most immediately, the DSNY was grappling with finding a home for refuse displaced by landfill closures in the city. In January 1981, the Hamilton Avenue incinerator in South Brooklyn, which had been operating for eighteen years, was shut down. Day by day, the relics of past disposal policy were fading away.[3] The Pelham Bay landfill in the Bronx, which had taken in about 2,700 tons of refuse each day, closed in 1979; Staten Island's Brookfield truck fill closed in the summer of 1980 (its eight hundred tons of waste diverted to Fresh Kills); and Brooklyn's Fountain Avenue landfill was slated for closure in 1985 (much of its eight to ten thousand tons a day also likely to be diverted to Fresh Kills). By default, Fresh Kills was the city's primary disposal option.[4]

In 1984 and 1985 Fresh Kills clearly was the world's largest landfill. It required a staff of 350, representing twenty occupational titles. The facility covered an area of about three thousand acres, 2,200 of which were available for fill. The operation, which ran six days a week (sometimes seven), was taking ten thousand or more tons per day (on busy days at least 12,000 tons). Plant #1 (occupying 1,200 acres) remained the chief unloading facility; while Plant #2 (980 acres) had been relatively inactive since 1978 and was being redesigned. At the time, sections 1 (263 acres) and 9 (358 acres) were the principal dumping areas. Since refuse was being or would be transferred from closed or closing landfills, sections 6 (266 acres) and section 7 (258 acres) also were open to dumping. Sections 3 (300 acres) and 4 (131 acres) were soon to be opened. Section 5—the 176-acre Davis Wildlife Preserve—was off limits to dumping. A landfill gas-recovery project began operation in the summer of 1982. Several new roads, ramps, and other construction projects were underway to relieve congestion, but some of the landfill's infrastructure was subpar or decaying (few floodlights at night, decrepit docks, equipment breakdowns) and needed attention.[5]

Daily operations at Fresh Kills were demanding, but the DSNY also had to find a way to fulfill the requirements of the 1980 consent decrees. The landfill's environmental shortcomings were ongoing: It faced a serious mosquito problem, rainwater falling on the mounds of refuse percolated into the groundwater as leachate (which contained microbes, metals, arsenic, and PCBs), and exposed garbage spread repulsive odors into the nearby residential communities. One resident in the Huguenot section called the smell "overpowering." Deputy Commissioner Paul Casowitz made the case that some progress had been made, such as walling off parts of the landfill with earthen berms, preventing litter, and using more dirt to cover the refuse.[6] But the department had not yet met state codes on a daily cover, and remediating leachate was costly and time consuming.[7] In a more cosmetic effort, highway workers began to install a three-thousand-foot-long fence to hide the landfill from the West Shore Expressway.[8]

During his tenure, however, Commissioner Norman Steisel had called for more attention to developing end-use plans for the landfill in anticipation of closing sections or even the entire facility.[9] But he stated with some concern, "Waste disposal has become the biggest problem facing the Sanitation Department.... If all these new waste disposal plans bomb out, we're going to have to look to Fresh Kills."[10]

In a November 1980 letter, Mayor Edward I. Koch tried to assure Senator Bill Bradley (a Democrat representing New Jersey) that the City of New York was doing everything it could to deal with complaints about floating refuse originating at Fresh Kills and polluting New Jersey. "I am aware of the history of problems related to refuse washing up on the Jersey Shore," he stated, "and can appreciate your concern about the pollution that has plagued the beaches of Woodbridge and Carteret." He added that in July New York City and Woodbridge had entered into an agreement through which the latter would receive payments for cleaning its beaches. Koch also mentioned that New York City was attempting to bring the landfill into compliance with state regulations, with the objective of "operating an environmentally sound landfill." As a gesture of goodwill, he thanked Bradley for his support of resource-recovery legislation.[11]

The letter may have been conciliatory and put the best spin possible on things, but it alone solved nothing. In remarks at a December 1980 meeting of the Department of Environmental Conservation (DEC) on the future of New York City's landfills, Senator John J. Marchi bemoaned the fact that under the mayors Robert Wagner Jr., John Lindsay, Abraham Beame, and now Koch, officials "were not prepared to contemplate a New York City without a Fresh Kills site." Marchi argued that for years, "the politics of doing nothing substantial about Fresh Kills proved beneficial to those in all boroughs save Staten Island." He added, "Resource recovery is an approach well worth advancing, but the rate at which the city is proceeding leaves little doubt that, unless there is a dramatic change in the intensity of the city's commitment, Fresh Kills will continue to be buried under the refuse of all our boroughs." Marchi's plan called for developing a resource-recovery plant at Fresh Kills within five years.[12] Other proposals also were on the table. But the resource-recovery technology that held the promise of replacing landfills had yet to prove itself on the environmental front, especially regarding air pollution. The only such plant operating in the area was located in Hempstead, Long Island, and it had been closed for several months because of emissions, bad smells, and labor problems.[13] Where to go next was in Mayor Koch's and the DSNY's hands.

PLANNING AND PROTEST OVER THE BROOKLYN NAVY YARD

The first tangible step (quickly blunted) was to construct a resource-recovery plant based on the recommendation of a DSNY site-selection team in 1979. In April 1980 the Board of Estimate gave the department approval to issue a Request for Proposals (RFP) for the design, construction, and operation of a resource-recovery facility at the Brooklyn Navy Yard, the first choice after a review of several sites. They hoped to develop a major public-private project.[14]

In April 1981 the department was granted the right to contract for an Environmental Impact Statement (EIS) for the site. In May it received four proposals in response to the RFP, and in December United Oil Products, Inc. (UOP, a subsidiary of The Signal Companies, Inc.), was chosen.[15] Opposition arose even before the work began, especially from nearby neighborhoods in the Greenpoint-Williamsburg area. Locals had suffered widespread asthma and lead poisoning in the past from other industrial development and opposed the inevitable truck traffic and congestion that would increase once the plant opened.[16]

Public resistance and differences of opinion over technical issues related to the plant led to a protracted battle over the site. Implementing the resource-recovery plan proved difficult, especially when a borough other than Staten Island faced the possibility of accommodating a new disposal facility. City Comptroller Harrison J. Goldin made it known that he could not support the proposed plant at the Brooklyn Navy Yard. In a January 21, 1983, letter he declared, "I oppose the plant. My reasons relate to the environmental and technical advice that I have received from independent consultants." Despite

FIGURE 12.1 Waterfront in Greenpoint, Brooklyn, June 2007.

Source: Wikimedia Commons.

FIGURE 12.2 Aerial view of Brooklyn Navy Yards, c. 1966.

Source: Wikimedia Commons.

supporting resource recovery as "the best available technology for solving our garbage disposal problem," Goldin thought that the proposed plant was the wrong size and to be set in the wrong place, and he believed that dioxins produced in the burning process "cause the greatest concern" (as did furans).[17] Despite claims that emissions from the resource-recovery plants would be minimal, Goldin concluded that "fears about the dangers of dioxins are not trivial and should not be dismissed lightly or ridiculed as parochial."[18]

Mayor Koch accused Goldin of trying to undermine the plan in order to appease Brooklyn residents who opposed it.[19] For the moment, Goldin's reservations were in the minority. Steisel and Casowitz downplayed the risk of dioxin by arguing that high combustion temperatures would mitigate against it.[20]

The eminent scientist Barry Commoner and his fellow researcher Karen Shapiro of the Queens College Center for the Biology of Natural Systems took a different view, and this resulted in a debate that became more animated as time passed. By the 1980s the intractable Commoner was a leader in the environmental movement, a scientist-activist with a long history of ecological advocacy. He described himself as a "visionary gadfly" with a deep commitment to the central role of science in society, and rightly so.[21] Brooklyn born (in 1917) and raised, the son of Russian Jewish immigrants, he grew up in a modest household during the Great Depression, during which time his parents lost their life savings. Commoner became an enthusiastic biology student in high school and went on to earn an undergraduate degree in zoology at Columbia University and advanced degrees in biology from Harvard University.

In 1933, when he enrolled at Columbia, the university was a center of social activism, which inspired a good deal of Commoner's future public interests. He also was keenly aware of the stigma placed on him because of his Jewish background (at a time of mounting anti-Semitism), which sharpened his criticism of the status quo. He served in the Navy during World War II, ironically involved in the spraying of Pacific islands with DDT (unaware of its properties), to prevent disease. His early experience with DDT made him a lifelong skeptic of introducing new chemicals and technologies into society if there was any reason to think they would be a health risk.[22]

Commoner eventually joined the faculty of Washington University in St. Louis in plant physiology and later became a university professor in environmental science. His study of strontium-90 (a radioisotope contained in nuclear fallout) took him out of the laboratory and permanently into the realm of social activism. He used his Committee for Nuclear Information to urge parents in St. Louis to provide him with their children's baby teeth for study, which led him to raise serious concerns about children ingesting milk tainted with strontium-90. (Dairy cows had grazed in areas that had been contaminated by the carcinogen.)

At Washington University he established the Center for the Biology of Natural Systems (CBNS) in 1966 and moved it to Queens College in 1981, which increased his interest in environmental problems faced by New York City. Commoner's causes always linked his view of the environmental crisis with what he believed were flaws in the economic (corporatist) and social system. He thought that scientists had an obligation to make information available and accessible to the public, and he stood up against a whole array of environmental risks, including nuclear fallout, pesticides, water and air pollution, toxic metals, and urban waste disposal. He became a noted participant in many grassroots environmental campaigns, including battles over New York City's refuse-disposal policies. He was a relentless opponent of incineration and was instrumental in shaping the dispute in New York City. Commoner even ran for president on the Citizen Party ticket to make his case.[23] He once said, "Environmental pollution is an incurable disease," stressing an inherent problem with human-led economic growth.[24]

Commoner and Shapiro argued that burning garbage without first removing polyvinylchloride (PVC) plastics would produce an unacceptable dioxin risk.[25] Early protests against the proposed facility were getting louder, but more because of NIMBY concerns than from fears of dioxin or other pollutants.[26]

The siting problems certainly hindered the DSNY's plans for constructing its first waste-to-energy-type resource-recovery plant. Also crucial was determining the effective life and efficient functioning of Fresh Kills Landfill as a complement to (instead of as an alternative to) resource recovery. Beginning in late 1981, Casowitz's Office of Resource Recovery and Waste Disposal Planning (in association with Wehran Engineering and Holzmacher, McLendon and Murrel) worked "to develop a plan for averting the crisis in waste disposal that confronts the City." An interim report was issued in January 1983, stating three fairly obvious objectives: upgrade Fresh Kills Landfill and related marine facilities to offset "imminent closure" of the remaining city landfills, ensure that Fresh Kills would meet current waste disposal needs, and bring Fresh Kills into compliance with regulations and laws.[27]

The team also developed three alternative "grading plans," each with different values on space or volume to be made available for dumping. The "Ultimate Grading Plan" (maximizing the amount of solid waste that could be placed on the site) projected eighteen more years of life for Fresh Kills. The "Modified End-Use Plan" would make portions of the site suitable for recreation uses, but at a cost, and would extend the life of the landfill by 17.8 years. The "End-Use Plan" gave greater attention to recreational land development but would extend the life of the fill only 13.5 years. The report concluded, "These plans graphically illustrate the need for accelerating the pace of efforts to develop alternative methods of waste disposal."[28] Developing the grading plans had fallen to the contractors, with a goal of pushing back the closing date for Fresh Kills a little further into the—still unspecified—future.[29]

The report looked at ways to relieve the transfer of refuse from the closing landfills. It regarded the marine transfer stations, because of their design and locations, to be able to handle a significant portion of the waste. But the value of the transfer stations was being undermined by the unreliability of the transport, unloading, and dumping operations. There had been reductions in the transport of materials by barge because of significant patterns of "downtime" and the age of the barges. Also, marine unloading at Fresh Kills was "breaking down with increasing frequency."[30] The report concluded that unless significant capital investment was made in the whole Fresh Kills operation, and if counterproductive work rules were not changed, the Staten Island facility "will not only be unable to meet future demand, but will probably fail, with increasing frequency, to meet existing disposal needs."[31]

A grading plan alone could not assure Fresh Kills' future viability as a disposal site. Rehabilitating existing incinerators or possibly extending the life of remaining landfills might postpone but not solve the disposal problems. Within the context it provided, the team considered how to plan for the construction of resource-recovery facilities but also to determine immediate steps "to weather the closure of three landfills and the transition to resource recovery without any prolonged or serious negative impacts." The report reaffirmed the dual-track approach for dealing with disposal.[32]

LANDFILLS AND TOXINS

As Steisel, Casowitz, and the DSNY were keenly aware, prolonging of the lifespan of Fresh Kills bought more time to implement resource recovery. Yet rising public awareness and heightened sensitivity over the disposal of toxic and hazardous materials in landfills intensified the need for their closure as soon as possible. The Love Canal controversy in the late 1970s became a cause célèbre for that issue. Love Canal, an incomplete canal dug near Niagara Falls in upstate New York, was the burial site for an estimated 21,800 tons of hazardous materials from Hooker Chemical's manufacture of pesticides and other chemicals between the early 1940s and 1953. Problems became public in the 1970s, when local residents complained about chemical odors and possible illness caused by the deposits in the canal near their homes and schools. The U.S. Environmental Protection Agency's (EPA) Superfund program was begun in 1980 to clean up some of the nation's most contaminated sites and to respond to a variety of environmental emergencies. Between 1979 and 1989 the State of New York and the federal government spent about $250 million on site remediation and relocation at Love Canal.[33]

One of the features of the hazardous waste controversy in New York City in the early 1980s was the legal and illegal dumping of hazardous and toxic materials into city landfills. On January 3, 1982, Assemblywoman Elizabeth A. Connelly announced that she was considering legal action against the DSNY to prevent the dumping of such wastes at Fresh Kills, demanding the closing of the site if necessary. "It seems unlikely, that if toxic wastes were dumped and buried at the (Fresh Kills) landfill," she stated, "it could be done without the knowledge and cooperation of someone within the Department of Sanitation." A possible source of illegal dumping under investigation concerned the possible disposal of industrial waste at either Fresh Kills or Brookfield.[34]

Two days later the New York State Senate Select Committee on Crime announced that a New Jersey company, the Hudson Oil Refining Company,[35] had dumped more than 150,000 gallons of highly toxic material in three city landfills between 1978 and 1979. Investigators followed up on evidence in another case that suggested that DSNY workers might have cooperated with the company's truck drivers in the illegal act. In their testimony, Hudson drivers admitted that they had been sent to the Pennsylvania Avenue landfill in Brooklyn, Pelham Bay landfill in the Bronx, and "Arthur Kill" (meaning either Fresh Kills or Brookfield). The waste included a variety of known or suspected carcinogens. Senator Ralph J. Marino, chair of the select committee, stated, "It looks like we're just scratching the surface at this point. There are other companies involved, and we haven't really dug it all out yet."[36]

Casowitz expressed deep concern that department employees were implicated in the scandal.[37] In February, Superintendent John Cassiliano, a nineteen-year veteran of the DSNY, was dismissed. He was charged with being complicit in the illegal dumping and paying off a subordinate to help him, and in November he was found guilty.[38] By the spring, the DSNY sought to establish a medical-screening program for sanitationmen working at landfills, after evidence of hydrocarbons and toxic waste were discovered at Brookfield. The findings also provided Representative Guy V. Molinari with an opportunity to reaffirm his opposition to the proposed electricity-generating plant on the Arthur Kill.[39]

An interagency report issued later in 1982 reaffirmed earlier findings (which would be questioned later) that no toxic contamination existed at the city landfill sites. A Health Department survey of people living near Brookfield uncovered a greater number of minor respiratory complaints from citizens but no serious illnesses. Members of IRATE (Islanders Against a Toxic Environment) were neither convinced nor reassured by the findings. Rose Pisciotta, who headed the group, stated, "There definitely is a problem here, but they're afraid of causing a panic."[40]

The latest toxic waste dispute added one more layer of opposition to landfilling. In March 1983, for example, the College of Staten Island chapter of the New York Public Interest Research Group (NYPIRG) planned an April "Toxics on Tour" bus trip that would visit Brookfield, Fresh Kills, the former Chelsea Terminal Tank Farm in Travis, the Rossville LNG tanks, and other sites to bring attention to current antipollution bills before the legislature.[41]

THE FIVE-HUNDRED-FOOT FUROR

While many Staten Islanders viewed resource-recovery plants as a welcome alternative to Fresh Kills, potential environmental risks there raised anxiety about the local impact of the site. Rumors circulated that the DSNY was considering elevating the refuse mounds at the landfill to an unheard-of five hundred feet, which also spoke to its long-term physical presence on the landscape.

The Republican assemblyman Robert A. Straniere (South Shore, Staten Island) introduced legislation on March 7, 1983, that would require all municipal landfills closed by June 30, 1990. The link between promoting the resource-recovery plants and closing Fresh Kills was obvious: "The people of Staten Island," Straniere declared, "are simply not going to accept 500 feet of garbage and unless the city is forced to develop alternatives to Fresh Kills that is exactly what we can look forward to."[42] In May, Assemblywoman Connelly wrote to Comptroller Goldin to express her deep concern about Fresh Kills becoming the sole refuse depository for the city and about the projections for continuing disposal there "for a minimum of twenty years." "I am sure you can understand the panic Staten Island residents are experiencing," she stated, "as they try to comprehend the potential impact of all the city's refuse being dumped in their borough. The prospect of creating mountains of garbage, reaching heights in excess of five hundred feet... is certainly cause for alarm." She requested an updated audit of the current operations at Fresh Kills, a DSNY study of the fiscal projections for the life of the landfill, and a review of "the managerial and technological assumptions made by Sanitation about the operation of the landfill under increased use."[43] She reiterated that the city's inaction was "destroying air quality and property values, damaging beaches, waterways, tidal wetlands and environment."[44]

On May 18, Borough President Anthony R. Gaeta, a Democrat, took members of the Board of Estimate and their representatives (including Mayor Koch, Commissioner Steisel, and Queens Borough President Donald R. Manes) on a tour of Fresh Kills. Gaeta stated, "We wanted to show all the members of the Board of Estimate first hand just what the term 'landfill' actually means."[45] A few days after the tour, Deputy Borough President Ralph J. Lamberti, a Democrat, and other Staten Island officials met with Commissioner Steisel to review the department's plans to raise the heaps of refuse to five hundred feet. "My fear," Lamberti stated, "is the 'mountain' is a stopgap measure."[46] Gaeta announced in early June that he intended to introduce a resolution before the Board of Estimate linking any future use of Fresh Kills to the development of resource-recovery plants in all five boroughs.[47] For Gaeta, such a resolution expressed a need to have all the boroughs bear the burden of disposal, not just Staten Island.

Local complaints were turning into intensified political action. Representative Molinari had "lost patience" with the handling of Brookfield landfill, and in May his office asked Jacqueline E. Schafer, the EPA's regional administrator, to review the data on claims that toxic materials were saturating the dumpsite. Molinari had serious reservations about the tests run by the departments of Health and Sanitation denying that levels of toxins were any higher there than at other landfills.[48] Area residents and local officials supported Molinari's position. Steisel stated that he was interested in performing additional health and environmental studies but had yet to identify the money to do so. Some skeptics suggested that it was more than money keeping the city from further probes.[49] On another front, City Councilman Frank Fossella planned to file suit to close Fresh Kills if the federal government phased out the Fountain Avenue landfill in Brooklyn (scheduled for closure in 1985).[50]

Eric N. Vitaliano, elected in 1982 as assemblyman from the Fifty-Ninth District (Democrat, East Shore, Staten Island), quickly became one of the new voices in opposition to Fresh Kills. (The landfill became part of his territory after reapportionment.) Elizabeth Connelly later remarked that once Fresh Kills was within Vitaliano's district, "You don't intrude on other members, so I left them to decide, to introduce bills to close it." At that point, she became a supporter of Vitaliano's efforts to deal with the disposal issue rather than taking a lead role. But, she added, that for the whole delegation Fresh Kills was "a cause celebre."[51]

Vitaliano was born in 1948 in the New Brighton section of Staten Island. He was a graduate of Fordham University and NYU School of Law (where he served on the school's Environmental Law Council). Vitaliano was admitted to the bar in 1972 and practiced law until 1979. Elected to the Assembly in 1982 as a Democrat, he was more conservative than many others in his party, which benefited him on Staten Island.[52] He was strongly identified with supporting the death penalty, lobbied for good transportation on Staten Island, and was a strong environmental advocate. (His efforts in addressing freshwater wetlands was of particular note.) In 1984 he was appointed to the Legislative Commission on Solid Waste Management, which was charged with exploring alternatives to landfills.[53] "The district I represent in the Assembly," he stated, "can point with pride to many wonderful places which we have." But Fresh Kills "would not make the highlight section of my Chamber of Commerce brochure for Staten Island."[54]

Even officials on the state level were getting into the act about the way New York City ran its dumpsites. The *Advance* observed that environmental administrators "haven't tried to hide their disgust with the way the city runs its landfills, including Fresh Kills, and the lack of aggressiveness in planning for resource recovery plants throughout the five boroughs."[55]

The flap with New Jersey over polluting its beaches with waste from Fresh Kills added to the criticism of New York City's sluggish response to the disposal problem. In June 1983 U.S. District Judge Herbert J. Stern ordered New York to build a planned multi-million-dollar barrier by 1985 meant to block refuse floating from the landfill to the beaches of Woodbridge. He also ordered New York to bear cleanup costs. He added, "If they fail, if the situation is not substantially abated, this court will order that that landfill be shut."[56] Medical waste remained a problem; hypodermic needles and other medical supplies regularly washed up on New Jersey beaches. Mayor Philip Cerria threatened to truck debris from the Woodbridge beaches and dump it on the steps of New York's City Hall if the medical waste was not contained. This was yet another battle in the ongoing garbage war between the two states.[57]

A CHALLENGING PLAN FOR MANAGING DISPOSAL

The DSNY's self-evaluation in 1984 was positive. Higher productivity in collections enabled the department to redirect resources to other functions such as street cleaning, which had been starved of funds during the early period of the financial crisis.[58] Using two major performance measures in the area of collection—how much garbage was picked up and how much was left behind—department officials claimed marked improvement from past years.[59] In the March-April issue of *Interfaces* (a journal focusing on the practices of operations research and management science) Assistant Commissioner for Operations Planning Lucius J. Riccio noted that before Steisel's appointment as commissioner, the DSNY "was a municipal embarrassment." The department now was "generally recognized as the finest in city government" and "one of the outstanding sanitation agencies in the country."[60]

The DSNY's inability to resolve the disposal problem was not praiseworthy. The Brooklyn Navy Yard construction project remained in limbo as the Signal Companies (the parent company of OUP, who won the plant contract) reorganized. Signal acquired Wheelbrator-Frye (an unsuccessful bidder for the incineration project) and saw the exit of UOP. Signal's new resource-recovery unit was called Signal Environmental Systems, Inc. In January 1983 the department undertook an independent study of health effects of dioxin emissions from proposed resource-recovery facilities, and in June it was instructed to produce a Draft Environmental Impact Statement (DEIS) for the Brooklyn Navy Yard and a general report on future disposal plans.

In September 1984 Steisel submitted three planning documents to the Board of Estimate: an assessment of potential public health impacts associated with dioxins and furans from the proposed Brooklyn Navy Yard facility, a draft EIS for the Brooklyn Navy Yard facility, and a supplement to the department's citywide action plan completed in April. The commissioner stated in his memorandum to the Board of Estimate, "I believe that the need for solid waste disposal is one of the most critical infrastructure problems facing our City."[61]

Fred C. Hart Associates conducted the dioxin and furan study. The primary finding was that the proposed plant was likely "to reduce the emissions of dioxin to the greatest extent possible given the current state of the art." Steisel added that eight of the nine reviewers "essentially agreed with the findings of the Hart report and its conservative approach." The ninth reviewer did not directly disagree with the findings but was concerned that the study was conducted under contract to the DSNY. The Hart report also was favorably reviewed by the commissioners of Health and Environmental Protection.[62]

The six-volume DEIS of the proposed Brooklyn Navy Yard facility outlined how the plant would dispose of three thousand tons of municipal solid waste (MSW) per day, recovering the equivalent of one barrel of oil for every ton of waste processed. Refuse would be barged to the site, and the resulting steam would be sold to Consolidated Edison.[63] The mass-burning water-wall technology chosen for the project represented "several significant improvements over existing facilities."[64] The air-quality analysis indicated that emissions of sulfur dioxide and other acids would be minimal. The DEIS also affirmed that the Brooklyn Navy Yard project would provide "an effective and proven means of disposing of our solid waste, recovering valuable energy, and minimizing any public health impacts to well below accepted standards."[65]

In *The Waste Disposal Problem in New York City: A Proposal for Action* (April 1984), the DSNY recommended that the city construct eight resource-recovery plants around the city (at least one in every borough) and extend the use of the Fountain Avenue landfill beyond 1985. It also called for placing greater emphasis upon recycling as a means of solid waste management. The Brooklyn Navy Yard facility was planned for completion in 1988 (the first new waste disposal facility in fifteen years). Construction debris, discarded appliances, and street dirt still would need to be dumped into landfills. Steisel also sought the participation of the board in clearly articulating the disposal problem and the need to take action, critically examining available alternatives, and involving the public in any decisions.[66] In the face of few remaining disposal facilities in operation in the city, the DSNY intended to "reverse this trend" by turning to waste-to-energy plants, starting with the first at the Brooklyn Navy Yard.[67]

The report noted the continuing drift toward "narrowing our waste disposal facilities and options," the limited capacity of Fresh Kills (at about sixteen to eighteen years), efforts to rehabilitate and expand the five existing incinerators, and the tenuousness of the barge transport system as exemplified by the 1979 tugboat strike. In summary: "the Department of Sanitation views the waste disposal problem in terms of the contraction of the disposal system and the depletion of disposal capacity." Landfills were "non-renewable resources," not parks in waiting as Robert Moses presumed.[68]

The DSNY objectives, therefore, were to ensure that there always was sufficient disposal capacity (reserving at least some landfill space), to implement new resource-recovery facilities, to promote waste prevention measures, and to reduce the need for landfilling as a singular disposal method. The commissioner realized that all of these measures would take time and require a major financial commitment. On the issue of siting, the general report restated what had been constantly repeated, that waste disposal facilities needed to be distributed throughout the five boroughs. Also, "in recognition of Staten Island's history for bearing a significant share of the City's waste disposal burden and the continued operation of Fresh Kills through the transition to new technologies, Staten Island will be the last borough in the geographic order of implementation for new facilities."[69]

While not overlooking a variety of alternatives to landfilling (including processing, exporting, and waste prevention), the department's policy hinged on the implementation and success of the waste-to-energy technology and an extended tipping capacity at Fresh Kills. It acknowledged that "in effect, we are already in a crisis situation" and that all of the department's good intentions needed to be realized in under two decades. The faith in resource recovery was quite definitive: "The real issues are not whether we will implement resource recovery, but rather when, where, and how."[70]

In the memorandum accompanying the studies, Steisel attempted to clarify what appeared to be "a perceived lack of commitment to recycling" in the April 1984 report. "I would like to recall that the Mayor and I successfully led the fight that culminated in the adoption of a returnable beverage container law by New York State." He asserted that 550 tons of material per day were being recycled as a result of the bottle law in its first year of implementation. He added, "Recycling and waste volume reduction are obviously the most desirable means of dealing with solid waste. While materials recycling is not a feasible solution for our total waste disposal needs, we intend to develop and exploit recycling opportunities more fully."[71]

Steisel's presentation to the Board of Estimate was well considered, rational, and concrete—if a little too optimistic about resource recovery. Such a serious and dogged issue as refuse disposal was salted with alternative assessments, local resistance, and a certain fear of the known and unknown, as the previous few years had demonstrated.[72] But some clarity was afoot in DSNY disposal policy, at least in tone and intention: resource recovery would become the primary refuse-disposal method for the city, the use of Fresh Kills would continue and its life extended as a complement to resource recovery, all boroughs would participate in disposal programs, and recycling (if not source reduction) would be actively pursued.

Senator Marchi had been on the Board of Estimate since at least 1980. Along with his support for the DSNY's plans was constant attention to the fate of Fresh Kills. He proposed legislation to appropriate $1 million to the DEC to begin detailed plans for a Staten Island resource-recovery plant. But it struck a nerve among Republicans and Democrats. Molinari called it "an asinine proposal that doesn't make sense." In his mind, it singled out Staten Island rather than spreading responsibility across all of the boroughs. Connelly agreed. Vitaliano declared, "We don't have to create a guinea-pig situation on Staten Island." Resource recovery, he argued, should begin in the other four boroughs "as soon as possible." Straniere was a little more conciliatory but stated, "The Senator is right in trying to get the city moving on implementing a resource recovery network, but I don't agree our borough should be first."[73]

Although Marchi's plan was pragmatic, it was a political hot potato even beyond Staten Island. Steisel told Marchi on January 5, 1984, "I am writing to express my appreciation for your announced support of a resource recovery facility at the Fresh Kills Landfill. I can only hope that your willingness to grapple with our city's waste disposal problem will set an example for elected officials in the other boroughs."[74]

Not dissuaded by resistance from his Staten Island colleagues and others, Marchi introduced the bill in the Senate on January 10. "If Staten Island is to be just one of the sites for resource recovery plants," he stated, "then we will have an equitable distribution of responsibility for handling of the city's refuse. And that is what I envision."[75] He continued to drive home this point as the bill worked through the legislature. Responding to criticism, Marchi stated, "If I had suggested that we build a mammoth plant designed to handle ALL of New York City's garbage well into the 21st century, I could understand the reaction. But what I have suggested is that we host one of *many* incinerators to be built throughout New York City, one of limited capacity." The senator trusted Mayor Koch to be true to his word about constructing the Brooklyn Navy Yard plant, thus making the proposed Staten Island plant second in line.[76] Yet there was no smooth sailing for the bill, some boroughs fearing a precedent that would affect them adversely. Nonetheless, the bill passed the Senate on May 16 and moved to the Assembly.[77] The project, however, was never completed.

Vitaliano took a slightly different tack than Marchi. In February 1984 he proposed legislation to contract design plans to convert the city's nine marine garbage-loading stations into resource-recovery facilities. "If neighborhood opposition is delaying construction of resource recovery facilities in our city, why not bring the resource recovery facilities to the garbage." Like Marchi, he wanted to move away quickly from relying on Fresh Kills: "The time for fiddling while Staten Island chokes on garbage at the Fresh Kills Landfill is long gone."[78] During the year he moved closer in support of the DSNY's resource-recovery plan. Vitaliano regarded the idea of extending the life of the Fountain Avenue landfill as welcome, but only as an "interim cushion." "We must," as the department suggested, "move to new alternatives."[79]

Developing resource-recovery plants was a race against time, but one that the DSNY thought they could win. Fresh Kills was a constant topic for discussion, and Brookfield, now long closed, was an ongoing source of community anxiety on Staten Island. About twenty-five neighbors of the Brookfield site met in January 1984 to discuss continuing problems and request a public hearing. There was a recent discovery of at least one industrial pollutant (pentachlorophenol or PCP) and several dozen other chemicals during test drillings. NYPIRG members charged that the city had not been doing enough testing at the landfill. One attendee stated, "At this time there is no apparent documentation that anything in the landfill is harming the community. But the city's position seems to be that until it is documented, they're not going to do anything about it."[80]

A staff liaison for Vitaliano attended the meeting and urged the citizens to keep pressure on the DSNY to inform the EPA of its findings. In response, the DSNY unveiled plans to seal the surface of the dumping site at Brookfield.[81] An air-quality monitoring study (November 1983) was released in March 1984, and a consent order issued in 1985 called for the DEC to make information about the site available, to develop a work plan for a remedial investigation, and to prepare a feasibility study.[82] A federal investigation determined that up to fifty thousand gallons a day of hazardous waste (oil, sludge, metal paint, lacquers and solvents) had been disposed of at Brookfield during its last six years of operation.[83]

NEW NIMBYISM

The DSNY's ability to assuage concerns over extending the life of Fresh Kills was badly strained in the early 1980s. Speculation on the life of the landfill after 1985 and talk of five-hundred-foot mounds on the site offered little hope of imminent closure. In the fall of 1984, one more potential indignity surfaced. The DSNY was seeking permission to open a new seventy-five-acre section of the landfill adjacent to an existing community in Greenridge–Arden Heights. The department claimed that the process of filling the site and then converting it into a park would take but four years. The *Advance* referred to it as "the creeping landfill." "We expect the Fresh Kills landfill to be with us for some time to come," it stated, "and we grudgingly accept that. But we do not have to accept the fact that it will be built right up to our very doors. That is where we draw the line."[84]

Writing to DEC Commissioner Henry G. Williams on October 5, Vitaliano warned of "a grave new threat to the environment and quality of life of Staten Islanders living in close proximity to the Fresh Kills garbage dump." Vitaliano reminded Williams that he had long advocated "the progressive reclamation" of the landfill, especially to constrict the site's perimeter, thus moving dumping operations "further and further away from people." "Tightening the noose around Fresh Kills," he stated, "will permit revegetation and reforestation of areas which have been (or still could be) used for landfilling operations and will keep stench, rodents and other health hazards further away from people." Vitaliano opposed extending the fill site to sections 2 and 8 (including the seventy-five-acre tract), which were behind the Jewish Community Center on Arthur Kill Road in the Greenridge area. In selecting the site, the department rejected using sections 1 and 9, which were remote spots on the other side of the West Shore Expressway and thus more costly to develop. Vitaliano urged Williams to halt the implementation of the dumping plan.[85] Williams's reply was vague and of little consolation.[86]

Somehow, the DSNY was having trouble limiting any expansion of the dumping sites at the very moment when it needed more support for its resource-recovery plan. Instead of conciliation, it pushed back against local criticism. Steisel argued that the DSNY had "to make the best possible use of [Fresh Kills'] remaining capacity" and in ways "that will have the least undesirable impact on the adjacent communities."[87] Casowitz added, "The capacity of Sections 2 and 8 is just too precious a resource to leave forever untapped."[88] But just like the imagery of five-hundred-foot mounds, dumping within two hundred feet of the JCC's children's playground and five hundred feet from a church in Greenridge–Arden Heights were graphic reminders of the encroachment of the landfill over more than thirty-five years. A protest letter to Mayor Koch from the Village Greens Residents Association (representing 725 homeowners) linked the proposed new site to Brookfield: "Please do not forget the tragedy that occurred at the Arthur Kill Road Brookfield Dump [the discovery of toxic wastes]. We are afraid that a similar tragedy will occur at this new site, since it is in such close proximity to neighboring homes."[89]

With the February 1, 1985, scheduled start date for developing the new site, protests became louder. One local wrote, "Isn't it bad enough we all have to live with the by-products of this dump, now they are literally throwing garbage in our faces."[90] Borough President Lamberti, who succeeded Anthony Gaeta, also reminded Commissioner Steisel that Borough Hall and the DSNY had allied over the resource-recovery plan. "In light of this history of cooperation and alliance, I am calling on you to show good faith in a related matter: garbage dumping in sections 2 and 8 of the landfill."[91]

The DSNY abandoned sections 2 and 8 for the moment. Locals had won a significant battle in their war against the landfill. The *Advance*, a moderate voice in many of the Fresh Kills debates, admonished island citizens at a community meeting who were "hooting and heckling the very sanitation officials who had agreed to their request." It added, "If there is anything worse than a sore loser, it is an obnoxious winner." The time had come to work with the department, it stated, for long-term change.[92]

But strong public reaction against the landfill did not ebb. In April 1985, one member of a citizens group generally upset with the quantities of refuse being dumped at Fresh Kills suggested that garbage be thrown in front of runners at the New York Marathon.[93] In a February 12, 1985, letter to a constituent, Vitaliano wrote that persuading DSNY to use a more remote section of Fresh Kills instead of sections 2 and 8 "will buy us some time to get resource recovery on stream, which is the long run solution to our landfill problem." But he concluded, "The fight [regarding Fresh Kills], however, is far from over."[94]

In November 1985, the Department of Sanitation reevaluated the possibility of using sections 2 and 8 for general dumping. Protests rose again, but they were not solely about those sections.[95] In late 1985 and early 1986 the possible dumping of asbestos at Fresh Kills put Vitaliano back on the defensive. The DSNY won out, but he continued to fight the ruling. Not being able to dump asbestos at either the Fountain Avenue or Pennsylvania Avenue landfills, the DSNY had turned to Fresh Kills.[96]

BATTLE OVER BROOKLYN

In the mid-1980s the DSNY not only found itself defending its actions at Fresh Kills but also saw its plans for developing resource-recovery plants beginning to unravel. By now, the plan was trimmed to have one incinerator per borough, especially because of disputes over the location of the facilities and budget limitations. On December 6, 1984, members of the Board of Estimate objected to the sites selected for the first new facilities during testimony over the Sanitation Department's disposal scenario. Several of the members threatened to vote down the resource-recovery plan unless changes were made (especially for sites in their particular boroughs). Queens Borough President Manes stated, "Everyone has reservations on the sites that he has." Staten Island officials kept pushing for the plan, along with mayoral aides and Commissioner Steisel. "If we don't solve this problem," stated Staten Island Borough President Lamberti, "you won't have anyplace to put your garbage. You will have to eat it."[97] Making every borough responsible for much of its own refuse was becoming a hornet's nest of an issue.

Complicating the already challenging turn of events, the U.S. Department of the Interior announced that the Fountain Avenue landfill would have to close by the following December. An *Advance* editorial put the issue simply but accurately: "That's terrible news for the city as a whole, and even worse news for Staten Island in particular."[98] After substantial political pressure, the Board of Estimate on December 20, 1984, approved by a narrow margin to continue planning for five resource-recovery facilities (with the Brooklyn Navy Yard not formally included in the decision). This was a nonbinding recommendation.[99] Like on Staten Island, citizens in Brooklyn mobilized against the possible imposition of a waste facility. The Citizens Advisory Committee for Resource Recovery (CAC) had formed there in 1981, with members appointed by Commissioner Steisel, Borough President Howard Golden, and Community Boards 1, 2, and 3. A major focus of their work involved unresolved questions regarding the Brooklyn Navy Yard incinerator. In an April 4, 1985, report, CAC concluded, "Environmental considerations were not paramount in the site selection, nor were those related to the community. The two major considerations . . . were those of speed of development and ease of garbage delivery." Thus, "the available information fails to demonstrate that the Brooklyn Navy Yard is the most appropriate site for the proposed plant." The concerns included air emissions, odor, noise, permitting, construction, operations, and aesthetics.[100]

Specific reservations to the DSNY's resource-recovery plans took a back seat to siting. The attorney Irving Scher filed legal paperwork in the Brooklyn Supreme Court in April to halt the Brooklyn Navy Yard project. He claimed that the facility was illegal because it violated environmental-planning laws.[101] On August 15, the Board of Estimate—after its seesaw deliberations over several years—approved by a vote of 6 to 5 construction of the Brooklyn Navy Yard plant, with the help of Commissioner Steisel, influential New York leaders, and Eric Vitaliano and Betty Connelly.[102] With the Board of Estimate vote in hand (and that of the City Council), officials turned to getting the necessary permits from the Department of Conservation. The first sanitation facility to be built by the city since 1961 (a mass-burn plant for Brooklyn) inspired renewed neighborhood opposition.[103]

The dispute over the incinerator had become increasingly messy. Borough leaders split over issues of self-interest. Environmentalist groups split as well. Established national groups, such as the National Resources Defense Council (NRDC) and the Environmental Defense Fund (EDF), supported waste-to-energy facilities if they were equipped with ample environmental controls and if the city agreed to implement permanent recycling programs in order to reduce the total amount of waste to be burned. In 1985 the EDF released a study emphasizing that a curbside recycling program with a recycling rate of 40 percent could be cost competitive with resource-recovery plants. Mayor Koch (upon Steisel's recommendation) decided to give recycling equal weight with the waste-to-energy facilities, and in June 1985 the city agreed to execute a voluntary recycling pilot program in five residential neighborhoods. The DSNY had a tepid response to the recycling plan, believing that EDF's 40 percent mark could not be achieved in the three-year projection.[104]

Resolve to develop curbside recycling in the 1980s for New York City fell in line with national trends but would not be easy in such a large and complex city. In 1968 Madison, Wisconsin, may have been the first city to begin curbside recycling (of newspapers). Recycling centers in several communities experienced modest success, but the inconvenience of dropping off materials at a centralized location severely limited participation. Curbside collection programs became the most effective method of gathering recyclables, achieving the highest diversion rates from the waste stream of any option available. From the modest beginnings in Madison and a few other locations, citywide and regularly routed curbside programs increased to 218 (located primarily in California and the Northeast) by 1978. In 1989 it was estimated that there were 1,600 full-scale and pilot curbside recycling programs in the United States, with participation rates estimated from 49 to 92 percent. Not until the 1990s, with markets for recovered materials on the rise, would recycling become a growth industry.

Recycling had developed strong political and social appeal in the 1980s. While not inexpensive, recycling put policy makers on the side of conservation of resources; it also gave concerned citizens a personal way to participate in confronting the solid waste dilemma. Recycling's link to waste reduction and waste minimization as ways to conserve resources and reduce pollution attracted new followers. A major goal of most communities and the nation in general was to increase the recycling rate, which stood at about 10 percent in the late 1980s. In 1988 the EPA called for a national recycling goal of 25 percent by 1992.[105] For a recycling program not yet in its infancy in New York City, Koch's willingness to balance curbside recycling with resource-recovery plants was an abstract idea with some political allure but no operational strategy.

Opponents of resource-recovery plants—with or without a recycling component—faced stiff opposition. Citizen activists were joined by NYPIRG, and especially Barry Commoner, in opposing the siting of incinerators in New York City. NYPIRG's own report in 1986 claimed that incinerators could have a consequential effect on health and advocated an expanded recycling plan as an alternative.[106] Commoner's voice rose above the crowd in condemning the proposed new plants. The dispute over resource-recovery plants for New York City was just the kind of issue Commoner could sink his teeth into—a grassroots issue, a questionable technology, and too pat a solution for a metropolitan area struggling with its disposal problems.

For a time, it seemed that Commoner and Steisel would be able to come to an agreement over a way to use the city's refuse for energy generation. Robert Randol of Smith Barney had extolled the virtues of resource recovery as a tempting new energy producer, a technology that could attract subsidies, and a way to turn waste into a marketable commodity. He passed along his message to Steisel, who became a leading advocate of a disposal technology that just might be the answer to the dying landfills. Commoner and Steisel had met at a conference in the spring of 1982, where the former gave a talk about biomass cogeneration, which could convert methane gas from garbage into heat and electricity and could be managed by small energy cooperatives rather than the city or a private company.[107] Steisel's publicly managed steam-generating incinerators were moving the resource-recovery idea in a different direction. The two men met again to discuss their differing views but ultimately could not agree over the production and risks of dioxin.[108]

In a 1985 article in *New York Affairs*, "Incinerators: The City's Half-Baked and Hazardous Solution to the Solid Waste Problem," Commoner concluded, "In sum, the City has gone to considerable effort to evade its legal responsibility to consider all the available evidence that bears on the potential effect of the proposed incinerator on public health. . . . This conclusion is a sufficient reason to reject the proposal [for resource-recovery plants]."[109] Steisel and Casowitz rebutted Commoner's claims in the same publication in 1986. They questioned Commoner's forecasting of the timing of a coming disposal crisis, stating that three years earlier he had viewed the DSNY's assessment of landfill depletion as "exaggerated." They accused Commoner of shifting his evaluation of dioxin risk "quite dramatically" and of being selective in his evidence.[110] In late 1985 the Board of Estimate supported Steisel's position. (It is unclear whether the magazine exchange influenced their decision or was even available to members of the board at the time.)[111]

Three months after the vote, Steisel resigned his post (February 1986) and moved on to join a bond firm.[112] Nevertheless, the commissioner's policy to move forward with resource recovery and thus replace landfills was in place—at least for the moment.

THE DSNY'S FRESH KILLS WOES

Dennis S. McKeon, president of Citizens for a Cleaner Staten Island, Inc., wrote to Vitaliano in May 1985: "Politics in the landfill matter stink since our health and the health of our children is at stake. Some people understand that but others do not."[113] When the citizen's group organized in January 1985, it had a goal of closing sections 2, 3, 4, 6, and 7 of the landfill, that is, the sections that bordered the residential communities of Greenridge and Travis and the Staten Island Mall in New Springville. McKeon noted that "as the problem keeps growing, our aims are growing as well." The next concern was hospital waste. DSNY whistleblowers had made it known that several bags of medical refuse had been found on barges headed for Fresh Kills in June. Disquiet rose over the possible spread of hepatitis and AIDS (at a time when this was a touch point for panic). As a result, two private hospitals (Mount Sinai Medical Center in Manhattan and Maimonides Medical Center in Brooklyn) accused of repeated dumping of potentially infectious waste were barred from city landfills, and the DSNY began considering fining first-time offenders.[114]

Islanders found themselves battling on several fronts to minimize the problems associated with Fresh Kills. Citizen activists demanded buffers between the growing fill sites and the residential neighborhoods in Greenridge, Arden Heights, and elsewhere, with the support of local politicians. Straniere called for setting strict boundaries at the landfill to protect nearby communities and thus reduce the size of the facility.[115] Leaders like McKeon had more faith in protests from the grassroots organizations than from the Fresh Kills Landfill Citizens Advisory Committee (set up under the consent decree), which he deemed to be a "phony committee designed to stroke the public and do politicians' bidding."[116]

Along with the long-term concerns over limiting the use of Fresh Kills was the threat of the imminent closing of the Fountain Avenue dumpsite. "If they close Fountain we'll get an additional 800 trucks per day," McKeon asserted at a public meeting in April. "There's already a backup at the Verrazano Bridge during rush hour in the morning, and it'll be worse with the additional trucks coming from Brooklyn and Queens. We're not ready for that, and we don't want it. It scares the hell out of me."[117]

While Fresh Kills increasingly bore the load of the city's refuse, failure to comply with all of the elements of the first consent order made it ill-equipped to do so. The DEC had been dealing with complaints about the Fresh Kills operations and other New York City landfills for some time, as well as with concerns about waste that found its way to neighboring beaches and waterways. In late 1984 Vincent J. Moore, regional director of the DEC, had been tracking refuse that was migrating from Fresh Kills into Raritan Bay adjacent to New Jersey and debris entering Fresh Kills Creek from barge unloading facilities.[118]

On October 18, 1984, DEC Commissioner Williams sent what were called "final" Orders of Consent to Commissioner Steisel. They included orders for three inactive sites (Brookfield, Pelham Bay, and Pennsylvania Avenue) as well as orders for Fountain Avenue, Edgemere, and Fresh Kills.[119] There were two orders for the Pennsylvania Avenue landfill, one for closure and the other for remediation of the area's illegally dumped hazardous waste. In an otherwise pro forma cover letter, Williams concluded that recent inspection reports on Fresh Kills "indicate continuing, flagrant violations of the Order of Consent executed by our respective agencies on October 23, 1980. The Department cannot permit these violations to continue."[120]

In response to the package of materials and especially to the cover letter, Steisel responded that "in view of the history of our negotiations, the content and tone of the letter is most disturbing. This Department has endeavored in good faith to cooperate with DEC and has been confronted with uncertainty in DEC as manifested by contradictory information from the regional and Albany offices."[121] Steisel and others in the DSNY were particularly concerned about the consent orders because they wanted the city to put a date on the closure of Fresh Kills. According to Casowitz, "when Koch wasn't [re]elected, I was urged to sign a consent decree, which would put a closure date on [Fresh Kills]. And I said, 'The next mayor can do that. I'm not going to do it,' because to me this was a substantial financial asset to the city."[122]

While the DSNY was trying to implement its resource-recovery plan, it also was engaged in a full-blown jurisdictional battle with the DEC. The new consent orders included the termination of dumping at Fountain Avenue by December 31; upgrading and improvements at Fresh Kills with an end to dumping of raw garbage by 1998 (the date being pushed back again from previous projections); investigation and remediation of hazardous waste at Fountain Avenue; and final closure after remediation of landfills at Brookfield (June 1, 1987), Pennsylvania Avenue (December 31, 1987), and Pelham Bay (December 31, 1989). The orders were not signed until December 1985.[123]

But as early as summer 1986 Vitaliano and Connelly were complaining to DEC that "operations of the Fresh Kills landfill are not following the details of the Consent Decree of December 1985."[124] Making matters even more difficult for the DSNY, Fresh Kills was receiving 29,000 tons per day (an all-time high) in 1986–1987, which included the transfer of refuse from Fountain Avenue. This more than doubled the amount it had been accepting before the closure of most of the other landfills in the city.[125]

A WAITING GAME

By early 1986, as Norman Steisel said his goodbyes at the DSNY, the disposal plan in place depended on maximizing the value of Fresh Kills while waiting for resource-recovery plants to come on line.[126] The new commissioner, Brendan J. Sexton, supported the direction Steisel had taken the department.[127] Sexton was the director of the mayor's Office of Operations. Born in Detroit in 1945, he studied experimental psychology at NYU. He was executive director of Encounter Incorporated (a counseling program for drug abusers in Greenwich Village) and also served for five years as a director of the Addiction Services Agency. More inclined to public service than to corporate life, he helped set up the Office of Operations and ultimately became its director. He had had some exposure to sanitation and snow-removal issues in the Office of Operations and was highly regarded by Mayor Koch. Sexton once stated that he considered the DSNY as "the NASA of city government." "We're moving something this municipality has never done before, resource recovery. In city government, you almost never get a chance to do an entirely new thing. So this is very exciting, very challenging."[128]

While outsiders could see little new in Sexton's support for resource recovery and recycling, he went out of his way to be more conciliatory with Staten Islanders. He promised to back away from dumping refuse at the edges of several sections of Fresh Kills closest to residential neighborhoods and planned buffers, berms, and screens to separate the mounds from the nearby homes and businesses.[129] Dredging along the waterways surrounding Fresh Kills also began in early 1986 in order to keep sunken refuse and sludge from blocking access to the marine disposal facilities.[130] However, Sexton sought compromise, not fundamental change, in dealing with Fresh Kills: his aim was to preserve waste disposal capacity but also lessen impacts on the community. As he told Staten Island civic leaders, "I hope that these actions demonstrate how seriously I take my responsibility to make the landfill operation a better neighbor to the community." He sought to develop an effective capping program and better leachate containment and collection, but he could not support a thousand-foot buffer around the whole boundary of Fresh Kills. Although he favored restrictions in elevation of mounds, some would reach the five-hundred-foot level. Like Steisel and others before him, Sexton saw Fresh Kills as too vital a facility simply to close down.[131]

Staten Islanders were not particularly inclined to view Sexton as their savior, although they welcomed the tone of his responses and the optimism of his promises. Neither would close Fresh Kills. Vitaliano worked aggressively to create a commission on the state level to fight against the continuation of the landfill. The Legislative Commission on Solid Waste Management called for a public hearing on Fresh Kills and published the results in September 1986. "This is music to my ears," Vitaliano said. "I have always believed that a fair and independent look at the City's plans for Fresh Kills was essential to guard the interests of Staten Islanders."[132]

While the testimony was deep and lengthy, the issues raised in past years echoed again—the fate of the Fountain Avenue landfill and its impact on Fresh Kills, the failure to comply with the consent order, the incessant odors, the five-hundred-foot mounds, the need for buffer zones, sections 2 and 8, and the demand for resource recovery.[133] Gordon Boyd, executive director of the commission, threw some cold water on the event when he stated, "I think most people would agree that Fresh Kills will never close."[134] That was not what most participants were hoping for.

The resource-recovery plan was still as much fantasy as reality in the mid-1980s and still subject to influence by all manner of economic, political, and environmental factors. By the start of 1986 the city had awarded a contract to Signal Environmental Systems to build a three-hundred-million-dollar incinerator at the Brooklyn Navy Yard, with construction to begin later in the year.[135] Fresh Kills, however, remained a battleground of protests. Efforts were underway in January to unite fifteen island civic groups in airing their concerns and heading back to the courts.[136]

Groups Against Garbage (GAG) pressured the city to reopen the Fountain Avenue landfill, among other things, to avoid more dumping at Fresh Kills.[137] In February, the DEC released a preliminary environmental impact study on Fresh Kills Landfill (DEIS 5-77), which spoke to changes in its topography, namely elevating mounds to 505 feet. The study concluded that the altered topography "would affect the microclimate of the area, the dispersion of pollutants from nearby industrial sources, and dispersion of pollutants from onsite landfill gas recovery facilities." Changes in the microclimate and pollution dispersion could potentially increase pollution levels "in communities adjoining the landfill."[138]

Was the city any closer to a solution to its disposal woes in 1986 than it was a few decades earlier? A staff report of the New York State Legislative Commission on Solid Waste Management published in 1986 stated, "While every community faces a slightly different mix of disposal problems and resource constraints, the state's solid waste problem is most critical on Long Island and in New York City.... New York City's problem is at least as urgent."[139] "If no other steps are taken," it added, "[Fresh Kills] will be full before the end of the century." The report outlined concerns about constructing mass-burn plants and advocated a recycling alternative. "Recycling cannot be the sole answer to the solid waste disposal problem," it concluded. "Ultimately, the solution must come from changes in the ways we produce and consume."[140] But this was a solution few in the city or elsewhere would carry out, let alone imagine, for many years.

By the late 1980s and into the 1990s (and because of the phasing out of Edgemere in 1991), the city was left with only one major landfill for its residential waste. Disposal options were more limited than ever before, especially in the short term. The gauge on the duel track was narrowing too.

The search for a refuse-disposal solution was not merely a battle over available technologies and practices (dumping, burning, and reusing). Nor was it a way to deal with Staten Island's obsession over closing Fresh Kills. But it was a potential reimagining of the disposal map that would focus on all the boroughs and eventually even beyond the city.

CHAPTER 13

BARGE TO NOWHERE

The saga of the garbage barge *Mobro*, in 1987, unveiled one of the quandaries of mass consumption: the adversity, perversity, and futility of dealing with its residue. As a story in a June 1988 issue of the *Poughkeepsie Journal* bluntly stated, "The drive to consume is exacting a price."[1] A positive consequence of the *Mobro* story was intensifying the public discussion about reuse and recycling. For the Department of Sanitation (DSNY) and city leadership in New York, however, a much stronger push for recycling now challenged the dual-track disposal policy contingent on Fresh Kills Landfill and the promotion of resource-recovery plants (to produce steam for electricity).[2]

A consent order issued in 1990 broadly displayed the limits of what had come to be seen as the reliance on and failures of Fresh Kills. Recycling would be pitted against landfilling as a kind of refuse-disposal savior. The DSNY's insistence on prolonging the life of Fresh Kills revealed the entangled problems of vast consumption, large-scale labor-intensive collection, and disposal practices that could not be easily resolved with "either/or" strategies. The eyesore and known health risk that was Fresh Kills was a dreadful blight on the landscape, an all-too-visible reminder of the limitations and social stigma of land dumping.

THE *MOBRO* STORY

The story of the *Mobro 4000*—the "gar-barge," the "barge to nowhere," the "pariah barge," "ship of fools," the "most watched load of garbage in the memory of man"—is virtually a thing of legend. The infamous garbage barge made national headlines because of its futile 112-day journey, which started in Islip, Long Island. The *Mobro* was attempting to locate a dumping ground for its unwelcome cargo many miles from home but ultimately had to return to New York with its load of waste intact.

The chain of events leading to the fateful trip of the *Mobro* began in 1983. In that year the New York State Legislature passed a law banning new landfills on Long Island and ordered the existing ones to close by the end of 1990. Islip's landfill was almost full, and in 1986 the facility was closed to refuse from schools and commercial establishments. Bids were advertised to haul garbage to other locations. A contract was let to Waste Alternatives, Inc., owned by Thomas Hroncich, the former environmental control commissioner for Islip. Hroncich allegedly had ties to the Lucchese and Gambino crime families. (Long Island's waste hauling businesses long had deep ties to the mob.) Hroncich and Thomas Gesuale, president of Review Avenue Enterprises, struck a deal with Lowell Harrelson, a heating and air-conditioning contractor (and enthusiastic entrepreneur) from Bay Minette, Alabama, who hoped to make money on producing methane from garbage.[3] Harrelson sought to acquire refuse from Islip or from anyone else willing to pay handsomely to have their commercial waste hauled away.[4]

Harrelson chartered a barge and tugboat,[5] and on March 22, 1987, the 240-foot *Mobro 4000* left Long Island City, Queens. It was carrying 3,206 tons of refuse from Islip and commercial waste from New York City and Nassau County. The *Mobro*—along with its tugboat *Break of Dawn*—was headed for a dumpsite in Morehead, North Carolina.[6] After four days on the water, the *Mobro* arrived at Morehead. At the insistence of state environmental experts, it was ordered out of North Carolina waters on April 3, and word spread that the barge was moving south. One account suggested that an inspector had found a bedpan among the refuse bales, a sign of possible medical waste. Harrelson told reporters, quite inaccurately, that the real destination all along had been Louisiana.

As the barge and tugboat neared Alabama, a court ordered them turned away. Officials from the Louisiana Department of Environmental Quality and state police met the barge and tug at the mouth of the Mississippi River and made clear that it could not unload at a landfill outside of New Orleans. The story was much the same in Mexico, Belize, and Florida. In May, the Environmental Protection Agency (EPA) sent inspectors to evaluate the *Mobro*'s load but found nothing dangerous or suspicious.[7] Much of the baled refuse was business waste—cardboard, paper, foam rubber—and some tires. There was no smell, but the crew had complained about horrendous numbers of flies. "We went to Morehead and got rid of all their flies," the tugboat captain was quoted as saying. "And then we went into Venice, Louisiana, and took all their flies out to sea."[8]

The journey of the "gar-barge" inspired a media frenzy, and the *Mobro* quickly became a symbol of the emerging garbage crisis in the United States. The barge also entered the realm of pop culture, with barge T-shirts, souvenirs, and cartoons, and received attention on both the Johnny Carson and Phil Donahue talkshows.[9] And in 2010, the children's author Jonah Winter published *Here Comes the Garbage Barge!*—a humorous fictional account of the gar-barge story for children illustrated by Red Nose Studio.[10]

After fifty-one days at sea, covering over six thousand miles, the *Mobro* had found nowhere to dump its refuse and headed back to the New York City area.[11] "We decided it would be in the best interest of all concerned to return the barge for offloading and disposition of the garbage," stated Robert Guidry, the president of Harvey Gulf International Marine, which owned the *Mobro* and *Break of Dawn*. Harrelson responded that "it was foolish of me to object unless I had a better plan. And right now I don't." Department of Environmental Conservation (DEC) officials and the Town of Islip reached an agreement that allowed for the increase in capacity of its landfill in Hauppauge. The ban on commercial refuse was rescinded, and the way seemed clear to unload the bales from the garbage barge at its point of origin.[12]

Yet the process of reentry into New York Harbor and final disposition of the refuse became a protracted affair. Islip officials were willing to take the refuse from the *Mobro* once the agreement to expand its landfill was accepted. Some critics charged that the town was exploiting the *Mobro* in order to get that agreement. On May 12, after a five-month soap opera, the Town of Islip agreed to accept the cargo. Harrelson said that he had yet to hear from town officials and was not interested in returning the refuse to Long Island. He was paying $6,000 a day to keep the barge at sea but seemed to balk at the idea that Islip would charge him $40 per ton ($124,000) to deposit his bales at its dumpsite.[13]

Several court orders filed by New York City Council members—fearful of political fallout from accepting the refuse—barred the *Mobro* from entering the harbor. Plans to unload the refuse were stalled again. First Deputy Mayor Stanley Brezenoff stated, "This is garbage without a home, and I don't want it to squat anywhere in New York City." Harrelson's response: "This is unbelievable. This is like a nightmare." A State Supreme Court decision barred the dumping in Queens until the refuse proved nontoxic.[14]

On May 17, the barge was anchored off Brooklyn, waiting for the outcome of the various legal battles and attracting rubberneckers. Mayor Edward I. Koch, for his part, would allow no unloading unless private, enclosed sanitation trucks were utilized. "We are treating garbage like Germany treated Lenin," he said. Lenin had to be in a sealed train while passing through Germany, Poland, and Finland, and so had the garbage to be likewise enclosed.[15] On May 28, a state judge declared the *Mobro* cargo nonhazardous, but city officials would not budge on allowing the barge to unload. In early June, the *Mobro* was seeking another federal anchorage, and the Town of Islip now agreed to accept only 40 percent of the barge's cargo.[16]

All kinds of subplots were playing out, including a familiar one. Senator John J. Marchi wrote to Mayor Koch on June 2: "There is, it appears, at least a chance that the now famous Islip garbage barge may ultimately deposit its contents at Fresh Kills. I am confident that you will move promptly to prevent such an occurrence." He went on to repeat the history of Fresh Kills indignities and how "outraged" Staten Islanders would become if forced to accept this refuse.[17] Various publications observed that the garbage-barge fiasco was drawing national and international attention to, as *Science* stated, "an approaching crisis in municipal waste management in this country."[18] More locally, it also pointed to the continuing tension over Fresh Kills.

On July 10, after 112 days of travel and conflict, the *Mobro* was to be allowed to unload its refuse and burn it at the Southwest incinerator, in Brooklyn. The ash would be carted to Islip. The decision was announced by Thomas C. Jorling, the new DEC commissioner, who said he spent the first two weeks of his term "finding a path through the thicket" of the barge situation.[19] Jorling could do little to influence the court or end the political wrangling that persisted, however, and so the plan to terminate the odyssey was blocked for the moment. Finally (finally!), after further inspections, the refuse was burned on September 1, and after several more efforts to bar it, the ash was unloaded in Islip, where the story had begun.[20]

Jorling had been appointed to his post in 1987 by Governor Mario Cuomo. Born in Cincinnati, he had degrees in law and forest ecology. He had taught science and environmental studies and also spent a good deal of his adult life as an attorney and government administrator—taking an active role in the development of landmark legislation such as the Clean Air Act, Clean Water Act, and Superfund. During his term as DEC commissioner (1987–1994), Jorling was an important voice in the debates over refuse disposal on the state level and in New York City, including those concerning the use and viability of Fresh Kills Landfill.[21]

Consensus opinion was that the implications of the *Mobro* debacle loomed large. The garbage barge was a symbol of the general apprehension and consternation about what to do with our waste and who should be responsible to accept it. The saga was buttressed by a similar but much longer voyage of the cargo ship *Khian Sea*. Loaded with 15,000 tons of Philadelphia incinerator ash, it searched for a landfill for sixteen years and was denied access in at least eleven countries and five states. The *Khian Sea* was barred from unloading in places as far away as West Africa. It illegally dumped thousands of tons of ash into the ocean and onto a beach in Haiti. A forced cleanup was followed by a return of about two thousand tons of its cargo to Pennsylvania in 2002. As dramatic as the *Mobro* and *Khian Sea* stories had been, waste could be legally exported by barge, in the United States at least, if the carrier registered with the U.S. EPA. In this sense, the voyages were not so much anomalies but emblems of the challenges of garbage disposal.[22]

While the *Mobro* did little to bring changes in consumption practices, it may well have stimulated national debate about the need for effective recycling efforts.[23] The "gar-barge" certainly influenced the recycling issue in the state of New York and in New York City in the late 1980s. Increasingly recycling and reuse were viewed as important counterweights to landfilling as a disposal panacea. To Staten Islanders, the *Mobro* was just one more garbage barge being trained on its bloated landfill. The story of the barge's rejection in port after port demonstrated how no one wanted to be a dumping ground for the waste of others, a role Staten Island was playing in a major way. Between 1987 and 1992, a critical decision over the future of refuse disposal in New York City, a decision that now included recycling as well as the fate of Fresh Kills, would have to be made. The noose was tightening.

THE STATE OF NEW YORK AND REFUSE

By the late 1980s, the state of New York was taking a much more active role in trying to confront its municipalities' problems with refuse. Its 1990/1991 Update of the State's ten-year Solid Waste Management Plan was its fourth, which contained information on the current status of solid waste management and regulations in New York.[24] The studies provided some valuable information for setting the context for New York City. Whereas disposal by landfill had been at 82 percent statewide in 1988, it dropped precipitously to 61 percent by 1990. Incineration remained relatively constant, while resource recovery of various types grew from 8 percent to 18 percent in those years. And out-of-state shipping now also stood at 18 percent. A total of 177 landfills were active in 1990, but only fifteen had valid permits. On average, New Yorkers were discarding about five pounds per day, which was above the national average.[25]

On the national level, solid waste output had quadrupled between 1960 and 1980. In 1960 approximately 88.1 million tons (2.68 pounds per capita) of municipal solid waste (MSW) was generated in the United States; in 1980 that had risen to 151.6 million tons (3.66 pounds per capita). Recycling rates in the same period were 6.4 percent (6.4 million tons) in 1960 and 9.6 per cent (9.6 million tons) in 1980.[26]

A 1989 report stated, "For local communities, confronted with the growing volume of trash and increasing difficulties in disposing of it, garbage, during the 1980s, had become an issue of economic survival." Solid waste collection and disposal costs were major components of total municipal budgets. Landfill costs, which had been cheap historically, began to rise, as did the cost of incineration plants. The cost of dumping more than doubled between 1984 and 1988, after remaining fairly steady from the 1950s to early 1980s. In 1988 tipping fees ranged from about $18 per ton to as high as $35 per ton in the first year of operation. Future fills were expected to have tipping fees exceeding $50 or $60 per ton. In 1988 New York doubled its fees at Fresh Kills, forcing private haulers to seek more distant but less expensive dumping sites.[27]

Financing landfills, negotiations with vendors, and local taxes all influenced the rise in fees. One additional reason for the jump in fees was the implementation of Subtitle D of the Resource Conservation and Recovery Act, which placed much stricter requirements on existing and future landfills. Not only were regulations stiffer and disposal space becoming dear in some parts of the country, but through the late 1980s many facilities were accepting a wide variety of hazardous materials and experiencing more illegal dumping, all of which raised grave concerns about the environmental risks of landfills. Leachate and noxious gases were now recognized as increasingly unwelcome outputs in places with few control systems.[28]

If nothing else, the *Mobro*'s journey in 1987 helped provoke a reaction. In that year, the DEC underwent a major reorganization, establishing the Division of Solid Waste as one of three new units.[29] Also in 1987 the DEC released its first State Solid Waste Management Plan. Getting to this point took almost seven years of preparation and the lobbying of many environmental groups. In 1986 and 1987 organizations like the New York Public Interest Research Group (NYPIRG) and some research projects were turning attention away from environmentally risky incineration. Recycling, reuse, and waste reduction were to be given priority over landfilling and incineration in the DEC plan.[30]

A statewide poll in 1988 indicated that 60 percent of the respondents favored recycling as the best way to deal with the regional refuse problem, but overall they did not view solid waste disposal as a very urgent issue.[31] Lawmakers in Albany were not as blasé, especially given the growing volumes of refuse in the state as a whole. New Yorkers were generating 14 percent more waste in 1988 than 1986 (more than 20 million tons per year). Between July 1988 and the end of 1989, per capita waste generation jumped 4.2 percent. While landfill life expectancy declined, the amount of refuse going to landfills increased. Over 52 percent of the state's annual capacity and 43 percent of all waste disposed went to three landfills, with Fresh Kills the most noteworthy (handling 40 percent of the state's solid waste stream by 1990). Landfilling accounted for 82 percent of disposal in 1988. Between 1986 and 1988, fifty-three landfills had closed (thirty-six more between July 1988 and January 1, 1990), and few of those remaining open had valid operating permits.[32]

Incineration was on the decline in the state; only five resource-recovery facilities had replaced them by 1988. But even then, resource recovery was the second most utilized disposal method (8 percent), with traditional incineration (non–energy generating) and out-of-state disposal each ranking third. Expectations about increasing waste-to-energy capacity, at least in the short term, were undermined, since the amount of waste processed through those facilities in 1988 had actually declined over the previous two years. Dependence on out-of-state disposal was increasing, despite the battles with New Jersey. New capacity of any kind within the state was likely to be meager. In a 1988/1989 update, the DEC reiterated its concern about a "crisis in capacity for disposal of solid waste."[33] Such a prognosis made recycling and waste reduction more urgent, not less.

The State Solid Waste Management Act of 1988 recognized the concept of integrated solid waste management as a way to guide its program—a practice that was becoming popular nationwide. The plan would blend various approaches to disposal and set a goal of 50 percent reduction in refuse by 1997, based on a projected 8 to 10 percent prevention of waste at the source and 40 to 42 percent recycling. Each municipality was required to set up source separation and recycling programs by September 1, 1992.[34] Every "planning unit" in the state had to adopt a twenty-year plan for managing all solid wastes (including sewage, harbor waste, municipal solid waste, and so on) within its boundaries. Until the plan was adopted, the DEC would not issue a permit for any waste management facility.[35] The new law and its amendments were among the most comprehensive in the nation. The DEC also was pressing New York City to change its disposal priorities again, this time away from landfilling and incineration and to a more aggressive—and some would say— unattainable dependence on recycling. Critics argued that the size and complexity of the city did not easily lend itself to recycling.[36] In February 1986 Deputy Sanitation Commissioner Paul Casowitz reflected the view of others in the DSNY when he stated in defense of the Brooklyn Navy Yard facility (supposedly to be operational in 1987 or 1988) that "recycling was terrific" but total recycling was not very practicable.[37]

OF RECYCLING AND LANDFILLS IN NYC

The news about the increasing generation of waste but fewer places to put it was grimmer for the city than for the state as a whole. In 1988 the city was producing 30 percent more refuse than it had been two years earlier. At Fresh Kills alone the amount increased from 5.8 million tons in 1985 to 7.63 million tons in 1987. Although Fresh Kills had received less refuse in 1990 than a year earlier (about 19,000 tons per day, compared to about 27,000), this could be attributed to high tipping fees especially for private commercial haulers (which encouraged more out-of-state shipping).[38] In 1990, 73.4 percent of the city's waste went to Fresh Kills. Municipal and apartment-house incinerators handled

12.5 percent; city-sponsored recycling and bottle and can returns amounted to only 6.5 percent.[39] The waste stream also was becoming more complex, with increasing amounts of nonbiodegradable plastics, especially in the form of polystyrene food containers and disposal diapers. Assemblyman Eric N. Vitaliano quipped, "The throwaway society, although more aware, is still a throwaway society."[40]

In 1989 New York City passed Local Law 19, requiring mandatory recycling (source separation of recyclables from household refuse) with a target of 25 percent by 1994.[41] Fledgling recycling efforts in New York City before Local Law 19, including pilot programs in curbside recycling, had not caught on very well.[42] The city had only opened a recycling office in 1985. Comptroller Harrison J. Goldin stated, "The Department of Sanitation has dragged its feet because it has not received the kind of direction from the top that it might have or should have." Sanitation Commissioner Brendan Sexton countered that the DSNY was moving as quickly as possible but that the logistics were staggering and the new law vague. "The department has been saying for some time that mandatory recycling was coming. Once you mandate recycling, that doesn't do anything. You still have to have somebody to go pick it up."[43]

The productivity of the DSNY's collection program had shown improvement, which suggested that progress in recycling was possible.[44] The operating expenditures for the department were close to $600 million for fiscal year 1989, of which $210 million went to capital equipment and facilities. The largest share of the budget remained devoted to collection of household refuse and street litter, with more than eight thousand uniformed workers engaged in those areas.[45] But disposal remained a quandary: the city disposed of more waste than it collected, especially because of private carters.[46]

Recycling evolved slowly. In 1988 New York City recycled only 2 percent of its waste (mostly beverage containers). The curbside program was at best "a step in the right direction." Joan Edwards, director of recycling for the DSNY, stated in March 1988, "We have a waste-disposal crisis that is only going to get worse. Fresh Kills is going to run out in the next 10 years. Recycling extends the life of landfill land that is priceless. And the cost of exporting will be so high 10 years from now that it will make the cost of recycling look like nothing."[47] In a promising move, recycling programs in the city budget for fiscal year 1990 were allocated $43 million, twice the amount allocated in 1989.[48]

The appeal of recycling and waste prevention did not erode the hopes of many city leaders that resource-recovery plants could replace landfills or at least preserve remaining capacity.[49] In 1987 a hearing was held on the proposed Brooklyn Navy Yard Resource Recovery Facility. The Interstate Sanitation Commission (a tri-state environmental agency) was granted party status at the beginning of the hearing, making New York City's disposal policy more of a regional issue.[50] However, little headway was made on the decision to open the plant by late 1988.[51] Carl Campanile, writing in the *Staten Island Advance*, characterized the decision as "a temporary blow not only to the Koch administration, but to a national movement toward using incinerators to solve a garbage capacity crisis."[52]

The *New York Times* supported Commissioner Jorling's decision to postpone the project, stating that he deserved praise "for insisting on tough safety standards." He also threatened to close Fresh Kills if the landfill did not meet the consent decree.[53] Groups such as NYPIRG were happy with any decision that delayed or, better yet, curtailed the plant schedule for Brooklyn.[54] Not everyone was pleased with Jorling's principled stances, as the push and pull of decision making over refuse disposal remained unbearably complicated.

Staten Island fought its own battles over the proposed resource-recovery plant at the Arthur Kill, which irritated already raw nerves over New York City's attempts to develop a new disposal policy. The Staten Island Citizens Advisory Committee on Resource Recovery (CAC) was fixated on that issue.[55] Speaking at Wagner College in June 1988, Barry Commoner asserted, "Staten Island is exactly the wrong place to build incinerators. You're bathed in carcinogens from northern New Jersey."[56] But opinion on Staten Island never went undisputed, especially with Marchi, Vitaliano, and others favoring a resource-recovery plant to offset dumping at the local landfill.[57] The future of Fresh Kills, of course, remained of primary interest.

In January 1987 the *Advance* published a three-part series by William Kleinknecht about Fresh Kills and the refuse problem. In the January 18 installment he stated that the landfill was more than just an eyesore: "At a time when there is wide agreement on the environmental hazards of landfilling, Fresh Kills remains a stubborn reality for a city that has no place else to put its garbage and won't for at least several years." The story emphasized the NIMBY problem at the site: "a nightmare for thousands of people settling on the West Shore, many of them Brooklyn natives in search of a cleaner, safer place to raise children. Instead of roasting hamburgers in the back yard, they are spending the entire year with their windows shut tight to escape the landfill's odors."[58]

Fresh Kills had become an environmental menace, particularly because of the unwelcome flow of leachate.[59] Walter Hang, the director of the NYPIRG Toxics Project, proclaimed that Fresh Kills was "an enormous pollution problem and one that has never been fully understood." Al O'Leary from the DSNY conceded, "People have this idea that we like Fresh Kills. We'd be happier if we didn't even have to use landfills." Yet he also expressed a sense of pride at the engineering feat of building five-hundred-foot mounds.[60]

Such satisfaction was not shared by others, including Congressman Guy V. Molinari, who believed that landfills ultimately were based on "unproven technology."[61] It was little consolation to locals that the DSNY employed some cosmetic approaches to "improving" the landfill site, such as avoiding dumping in some areas, landscaping with trees, installing wire mesh to discourage seagulls, and constructing dirt berms between landfill sites and residential neighborhoods.[62] The DSNY also responded to chronic complaints about odors emanating from the landfill by beginning a deodorant test in the summer of 1989.[63] But one thing the DSNY could not do was up and move the giant dumpsite.

In Kleinknecht's second story, he took on the closing of the Fountain Avenue landfill in Brooklyn (on National Park Service land, closed when the city's lease ran out), which raised the intake at Fresh Kills to 22,000 tons and daily brought roughly 850 garbage trucks across the Verrazzano-Narrows Bridge. Representative Guy Molinari and others fought to reopen the Brooklyn landfill, but to no avail. Kleinknecht added prophetically, "But on Staten Island, where discussions of city government often turn into angry calls for secession, many still regard the closing of Fountain Avenue as yet another indignity foisted on this beleaguered borough."[64]

Kleinknecht rightly sensed the passion building up around the Fresh Kills issue. In part 3, he made the story more personal, turning to the reaction of homeowners adjacent to the site. Barbara Savino and her husband bought a condominium in Arden Heights in the summer of 1986, aware of its proximity to the Fresh Kills facility but not fully conscious of what that might mean to their lifestyle. The persistent odors and the seagull droppings on their skylight became constant reminders of what they were up against. "When I realized how big the landfill really was, I panicked," said Mrs. Savino. "This is our first home, and we don't even know if we're going to be able to sit outside in the yard."[65] Like other exposés, the series was somewhat anecdotal. But taken as a whole, it revealed the range of issues that were intensifying Staten Island's distaste for what was becoming the lone source of disposal for New York City's residential solid waste.

Stories like the Savinos' were not uncommon. Susan Chew sent a letter to the Natural Resources Protective Association (with copies to legislators and environmental agencies) stating,

> Last night [February 1, 1987] . . . I smelled garbage as I was crossing the Verrazano Bridge into Staten Island. When I reached my home, the smell was very strong and there was no doubt that it *was* the smell of garbage. I live on the North Shore where we ordinarily do not get these odors; however, we are now! I understand that the Arthur Kill [Fresh Kills] Landfill is not being run up to New York State standards, and I'm sure that things must be getting worse there for these garbage odors to reach the North Shore as far as the Verrazano Bridge.[66]

Given the circumstances that Kleinknecht outlined, comments from Commissioner Jorling at a press conference in May 1988 were downright ironic and peculiar: "In defense of Fresh Kills, it is one of the best-run landfills in the state." He hedged a bit by adding that Fresh Kills "certainly has been upgraded from what it was several years ago. It probably is one of the best maintained, but it isn't perfect by any means. That's why landfills are so in disfavor. Even when they are managed well, they are eyesores or nose sores."[67] Jorling's statements clarified nothing. Was this simply politician talk?

Nevertheless, there was a direct correlation between Fresh Kills becoming New York City's only dumping site and the rising temperature of protest, with things getting substantially worse by 1987. Attention given to the Fountain Avenue closing now turned to Edgemere. The landfill in Queens accepted only six hundred tons of refuse per day, compared with the more than 22,000 at Fresh Kills, but its closure, which was being bruited about, was another ominous sign for Staten Island. Once closed, only Staten Island would possess a landfill. As Assemblywoman Elizabeth Connelly believed, "A borough without a landfill is a borough without a garbage problem, and a borough without a garbage problem is one whose leaders will feel much less pressure about accepting a resource recovery plant, or two."[68]

Concerns about Edgemere were intensified (and disposal choices narrowed) when in June New Jersey legislation banned New York private haulers from dumping refuse in the Edgeboro landfill in East Brunswick. The refuse of those private haulers would be diverted to Fresh Kills. The *Advance* admonished, "The wandering garbage Barge *Mobro*, and the sudden inhospitality of New Jersey to New York garbage are very clear signs that we had better get a plan—a long-term plan—for solid waste disposal in place, and soon."[69] Unless New York City had no other place for its refuse, a city-state agreement affirmed that Fresh Kills would need to close by 1998. Edgemere, under study by the DSNY, was to close in 1991. All of this was quite tentative and made more ambiguous by a statement from Casowitz that even with proposed incinerators in place, "not all waste can go to an incinerator": space would need to be found for ash residue.[70] Robert Moses and others had made this point privately some years before, but they had not uttered it publicly, at least not loudly.

ASHES, ASHES . . .

Concerns over incinerator ash being sent to the landfill were no idle musings. Combustion residue had been going to Fresh Kills since its opening in 1948.[71] In April 1988 the DEC recommended that all new landfills built in New York have state-of-the-art double linings to prevent leaks and have separate spaces for incineration ash that exceeded the federal limits for hazardous waste.[72] Neither the EPA nor state agencies had paid much attention to hazardous wastes until publicity over Love Canal in 1978 created a national stir, but now they were extremely sensitive to it.[73] One news story stated, "The region that includes the infamous Love Canal is the only area in New York that causes the state Department of Environmental Conservation . . . more headaches than Staten Island."[74]

Legislators and citizen groups on Staten Island knew what was coming and opposed a separate ash dump at Fresh Kills. But Commissioner Jorling stated publicly that New York City would need to have such a facility and that Fresh Kills would have to be the first choice. "We prefer landfill operations concentrated in one area," he stated, "rather than dispersed throughout the region." At the time, six hundred tons of ash daily from city incinerators and about two thousand apartment buildings already were being dumped at the Staten Island site, but the ash was not regarded as toxic.[75] A 1987 survey commissioned by the state of New York, however, had found lead and cadmium levels exceeding federal guidelines in six existing incinerators. The report set off alarm bells.[76] The *Advance* spoke for its neighbors when it proclaimed, "The suggestion that Staten Island play host to the largest toxic ash dump on the planet is an insult and an outrage. Such a dump is unacceptable here and must not be built."[77]

Locals got support (of sorts) from the chair of the State Assembly's influential Environmental Conservation Committee, Maurice Hinchey, a Democrat from Ulster, who stated that if an ash dump were located on Staten Island it should be double lined. The Koch administration simply wanted an exemption from existing regulations, which would allow it to build a separate ash pit without a liner.[78] Assemblyman Robert A. Straniere and other Staten Island political leaders strenuously opposed any ash landfills in New York, especially at Fresh Kills. This newest battle was doubly intense because the DSNY already had decided that ash from the proposed Brooklyn Navy Yard incinerator would be dumped at Fresh Kills, and Casowitz acknowledged that ash from other new incinerators would end up on Staten Island as well.

Concern over hazardous waste was coupled with the realization that if it became a site for ash disposal, the Fresh Kills Landfill would be open in perpetuity. The debate also raised a variety of other issues: As it would be constructed on old fill, was lining an ash mound even feasible at Fresh Kills? Could nontoxic and hazardous ash be separated? Should incinerator ash itself be considered hazardous waste?[79]

A distressed citizen wrote to Vitaliano:

I can certainly appreciate your views against a toxic ash dump in Fresh Kills . . . but you still seem to be missing the real point. We are quickly running out of places to dump ANYTHING much less toxic ash. This is the real nature of the garbage crisis. By supporting the burning of garbage you are helping to create an even worse problem when there is no place at all for the ash. You will also be involved in endless battles to keep it out of fresh kills, as the push to dump it there will be constant.[80]

Vitaliano, among others, was finding himself in a bind by supporting resource-recovery plants as an alternative to dumping at Fresh Kills. The two disposal methods could not be so easily separated. He was offended at being called a "knee jerk" environmentalist; he much preferred "reflective environmentalist."[81]

In late May 1988 much of the battle over dumping ash was being waged in Albany. The Legislative Commission on Solid Waste conducted hearings that "raised more questions than answers," but it was clear that Fresh Kills was the focus of attention for New York City's ash-disposal problem. Debate also centered on the DEC wanting to treat ash as "special waste" as opposed to "hazardous waste."[82] But momentum was heading toward a "transitional rule" that would allow the city to dispose waste from the proposed Brooklyn Navy Yard plant at a dormant location within Fresh Kills, which prompted immediate protest and legislative action from Representative Guy Molinari.[83] A state decision not to grant a permit for the Brooklyn Navy Yard plant complicated the ash debate. Jorling had ruled against the permit (at least temporarily) because of inadequate plans for recycling and for dumping ash at Fresh Kills. This seemed like a catch-22 situation, although Commissioner Sexton asserted that the decision would not keep the city from dumping ash at Fresh Kills.[84] The current dumping would go on, even as the EPA was discovering that ash from garbage burning itself included toxins.[85]

Beyond its focus on being a potentially serious pollutant, the questions about incinerator ash intensified the long-standing fight over closing the Staten

FIGURE 13.1 Gansevoort MTS, 1989.

Source: Courtesy of Steven Corey.

Island landfill.[86] Other challenges focused on the ash's impact on New York City's neighbors. In the late 1980s pollution was part of a many-front environmental war in and around the city. Earlier in the decade New Jersey authorities had complained about garbage and trash floating from the landfill to the beaches of Woodbridge. A judge had called for New York City to pay the cleanup costs or face the closure of Fresh Kills.[87] Medical waste had been of particular concern, as stated earlier, as hypodermic needles and other medical supplies washed up on New Jersey beaches.[88]

In 1987 and 1988 medical waste and other refuse continued to be found on the New Jersey shore, and what had become an eight-year scuffle between Woodbridge and New York City continued.[89] In October 1987 a U.S. district judge found New York City in contempt of court, stating she was "outraged" at the city's "Rip Van Winkle" delays in keeping waste from the Jersey beaches.[90] The next month a tentative pact was reached, calling for the installation of a floating boom around the Southwest Brooklyn marine transfer station to contain trash and another to be installed at the landfill.[91]

New Jersey officials filed a second suit, claiming that waste from Fresh Kills was spilling into the Arthur Kill and thus posing a health risk to their citizens.[92] Needless to say, Staten Islanders felt set upon by the lawsuits. Vitaliano and others went so far as to suggest that the agreement reached in November had dodged the issue of an enclosed barge unloading facility, which meant that the persistent odors from the rotting garbage would not be abated.[93] In December

FIGURE 13.2 Fresh Kills Plant 1, overloaded Athey wagon, 1988.

Source: Courtesy of Steven Corey.

FIGURE 13.3 Hydraulic crane unloading barge at Fresh Kills Plant 2, 1989.

Source: Courtesy of Steven Corey.

the announcement of a court-supervised consent order regarding compensation for Woodridge and stricter disposal standards at Fresh Kills and at New York City's marine stations formalized the tentative November pact.[94]

The parties to the consent order met on a regular basis in 1988. New York City hired an independent monitor to evaluate landfill operations and an independent consultant to determine short- and long-term measures to avoid future debris problems. For the moment, floating material had tapered off in the area.[95] There were rumblings that New York City had been let off easy in this case and that some waste might still be washing up on shore.[96] Similar problems with medical waste on Long Island beaches and recurring issues about sludge discharge at sea (and possible deposits at Fresh Kills) kept the debate over marine pollution alive, as did criticism of the EPA for not getting more directly involved in finding solutions.[97]

THE WETLANDS ACT

At this point, Staten Islanders were taking as a slight any environmental decision in the state or city. The New York State Freshwater Wetlands Act of 1975 required that the DEC map wetlands larger than 12.4 acres as well as any other wetlands of local importance. The goal was to protect the public and natural environment in wetland areas from private development. This was a far cry from the days when marshes and swamps were viewed as waste land. Modern scientific thinking had gone a long way in demonstrating the ecological value of marshland and the need for their protection.

While the law did not expressly treat salt marshes and wetlands relevant to the Fresh Kills site, in a substantial way it did affect any further expansion of the landfill on Staten Island. And, in the long run, new attention to wetlands would focus on flyways, animal habitats, pollution control, and marshland preservation that could be connected to Fresh Kills. Most immediately in the 1980s, however, the law's effect on local property rights spurred the most controversy on Staten Island.

Wetland property owners now needed DEC approval to make substantial changes to their land. Although the law was enacted in 1975, the first maps were not filed until 1984. Senator Marchi's Legislative Commission on Expenditure Review released an audit of the mapping activity in 1987 (only 58 percent of the required maps had been filed). "This unacceptably long delay in completing the mapping in certain areas of the State, particularly on Staten Island," Marchi stated, "has caused public opposition to the methods by which the program has been administered." Many property owners had complained that the law interfered with their rights, and Marchi's interest clearly sided with the landowners on Staten Island over preservation.[98]

Marchi then introduced a bill that would make the DEC's 1981 tentative map of the island official, resulting in seven hundred acres of property subject to protection, rather than an estimated six thousand parcels. He believed that such a bill would help property owners who were "suffering economic hardship because they are in a kind of bureaucratic and legal limbo."[99] Marchi's bill passed the State Senate in March, which removed about six hundred acres (designated as wetlands) from state control for five years, allowing people to build on or sell their land.[100] Assemblyman Vitaliano and City Councilwoman Susan Molinari (the daughter of Representative Guy Molinari) also supported compensation for condemned property in marshlands.[101]

A group of developers challenged (unsuccessfully) the methods of the DEC in determining wetlands designations.[102] Some property owners worried that designation of a field adjacent to their homes as marshland might create a fire risk. "Some people might think it's one of life's greatest joys to gaze upon a field of cattails... then turn and go home," one man stated. "I and my neighbors see it somewhat differently. That field is a source of fire."[103] At one demonstration, a "patriotic protest" accompanied by a colonial-style fife-and-drum corps denounced condemnation without compensation. Vitaliano and Connelly wanted more stringent criteria for wetlands designation.[104]

While some groups were concerned about this most recent threat to wetlands preservation, NIMBYism appeared to be winning out. It was not acceptable to turn marshland into a giant landfill, but it was quite another matter when how private wetland property might be exploited was affected by eminent domain.[105] Staten Island also was becoming a test case of sorts. A bill in the New York State Senate in early 1989 would have the law covering Staten Island wetlands designations applied statewide.[106] As far back as the burning of the Quarantine and continuing through to the ongoing Fresh Kills saga, Staten Islanders did not like any government encroachment on their property, no matter the reason. Nonetheless, in a 6–0 decision in March 1990 the New York State Court of Appeals upheld the DEC's practice of using temporary maps as guides before making a final judgment as to whether land would or would not be designated as wetlands. The court ruled against a group of developers who had planned to build private homes on Staten Island. They in turn appealed the ruling, and the dispute continued.[107]

THE CONSENT ORDER OF 1990: A TURNING POINT

Rarely do key events so neatly coincide with the beginning of a new decade, but the third consent order on Fresh Kills in 1990—if not in and of itself successful in determining the landfill's fate—set the stage for a convergence of several issues relevant to Fresh Kills' future and the general debate over the disposal dilemma. Particularly important at this point in time was the triangular relationship on the refuse issue among Albany, New York City, and Staten Island. Albany had been a player on and off, especially beginning in the modern environmental era and with respect to the two previous Fresh Kills consent orders. But its role in helping determine the future of refuse disposal locally and statewide proved central and challenging.[108]

On May 27, 1988, Assemblyman Vitaliano met with DSNY Commissioner Sexton, Deputy Commissioner Casowitz, and other staff members. Department officials acknowledged that odors were a continuing problem at the landfill, but they more narrowly focused on carting waste on the site's garbage trails as a worse offender than barge covers. An update on the Woodbridge consent order was discussed, as were other water-quality issues. The DSNY was reviewing a proposal for methane recovery, developing a new chemical formula to reduce odors, and evaluating measures to control litter and leachate. Officials reiterated their view that neither raw garbage nor ash posed a leachate problem. And they were emphatic that Fresh Kills' life "will be maximized." The department was moving to close some sections of the landfill in two years, but a target of 1996 for total closure was only possible if alternative disposal options were available. The place of Fresh Kills in New York City's disposal plans was well entrenched, to the consternation of Vitaliano and a majority of Staten Islanders.[109]

On September 6, Assemblywoman Connelly wrote to Deputy Secretary to the Governor Francis J. Murray Jr., inquiring about negotiations regarding a revised consent order for Fresh Kills, since the DSNY was "unable to comply" with the New York State Environmental Conservation Law regulating permits for continued operation. Connelly was particularly concerned that "the outcome of such negotiations will have a tremendous impact on the residents of Staten Island" and would require "some [public] input in the decision making process." "Unfortunately," she stated, "these discussions have been fruitless and there seems to be little interest in revealing any part of the proposal under consideration." As a result, Connelly sought Murray's intercession.[110]

Throughout 1989 complaints mounted about pollution at Fresh Kills and in the surrounding waters of the bay. The newspapers exposed that engineers quietly were creating a 399-acre, 505-foot "pyramid of garbage" at the site.[111] An insightful "Opinion" in the *Advance* on June 4 noted, "When people look at the Fresh Kills landfill, they see many different things." For residents of other boroughs it is "some vague, boundless, far-off repository for their garbage, as safely removed from them as if it were on Mars." "Staten Islanders, for the most part," he added, "would like nothing better than to see the thing padlocked and piled over with dirt." The state government had another view: "More than a simple dump for 25,000 tons of all of the nation's largest city's garbage every day, they claim to see the landfill as some vast, untapped West Virginia of a place, full of good things to be extracted (although this had yet to be done other than for methane)." The editorial exaggerated how state officials valued the site for reclamation and redevelopment (several had no such optimism), but observations that Fresh Kills was a place that courted divergent views was certainly accurate and validated by its history.[112]

By October someone leaked to the press that state and city officials were negotiating a new consent order for Fresh Kills (the third) at the same time that Commissioner Jorling was considering permits to construct the incinerator at the Brooklyn Navy Yard.[113] The situation turned intense on November 8, when Jorling filed a civil complaint against New York City, stating that it had ignored his orders to clean up Fresh Kills, and he threatened to obtain a court order to close the facility by the summer of 1991. Jorling wanted to pin down the city on a firm closing date for Fresh Kills, and the court decision came on the heels of what a DEC official called a "dead spot" in negotiations over a new consent order.[114]

Jorling's complaint about the need for cleanup at Fresh Kills focused on the millions of gallons of leachate running into groundwater, garbage spills on the waterways, and uncontrolled odors and gases. According to the *New York Times*, "Vowing to fight any effort by New York State to close the Fresh Kills landfill on Staten Island, Mayor Edward I. Koch yesterday denounced the state's environmental chief as the 'Colossus in Albany,' a 'hanging judge' and 'zealot' whose tactics were 'insane' and 'most vicious.'"[115] Commissioner Sexton uttered "astonishment" at Jorling's action. "He can't be serious. We're talking about half the trash in New York State." Sexton added, "We've given him everything he asked for and lots of stuff he never even thought of." Jorling's believed that the city had defied the previous consent orders; he also was unwilling to issue a permit for the Brooklyn Navy Yard plant without a better city recycling plan.[116] The stakes were greater than just the dispute over Fresh Kills.[117]

The issue at hand was how long the Staten Island landfill would remain open. An "optimistic" estimate, according to Sexton, was that Fresh Kills could still accept garbage until 2029 if an effective recycling program was implemented and five incinerators were brought on line. One could imagine how much these comments inflamed Staten Islanders.[118]

Jorling was wielding a big stick by trying to get city compliance on refuse disposal or face a premature closing (by city calculations, at least) of Fresh Kills. Some environmental-advocacy groups applauded his stance, but Sexton questioned its timing. "Maybe this is his way of trying to get the new administration [that of Mayor David N. Dinkins] on the defensive." DSNY officials, however, had used the mayoral election to delay specifying a closing date for the landfill.[119] A January 22, 1990, story in the *Advance* may have summarized the department's sentiments when it stated, "No matter what the future of garbage disposal is, experts say, it is inevitable that there will be some need for dumps."[120]

It was difficult for the DSNY to contemplate—or anyone else for that matter—where the refuse would go after closure and how to do it without costing millions of extra dollars. The department also faced other immediate problems and pressures related to Fresh Kills. Sanitation workers, for example, requested studies concerning health risks to employees at the Staten Island facility. Attorney General Robert Abrams sought legislation to allow citizens' groups to sue polluters for violating state environmental laws.[121] Aside from Jorling's actions, Senator Marchi proposed legislation to have New York City set aside money each year to pay for the cost of closing Fresh Kills.[122]

Some kind of showdown was imminent. Jorling's threats could not, in and of themselves, force the shuttering of Fresh Kills.[123] They did move the negotiations forward on the new consent order, which the parties entered into on April 24, 1990. The order touched all of the familiar points: improving and remediating the landfill according to state regulations, developing interim operating requirements, and progress reporting on compliance. It also made accommodations for public access to information, stipulated penalties for noncompliance (although the city denied any liability, wrongdoing, or violation of any statutes or regulations), and, quite significantly, outlined DEC's role in monitoring that compliance.

The consent order called for spending $600 million to clean up the site as well as four other (inactive) city landfills where hazardous waste had been dumped illegally (one of the largest environmental commitments made by the city). The city also was required to submit permit applications by March 15, 1995, to meet state regulations. One enormous sticking point was recognition of the continued interest in developing incinerator-ash capacity at Fresh Kills. A public meeting would be held on Staten Island on May 30, 1990, to respond to concerns over the order's provisions, and the DSNY also would prepare testimony and comment. Within sixty days after the end of public comment, the DEC could propose to modify the consent order to reflect concerns raised in the meetings.[124]

Under the 1990 Consent Order (between the DEC and the city), the state allowed the DSNY to continue operating the landfill while the city made environmental and operations improvements to bring the site into compliance with state requirements under the new Part 360 solid waste regulations. Several contemporary landfill designs were to be incorporated as an alternative to the lack of a protective liner present in newer landfills. A DSNY document summarizing the consent order was not so enthusiastic about the upgrading process and suggested that "[the consent order] allows for continued operation of the facility after 1998 if certain vague criteria are met." "While it is understandable that the closure of the Fresh Kills Landfill presents a serious dilemma," it added, "it seems quite unlikely that the facility could be remediated to such an extent that it provides no threat of contamination or health impacts to the public health or the environment."[125] This was an assessment the department did not want widely circulated among the public or in Albany.

In late June, the Board of Estimate's decision to lower tipping fees for private carters (if they removed recyclable materials from their refuse) was another indication that the city had little intention of closing Fresh Kills anytime soon. While the action was meant to encourage more recycling, the lower fees sent a contrary message. Eric A. Goldstein, a senior lawyer for the Natural Resources Defense Council, stated in a letter to Mayor Dinkins: "This is one of the nuttiest ideas to emerge from City Hall in years."[126] Senator Marchi commented, "The Board of Estimate's action . . . leaves me aghast. The decision is so bad as to be incredible."[127] Faced with vehement criticism of the lower fees for private carters, the board backed away from its initial decision when a key swing vote balked, but not before riling Staten Islanders again with its misguided plan.[128]

City Council Member Jerome X. O'Donovan (Staten Island) summed up the local response to the consent order when he flatly stated, "It is particularly distasteful to me to come to these annual 'Consent Order' meetings because the real purpose is to facilitate continued dumping at Fresh Kills." He added, "It is also insulting to continue to ask the people of Staten Island for participation or 'consent' in operating the dump when the City Administration cuts back funding on recycling, fails to start any meaningful waste prevention programs, doesn't pursue refuse exportation possibilities and worst of all, stalls on incineration for blatant political reasons."[129] The new Staten Island borough president, Guy Molinari, was curt: the consent order was "a scam." The city had "bought the time they wanted."[130] Molinari added, "It's like a mating ritual. If anybody thinks there will be any difference in landfill operations, they're out of their heads. The garbage there will stink just as bad next year as it did this. The debris will be flying around, only in increased quantities because the mountain of refuse will be higher."[131]

Assemblyman Vitaliano echoed those sentiments (albeit a little less stridently) in his public-hearing testimony on the consent order. He too emphasized the failure of the document to include "no date certain" for the closure of the landfill. "For the citizens of Staten Island, it is intolerable that this Consent Order is not a Consent Order to Close" as he hoped it would be. He also noted the "the trail of broken promises" made by the city concerning past legal commitments and the lack of provision for closure for Fresh Kills as required by the DEC for all new landfills (as stated in the Part 360 regulations). Unlike O'Donovan, Vitaliano conceded that not all of the 1990 order was bad, emphasizing how his "full court press" for legislative and administrative remedies were paying off.

What followed was an extended discussion of the "history of noncompliance" going back at least to 1979, with particular concern over leachate migration, groundwater quality, and methane-gas emissions. He concluded:

> In the absence of real world commitments from both [the new] Dinkins Administration and Commissioner Jorling, and in the face of a retreat from the 1985 binding order to close Fresh Kills by a date certain, it is once again necessary to take our needs to the State legislature, to the City Council, and to the Congress to force the City and this State to commit specifically to an end to an almost half century's old addiction to Fresh Kills, which in and of itself, provides an expedient and unacceptable excuse for our failure as a society to come to grips with the need to reduce and recycle, rather than continually dump, our garbage.[132]

The *Mobro* may have highlighted the need to rethink refuse disposal in the country, and it gave recycling a serious nudge forward. But at the turn of a new decade, just how solid-waste disposal practices were to change in Gotham now depended on the interaction of Albany, Manhattan, and Staten Island collectively. Could the three entities cooperate in some meaningful way? What would be the role of a fourth player—the federal government—especially through the actions of the Environmental Protection Agency? Did the public have any role to play? And finally, would rational planning, politics, or some unforeseen force lead New York City away from its entrenched dual-track disposal plan?

A 1992 report by NYPIRG's Toxic Project's Arthur S. Kell made a strong case for its Recycle First plan for the City of New York and criticized the DSNY's claims that recycling was much more expensive than landfilling and incineration.[133] As had become readily apparent over the years, a solution to the disposal dilemma intermixed economics, administration, environmental factors, available technology, politics—and a dash of public pressure. The cycle of action (and debate) around dumping, burning, and reuse continued and was complicated by all these factors. What disposal option(s) ultimately would come out on top going forward was a constant source of speculation, and how Fresh Kills would fit in the mix remained unresolved.

CHAPTER 14

A NEW PLAN

The 1990 Consent Order accomplished very little in the eyes of many. It stipulated and even fostered a range of necessary technical studies, but it perpetuated existing tensions among the three principal parties: Staten Island, Manhattan, and Albany.[1] The city was completing a Comprehensive Solid Waste Management Plan and was keeping the Department of Environmental Conservation (DEC) informed of its progress in formulating and assessing alternative scenarios moving into 1991.[2] After a protracted statewide fracas, the City of New York developed a new plan of action in 1992 for disposing of its copious wastes, including a renewed commitment to recycling. It was a complicated plan with considerable contingency built into it (including a role for resource-recovery plants).

The elephant in the room was whether to close Fresh Kills. But heated debate also built up around the proposed ashfill for Staten Island, its relationship to the Brooklyn Navy Yard project, the form and scale of the city's recycling efforts, and the larger issue of how to implement the city's future plans for disposal. The back and forth over compliance or noncompliance was to be expected, especially on whether the city was meeting "milestone" designations.[3] Public meetings on compliance at Staten Island venues failed to quell the citizenry's frustrations.[4] Exasperation and weariness soon spilled over into the most significant local secessionist effort in the nation's history. It also fueled the unwavering demand for closing New York City's gigantic land sink. Fresh Kills and the issues surrounding it were becoming wearisome in Gotham.

MAYOR DINKINS

Enter David Norman Dinkins, New York City's first and only African American mayor. His time in office was wedged between two iconic figures—Edward I. Koch before him and Rudolph W. L. Giuliani afterward. Who Dinkins was as a leader and what he represented created a mixed legacy during his single term.

In 1989 Dinkins defeated Koch in the Democratic primary (running strongest in Manhattan, the Bronx, and Brooklyn), in what should not have been a surprising political end for the fading local institution referred to as "Hizzoner." Dinkins played upon what he believed to be Koch's lack of responsiveness to minority issues. Dinkins then went on to win the general election by a narrow margin (3 percent) over the Republican-Liberal Giuliani largely thanks to substantial backing from racial groups (African Americans and Latinos), labor, and some liberal white voters. Many white Democrats who had supported Koch in the primary moved to Giuliani in the general election.[5]

Dinkins was born in Trenton, New Jersey, in 1927. After high school he tried to join the U.S. Marines but was told that the "Negro quota" already had been met. After being drafted he was able to serve with great honor as a Marine. The GI Bill of Rights was his ticket to entering Howard University, where he completed his degree in mathematics in 1950. He received his law degree from Brooklyn Law School in 1956 and began a private law practice. A founding member of the organization 100 Black Men (1963), he became Harlem's representative to the New York State Assembly in 1966. In 1972 he was appointed chair of the Board of Elections and became city clerk from 1975 to 1985. Dinkins was elected borough president of Manhattan in 1985.[6]

The mayoral election in 1989 was a day that saw several African Americans (including Dinkins) elected to key city and state offices across the nation. A story in the *New York Times* stated, "In New York, Mr. Dinkins invoked the idea of unity in a city that has known very little of it lately, hinting ever so subtly that a member of his race was perhaps better positioned to bring people back together."[7] In his victory speech, Dinkins reflected: "My father remembered when he was young, talking with neighbors who themselves remembered the days of slavery. Tonight, we've forged a new link in that chain of memory. One of those folks my dad knew never thought America would see ... someone like them be elected to the highest office of the greatest city in the world."[8] Observers took away varied feelings from the election results. A woman in Harlem proclaimed, "Now we have a compassionate man in Gracie Mansion. That's what we need—someone with a heart. He's a great person." A young accountant was less optimistic: "I'm pretty hopeless that any one man, white or black, can make a difference." A Democrat who crossed party lines to vote for Giuliani stated, "Dinkins is just a high-priced secretary." A Queens voter confessed, "I wanted

Mayor Koch again." And, sadly, there was sometimes the ugly reaction: "He looks like [Rev. Jesse] Jackson. Then they all look alike."[9]

The new mayor faced many of the same challenges as those who came before—and some additional ones. Racial and ethnic tensions were intense, despite his pleas for unity. White flight from the city continued. The AIDS crisis deepened, as did homelessness. The crack epidemic caused drug wars and a crime wave that strained race relations even more. Wall Street woes and a real estate collapse plunged the city into another recession, leading to higher taxes, layoffs, and cut services. Despite the high expectations of supporters, Dinkins could not shake his image as an ineffectual manager, which led to his fate as a one-term mayor. Those focusing on conditions at the time may have judged him too harshly, and future leaders inherited the momentum of some worthwhile projects that got them the credit and approval Dinkins was denied. He ultimately achieved a balanced budget, expanded the police force, rebuilt neighborhoods, witnessed a decline in homicides, and began cleaning up Broadway.[10]

From the start, Mayor Dinkins was extremely unpopular and downright disliked by many Staten Islanders. The island's population at the time was under 400,000, but the 1990s would see the largest percentage increase (17 percent) since the 1970s (19 percent). The growth rate on Staten Island would amplify its voting presence sufficiently to help conservative candidates in a Republican-leaning borough where the population was still overwhelmingly non-Hispanic white (80 percent).[11] Not surprisingly, Dinkins's standing on the island was largely influenced by a mixture of race and political ideology. The secession movement in the 1990s would heighten that antipathy.[12]

Economic woes identified with the beginning of Dinkins's term in office affected the mayor's popularity not only on Staten Island but also in the other boroughs. As early as January 3, 1990—Dinkins had been in office only three days—the mayor had to delay the hiring of 1,800 police officers, call for spending cuts, and propose a property-tax increase. Retail sales were declining rapidly, and thus sales-tax revenue had shrunk. The budget deficit for 1991 was estimated at $1.5 to $2.5 billion. The gross city product fell throughout 1990, job losses in the third quarter exceeded 34,000, and construction in many places was halted. During Dinkins's term 400,000 jobs were lost, reversing the gains of the Koch years. While this economic setback would be shorter and less severe than the mid-1970s crisis, rumors of another state takeover were in the air.[13]

Brendan Sexton remained commissioner of the Department of Sanitation (DSNY) during the first four months of the new mayoral administration. But like many mayors before him, Dinkins wanted his own team in place. He selected yet another administrative type in the youngish Steven M. Polan, a thirty-nine-year-old attorney serving as general counsel for the Metropolitan Transportation Authority (MTA). Polan was a protégé of Richard Ravitch, the former chair of the MTA; he had worked for the authority since 1981. He was credited with being "one of the architects of the M.T.A.'s $16 billion capital rebuilding program" and, according to the current chair, Robert R. Kiley, "helped to translate the vision of rescuing our massive transportation system into a reality." Polan came to the DSNY position having no experience in solid waste management but with a full understanding of the coming citywide budget cuts. He was convinced that he could handle the job given his background as a manager and as a lawyer.[14]

ASHES AGAIN

The proposal to site an ashfill on Staten Island linked several controversial issues: the closure of Fresh Kills, the fate of the Brooklyn Navy Yard incinerator, the recycling program, and concerns over pollution. The Department of Environmental Conservation's (DEC) decision in 1988 to build an ashfill on the Fresh Kills property appeared even more questionable to Staten Islanders after the new consent order was struck. In June 1990 DEC Commissioner Thomas C. Jorling stated in a letter to Assemblywoman Elizabeth A. Connelly that now that the consent order was signed, "among other items, ash disposal practices must be evaluated and addressed."[15] Soon thereafter, Assemblyman Eric N. Vitaliano and others sponsored a bill requiring incinerator ash to be "monofilled" in ash-only dumpsites to segregate it from other wastes. Fly and bottom ash would be separated and tested quarterly. The most telling part of the bill was the requirement that monofills under most conditions be located in the same communities where the incineration plants of any type were found.[16]

A technical report released in October 1990 essentially stated that the Staten Island ashfill was a fait accompli. The report acknowledged the construction of a thirty-acre dumpsite for approximately 900 to 1,800 tons per day of incinerator ash/residue "generated by the existing and proposed municipal solid waste (MSW) incineration facilities of the NYCDOS." The project would be designed to have a maximum useful life of ten years if the Brooklyn Navy Yard incinerator was built, or twenty years if only existing incinerators were in operation.[17]

That same month, the DSNY applied for a permit that would allow Fresh Kills Landfill to accept incinerator ash. Public protests arose again, quoted in the newspapers and finding their way to letters sent out to any authority who would listen. "This is really outrageous," stated Ellen Pratt of the Protectors of Pine Oak Woods. Councilman Alfred Cerullo, a Republican from Staten Island who had fought the ashfill for some time, declared, "This is truly the straw that will break the camel's back." It was clearly understood that the designated location for the ashfill was the one remaining impediment to having the DEC sanction the Brooklyn Navy Yard resource-recovery (waste-to-energy incinerator) facility.[18]

In November, Wehran EnviroTech published its site investigation plan, which located the proposed ashfill on a seventy-five-acre parcel at Fresh Kills between the West Shore Expressway and the Arthur Kill.[19] By year's end, DSNY Commissioner Polan had made no attempts to cover up that he believed the city needed at least one high-tech resource-recovery facility and that he did not expect recycling to provide the single solution to the city's disposal needs. Thus it was best to build "a modern, lined ashfill" to handle the residue. For the time being, however, Mayor Dinkins, who was on record as opposing incineration, had put the Brooklyn Navy Yard project on hold. That decision further complicated the give and take over resource recovery.[20]

While Dinkins had run for mayor on an anti-incineration platform, environmental groups, some city officials, and Borough President Guy V. Molinari remained skeptical of his actual views. Dinkins's plan to delay expanding the mandatory recycling program seemed to offer evidence of a possible shift in policy—maybe back to incineration? It was anybody's guess at this point.[21]

The ashfill controversy became a cause célèbre for opponents. Calls went out to investigate the DEC's actions on the matter. A coalition of environmental and community organizations formed "Say No to Ash." The attorney and activist Matthew Pavis stated, "This issue ... is a lightning rod that coalesces everything all of the other groups have been trying to do."[22] On March 25, 1991, Representative Susan Molinari and several other members of New York's congressional delegation wrote to Governor Mario M. Cuomo to urge him to intercede with the DEC to prevent the city's permit application relative to the ashfill.[23] Patrick Joyce of the *Advance* observed, "Environmentalists fear state approval of the ashfill would be the first domino to fall in a series of policy moves leading to a greater reliance on incinerators, some of which burn garbage for energy."[24]

Yet the DSNY defended its current position on handling ash (which it claimed was not hazardous) as an indicator of how it would handle future disposal practices.[25] Assemblyman Vitaliano, who organized a coalition of twenty-five state lawmakers to oppose the ashfill, called on Governor Cuomo to block the city from building it. Without a comprehensive solid-waste management plan or any credible alternatives to land disposal (both required under state law), "the city remains unprepared to state how much waste they hope to recycle, how much they hope to burn and where, or how much raw garbage or incinerator ash is to be buried over the next decade." Authorizing the ashfill before the city decided to go forward with the Brooklyn Navy Yard facility, he added, was "putting the cart before the horse."[26]

Cuomo was regarded as a friend on environmental issues, but throughout the entire debate over New York City's disposal problems, his involvement varied depending on which issues were at the forefront. A native of Queens from an immigrant Italian family, he received a law degree from St. John's University in 1956 and made a name in private law practice over a public housing dispute in Queens. His first important venture into politics was as an unsuccessful candidate for lieutenant governor in 1974, which got him attention outside of New York City. He went on to serve as secretary of state from 1975 to 1978. During that time, he ran for the Democratic nomination for mayor of New York City in 1977 but lost. He became lieutenant governor the following year, which proved to be a springboard to his election as governor in 1982. Cuomo's greatest support came from New York City. In 1986 he was reelected by a wide margin, and again in 1990. Although there was little doubt of his liberal Democrat credentials and leanings, he strove to be "inclusive" and to avoid partisanship. Hard working, intelligent, and some would say abrasive, he was not particularly good at delegating authority, but he was a spellbinding speaker.[27]

Despite the flurry of criticism, the DEC tentatively approved the DSNY's ashfill permit on March 29, 1991 (although there would be a public hearing scheduled). The decision was contingent on the DSNY filing by year's end a draft environmental-impact statement (which it had done) and completing a solid waste management plan—the linchpin of establishing clear future disposal goals. City Comptroller Elizabeth Holtzman had written to Cuomo, expressing concern that New York City had not actually completed its application for the ashfill because the city's solid waste management plan had yet to be submitted.[28]

Representative Susan Molinari called the ashfill "another excuse for the city not to fulfill its recycling obligations." New York Public Interest Research Groups' (NYPIRG) Arthur Kell was not hopeful that the governor would intervene in stopping the DEC's action, calling it a "cowardly move on the part of the Cuomo administration." In an "eleventh-hour decision" on April 1, the DEC accepted as completed the DSNY application to site an ashfill at Fresh Kills.[29] On May 5, more than 150 environmental activists, including Say No to Ash, staged a rally to dispute the decision. Many of the protesters wore white face paint to stress their claim that the ashfill would be a major health risk.[30] A few days later Commissioner Jorling stated that there was "wide room for doubt" whether the state would issue the city permits to build an ashfill (and an incinerator) if the city's recycling plans remained in limbo. Confusion reigned.[31]

THE PENDULUM SWINGS

Scaling back the recycling program could be a quick budget fix for the Dinkins administration. But recycling was so deeply enmeshed in the disposal debate that such a decision would have immediate political and policy repercussions. On January 7, 1991, Council Member Stephen Di Brienza from the

Thirtieth District in Brooklyn wrote to Deputy Mayor Norman Steisel: "Although I am cognizant of the City's need to dramatically reduce expenditures, I believe that there are certain proposed cuts which make little sense from both policy and financial viewpoints. And while there are certainly numerous programs which deserve to remain intact, I will limit my comments in this letter to Department of Sanitation recycling programs."[32] Di Brienza went on to remind Steisel that the city was only recycling about 6 percent of its solid waste even though Local Law 19 called for 25 percent by 1994. He also raised the long-standing issue of shrinking landfill space. "As landfill space becomes scarcer and the cost of solid waste disposal continued to increase dramatically, reducing the City's funding for recycling is, in the long-term, short-sighted."[33]

The proposed cutbacks in recycling were a public acknowledgment of privately held feelings among city officials that recycling was more expensive than they had expected. In an October 1990 report, Polan stated that the cost of recycling by 1994 would be $198 to $273 per ton, instead of previous estimates of $65 per ton just two years before. The cost of collection prove to be the biggest miscalculation. The *New York Times* reported in March 1991 that the recycling program had fallen behind the annual tonnage requirements. The DSNY should have been collecting about 1,400 tons of household recyclable material each day but was falling short by 250 tons. Polan stated that the goal had not been reached because the expansion of the program had been postponed. He added that public participation had been less than expected, the costs were high, and market demand for recyclables was inadequate. Thus the timetable set by the law was "unrealistic."[34]

Polan understood the political implications of a cutback in recycling. In a confidential memorandum to Steisel on May 3, he wrote, "I can only assume from yesterday's meeting that the Mayor's Executive Budget will propose the total elimination of the recycling program, principally as a union and/or political bargaining strategy." Polan strongly objected to the approach, believing that Local 831 would not bend on route extensions as an alternative to the elimination of recycling. Better than threatening layoffs, he argued, would be to privatize recycling. "We can obtain the same advantage vis-à-vis the union without undermining the program or incurring the wrath of all the people who support it." He also believed that the program would be "hurt very seriously by threatening to kill it." He concluded, "Some may take the position that the program is indeed frivolous as a waste management strategy, but it is most certainly not as a regulatory matter. There will be no permit for the Navy Yard or any other facility we need if the program is eliminated. The Solid Waste Plan [currently being devised] will be dismissed as an irrelevant endeavor."[35] Publicly Polan had been positive about recycling, but he was keeping his eye directly on the main objective—sustaining the waste-to-energy strategy he had inherited. Recycling was not a side issue. It was central to the city's disposal plans going forward.

Polan's public remarks suggested his advocacy for recycling and concern that the suspension could erode community trust in the city's disposal program. But he also was trying to protect landfill space and the Brooklyn Navy Yard facility.[36] During a press conference on April 1, Polan stated, "There is no conspiracy to push incineration [over recycling]," which belied his private utterances.[37] For many New Yorkers, cutting recycling was vastly unpopular, especially because recycling was a visible sign that the city was making progress in modernizing disposal operations. Killing the recycling plan never got political traction, but a tentative decision was reached to suspend mandatory recycling for a year. For some officials this was a means to a budgetary end; to others it was a serious threat to any viable new disposal plan for the city.[38] In late October 1991 Dinkins abruptly changed course, announcing that funds had been found to keep the recycling program "at current levels" in the new financial proposals.[39]

This latest decision came on the heels of Dinkins's flip to push forward three "state-of-the-art" incinerators, contradicting his previous stance. (Five plants initially had been proposed before Dinkins's election.)[40] A detailed memorandum from Polan to Dinkins on March 21 laid out the reasons for the shift. The DSNY was soon to present its newest draft plan for dealing with the refuse problem, "and it is *essential* that I now begin to lay the groundwork for Plan acceptance by publicly building a case for the general approaches that will be detailed in the plan. I expect that our Plan . . . will rely more heavily on recycling and less on resource recovery [than previous plans], but it will still include construction of some new waste-to-energy facilities," Polan stated.[41]

Polan suggested that the DSNY was showing "systematic progress" in building a "legitimate recycling program" but viewed it as only "part of the solution" to disposal. The entire solution, he believed, included resource recovery (his long-held position). "Among the real world options, it is my view that resource recovery . . . must be a substantial element of the solution." And he added, "The likelihood of success in siting, permitting, and building all the capacity that is required to augment the remaining capacity at Fresh Kills—including recycling facilities, composting facilities, and resource recovery facilities—is dependent upon the credibility of the Department and of our Plan." The plan would be "greatly enhanced" if he began building the public case immediately. A major concern was that he felt "hampered" until he was able to explain "the entire solution" to the public.[42]

Polan expected the new program would include the Brooklyn Navy Yard incinerator. He warned the mayor that "to the extent that we propose *any* new resource recovery capacity, we will certainly be criticized by the recycling advocacy community and the many groups who have opposed the Navy Yard project.

It will be painted as a lack of commitment to recycling, when in fact recycling is one of the few areas of the budget that has grown under your Administration." Polan made clear that Dinkins would be accused of reversing a campaign pledge, even "though you only agreed to defer a decision until a Solid Waste Plan was developed." Polan stressed that to gain the support of the "opinion makers" and the public at large, they would have to state the "dire nature of the capacity problem" and make a convincing argument for their chosen path. This also would mean progress in "building a legitimate recycling program" rather than the "totally unrealistic one contained in Local Law 19."[43]

Polan outlined the necessary shape that their argument about refuse disposal should take, underscoring realistic outcomes. He stressed that "the critical question that we need to address now" is what to do with nonrecyclable waste. "The choice is *not* between recycling and resource recovery, as the issue is conventionally posed, but rather one between resource recovery and landfilling [at Fresh Kills or through export]." Polan was concerned that while continuing to use Fresh Kills would delay having to depend on other options, it would "expose the City to extraordinary risk and expense, perhaps by the end of the decade."[44]

Dinkins appeared convinced by Polan's arguments, and he began turning New York City back to resource recovery as a disposal priority, just as his immediate predecessor administrations had. Controlling the public debate over disposal proved difficult for the DSNY, however, not only in terms of getting out its own message but also in carrying out its daily responsibilities, confronting budget shortfalls, giving attention to compliance with state regulations, and addressing the incessant public protests, especially from Staten Island but also from Brooklyn. Complicating matters was a federal health investigation of Fresh Kills prompted by a petition from Staten Island Citizens for Clean Air (SICCA).[45] Senator John J. Marchi's bill to have the city set aside money for the eventual closing of Fresh Kills also made its way through the Republican-controlled Senate and was now moving to the Democratic-held Assembly, but current budget woes were likely to stall it.[46]

In May 1991 the DSNY released a document describing the scope and process for a comprehensive solid waste management plan and a generic environmental-impact statement called for by the state. Benjamin Miller, director of policy planning for the Office of Waste Management and Facilities Development, described the major elements of the planning process currently underway: "The challenge the City faces . . . is to develop a near- and long-term plan to guide implementation of cost-efficient, environmentally sound programs and facilities that will reliably meet the City's waste management needs into the next century." Significantly, he stated, "Despite continuing efforts, the City has not constructed any new waste-disposal facilities in 30 years" and instead "depends almost exclusively on only one major landfill—Fresh Kills on Staten Island—to meet its huge waste-disposal needs." Miller reviewed the current legislative mandates and made the case for public scrutiny in the planning process. The plan would be intended to "guide the development of new programs and facilities that will be needed over the next 20 years to ensure that all of the solid waste generated in the City *can* be managed effectively."[47]

Miller then outlined a sophisticated methodology that evaluated all the different waste streams, all available disposal techniques, baseline conditions, and possible solid waste management scenarios and options. The plan was being developed by multiple city agencies under the direction of the deputy mayor for planning and development, Barbara J. Fife, but Miller and his team would do the heavy lifting. DSNY personnel would prepare the document with technical assistance from sixteen consulting firms.[48] Although Polan certainly was hopeful that the study would reconfirm his own assessments of the city's waste disposal needs, this was not going to be a whitewash of the DSNY. It would be the most thorough study of the department and its future ever completed, clearly within the objectives of developing an integrated solid waste management system as envisioned in Albany. Given the times, it was likely to be controversial no matter what it concluded.

INCINERATION VERSUS RECYCLING

Complicating the story was the resignation of Commissioner Polan. He submitted a formal request to Mayor Dinkins on November 26, 1991, effective January 1992. Polan thanked Dinkins in the customary fashion for "the confidence you have placed in me and for having given me the opportunity to serve the people of New York." He added that "the progress the Department has made is due in no small measure to your support."[49] Polan was replaced by Emily Lloyd in February 1992 (with Steisel's intercession), who served through 1994. Lloyd had a background in planning, transportation, and environmental issues. The former Boston transportation commissioner spent several years with the New York–New Jersey Port Authority as general manager of the Aviation Customer and Public Services Division and then as director of the Office of Business Development. The first woman appointed to the post, Lloyd was regarded as one of the ablest commissioners in the Dinkins administration. She hoped to develop new recycling markets and to continue neighborhood-cleanup programs sponsored by businesses and community groups. Environmental organizations praised the appointment even though she had yet to make a commitment for or against waste-to-energy incineration.[50]

Shortly before stepping down in early January (to join an international construction company), and contrary to his formal resignation letter, Polan stated that he was leaving because he was unable to get the Dinkins administration to address disposal more aggressively. He told a *New York Times* reporter that while Dinkins offered "tepid" support for incinerators, the mayor took no steps to move forward to build them, nor did he fully support recycling. "Every elected official in this city and state that I have spoken to in the last two years has told me privately that we need to build incinerators here," he stated. "But none will say so publicly. If the problem doesn't need to be solved today, in their terms in office, why should they bother? We haven't built a new solid-waste facility in this city in 30 years. The obvious question here is, are these people leaders or followers?" On recycling, he called the original idea for the program "ridiculously naïve." "The underlying premise was that the city would embrace it. That didn't happen. We have not met our goals. Markets have not developed. And right now we are spending money we don't have."[51]

Polan's comments came as the DSNY was completing its twenty-year plan, which had been under development for a year. The draft was due to the DEC on March 31, with the final version due by July 30. Between April 1 and July 30, the city would conduct its environmental review of the plan, along with public hearings.[52] The disgruntled retiring commissioner asserted, "After all this time and effort I think there is a real question whether it [the plan] will mean a thing. It may become just another volume for the bookshelf, and the city cannot afford that."[53]

Albany officials were anxiously and impatiently waiting for New York City to frame its future disposal policy. City Comptroller Holtzman, in particular, went on the attack against incineration—something that had been largely the province of local protesters until 1992. In her January report *Burn, Baby, Burn: How to Dispose of Garbage by Polluting Land, Sea, and Air at Enormous Cost*, she declared, "The urgent need to develop alternatives [to current disposal practices] has been known for over ten years, but the New York City Department of Sanitation . . . has yet to present an acceptable and feasible plan of action." Holtzman added that the DSNY had "ignored legal mandates" and instead had "simply declared an intent" to rely "more heavily on incineration, a technology which is costly and poses very serious public health hazards." Time was ticking away on Fresh Kills, the city had missed the April 1, 1991, target date (an extension of the original deadline) for its solid waste management plan, and the DSNY was continuing to pursue "incineration aggressively instead of carrying out its recycling mandate."[54] She concluded:

> What is wrong with any waste program that relies heavily on incineration can be summed up very simply: it is environmentally hazardous and bad for the health of New York residents; it will not save nearly as much scarce landfill space as DOS thinks—nor would it save much more than a program based on 50% recycling would save; and it will cost much more than DOS has projected—perhaps more than safer alternatives, both in budgetary terms and otherwise.[55]

Holtzman's missive had something for every critic of Dinkins and Polan, New York City, and its Department of Sanitation—from Commoner and Kell to Jorling, from Staten Islanders to recycling advocates. Her March 1992 follow-up, *Fire and Ice: How Garbage Incineration Contributes to Global Warming*, and her May 1992 *A Tale of Two Incinerators: How New York City Opposes Incineration in New Jersey While Supporting It at Home*, poured on more condemnation.[56]

All the while, the drumbeat to close Fresh Kills, to fight the ashfill, and to protest the Brooklyn Navy Yard incinerator went on. The DSNY's February 10 request to raise the maximum height of sections 3 and 4 of the landfill to 191 feet only inflamed the situation.[57] Nor were locals pleased when, on February 27, 1992, DSNY Deputy Commissioner Robert P. Lemieux told a crowd of about 250 at the College of Staten Island, "You're not going to like what you hear. We have 100 million yards of capacity in the landfill and we plan to use all of it." Asked when Fresh Kills would close, Lemieux flatly stated, "In about 12 to 20 years."[58] The DSNY had developed "A Final Closure Plan," but only for Sections 2/8 and 3/4 of the landfill.[59] Despite the ambiguity of the department's goal for Fresh Kills as a whole (or maybe because of it), protesters continued to fight the ashfill. The state decision to allow for its construction would be challenged in the courts until the review procedure was completed.[60]

THE 1992 PLAN ON ITS OWN TERMS

The Comprehensive Solid Waste Management Plan and Draft Generic Environmental Impact Statement became public in the spring of 1992. The executive summary provided a good rendering of what the DSNY's future policies might be. It began with a blunt statement of the bewildering reality: "Every day, New York City receives and consumes tens of thousands of tons of materials extracted and transported from all over the world, and in turn, produces tens of thousands of tons of solid waste. These millions of tons every year make it the largest 'materials sink' on the planet."[61] While the Fresh Kills Landfill, the report went on, seemed to be the cheapest method of disposal, "such a shortsighted analysis ignores the environmental damage created by landfills and the economic and environmental benefits that alternative waste management options could provide." The report did not deny the current need for the use of Fresh Kills despite its possible shortcomings, but it recognized that the facility "will not last forever, and when it is gone, it is inconceivable that a replacement could be sited within the City's limits."[62]

The city's residential waste and ashes from incineration, in particular, were sent to Fresh Kills, but other refuse had different disposal outcomes. Private carters, for example, hauled most commercial solid waste generated by businesses to landfills outside of the city. After July 1, 1992, sewage sludge (coming from different sources than solid waste) could no longer be deposited in the ocean and would need to be dealt with in some other way. More than half of medical waste (which was potentially more hazardous than municipal solid waste) was exported. Because onsite hospital incinerators were to be shuttered, more medical waste would have to be removed to remote facilities. Dredged material—too bulky for a landfill and requiring expensive transportation to it—was often burned at sea, but in the future it was more likely to be dumped on land, as any kind of sea dumping was now frowned upon. And very little recycling was being carried out, given the inadequate system in place, and it played only a small role in taking material out of the waste stream.[63]

Given the complex nature of the makeup of waste, the increased costs, and the dwindling disposal options, the report argued that "this historic trend, though it has been obvious for decades, has never been adequately addressed by prior planning and facility implementation efforts." Thus, "no major new facility has been developed in 30 years," which the report blamed on the government's inability to carry out needed projects through the public and regulatory approval processes. Inadequate planning was a major cause of poorly executed disposal. Additionally, depletion costs were not reflected in the city's budget, adding to the lingering perception that landfilling at Fresh Kills had been "free." The DSNY's new approach was meant to overcome both of these difficulties "by fundamentally rethinking the premises on which the waste-management system is based."[64]

The report outlined eight premises to guide the planning process: (1) the waste stream needed to be understood as encompassing "many discrete material components" from various sources, thus requiring different types of management techniques; (2) the entire waste management process had to be treated as an "integrated whole"; (3) the DSNY needed to do what is feasible to prevent waste generation, to reuse, recycle, or compost waste, and to recover energy from waste (and landfill what cannot be managed in "any of these preferred ways"); (4) the management of wastes from different sources needed to be combined; (5) a long-term perspective (as a "decision tree") to maximize achieving goals and build in flexibility needed to be adopted; (6) an "expansive analysis of the full universe of potentially feasible options" had to be executed; (7) technical and policy expertise from a wide variety of disciplines and perspectives had to be mobilized; and (8) sophisticated data-analysis capability needed to be employed.[65]

Based on the policy approach chosen and the massive amount of data gathered, the report discussed in some detail its conclusions. Out of hand it rejected continuing the current waste management system unchanged into the future, which was "the most costly, the most environmentally degrading, and would put the City at greatest risk of having to depend on out-of-City landfills to meet a substantial portion of its future waste disposal need."[66] Two quite similar "finalist" systems, which optimized recycling and composting, were not chosen. Instead, the plan proposed a "real world" plan somewhere between the extremes but focused on the immediate decisions to be made.

The path chosen grew out of the assumption that there was a high level of uncertainty in every aspect of the solid waste management planning process: "The plan can best be envisioned as a decision tree which identifies a set of immediate decisions that must be made, the additional information that will be available to inform the second round of decisions, and finally, the likely timeframe when the last, long-term decisions will have to be made to assure an adequate system to handle the City's solid waste at the 20-year mark."

The city, therefore, had not decided to proceed with any particular twenty-year program at this time but to concentrate on a near-term five-year implementation plan. This would include waste prevention legislative proposals, proposals to reform the mandatory recycling program, supplemental techniques for diverting constituents for recycling, a secondary-materials marketing strategy, a composting program, continued maintenance and upgrading of waste management infrastructure (including marine transfer stations, the Southwest

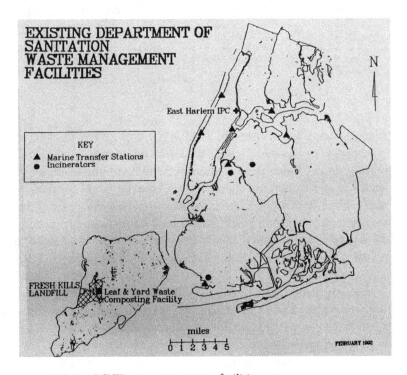

FIGURE 14.1 Existing DSNY waste management facilities, 1992.

Source: A Comprehensive Solid Waste Plan for New York City and Final Generic Environmental Impact Statement, August 1992.

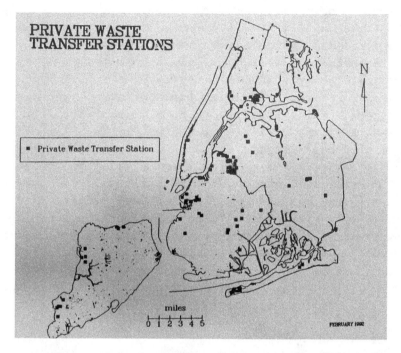

FIGURE 14.2 Private waste transfer stations, 1992.

Source: *A Comprehensive Solid Waste Plan for New York City and Final Generic Environmental Impact Statement*, August 1992.

Brooklyn municipal incinerator, and Fresh Kills), a new waste-to-energy processing capacity, ash-residue disposal, exploring future disposal options, and additional technical fixes.[67] In its assessment of the environmental risk factors, it defended its approach as not appreciably different from other options.[68] The plan also committed to a long-term implementation phase that combined reuse and recycling measures with possible additional waste-to-energy capacity. As noted in a supplemental report: "The City must take a long-range view and have the ability to handle its waste, but must also have the flexibility to adapt to changing conditions."[69]

It is not surprising that skeptics regarded the plan as business as usual. Resource-recovery remained. Fresh Kills stayed open. But read more carefully, the plan laid the groundwork not only to improve existing practices but more importantly to commit the city to a careful assessment of future needs. The 1992 plan was notably different from past policy stances that simply chose landfilling over ocean dumping, incineration over landfilling, and recycling over everything else. Ideas discussed and debated nationally with respect to integrated solid waste management were now to be employed by the DSNY for the first time. The new tone and perspective, of course, did not guarantee a vastly improved disposal system, but it was a significant indicator of the department's realization that simple solutions to complex problems were unworkable.

REACTIONS TO THE COMPREHENSIVE PLAN

The plan received considerable positive support. Taken as a whole, however, the arguments and charges against it were a recapitulation of the ongoing debate over New York City's disposal policies—understandable given the years of frustration and ineffective solutions. Critics of the ten-volume report and opponents of the New York City government did not acknowledge the depth of research invested in the plan or recognize the change in perspective and tone it represented. Even if they heard the words, many did not trust the DSNY's sincerity.

Negative reaction was immediate from Staten Islanders fed up with Fresh Kills and from Albany officials wary of New York City's slow response to regulatory requirements. One of the harshest criticisms came from an editorial in the *State Island Register* on April 21: "Despite motive, opportunity and outrage, Staten Islanders killed no Sanitation officials at [the] Community Board's Environmental Committee hearing on the Fresh Kills landfill last week." It charged that despite the new study, the DSNY had no long-range management plan for Fresh Kills and "obfuscated on the issue of closure."[70] Such feelings were in stark contrast with the DSNY's objectives, which meant to confront broadly and widely Gotham's long-standing refuse problems. The controversy clearly was a clash between short-term desires and long-term goals, all expressed through a loss of trust.

At a May 18 Staten Island hearing on the plan, several local political leaders aired their disfavor. Assemblyman Vitaliano, who had openly supported the Brooklyn Navy Yard project but fought the ashfill on Staten Island, expressed "great concern" about the plan:

> My concern starts with the obvious—This plan perpetuates the long standing city perspective that it can thoughtlessly continue to heap its waste into an unlined, grossly polluting, unpermitted garbage dump. This plan continues reliance on the Fresh Kills Landfill, and allows the city to continue to procrastinate in developing much needed, long overdue waste management alternatives.[71]

In a release for the meeting, the Staten Island Coalition for Survival, representing six local community organizations, voiced support for recycling over incineration. The Staten Island Environmental Coalition, in its own release, supported recycling and also focused on landfill pollution: "The bottom line is—Staten Island has become one big toxic waste dump and we want it cleaned up. We want the landfill not just closed, but cleaned up so we can breathe again."[72]

Congresswoman Susan Molinari broached similar views about Fresh Kills and the DSNY: "Operating such a facility in a City that traditionally has the biggest and best of everything, is in my estimate a disgrace." She also stressed that her dissatisfaction with the plan was based largely on her efforts to get federal authorities to conduct pollution tests at Fresh Kills. She claimed that the plan failed to confront the levels and concentration of pollutants relative to waterways surrounding the landfill, the rural terrain of Staten Island, and soil contamination. Molinari also took a swipe at the ashfill and inadequate recycling.[73]

Council Member Cerullo emphasized how Staten Islanders "have shouldered an unfair and basically criminal burden of the city's solid waste disposal responsibilities." He added, "Unfortunately, the details of this draft plan do not echo the sympathy that Commissioner Lloyd expressed for the residents of this borough. In fact, if certain scenarios of this plan are implemented, we as Staten Islanders will find ourselves with an even greater burden of the city's solid waste disposal responsibilities in the next century."[74] Assemblywoman Connelly, before turning to Fresh Kills, broadened the discussion: "Many aspects of the plan are quite disturbing, including the persistent over-reliance on planned resource recovery and the failure to get serious about recycling."[75]

The hearing, a joint meeting of the City Council and the DSNY, attracted about five hundred people to the United Artists Movie Theater in Travis, with about four hundred more in the parking lot listening on speakers. It began at 6:30 p.m. and continued until midnight. The *Advance* observed, "Fueled by health fears, Staten Islanders flung curses, threats and promises of political retribution at Sanitation Department officials and City Council members during a fiery hearing last night on the mayor's draft Solid Waste Management Plan."[76]

Local community and environmental groups piled on. Staten Island Citizens for Clean Air criticized the format of the plan: "The discussions are left vague and it is impossible to understand what the analyses revealed." SICCA and some others wanted a sixty-day extension to review and comment on a variety of issues.[77] On May 26, Barbara Warren Chinitz, secretary of SICCA and a member of the Staten Island Solid Waste Advisory Board, wrote to City Council Speaker Peter F. Vallone, calling for the council to reject the plan. Among its weaknesses, she charged, it had "grossly deficient information on the current crisis nature of the city's solid waste problem, i.e. the Fresh Kills landfill." The plan also was a "dishonest presentation of the many important issues central to what the city chooses to do. The information does not reflect the REAL WORLD. Without special access to additional information, anyone reading the plan cannot possibly know what the lies are."[78]

Holtzman immediately went on the offensive. In a statement to the Committee on Environmental Protection of the New York City Council, she criticized her number-one target: "Today I am here to tell the City Council the stark truth about the city's proposed solid waste plan. The truth is that this plan promotes incineration. The truth is that incineration is a bad idea for New York City. Incineration is bad for health. Incineration is bad for the environment. Incineration is bad for the budget." She charged that the plan "is a smokescreen for an incineration blueprint."[79]

The Center for the Biology of Natural Systems (including Barry Commoner) and Mirah Beck, from the Bronx borough president's staff, offered their own assessment on May 28. The plan, they argued, failed to meet requirements established by state and local laws, chose to implement a system inferior in environmental and economic impacts, and used faulty data and assumptions.[80] The Center and Beck also charged that the plan was undermining the city's recycling program. They restated their criticism of incineration and called Fresh Kills "more than an eyesore and harshly offensive odor."[81]

Modifications to the report were released on August 26, 1992, with a bombshell. The DSNY was going to scrap the ashfill at Fresh Kills. This obviously was politically motivated but surprising nonetheless. "The Department," the revised report stated, "does not intend to proceed with [the ashfill] and will withdraw the permit application when this plan is submitted to State DEC. Out-of-City ash-disposal capacity, instead, will be sought through an RFP process."[82] Hitherto often scoffed at, out-of-state disposal was increasingly being considered as a solution to some of New York City's internal waste problems, especially as the days of Fresh Kills were numbered.[83]

Pressure from City Council leadership played a role in the ashfill decision, and it also helped convince Mayor Dinkins to defer building the Brooklyn Navy Yard facility until 1996.[84] Dinkins wanted to drop the incineration issue so that it did not haunt him in the 1993 mayoral race. Commissioner Lloyd, clouding the issue, added that the plan recommended "we commit to beginning the site-selection process for three major waste-management facilities, without specifying the technology."[85] Whether critics liked to acknowledge it or not, debate over the plan had brought them substantive victories, even if the results only led to deferring some unpopular decisions.

FROM HERE, WHERE?

The filing of the 1992 comprehensive plan was a pause—albeit a short one—in the intense debate over the refuse dilemma in New York City. It marked a reflective moment for the DSNY, which had been trying to fend off its detractors and also trying to figure a way forward in dealing with the perplexing challenges of disposal. The filing also marked a moment when critics of the department and the Dinkins administration had to ask themselves: "Where do we go from here?" Was the report a smokescreen simply to keep the wolves at bay? Would it simply inspire another feeding frenzy of criticism over the symbols of controversy—Fresh Kills, the Brooklyn Navy Yard, and the trampled recycling program? For the remainder of 1992, these contentious issues were constant subjects of dispute.[86] But finally beginning to be heard were those pondering the likelihood of making over Fresh Kills (reclaiming it) after closure—whenever that might be. The report had not made the timetable any clearer than numerous earlier pronouncements.[87]

There was some room for an inkling of hope on Staten Island in mid-to-late 1992. The State Assembly passed a Vitaliano bill on June 22 that would bar New York City from dumping refuse in buffer areas of the landfill near residential neighborhoods. This was merely a stopgap measure, but it carried much symbolic weight. Mayor Dinkins opposed the bill, arguing that it would put undo constraints on the DSNY in dormant and unused portions of the facility, closing off 75 percent of the landfill's remaining capacity. Marchi shepherded the legislation through the Senate, and it passed on June 26, despite opposition from ten Manhattan Democrats who believed that the action would trigger the building of incinerators throughout the five boroughs. Vitaliano withheld sending the bill to Governor Cuomo, hoping to use it in obtaining some political leverage through a negotiated deal.[88]

Cuomo, like other government officials, was learning how much Fresh Kills had become a political liability by antagonizing Staten Island voters.[89] After some discussion, Commissioner Lloyd agreed to "cease future landfilling outside the existing footprint of the Fresh Kills Landfill" if Vitaliano agreed not to submit his bill to the governor.[90] In an agreement announced in December, the DSNY pledged not to extend landfill operations within a thousand-foot buffer area bordering residential neighborhoods, with some exemptions given for existing operations. The one caveat was that the department could dump almost anywhere during an emergency.[91] It was not perfect, but it was a start.

Another quasi-victory was a watered-down measure that would have the EPA conduct a two-year study of environmental conditions at Fresh Kills and audit compliance with federal standards. Representative Susan Molinari fought for the EPA to have the authority to close down the landfill if the city and state did not take steps to upgrade it. Governor Cuomo moved to derail the effort, preferring to sustain state control and deflect federal involvement in the matter. Molinari charged that Cuomo's argument against her proposal was "the same old line we have been hearing for 12 years."[92]

In November the DSNY announced that a new bridge would soon be in operation, allowing the closure of two sections (3 and 4) of the landfill. Sections 2 and 8, near Arden Heights and Greenridge and closest to heavily populated

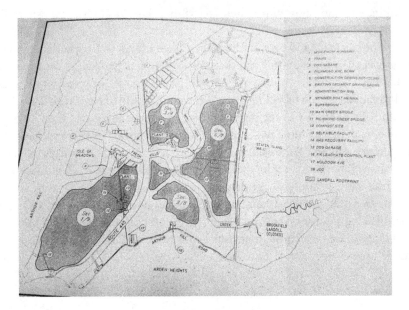

FIGURE 14.3 Fresh Kills' sections, 1994.

Source: *DOS Fact Sheet: Fresh Kills Landfill Closure Plans*, February 1994.

areas, were to follow by the end of 1993. Sections 1/9 and 6/7 would close too. This was good news or bad news, depending on one's perspective. It was a positive sign of "shrinking" Fresh Kills. Yet it also was an indicator of the relentless use and vast size of the facility and insinuated the possible shift of dumping operations to other vulnerable spots.[93]

What was the way forward for Fresh Kills? Staten Islanders in particular saw little in the 1992 plan to give them comfort about an end to their dreaded landfill. Fresh Kills remained an artifact and symbol of all things that made the disposal problems so unwieldy and so discouraging. Could a rational planning process cure the ills of generations of trial-and-error disposal methods? Would such a plan be subverted, diverted, or simply ignored? Yet the end for Fresh Kills did not come with a systematic phasing out or through the embrace of well-contemplated alternatives. Its end was strictly political.

On a trip to Staten Island in late July, 1992, Mayor Dinkins had tried to assure Staten Island residents that he cared about their problems, be it transportation or the landfill. He spent nearly two hours discussing environmental issues with local business leaders and community representatives. Chinitz accused Dinkins of breaking his 1989 campaign promise not to build more incinerators. She added that this reversal would lose him support among people of color. Dinkins was clearly on the defensive. The visit did little to smooth the way for

Dinkins's relationship with Staten Island or pacify local citizens, especially about Fresh Kills.[94]

In the late 1980s and early in the 1990s, Staten Islanders' frustration and outrage over their megalandfill and the endless stream of garbage scows from the other boroughs finally boiled over. Staten Island's move for secession was the culmination of years of feeling outside the mainstream of New York City. And Fresh Kills Landfill was the biggest thorn. In a letter to a constituent in September 1991, Betty Connelly stated what was unvarnished truth for those in the forgotten borough: "Each well-fought win for a Queens community, or a Manhattan neighborhood, is a loss for Staten Island . . . because it all ends up here in Fresh Kills."[95] About the same time, John Holusha, in the *New York Times*, began an article about national dumping issues by stating: "No one likes the idea of being someone else's trash can."[96] The question about who bore the burden of disposal—my borough or yours—always flowed in the undercurrents of the refuse debate.

PART V
THE ROAD TO CLOSURE

CHAPTER 15

SECESSION

The 1993 vote on Staten Island's secession from New York City was a poignant event, possibly not as significant as the 1898 consolidation but certainly the latest reminder that the potential benefits of uniting the boroughs did not always translate into a harmonious relationship. There were several reasons why Staten Island felt ill-served within the consolidated metropolis, the nagging presence of Fresh Kills Landfill chief among them. A staff member for State Senator John J. Marchi prepared a background paper for an October 1993 *Larry King Live* show that was to feature the senator. It stated in part: "The most visible issue [with respect to secession], symbolic of the borough's perceived step-child status, is the Fresh Kills Landfill, the world's largest dump, located on Staten Island."[1] The landfill was too big of an area to be ignored on the relatively small island—its odor, its rats and seagulls, and its leachate and methane were constant reminders of its existence. The uncertainty of New York City's disposal policy in the early 1990s only made Fresh Kills stand out more as a contentious issue.

Fresh Kills, of course, employed many workers across the island's several communities, but that did not offset how much the landfill had shaped Staten Island's identity since its opening in 1948—and not for the better. Resentment ran deep about the landfill, bolstered by memories of the Quarantine and the World War I–era Garbage War.

SECESSION CYCLES

To understand the significance of the 1993 secession vote and the role Fresh Kills played requires first going back to earlier secessionist uprisings on Staten Island. Some have argued that Staten Islanders have threatened to secede from New York City every decade. This is surely an exaggeration, but during times of extreme aggravation separation seemed preferable to marriage. Staten Island voted to consolidate in 1898 based on the assumption that Richmond Borough would benefit financially—better transit, new infrastructure. In the wake of the Garbage War of 1916–1918, the reduction plant on Lake's Island threatened to make Staten Island a "garbageland" for the city as a whole. William Wirt Mills, chair of the Staten Island Separation Committee, at that time proposed a legislative secession bill, but it did not get very far. It was purportedly the first such effort to regain the island's independence.[2]

In the mid-1930s, periodic declarations appeared in the newspapers calling for secession for one reason or another. A man from Brooklyn in a March 1935 *New York Sun* letter to the editor was "startled" that someone had advocated for "the secession of Staten Island from New York, both city and State, and [applied] to New Jersey for admission to that State." The writer was not shocked by the suggestion but because he too had made the same proposal over twenty years earlier. He had gone even further by suggesting that if New Jersey gave Staten Island "the cold shoulder" it should apply to the federal government to be named a federal district. As such, Staten Island "would long ago have had the rapid transit of which it has been deliberately deprived by the other boroughs but taxed never the less to build miles of track into the swamps of Queens and to supply double and triple service for Manhattan, Kings and the Bronx."[3] There were other calls for Staten Island to join New Jersey. A letter from a reader in April 1940 stated what many believed: "Even the strongest proponent for New York admits that [Staten Island] is geographically New Jersey."[4]

The Port Richmond lawyer Tom Garrett advocated secession in 1940 to give Staten Island "a fresh start." The borough needed tunnels and bridges, waterfront industry, and more. The plight of the island, he argued, was akin to Jonah being swallowed by the whale.[5] An editorial in the *Advance*, foreshadowing what was soon to come in the case of Fresh Kills, proclaimed that Mayor John Mitchel "did not hesitate to plan a monster garbage dump here [in the late 1910s], to absorb the city's refuse, and to exert his influence at Albany in order to deprive us of such protection from exploitation as the legislature was willing and able to give us."[6]

The Republican assemblyman Edward P. Radigan's failed efforts to introduce a secessionist bill to the legislature in 1947 was the closest thing to a real secession movement until the late twentieth century. His concerns focused on eliminating payment of the city sales tax, the need to attract new industries, and—quite importantly—an effort to stave off plans for the placement of a new city garbage dump on Staten Island. What would become Fresh Kills Landfill was a rallying point for secessionist sentiment even then.[7]

Radigan's campaign continued to be a source of conversation well into the 1950s and after.[8] Periodically through the 1970s talk of secession (and possible annexation by New Jersey) surfaced at least informally. Some viewed the call for independence as largely political; it was a way to highlight the problems of Staten Island in a symbolic or general way.[9]

An event in 1981 directly related to the concerns of Staten Islanders was the first step in the most serious and potentially successful effort at secession in New York City history. That year, the New York Civil Liberties Union (NYCLU) filed a lawsuit on behalf of three Brooklyn (Kings County) residents challenging the constitutionality of membership on the Board of Estimate. Unlike the City Council, where apportionment was based on population, board representation was not. Each borough president had one vote on the board, while the mayor, comptroller, and City Council president had two votes apiece, for a total of eight votes. Brooklyn, for example, with six times the population of Staten Island, had a single vote. Since the Board of Estimate negotiated and approved city contracts, supervised land-use regulations, and negotiated the city budget, borough participation was crucial.[10]

Of all the boroughs, Staten Island benefited most from the current bureau representation because it had the smallest population. It was not surprising, therefore, that an effort to determine representation by population was contrary to the forgotten borough's interests.[11] Wrangling in the courts about one person, one vote and its possible application to the Board of Estimate got messy.[12] In the meantime, the debate over representation went public. A *New York Times* editorial on May 20, 1983, observed that trying to alter the board according to population "would be immensely complicated, with no obvious benefit for democracy." It added, somewhat ominously, "Now to annul this vehicle of unity might well tempt people in Staten Island, the smallest but fastest-growing borough, to see advantage in independence."[13]

On May 16, 1983, the U.S. Circuit Court of Appeals determined that one person, one vote should apply to the Board of Estimate. This prompted a strong response from Staten Island leaders. On May 23, Senator Marchi wrote to Mayor Edward I. Koch that the court decision "poses very serious and negative consequences for Staten Island, if the ruling ultimately stands." Marchi added rather adamantly:

> To make a fractional vote system law would be to make this borough even more of a governmental "poor relation" than it already is. Realistically, the result would be the virtual elimination of Staten Island from the governmental and political map. We would . . . become a "Ground Zero" for every obnoxious idea or undesirable facility from entities like the Power Authority.[14]

For many prosecessionists, the issue was never really one person, one vote but one borough, one vote.[15]

TENTATIVE STEPS

In the summer of 1983, Marchi (who became known as the "Father of Secession") was reluctant to scream "secession!" as some of his colleagues began to do. But it was an option for him if all else failed.[16] He nevertheless began exploring "the pluses and minuses" of setting up a separate municipality.[17] In a letter on June 17, Mayor Koch told Marchi that his concern was "premature." The Court of Appeals had not ruled that the Board of Estimate as constituted was unconstitutional but simply remanded the case to the district court for further study. Koch assured Marchi, "The City will defend the present structure of the Board" and will appeal the decision if the court finds it unconstitutional.[18]

Marchi was not convinced that the mayor's office would carry the flag to protect Staten Island's interests. On June 20, he made public a preliminary report that found that Staten Island could survive financially as an independent governmental unit.[19] Among other things, it stated that apportioning payments between Staten Island and the other boroughs for capital assets (such as infrastructure and equipment) would be "complex and arduous." But in Staten Island's favor the city would be under pressure to negotiate over disposal of its garbage, since Fresh Kills Landfill would become a valuable hostage on the island.[20] A big question was whether local authorities would assume control of it or cede it to New York City.

The *Advance* conducted an unscientific survey and published it during the first week of July. It showed 96 percent of those polled on the island favored secession.[21] As yet, however, secession was not a settled issue on or off Staten Island. For example, an unofficial report described secession as "not the answer" and "a financial disaster."[22] The principal voice of opposition was Congressman Guy V. Molinari. He did not believe that Staten Island was well served within the City of New York, but "at this point," he stated, he was not prepared to favor secession. Molinari was concerned, however, about two giant liquefied natural gas tanks placed on the island against its will, the plans to build a park rejected by Queens, an almost nine-year battle over building an unwelcome power plant in the borough, and most especially the presence of Fresh Kills Landfill.[23] In his memoirs, he reflected that Albany would never let Staten Island secede. "I knew that Staten Island was more important to 'The City' than 'The City' was to Staten Island. In those days we had the God forsaken garbage dump. If for no other reason, that was reason enough for Albany to deny secession to go through. Additionally, the state did not want to see New York City reduced in size."[24]

Marchi kept putting on the pressure. On July 12 he wrote Governor Mario M. Cuomo, taking issue with the governor's opposition to "reconfiguring New York City." The senator declared, "In my view [Cuomo's opposition to secession], assuming you are quoted accurately, expresses euphemistically the idea that might, indeed, does make right." Marchi reminded Cuomo that the governor's credentials with Staten Islanders—and with Marchi—"are solid" and that Cuomo and Mayor Koch had generated "genuine compassion and sympathy" for the community. But, he cautioned, "Please, Governor, do not misread us." While Staten Island wanted to remain part of New York City, as the potentially second-largest city in the state, "we will brook no judicial violence and we will not submit to a servile fawning dependency on benign mayoral or gubernatorial paternalism."[25]

On August 2 Marchi's stance on secession was hardening: "I am now prepared to say that secession is not only feasible [economically], but that it is also legally and politically viable." He added, "We will not abide passively the semi-colonialization of upwards of 400,000 men, women and children in a society that professes to condemn colonialization."[26] Marchi particularly was troubled about how the recently formed New York City Charter Revision Commission, which was taking up representation on the Board of Estimate, among other things, would affect Staten Island.[27]

For many people off the island, the threat of secession seemed foolhardy at best. But as an editorial in the *New York Times* observed:

> For Staten Island to persist in its secession threats... sounds like exaggeration, even folly. But it would be equally mistaken to minimize the local passions that underlie the secessionist idea, and they are not limited to Staten Island. If it left, parts of other boroughs would try to follow, hoping so to preserve their character and to hide from change. They could not succeed for long—but would destroy much in the process.[28]

On November 17, Marchi edged closer (but cautiously) toward testing secessionist sentiment in Albany. He offered three bill options in the pre-filing period before the opening of the 1984 legislative session on January 4, 1984. Assemblywoman Elizabeth A. Connelly and Assemblymen Robert A. Straniere and Eric N. Vitaliano provided "bipartisan support" (from Staten Island at least) in the lower chamber.[29]

Vitaliano used the occasion of the thirty-seventh anniversary of Radigan's call for secession to write Governor Cuomo about the garbage crisis facing New York City. He stated that Radigan "saw the city's proposal to convert the Fresh Kills wetlands into New York's garbage pail as the 'last straw' in a long series of abuses Staten Island had suffered through remote control." "My message to you," he continued, "is not about secession or equal municipal representations for Staten Island. It is, rather, about the frustration articulated by Assemblyman Radigan on this day 37 years ago. It is about the true passion that faces a new move for Staten Island to secede. It is about what makes secession a Staten Islander's code for dealing with Fresh Kills and garbage." Truly, the letter was about Fresh Kills *and* secession.[30]

In late 1984 and early 1985 Marchi had yet to commit to secession and thus was withholding such a bill from legislative action. The senator's stance was guided by his view that secession was the "last resort" for solving the current representation crisis and part of a waiting game as the courts wrestled with one person, one vote and home-rule questions.[31] Increasingly, parties outside of the courts were weighing in on the possible outcome of the court deliberations.[32] The Citizens Union of the City of New York (filing an amicus brief in *Morris v. Board of Estimate*), for example, did not favor the Board of Estimate as constituted. It did support state legislation to create a charter commission to make recommendations about governmental restructuring in cases where bodies did not meet constitutional requirements.[33]

THE SECESSION BILL

In 1986 Judge Edward R. Neaher (under orders to review *Morris* again) ruled that the Board of Estimate had to follow the doctrine of one person, one vote but that the current board could continue to function while the city brought it into compliance. Few favored this stopgap measure.[34] While the case lingered, Staten Island Borough President Ralph J. Lamberti attempted to shape the secession issue locally. Unequivocal support for secession among locals was not yet in the air. On December 15, 1986, he created the Committee for the Review of Secession with some of the most prominent leaders in the community. After about ten months of work, Chair Paul E. Proske told Lamberti that "the data compiled by this Committee clearly indicates that secession is feasible. That is not to say that it is desirable nor does it mean it would be easily achievable."[35] The so-called Lamberti Report emphatically stated that "self-determination" was the issue at hand.[36]

Others on the island voiced concern about the borough's capacity to govern itself.[37] In early 1987 Lamberti continued to support the retention of the Board of Estimate and its budgeting powers with the backing of Senator Marchi; he was getting resistance from the City Council members Susan Molinari and Jerome O'Donovan. In a case of self-interest, they had endorsed Council Majority Leader Peter F. Vallone's call to give the board's budget powers to the City Council.[38]

The legal roller-coaster ride continued. On October 9, 1987, the federal appeals court sustained the lower court's ruling that the Board of Estimate was unconstitutional in its composition. The very next day, Marchi came out unequivocally for secession.[39] Marchi and others were buoyed temporarily when they learned that the U.S. Supreme Court would take up the constitutionality of the Board of Estimate's apportionment.[40] During the court hearings, Associate Justice John Paul Stevens asked, "Is Staten Island going to secede?" The city's corporation counsel responded that the Board of Estimate was "the glue which has held the City of New York together."[41]

Frustrations on Staten Island ran deeper than the administrative structure of the board. Theresa Grey, a clerk at a coin-operated laundry on the island, flatly stated, "I dislike New York." James Barron, reporting for the *New York Times*, added, "[Staten Islanders] are tired of paying city taxes. Tired of smelling the city's huge Fresh Kills landfill. . . . Tired of waiting for city buses that are late, or never come. And tired of complaining about it all."[42]

On February 21, Marchi introduced a bill to the Senate "to provide for a referendum of the people of the borough of Staten Island on the question: "Shall the borough of Staten Island separate from the city of New York to become the city of Staten Island? And creating a charter commission for such purpose." Connelly, Vitaliano, and Straniere had submitted the same bill in the Assembly a few days earlier.[43]

In March, the Supreme Court, which was now hearing the *Morris* case, declared the Board of Estimate unconstitutional. "Today's decision by the U.S. Supreme Court," Marchi stated with imagery he used many times before, "drops 400,000 Staten Islanders into a colonial state by effectively separating them from the governance of the larger community in which we live." He added, "This turn of events impels me to move forward with my legislation seeking secession of Staten Island from the City of New York." Marchi requested that his bill be reported to the Senate calendar for quick action.[44]

In the last minutes of the legislative session on June 30, Marchi's bill authorizing secession if Staten Island voters approved it passed the Democratic-controlled Assembly by a vote of 117 to 21 (some said 116 to 22) the day before it passed the Republican Senate with only one dissenting vote—43 to 1.[45] Secession was moving much closer to reality than many had expected or could have envisioned.[46] Marchi stated, "My pleasure, however, is tempered by an ever lingering sense of regret that such a wrenching move as secession has to be considered at all. It is only with the greatest reluctance that we would break away from New York City. But the turn of recent events gives us no choice."[47]

Many in New York State now began to see the threat of secession as very real. Concerning the referendum, Assemblyman Anthony J. Genovesi (a Democrat from Brooklyn) stated, "It's fundamentally unfair to all the residents of the city of New York. I really can't believe we're voting on this. I don't know that anyone knows the implications of such a referendum."[48]

Before the Staten Island citizens could vote on the measure in November 1990, Governor Cuomo had to sign the bill. This was no foregone conclusion. Several city and state agencies and administrative bodies recognized the bill as destabilizing. Executive Deputy Secretary of State James N. Baldwin feared that the bill "set in motion a process that is less cognizant of and responsive to the State's interest than that which has been used historically to establish new city governments."[49] In other words, the bill would leave too much power in local hands, which was a battle as old as the nation itself. Edward C. Farrell, executive director of the New York State Conference of Mayors, asserted that his organization was not opposed to the establishment of new local governments but that "we strongly believe that the method of creation set forth in this proposal would be inequitable and unduly detrimental to the currently existing City of New York."[50] Peter Vallone, vice chair of the City Council, was particularly adamant that the governor veto the bill, which he believed was "constitutionally flawed."[51]

RACE AND SECESSION

Staten Islanders had mixed feelings about the bill, which began with initial differences over the value of secession as expressed most graphically by Marchi and Molinari. Congressman Guy Molinari was no fan of the bill, and his relationship with Marchi continued to sour.[52] Marchi transmitted the bill to Cuomo (along with a memorandum explaining the provisions), ending his letter thus: "The Iron Curtain is lifting, the Berlin Wall crumbling. If the Cold War is indeed ending over in Europe, do we deserve a Cold Shoulder here on Staten Island?"[53]

Assemblyman Vitaliano made a strong case for the prosecessionist forces, urging Cuomo's approval of the bill, dismissing the notion that he and others in favor were engaging in "political pandering," and emphasizing that Staten Islanders should have a choice on the issue.[54] Assemblywoman Connelly took another tack, reminding the governor that in a recent election "68 percent of nearly 75,000 voting Islanders said 'NO' to the new City Charter [which changed the status of municipal representation]." She added, "Their rejection was responding to a new Charter which gives them colonial status, strips away their ability to impact on their own affairs, and grants them a mere two votes among a Council of 35 members currently."[55]

The attorney David Goldfarb, former president of the St. George Civic Association and a founder of Staten Islanders for a Unified New York (SIUNY), argued contrarily that no community should have the right to decide to secede unilaterally. Plus, some sections of the borough had closer ties to Greater New York City than others. "These are the older well integrated neighborhoods," he argued. "These are the neighborhoods that have resisted racial steering by real estate interests, while other communities have been controlled by the mentality of 'white flight' fears." Some North Shore areas (the most economically and ethnically diverse on the island), therefore, "will become hostage to the rest of Staten Island, much as Staten Island feels it is put upon by New York City."[56]

Goldfarb, a Democrat who had challenged Assemblyman Connelly for her seat unsuccessfully, was not part of the Staten Island political establishment. From the start of the secession battle, he believed that the borough representatives (especially Connelly and Vitaliano) were promoting a political scheme that might score them points but never pass the Assembly. He also believed that the *Advance* was pushing a secessionist agenda at the expense of presenting the opposite viewpoint.

Goldfarb conceded that Staten Islanders felt unfairly beset by problems such as Fresh Kills and a serious homelessness problem.[57] But he also sensed an underlying racial element to secession, including negative local feelings about newly elected Mayor David N. Dinkins that might be construed as racially motivated. In an oral interview conducted some years later, Goldfarb stated that there was considerable hostility to the antisecession position. "People thought we were robbing them of their right to be free. Some of the terms that it was put into were pretty nasty. I mean, there was to some degree, in some of the areas, a real degree of racism expressed." He went even farther in suggesting that some prosecession officials were at a meeting, "when people talked about, you know, 'We want to keep the garbage out of Staten Island and we're not referring to the dump.' They were talking about welfare, and welfare and race were often lumped together on Staten Island." Goldfarb added that because of his views he sometimes felt like "the most hated man on Staten Island."[58]

While Staten Island remained the mainstay of conservative politics in New York City and predominantly white, its racial and ethnic composition nevertheless was undergoing some changes in the 1980s and 1990s, which had not settled well with many white residents since at least the completion of the Verrazzano-Narrows Bridge. Part of the reason for the island's demographic changes was its notable rise in overall population. Between 1960 and 1990, Staten Island's rate of growth far exceeded the other boroughs; Brooklyn and Manhattan actually lost population between 1960 and 1980. In the 1960s Staten Island population increased by a little over 33 percent. The closest rival was Queens, with 9.8 percent. In the 1970s and 1980s Staten Island's rate of increase slowed (to 19.2 and 7.6 percent respectively) but grew nonetheless. The island population in 1990 was 378,977.[59]

The Staten Island of this period continued to be perceived as an Italian American haven.[60] However, the borough's identity as a white, racially homogenous borough of European ancestry had begun to change in the 1960s, with an influx of African Americans especially after the opening of the Verrazzano-Narrows Bridge. A second important wave began in the 1980s, with immigrants from developing countries. Nonwhite population increase was slow but steady since the 1950s, at only 3 percent in that decade, rising to 6 percent in 1970,

12 percent in 1980, and 15 percent in 1990. Until the 1980s, blacks were the primary minority, but after that decade the census characterization of people "other than black" jumped to twelve different categories or racial and ethnic groups.[61]

Between 1960 and 1990, racial distribution for "White," "Black," and "Other" changed from 95.4, 4.3, and .3 percent to 85, 8, and 6.9 percent respectively. For racial and ethnic groups, the percentage change in population growth was competitive with the other boroughs. Still, by 1990, the percentage of whites outnumbered nonwhites by a wide margin (wider than in any other borough). A study on ethnicity and race written in 1991 stated, "Relative to the racial and ethnic composition of the other city boroughs, the shift toward increased heterogeneity in Staten Island may appear as minimal. It assumes significance, however, in the context of Staten Island itself because this county has remained insulated, until recently, from the changes in population composition taking place in New York City."[62] When racial issues came up during the secession battle, it was a reminder of the differences between Staten Island and the other boroughs. At the very least, there was an undercurrent that buttressed a sense of estrangement or more clearly reinforced the desire to maintain the island's heritage as largely a bastion for the descendants of white Europeans.

That Goldfarb questioned secession on the grounds that it demonstrated hostility to diversity contrasted with the belief supported by conservative leaders that Staten Island was simply a stable, conventional, suburban place needing its autonomy. Senator Marchi stated (clearly overlooking the changing demographics in the borough especially in the 1980s and 1990s), "Staten Island doesn't have the crime or welfare problem, and [has a] good bourgeois middle class."[63] The implication was clear: the island was (or should be) a homogenous community with a singular purpose and common goals.

Joseph P. Viteritti, a New York University associate research professor in public affairs and executive director of the Staten Island Charter Commission, argued that Marchi never allowed the commission to build its case on race. "I mean, there was a racial overtone to the debate when David Dinkins became mayor, and it could have really gotten ugly. Having John as the chairman really threw a lot of water on that because John was respected by everyone."[64] Viteritti reiterated what several secession advocates argued, that "a lot of New Yorkers" saw secession "as a racial issue because they didn't understand the issues."[65] There were people who wanted to "racialize the issue and make it seem like, oh, here's Staten Island, you know, those conservative folks out there want to leave the city because we have a black mayor."[66]

Intentional or otherwise, the secession debate could never quite shake the implication that Staten Island's homogeneity set it apart from the rest of New York City—even though the island was undergoing real demographic changes. These changes, however, helped make the case that many locals saw the loss of homogeneity as another reason to seek independence.

Among people of color, secession views varied according to vantage point. Organized African American groups on Staten Island, like the NAACP and the Urban League, opposed secession. Evelyn King, vice president of the local NAACP, stated, "It would be a dangerous step and detrimental to minorities. It's bad enough as it is [for blacks] in areas such as housing and employment, but I feel we wouldn't have a chance if we secede." For many years, black leaders had been very concerned with what they called "blatant segregation" in housing on the island.[67] They believed that their interests were best served in a unified city, where racial minorities constituted a majority of the population (57 percent in 1990). African American and Latino political leaders in other boroughs were more receptive to secession. Some were sympathetic to the idea of self-determination, while other more pragmatic leaders were not unhappy that a predominantly white borough might leave the city.[68]

Connelly reflected some years later:

> One of the arguments in conference, from some of our Black and Puerto Rican members, they felt that the goal was we didn't want "their people" coming to Staten Island. It was a fight I had to fight often because my district, the whole north shore, was and is a very diversified area. I couldn't make them understand, these are people that I represent, that need the same kind of services that you're fighting for in your district, and you're not allowing them to be shared with me.[69]

Connelly's district, if what she said was true, was not necessarily typical of the whole island.

Race and to some degree class were factors in the secessionist battle. Unease on Staten Island because of changes in its population makeup and the mayoralty of David Dinkins cannot be overlooked. Not only Dinkins's Democratic policies but also his race had made him unpopular on the island, especially among its white working class. Such issues were prevalent in the drawn-out independence effort, even though concerns over the borough's political emasculation (ignited by the ending of the Board of Estimate) and especially a sense of alienation symbolized by Fresh Kills Landfill dominated the public debate.[70]

FRESH KILLS AND SECESSION

The presence of Fresh Kills Landfill on Staten Island was never very far from the surface of the secessionist debate as a cause célèbre, as a problem to be resolved, and also as a bargaining chip. In November 1989 Hilary David Ring of the Office of Fiscal Studies of the Senate Finance Committee wrote to Senator Marchi, raising an important issue: "One of the major questions surrounding the potential secession . . . revolves around the manner in which two basic municipal services would be provided: the disposal of solid waste and the provision of fresh water." On the matter of Fresh Kills, Ring stated:

> If this landfill accepted solid waste from jurisdictions other than the independent City of Staten Island . . . a great deal of money could potentially be generated [for Staten Island] that could be used to support municipal services or reduce the need to levy some local taxes. . . . On the other hand, however, public health and environmental officials have repeatedly warned that long term garbage dumping on Staten Island constitutes a clear and present health hazard to Staten Islanders.

An independent Staten Island, therefore, would have a choice between managing the landfill for profit or attempting to close the landfill as soon as possible to protect the island's health.[71]

The consulting firm Schillinger, Salerni and Boyd, Inc., reviewed Ring's memorandum concerning project revenues that could be generated by Fresh Kills Landfill should Staten Island secede. Although they did not have "ready access to the presumably accurate and up to date information" maintained by the Department of Sanitation (DSNY), the firm had some relevant observations (based on the memo's assumptions and calculations) about Staten Island managing the landfill: Although Fresh Kills was the city's only landfill, the potential profits from such management were unclear, based on uncertain tonnage estimates of disposed waste, the viability of current tipping fees, and the cost of remediating and closing the site.[72] Projections for a Staten Island takeover of Fresh Kills were not so rosy but nevertheless intriguing.

Voter approval of the Charter Revision Commission's recommendations in November 1989 added fuel to the intensity of the secession debate. In early 1988 Mayor Edward I. Koch recommended that the commission defer making a recommendation to the voters regarding the structure of the Board of Estimate until the issue "is resolved by the courts." Since the commission's authority would expire with the 1988 general election, he was appointing a new commission (the same body) to take up the board's structure pending the courts' "final disposition of the matter."[73] Voters approved abolishing the Board of Estimate—a major change in the governance of the city and a reduction in the role of Staten Island. The shift of political and economic power to the mayor and city council further inflamed Staten Islanders.[74]

CUOMO SIGNS!

In a surprise move, Governor Cuomo signed the Marchi-Connelly secession bill on December 15, 1989. Senator Marchi was "elated," in what culminated "a six-year effort begun in a climate of skepticism and scoffing." Recognizing that the bill simply provided a referendum for the voters of Staten Island and not automatic liberation, Marchi accepted the "early Christmas present" nevertheless.[75] Other secession proponents were stunned but overjoyed, including a bevy of Staten Island officials and Les Trautmann, the editor of the *Staten Island Advance*, who asserted, "Staten Islanders should thank Gov. Cuomo for giving them an opportunity to speak on secession. He has shown that democracy still exists."[76] Mayor Koch was not so cheery: "[Cuomo] is plunging a dagger into the city's heart."[77] Incoming mayor David Dinkins also opposed secession and would continue the legal battle over the home-rule issue initiated by Koch. Dinkins's election was regarded by prosecessionists as an additional obstacle to Staten Island's aspirations.[78]

Cuomo was careful about how he couched his views on secession. He wrote in a memorandum that the bill provided that the question "Shall the borough of Staten Island separate from the City of New York to become the City of Staten Island?" be submitted to the voters of Staten Island in a general election in November 1990. Also, "If, after the intervening year and the discussion and education process that is sure to follow, the people reject the idea, the secession movement will end." If the people approve, a charter commission would be created, advisory committees established, and a commission of state legislators set up for apportionment purposes. Within thirty months, the charter commission was to submit its work to the governor, legislature, and borough president and then conduct public hearings for at least six months.

The next step would be to submit the charter to the voters of Staten Island. Two questions would be on the ballot: "Shall the charter of the City of Staten Island be adopted?" If the vote was negative, a second question would need to be addressed as to whether the commission should continue "for the purpose of drafting an alternative proposed charter for the City of Staten Island." If both questions received a "no" vote, the attempt to secede would end. The whole process would take a minimum of three years for the people of Staten Island "to be educated on all of the issues involving secession before they are required to actually cast their vote on whether or not Staten Island will secede from the City of New York."[79]

Since the potential separation affected everyone in New York, the state would "bear all of the expenses and requiring the proposed charter to be served on both the Governor and the Legislature." Cuomo also called for a separate bill obliging the Legislature to adopt "any new charter after it has been approved by the voters of Staten Island." He concluded, "I am pleased to give Staten Island, and the whole State, the fair chance to be heard on this important question."[80]

There was much speculation as to why Cuomo signed the bill. He took no public stand for or against secession; he may have believed that securing it would be so arduous for Staten Islanders that they would choose to remain in New

> # Charter for the
> # City of Staten Island
>
> August 2, 1993

FIGURE 15.1 Proposed charter for the City of Staten Island at the completion of public hearings, August 2, 1993.

Source: College of Staten Island Archives and Special Collections.

York City. He liked and trusted Marchi, and facing reelection Cuomo may also have wanted to cultivate additional support from the renegade borough. But as a New York City native, Cuomo stated publicly that "we need to keep Staten Island in the city of New York."[81]

Another clue factored into Cuomo's decision. As a condition of his support for the bill, the governor offered an amendment in March 1990 (pursued surreptitiously) that gave the state legislature the final decision on secession, not Staten Island voters. Under the revision, state approval would be necessary for independence to be implemented.[82] Vitaliano initially opposed the amendment but carried it in the Assembly, believing that it would strengthen the state's ability to undercut the city's home-rule claim. The measure passed by 51 to 6 in the Senate and 94 to 36 in the Assembly.

The governor signed the March-Connelly bill at a ceremony held at the College of Staten Island, St. George campus. Vitaliano, Marchi, Connelly, Straniere, and Senator Martin Connor stood by smiling. While Cuomo claimed that the bill would help secure independence for Staten Island by ensuring that the courts would not overturn the Marchi-Connelly bill, it actually would undermine it.[83] Even after the 1993 referendum on secession, Cuomo continued to deny that backing the decision to allow Staten Island to vote on a separation path was done with the expectation of the legislature upending secession.[84]

The *New York Times* was livid with Cuomo for signing the Marchi-Connelly "flawed legislation." "Given the risks, the least Governor Cuomo could do is show some belated leadership by saying whether he's for or against secession, and working to keep the process he has so unwisely let loose focused on substance. That is the only way the people of Staten Island can vote intelligently on their fate, and the city's."[85] To some, the amendment simply was "lawsuit insurance" that would protect Albany from charges that it had violated home-rule provisions of the state constitution.[86]

What they were yet to discover was that with the March 1990 amendment the governor had given the legislature a tool to defeat secession. Whether he intentionally or unintentionally wanted secession to fail, the amendment gave political cover to Cuomo. Borough President Guy Molinari understood (even before it was signed) that the legislature would not support secession, because the other New York City boroughs held more sway than Staten Island. He had regarded the Marchi-Connelly bill as "legally deficient" and came to believe that secession had become a "kind of a tragic hoax."[87]

THE FIRST REFERENDUM AND THE COMMISSION

While preparations for the November 1990 vote were underway, opposing sides continued to maneuver. The City of New York, through Corporate Counsel Peter L. Zimroth, filed a complaint for summary judgment with the New York Supreme Court, in order to stop the referendum on "whether the City of New York ... shall be dismembered—i.e., whether the geographic, political, economic and social structure of the City, which has existed for nearly a century, shall be destroyed." It argued, among other things, that only Staten Island voters were allowed to cast ballots in the referendum, thus denying equal protection under the law for voters in the Bronx, Brooklyn, Manhattan, and Queens.[88]

Senator Marchi objected to the lawsuit, declaring, "I feel certain that no American deserving of the respect of a free society would accept what is being proposed for us without a total loss of their self-respect."[89] The courts thwarted efforts by New York City to block the 1990 referendum. While the Court of Appeals affirmed the rulings of the lower courts about the constitutionality of the secession legislation in September 1990, it did not approve secession or authorize the voters of Staten Island to decide the issue. Significantly, it also did not preclude a final decision by the legislature about whether to allow Staten Island to break away from New York City.[90] The intent of Cuomo's amendment would stand.

The referendum held in 1990 was a resounding step forward for Staten Island's independence. Its passage set up a Staten Island Charter Commission by a decisive vote of 82 percent (some claim 83 percent) in favor, although only half of the registered voters cast ballots. (Marchi, Straniere, Connelly, and Vitaliano all won reelection, as did Governor Cuomo, in that year as well.)[91] One secessionist supporter noted, "Living on Staten Island is like getting a phone bill without having a phone."[92]

In 1991 the thirteen-member Charter Commission was sworn in. They were a moderately conservative–to–liberal group of people with good ties to city affairs. Two of the members were women; only one was African American. Chaired by Senator Marchi, it included five Staten Island legislators, five of their appointees, and one member each selected by the governor, speaker of the Assembly, and the president pro tem of the Senate. Viteritti served as executive director.[93]

Viteritti had been an advisor to the New York City Charter Commission, which is how the current commission found him. Aside from his academic credentials, Viteritti had more than twenty years of government experience. He was currently executive director of the Center for Management in the Robert F. Wagner Graduate School of Public Service at NYU. He had not favored the dissolution of the Board of Estimate, which he regarded as a major loss to the city. A Brooklyn Heights resident, Viteritti stated publicly that he was not predisposed to support or reject secession. This stance did not affect his association with Marchi, and they developed a good working relationship.[94]

The commission worked earnestly, holding twelve public hearings (all but one on Staten Island) between October 1991 and October 1993 and also convening monthly open meetings.[95] Borough President Molinari remained skeptical of the commission's work. "If the majority of the commission members believe the cost analysis cannot be done, then it might be fruitless and irresponsible to proceed any further," he stated.[96] A variety of opinions and documents—for or against secession—made their way to the commission at this time. Antisecession responses often were vigorous and detailed. SIUNY presented such a document to Marchi and Viteritti in May 1992. "Why We Oppose Secession Now" argued that by its very nature, the commission was not going to produce a balanced independent study. And most significantly, "The economic realities are being ignored."[97]

In July the mayor's Task Force on Staten Island Secession put out its report. Financial feasibility was the central issue: "Staten Island's fiscal health is closely bound with that of the City of New York. Rather than being in an improved economic position, as an independent municipality, Staten Island would face a large budget deficit with profound implications for its residents, businesses, and economic future."[98] The report concluded that as an independent city, Staten Island "would be forced to either raise its taxes sharply or cut its spending on services" and that New York City would not accrue an "automatic savings."[99]

On Fresh Kills, the report stated: "The Fresh Kills landfill has been at the center of earlier studies on the feasibility of Staten Island secession. Observers and advocates have argued that in the event of secession, the landfill could be a major source of revenue for Staten Island since it could change prevailing rates for the disposal of New York City's garbage." The issue was a question of "the ownership and operation of the landfill if secession were to come about." The task force came down predictably and squarely on the side of New York City: "Our financial analysis assumes the Fresh Kills landfill would be kept in its current use for the disposal of the non-commercial waste of New York City (and Staten Island) on *a non-profit basis* [emphasis added]." The report raised, not so subtly, the prospect of "potential legal problems" if Staten Islanders thought otherwise.

The report stressed the importance of Fresh Kills to the recent history of the island but offered little hope of its economic value if secession was achieved:

> The Fresh Kills landfill has been at the heart of the controversy on the island since its opening. An independent Staten Island may be faced with great tension between the demands of its constituents to close the landfill and the demand to maximize revenue—without the financial stability offered by a city the size of New York. Given community opposition to the landfill and the probable cost of compensation, Staten Island is unlikely to realize any substantial net income from Fresh Kills.[100]

In February 1992 the commission hired a consultant to conduct a poll on secession among Staten Islanders. The results, released in March, found 58 percent favoring secession, 25 percent opposed, and 16 percent undecided; 49 percent stated that they would be willing to pay higher taxes as a consequence of secession. The consultant declared, "Support for secession among statistically meaningful groups cuts across demographic, socio-economic, partisan, attitudinal, and residential lines. People of Staten Island support secession because they are Staten Islanders.... Their support for secession transcends other characteristics and concerns." He added: "They feel 'dumped on' both figuratively and literally; Fresh Kills appears to be both a tangible example and potent symbol of Staten Islander's disaffection."[101] Fresh Kills was most frequently cited as "the most important problem" along the South Shore, North Shore, and Mid-Island. Many polltakers placed little confidence in New York City's government.[102]

Assemblyman Straniere, a member of the commission, presented an independent poll showing 73 percent of Staten Islanders favoring secession. The commission also reviewed several studies dealing with the economic viability of independence. A group of NYU economists addressed the revenue-generating potential of the borough, and in a second study they evaluated the strength of the revenue base. They concluded that the overall economic health of Staten Island was sound and that the rate of growth exceeded the other boroughs. Estimating expenditures was more difficult. Various studies were at odds as to whether Staten Island would face a deficit or not. Based on its own analysis, a subcommittee of the commission declared that the establishment of a separate city was feasible. A statutory panel set up by the legislature and the governor concurred, as did the full membership of the commission.

For the City of New York, such a break would mean the loss of revenue, population, infrastructure, and land, but the metropolis would survive nonetheless. Some were concerned that the secession of Staten Island could have a ripple effect, leading to other attempts at separation within the city. In April 1992, for example, a bill passed the State Senate that would allow Queens to follow a similar path as Staten Island. This was considered highly unlikely, given that Queens did not possess the resources to achieve and sustain a successful break.[103]

THE SECOND REFERENDUM

The charter commission completed its report on February 2, 1993. It laid out the work that it had done over the course of sixteen months, including conducting business at thirty-four public meetings, holding seven public hearings, and commissioning five background briefing papers, five consulting reports, a report on civil service rights and retirement benefits, a report of the Working Group on Education, and a report of the Working Group on the Budget.[104]

The polls conducted demonstrated "widespread support for secession" and concern about taxes, economics, and representation. The commission's two major findings on economic matters were that Staten Island had a strong revenue base and that it was possible to provide services at the current level of revenue. After considering a variety of options, it recommended a mayor-council form of government for the proposed new city. This was consistent with "local tradition" and provided "maximum effective representation for the Island's diverse population"—a tip of the hat to the rising minority growth on Staten Island.[105] The path to the November 1993 referendum, Proposition 5, was now made ready.

Antisecessionists clearly were dissatisfied with the report. Mayor Dinkins wrote to Governor Cuomo, "I understand the frustration that many Staten Islanders feel, and I share their frustration." But he did not truly seem to understand it, sounding a partisan note rather than acknowledging what actually undergirded the secessionist fervor: "The combined effects of a prolonged recession and twelve years of Republican administration in Washington have hurt all of us." Dismissing the conclusions of the report, he expressed an array of views against secession: "I therefore seek your support for keeping New York City one—for financial reasons, but also for moral ones." He closed, "We must deal with our problems, and celebrate our triumphs together. Please join me in protecting and sustaining the principle and promise of democracy."[106]

SIUNY, a natural ally of the mayor on secession, branded Dinkins's efforts as weak. "The mayor's stuff has been superficial at best," asserted Goldfarb, and charged that Dinkins had not made the independence issue a priority, especially through his own task force.[107] SIUNY and others continued to voice strong opposition to Proposition 5, including city officials from the various boroughs and state senators and assemblymen. They were joined by a minority of private citizens from Staten Island, several of whom were on fixed incomes and were leery of higher taxes; city workers who were concerned that they would lose their jobs; and even environmentalists who feared that the new government would favor business growth over nature protection.[108] The antisecessionists, however, were not well organized.

In April, well before the vote on Proposition 5 in November, Guy Molinari distributed a report to voters (conducted by the state economist) making the claim that Staten Island could survive separation but also questioning why and how that should or could be accomplished. The report stated that Staten Island did not generate enough revenue to pay for services currently provided by New York City and that the proposed new entity would not become a "Mini–New York City." "The original fiscal argument for secession," it argued, "as has been shown by virtually every study to date, is false. The residents of Staten Island must base their decision to secede on arguments other than the fiscal argument." Those issues included whether the city would pay to use Fresh Kills, if Staten Island would own its land, and what the fiscal consequences of a new government might be. This was a cautionary document that indicated how carefully Molinari had eased into the secessionist camp before the election.[109]

On November 2, 1993, Staten Island voters supported secession by a 2 to 1 margin (65 percent). The final decision now rested with Albany.[110] Well before the vote, rumblings appeared in newspapers about the "tough road ahead" for secession. In late October, Gregg Birnbaum in the *New York Post* wrote, "The secession steamroller from Staten Island could run smack into a wall when it arrives in Albany." The Republican-controlled Senate would very likely bow to Senator Marchi's wishes, but in the Democratic-held Assembly "an uphill battle" waited. Speaker Saul Weprin (a Democrat) favored keeping the city together, but no clear outcome in the Assembly could be counted on at the moment.[111]

Guy Molinari's significant turnaround before the referendum passed unified Staten Island officials over secession. In a letter to Assemblywoman Connelly, Molinari stated that he stood foursquare in favor of secession: "After the long and torturous road that brought us to this point, I am now directly seeking your affirmative vote to allow Staten Islanders to secede." He finally "embraced secession" after hiring three consultants over the past year and "after studying the issue for almost five years." In voicing this deliberative approach, he left out the political momentum he was up against as the referendum reached the voters in 1993.[112]

Molinari announced plans for a victory rally during which a Civil War–era cannon would be fired in the direction of Manhattan after the vote count. "The cannon fire," Molinari proclaimed, "symbolizes the spirit of revolution emerging from our 'forgotten borough'—four shots signifying a message for the other four boroughs."[113] He told *Newsday* on November 1: "We are small town U.S.A. We pride ourselves on our difference. We would like to close the garbage dump down and have the other boroughs deal with their own garbage for a change. We've known for a long time that the other boroughs don't give a damn about Staten Island."[114]

It was clear that the separation movement that had begun as a "sitcom secession" had emerged as a tangible and serious threat to New York City.[115]

SECESSION DIES

On December 15, the unprecedented process of crafting secession legislation began in Albany. The Republican head of the Legislative Bill Drafting Commission stated flatly, "It's a monstrous project."[116] Adhering to its mandate, the charter commission submitted legislation to both houses on March 1, 1994, that would create the City of Staten Island. On the following day, the new speaker, Sheldon Silver (a Democrat from Lower Manhattan), on the advice of his home-rule counsel, announced that the bill could not be brought up in the Assembly without a home-rule message from the City of New York. This meant that New York City would have to agree to let Staten Island go. Shockwaves!

A negative outcome for secession now was a foregone conclusion. Counsel to the governor weighed in that a home-rule message was not required, in keeping with his support for Chapter 773 (and tacit agreement with Marchi). But the corporation counsel for New York City prepared an opinion for the new mayor, Rudolph W. L. "Rudy" Giuliani (a Republican), arguing that the home-rule request was necessary under the state constitution. Marchi and his supporters were livid. At this impasse, the three members of the Staten Island delegation to the Assembly (all members of the charter commission) initiated a lawsuit against Silver, claiming that his action was unconstitutional.[117]

Bitterness stretched beyond the usual political squabbles. Staten Island officials had learned in early March that a secret memorandum on secession (unsigned and not printed on official stationary) had been sent from Assembly leaders to City Council officials on January 10. The memo, based on Assembly legal-staff opinion, stated that secession required a home-rule message from the New York City government before the legislature could act. This was in opposition to what had been stated by the governor's office and by key Staten Island officials. Interestingly, Weprin, not Silver, was speaker at the time. City Council leader Vallone had even kept the memo away from Councilman O'Donovan from Staten Island. Guy Molinari's reaction was that "there was a plan afoot to stop secession the fastest way possible." Assemblyman Straniere called the memo an attempt to "kill" secession. Marchi's response: "A mugging has taken place here."[118]

On January 17, 1995, a trial court dismissed the case against Silver. Judge Robert Williams ruled that the speaker could report a home-rule request out of committee. He did not decide on the larger issue of home rule, and the case faced appeal. At this point, the Senate already had passed a secession bill without the stumbling block of home rule, and the new governor, George E. Pataki, a Republican, had agreed to sign it.[119]

The bulldog in Senator Marchi came out with the January 17 decision. In a memorandum to the charter commission's members on January 19, he stated, "It is most important to note that the court did not decide that a home rule message is required on the secession bill; rather it was held that the resolution of same is a legislative not a judicial prerogative at this stage. Obviously, this places a heavy burden on Assembly Speaker Sheldon Silver." Marchi intended to introduce a bill after the commission had met on February 3.[120]

The final blow to secessionist dreams (for the near future at least) came on February 22, 1996. The five-judge Appellate Division of the State Supreme Court let stand the lower court's decision, which effectively blocked the Assembly from voting on secession without prior approval on home rule from New York City. The speaker's judgment, it stated, was "part and parcel of the legislative process" and thus outside of judicial review. Supporters talked about moving forward. Assemblyman Vitaliano declared, "We'll take it as far as we can go." Instead, Guy Molinari recommended trying to get Silver to change his mind.[121] But Straniere and others once more went off to court, hoping that the highest judicial body in the state would rule in their favor. But there was no change in direction for the court.[122] Secession was dead.

The bitterness on Staten Island would last a long time. And some would question whether the acrimonious fight over independence had been worth it. The City of New York did not want to let go of its sometimes overlooked borough. Was it pride? Economics? Political expediency? And the state of New York, especially its legislature and governor, too seemed unwilling to have a peaceful breakup of its powerful metropolis.

But when the reality hit and the battle was surely lost, it was uncertain what would be done to address the borough's grievances. It was quite clear, however, now that the revolt had died out, that Fresh Kills—as a source of secessionist fervor—would remain the mainstay of the city's disposal needs. This was a certainty, unless something else derailed the long-time dependence on the site . . .

CHAPTER 16

CLOSURE

Fresh Kills was a scab that kept getting pulled back, exposing a festering resentment on Staten Island. In one sense, the attempt to secede from New York City was a way for the powerless to assert some power. But the secession effort failed, and the landfill remained. The fleet of scows still relentlessly trekked from the other boroughs to the dumpsite on Staten Island, day after day.

While the passage of Proposition 5 failed to be the mandate that the rebels sought, that same 1993 election changed Staten Island's future nonetheless. What tempered the pain and riveted the attention of the citizenry was not the "good fight" but the election of Rudolph W. L. "Rudy" Giuliani as mayor. State Senator John J. Marchi noted in his May 9, 1994 newsletter: "We now see the value of assurances made some time ago by Mr. Giuliani that, while as mayor he would oppose secession, he would make no effort to block legislative consideration of Staten Island independence efforts in Albany."[1]

Giuliani's election as mayor and George E. Pataki's as governor coincided with Guy V. Molinari's term as borough president, producing a Republican trifecta. This perfect political storm ended the secessionist surge and opened the way for the long-awaited closure of Fresh Kills.[2] The closure decision was ultimately political, despite protestations that the environmental risks of the landfill had driven the decision. The welcomed declaration to the long-suffering Staten Islanders, however, did not come with clear, long-term plans for solving the city's disposal dilemma.

MAYOR GIULIANI

The mayor's race of 1993 was the mirror image of 1989. David N. Dinkins beat Giuliani by a mere 47,080 votes (2 percent of the vote) in 1989; Giuliani edged Dinkins by 53,340 votes (3 percent of the vote) in 1993. While not the sole reason for Giuliani's turnaround in 1993, Staten Island votes clearly mattered—particularly because the total number of island votes had been steadily rising since the late 1970s. In 1977, 59,246 Staten Islanders cast votes in the election that brought Edward I. Koch to office. In the first showdown between Dinkins and Giuliani the total Staten Island vote was 113,368; it jumped to 136,923 in 1993.[3]

Many attribute to Prop 5 the reason for the healthy increase in voter turnout in 1993. Guy Molinari, who played a hefty role in getting out the vote for the Republican candidate, stated so in his memoirs: "As Staten Islanders flocked to the polls that day, they were essentially casting two votes against Dinkins. The issue of secession brought them to the polls, but every vote for secession also included a vote for Rudy Giuliani to be the next mayor. It was a great political strategy to rally the citizens to make a difference for the entire city."[4]

Giuliani carried Staten Island in both elections, 90,380 (79.7 percent) in 1989 and 115,416 (84.3 percent) in 1993.[5] For Giuliani, the vote differential for Staten Island alone in 1993 (93,909 votes) was impressive and significant.[6] The election, however, hinged on more than winning Staten Island. Giuliani obviously benefited from the large turnout there, but Dinkins only managed to match (or in some cases slightly lose) his 1989 numbers in key African American communities, even with a rise in voter registration. Getting out his supporters was more of a problem for Dinkins than Giuliani. The pollster Mark Penn did not think that the difference in the campaigns simply came down to turning out the vote: "It was an '89 performance in a '93 environment."[7]

There were subtle and not-so-subtle issues at play in 1993. Dinkins had not been able to reverse many problems associated with the ongoing recession or convince many voters that he had the administrative skills to do so. Welfare, crime, racial discontent, bigotry, and jobs all became issues, reflecting local as well as national frustrations.[8] The 1993 election also had "racially charged" aspects.[9] Giuliani's advisers were divided over portraying their candidate as the tough former prosecutor or as the statesman who could rise above the fray. In either case, Giuliani's law-and-order platform clearly appealed to his constituency and carried implicit racial undertones.[10] It comes as little surprise that race played a significant role in the campaign. In the end, no one issue highlighted in the campaign effectively moved voters to take an alternative position.[11]

Giuliani's victory meant that the two largest cities in the country (New York and Los Angeles) now had Republican mayors.[12] Polarizing yet popular, Rudy Giuliani was the first Republican to hold New York City's mayor's office in twenty years, despite the fact that Democrats outnumbered Republicans fivefold. The tough-minded former prosecutor was born in May 1944 in East Flatbush, Brooklyn, to a working-class Italian American Catholic family. When he was seven years old, the family moved to Long Island. His father Harold had robbed a milkman at gunpoint in 1943 (which got him jail time) but wanted his son to grow up in a safe environment where he respected authority. Despite his father's wrongdoing, Giuliani was surrounded by family members who were police officers and firefighters.[13]

In 1965 Giuliani graduated from Manhattan College (the Bronx), where he studied political science and philosophy. He achieved magna cum laude at NYU Law School in 1968. He then accepted a prestigious clerkship with Judge Lloyd MacMahon, a U.S. district court judge for the Southern District of New York. This experience propelled Giuliani into the U.S. Attorney's Office in Washington, DC. In 1973, at the age of twenty-nine, he was named chief of the Narcotics Unit and then rose to executive U.S. attorney. After four years of private practice between 1977 and 1981, he was named associate attorney general in the Reagan administration and in 1983 was appointed U.S. attorney for the Southern District of New York. There he battled drug dealers, white-collar criminals, and organized crime.[14]

Giuliani's election as mayor brought a change in tone and new priorities to the office. He exhibited the aggressiveness, ambition, and self-promotion that had made him an iconic prosecutor. He attempted to apply the same political philosophy on the local level that Ronald W. Reagan had done nationwide, especially an emphasis on limited government (although his administration centralized several municipal agencies without drastically restructuring them). Giuliani focused on cutting taxes and reducing the total workforce, and by 1997 the city's economy showed significant improvement.

The mayor was interested in results more than process. He favored privatizing key agencies, such as housing, parks, and homeless services, although whether these efforts yielded positive results is debatable. He was unsuccessful in privatizing the largest service—municipal hospitals. And as we will see, his plan for closing Fresh Kills emphasized private solutions.

Many of Giuliani's initiatives garnered public support as economic conditions improved, but after ending his first year with a surplus he applied most of it to the following year's expenditures rather than increasing services. The mayor talked tough with unions and avoided layoffs but signed few contracts that lessened inefficient union work rules. The longer in office, the more he acted like previous mayors had, by using revenues to satisfy aggrieved groups such as the unions. Giuliani provided severance packages with financial incentives to reduce the workforce rather than resort to layoffs.[15]

More than management style, the things that people could see and feel brought rave reviews—or condemnation. The Giuliani administration initiated the country's largest "workfare" program, which during eight years in office transferred 691,000 people from the welfare rolls to jobs. It adopted the "broken windows theory," whereby even the smallest misdemeanor (such as graffiti or vandalism) was treated as unacceptable and subject to police action. The goal was to alter the perception of crime, starting at the bottom. Crime rates dropped significantly in the city, and income and property values rose. Some attributed the change to Giuliani's forceful programs; others viewed it as an extension of a trend started under Dinkins, an improving national economy, outmigration from the city, and the decline of overall crime rates.[16]

Giuliani's attention to "quality of life" issues (including cleaning up Time Square) and fighting crime got him the moniker "the Nanny of New York." Those less enthused by the mayor's style claimed he often took credit for changes not wholly attributable to his actions, regularly defended police officers' transgressions, and sometimes acted downright petty. Minority leaders loathed his reliance on profiling in law enforcement, which aggravated race relations, and they criticized his dismantling of an affirmative-action program for minority and women contractors. Nonetheless, Giuliani was able to maintain high approval ratings throughout much of his tenure in office.[17] He especially got high marks with the people of Staten Island.

FRESH KILLS BLUES

Fresh Kills had not been an explicit campaign issue in the 1993 election. In that year and through the first two years of Giuliani's mayoral term, push-pull over the fate of the landfill continued, without any clear resolution. Policy shifts with respect to collection and disposal in general, however, were becoming apparent as early as 1994.[18]

Concern over the environmental status of Fresh Kills was ongoing. The Department of Sanitation (DSNY) and the State Department of Environmental Conservation (DEC) generated several technical reports as per the consent order (and given the constant public pressure), focusing especially but not exclusively on leachate mitigation.[19] A lawsuit brought by the Staten Island Citizens for Clean Air (SICCA) and other parties triggered the investigation of leachate. In March 1996 a federal judge named an environmental attorney to confirm the charges that leachate pollution from Fresh Kills was violating the Clean Water Act. Under the consent order, the city was given the opportunity to demonstrate that it was making progress, and it promised that a new leachate-treatment plant was to be completed by the end of the year. The attorney for the plaintiffs, Matthew S. Pavis, argued that the DEC in allowing the leachate problem to continue "has been lax in putting the city's feet to the fire in remedying the violations." He added that the DSNY had been aware of the problems "back in 1980."[20]

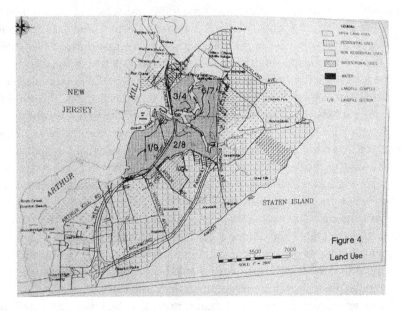

FIGURE 16.1 Sections of Fresh Kills Landfill, 1994.

Source: City of New York, Department of Sanitation, *Executive Summary: Project Information and Site Assessment Document, Fresh Kills Landfill*, September 23, 1994.

Complicating the situation was an alleged organized-crime link to a contractor from Long Island who initially had signed up to build the leachate-control plant.[21] (Ending the role of organized crime in the hauling business, in particular, would become an issue during the Giuliani administration.) The DSNY defended the capabilities of the new leachate-treatment plant after it was built and announced an additional plant to be constructed by 1997. SICCA and others remained skeptical, wanting more involvement in the completion of the new facility.[22]

Criticism over air pollution and odor problems at Fresh Kills also were aimed at the DEC and the city. In August, Borough President Guy Molinari charged that the DEC might not complete an air-pollution study because it was spending most of the $1 million earmarked on equipment and not on analysis. The DEC denied the charge.[23] The *Staten Island Register*'s Joan Gerstel queried, "Will a state study of air toxins and odors at the Fresh Kills landfill . . . give Staten Islanders a breath of fresh air? Or is it just another spending scheme designed to placate those worried about sharing their breathing space with the world's largest source of pollution?"[24]

Despite more funding for capital projects and efforts to address environmental problems at Fresh Kills, the DSNY continued to come under criticism for its management of the landfill and its modest (some would say substandard) recycling efforts.[25] In a January letter to Raymond Horton, head of the Citizens Budget Commission (CBC, a private, nonprofit civic organization), the DSNY Commissioner Emily Lloyd defended her department's actions against criticism on the recycling program, waste prevention, and disposal. She conceded that recycling had "experienced setbacks" as a result of the city's fiscal condition "over the past few years" and also suffered because of "the ad hoc way in which the program was initially implemented." But she believed the CBC report failed to acknowledge the "major accomplishments achieved this year," including the expansion of curbside recycling, the department's public-education strategy, and targeting new materials for collection. She added that the department had reason to be optimistic because the city's recycling diversion rate "will increase dramatically over the next two years."[26] Lloyd also questioned the report's charge that the DSNY did not have "an aggressive waste reduction program," asserting that it was proud of its emerging waste prevention program and was convinced that the DSNY "ultimately would be recognized as the architect of the premier Waste Prevention Program in the United States."[27]

Lloyd defended the department against the charge that it was making little progress in promoting new disposal approaches. She fell back on the 1992 plan, arguing that the city would pursue an "aggressive" waste reduction and recycling effort, and repeated the argument that the DSNY would emphasize flexibility in choosing disposal options, including waste-to-energy plants. While the DSNY worked to reduce its own exports of commercial waste, it was "prudent to encourage the private carters to continue to export waste and preserve as much of the capacity at Fresh Kills as possible." Lloyd added that there were many potential advantages to privatizing "certain functions" (the operation of the Brooklyn Navy Yard incinerator and disposal of ash were planned to be privatized), but she made no new concessions.[28] No signs pointed to Fresh Kills being shut down at this juncture.[29]

Immediate problems, including some at Fresh Kills, drew much of the DSNY's attention in 1993. In January seven families living near the landfill planned to file a negligence suit against the department, claiming a reduction in the fair market value of their homes and "substantial impairment" of the use and enjoyment of their property because of the negligence of the DSNY and city in operating Fresh Kills. Anthony Cantalupo, who lived behind the Staten Island Mall, claimed that his wife had been suffering from a unexplained cough and that he had not been able to open his windows in the summer "for the last four or five years" because his condominium was "inundated with stomach turning odors."[30] A story appeared in June that "formerly" radioactive waste was found at the landfill site, raising additional suspicions about monitoring.[31]

The battle over dumping ashes at Fresh Kills, which seemingly had ended, arose again. Although the DEC had withdrawn its plan to locate an ashfill on

Staten Island, it now sought to reclassify incinerator ash so that it could be used as landfill cover. Andrew Stein, the president of the New York City Council, stated in a January 28 letter, "Yet again the citizens of Staten Island have been misled. When the city passed the Solid Waste Management Plan, the proposed ash disposal site at Fresh Kills was removed. However, with this change in regulation, the City could dump the ash there anyway under the guise of beneficial use."[32] SICCA and Staten Island officials also denounced the proposal.[33] Hoping to defuse the controversy, Commissioner Lloyd announced that incinerator ash would not be placed at Fresh Kills even if state regulations were changed to allow it.[34] A ruling by the U.S. Supreme Court in May 1994 declared that incinerator ash was not exempt from the hazardous waste provisions of the 1976 Resource Conservation and Recovery Act. The decision simply made more complicated what could and could not be dumped at Fresh Kills.[35]

Like the ash debate, some other issues did not to go away quietly. Don Gross of the *Advance* reported in May 1993: "While most Staten Islanders wait impatiently for a long list of health studies to tell them whether or not something in the Fresh Kills landfill is causing cancer, other residents look a little to the southeast toward the closed Brookfield landfill in Great Kills as a possible source of rumored health problems."[36] Closed for almost thirteen years, little had been done to remediate Brookfield's toxic heritage. For at least eight years, the landfill's supervisor John Cassilano ("Johnny Cash") had taken bribes to allow trucks to dump toxic materials there in the dead of night. Rumors abounded that some drivers, tired of waiting in long lines to deposit their refuse, had simply dumped their loads in what became the Staten Island Mall and Heartland Village. The *Advance* had been reporting the presence of hazardous materials such as benzene, vinyl chloride, and trichloroethane at Brookfield since 1982. Wanting action, locals demanded independent testing of the air, soil, and water there.[37]

Representative Susan Molinari, who had urged local testing in the past, looked to the U.S. Environmental Protection Agency (EPA) to assure the public that there was not an immediate hazardous-substances risk at Brookfield.[38] In July the EPA conducted an initial air screening at Colonial Square homes near Arthur Kill Road in the Great Kills area and declared a provisional "all clear." Citizens groups remained active nonetheless.[39] In August Colonial Square homeowners decided to take legal action against the city, a form of recourse that seemed to be happening more often with each passing year. "Residents of these homes were not informed that their houses were built on top of and next to a former landfill," asserted the environmental attorney Michael B. Gerrard. "The developer should have told them, the city should have told them, and the city should not have granted building permits," he added.[40] The EPA continued to argue that their testing had not turned up evidence of chemical or heavy-metal contamination adjacent to the landfill site.[41]

THE BROOKLYN NAVY YARD AND ENVIRONMENTAL JUSTICE

In December, the long-awaited cleanup of the Brookfield landfill had its kick-off ceremony.[42] An opinion column in the *Staten Island Advance* on April 12, 1994, made the case that residents of Colonial Square townhouses faced an environmental-justice issue (a proposed civil rights violation): "At least one island bank, Staten Island Savings Bank, seems not to believe the clean bill of health [issued by the EPA]. It has acknowledged a policy of rejecting applications for first and second mortgages in the development based on a review of state and city reports by private consultants [about environmental conditions in the area]." The bank's chief lending officer stated, "Our review of these reports left many unanswered questions."[43]

Raising questions of environmental justice over Brookfield was problematic. The community was predominantly white, and charges of environmental-justice violations primarily focused on the plight of other racial groups at the time. The Brooklyn Navy Yard incinerator case, for example, generated a public debate concerning ethnic and minority communities.

Beginning in the 1980s, environmental-justice advocates (including activists, academics, ministers, and political leaders) reinvigorated the demand for civil rights in a new political arena. They focused on the disproportionate exposure to health and pollution risks faced by people of color at home and in the workplace, with an initial emphasis on African Americans and to a somewhat lesser extent Latinos. These risks, they argued, could result from unintentional or intentional efforts to site polluting industries and waste disposal facilities almost exclusively in minority neighborhoods or to place minority workers in harm's way on the job. The goal of the environmental-justice movement was urban focused and essentially political. It was directed at the government, private industry, and what it believed to be a white-middle-class-dominated environmental movement more interested in nature preservation than human health and well-being. The earliest environmental-justice literature reflected these goals and was clearly polemical. Incendiary terms such as "environmental racism" and more clinical academic definitions such as "environmental equity" were set aside and replaced by the concept of "environmental justice." The latter term offered a more positive, more inclusive concept to inject into the political discourse and suggested a broader objective (that is, justice) than simply the eradication of racist actions.

The efforts of environmental-justice advocates met with skepticism and resistance, but once the issue was on the table it could not be ignored. While the convergence of civil rights protest and environmentalism opened a new phase in the environmental history of the nation, the political objectives of the environmental-justice movement quickly exposed limitations in addressing broader issues than race, especially class and gender. Environmental-justice advocates did not dismiss class and gender as variables in environmental risk, but many (initially at least) accorded them a lesser status. From a political perspective, a broader definition offered an opportunity to gain new supporters (such as Native Americans and Pacific Islanders) but also threatened to weaken the powerful images of vulnerable urban neighborhoods that had put environmental justice in the political arena in the first place. A case would need to be made that rural villages, reservations, and even farms also were sites of potential injustice.

Among the first efforts to deal with environmental justice on the federal level came with a June 1992 report of the EPA (*Environmental Equity: Reducing Risk for All Communities*). It linked race and class but was not as emphatic about environmental racism as advocates liked. In February 1994 President William Jefferson "Bill" Clinton signed an Executive Order on Federal Actions to Address Environmental Justice in Minority Populations and Low-Income Populations "to focus Federal attention on the environmental and human health conditions in minority communities and low-income communities with the goal of achieving environmental justice." Also in 1994 the EPA renamed its Office of Environmental Equity to Office of Environmental Justice, reflecting the changing interpretation of environmental racism.[44]

The controversy over the Brooklyn Navy Yard would become a textbook environmental-justice case, especially as an industrial area bordering working-class and minority neighborhoods in Greenpoint-Williamsburg. The proposed site of the city's first resource-recovery plant had been mired in controversy. It was a project that remained on the table in 1993, especially with support within the DSNY, even though Mayor Dinkins had deferred its construction until 1996. DEC Commissioner Thomas C. Jorling announced in April 1993, however, that he was ready to issue the permit for the incinerator, despite local resistance. He stated that the city now had complied with his requirement to produce a suitable ash-disposal plan. Ash would be transported to the Charles City County landfill in Virginia. The city also would be required to separate recyclables from trash headed to the incinerator.[45]

Opposition to the incinerator took a decidedly meaningful turn by the summer of 1993. Anti-incinerator forces at the federal level, led by Representative Edolphus Towns, the Democrat whose district contained the Brooklyn Navy Yard, sought a moratorium on incinerator construction. Towns also acknowledged the significance of the Environmental Justice Act sponsored by Representative John R. Lewis, a Democrat from Georgia, former civil rights activist, chief deputy majority whip, and key member of the House Black Caucus. The act sought to address environmental problems unequally affecting people of color.[46] While few gave the moratorium a chance of succeeding, the raising of racial and civil rights issues elevated the debate over the Brooklyn Navy Yard and other sites beyond mere Not in My Backyard (NIMBY) claims. The act, however, never passed.

Commissioner Jorling stated in the fall of 1993, "In the more than seven years since the start of the review process for this facility, the solid waste crisis in New York City has heightened steadily." The link between the proposed Brooklyn facility and alleviating New York City's disposal woes was at the heart of the pro-incinerator argument. Brooklyn Borough President Howard Golden, who had called the Brooklyn Navy Yard incinerator a "monstrosity," vowed to continue the fight against it.[47] Writing to Mayor Dinkins on September 30, Golden pressed for a clearer rendering of the city's position on the fiscal implications of the incinerator. He questioned at length why the city had not given deeper consideration to exploring "the export option" as opposed to burning waste. "I remain skeptical of your contention," he added, "that export is not a viable, long-term solution for the disposal of the non-recyclable portion of the municipally collected waste stream."[48] Golden was raising what soon became New York City's most recent "new" solution to its disposal woes—exporting waste—which looked beyond the long-standing give-and-take over landfilling, incineration, and recycling within its borders. The debate over the Brooklyn Navy Yard also complicated the future use of Fresh Kills, which was affected by the city failing to jumpstart its waste-to-energy plan.

THE END OF AN ERA COMES CLOSER

Controversies over Fresh Kills, the Brookfield landfill, and the Brooklyn Navy Yard sounded like a broken record between 1993 and 1995.[49] At the annual public meeting on the Fresh Kills Consent Order on February 15, 1994, secession issues were still in the air. The DSNY also announced the postponement of the opening of the leachate-treatment plant. To allay concerns in general about the disposal site, Commissioner Lloyd offered tours to city officials and other interested parties.[50] The landfill still operated twenty-four hours a day, six days a week. Scows delivered approximately 13,000 tons per day (tpd) there; trucks delivered an additional 1,500 tpd. The enormous amounts of refuse were steadily shrinking space at Fresh Kills. In the late 1980s, the landfill had four distinct sections. In November 1992, Section 3/4 (131 acres on the northern portion of the property) was closed, and in June 1993 Section 2/8 (147 acres to the south) was closed. In mid-1993 landfilling was restricted to approximately 440 acres at Section 1/9 (to the west) and 336 acres at Section 6/7 (to the east).[51]

The targeting of certain sections of the landfill for dumping first and foremost meant bringing refuse to where space was available and attempting to comply with the various environmental-impact requirements. The DSNY also was concerned with having Fresh Kills fit within the integrated solid waste management plan reflected in its 1992 report. It did not believe that the operation of the landfill had negatively affected nearby neighborhoods as residents and their supporters contended.[52] Nevertheless, the DSNY formally agreed to place a permanent ban on dumping in sections bordering residential areas, such as in Travis, Annadale, Arden Heights, and Greenridge.

Confining the landfill had been a long-standing point of debate, most recently advocated for by Assemblyman Eric N. Vitaliano. "I believe that for the first time in the 45-year history of Fresh Kills, a written agreement puts Staten Island's quality of life first," he declared in January 1993. This was important, because the DSNY now was responding in absolute terms about dumping locations whereas before, such promises were tentative or contingent on other factors. Commissioner Lloyd stated that the buffer zone was "a terrific idea" and would help improve community relations. But most telling were her remarks concerning efforts to increase recycling operations to produce less refuse for disposal and the determination to shrink the use of the landfill "from 53 percent of our disposal capacity to 20 percent by the year 2000."[53]

The DSNY now viewed the use of Fresh Kills, and landfilling in general, as having definite limits. Yet this did not mean immediate closure. Early in 1993 Senator Marchi and Assemblywoman Elizabeth A. Connelly renewed their efforts to have the city fund the eventual closure of Fresh Kills with an eye toward studying the landfill's useful life and its closing costs. Marchi presented a similar bill to the Senate a year before, but it was never voted on in the Assembly. DSNY officials criticized the newest bill as duplicating their efforts, since the city already had started to retire sections of the landfill. The DSNY spokesperson Anne Canty stated firmly that Fresh Kills would operate "into the next century definitely." She added, "We already have annual closure costs budgeted and we'll continue to do so." Connelly responded that the "city has got to start actively working for closure. The longer we put it off, the longer we'll have reason to continue using [Fresh Kills]."[54] The sides were talking around each other in terms of a timetable.

Discussions about how the space might be altered after closure indicated that the prevailing view in the department was that the landfill could not continue for that many more years. In March 1993 the environmental committee of Staten Island's Community Board 2 invited the DSNY landscape architect Bill Young to discuss beautification of the site. His talk, "The Greening of Fresh Kills Landfill: Landscaping and End Use," ranged over various choices available for retired mounds (such as sections 3 and 4), an overview of potential uses for the landfill once shut down, and other projects underway at former landfills used as golf courses, athletic fields, and parks. The motivation to beautify the area was heightened by its proximity to the Greenbelt, wetlands, and other open spaces on Staten Island. Native species would be used when and if the plans were accepted, and birds and wildlife could disperse seeds to accelerate plant growth. Given the scale of the landfill, Young believed that the possible uses for the site were limitless. Young envisioned the landfill as a wildlife sanctuary three times the size of Central Park or at least as an extension of the Greenbelt.[55]

Young's article in the summer 1994 issue of *Garbage* stated that the DSNY had hired him in 1989 to demonstrate "the benefits of planting a more diverse vegetative cover, and restore entire plant communities as they once existed in the City's coastal areas." He conducted a trial on a capped six-acre nook at Fresh Kills. What his team ultimately produced was a "synthetic ecology." As such, Young hoped for "a restoration and stewardship ethic" to be incorporated into the landfill-closure regulations. He believed that returning the land to "created habitats" and linking them with nearby natural areas "taps the tremendous potential for secured landfills to become ecological oases."[56] On a less visionary note, some citizens wondered where the rats and seagulls would go once they lost their source of food in the landfill.[57]

The state of New York had seen a significant decrease in the percentage of refuse being landfilled. In 1994 almost 47 percent of the waste stream found its way to landfills, down from 82 percent in 1987. Between 1986 and 1993 there was a net loss of 244 such facilities (83 percent); in 1994 alone sixteen municipal solid waste (MSW) landfills closed. By the end of 1994 there were only thirty-nine MSW landfills left in New York State, with only twenty with valid operating permits. Three new fills opened in 1994. Other disposal options accounted for the decline in landfills on the state level. Exporting had increased from 5.35 percent of the total state MSW in 1988 to 19.38 percent in 1993, with over 4,199,448 tons exported in 1994 (a 5.46 percent increase in one year). This represented the second-most-popular method in 1994, with waste-to-energy incineration and recycling following. Recycling was on the rise, with 3,188,671 tons of secondary materials marketed during 1994 (16 percent of the total waste stream).[58] Statewide trends (especially the decline in landfilling and the rise of exporting waste) soon would be reflected in New York City, but not recycling.

At the February 15, 1995, meeting on consent-order compliance and milestone updates, it was clear that patience had worn thin (thinner?) on Fresh Kills' closure. The *Advance* reported, "With anger and disgust, speakers at the public meeting last night accused the city of deception and dragging its feet in the operation of the Fresh Kills landfill. The Sanitation Department was bitterly criticized for continually breaking its promise to set a closure date for one of the world's largest garbage dumps."[59] John Doherty, the new sanitation commissioner, repeated the DSNY's most recent defense, "to try and make the landfill a better neighbor," which, he admitted, "is very difficult."[60] Good at getting to the point, Molinari bluntly told Doherty:

> What we want, at this point, is a commitment on closure of the dump. After creating, in the least populated borough of the City, the largest garbage dump on the planet, we deserve no less. It is our intent to hold you accountable to

FIGURE 16.2 Aerial view of Fresh Kills, 1995.

Source: Courtesy of Jose Mario Lopez, Hines School of Architecture, University of Houston.

change the pattern of failure and irresponsibility of incompetence and mismanagement which has earned your agency and others the distrust of our community.[61]

Pressure for an explicit closure date was coming from elsewhere, too. Representative Susan Molinari reported that the EPA had asked the city to set a closing date rather than continue with uncertainty about its future disposal plans. Based on a two-year study, the EPA concluded that the landfill generally received "passing grades" on standards established by Congress in 1992, but investigators had yet to rate Fresh Kills on several significant areas because of the lack of data forthcoming from the city on issues such as leachate. Air testing had yet to be done as well.[62] Few reports were available despite the fact that a variety of health and environmental studies, including a city Health Department assessment of cancer among those close to the landfill, was underway.[63] The city's immediate response to closure was to earmark $750 million for the eventual shutdown. This was part of an effort to obtain a state permit to continue using the landfill into the next decade.[64]

GIULIANI AND SANITATION POLICY

At the heart of Mayor Giuliani's views on sanitation was his fiscal conservatism, particularly his attraction to privatization in the wake of the Reagan Revolution.[65] Like most new mayors, Giuliani began to change his administrative team. In the DSNY he had replaced Commissioner Lloyd with John J. Doherty, who had worked his way up the ranks from garbage collector. Doherty was the first commissioner in more than thirty years actually to have picked up refuse as a uniformed employee. His selection broke a more recent trend of selecting seasoned bureaucrats with little sanitation experience. The tradeoff was inside knowledge in exchange for broader administrative experience. Lloyd resigned to become a vice president at Columbia University, and Doherty entered at a time when Giuliani was not only facing changes in the recycling program but confronting a new union contract that would have sanitation workers collecting more refuse on longer routes. The mayor proposed cutting 1,400 positions of the department's 10,500 (13 percent), but some feared that this plan would weaken the DSNY.

Doherty, fifty-six years old, had been deputy commissioner since 1988 and a lifelong Staten Island resident. The child of an Irish immigrant father, he had chosen civil service over college. By 1981 he had been placed in charge of the Bureau of Cleaning and Collection (the department's largest division) and had overseen productivity programs including the reduction of sanitation workers assigned to each truck from three to two. Doherty's appointment was viewed as a way to make the changes in the department more palatable because he was looked upon favorably by the union and environmental groups alike. Eric A. Goldstein, a lawyer for the Natural Resources Defense Council, said that Doherty was more experienced than an outsider and "doesn't need directions to the Fresh Kills landfill."[66]

The new administration faced many of the same problems previous ones had, including inefficiency and rising costs.[67] High expectations for recycling under Lloyd would not be reflected in the new administration. With budget problems in 1994 (and an aversion to expanding government programs, especially those he did not like), Giuliani sought to cut recycling research, recycling education programs, and enforcement practices over three years. He also wanted to reduce the frequency of recycling collections from once per week to once every two weeks in Manhattan, Brooklyn, and the Bronx. A challenge from the City Council blunted that effort. Bronx Borough President Fernando J. Ferrer asserted that unless garbage was recycled, "we'll spend billions of dollars for filtration, river cleanup, beach cleanup and sending it to states that don't want it anymore."

Despite the fact that New Yorkers were recycling about half of all bottles, cans, jars, foil, newspapers, and magazines that had been discarded—and that markets for used paper were strong—advisers to the mayor argued that recycling programs were just too expensive. Deputy Mayor Peter J. Powers stated that "recycling does not come cheaply" and that just to meet the demands of the current law "would cost us $100 million." The budget cuts in particular would affect plans to start recycling mixed paper, such as junk mail and computer printouts. (Barry Commoner and others had predicted that collecting mixed paper would increase recycling rates to 25 percent, actually reduce the cost of collection of nonrecyclable materials, and provide additional funds from the sale of the paper.) Lack of enforcement of recycling laws also might mean that the activities of private carters would not be monitored.[68]

The City Council set the recycling program's goal at 25 percent when it passed its recycling law in 1989. The original deadline for meeting that goal was July 1994, but Mayor Dinkins's plan to suspend the program and the ensuing court battle further delayed the possibility of compliance. The Giuliani administration was ordered by a state judge in 1994 to comply with the law, with a new deadline set at July 14, 1996. But the administration argued that it would not have the funds available to meet the deadline and urged the City Council to change the law. Powers stated, "We have to amend the law; we have to. You've got to step back and look at your goals. It's not tonnage, but recycling as much as you can. You have to have a realistic sense of progress." But Stanley E. Michaels, the chair of the Council's Environmental Protection Committee, was not inclined to recommend a change in a program he regarded as "an absolute necessity." "We're running out of space in Fresh Kills," he added. "We don't have incinerators. We've got to get up to 40 or 50 percent or drown in our own garbage."[69]

The complicating factor in this story is how everything is connected to everything else—recycling, land disposal, exporting waste, and more. In keeping with his objective in promoting private solutions to public problems, Giuliani teamed up with Staten Island Borough President Molinari and Governor Pataki to support the development of a private paper-recycling plant and board mill on Staten Island (near Fresh Kills). In 1995 Pratt Industries of Australia began work on the Visy Paper Mill, using the existing city-owned marine transfer station system. It became the city's largest contractor for recycled residential paper, producing linerboard for cardboard-box manufacturing.[70]

The Republican George E. Pataki had been elected governor in 1994, stunning the political world by defeating the powerful and influential Mario Cuomo. Born in 1945 in Peekskill, Westchester County, he was the son of a farmer and postmaster. Pataki graduated from Yale University and Columbia Law School, practiced law, and soon found his way into Republican politics. In 1981 he was elected mayor of Peekskill, and in 1984 he joined the state Assembly. In 1992 he moved up to the State Senate, beating Cuomo for governor two years later.[71] Pataki was the first governor elected with Conservative Party backing. He shared with Mayor Giuliani and others in his party a fiscal conservatism. In 1995 he fulfilled a campaign promise by pushing through the legislature a large tax cut.

No ideologue or rigid social conservative, he appealed to moderates by supporting abortion rights, and he worked against discrimination of gays. Pataki was a pragmatic politician who simply wanted to get things done. "It's very important to stand on your own feet to the extent that you can," he told an interviewer.[72] The combination of Republicans in the statehouse and in City Hall in 1995 would have a far-reaching effect on Fresh Kills.

Unlike the objectives of the 1982 solid waste management plan, the approach of the new city administration was moving away from a multifaceted disposal program. All remaining city incinerators were to be closed by 1994. The mayor showed no signs of wanting to reopen the Gravesend Bay incinerator in Brooklyn (closed in 1991). "Mayor Giuliani's decision not to proceed with the [reopening] of the southwest Brooklyn incinerator, at least until after an environmental impact statement is prepared," stated Larry Shapiro, an environmental attorney for the New York Public Interest Research Group (NYPIRG), "is a tremendous victory for all of us who are concerned with public health."[73] More likely, for the administration at least, it was a cost-saving measure. Interest in the waste-to-energy Brooklyn Navy Yard plant received only a lukewarm response. And after it was discovered that contamination at the site had been left out of a supplemental environmental-impact statement (EIS), the administration decided to delay funding until 1999, pending a new EIS.[74]

A harbinger of things to come, Giuliani began to explore placing a greater emphasis on exporting refuse. In 1992 the total DSNY-managed waste (disposed and recycled) was 14,733 tpd. That figure rose to 15,236 tpd the next year but fell to 14,876 in 1994. This amount still was considerable, even though curbside-collection figures were flat between 1992 and 1994 (and would remain so through 2000).[75] Exporting waste was an approach that fit congenially in the political and economic philosophy of Giuliani's administration, which emphasized privatizing services and reducing the responsibilities of the city. In November 1993 the DSNY announced that as a way of relieving some of the burden on Fresh Kills, it planned to export more than one thousand tons of garbage each day over a five-year period from the Bronx (while the marine transfer station in the South Bronx was being rebuilt).

When the department's management plan was being negotiated in 1992, the Dinkins administration agreed to let the department seek out-of-state sites to handle some of the city's waste. As a small indicator of changing views on scaling down Fresh Kills, Councilman Alfred C. Cerullo stated, "The goal is to find ways for the city to stop relying on Fresh Kills as the sole repository for the city's garbage." The 1992 report had stated that by 2000 Fresh Kills was likely to be handling only 20 percent of the daily refuse disposed by the city.[76]

In 1994 the administration began searching for out-of-state landfills that were willing to accept about one-third of the city's refuse. This was a substantial change in policy, since residential and institutional waste collected by city trucks had gone to Fresh Kills; commercial garbage hauled by private carters could and had gone out of state. Canty stated, "We are looking for a company that has permanent landfill space outside the city." The period of the proposed contract would be twenty years. The timing seemed right: landfill prices had dropped because several new landfills had opened outside the city.[77]

The national trends concerning interstate movement of MSW were favorable. A *Waste Age* article published in January 1994 found a widespread and complex network of interstate refuse movement in North America. The District of Columbia, forty-seven states, some Canadian provinces, and Mexico exported some of their refuse. There were forty-four importing states in 1992, up from forty-two in 1990. The magazine estimated that 19 million tons of MSW moved between states during 1992, which was about 9 percent of the total refuse generated.[78] A variety of private solid waste companies made inquiries to the city and various officials about the possibility of exporting waste.[79]

A potential problem facing New York City at this time was the issue of flow control, but the city had no policy in place. Flow-control ordinances specified where MSW had to be processed, treated, or disposed. The imposition of flow-control ordinances was meant to secure financial support for public disposal facilities and to ensure that refuse was disposed of in an environmentally safe way. In a landmark Supreme Court case in 1994, *C&A Carbone, Inc. v. Town of Clarkstown*, the justices overturned a municipality's authority over waste in its jurisdiction. They claimed it violated the Commerce Clause of the U.S. Constitution, because in 1978 the court had determined that garbage was an entity within the meaning of the clause. Private haulers were overjoyed by the 1994 decision (especially large integrated waste firms); cities and other governmental entities were not. What ensued was a major controversy over the control of refuse, which threatened to influence decisions about its interstate transfer.[80]

For several towns and cities, the dilemma of the *Carbone* case was whether to construct new disposal facilities under their jurisdiction (and reap supposed profits) or to send the waste out of state. For New York City this was not the choice, but exactly how the court case would play out in the wake of current and future challenges added a layer of instability to the commitment to interstate transfer of refuse. Giuliani and New York City would face such challenges in the late 1990s. Disputes with the commonwealth of Virginia and others concerning the acceptance of Gotham's refuse, who could do the shipping, and how the refuse would be handled before shipping muddled future exporting plans.[81]

WINDS OF CHANGE

Everyone with an interest in Fresh Kills (the DSNY, the citizens and officials of Staten Island, the state of New York, and the mayor) was contemplating how much time remained to it, but with different aspirations or apprehensions. In what would be one bill of many in an avalanche of closure legislation, the South Shore Republican councilman Vito J. Fossella Jr. submitted his proposal on February 15, 1996, to shut down Fresh Kills Landfill by 2002. Such an effort drew criticism from Mayor Giuliani's office.[82] The hard-hitting *Staten Island Register* did not view Fossella's action as "grandstanding" but as an effort to "force the City of New York to confront the unthinkable—the loss of its only remaining landfill and its sole repository of 14,000 tons daily of residential solid waste." The news organ recognized the legislation as "the longest of long shots." But so was secession legislation, it added. "The current sorry state of the city's long-range plan for solid waste-disposal and the dismal lack of political will to make the tough decisions about it . . . make closure of Fresh Kills the city equivalent of nuclear war."[83] Borough President Molinari stated that he would meet with Giuliani to argue the case for the bill.[84]

Also in February the City Council was "preparing to shake hands with City Hall" on a revised solid waste management plan that would dump more garbage (and possibly hazardous materials) at Fresh Kills. The decision came as a result of months of "behind-the-scenes negotiations" and included "a major administration pull-back on recycling."[85] The February 15, 1996, *Comprehensive Solid Waste Management Plan* (released on the same day as Fossella's announcement) claimed that the city had "implemented one of the most comprehensive and successful recycling efforts in the United States." In a telling sentence, however, it added that "the goal was achieved and, *except for landfilling*, more MSW is now managed via recycling than any other MSW management option."[86]

It was landfilling that continued to be the crux of conflict over the city's disposal plan, which the recycling claim was merely camouflaging. The DSNY continued to insist that its various waste reduction efforts and out-of-city disposal opportunities were attempts to preserve Fresh Kills' capacity—"a depleting resource"—and to extend its life as long as possible.[87] Not surprisingly, the Staten Island delegation to the council roundly criticized the plan for its limited focus on recycling, its failure to address the Brooklyn Navy Yard, and its inability to set a closure date for Fresh Kills.[88]

The annual Fresh Kills Consent Hearing on February 28 proved to be the most imposing public showdown over the landfill in recent memory, providing a forum for many of Staten Island's political heavyweights. The mood of the approximately two hundred citizens attending the four-hour meeting was angry, with shouts of "close the dump" and "you're killing our children."[89] In a wide-ranging attack on the city's disposal efforts and its latest plan, Guy Molinari argued with some hyperbole that conditions at the landfill "have become immeasurably worse in the last six months." He concluded,

> I am stating emphatically and unequivocally that I will not stand by and allow the Fresh Kills Landfill to be "permitted" until reasonable and responsible reports on the health impacts are completed, released and adequate response enacted. Both the City and the State are hereby warned that this need not be a protracted battle—it should not be a battle at all—but the lines has [sic] certainly been drawn.[90]

Senator Marchi also focused on what had been a refrain throughout 1996: *close the landfill now.* "It is past time," he stated, "for the state government to force the city to do the right thing and close Fresh Kills in this century."[91] Assemblyman Vitaliano announced support for the Fossella measure but warned, "It is obvious that without some commitment in the upcoming City Budget right now to develop alternative disposal capacity outside of Staten Island [in terms of exporting waste] a closure law will not be worth the paper it's printed on."[92] Fossella, promoting his bill, stated that he had proposed the legislation because "through its action the Department of Sanitation has in no uncertain terms implied that there is no alternative besides the Fresh Kills Landfill."[93] Councilman John A. Fusco—piling on—was even more provocative: "Staten Island has been the victim of the cruelest environmental hoax in this City's history."[94]

The New York State Solid Waste Management Plan, 1995/1996 Update struck a more optimistic note than the February 28 meeting. It stated that actions taken in the early 1990s, including the implementation of local solid waste management plans and new recycling and disposal facilities, had "helped avert [the solid waste disposal capacity] crisis."[95] Speaking about Fresh Kills, the report added that the DEC had worked closely with New York City to provide technical assistance in developing operation and closure plans for the landfill. These efforts resulted in "significant improvements" in operations and in the closure of sections 2/8 and 3/4, which was anticipated completed in 1996. If all of this were true, it was not an assessment shared on Staten Island. The DEC made no friends there by adding that landfilling "will continue to be a necessary part of integrated solid waste management systems," even though it had started a Landfill Closure State Assistance Program.[96]

The Giuliani administration's stance on recycling and disposal in general confused Staten Islanders and others hoping to see the city move away from landfilling. Arthur Kell, the director of Toxic Projects for NYPIRG and a tenacious critic, accused the city government of retreating from recycling and giving insufficient attention to other disposal options. Despite the fact that the administration was floating the idea of refuse exports, Commissioner Doherty responded that exporting waste was too expensive and community resistance to resource-recovery plants too great.[97] He noted in a March 25 letter to the DEC, "While we continue to strive for less reliance on the Fresh Kills landfill, it remains an integral part of the city's waste disposal infrastructure. The goals of the DEIS and permit application are to demonstrate the landfill can be operated safely, responsibly and in a manner protective of public health and safety."[98]

Staten Islanders and others opposed to Fresh Kills were keeping up the pressure to secure a closing date. In March a group called Concerned Staten Island Realtors urged legislators in Albany to find a way to shut down the landfill. This was the first time island realtors had gone to the legislature about this issue.[99] Fossella's closure bill also was making slow progress through the City Council, garnering support of more than half of the members. Some council members sympathized with Staten Island's fate; others feared what might happen in their districts once the landfill was closed without a disposal plan in place. But apart from the City Council, Fossella still needed seven additional votes. The mayor, for example, was opposed to the bill. A spokesperson for Giuliani stated, "We fully recognize the concerns Staten Islanders have about the landfill. But we do need other [refuse disposal] outlets." The mayor clearly did not want a decision imposed on him.[100] In addition, no definitive reports had been released indicating the extent to which the landfill was causing health risks.[101]

Another tack taken was the decision by Guy Molinari, Susan Molinari, Fusco, and Fossella to file a sixty-day notice of intent to file a lawsuit in Brooklyn federal court under the Clean Air Act and the fair-share provisions of the City Charter.[102] The goal was to force the city to set "a date certain" for closure of the landfill.[103] South Shore Assemblyman Robert A. Straniere, Vitaliano, and Connelly also were moving to present a closure bill to the state Assembly in late March, as did Senator Marchi.[104]

Vitaliano's approach, in a revised version of his bill in April, built on strongly criticizing current practices in New York City by not only seeking to close Fresh Kills by January 1, 2002, but also prohibiting the building of the Brooklyn Navy Yard incinerator, enforcing recycling laws, and identifying in-state disposal capacity. Vitaliano described his legislation in his newsletter to constituents as Mayor Giuliani's "worst nightmare." This was clearly a change from his earlier, tempered approaches to the city's refuse-disposal problems.[105] Vitaliano stated, "We are putting together a coalition that goes beyond the concerns of Fresh Kills. We are reaching out to the opponents of incineration, the advocates of recycling and our upstate friends who might be interested in selling their excess disposal capacity."[106]

About the same time, Councilman Jerome X. O'Donovan (a Democrat) submitted a closure bill (an extension of Fossella's bill) calling for Fresh Kills' closure by December 31, 2002, and for the city to meet tonnage targets for reduction, recycling, and exporting waste.[107] On February 20 O'Donovan had written the mayor seeking support for his bill, asserting that "after 50 years of living with the world's largest garbage dump Staten Islanders' patience waiting for some relief from City Hall has been exhausted.... Two years into your term we now look to you for leadership on this issue." But seeking an endorsement on what he called "a non partisan, good government effort on the part of both the executive and legislative branches of local government" was not forthcoming.[108] In a much delayed April 29 reply to O'Donovan's February 20 letter, Deputy Mayor Powers shot back, "Contrary to the statement contained in your letter, the Giuliani Administration is vigorously pursuing alternative waste management programs." He concluded, "Mayor Giuliani is very sensitive to the concerns of Staten Island residents and will make all efforts toward coming up with a rational solution."[109]

Giuliani viewed his own plans as more "realistic." As he stated on May 20, "We're working very hard to come up with a realistic plan that we can stick with, rather than just a date that somebody has passed."[110] He also met privately with senior advisers to assess the focus and timing of potential actions.[111] As the various bills came closer to passing their respective bodies by late May, however, the window of opportunity for the mayor to act independently of the legislators appeared to be slipping away.

A CLOSING DATE IN SIGHT

The deluge of bills and the potential lawsuit in March and April 1996 were turning the screws on officials to find a closing date for Fresh Kills. In late March Governor Pataki stated that he would review the closure plan, but he remained silent about giving support to pending legislation. He added, "We want to work with the borough president and also the mayor and try to resolve the problem so that Staten Island is no longer burdened by this." Yet Pataki spoke cautiously, emphasizing that a closure plan could not "impose any undue pressures on the city." He preferred that the issue be settled locally given its politically explosive character.[112] A Republican governor and a Republican mayor would be responsive to Staten Island's plight, especially since that borough was the only one of the five Pataki carried in his political battle with Mario Cuomo. Neither could Giuliani's "razor-thin" victory over David Dinkins be forgotten, given his upcoming reelection campaign in 1997.

Some sensed a change in tone in Albany that might bode well for Staten Island. NYPIRG's Larry Shapiro was one of those people. Carl Campanile of the *Staten Island Advance* wrote on March 31: "The stars in the political galaxy might be aligned just right, Shapiro said, to get action of the dump." Shapiro also conjectured that a closure bill passing the Senate was a "reasonable assumption" if Marchi got behind it. The Assembly was much tougher, since it was dominated by representatives from the other four boroughs not looking to have garbage dumped in their territories. State Assembly Speaker Sheldon Silver (a Democrat from Manhattan) had opposed secession, but, Shapiro asserted, "Silver may be able to use Fresh Kills as an issue to get into the good graces of Staten Islanders." He added, "The time is right. This is the right moment in history for Staten Islanders to do what they can to get the landfill closed."[113]

Shapiro's speculation made sense, but the Staten Island delegation in Albany had yet to form a united front. Tension existed, for example, between the Marchi forces and Straniere over their versions of a closure law.[114] There also were some unresolved issues related to a state permit for keeping refuse flowing to Fresh Kills. In a telling response, Governor Pataki wanted to see New York City set a closure date for Fresh Kills before the permit was issued.[115] His public position on Fresh Kills now became crystal clear—he supported closure, and soon.

In early April Speaker Silver came around. When Vitaliano submitted his revised version of his bill on April 11, Silver became a cosponsor. The inclusion of a provision not to build the Brooklyn Navy Yard clearly had been directed at Silver, who had long opposed the incinerator.[116] Marchi also got behind Vitaliano's effort, but he balked at the way the assemblyman was politicizing the issue by referring to his bill as Giuliani's "worst nightmare."[117] Partisanship had yet to be unfastened from the controversy.

On April 23 the *Advance* reported, "A flurry of recent high-level meetings in the City Hall office of Deputy Mayor Peter Powers has focused on what in the past would have seemed an unlikely subject: Putting together a practical plan for shutting down the Fresh Kills landfill." Commissioner Doherty had met privately with Powers concerning the feasibility of ending reliance on Fresh Kills in the coming years. An anonymous official stated, "They're seriously crunching the numbers. The bottom line question is: Can the city survive without the landfill?" Exporting waste, incineration, and recycling all were being considered as alternatives.[118] Obscuring the picture was ongoing debate in the U.S. Congress over exporting wastes and flow-control measures, which would affect New York City's future choices.[119]

Some smelled blood in the water. The closure effort was strengthening in late April. Hundreds of environmental activists from across the state converged on the grounds of the State Capitol on April 22, Earth Day. A group of twenty-five citizens from Staten Island were most passionate in their call for closing the landfill. "Dump Fresh Kills," a black banner with lavender and green letters proclaimed. Other placards read "Our Children Are Counting on You, Support 2002" and "Close by 2002, a Recycling Incentive." Some of the protesters declared that they were hopeful about current legislation, but Governor Pataki was not ready to say that he would sign it. He and the New York congressional delegation were busy fighting off attempts at the national level to curtail the export of refuse across state lines. "Obviously," Pataki stated, "[an alternative disposal plan] is an important part of the entire Fresh Kills equation."[120]

Mayor Giuliani, also feeling the pressure, regarded Fresh Kills as "a big legitimate concern for Staten Island." At a town-hall meeting at Port Richmond High School on April 24, he stated, "We're going to try very hard to find a solution for you quickly, but I run an entire city, not just one part of the city." When several attendees demanded a definite closure date, he declared, "Don't try to put me on the spot to say when. You should trust me to solve it."[121]

Fresh Kills was not only a source of great controversy but a curiosity as well. In the spring, the DSNY had been planning a program to conduct regular tours of the landfill for public-relations purposes. Lucian Chalfen, assistant commissioner for public affairs, stated, "The tour is meant for people to get a broader understanding of the landfill. And we encourage people who live around it to tour it also. One day it will close, but the point is, it's here now." The tour would attract school groups, government officials, and international visitors. He added, "There is nothing so unusual about interest in heavy landfill equipment. A couple of people asked if they could picnic there. I guess the point is you don't have to pick up after yourself."[122]

Mayor Giuliani did not take the project so lightly, especially since Borough President Molinari and others vociferously complained about the plan. He sensed how bad the timing of these tours might be, given the current uproar over the landfill. On April 2 the mayor canceled the tours. "Except for educational purposes with regard to recycling or things that would have a purpose like that, there really is no point in having tours there," he stated. "And I think it was offensive to the people that live near there."[123]

Events surrounding debate over the landfill closure were coming to a head in May. At a seven-hour meeting of the City Council to consider two bills to close Fresh Kills, more than two hundred angry Staten Islanders booed Commissioner Doherty. After the meeting, Doherty stated that he expected city officials to unveil a plan for closure in "six months, maybe quicker." Earlier that day, Giuliani said that a plan could be ready in three to four months.[124] Pataki repeated his opinion that the closure could be done, but in such a way that did not mortgage the city's financial future.[125] City officials shared this concern, especially in considering the price of closure and the cost of possible alternatives to Fresh Kills once it closed.[126] A typical local response came in a letter to the *Advance*: "Within recent months several political leaders have made proposals to close the Fresh Kills landfill. I hope that these proposals are not just smokescreens to temporarily calm the nerves of Staten Islanders who have borne the brunt of storing the city's garbage for so many years. Will these proposals actually result in closing the landfill on Staten Island? I hope so, but I think not."[127]

THE SURPRISE ANNOUNCEMENT

Storm clouds cleared at a momentous press conference held on May 28 at the borough president's office on Staten Island. Mayor Giuliani, Governor Pataki, and Borough President Molinari appeared together. They announced what Staten Islanders had been hoping to hear for years: Fresh Kills would be closed to future shipments of refuse on December 31, 2001.

In fact, the December 31, 2001, date to close Fresh Kills had apparently been agreed upon by the governor, the mayor, and the borough president several weeks before, with negotiations in general having gone on for months.[128] The closing date coincided with the end of Giuliani's term in office.[129]

At the press conference the three Republican leaders outlined the process for closure. A task force composed of federal, state, city, and Staten Island officials and representatives from the environmental community soon would be formed to develop a master plan to deal with disposal. "It has become increasingly evident that the burden this facility now imposes on the citizens of Staten Island must be lifted," Giuliani stated. Pataki added, "the People of Staten Island have suffered long enough. The time has come to close Fresh Kills and put an end to one of the worst environmental nightmares this state and city have ever witnessed." He planned to work with the mayor, the borough president, the federal government, and the environmental community "to ensure that we lock the door on Fresh Kills. We must move decisively to shut Fresh Kills and develop a comprehensive plan to reduce New York City's waste stream, increase recycling and find alternatives for disposal."

Echoing his colleagues, Molinari pronounced this as "a great day for the people of Staten Island and for the future of Staten Island" and repeated his perennial assertion that the citizens of the borough have done "more than their fair share and have borne the brunt of a waste disposal nightmare." He praised Pataki and Giuliani for allowing Staten Island now to "see the light at the end of the tunnel."[130]

After such a protracted battle, the reactions to the closure ranged from elation to skepticism, from "it's about time" to "we've heard this before."[131] Guy Molinari promised a parade. A variety of legislators vowed to push for their bills to turn the pledge into law.[132] Making the closure declaration (and the development of a new disposal plan) functional would not be simple.

How Pataki, Giuliani, and Molinari arrived at a closure agreement after the city lived almost fifty years with Fresh Kills Landfill is at once easily and opaquely explained. The DSNY's lack of convincing steps forward in finding alternatives to Fresh Kills did not help the transition process. Public protests were intense and feverish (and getting more so), but also important were the closure bills presented to the legislature in Albany and to the City Council. Molinari's lawsuit kept the story alive and in the news on a daily basis.[133] As

Shapiro noted at the end of March (and many others mentioned subsequent to the closure announcement), the political alignment at the time favored an agreement. Three Republicans were in crucial positions of authority at the time, making united action possible.[134]

Michael E. McMahon, who later served as a Democratic member of the City Council representing Staten Island (also eventually serving in the U.S. House of Representatives), stated in an interview that the landfill was closed because

> it really comes down to political alignment, or alignment of the stars.... Because [Giuliani] won by so few votes [in 1993], certainly not as close as Giuliani's first election, but I think that that made Giuliani pay very close attention to Staten Island. The number one issue was the landfill. And Molinari and he were close, and with Pataki as the governor, lightning struck, and the deal was done.[135]

Governor Pataki was persuaded that a firm decision was necessary given the environmental mess that the landfill had become; he also appreciated the political advantages of taking up Staten Island's cause (especially in seeking its support for his reelection bid in 1998). In fact, when he ran for office in 1994 he said he would like to see the landfill closed. His efforts were crucial in pushing the process forward. But he also knew that acting without Mayor Giuliani's concurrence looked too much like meddling, which was an ongoing issue between city and state governments.[136]

Giuliani, for his part, wanted to appease Staten Islanders and curry their support for the 1997 election. He also was happy to lighten the financial burden that Fresh Kills represented for the city for some of the same reasons that he had favored cutting back on the recycling program. This was much in line with his Reaganite fiscal views. Giuliani's previous efforts to investigate exporting waste as a disposal option offered a potential private-sector solution not possible as long as the giant landfill remained open. Still, he balked at supporting any of the many closure bills (particularly Democratic ones) that would usurp his flexibility in dealing with the issue. In fact, when the public closure announcement was made, no Democratic legislators were invited to the event.[137]

Perhaps the true catalyst of the closure agreement was Borough President Guy Molinari. Since he was important in the election victories of Giuliani and Pataki, he was not afraid of leveraging it for his advantage. A longtime outspoken critic of Fresh Kills in almost every fight, he had kept the pressure on his Republican colleagues, the city, and the state.[138] This was represented most graphically by his federal lawsuit (it was not withdrawn after the announcement), which now acted as a way of protecting Staten Island's position on Fresh Kills. He also had the ear and the confidence of both Pataki and Giuliani, adding an essential personal touch to resolving the controversy. Daniel L. Master Jr., legal counsel to Borough President Molinari, had pushed for the federal lawsuit that played some part (although it is unclear what part) in urging the city to seek closure of the landfill. But first came Molinari's relationship with the mayor. They had become friendly in the 1980s when Giuliani was a federal prosecutor. They shared conservative values, including support for "law-and-order" programs, and were proud of their Italian heritage.[139]

Master stated that Molinari was convinced that Giuliani would win the 1993 election and that the borough president's enthusiasm for the new mayor was "infectious." He added, "Rudy was certainly very close to Guy. Guy having been the person who kind of drafted him to run for the mayoralty [in 1989]."[140] Molinari would serve as Giuliani's chief campaign adviser in 1989, 1993, and 1997. Several years later, he and former deputy mayor and state Republican Party chair Bill Powers led the statewide effort for Giuliani's run for president.[141]

Giuliani purportedly requested that Guy Molinari run for the borough presidency in 1989. Molinari also was on good terms with Pataki, whom he helped elect through their connection with the Conservative Party.[142] In early 1997 Molinari stated, "Staten Island's 'impossible dream' is coming true.... Last May, I forged an historic agreement with Mayor Rudy Giuliani and Governor George Pataki to close the Fresh Kills landfill by December 31, 2001. That's years before the Sanitation Department planned to cease dumping."[143] According to Molinari, he had been enraged by the DSNY's announcement of possible tours

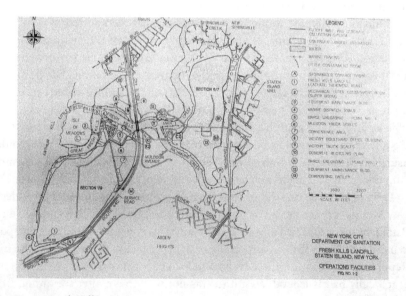

FIGURE 16.3 Fresh Kills, 1997.

Source: NYC Department of Sanitation, *Final: Operations and Maintenance Plan for the Fresh Kills Landfill*, vol. 1, January 31, 1997.

at Fresh Kills. That and the lawsuit against the mayor were immediate events prompting the May 28 closure announcement.[144]

A firm closure date was what every Staten Islander had been waiting for. Molinari, Giuliani, and Pataki delivered it. But the citizens of New York City were cautious. Would this be another false promise? For state and city legislators, the question was whether the announcement, in and of itself, was legally binding to current and future administrations. In some respects, the announcement of a closure date was the easy part. Big issues remained. How—and at what cost—could the landfill be shut down? And what would replace it? The answers to these questions were not fully developed, and the problems remained enormously complex, despite the significant step just taken to end the landfill's long and inglorious life.

At a public hearing of the Committee on Environmental Protection of the New York City Council on May 1, Assemblywoman Connelly made some sobering and prophetic remarks:

> I think it's fair to say that my colleagues and I will see the landfill's closure as a hollow victory without serious methods in place to reduce the city's garbage and to handle the tonnage left over in a way that does not impact on people's lives the way the Fresh Kills Landfill has strangled our Island's citizens for decades. Even with a defunct landfill, Staten Islanders realize that they will still require disposal methods to handle their own wastes. . . . The city has hidden behind the "luxurious" explanation that garbage was only the problem in one of its boroughs . . . and there were extreme other pressures in other boroughs that required attention first. With Fresh Kills' closing, garbage becomes every borough's . . . every New Yorker's . . . problem now.[145]

Moving from a one-borough solution to disposal to a multiborough approach would be extremely challenging, as would changing the dumping, burning, reuse cycle.

CHAPTER 17

NOW WHAT?

An August 30, 1999, article in the *New York Times* reviewed the aftermath of the May 1996 announcement to close Fresh Kills.[1] Until forty-eight hours before the "splendid day for celebration and champagne on Staten Island," no one in the mayor's office had told the Department of Sanitation (DSNY) about the closure timetable. On December 31, 2001, "not a single old shoe, not a single 'I Love New York' coffee cup, not a single piece of rotting refuse would ever again be buried at Fresh Kills. No exceptions." The DSNY "suddenly had to figure out what to do with the 13,000 tons of household garbage added each day to the artificial mountain." The challenge, according to one frustrated official, "seemed like skiing down a slope with a lot of trees, on one leg." Eric A. Goldstein, an attorney for the National Resources Defense Council, asserted, "Everyone agrees closing Fresh Kills makes sense, but it is wrong to set a date before you have a realistic plan."[2]

The announcement to close Fresh Kills with a date certain was a political choice, pure and simple.[3] Moving from a celebration to a new disposal plan had neither been adequately deliberated nor designed. The May 29 decision arose out of converging circumstances: Republican control of local and state government, increasingly restive Staten Islanders, pending closure legislation, a possible lawsuit, and an upcoming mayoral election in 1997.[4]

Skeptical Staten Islanders wanted to believe that the end was near for the landfill, but they pleaded "show me."[5] The reaction of the *Staten Island Register* was scathing, labeling the announcement of the Giuliani-Pataki-Molinari pact a "performance." The unanswered question "throughout the exercise in propaganda is the only one that matters—what will be done with the 26 million pounds of trash produced by the residents of New York City each day?" It accused Mayor Rudolph W. L. Giuliani and Borough President Guy V. Molinari of "gutting" the recycling program and added, "by their own hand and through forces beyond their control but of which they're perfectly aware, the politicians who announced the closure of the dump last week are caught in a web of contradictions that renders the whole business a hoax."[6]

The joint city and state Waste Disposal Task Force formed to examine alternative disposal options was asked to deal with long-standing, complicated issues under an impossible deadline. On the day of the closure announcement, Giuliani stated (and repeated on several occasions), "It has become increasingly evident that the burden this facility now imposes on the citizens of Staten Island must be lifted. We have worked closely with State and Staten Island officials to ensure that we develop a plan to address the waste disposal issues that the closing of Fresh Kills presents."[7] A modest recycling program was underway in New York City, hardly sufficient to deal fully with the tons of refuse generated every day. Exporting waste had strong support locally (especially from private companies), and the Giuliani administration claimed to have spent a couple of years trying to devise alternatives to Fresh Kills.[8] But implementing a comprehensive refuse program still had many pitfalls in its path.

Giuliani clearly favored privatizing refuse disposal as much as possible and supported exporting over any other method. This stance sparked resistance from advocates of recycling and recovery and from states that could receive the city's refuse. Each borough, in addition, was fearful that any new plan would mean new responsibilities at its doorstep that could adversely affect its neighborhoods. On top of everything else, the high cost of closing the landfill fell heavily on the city and state.

ANNOUNCEMENT AFTERMATH

The closure announcement did not, in and of itself, secure the termination of Fresh Kills. Staten Island officials, in particular, sought ironclad assurances and legislation to make sure that future city governments did not renege on the decision or postpone it beyond the deadline. Hours after the May 29 ceremony, the New York State Senate passed Senator John J. Marchi's bipartisan bill requiring that Fresh Kills be closed by December 31, 2001. The bill also barred operation of the proposed Brooklyn Navy Yards incinerator, which now was considered environmentally unsound and too costly. The governor's office regarded the bill as unnecessary but also understood that it could never be repealed.[9] Giuliani's legislative representative wrote to the governor that the mayor supported the bill as being "consistent with [his] decision to expedite closure of the landfill and to consider other waste disposal options besides incineration for the City of New York."[10]

On May 30, 1996, the closure bill "breezed through" the Assembly and was forwarded to the governor. Eric N. Vitaliano, the sponsor of the bill in the Assembly (along with Speaker Sheldon Silver), was not satisfied that the existing Fresh Kills consent order was sufficient to close the landfill. "Let's finally write into law the epitaph on the tombstone of the Fresh Kills landfill," he proclaimed.[11] And along with Marchi's bill and similar legislation in the City Council,[12] Borough President Guy Molinari and Representative Susan Molinari continued to push ahead with their federal lawsuit (on violations of the Clean Air Act), as added pressure to close Fresh Kills.[13]

Everyone now awaited the appointment of the task force. Giuliani stated that any findings about alternatives to the landfill would be forwarded to the task force, but he declined to disclose what those findings were.[14] Governor George E. Pataki wanted to see "a comprehensive plan to reduce the city's waste stream, to increase recycling and find alternatives for disposal, including the exportation of garbage." The mayor sided with those who opposed incinerators, asserting that there was no shortage of export markets. But with New York City only recycling 15 percent of its refuse at the time, it was difficult to tell how strong a commitment the mayor was willing to make to that option.[15]

Drowned out by the closure celebration was the anxiety of workers at the landfill site who might lose their jobs. Allen J. Adams, a tractor operator, wrote to Assemblywoman Elizabeth A. Connelly:

> Now that all the "concerned citizens" have gotten their way, and the politicians have joined the bandwagon and are closing the dump, there is one question that no one has seemed to address. What is going to happen to the 152 tractor operators and 35 crane operators who work at the landfill? . . . Everyone keeps saying that the dump is a health hazard, even though this has never been proven. If all our elected officials think that this is true, why hasn't anyone done a health study of the men who work there every day? Don't we matter, or is it easier to just forget about us? Perhaps you are all afraid to find out that it isn't really as bad as everyone wants to believe.[16]

Responding to Adams's letter, Connelly stated that she was skeptical that the landfill would close on time; she also was "not prepared to say that there will be no further use for employees currently working at the Fresh Kills landfill." She was willing to sit down with the union, DSNY, and local officials to discuss "future options" and to pass along the tractor operator's concerns to the task force.[17]

There was no denying the sickening odors that frequently emanated from the landfill. Staten Island also suffered from its proximity to other sources of air and water pollution, especially from factories and industrial facilities along both the New York and New Jersey side of the Arthur Kill.[18] On health, Connelly wrote to Adams:

> There have been and continue to be studies on the possible ill-effects the landfill may have on the citizens of the borough. However, regardless of what these studies reveal, the issue that cannot be argued is the smell. It is the most intrusive, destructive aspect of the quality of life for people living near the landfill and for those of us living in the wind's path. The future economic strength of the Island depends more, I believe, on dissipating the odor.

Connelly was sidestepping Adams's concerns and focusing more narrowly on NIMBY issues and economics.[19] Nonetheless, Connelly made good on her promise to Adams, conveying to the task force that "there must be a plan in place to address the workers' justifiable concerns."[20] She was assured that Giuliani promised to redeploy, not fire, DSNY workers who were not needed on Staten Island.[21]

For most Staten Islanders, however, the ultimate fate of the workers at Fresh Kills was not as pressing as seeing clear evidence of the landfill's closure. At Borough Hall on June 9, Governor Pataki signed the Marchi-Vitaliano bill into law.[22] This was a nonpartisan event, unlike the May 29 announcement. "Today," Pataki declared, "we put an end to all the doubts and deliver on our pledge by setting in stone the day when this environmental nightmare will close for good."[23]

THE BREWING DISPOSAL DEBATE

In May 1996 Mayor Giuliani named Marilyn Haggerty-Blohm, former chief of staff for Borough President Molinari, to head the twelve-member state-city task force. Its job was to recommend a new disposal plan for New York City.[24] By June, the members of the task force had been selected, including staff members from the mayor's office, the Department of Environmental Conservation (DEC), the State Department of Health, the secretary of state's office, the city departments of Environmental Protection and Health, the U.S. Environmental Protection Agency (EPA), and Susan Molinari's office. Also included were John Doherty from the DSNY, Borough President Molinari, and a citizens' environmental advocate.[25] Conspicuously missing were representatives from the boroughs of Manhattan, the Bronx, Brooklyn, and Queens. The mayor received letters from borough leaders asking to be placed on the task force, but he did not alter its composition.[26] Task force meetings were held behind closed doors.[27]

The *Staten Island Advance* observed that the task force had been given only a few months to come up with a plan (a "mission impossible"), after the landfill had "blighted this borough for half a century and proved itself a deathless dump."[28] As the body deliberated, Borough President Molinari hired a consultant to produce a separate report, stressing that he was "protecting Staten Island's future and the health and well-being of the people who live and work here."[29]

The central question for the task force was: What would replace landfilling? Over its history, New York City had jumped from one lily pad to the next, embracing one viable disposal option, then another. In the late 1990s, the idea of integrated solid waste management was becoming orthodoxy nationwide, but in practice the exact mix of options varied from place to place. In New York City, emphasis on private-sector exporting of waste, favored by the mayor and his supporters, seemed to be pitted against the recycling and recovery preferred by environmental groups and their supporters. But even seeing the situation in terms of this rift overly simplified matters, because any ultimate decisions would need to take into account short- and long-term goals, local self-interest, the art of the possible, and the political need to compromise.

Commercial refuse from the city already was being exported, but resistance to expanding the practice came from several quarters: those against privatizing disposal, those dreading prohibitive costs, New York communities and states that would or would not accept refuse, and a movement afoot in the U.S. Congress to ban or constrain shipping solid waste across state lines. The debate over the right of cities to enforce flow-control laws versus the right of private haulers to transport refuse across state lines remained a slugfest in Washington.[30] There was no pending resolution in sight despite the fact that thirty importing states demanded passage of federal legislation to stop or slow down trans-state movement of refuse. The closing of Fresh Kills provided a convenient "hot potato" to spur on those who believed exporting refuse was now likely to escalate in the nation's largest city.[31]

Mayor Giuliani's office favored exporting refuse across state lines and opposed bills to limit such activity on the state and federal levels. Deputy Mayor Mastro, testifying before the U.S. Senate Environment and Public Works Committee in March 1997, asserted that it would not be fair to cite the closing of Fresh Kills as justification for passage of interstate waste legislation, since the debate over such legislation predated the announcement of the closure.[32]

In one of many salvos over exporting refuse, Pennsylvania's Republican governor Thomas Ridge told Governor Pataki in the spring of 1996, "New York City should not be looking to export more waste to Pennsylvania unless our communities with disposal facilities have agreed to accept waste from other states." At the time, private haulers from New York were depositing about 2.7 million tons of commercial waste in privately owned landfills in Pennsylvania. Ridge accused Pataki of not following through with negotiations carried out with several importing states to secure compromise legislation. Pataki responded that he needed to wait for the task force's report before taking any steps to work out agreements with the importing states.[33]

Ridge's testiness was understandable but also a little confusing. Exporting refuse was hardly a new issue in 1996. A 1995 *Waste Age* survey confirmed "the existence of an extensive and intricate web of interstate solid waste movements."[34] In that year New York State exported waste not only to Pennsylvania but to Connecticut, Illinois, Indiana, Kentucky, Maryland, Massachusetts, New Jersey, Ohio, Vermont, Virginia, and West Virginia. It also imported refuse from Pennsylvania, Connecticut, Massachusetts, New Hampshire, New Jersey, Ohio, Vermont, and Canada. Indeed, in 1995 every state either imported or exported solid waste—or both. Four states—Maryland, Missouri, New Jersey, and New York—exported more than 1 million tons, which was greater than 50 percent of all refuse exported in that year. Among them, New York exported the most, with 3.8 million tons.[35] The major reasons for the interdependence for disposal among neighboring states had to do with the availability of low-cost disposal capacity, a trend toward regionalizing landfills, and the consolidation of the solid waste management industry.[36] This is why Giuliani's disposal plan emphasized taking advantage of exporting refuse.

Recycling as an alternative to landfilling, however, had gained strong support among environmentally minded citizens and officials in New York City. Ruth W. Messinger, the liberal Democratic borough president of Manhattan, was a particularly strong proponent of recycling. Presenting testimony before the City Council on her behalf on May 1, Timothy Forker linked the closure of Fresh Kills to recycling. Messinger, he stated, supported "rapid and profound increases in waste prevention and recycling throughout the City's residential, institutional and governmental sectors." She also favored "careful planning of the City's waste export infrastructure and its regulation." The local impacts in the boroughs post–Fresh Kill was a clear reminder that "garbage cannot just be thrown away. 'Away' is always someplace—and in New York City, someplace is usually in somebody's backyard."[37]

Writing to Giuliani, Messinger repeated her views on recycling (and exporting). In light of the permanent closure of Fresh Kills, she was "dismayed" that the "majority of public discussion has centered not on waste reduction, but on exporting the approximately 13,000 tons per day now dumped at the landfill." Her fear was that "export on this scale would move the solid waste problem from Staten Island to Brooklyn and the Bronx," where enormous amounts of waste—residential and commercial—would go through private transfer stations before being shipped out. At the time, the private stations were used primarily for commercial waste.[38]

Once regarded as a grassroots approach to source reduction and a reaction against overconsumption, recycling now was becoming widely accepted as an alternative disposal strategy. As the 1990s began, recycling was a growth industry; by mid-decade markets for recovered materials were on the rise. The number of programs reached more than 2,700 in 1991, and forty-seven of the fifty largest cities had recycling programs in 1992. High diversion rates became the measure of success for recycling. In the late 1990s the EPA raised its recycling goal from 25 to 35 percent, although this was difficult for many cities.[39] New York City's own efforts fell far short.

Mayor Giuliani was no fan of recycling and had not supported full funding for it throughout his first term. The City Council approved his budget plan for fiscal year 1997, which called for a $29.8 million cut in the program. (The city then was recycling about 1,700 tons daily.) The budget was part of a $1 billion service-reduction package that also affected city cultural agencies, AIDS outreach, and more. Some regarded the roughly 38 percent cut in recycling and waste prevention as essentially dismantling the programs entirely. But Giuliani stated that recycling would play a larger role in the city once Fresh Kills was closed.[40] By then, he would be at the end of his term.

Among those criticizing the decision to cut recycling was Thomas Outerbridge, chair of the Manhattan Citizen's Solid Waste Advisory Board, who recommended holding off thinking about "dramatic changes" in the recycling program until the task force issued its report.[41] In responding to Outerbridge, Powers asserted that reducing the recycling budget was "essentially a fiscal decision." "While there are many important programs and services provided by the City, many of these, while compelling, are also competing for limited dollars. Recycling is no exception."[42] This was not the first cut that the recycling program had endured under Giuliani.

Messinger was "appalled" that the 1997 budget "lops off approximately $30 million in funds for the City's recycling program," including the elimination of mixed-paper recycling, education in recycling collection frequency, and waste prevention programs.[43] On June 20, Arthur S. Kell of the New York Public Interest Research Group (NYPIRG) repeated his arguments against Giuliani's recycling views: "Now there isn't even a pretense of complying with the city's 1989 recycling law. It certainly doesn't bode well for the timely closing of the Fresh Kills landfill, since recycling should be the primary method of disposing of New York City's mountain of waste." He added that the mayor "is putting the city at considerable risk with such a strategy. He's also providing potent arguments for out-of-city residents who don't want to be dumped on just because Giuliani can't be bothered with what he calls 'unrealistic' recycling goals."[44]

To Kell, the mayor was painting himself into a corner by closing the landfill, abandoning incineration, and questioning recycling. Responding to Kell's earlier criticism that substantial reliance on exporting and less on recycling was bad policy, Giuliani stated, "I think we're being criticized by people who just don't understand that very, very unrealistic goals were being set" about recycling.[45] Recycling had to be propped up by reliable markets for the mayor to view it as essential. In addition, he had all but ignored Local Law 19, which mandated what materials, how much, and by what date the residents of New York must recycle.[46] Estimates for fiscal year 1997 suggested that the recycling rate for the city was likely to rise only 1 percent over the previous year's very modest 13.8 percent of household waste.[47]

In a July 12 letter to Giuliani, Comptroller Alan G. Hevesi (a former Democratic assemblyman) also chimed in about the mayor's post–Fresh Kills strategy: "I am concerned that you are making a succession of far-reaching policy pronouncements limiting the City's options for dealing with its garbage without any meaningful attempt to create a complete and realistic plan." Hevesi was emphatic that "recycling must be an important part of New York City's program for dealing with its garbage" and that the FY 1997 Adopted Budget will result "in the total elimination of the [recycling] education and outreach program."[48]

Along with the mayor's tepid feelings about (or animus toward) recycling was a chorus of critics nationwide who also believed that the faith in recycling was overstated. Skeptics of recycling were concerned about cost, lack of stable markets, and what they believed to be an inaccurate portrayal of the success of many recycling programs. In 1996 the libertarian journalist and commentator John Tierney wrote an article for the *New York Times Magazine* stating that as a result of the *Mobro* incident in 1987, "The citizens of the richest society in the history of the planet suddenly became obsessed with personally handling their own waste." As a result of concern over landfilling, Americans saw recycling as their only option but failed to recognize that the 1987 crisis was overblown and that mandatory recycling programs "aren't good for posterity," diverting money from "genuine social and environmental problems." He concluded that "Americans have embraced recycling as a Transcendental experience, an act of moral redemption."[49]

Recycling supporters were outraged. The piece was selective in its use of data, and Tierney had thrown out the proverbial baby with the bathwater for the sake of a story. He characterized the recycling craze as new, and therefore impulsive, rather than as part of a long-standing practice. He criticized recyclers for not being grounded in the economic realities of cost-benefit but did not provide careful documentation of recycling's alleged failures. He depicted advocates as representing a clearly identifiable social or political fringe rather than more accurately as a diverse group. He regarded recycling as independent of other waste disposal issues instead of viewing it as a component in a large and more complex system. And he painted the history of recycling as if it already had played itself out.[50]

While critics of recycling often were guilty of overstating their case, proponents too were sometimes overly optimistic about its successes. Just as the sanitary landfill came to be viewed as a disposal panacea, recycling would need to demonstrate a strong record of tangible results. The local recycling dispute in New York City mirrored the national debates. Proponents emphasized the environmental benefits and saving virgin materials; opponents tended to stress the economic costs.[51]

The expense of closing Fresh Kills loomed in the background. Officials estimated that over a thirty-year period the total cost of closure would be $773 million. The mayor stated that funding would be found. Assemblyman Vitaliano insisted that "it would be far more expensive to bring Fresh Kills up to environmental compliance than to export the city's trash."[52] The landfill mounds had to be covered and capped, air pollution monitored and eliminated, and millions of gallons of leachate collected. And in the background was another question: What would the vast landscape be used for next?[53]

Now that closure was policy, several officials underplayed the cost. Not everyone agreed, but the exhilaration over closing the landfill masked the assortment of issues the city now faced in making such a dramatic break with its venerable disposal site.[54]

CONFRONTING THE MOB

Before and after the task force issued its report, Giuliani pressed forward with an export stratagem (dependent on a much greater emphasis on privatizing disposal) as the centerpiece of his post–Fresh Kills disposal plans. A significant problem, however, stood in the way: a private commercial waste-collection market dominated by organized crime. The mayor faced the conundrum of how to extract the mob from commercial collection and replace it with large national waste companies who could collect refuse and help with disposal.[55]

In the 1990s the commercial waste hauling business was controlled by two formidable Mafia cartels—the Lucchese crime family in Long Island and the Gambino crime family in New York City. The two families also controlled the trade associations and the unions. The city had been receiving complaints about illegal activity in the carting industry as early as 1947. The New York City Commissioner of Investigation ordered the dissolution of various trade associations, but they soon resurfaced under different names. In the 1950s Teamsters Local 813 (representing the drivers employed by the cartel) came under heavy scrutiny, along with the hauling companies themselves. In 1956 Mayor Robert Wagner Jr. established the Department of Consumer Affairs (DCA) to monitor the waste business by setting maximum rates and establishing licensing requirements as a way of controlling entry into the carting business. It did neither, virtually granting licenses without limits.[56]

In 1990, under the directorship of Commissioner Mark Green, the DCA undertook measures to challenge corruption and racketeering in the carting industry. During the next several years, criminal prosecutions, civil racketeering lawsuits, and the entry into the market of a few large waste disposal companies helped squeeze out mob business both in Long Island and the rest of New York City. Browning-Ferris Industries (BFI, Houston, Texas) and Waste Management, Inc. (Oak Brook, Illinois), announced that they might enter the Long Island market. After several strenuous court battles, competition among carting companies broke the stranglehold of the mob in Long Island. BFI decided to test entry into carting in New York City, obtaining its first contract in 1991. The Mafia responded by leaving the severed head of a German shepherd for them with a note: "Welcome to New York."[57]

BFI soon realized that the task of entering the market would be difficult, and in 1993 it actively cooperated with the Manhattan District Attorney's Office in an undercover sting. Federal prosecutors mobilized, indictments followed, and the New York City Police Department gathered additional evidence. Efforts to purge the mob from Local 813 intensified.[58]

To help break the mob's control of the carting industry, the Giuliani administration established a new regulatory agency—the Trade Waste Commission (TWC)—which focused on licensing, cutting rates, and increasing market competition. The mayor wanted to remove the "smell" from the industry.[59] While the DCA's licensing efforts had been tried—and had failed—they proved more successful in other city actions, such as fighting corruption in Fulton Fish Market. On September 30, 1996, the TWC issued new, wide-scale rules.[60] Actions of local, state, and federal officials, plus the establishment of the TWC, essentially rid the waste hauling industry of mob influence.[61] The question remained as to whether industry competition would endure or face new problems associated with business consolidations.

THE TASK FORCE REPORT

After a postponement, the task force's report was completed in November 1996 and released to the public in early December.[62] Haggerty-Blohm was convinced that city and state officials were close to an agreement over how to move forward. Some were not so generous. The Staten Island environmentalist Barbara Warren had not been placed on the task force until the first half of September (as was the Environmental Defense Fund's general counsel James T. B. Tripp). She asserted, "I don't think the closure deadline is too secure right now. What the panel is doing has added more fuel to my belief that this is a joke, a charade."[63] Equally dramatic, during the first meeting of the task force in the summer, Guy Molinari "stormed out," complaining that the group only allowed him a marginal role in the work. But, as might be expected, he worked his way back into the deliberations.[64]

Guy Molinari had released a study of his own in October, before the much-anticipated task force's report came out. Sadat Associates, Inc. (Princeton, New Jersey), prepared a study similar in scope to the work of the task force, but it allegedly never shared any of its work with the task force until the final product was released.[65] The objective of the Sadat Report was "to provide an analysis of alternatives for the phasing out of the Fresh Kills Landfill." In particular, it favored developing "in-City and in-state" treatment and disposal programs "to encourage the State of New York to cease being a net exporter of solid waste within ten years."[66]

The Sadat Report called for the private sector to identify communities in the state where recovery, transfer, treatment, and disposal facilities could be located; it also favored utilizing "former contaminated sites" (brownfields) for disposal. Such a proposal was deemed economically beneficial to those communities that opted to accept such facilities. As for in-city treatment, the report backed material-recovery facilities (MRFs) and composting.[67] Reflecting a Staten Island perspective, the report also proposed a "borough-specific" approach in which each borough would be responsible for its own refuse disposal.[68]

It remained to be seen if the Sadat Report (completed in October) would contradict what the task force might propose or possibly steal its thunder. Clearly, the report deviated from the ongoing debate of exporting waste versus recycling by bringing state planning into the equation, downplaying out-of-state exporting, redefining reduction and recycling issues, and placing responsibility clearly at the borough level.

A month later, the task force's own report was made public. It acknowledged that Molinari's Sadat Report recommended "many initiatives that the Task Force adopted" but assumed "a more aggressive stand on reducing the amount of waste going to Fresh Kills." The heart of the report placed an emphasis on developing borough-specific recommendations for waste management and out-of-the-city disposal as a "safety net" to meet its targets for phasing out Fresh Kills. Transfer stations would be the destination for borough "unrecycled waste," and exportation would be defined as "the disposal of waste outside the City of New York (within the state or in other states) under transport contracts with the private sector waste industry." The destinations would be "either public or private landfills, incinerators, and/or waste-to-energy facilities," but the report did not specify *how* waste was disposed of.[69]

The task force suggested several approaches to relying on transfer stations but did not favor one in particular. They included retrofitting Fresh Kills to serve as a citywide marine transfer station, a single island-based transfer station, a single waterfront transfer station, exporting waste from existing facilities, and borough-based transfer stations. The report also devoted substantial attention to the operation, closure, and post-closure of Fresh Kills.[70]

The report announced that city officials—at the City Council's urging—had agreed to restore $6 million in funding in FY 1997 for mixed-paper and bulk-metal recycling collection in Staten Island and the Bronx and to expand it in Manhattan by March 1, 1997. The city would increase recycling resources, begin a pilot program to collect and recycle chlorofluorocarbons from discarded appliances, provide funding for borough self-help bulk sites, and restore funding for waste prevention research. If carried out, New York City would be more in line with national trends in recycling and waste reduction.[71]

The task force's report concluded that it had brought together "key stakeholders," but it failed to mention the lack of representation from the boroughs other than Staten Island. It ignored responding directly to the in-state exporting of waste and essentially dodged any in-depth analysis of private out-of-state exporting. The task force deliberations may have influenced (or nudged) the Giuliani administration into taking recycling more seriously, but it was emphatic about accepting the inevitability of exporting waste.[72]

Both reports recommended a borough-based waste management system that would satisfy Staten Island but not please the other boroughs. Now began the tortuous fight over implementing a new system that was sure to generate a wider circle of discontent than Fresh Kills protesters.

BOROUGH-BASED RESISTANCE

Well before the task force report, inquiries from private companies wanting to cash in on various disposal opportunities flooded the mayor's office.[73] Release of the task force's report alone solicited proposal from more than two hundred waste haulers. In January 1997 there was a standing-room-only meeting at the offices of the Economic Development Corporation concerning a city contract to export refuse from the Bronx (the first of such borough contracts).[74]

Governor Pataki was among the first to praise the "sound and balanced" task force report. "I think that it provides the framework by which we can go forward and make sure Fresh Kills is closed in accordance with the dates [for closure]." He also was pleased with how the report handled the way that the city could cooperate with other communities willing to accept its refuse.[75] Pataki immediately sought funds for the landfill's closure. Somewhat in jest, he stated that he wanted the Los Angeles Dodgers to move back to Brooklyn, but if that did not happen, "We'll close Fresh Kills, level it and put the Dodgers right here."[76]

Staten Island became the first borough to establish a working group to develop a plan for its residential waste after Fresh Kills closed. Guy Molinari, who set up the borough-based Closure Panel, stated, "I believe it is important for Staten Island to set an example for the rest of New York City by being the first borough to appoint a working group and the first borough to release alternatives to dumping at Fresh Kills."[77] He and other locals, however, looked with displeasure on a scheme to build a solid waste management industrial park on the site of the old Proctor & Gamble Company plant at Port Ivory and on a later suggestion for a garbage transfer station there.[78] In April Connelly and Marchi submitted a bill to the legislature that would prevent the city from building large transfer stations or refuse-processing facilities on Staten Island after Fresh Kills closed.[79]

With reliance on Fresh Kills Landfill winding down, a boroughcentric disposal plan clearly meant new (and unwanted) responsibilities across New York City. Haggerty-Blohm asked the borough-based working groups to recommend how each borough would handle its solid waste. In response, Manhattan Borough President Messinger (through Timothy Forker) thanked the task force chair for the offer to participate but added that "the exclusion of so many interested and informed parties [on the task force] is a fundamental reason why the . . . report does not address in detail many of the most difficult solid waste issues facing the City, such as the preferred methods for waste reduction and exportation." Eventually, however, Manhattan would set up a working group.[80]

By mid-December 1996, Queens Borough President Claire Schulman had signed on to develop a borough plan, but she had reservations about what would happen once Fresh Kills closed.[81] The Bronx convened a working group in January 1997. Under Borough President Fernando J. Ferrer the panel was particularly wary of permitting new facilities. Despite efforts to block it, the Bronx was the first borough to export its residential waste, beginning in July 1997. This was the first time since the 1930s that New York City had deposited residential refuse outside its city limits. Waste Management, Inc., had a three-year contract to export the borough's 1,750 tpd to a landfill in Waverly, Virginia.[82]

For the mayor, the Waste Management contract was an example of how the administration was trying to take the waste burden off of Staten Island immediately. "We are addressing the problem now because we understand that this is a matter of urgency for the people of [Staten Island]," Giuliani stated, "and really for the city as a whole."[83] Others worried that exporting refuse out of state could nearly double the city's disposal costs (about $75 per ton as opposed to $41.50 per ton at Fresh Kills). In the case of this first contract, the rate was $51.72.[84] Environmentalists and some experts in the waste industry cautioned that while this first round of bidding offered good rates, the charges were most likely to rise when the city became increasingly reliant on the contracted companies.[85] Advocates of recycling and waste reduction clamored even louder for new programs and increased diversion rates.[86]

The borough-based approach to refuse disposal raised more than NIMBY concerns. Testifying before the City Council's Environmental Committee in mid-December 1996, Brooklyn Borough President Howard Golden stated, "Though the Task Force raises the issues of fair share in the siting of waste processing facilities, it does not commit the city to developing these standards." He was referring to existing transfer stations and new facilities and to what extent they were equitably placed. More than 25 percent of the private transfer stations in the city were located in the minority community of Greenpoint-Williamsburg. "Similar conditions," he said, "exist in Red Hook and East New York. This recommendation reflects the administration's position that it has no responsibility to develop siting standards for waste facilities to ensure that new 'sacrifice zones' will not be created by the closing of Fresh Kills." He concluded that Brooklyn, "which already processes the majority of commercially collected waste, stands to suffer an even greater disproportionate burden of managing the city's garbage through the increased processing of residential waste."[87]

The task force's report raised many questions but did not provide all the answers.[88] Borough President Molinari was concerned that criticism of the report in some ways could undermine the closure of Fresh Kills. At one point Messinger (now understood to be running for mayor) quipped, "A plan to make a plan is not a plan."[89]

A January 6 *New York Times* commentary echoed Messinger's pronouncement:

> Though the Mayor and the Governor have suggested that their plan to shut Fresh Kills was both a bold act and one that took years of preparation, in fact it was neither . . .
> Their plan was not so much a plan as a plan to have a plan. . . . As a result, it has become clear only over the last month or so just how complicated closing Fresh Kills will be.[90]

TRANSFER-STATION CONUNDRUM

While long-term solutions to the disposal problem were under review, Mayor Giuliani needed an immediate strategy for disposal lest he renege on his promise to close Fresh Kills by the end of 2001. His "interim plan" would rely on existing privately owned waste transfer stations in operation to house the city's refuse before it was exported. In late January 1997 he proposed approximately $321 million over a five-year period to pay for the closing of Fresh Kills, $254.5 in new funds and $66.5 million in allocated funds. Of that amount, $76 million would go to recycling programs, while the lion's share ($245 million) would be dedicated to exporting and landfill closure. The budgeting gave more than a hint of where his interests lay.[91]

There was no talk about replacing Fresh Kills Landfill with any kind of a public, centralized disposal facility, but there remained deep concern in Brooklyn nonetheless. Environmentalists and government officials there were troubled that an alternative disposal plan would shift much of the burden to its borough, especially because of existing installations already handling great amounts of commercial refuse. Three of the biggest carting companies in the country operated transfer stations in Brooklyn and might be in line for more business. The activist Deborah Masters stated, "I'm afraid that a third of the garbage that is going to Staten Island will go to Brooklyn."[92]

The mayor did not bargain for the storm of criticism that soon followed his initial plans to use the private transfer stations. Community activists had protested against them for years. Many of the facilities did not have proper licenses and often violated health and safety codes. Most transfer stations—serviced by heavy truck traffic—were located in industrial areas with salvage yards, repair shops, and recycling warehouses and near neighborhoods populated by large numbers of minorities in Brooklyn, Queens, and the South Bronx. Brooklyn alone had about 40 percent of all the transfer stations in the city, most located near a gigantic sewage-treatment plant, a nuclear waste facility, and a huge underground oil-storage tank. Hunts Point in the South Bronx, with about ten thousand people, was home to numerous manufacturing facilities, automobile salvage yards, food markets, and transfer stations. The severe asthma problem there was a testament to the kind of "sacrifice zone" Howard Golden had noted.[93]

By February the Giuliani administration had a modified plan in mind for post-closure disposal, which would not rely strictly on private transfer stations. In addition to private stations, DSNY trucks would dump as much residential waste as possible from Brooklyn, Manhattan, and Queens in barges at the city's eight existing marine transfer stations (MTS) distributed throughout the city.[94] Such a system reduced congestion from vehicle traffic, since the refuse traveled by water. Bypassing Fresh Kills, the barges would take the waste to three new enclosed barge-unloading waste transfer stations (in the metropolitan area), containerize the refuse, and then ship it by truck, railcar, or barge out of the city. A new Fresh Kills transfer station would handle Staten Island's household waste. The city also intended to negotiate a new contract to export refuse from the Bronx by rail or barge. Possible locations for the new facilities included Red Hook, Hunts Point, Sunset Park, and sites in New Jersey.[95]

An article in *City Limits* stated, "[The] nominees selected for this honor [as new waste-facility sites] are among the poorest and the most environmentally blighted communities in Brooklyn and the Bronx."[96] Several neighborhoods came to believe that the mayor placed more stock in law-and-order issues than in environmental and health concerns in their communities.[97] In Hunts Point, Red Hook, Greenpoint, and Williamsburg, protest against transfer stations had intensified after the announcement to close Fresh Kills. In response to the mayor's proposed siting plans, organizations in Red Hook such as Groups Against Garbage (GAGS) and Organization of Waterfront Neighborhoods (OWN, established by the New York City Environmental Justice Alliance) stood alongside other community groups, some of which had formed as early as 1982. Yolanda Garcia, executive director of Nos Quedamos (a Bronx-based community-planning group) stated, "Everybody's organizing now, spreading the word. We cannot be the sacrificial lambs for the rest of the region."[98]

Memories of the Brooklyn Navy Yard battles were fresh in the minds of the protesters. They now pleaded with Giuliani to take a "fair-share" approach to distributing new transfer stations, not concentrating them in poor and minority communities. The pushback from the mayor and his supporters did not help diffuse the protests. For example, an EPA civil rights investigation looking into allegations of environmental racism in the location of the transfer stations in 1998 drew an editorial scolding from the *New York Post*. The conservative, pro-Giuliani newspaper called environmental racism a "crunchy-granola-y diversity-crazed notion." Citing Fresh Kills, located in a community which was 85 percent white, it argued that "there is simply no evidence that such a thing as 'environmental racism' even exists." Such facilities as transfer stations, it added, would create jobs in poor areas.[99]

What was and what was not "environmental racism" was a point of intense debate. Some community activists threatened to sue the city because they believed that the policy that had led to the closure of Fresh Kills discriminated against poor and minority communities in Brooklyn and the Bronx. In early March 1998 Representative Nydia M. Velasquez (a Democrat representing parts of Brooklyn, Manhattan, and Queens) called for a lawsuit, and she was joined by about two hundred residents and several other elected officials in a protest at a DSNY office. Giuliani's response was that Velasquez's reaction was the worst kind of "not in my backyardism." "We are not asking any one of the boroughs to take on more of a burden than the burden itself creates by virtue of its own waste," he argued.[100] But the mayor and the commissioner were missing the point: the transfer facilities would not be affecting each borough as a whole but selectively targeting those same neighborhoods that had suffered environmental inequality before.

In an effort to placate communities facing the installation of new waste facilities, the administration proposed rules that set distances (three hundred feet) separating waste transfer facilities in light manufacturing zones from residences, schools, and parks. This policy mirrored the efforts to create buffer zones at Fresh Kills. Not surprisingly, borough officials and environmentalists criticized the standards because they did not protect people living in heavy-manufacturing districts in Brooklyn and the Bronx. Threats of legal action arose, and those anxiously awaiting the closure of Fresh Kills tensed up again.[101] The *New York Times* observed sometime later that the three-hundred-foot rule for transfer stations was two hundred feet lower than the buffer for pornography shops.[102] A DSNY proposal on April 29 called for extending the buffer distance for transfer stations from three hundred to four hundred feet.[103]

Almost one year after "the announcement," euphoria was replaced by endless tension. "Where will our garbage go?" had yet to be clearly answered. An exasperated Deputy Mayor Mastro stated in April, "Sometimes individuals get so provincial they forget that the city is dealing with the trash in a much more environmentally sound way than dumping on Fresh Kills."[104] The administration did not waver in claiming that closure was near, but the "greater-good" argument was not selling in Manhattan, Brooklyn, Queens, and the Bronx. City leaders often failed to appreciate that a borough approach to disposal was the change agent for many communities, not the closure of Fresh Kills itself.

On April 30, the deadline for submission of borough plans, Manhattan presented its report.[105] It stated, "The real challenge for the city as a whole—a challenge that must be taken up by every home, business, agency, neighborhood, and borough—is to select and develop a waste management system that avoids adverse environmental impacts, provides long-term stability, and captures the economic and environmental benefits inherent in waste prevention and recycling."[106] The report called for citywide planning, declaring that the notion that "each borough is or can be an island unto itself is simplistic and ignores existing and developing recycling and waste disposal markets."[107] But it complained that formulating its strategy "was severely hampered by the limited access allowed to information and expert agency personnel."[108] Another dig: "Too often the Administration treated the development of alternatives to Fresh Kills like Fort Knox," thus undermining collaboration.[109]

CITY COUNCIL'S CASE FOR RECYCLING

On October 27, 1997, almost one year after the task force's report had been made public, the City Council produced *Without Fresh Kills: A Blueprint for Solid Waste Management*. A few days before its release, Stanley E. Michaels, the chair of the Committee on Environmental Protection, called it "one of the most significant documents to be issued by the Council in recent years." He did not regard the document as "an attack on anyone but as constructive criticism," although it stood in stark contrast with Giuliani's plans.

Recycling was at its core. "The [disposal] program cannot work if the public gets the message that recycling is a frill," Michaels stated. "Let no one be distracted by the argument that it will cost too much. Especially in light of the pending Fresh Kills shutdown. Recycling must finally be viewed as an essential service."[110] Speaker Peter F. Vallone, also pushed the recycling program: "There is only one long-term viable solution, and the City Council knows what it is. It's called recycling—and it's the law of New York City.... We cannot burn our way out of this crisis, or bury our way out of it, or dump our way out of it, or ship our way out of it. We have to recycle—period. We have to recycle—now."[111]

The report called for the development of "a comprehensive economic development policy for recycling." It added forcefully:

> The Council is outraged that eight years after the passage of Local Law 19, after seven State Court decisions upholding the mandates of the recycling Law, the development of the 1992 Comprehensive Solid Waste Management Plan, the 1996 Update and Modification, and countless Council oversight hearings on the City's recycling Program, that DOS and the Mayor are essentially declaring that DOS does not possess the in-house expertise to maximize recycling participation and diversion.

Wiser use of funds could be directed at public education in areas with low levels of recycling and waste prevention, and it wanted to enhance and expand the city's procurement of products, material, and equipment with recycled content.[112]

On the management of unrecycled waste, the report favored utilization of the MTS system but viewed exporting as a last resort. The council also called for a moratorium on new permits for private transfer stations. This objective stood in the path of the Giuliani administration's reliance on these facilities to help meet its closure deadline for Fresh Kills.[113]

THE MORE THINGS CHANGE...

By the end of 1997 uncertainty remained as to whether the complex task of closing Fresh Kills could be met by the imposed deadline. Equally important questions were whether Giuliani's export plans would win out over enhanced recycling and waste reduction and to what degree a borough-based disposal program could actually work. A December 21 editorial in the *Advance* did not question the intention of the mayor and the governor to live up to their promises about the landfill closure. "Beyond that, however, problems lay. It is one thing—and far easier—to promise to close the landfill than to actually accomplish that."[114]

With Giuliani's mayoral victory in November 1997, it was now clear that the sitting government would shepherd Fresh Kills to its closure. There had not been much question that Giuliani would win. Running on the Republican-Liberal ticket, he beat Manhattan Borough President Ruth Messinger by sixteen points, including taking her home borough of Manhattan, as well as Brooklyn, Queens, and Staten Island.[115]

A minor flare-up occurred before the primary in the summer of 1997. Complaints arose about Giuliani using taxpayer funds to appear in television and radio commercials. In one case, the DSNY put together an expensive television ad featuring the mayor as a spokesperson for the value of recycling. Giuliani was shown throwing an empty water bottle into a recycling bin, at the prompting of the New York Yankees' manager Joe Torre. Since Giuliani was not associated with any active promotion of recycling, the ad was most curious.[116]

The mayor's supporters continued to defend his actions in dealing with the disposal problems well after the election. Deputy Mayor Mastro criticized Howard Golden and others for continually dragging out NIMBY arguments about refuse issues in their boroughs. Mastro emphasized that the administration was implementing its closure plan ahead of schedule.[117] Ever faithful to Giuliani, the *Staten Island Advance* asserted that his administration was "the only one in history that was bold enough to propose closing Fresh Kills and then bold enough to actually plunge into the hard work of accomplishing that goal."[118]

CLOSURE OPERATIONS

Despite the political back and forth, the DSNY had been hard at work implementing final closing specifications. Recent estimates had the city spending more than $1 billion to close and monitor Fresh Kills. According to Assistant Commissioner Lucian F. Chalfen, "The cost of closure is larger than the annual budgets of many countries."[119] Two sections at the landfill still remained active: section 1/9 (462 acres) and section 6/7 (309 acres).[120] With the last barge of refuse, section 1/9 near Arden Avenue (the highest point at Fresh Kills) had a mound 270 feet tall—higher than the Statue of Liberty.[121]

Closing the landfill required a major effort, estimated to take about five years. Among the tasks to be completed were constructing slurry walls to contain leachate; building a collection system and treatment plant for contaminants; finishing wells, pipelines, and process plants for methane; implementing a storm-water drainage system; and completing vegetative and geosynthetic covers. Monitoring and maintaining the closed landfill would last for thirty years under federal law. Hundreds of DSNY employees and private contractors would need to be employed in the closure work.[122]

Away from the landfill, discussion continued about modifying the existing consent order to make it, as Assemblyman Vitaliano stated, "a legal protocol to implement the closure law."[123] The DSNY also faced challenges of responding quickly—with little preparation—to new disposal policies being designed by the Giuliani administration and the City Council, including problems related to exporting waste.

On April 3, 1998, the DSNY released a draft modification of the Comprehensive Solid Waste Management Plan, putting its best face on post–Fresh Kills operations and goals. The modification was intended to augment the broad framework of the original 1992 plan with the changes made in 1996. The changes were in alignment with Giuliani's objectives. The report praised the mayor (working with Pataki and Molinari) for announcing "a fundamental change in the management of the City's Solid Waste" and pledging "that no incinerator would be built or renovated in the City." The modification also asserted that the mayor's leadership was demonstrated in shifting the solid waste management system away from the landfill "to increase recycling, waste-prevention and waste export," reflected in his budget priorities.[124]

These statements shaded the context in which the new disposal options had been made. Not surprisingly, they failed to mention the political pressure necessary to reinstate funds for recycling into a budget that had initially favored exporting waste. In the report's "Financial Plan Highlights," six-year projections combined recycling and waste export, showing FY1997 at $17.2 million, increasing to $117.5 million by FY 2002. This was misleading because it was an aggregate total. When broken down, $373.2 million of a total of $485.6 million for the six-year period went to waste export. The report added an optimistic portrayal of the curbside recycling program after 1996 and also praised the mayor for dramatically improving the commercial solid waste management practices in the city.[125]

The key plan-modification elements essentially included activity underway and decisions already made.[126] The report reconfirmed the decision to close Fresh Kills on December 31, 2001 (with a phasedown from 13,000 tons/day in 1996 to zero in 2001) and not to replace the landfill with incinerators or other waste disposal facilities.[127] The city's export program would rely "to the maximum extent possible on marine and rail-based transfer capacity." It stressed that it would not depend on "in-City, land-based truck-to-truck transfer stations."[128] The DSNY looked to grow the recycling program, with a commitment to a 25 percent diversion rate by 2001 and to expand the waste-prevention programs.[129] Such tasks would be a tall order, especially since tonnage for the curbside program remained essentially flat since 1992 and would continue to be so through 2000.[130]

The year 1998 marked the fiftieth anniversary of Fresh Kills Landfill. The activist Linda Angelone, executive director of Rainbows Hope, stated, "Are we celebrating because we honestly believe all of this is going to come to the closing of the dump?" Representative Vito Fossella Jr. asserted, "I will continue to work so that our children will celebrate the golden jubilee of landfill closure in 2051." The *Advance* reporter Maureen Seaberg called it "a molden anniversary. A real jubi-leachate."[131] The *New York Times* declared on May 2, 1998: "Dealing with the waste stream in the 21st century is probably New York's most important unsolved environmental problem."[132]

Public advocates such as Mark Green wanted a full disclosure of the environmental risks associated with the landfill before the termination deadline.

Several borough leaders continued to question the effect of closure on their neighborhoods.[133] The *Advance*, in early May, reminded citizens of the "hidden toxic powers" of the landfill, "when nine Sanitation workers got ill, some violently, in a sudden wave. Then, when ambulances showed up to give them aid, 10 paramedics got sick." But no one knew the cause of the outbreak.[134] Some protesters called for compensation to Staten Islanders for bearing the burden of the dumpsite for so long.[135] Standing on the steps of St. John Neumann Church on Arthur Kill Road, Sister Mary Michael could see the mounds at Fresh Kill. "It's evil," she said. "It's how I imagine hell would be, souls rotting away for eternity but without ever fully decomposing."[136]

THE SECOND FRONT OVER CLOSURE

The closure decision opened up two fronts. The first was a hopeful but also skeptical Staten Island. The second resided in the other boroughs (especially Brooklyn, Queens, and the Bronx), which faced the unwanted responsibility of dealing directly with disposal of their refuse. Since the closure declaration, and even before, the focus was unreservedly on the transfer stations. In 1998 New York City had eighty-five private waste-transfer stations and nine public marine transfer stations (MTSs).[137]

Racial politics reared its head again during the latest debate on the transfer stations. Poor and minority neighborhoods, especially in Brooklyn and the Bronx, renewed claims of unfair treatment and possible environmental racism over plans to rely on existing transfer stations or build new ones. A story in the November 18, 1998, *Advance* stated that Staten Island officials and civic leaders strongly denied that the mayor and his administration "in closing the Fresh Kills landfill, is sacrificing minority communities and placing objectionable garbage facilities within them to pacify white Staten Islanders."[138] But there is little doubt that as the city sought alternatives to the landfill, numerous groups (especially in neighborhoods that had suffered unequal treatment in the past) were feeling threatened again.

In May 1998 James T. B. Tripp (co-chair of the citywide Recycling Advisory Board) met with David Jaffe (counsel to Senator Marchi). In a letter to Jaffe on May 13, Tripp and co-chair Thomas Outerbridge[139]—both with ties to Staten Island—laid out their views on the transfer stations and the "failure of adequate planning" for the Fresh Kills closure, hoping to persuade Marchi to support a bill "that provides ironclad enforcement of the Fresh Kills closure date as well as solves some of the transfer station problems."[140]

Tripp had been appointed to the mayor's task force, but he believed that it had not been "a satisfactory process," especially since Giuliani "ignored at least 80% of their own recommendations, including very important waste reduction and recycling ones." The failure of planning, Tripp added, could jeopardize the closing of the landfill but also affect those with transfer stations in their neighborhoods. He added, "It is in Staten Island's own interest to look realistically at what the situation is and not just assume that an entire borough does not want to manage its fair share of waste."[141] This perspective was not likely shared among New Yorkers, who were looking out for their own neighborhoods.

"The same organized crime outfits that operated the carting companies were operating the transfer stations," Tripp charged. He also claimed that the mayor and the DSNY "ignored the problem of transfer stations." This was an exaggeration, but his point was that closing Fresh Kills required a more serious and all-encompassing look at the existing refuse infrastructure—including recycling.[142] None of Tripp's suggestions were new, but his desire to avoid direct tensions between Staten Island and the other boroughs was insightful.

It was a difficult task for residents of a particular borough to view the disposal crisis as more than a local threat. Protests became increasingly forceful, especially in opposition to new transfer stations and the potential for increased truck traffic. For example, hundreds of people (including Borough President Golden) turned out to condemn a plan by USA Waste to expand and rebuild a truck-only transfer station on the Williamsburg (Brooklyn) waterfront. Other proposals were being developed for Red Hook in Brooklyn and Hunts Point in the Bronx.[143] Environmentalists on Staten Island continued to criticize a proposed transfer station on the former Proctor & Gamble site at Port Ivory.[144]

In the fall of 1998, the city government reviewed thirteen proposals from companies interested in handling residential waste and in upgrading the eight existing MTSs.[145] Much of the current controversy over transfer stations centered on Brooklyn. Staten Islanders viewed the protests and possible lawsuits over the use of existing or new transfer stations there as efforts to stall the closing of the giant landfill or at least obstruct the city's plan for an alternative. To citizens of Brooklyn, deep concern grew around the potential added burden of new waste disposal facilities and the transporting of tons of refuse through the borough.[146]

Howard Golden, a major proponent of a fair-share approach to refuse, challenged the contract to ship 2,400 tons of residential garbage out of Brooklyn, arguing that it violated state environmental law.[147] Although the City Council was considering not approving the latest solid waste management plan, Mayor Giuliani repeated that each borough had to handle its own solid waste and that transfer stations would be placed throughout the city in areas where they would have the least impact on the neighborhoods. He added that while Fresh Kills had been a long-term depository for the city's refuse, the transfer stations would hold waste only temporarily.[148]

Little skirmishes kept the big issues alive. A Brooklyn judge issued a temporary restraining order in late October 1998 that halted the export of local residential waste. Deputy Mayor for Operations Joseph Lhota responded, "This is an effort to block us from keeping our promise to close Fresh Kills landfill. We're not going to let them get away with this."[149] The order was soon overturned, but the tension continued when the state withdrew permits intended to allow two transfer stations in Brooklyn to process residential garbage for shipment. Soon thereafter the permits were reissued, and the courts dismissed a lawsuit by Brooklyn residents to stop implementation of the city's solid waste disposal plan.[150]

THE MAYOR'S "PERMANENT" DISPOSAL PLAN

In December the mayor unwrapped what he believed would be part of a permanent solution to the disposal issue, announcing three large enclosed barge-unloading facilities in Carteret and Newark, New Jersey, for export of Manhattan and Queens waste, and a third in Red Hook, Brooklyn, for export of Brooklyn refuse. The plan also called for a Staten Island and a Bronx barge- or rail-export facility.[151] Praised by some, the unveiling nevertheless set off a variety of negative responses locally, especially since city officials had not given New Jersey Governor Christine Todd Whitman advance notice.[152] "The proposal being promoted by the city of New York is a direct assault on the beaches of New Jersey and a very real threat to our quality of life," Whitman declared.[153] Her office issued a press release with the headline: "Whitman to New York's Garbage Plan: Drop Dead."[154] Peter McDonough, the governor's director of communication, added, "New Yorkers think they can buy anything. The level of arrogance coming from New York City is just beyond the pale."[155]

Deputy Mayor Lhota took responsibility for the miscommunication and tried to clear up the mess.[156] For his part, Giuliani reacted to the negative responses from the Garden State by arguing that "New Jersey is not the only alternative." He added that, given the potential economic benefits of the disposal business, there were a number of places that would take New York City's refuse. He claimed that it was "easy to get people to take our garbage" but provided no specifics. Giuliani realized that not keeping Whitman apprised of New York City's objectives was a mistake, but his pronouncements still came across as arrogant and self-serving. Officials in Carteret and Newark also felt slighted. Mayor-elect James Failace of Carteret asserted, "New York City apologized for not briefing Gov. Whitman, but no one apologized for not informing us."[157]

The Giuliani solid waste plan was under criticism from many sides, not just New Jersey and the minority neighborhoods of Brooklyn and the Bronx. In mid-December, a citywide alliance of environmental groups pushed for the DSNY to use already existing facilities, particularly marine transfer stations,

FIGURE 17.1 Arthur Kill is a tidal strait separating New Jersey (above) from Staten Island (below). The water body perpendicular to the Arthur Kill in New Jersey is the Rahway River, which divides Carteret on the south (left) from Linden on the north. (The island in Arthur Kill is Prall's Island.) Mayor Giuliani hoped to export New York City's refuse to New Jersey after the closure of Fresh Kills Landfill. Photo taken on February 22, 2010.

Source: Wikimedia Commons.

and also raised familiar environmental-justice concerns. A letter to Lhota stated that the city had "failed to explore fully, in an open manner," the advantages of direct containerization of waste at any of the city's marine transfer stations. This would avoid the problems associated with the barge-unloading facilities. Lhota was not convinced that a conversion of MTS facilities was viable but asked the DSNY to look into it.[158]

At a town meeting on December 14, Giuliani told Staten Islanders, "You've got to show up for meetings, make sure people understand how important [the city plan for closure] is. . . . There are some politicians out there who are trying to frighten the public for absolutely no reason. There's an opposition out there, some politicians, who have tried to make the garbage issue into a nonsensical discussion." He was speaking broadly about opposition to his plan, but he also directly addressed the New Jersey issue. "You've got to understand that garbage means money, lots of money and jobs."[159] In New York City the story was different. As Neil Cohen, a board member of the community group Bay Ridge Against Garbage Sites (BRAGS), astutely noted, "The unstated purpose of the plan is to privatize the city's garbage, to have the facilities built and get the city out of the garbage business."[160]

Not surprisingly, recycling remained outside the major thrust of the plan. Proponents of recycling continued to proclaim that the city's program was undersupported and thus not working very well.[161] A most strident attack came from the *Staten Island Register* in November. It declared that the closure plan "is a hoax, a political sham meant only to garner publicity and votes for people who will be out of office by time [sic] the alleged 'deadline' for closure arrives." The City Council accused Giuliani of interrupting a hearing on recycling to state that he intended to veto its proposed bill. "In what is becoming a disturbingly typical example of Giuliani insults to people with whom he disagrees, the mayor snarled to the Council that citizens who support recycling regard it as 'a religion or supernatural thing.'" The editorial further argued that it was no secret that the mayor opposed the 1989 recycling law and had not been a friend of maintaining or expanding the recycling program.[162] Faced with pressure from several quarters to sustain the recycling program (and to garner support for his disposal plan), Giuliani signed a bill on December 22 that mandated weekly recycling pickups throughout the city within sixteen months.[163]

Despite the heated public controversy over the refuse disposal plan, the DSNY's 1998 *Annual Report* painted a rosy picture about the handling of the department's responsibilities.[164] Mayor Giuliani's introduction proclaimed, "The Department of Sanitation continues to be a source of pride for the City of New York, cleaner today than it has been in many, many years." He especially noted that there had been "milestones" in the reduction of tonnage delivered to Fresh Kills through the export program in the Bronx and Brooklyn, a reduction of odors at the landfill, and the expansion of the recycling program.[165] Kevin P. Farrell, who had replaced Commissioner John Doherty, seconded those remarks.[166]

There was no doubt in the DSNY report that the private-company, export-heavy plan would be implemented. In fiscal year 1998, the Waste Management facility in the Bronx exported 565,508 tons of refuse (1,897 tons per day). In Brooklyn, the DSNY awarded Waste Management a contract in the Williamsburg-Greenpoint area, with exporting to begin in December 1998. The mayor's office also began negotiations with two vendors to use existing marine transfer stations for receipt and transfer of waste. In spring, the DSNY completed the design and permitting of its Fresh Kills Composting Facility relocation. Closing Fresh Kills was moving apace.[167]

EXPORTING WOES AND TRASH TALK

Promises to the contrary, Staten Islanders and others did not have full confidence in the mayor's solid waste plan.[168] This feeling persisted despite the rapid approach of the deadline (or maybe because of it). On March 18, 1999, the Staten

Island Chamber of Commerce displayed a sign in the window of its St. George office: "1,019 days to landfill closure." It planned to count off the time until the glorious end occurred.[169] On April 5, after less than one month on the job, Commissioner Farrell was emphatic that only a meteorite falling on the city would prevent the landfill's closure on December 31, 2001.[170]

A political storm over exporting was brewing in January 1999. New York State already was the largest exporter of waste in the country.[171] In addition, the fight over flow control in Washington was tethered to New York City's future exporting plans and to the closing of Fresh Kills. In an urgent January 14 fax to Deputy Mayor Jake Menges, State Representative Patrick Lally noted, "Things are really heating up here."[172] With the new Congress ready to assemble, Representative Fossella, for example, continued to push against efforts to restrict interstate movement of refuse. "It's the most important issue for Staten Island," he stated, "environmental or otherwise."[173]

More ominous was recent news from Virginia. In his State of the Commonwealth address, Governor James S. Gilmore III stated opposition (supported by state environmentalists) to accepting more refuse from New York City with the pending closure of Fresh Kills. New York City was determined to barge 3,900 tons of waste per day from Brooklyn to Virginia—an increase of 30 percent in total refuse exports. A spokesperson for Gilmore stated, "As governor of Virginia, he will not sit idly by and allow out-of-state garbage to be randomly dumped in Virginia."[174]

In response, Giuliani cited the Interstate Commerce Clause of the Constitution, which had been interpreted to protect the right to ship refuse across state lines. He added that endorsing import limitations was "a very popular political thing to say." Then he went too far. The mayor, in what soon would be taken as a condescending rejoinder, was quoted as saying that Virginia and other states needed to help New York City in disposing of its waste because of the city's status as a national cultural and business center. Giuliani also repeated what he had mentioned on other occasions, that the waste importing business made good economic sense. "So I think when people move past the knee jerk political issue here, there is [sic] whole industry out there."[175]

The *New York Post* reacted immediately to the mayor's alleged response to Virginia officials: "Gotham's garbage problem cannot be solved exclusively by exportation, which is the current plan. Giuliani would do well to acknowledge that. Instead, he has made an argument for exportation that it's hard to believe he actually believes: 'People in Virginia like to utilize New York City because we're a cultural center.' "[176] Several days later, the *Westchester Journal News* chimed in: "New York City Mayor Rudolph Giuliani should take a break from hawking the Big Apple's garbage-export business. Really. He's embarrassing the rest of the state."[177] Believing he was misquoted or misinterpreted, Giuliani attempted to clarify what was regarded as an inflammatory statement: "What goes with being a cultural center is [we're] very crowded. We don't have the room to handle the garbage that's produced, not just by New Yorkers, but by the 3 million more people that come here. So this is a reciprocal relationship. The city . . . brings great benefits to the American economy." Virginians and others heard something a little more chauvinistic in his words.[178]

The mayor soon backtracked, suggesting that Virginia was not obliged to take New York City's refuse. "We'll pay them for it. This is a business. It isn't a question of obligation, it's a very big business. It supplies lots of jobs for people. We have no problem finding people who want to take our garbage in exchange for the huge amount of money we give them for it." Still, he sounded cavalier. "There are many more places that want to do business with us than you could possibly imagine," he added. "I mean, every time somebody makes these [anti-export] statements—some politician trying to get on television—we get five mayors and 10 communities to call us to say don't listen to him, we need your business." He apparently could not resist another dig at Gilmore.[179]

City officials attempted damage control, considering the furor in Brooklyn, New Jersey, and Virginia. The DSNY scheduled a community meeting to discuss the use of marine transfer stations. But without making a commitment beyond the December 1998 announcement to continue using them, Deputy Commissioner Martha Hirst stated, "There's a lot of room for conversation here."[180] Governor Gilmore wrote to Giuliani, taking umbrage with the comments about any kind of obligation to take the city's refuse. Giuliani wrote back, "There is no reason for you to be offended. No one is obligated to accept New York City's garbage. It is a relationship of mutual benefit, entered into freely and voluntarily." In a news conference on January 19, the mayor stated, "I didn't say they were obligated to take the garbage. Somebody put that spin on it, so now their spinning based on the spin and then everything is spun around."[181]

The hole was getting deeper, and the controversy seemed to be giving lawmakers in Washington some ammunition to move the flow-control debate further along.[182] In the meantime, New York City was taking steps to ship waste from Manhattan and Queens to transfer stations, landfills, and incinerators in New Jersey, Long Island, and Westchester County.[183] In Brooklyn's East Williamsburg, the hauling company Eastern Transfer, Inc., called a dumping facility a "shed," while locals called it a "stealth" waste transfer station. The debate over what actually was transpiring in the community now was mired in the courts after an eight-month standoff.[184]

In early February 1999 environmental commissioners in New Jersey, Virginia, Pennsylvania, Maryland, and West Virginia sent a letter to Giuliani, calling his plan "an unacceptable policy." The commissioners wanted the city to "bring balance to the waste marketplace."[185] On February 6, Giuliani gave the keynote address at the Pennsylvania Republican Committee's annual Lincoln Day Dinner. About a dozen protesters (someone dressed as Benjamin

Franklin) waited outside the Harrisburg Hilton and Towers. Jeff Schmidt, speaking for the Pennsylvania chapter of the Sierra Club declared, "Mayor Giuliani should stop exporting his problems to other states. When this Mayor decided to close New York's landfill, he bowed to local political pressure. His answer has been to dump on his neighbors." Earlier, a Democratic state representative led a demonstration at the State Capitol, where fifty pounds of workplace trash had been collected to ship to Giuliani's Manhattan office.[186]

A few days later the Virginia governor reported that illegal medical waste from Brooklyn had been found in one of the state's landfills, prompting the legislature to pass a bill (signed by the governor) preventing the barging of waste to Virginia.[187] This was a more serious hit for New York City's trash plan than the war of words. At this time, Governor Pataki's administration was in quiet discussions with refuse-importing states to avoid a major impasse.[188] The *Rome Daily Sentinel* poignantly declared, "Metropolitan New York's scramble for a location for its trash is reminiscent of the 'garbage barge' in the late 1980s."[189]

Giuliani's ill-advised comments about importing waste to Virginia stirred the pot of what was becoming a full-blown controversy. For the James River Environmental Group and their ally, the Virginia chapter of the Sierra Club, the comments served their ends. "Giuliani was our secret weapon," a local Sierra Club director stated. "His comments provided impetus for our elected officials, although we had been fighting this issue for some time." The Sierra Club gave "I LOVE NY" T-shirts to waste industry lobbyists and sponsors of the legislation.[190]

Despite such a clever demonstration, Virginians and others in waste-importing states did not hold a unanimous opinion about the value of accepting New York's refuse, which in the long run would favor the net exporter. In Sussex County, Virginia, the local landfill that accepted New York refuse accounted for a third of all local revenue. While citizens might be unhappy with the facility as a neighbor, the amount of funds coming to the county was difficult to turn down. Pending legislation over barge traffic could jeopardize it.[191]

In the long run, Mayor Giuliani was correct: New York City would find suitors to take its refuse. An announcement in mid-March declared that the City of Harrisburg, Pennsylvania, would be willing to take a major portion of New York City's waste once Fresh Kills was closed. The outcome still depended on flow-control legislation being debated in Washington, however.[192]

In April the DSNY published a feasibility report on MTS conversion, which focused on alternatives for converting the existing stations for the purpose of containerizing waste for direct export.[193] In what was a surprise announcement during a hearing of the City Council's Environmental Protection Committee on April 29, Commissioner Farrell stated that the DSNY would be seeking new ideas about reconfiguring the MTSs or building new facilities to help with the closure of Fresh Kills. This decision was in direct contrast with the DSNY's stated position that it would not retrofit MTSs because it was uneconomical and time consuming. While some committee members welcomed the apparent change in policy, others wanted more than a solicitation of ideas and demanded a formal request.[194]

On May 28 the city issued a document outlining the scope of the Draft Environmental Impact Statement (Plan EIS).[195] While providing a further commitment to utilize the MTSs, the Plan EIS expanded the area for refuse facilities and thus increased the NIMBY concerns in the boroughs. At an April 13 public hearing on the scoping document at City Hall, eight hundred people showed up to testify (many were from Brooklyn). Questions abounded on why the refuse plan emphasized exporting over recycling and waste reduction. One comment seemed to encapsulate neighborhood feelings: "One conclusion is clear. The New York City Department of Sanitation must go back to square one and develop a waste management plan that is more concerned with the residents of the City of New York than it is with corporations engaged in the business of waste management."[196] A crowd of two hundred at a Brooklyn meeting on April 6, according to the *Advance*, "overflowed out into the halls, as the assembled listened intently to the organizers' agenda: Trashing the Giuliani administrator's plan to build a waste facility in their borough to close the Fresh Kills landfill."[197]

CRACK IN THE REFUSE PLAN

Out-of-state disposal was not the only option some New Yorkers envisioned. Upstate New York seemed to open up possibilities. In March, Guy Molinari wrote to the DEC's John P. Cahill asking for assistance in locating in-state disposal sites. In his mind, these sites avoided the problems the city was facing with other states and, most importantly, would help ensure that the city met the Fresh Kills closure date.[198] Molinari took the lead on this initiative, which was linked to his reputation as New York City's unofficial "trash czar."[199]

DEC records indicated that there were more than thirty municipal and private landfills upstate with sufficient capacity to accept more refuse. There also were ten resource-recovery plants available. Even Mayor Giuliani saw an opportunity to court upstate towns, which might benefit economically from receiving Gotham's refuse (especially if he sought the Republican nomination for the U.S. Senate).[200] Deputy Mayor Lhota told the *New York Daily News*, "The more upstate is involved, it deflates some of the arguments of Pennsylvania and Virginia."[201]

The shipyard on Newburgh's waterfront was being considered as a barge transfer station from where refuse could be taken by rail to various landfills.[202]

Two companies already were trucking small amounts of solid waste from Brooklyn and the Bronx to an incinerator in Peekskill, and Niagara Falls was looking for its share of refuse for its incinerator. Not all places were equally enthusiastic. Attorney Michael B. Gerrard was fighting against the construction of a landfill near Buffalo, where a rural community was trying to maintain its hunting and fishing sites. "We do not want the intrusion of hundreds of thousands of New York City garbage trucks despoiling the natural environment," he argued, "and we are not a regional dump."[203]

The current solid waste plan, despite the positive spin of the Giuliani administration, was springing leaks in mid-1999. Ed Koch launched a particularly vicious verbal assault on the current mayor in a June 25 op-ed piece appearing in the *New York Daily News*:

> Critics of Mayor Giuliani point to his history of curtailing civil liberties, threatening the independence of the judiciary, demonizing opponents and taking credit for other people's achievements.
>
> But the impact of these actions is modest compared with the economic and environmental devastation that will ensue from his support of the 1996 legislation ending the construction of incinerator plants in New York City and authorizing the closing of the Fresh Kills landfill on Staten Island . . .
>
> This ill-advised decision was clearly based on his desire to win votes from Staten Islanders and residents of Williamsburg, Brooklyn, rather than to do what was best for the city. But Rudy made certain that the political fallout from this decision would be delayed until after he left office. *Apres mois le gar-bage*.[204]

Koch's screed was somewhat apocalyptic. While few argued that New York City should bring back incinerators, he more than intimated that unrest over the garbage plan might occur.[205]

The drumbeat continued among those who believed the closure was an impractical dream. An *Advance* editorial noted, "The four-and-one-half-year timetable to close the Fresh Kills landfill has usually been referred to as a process, but the truth is, it has often been more like a minefield."[206] In late June the Giuliani administration announced that all parts of the permanent disposal plan may not be in place by the closure deadline and that emergency powers to stop refuse from heading to Fresh Kills might have to be invoked.[207] Senator Marchi was persistent: "As long as I am in the Senate, I will not allow a bill to pass the Senate which would extend the closure date mandated by law."[208] In *Waste Age*, Kim O'Connell concluded, "One thing is certain. As long as interstate battles continue in Virginia, New York, or anywhere the amount of waste exceeds capacity, Fresh Kills Landfill—or at least the issues surrounding it—never will be fully closed."[209]

Giuliani continued to defend his plan, but he did soften it somewhat out of political necessity. At the end of June, weekly curbside recycling pickups would phase in, replacing biweekly pickups in South Shore and mid-Island communities on Staten Island and in nineteen other districts in Brooklyn and Manhattan. By the fall, collection rates were rising.[210] The DSNY continued to rethink the use of MTSs. The city comptroller also investigated new garbage trucks with removable refuse containers that could reduce or eliminate transfers.[211]

The export battles began shifting in New York's favor by the summer of 1999. In late June a federal judge temporarily blocked two Virginia laws meant to stop refuse imports. The judge also barred an attempt by Virginia officials to include New York City in the hauler's lawsuit. Another court decision struck down the Virginia state statutes blocking imports of refuse in February 2000.[212] And a Kings County Supreme Court judge ruled against a claim by Brooklyn that New York City's contract to ship that borough's refuse out of state rather than to Fresh Kills violated state law.[213]

Things were not so rosy in New Jersey. Officials in Bayonne welcomed a proposal for a transfer station to be located there, although many residents opposed it. Governor Whitman and others continued to express opposition to any cooperation with New York over disposal.[214] In fact, New Jersey police planned to set up a temporary inspection station near Goethals Bridge to halt garbage trucks from entering the state.[215] The conflict escalated when Molinari threatened to start a "borough war" by seeking federal action and suggesting a shopping boycott against malls in Elizabeth and elsewhere in New Jersey if the police enforced the inspections. Elizabeth Mayor J. Christian Bollwage quipped, "I can't imagine any resident of Staten Island giving up a 40-percent discount that they get if they shop in our stores. If I tried to tell my constituents what Guy Molinari is saying, they would laugh at me."[216] The City of Elizabeth then marched to court, but a Union County judge denied an injunction to stop the hauling of trash across the state border.[217] Sounding like a Staten Islander, Mayor Bollwage asserted, "New York is trying to say that we are still a town that can be dumped on."[218]

Neither New Jersey nor Virginia got much relief from Washington's ongoing skirmishes with New York City. For their part, Mayor Giuliani and Governor Pataki chose not to testify on the flow-control bills, stating scheduling conflicts. Spokespersons from refuse-importing states, however, were quick to participate. Senator Charles S. Robb (a Democrat from Virginia) and others pleaded with Congress to act before Fresh Kills closed.[219] In July about fifty House members wanted hearings on a bill to put obstacles in the way of New York's exporting plans, which would possibly threaten the closing of the gigantic landfill.[220] The Staten Island Chamber of Commerce organized a trip to Washington to bring such concerns to Congress.[221] By August a flow-control bill in the House was dead, with no immediate prospects in the new year.[222]

ENVIRONMENTAL RACISM AND ENVIRONMENTAL JUSTICE

Much of the battle over the mayor's solid waste plan revolved around a firm closing date for Fresh Kills, exporting versus recycling, and the siting of transfer stations. This third issue was a major community concern, especially in Brooklyn and the Bronx. It went to the core of who should bear the burden of living with refuse as an ongoing part of their lives (at least within the city borders). The concentration of transfer stations in poor and minority communities was a symbol of neighborhood degradation and a source of environmental risk from never-ending truck traffic and diesel fumes. More than half of the city's private transfer stations were in two neighborhoods: the South Bronx had thirteen and Williamsburg-Greenpoint seventeen. Sunset Park–Red Hook in Brooklyn also had several.

The issue was not new. In the late 1980s tipping fees at Fresh Kills increased for private carters, which put additional pressure on the transfer stations. A presumed step in the right direction occurred in 1990 when the city passed Local Law 40, mandating that the DSNY develop siting regulations. The existing stations, however, still were heavily concentrated in too few communities. With the announcement of the closure of Fresh Kills, OWN formed, and it planned many demonstrations and public events and testified at numerous hearings.[223]

For communities like Red Hook, which were trying to make an economic comeback, such problems were truly frustrating. In 1998 the real estate market was picking up. Light industry was moving to old restored piers. New retail stores were opening. One block from the proposed transfer facility was a new esplanade. A local resident stated, "It seems shameful that the city would spend all this money to lift up the community and make people feel good and then hit them with a lightning bolt like this."[224]

In March 1999 White House and other federal officials came to New York City to determine if transfer stations located in the South Bronx (blamed for causing widespread respiratory and other health problems) offered evidence of environmental racism in the predominantly African American and Latino neighborhoods. The federal action was in response to Representative Jose E. Serrano (a Democrat from the South Bronx), who was frustrated with the EPA's delays and dismissals in reacting to previous complaints. The transfer stations, sewage-treatment plants, and automotive shops in the South Bronx, he asserted, increased truck traffic and diesel fumes in the area. The federal government ultimately acted on Serrano's grievance based on a civil rights interpretation of recent environmental-justice legislation.[225]

This was the first inquiry of its kind conducted in New York State. About fifty officials, sponsored by the White House Council on Environmental Quality (CEQ) and led by representatives of the EPA and the Department of Justice, took a bus tour of the South Bronx and parts of Brooklyn, visited transfer stations, and talked with locals.[226] In Red Hook, a protester held up a sign—"Stop the Garbage"—which summed up the communities' feelings. About the same time as the tour, a hearing on the issues was underway at Columbia University's School of Public Health, to permit locals to air their complaints.[227]

The federal investigation began in September and required a few months to complete. Similar inquiries had taken place in Louisiana, Michigan, and elsewhere. Bonnie Bellow, speaking for the EPA, stated, "The waste-transfer issue is difficult because we don't have a full understanding of their impact. This is kind of new territory."[228]

Congressman Fossella was neither pleased with the tour nor with the hearing at Columbia. "I find it outrageous that they convened a meeting with all these high-level officials and they didn't come to see the city's worst environmental nightmare, the Fresh Kills Landfill." He also rejected the notion that environmental justice should focus only on communities of color, adding, "I believe that this country stands for equality for all. If something adversely affects someone, it doesn't matter if they are black, Hispanic or white. If it is bad for one, it is bad for all."[229] Guy Molinari, defending Fossella's stance, described the environmental-racism position of the federal government as lacking and was concerned that the recent query threatened the closure of the landfill.

In response to the federal probe in the Bronx and Brooklyn, Representative Fossella filed a federal countercomplaint charging that the landfill operations at Fresh Kills violated the civil rights of Staten Islanders. "The very existence of Fresh Kills," he stated, "is the best proof that Staten Islanders have been discriminated against in connection with New York City's disposal of residential trash."[230] While Fossella had a reasonable point about the impact of Fresh Kills on locals, he also pushed his claims in a partisan direction at a March 10 press interview, suggesting that the federal probe in the Bronx was a "political escapade" designed to boost the presidential campaign of Vice President Albert A. Gore, Jr. "Where has the EPA been for 50 years," Fossella stated, "when we've been raging on Staten Island about the abuses we've suffered and the environmental injustices that have resulted from the Fresh Kills landfill?"[231]

Fossella then invited Gore to tour the landfill. In response, the White House supported a federal hearing on an environmental-justice investigation of Fresh Kills. There was some question if such an investigation would take place, and ultimately the White House denied that a promise was ever made.[232]

Emotional pleas from the Staten Island Chamber of Commerce led to a visit to the landfill by Bradley A. Campbell, the head of the CEQ.[233] But Staten Island was never made part of the federal environmental-justice probe in early 1999. The CEQ chose cities and communities that had large concentrations of minorities who had borne serious environmental burdens.[234]

In a January 2000 letter to Fossella, Anne Goode, the chief of the EPA's civil rights office, told the congressman that she viewed his call for a probe inadequate. It did not, she argued, suggest how the continuing operation of the landfill constituted discrimination against a specific minority group.[235] In the 1990s, environmental justice was interpreted by its supporters and federal executive-branch authorities as essentially a civil rights issue. For good reason, it was so. Fairly or unfairly, it was difficult to make an argument that siting the landfill on predominantly white Staten Island could violate the civil rights of the majority population.

It is unfortunate that Fossella's spotlighting of Fresh Kills as an example of environmental injustice devolved into partisan politics and was wrapped up in the desire to close the facility on schedule. About fifty thousand Staten Islanders lived within one mile of the landfill.[236] Fossella clearly was standing up for his constituency, but he did not dismiss the problems of the Bronx and Brooklyn in doing so. He did believe that it was hypocritical to take the grievances of those in Red Hook seriously but not similar ones on Staten Island. He also argued that the application of environmental justice should be color blind, but there was no precedent for this position at the time.[237] It was an unresolved quandary then and now.

The emphasis on environmental justice at the level of the federal government was focused on the intersection of environmental racism, civil rights, and poverty. In this context, Fresh Kills was an anomaly. A more inclusive definition of environmental justice and how it might operate more widely in society was just not a national concern in 1999. Fresh Kills was an environmental problem, but just how it might be linked to the citizens of Staten Island did not extend much beyond NIMBY considerations.

A NEW CENTURY

Mayor Giuliani was right when he stated in the DSNY's 1999 *Annual Report*, "Once the Landfill closes, a new era in your Department will begin." In October 1999 the "third phase" of the Fresh Kills closure began. For the first time, DSNY collection trucks were used to haul refuse to stations outside the city limits, and the department was formulating plans for post–Fresh Kills disposal. In 1999 the Staten Island landfill received only 9,272 tons of refuse per day (sizable, but shrinking), and attention continued to focus on gas control and recovery, leachate prevention, composting, and dredging for barge mobility.[238] Section 6/7 was closed ahead of schedule on June 18, 1999, and solid waste was no longer being accepted at the Plant 2 Unloading Facility, with only 150 acres of the landfill continuing to receive waste.[239]

There was a new solid waste management plan—again.[240] Coming a year and a half after the last one, the newest version still emphasized exporting waste. But problems remained in dealing with importing states such as Virginia, Pennsylvania, Ohio, and towns in upstate New York.[241] About seven thousand tons per day were being dumped in Virginia landfills, at an incinerator in Newark, and elsewhere.

The Giuliani administration gave up on trying to sell its exporting plan as economical. In the long term it would cost the city about $323 million per year, which was $90 million more than projected. Some critics charged that setting a firm date for the closure of Fresh Kills meant that the mayor had lost negotiating leverage with private haulers. But Deputy Mayor Lhota argued that the latest plan was being managed differently than the earlier one and relied upon keeping the parties informed of what they intended to accomplish.[242] Despite opposition from the Bronx, final approval for the new plan passed the City Council on November 29, 2000.[243]

The proposed long-term export plan—which applied to the 40 percent of the city's waste managed by the DSNY—included several elements. Exporting refuse would rely on rail and barge transport, rather than trucks. Several of the city's marine transfer stations would be used in transfer operations. Focus remained on borough self-sufficiency. The projection for recycling was a 25 percent diversion rate of the total residential waste stream by 2002 (estimated at about 20 percent in 2000). Nonrecycled waste that was exported would be handled through twenty-year contracts with private vendors.[244]

Raising diversion rates was no easy task, especially given the uneven performance of recycling throughout the boroughs. Estimates for June 1998 pegged the citywide diversion at 17 percent. In Manhattan, it was 21.6 percent; on Staten Island, 21.2 percent; Queens, 19.2 percent; Brooklyn, 16.5 percent; and the Bronx a low of 14.8 percent.[245] In all, there were fifty-nine community districts, some that were lower-diversion areas and others that performed much better. In Manhattan's District 1, the diversion rate was 31.1 percent; the rate was only 6.8 percent in District 3 in the Bronx.[246] Because the districts were unique places in many ways, a one-size-fits-all strategy was doomed to fail. The DSNY had to struggle with that reality in moving forward with its recycling and waste prevention programs, even though its public statements about recycling reflected an optimistic outlook.[247]

For people in Brooklyn and the Bronx, the most popular decision made under the plan was to abandon construction of new transfer stations in Red Hook and Hunts Point. Local pressure was a major factor. The project in Carteret, New Jersey, also was junked. The city instead intended to rely on five of its smaller MTSs in Manhattan, Brooklyn, and Queens and to build an enclosed barge-unloading facility (EBUF) in the blue-collar town of Linden, New Jersey (population of 37,000 and not far from Fresh Kills).[248]

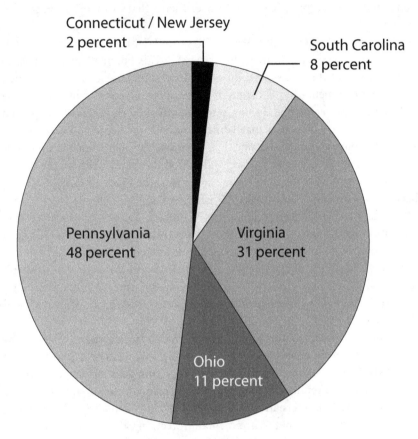

FIGURE 17.2 Where New York City refuse would go by 2010.

Source: City of New York, Department of Sanitation, *NYC Recycles, NYCHA-wide Residential Recycling Program* (May 2015).

After being placed in sealed containers, the refuse would be shipped by rail to landfills in Illinois, South Carolina, Georgia, and other states, under the supervision of Allied Waste Industries. Two other MTSs would be converted into compacting and shipment centers. Unlike the slight over the Carteret facility, New York officials consulted Linden in advance. Although some citizens were dubious of the decision, supporters looked forward to the sizable revenue and to the fact that the facility would be barge-to-rail instead of a trucking site. Mayor John T. Gregorio, noting that the port would be about two miles from the closest residential neighborhood, stated, "You won't see a truck. You won't smell it. I think we can ensure that our taxes won't be going up for years, and maybe they'll decrease."[249] "The area we're talking about," Councilman Richard Gerbounker stated, "has been contaminated by heavy industries.

You're not going to put pristine industries there. That's why the garbage is so attractive."[250]

Those most affected by the new facilities (the Tremley Point neighborhood) were the most emphatic in their disapproval.[251] Talk about a hazardous waste incinerator and the transfer facility found no support. Concerns remained among environmentalists, some property owners, and communities along the rail path in New Jersey. The Union County Board of Freeholders (who approved the transfer station) stated that while they had listened to objections, the interstate-commerce rules dictated their decision. The board's chair Daniel Sullivan asserted that a glut of trucks would flood the streets without the barge-to-rail system: "We're being told you can't reject [the Linden facility] basically because you don't like New York State garbage."

Officials and residents in neighboring Middlesex County opposed the transfer station. It would receive no revenue, but trains with refuse would travel past homes in a number of towns, including Carteret, Woodbridge, Edison, Metuchen, South Plainfield, Piscataway, and Middlesex Borough.[252] Metuchen's mayor Ed O'Brien queried: "Why doesn't Giuliani send his garbage through New York State? Why not barge it all the way down to Virginia? They say it will cost too much? Well, that's not our problem."[253]

While the transfer stations were to be moved to largely receptive new locations, the city's solid waste plan remained heavily dependent on exporting waste elsewhere.[254] While some praised the new recycling goals at public hearings, others remained skeptical.[255] Some wanted the plan junked altogether, but this was never a credible alternative. In late October, City Council Speaker Vallone, along with Committee on Environmental Protection Chair Michels, presented council members with a laundry list of ideas to help devise a plan that managed the city's waste "in the most cost-effective, efficient and environmental sound way." The report outlined a range of issues, including commercial waste management, recycling and waste prevention, and the city ownership of solid waste facilities. Commercial waste management seemed to be the largest question not fully addressed.[256]

The DEC provided detailed comments about the new plan, rather than suggesting big-picture revisions.[257] Some of the harshest criticisms came from those fearful of seeing a public-sector service become privatized, worried about what this meant especially for the city's DSNY employees. An overview prepared by Program Planners, Inc. (apparently speaking on behalf of the Uniformed Sanitationmen's Association, Local 831 IBT), concluded, "New York City is clearly putting itself in the grasping control of private sector entities accused over the years of environmental degradation, price fixing, restraint of trade and other illegal practices."[258]

A key concern was the overall cost of the city's plan. Benjamin Miller, the director of the DSNY's 1992 plan and author of *Fat of the Land* (on New York

City's refuse history), warned that the cost of disposal could rival the cost of collection—a new phenomenon. "Disposal costs," he said, "are going to dominate the budget."[259] He was essentially asking if closing Fresh Kills was worth it. Others feared that the long-term contracts offered private firms for construction or exporting services would lock the city into high payments.[260]

A tough critique of the new solid waste plan came from a report written by Barbara Warren under the auspices of OWN and the Consumer Policy Institute/Consumer Union. A longtime activist, Warren was currently director of the New York Toxics Project at the Consumer Policy Institute/Consumers Union. She had been involved in waste management issues in New York City since 1980, especially on Staten Island. "This closure [of Fresh Kills]," she stated,

> presents the City with both an opportunity and a danger. The City has an opportunity to rethink and restructure the City's solid waste system in an equitable and sustainable manner, which reuses and recycles valuable resources creating economic development and jobs. The danger is that the choices made now are ones that will impact the City for decades, and the wrong choices could lead to financial, social and environmental difficulties which could be severe in the years ahead.[261]

Complaints over the new plan were well understood by Staten Islanders but firmly within the context of their own historic interests. Their mantra remained the same: Will Fresh Kills close on time, and to what extent will the new solid waste plan foster that closure? On May 23, the *Advance* quipped: "Staten Islanders have lost count of the number of times politicians have summoned the press to announce that the Fresh Kills Landfill will close on Jan. 1, 2002." It was responding to the agreement struck by Mayor Giuliani and Governor Pataki the previous week for the City of New York to withdraw its permit to operate the landfill (under the initial consent order) and to draw up a "less stringent" set of milestones and a system of weak fines if the city failed to meet the goals.[262]

Senator Marchi praised the agreement but in a letter exhorted Pataki that in future declarations "you give yourself credit for the *law* you signed in 1996 to make that Fresh Kills closing *mandatory*." "Your approval of that law," he added, "has to be rated one of the most important environmental protection actions ever undertaken by a governor of New York, or, for the matter, any governor in American history."[263] For supporters like Marchi, the agreement signaled one more affirmation of the closing of the landfill—"the final nail in the coffin," noted DEC Commissioner Cahill. To skeptics it signaled possible lax standards in maintaining the disposal facilities, and rerouting refuse disposal created problems such as noise and congestion along truck routes and stirred new fears about air pollution.[264]

Nonetheless, there was great anticipation in the air for the closing. The *Advance* led the way: "Is it possible? Could the Fresh Kills landfill, the enormous city garbage dump that has plagued Staten Island for 50 years, actually be closer—perhaps much closer—to shutting its gates for good than anyone could have hoped?" Fossella, Straniere, and others expected an early closing date. The DSNY stated they were ahead of schedule. Deputy Mayor Lhota was more cautious.[265] By December 1999, the amount of refuse diverted to Fresh Kills had declined to five thousand tons per day (down from 8,500 a year earlier).[266] The countdown was on.

Yet the *New York Observer* raised the question (not unlike Ben Miller): What's the rush? "The state and city are desperate to please that Republican outpost known as Staten Island. . . . Instead of floating the junk on giant barges to Fresh Kills, we're now using trucks to take it elsewhere. Ironically, the result has been an environmental and traffic disaster," it stated. "Let's stop the clock and figure out exactly what we're doing. Otherwise, Staten Island's pain will not have been eliminated. It simply will have been transferred."[267]

The momentum to close Fresh Kills was just too great for this kind of reflection to be taken seriously, despite the possible future implications. Steve Violetta, who had worked for thirty years at the DSNY (the last eleven at Fresh Kills), had a different take: "I will definitely miss the place. Once Fresh Kills goes, it's the end of an era."[268] Oh, the contradictions that the landfill inspired.

PART VI
THE POST-CLOSURE ERA

CHAPTER 18

9/11

At 8:46 a.m. and again at 9:02 a.m. hijacked jetliners originally bound for Los Angeles deliberately slammed into the North and South Towers of the World Trade Center (WTC), located in Lower Manhattan. A third plane crashed into the Pentagon in Washington, DC, at 9:40 a.m. A fourth crashed near Shanksville, Pennsylvania, without striking a targeted building, thanks to the heroism of several brave passengers.[1]

In testimony given later, Chief Joseph Pfeiffer, Battalion 1 of the New York City Fire Department (FDNY), stated,

> In Manhattan, you rarely hear planes because of the high buildings. So we all looked up. In almost disbelief, we see the plane pass, and it's flying so low. Our eyes followed it as it passed behind the buildings, and then it reappeared, and it appeared to me that it aimed right into the building. It smashed into the building. There was a large fireball. And then a couple of seconds later you heard the sound of the explosion. I told everybody to get in the rigs because we're going down there, to the Trade Center.[2]

The total number killed in the New York attack: approximately 2,750. The number of firefighters, paramedics, and police killed: more than four hundred. Passengers and crew killed on American Airlines Flight 11 and United Airlines Flight 175, which hit the Twin Towers: 157. Pentagon personnel killed: 128. Passengers and crew of American Airlines Flight 77, which hit the Pentagon: 64. Passengers and crew of United Airlines Flight 93, which crashed in the Pennsylvania countryside: 44. The number of nations whose citizens were killed in the attacks: 115.[3]

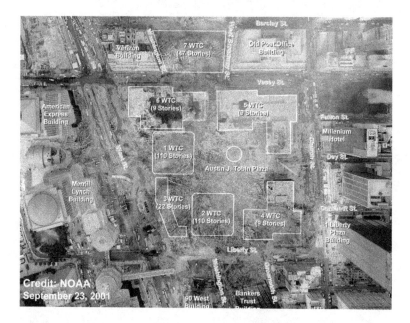

FIGURE 18.1 Ground Zero with overlay showing the original building locations.

Source: Wikimedia Commons.

After al-Qaeda struck the towers, they stood for only fifty-six and 102 minutes more, respectively. They then collapsed in seconds.[4] While the shock, grief, and anger would linger for much longer, reality set in almost immediately. Local authorities were confronted with an urgent search for survivors, the horrific task of recovering human remains, assessment of the crime scene, and a massive cleanup. Little did anyone realize when the last barge docked at Fresh Kills in March that an unforeseen catastrophic event a few months later would force the landfill's further service.

THE LAST BARGE

Six months earlier, on March 22, 2001, the final barge of refuse embarked for Fresh Kills Landfill. On a gray Thursday morning at about 8 a.m., D.S. 123, loaded with 505 tons of refuse, started its five-hour trek from its berth at the Marine Transfer Station in Flushing Bay, Queens. The spruced-up blue barge, measuring twenty-five by one hundred feet, was draped in colored ribbon and sported a sign: "Last Barge." Five tugboats escorted D.S. 123, led by one crewed by Jerry Joyce, who had been transporting refuse across the harbor for twenty-five years. In a separate tugboat three newspaper reporters, two photographers, and a television cameraman recorded the event for posterity. As the flotilla passed the mayor's home at Gracie Mansion on Manhattan's East Side, Rudolph W. L. Giuliani and Deputy Borough President James P. Molinaro waved in approval.[5] The mayor later said, "We waved to it and they signaled to us and it was really absolutely terrific." Staten Islander Tommy Napoleone, working on the escort tug *Justine McAllister*, was less enthusiastic: "I'd rather not see it [Fresh Kills] close—for the jobs."[6]

In dramatic prose, Heidi Singer of the *Advance* recounted D.S. 123's journey:

Exhaling methane, she pushed past the Statue of Liberty, playing chicken with the Staten Island Ferry in their tub of sloshing bay water and getting a salute from the Fireboat Firefighter, ceremonially spraying water as she passes. She creeps along the New Brighton shore, past burnt-out houses and mounds of dirty white salt. She slides past bare tree trunks, the Greek columns of the Snug Harbor Cultural Center and the neon of a waterside bar.[7]

Standing together at the end of Fresh Kills Creek to witness the arrival of the barge, Mayor Giuliani (who had come from Gracie Mansion), Governor George E. Pataki, and Borough President Guy V. Molinari prepared to commemorate the event.[8] Also in attendance were Deputy Mayor Joseph J. Lhota, Department of Sanitation (DSNY) Commissioner John J. Doherty, and Department of Environmental Conservation (DEC) Acting Commissioner Erin M. Crotty.

Over its fifty-three-year history, more than four hundred thousand barges had deposited their contents at Fresh Kills. "Five years ago, I promised the residents of Staten Island that the City would stop shipping its garbage to this landfill," Giuliani intoned. "Today, I'm proud to announce the fulfillment of that promise. As Staten Island continues to thrive and grow, its residents can take pleasure in knowing that the world's largest landfill has taken on its last shipment. Staten Island's future has never looked brighter." Governor Pataki added, "In 1996 we promised New Yorkers and the residents of Staten Island that we would close the Fresh Kills landfill and end the environmental nightmare that has burdened this area for more than 50 years." Thanking Giuliani and Pataki for helping "slam the door on what has been the most notorious environmental burden in Staten Island's history," Molinari proclaimed, "This is a glorious day for Staten Island." He added that the closure was "by far, the highlight of my 27-year political career, and perhaps the greatest victory in our Borough's history."[9]

A few days earlier, Kirk Johnson of the *New York Times* declared: "The scarred landscape of the Fresh Kills landfill in Staten Island is at rest." Calling it "a baby boomer's landfill," viewed by environmentalists as "a giant accident," he also asserted, "Few will mourn the passing of Fresh Kills. Its

FIGURE 18.2 Last load to Fresh Kills, March 22, 2001.

Source: NYC Department of Sanitation.

name on too many 10-worst lists, its stench too great a symbol of all that Staten Islanders endured."[10]

FRESH KILLS' LAST DAYS

Fresh Kills closed with an estimated 150 million cubic yards of refuse buried in its four main sections. Two sections had been permanently capped, one was being prepared for its final cover, and the fourth received the last shipment from D.S. 123. (A proposed transfer station in Travis was to continue to process the island's refuse there.)[11]

The DSNY's budget jumped from $590.5 million in 1997 to $838.3 million in 2000. Closing Fresh Kills was likely to raise it to $1.1 billion by 2005. Political promises had been fulfilled to terminate Fresh Kills, but this alone did not guarantee success for a new disposal program. As a temporary measure, until the city implemented a proposed long-term refuse-export plan, the DSNY began an interim export program beginning in July 1997 by trucking waste to incinerators in Long Island and New Jersey or to transfer stations in the city and New Jersey. From the stations, private vendors using trailer trucks carried the refuse to incinerators or landfills in Connecticut, Ohio, Pennsylvania, South Carolina, and Virginia. Former DSNY director of policy planning Benjamin Miller suggested that the city should keep part of Fresh Kills open in case hauler contracts dissolved. "We could effectively be held hostage by the waste management industry," he cautioned. Giuliani's response: "It's closed. Gone. Over. Environmentally, it's a disaster and has been for some time."[12]

Miller's comments were an elaboration of those he had made in January, which acknowledged that Fresh Kills had more remaining space for dumping than all but five other landfills in the United States. "The loss of New York City's last waste disposal facility," he argued, "will make the city dependent on the kindness (or greed) of strangers for the disposal of its 13,000 tons of garbage a day, perhaps forever." Echoing other critics, he added that the city had lost bargaining leverage for out-of-state disposal when the New York legislature passed a law "making Fresh Kills illegal" and "outlawed" the Brooklyn Navy Yard incinerator already approved by the City Council. He surmised:

> At a cost of $6 billion—the price of closing Fresh Kills 20 to 50 years before it would be filled—Staten Islanders will no longer bear the embarrassment of being associated with the world's largest landfill. It is not clear, however, how much the quality of their air, water or land will be actually affected, since the mounds in their midst will still produce landfill gas and leachate for another generation or two.[13]

Staten Islanders, flush with the promise of the landfill closure, could not or would not accept Miller's assessment. However rational his argument might have been, the reopening of Fresh Kills was unthinkable.[14] Hope was high that the stench that lingered would disappear eventually, but uncertainty remained as to what closing the landfill actually would mean to Staten Island. Skeptics were plentiful. Others, however, anticipated the possibility of a real estate windfall in areas around the facility.[15] Cesar Claro, head of the Staten Island Economic Development Corporation, asserted, "The bottom line here is that without the dump Staten Island's image is greatly improved. That's important to businesses that want to set up shop here." Lawrence DeMaria, president of the Staten Island Chamber of Commerce, added, "The dump has always been the background noise that confused people. Without the background noise, things look a lot better to people."[16]

News of an early shutdown of the facility (in July 2001 rather than on December 31, 2001) also pushed aside talk of a reopening. *Waste News* reported on January 15, 2001: "A near-crisis mood has surrounded the closing of New York City's Fresh Kills landfill for some time now. So it's a pleasant irony that what was once the United States' largest landfill now apparently will close well ahead of schedule." In his State of the City address the week before, Mayor Giuliani announced that Fresh Kills would close July 4.[17]

Implementing a new city disposal plan would not be easy. Many of the tensions over exporting, flow-control legislation, neighborhood pollution, and recycling lingered. Heavy truck traffic moving refuse throughout the boroughs was a reminder of what might become more typical after Fresh Kills.[18] Initially, Giuliani's so-called interim plan would depend on truck traffic to public (marine) and private (land-based) transfer stations before export to New Jersey, Pennsylvania, and Virginia. There also would be some enhancements in recycling but not an elaborate program. By 2005 the city intended to have ready five public marine transfer stations in Manhattan, Brooklyn, and Queens (equipped with compactors) to ship by barge about half of its refuse (6,500 tons/day) to the private waste transfer station in Linden, New Jersey, and to ship by rail or barge the remaining half to out-of-state refuse disposal facilities.[19] Community groups in the affected borough neighborhoods commented, watched, and waited.[20]

The Linden deal was under investigation (Mayor John Gregorio purportedly used undue influence in securing the waste transfer station's land, which was owned by his son-in-law) by the New Jersey attorney general, and there was concern that it would drag on for some time. Browning-Ferris Industries (BFI) was to construct the enclosed barge-unloading facility (EBUF), but not without facing its own legal disputes. In September Garden State officials formally delayed plans for the facility.[21] The flow-control debate droned on in Washington, while tensions over exporting New York City refuse had yet to ease, let alone be resolved.[22] Nevertheless, as Susanna Duff of *Waste News* stated in August 2001: "Municipal solid waste shipments crossing state borders continues to rise with no signs of stopping." In mid-2001, New York State was the leading net exporter of MSW in the country by a large margin.[23]

WHAT TO DO WITH THE WASTESCAPE

Well before the last barge marked the end of Fresh Kills Landfill—and before the events of September 11 complicated the question of its closing—what to do with the gigantic wastescape most immediately and into the future was a knotty problem. Most Staten Islanders wanted to blot the landfill from their memories. There were practical concerns to deal with, including the covering of the last-used mounds, leachate control, methane monitoring, and a wide array of shutdown requirements. Looking farther down the road, nothing had been decided on whether or how the vast acreage would be reclaimed and utilized. In March 1993, as noted in chapter 16, the environmental committee of Staten Island's Community Board 2 invited Bill Young, a DSNY landscape architect, to discuss the beautification of the waste complex and adjoining land. He discussed landscaping choices for retired mounds, other projects underway at former landfills, and potential uses for Fresh Kills once it was shut down completely. Potential uses were limited, given the scale of the facility, but Young promoted the idea of the space as a wildlife sanctuary or as an extension of the Greenbelt.[24]

The DSNY experimented with a variety of beautification efforts and barrier zones over the years and also regularly considered future uses for the landfill space (with some end-use plans going back as far as the 1980s). In 1998, two years after the closure announcement, sanitation officials notified the mayor's office about the "Fresh Kills Landfill Study." The New York Department of State had awarded a grant of $200,000 (under Title II of the Environmental Protection Fund) for the two-year study to be conducted jointly by the DSNY and the City Planning Department. Once the landfill closed, the DSNY would be required to manage the site for thirty more years to monitor environmental conditions.[25]

It was clear that the entire landfill would not be ready for reuse for many decades, but parts that had been closed and capped might be suitable for "interim uses." "More important," a December 30 DSNY memorandum to the mayor's office stated, "a conceptual reuse plan may influence engineering plans for closure of the landfill so that it will be suitable for proposed uses." It made sense, therefore, to consider long-range plans for the site while the closure process was being finalized. On an optimistic note, the memorandum added:

> The Fresh Kills Landfill represents a unique opportunity to create an asset of great value to Staten Island and the region. Terrain that was once considered a community blight could be transformed from an eyesore into a beautiful landscape of native vegetation providing a myriad of experiences. These might include sweeping views from the tops of the hills extending out to Raritan Bay; trails for hikers, birdwatchers, bicyclists and horseback riders; playing fields, courts and amusement areas; boating in the creeks and tributaries; and diverse ecological communities in restored wetlands and streams. New industrial and commercial enterprises could be developed in nodes along the periphery.

The DSNY and City Planning eventually included several other departments and agencies in the project and set up an advisory committee. The completed study also would provide "a sound foundation" for a design competition that the nonprofit Municipal Art Society had proposed in partnership with the borough president, surrounding communities, and other groups and organizations.[26]

Plans for a "second career" for landfills had been and were being considered elsewhere in the country. Kim O'Connell from *Waste Age* suggested, "Increasingly, closed landfills are being converted into recreational areas such as parks and golf courses." For example, the Danehy Park landfill in Cambridge, Massachusetts, was opened as a recreation area in 1990, with playing fields, a fifty-acre park, and open areas. The conversion was not easy, however. The ventilation of gas, the unpredictable nature of waste decomposition, the settlement of the landfill's soil, and other issues had to be addressed in order to repurpose the site. O'Connell concluded that the Danehy experience could be a model for Fresh Kills but that the gigantic scale of the Staten Island facility posed its own challenges.[27]

For his part, Staten Island Borough President Guy Molinari set up a "Landfill Fresh Start" committee in March 2000, to investigate redeveloping the landfill after it closed.[28] As a way of easing congestion, Molinari also instructed traffic consultants to develop plans for a new highway that would connect the existing roads used for garbage trucks. Talking with people on the street elicited a variety of other opinions about the future use of the landfill, including such things as "organic teaching gardens," "a giant piece of land art," "a fabulous zoo," "a really old-style park," "a giant new museum," and a new golf course. But "no more tract homes. Staten Island has enough." More facetiously, one person called for a day prison for cell phone abusers, and the singer Bette Midler chimed in, "My first thought is that they should build a huge municipal building to house all the politicians who are not friendly to the environment."[29]

In the fall of 2000, Executive Director of City Planning Andrew S. Lynn announced that an international competition for landscape-design experts would be held. The first step was to solicit individuals to be the professional advisers overseeing the event.[30] As closure approached, the Department of City Planning announced preliminary procedures for the competition. Soon thereafter, William H. Liskamm, an architect and environmental planner from the San Francisco area, was named coordinator of the "Fresh Kills Landfill End Use Conceptual Master Plan Design Competition," to be carried out over the following eight months. Liskamm had led more than forty such contests, many for the National Endowment for the Arts. After the proposals were submitted, a technical advisory panel would evaluate them for engineering, environmental, and financial feasibility. The findings would then go to a design jury. The announcement of the winner was anticipated in mid-November 2001.[31]

City Planning held public sessions and forums on Staten Island beginning in late March 2001 to get input for the eventual design of the post-closure project.[32] Councilman James S. Oddo (a Republican from mid-Island) was selected to provide opening remarks at a panel discussion held at the College of Staten Island in May. He observed, "It is a little irony that for the longest time no one outside of Staten Island cared about Fresh Kills. Now folks all over the country are coming up with ideas for its post use. While I welcome this interest, Staten Islanders should have a say in what becomes of it. I encourage the community to come out so Staten Island has a clear voice in the process."[33] Throughout the post-closure period, this issue of Fresh Kills' ownership among Staten Islanders, with its odd love/hate quality, would show up repeatedly and permeate the discussion of the gigantic wastespace's future.

GROUND ZERO

The momentum and euphoria around the closure of Fresh Kills was sidelined after the attacks of September 11. In late spring 2002, a committee of the American Society of Civil Engineers (ASCE) issued a report on the reasons for the collapse of the World Trade Center building complex (seven buildings completed in 1973).[34] The Twin Towers (110 stories each) were designed for lightness. They essentially were cylinders reaching 1,362 feet (the South Tower) and 1,368 feet (the North Tower). Both were built on a sixteen-acre site and anchored in the island bedrock for strength. Architects designed them to withstand winds of 150 miles per hour, and they could sway slightly because some give was engineered into the structures. The total weight of each tower was approximately 600,000 tons. The South Tower might have survived the crash had it not been for the intense fire caused by the aviation fuel of the crashing Boeing 767, which ignited vast amounts of flammable material and degraded the steel supports. Investigators had less precise information about the collapse of the North Tower, but it suffered from a colossal fire as well. Other buildings in the complex also collapsed.[35]

The rubble from the destruction was strewn all over Lower Manhattan—from concrete to twisted steel, insulation to broken glass. Ash and dust covered everything. Fires broke out in many places in the general vicinity. *Atlantic Monthly*'s national correspondent William Langewiesche wrote about collateral damage beyond the towers:

> There was wider damage, of course, and on the scale of ordinary disasters it was heavy.... Massive steel beams flew through the neighborhoods like gargantuan spears, penetrating subway lines and underground passages to depths of thirty feet, crushing them, rupturing water mains and gas lines, and stabbing high into the sides of nearby office towers, where they lodged. The phone system, the fiber-optic network, and the electric power grid were knocked out. Ambulances, cars, and fire trucks were smashed flat by falling debris, and some were hammered five floors down from the street into the insane turmoil erupting inside the World Trade Center's immense "bathtub"—a ten-acre foundation hole, seventy feet deep, that was suffering unimaginable violence as it absorbed the brunt of each tower's collapse.[36]

There was profound unease about potential environmental risks caused by burning paint, solvents, lubricants, plastics, and fireproofing materials. Asbestos was not used as insulation for the steel beams, but elevated levels of asbestos were discovered at the site, possibly from flooring. There also was worry that the intense heat could generate dioxins from the burning of PCBs found in light fixtures and other equipment. The scale of the disaster was unprecedented. Eduardo Kausel, a civil engineering professor at MIT, stated, "No building this high has ever collapsed, so there really is no comparison."[37]

RECOVERY AND CLEANUP

Initially chaos ruled at Ground Zero (sometimes referred to as "the Pile," "the Hole," "the Crater," or "the reckoning").[38] Cleanup went slowly and without much coordination at first because of the search for survivors. Firefighters, police, federal-agency personnel, and numerous volunteers worked through the nightmarish landscape by hand and in bucket brigades, knowing that time was of the essence. Twenty-one people were found alive on the first day, but none thereafter. Saying that "only" so few were rescued, however, diminishes the importance of saving that fortunate group.[39] A "mole" (a rescue worker who climbed into the rubble looking for bodies) marveled at the size of the debris piles but was annoyed with reporters looking for stories. "They asked me if I found any bodies, and what do they want me to say? I'm not going to talk about that.... If you see me on TV, talking to a reporter, you can be sure I'm making everything up."[40] Sanitation Commissioner Doherty observed that some of the workers at Ground Zero were "seriously traumatized." "That was a horrible scene down there," he stated. "There were things that people don't normally see unless they're in combat."[41]

The debris was separated into four different sectors at the disaster site. The sectors were cut off from one another and initially difficult to reach by heavy equipment. Cleanup was slowed not only because of the recovery mission but because workers had to reinforce a wall that kept out water from the Hudson River. The wall was necessary because half of the World Trade Center site was located on fill land extending into the river, and it had been weakened by the collapse of the buildings.[42] As a story in the *New York Times* ominously noted:

> In a part of New York where much of the land near the Hudson and East Rivers was created with landfill over the last three centuries, and where the ground below the streets and buildings is a latticework of train tunnels, telecommunications cables, and sewer and power lines—the city simply does not know yet whether it is safe to allow anything like a normal return to the tumult of trucks and trains that was commonplace before.[43]

Within days, more than one thousand emergency and construction workers began clearing the wreckage. Private companies (Bovis Company, Turner Construction, Tully Construction, and Amec Construction Management) were hired, and they mobilized a substantial amount of heavy equipment for the job.⁴⁴ "Because it's a rescue operation, everything is being done carefully," stated AMEC Vice President Lee Benish on September 13. "You have to be sure you're moving things in the right way and at the right time."⁴⁵ Mark Loizeaux, president of Controlled Demolition, Inc. (which prepared the preliminary demolition plan), asserted, "How do you eat an elephant? Carefully and in small bites. They have an elephant."⁴⁶

The grim task of gathering body parts began at once, and they were brought to four temporary tent morgues. Later, the remains were taken to a central morgue. The construction crews could not begin their work in the various sectors until a targeted area was examined for bodies. (This did not always go smoothly, with firefighters primarily interested in recovery efforts and construction workers determined to clear the site of rubble.) Structural steel went to scrap yards, then melted down or reused abroad without further examination as fire-scene evidence (which concerned those who thought the investigation into the collapse of the towers was conducted too hastily). Each of the towers contained 78,000 tons of recyclable steel.⁴⁷

Within a few hours of the attack, the governor's office called Borough President Guy Molinari to discuss using Fresh Kills for the WTC debris. Local officials quickly complied. The material was primarily structural, lacking the trees and residential material normally produced after a natural disaster. Borough Environmental Engineer Nicholas Dmytryszyn stated, "It's not really an opening of the landfill. I think that term is inaccurate and incorrect. The stuff that's going to be there is strictly construction and demolition debris. This is not regular garbage."⁴⁸

By late in the day on September 11, heavily loaded trucks headed to either permanent or temporary MTSs or directly to the landfill. Until barges were utilized for transport, trucks crossed the Verrazzano-Narrows Bridge, heading for Muldoon Hill in Fresh Kills. By Friday, September 14, 1,500 truckloads (about nine thousand tons) of debris had made the journey. Eventually material loaded on barges at the MTSs was segregated by type for efficient handling.⁴⁹

At the landfill site, DSNY personnel unloaded, stored, and separated the material. According to Martin J. Bellew, DSNY's director of the Bureau of Waste Disposal and the person in charge of disposal of the WTC material, "the top of the landfill was relatively flat, which provided a good area for screening all the material."⁵⁰ The New York City Police Department and the FBI handled the sorting and investigative actions. Even during transit to Staten Island, the material was being prepared for screening and disposal. Upon arrival, it was stockpiled in two areas on a 135-acre plot called the Staten Island Landfill World Trade Center Recovery Site. A metal vendor took away the steel beams; other metals were removed after they had been screened and cleared for loading. Crews stacked about 1,400 damaged vehicles near the edge of the landfill; most were recycled.[51] According to Tom Harnedy, with the U.S. Army Corps of Engineers' (USACE) North Atlantic Division,

> Working the World Trade Center debris recovery mission at the Fresh Kills Landfill was a challenging experience for me both professionally and personally. As a Staten Islander, having witnessed the tragedy, as well as having family friends who lost their loved ones, I became very aware of the sensitivity of the work the Corps was accomplishing in support of the recovery and the importance of my role as the contract manager for the Corps in this effort.[52]

Under tight security, the police and FBI initially set up a manual screening operation close to the unloaded debris (eventually mechanical sorters were employed), searching for black boxes from the airplanes, human remains, and personal or identificatory items. The first loads arrived at 2:00 a.m. on September 12. This area (the "Hill") may have been the largest crime scene in history. Hundreds of agents for the FBI worked at the 174-acre site. During the operation's peak, ten thousand tons of material were delivered to the site daily (with piles reaching ten to fifteen feet in height), and more than 55,000 pieces of evidence were recovered. Aside from the DSNY, NYPD, and FBI, working space was made available for the Department of Environmental Conservation (DEC), the U.S. Environmental Protection Agency (EPA), the FDNY, the Port Authority Police, the Corrections Department, and the Red Cross. The Army Corps signed a contract with Phillips & Jordan of Zephyrhills, Florida, to serve as liaison among the various agencies. Evans Environmental developed the health and safety plan for Fresh Kills.[53]

Crows, gulls, and turkey vultures foraging through the debris complicated the grisly, dangerous work of the crews, who were dressed in protective white jumpsuits and examining and clearing the rubble. Chief of Detectives William H. Allee (a Staten Islander) was one of the many who were happy that Fresh Kills had closed in July. "But," he stated, "I think everybody understands the reasons why we're doing this here." A Rossville resident interjected that he was glad that the WTC material remained in the city, where it could receive proper burial. "I hope to God they put all of it there and make it a memorial."[54]

The little-known Department of Design and Construction (DDC) was the city agency that oversaw the cleanup at Ground Zero. Created in 1996, its primary task before the attack had been dealing with municipal capital construction contracts. It had been given the 9/11 duty largely on the strength of its two top officials, Kenneth Holden and Michael J. Burton. They had been effective early responders and actually had begun directing operations on their own, before receiving formal authority to do so. Holden and Burton first surveyed and mapped the site and then replaced the volunteers primarily with a professional labor force made up of firefighters, police, engineers, and as many as three thousand union construction workers. Crews labored twenty-four hours each day for at least nine months. The DDC's control of the operations was hampered because it only represented one vote on a committee made up of twenty-six federal, state, regional, and city agencies and departments—some with more sway than others. Decision making was never easy throughout the cleanup process.[55]

Many groups, offices, and agencies took part in the operation on one level or another. Some nongovernmental groups provided help as well, including the National Solid Wastes Management Association. Within the first twenty-four hours of the attack, about 1,500 sanitation workers were removing debris.[56] USACE worked in various capacities in the response-and-recovery effort, including interfacing with the Federal Emergency Management Agency.[57] FEMA's mandate was to assist directly those affected by natural or human-made disasters. Emergency services personnel from eighteen states were on the scene almost immediately, including twenty-eight search-and-rescue units employing sniffer dogs, life-detector sensors, and cameras.[58] Yet for all the good that FEMA performed at the site, they were criticized for their handling of disaster-relief funds for those who had lost their jobs.[59]

The DDC may have overseen the cleanup, but it was Giuliani (in his last months as mayor) who proved to be at the center of the city's response to September 11. By this time, Giuliani's administration was faltering; he was involved in a highly publicized divorce, had pulled out of the U.S. Senate race against Hillary Clinton, and had announced that he had prostate cancer (which he survived). But nothing deterred him. When the first plane hit, Giuliani rushed to his emergency command center at No. 7 World Trade Center, but he, his bodyguards, and his staff were forced to flee it when the second tower collapsed. The new command post was set up at the Police Academy, and the mayor immediately began getting information and assurances to the public and rallying city workers and agencies to meet the crisis.[60] He announced to New Yorkers, "Tomorrow New York is going to be here. And we're going to rebuild, and we're going to be stronger than we were before. . . . I want the people of New York to be an example to the rest of the country, and the rest of the world, that terrorism can't stop us."[61]

Giuliani's constant public presence during the disaster and his frenetic efforts to get the city back on its feet earned him international acclaim. Queen Elizabeth awarded the mayor an honorary knighthood. The president of France called him "Maire-heros"—"Rudy the Rock." *Time* named him Person of the Year. On the streets of New York people chanted "Four more years!"[62] Even his old nemesis, Ed Koch, praised him: "He rose to the occasion. He set the standard for what a mayor should do in a time of crisis. People want you to lead."[63]

Not everyone praised the mayor. Some critics referred to the command center as "Rudy's bunker." Others suggested that Giuliani had failed to prepare the city for another attack after the 1993 WTC bombing and that he underplayed the health risks around Ground Zero after September 11. Firefighter and police unions blamed him for the lack of sufficient protective gear for responders and thus for illnesses they contracted after the attack. Firefighters in particular did not like being pulled off the frontlines while they searched for their comrades' bodies. Some family members of the victims lodged complaints. Later, Giuliani was disparaged for his attempts to profit personally from his actions after the attack.[64]

The onerous cleanup process continued through the rest of the year. The Giuliani administration was rumored to be preparing a plan to replace the DDC with a private construction firm (in this case the Bechtel Group) to oversee the operations. Tensions also grew because the companies clearing the rubble were working without contracts. This arose largely because of disagreements over who was liable for damages concerning toxic materials at the site.[65] Yet the laborious work did not go unrecognized. *Waste News* gave its Newsmaker of the Year Award for 2001 to the people working to clean up the WTC area. It was estimated that the collapse of the towers had created anywhere from 1.2 million to 1.6 million tons of debris, more than 700,000 tons of which had been removed by December. Another 121,000 tons of steel was shipped to recyclers.[66]

THE DUST

Attention to the feverish rescue and cleanup tasks at hand could not hide the potential environmental risks caused by the collapse of the towers. These risks threatened workers and local citizens at and around the attack site and at Fresh Kills Landfill. Initially, clouds of dirt, soot, and fumes inundated the area, and the dust settled on almost everything within reach. It consisted mostly of pulverized cement and particles with glass, fibers, asbestos, lead, hydrocarbons, PCBs, furons, dioxin, and human remains. Everyone involved in the recovery or cleanup was breathing the air, and many experienced hacking soon labeled "the World Trade Center cough." Concern that freon gas was escaping from refrigeration units was one of many possible threats that had to be met. Unexploded ammunition belonging to the U.S. Customs Service also had to be dealt with, but this led to no injuries or fatalities.

More insidious, if less dramatic, were the chronic respiratory problems that a number of workers developed. Although many at the site were supplied with personal protective equipment, not everyone chose to wear their respirators. Firefighters in particular argued that they had to be able to smell to find dead bodies. The Occupational Safety and Health Administration (OSHA) and other safety officials did not have the authority to force the use of the equipment.[67]

Initial newspaper accounts often played down the danger to workers at the attack site. At the time, experts were not in a position to provide definitive answers without further study. A professor of environmental medicine from NYU's School of Medicine noted in mid-October 2001: "We have not yet seen anything that would tell you [dust and soot is] dangerous, but we don't know everything yet."[68] According to an OSHA national news release provided in that same month, OSHA and EPA "found no evidence of any significant public health hazards to residents, visitors or workers beyond the immediate World Trade Center area." This did not include Fresh Kills.[69] There were skeptics, nevertheless, of any offhanded assurances of worker safety or of no long-term environmental pollution.[70]

Years after the end of the cleanup, the health and environmental risks from the operation among "the forgotten victims" were better understood.[71] NYPD detective James Zadroga died in 2006 of a lung disease directly attributable to his many hours of work at Ground Zero.[72] Reports of more than twenty additional deaths likely from pulmonary disease or possibly cancer were reported. About four hundred NYPD detectives suffered from health problems likely associated with their work at the attack location. By the fifteenth anniversary of the World Trade Center disaster, the Health Department of New York State announced the latest findings on health impacts at the site. Studies revealed more cancer cases than expected and provided a more precise understanding of post–traumatic stress disorder (PTSD) experienced among workers. Around 12,000 of the approximately thirty thousand individuals who worked at Ground Zero and Fresh Kills filed health claims by 2006, and many others complained of breathing problems. Several workers filed lawsuits, with litigation extending into the 2010s.[73]

Health risks and the presence of hazardous materials at the Fresh Kills recovery area were similar to those at Ground Zero. The decision to hold tons of debris from the WTC site at the landfill complex reopened old wounds on Staten Island. Locals of course were grieving about the attack in general, plus 274 of their family and neighbors had died (a disproportionate number for the borough compared to the overall population). But neither the possibility of resurrecting the landfill nor the health dangers experienced by local workers during the cleanup and investigation sat well.[74]

As late as January 2002, EPA and DEC officials claimed that no significant amounts of hazardous waste would enter the landfill. The DEC's spokesperson Jennifer Post stated, "The material is no different than what is put at any other municipal landfill."[75] The environmental attorney Michael B. Gerrard did not agree: "Scattered among the debris, is material that ordinarily could not lawfully be disposed at a C&D [construction and demolition] or MSW landfill and would need to go to facilities specially designed for hazardous, asbestos, or petroleum wastes."[76] There were obvious signs of health issues among workers at the Fresh Kills recovery site and on the transport barges, such as an increased risk of new-onset asthma. One study concluded that "post-9/11 asthma cumulative incidence among Staten Island landfill/barge workers was similar to that of other WTC disaster rescue and recovery workers."[77]

As early as September 13, 2001, EPA Administrator Christie Todd Whitman announced that her agency would begin taking steps to monitor and protect the safety of workers and the public (and the environment) at both the World Trade Center and the Pentagon disaster sites. Whitman stated, "EPA is greatly relieved to have learned that there appears to be no significant levels of asbestos dust in the air in New York City. We are working closely with rescue crews to ensure that all appropriate precautions are taken. We will continue to monitor closely."[78] The EPA coordinated its activities with various levels and branches of government, leading or participating in monitoring air, water, and dust; vacuuming of debris and dust from outdoor spaces; disposing of hazardous wastes; setting up wash stations and providing protective equipment; and developing testing programs for indoor residences.[79]

A Natural Resources Defense Council (NRDC) study conducted in early 2002 presented a preliminary assessment of the overall environmental impact of September 11 and a somewhat critical appraisal of its handling. It concluded: "The terror attacks on the World Trade Center, in addition to their heart-wrenching toll on human life and wide-ranging economic impacts, constituted an unprecedented environmental assault for Lower Manhattan. At least 10,000 New Yorkers have suffered short-term health ailments from Trade Center–generated air contaminants." The report noted "good news" about the quality of outdoor air in the vicinity days after the attack but remained concerned about "indoor pollution threats" in some residences and offices, which it argued was "manageable." While rating positive much of the overall government response to environmental problems posed by the disaster, the report asserted that it "fell short in several crucial areas," namely, gaps in coordination and leadership, failures to communicate environmental information to the public, and lapses in occupational and safety actions. The New York City Department of Environmental Protection and OSHA received the harshest criticism. That being said, the NRDC report praised the debris-removal efforts, which "advanced swiftly and without major environmental problems."[80]

The report addressed several of the structural and organizational issues of the post-attack recovery and cleanup and the environmental impacts on air and water pollution. Yet it gave scant attention to worker health issues, which proved much more serious as time went by. In assessing what transpired at Fresh Kills, it made the pointed observation that "additional attention is warranted concerning the burial of potentially contaminated waste at the Fresh Kills landfill and the final waste cleanup plan at Ground Zero."[81] Such observations pointed to the central role of Fresh Kills in the World Trade Center tragedy, along with its function as an extension of the crime-scene evaluation and the larger implications of the landfill—now understood to be hallowed ground.

CRIME SCENE AND CEMETERY

From a strictly logistical standpoint, reopening Fresh Kills to take the debris from the Twin Towers made sense. That decision, however, was complicated by controversy and symbolism. Fresh Kills was not only an extension of the crime scene and a final dumping ground for WTC debris. It also became a cemetery—and, in some strange way, a memorial. The impossibility of separating all human remains from the rubble fueled an inevitable dispute.

In late September, Mark Schoofs, a staff reporter for the *Wall Street Journal*, observed,

> In a round-the-clock procession, trucks bearing the remains of the World Trade Center leave Manhattan by the Brooklyn Battery Tunnel and make their way across the Verrazano Narrows Bridge. Barges with debris pass through New York Harbor, skirting the New Jersey shore, then up Arthur Kill inlet. They come together on Staten Island to form the last chapter of a storied place that has housed the cast-offs of New York City for half a century: the 3,000-acre Fresh Kills Landfill.[82]

A backlog of debris kept the operation at Fresh Kills open until after the work had ceased at Ground Zero. On May 30, 2002, an emotional ceremony brought the recovery effort to a formal end.[83] A somber closing event took place at the Hill at the end of the sorting process on July 15, 2002. Port Authority Police Lieutenant Brian Tierney stated, "In a lot of ways, it will be hard to stop, because you know you haven't recovered everyone. It'll be hard to say what we've done is as much as we can do—it'll never be enough."[84] Of the approximately 1.2 million to 1.6 million tons of material brought to Staten Island, more than one million tons were screened and sifted down to a quarter-inch in diameter; the remaining WTC material was dumped in a forty-eight-acre area adjacent to the recovery site on the West Mound in Section 1/9.[85]

Dennis Diggins, who directed the landfill operations, said that the work at the site was so important that there were days when workers felt guilty for taking time off. "This is a project that you will be thinking about for the rest of your life," he declared.[86] As numerous other comments and memoirs attest, many workers and observers at the post-9/11 site at Fresh Kills were keenly aware that the landfill and environs had become something other than a dumping place for society's discards. In a conversation with Chief of Detectives Allee on January 15, 2002, CNN correspondent Jason Carroll noted, "The Fresh

FIGURE 18.3 Debris from Ground Zero to be sorted at Fresh Kills Landfill, October 3, 2001.

Source: Wikimedia Commons.

Kill landfill in Staten Island has become more than it was originally intended for." Allee: "Listen, this isn't a dump. This is sacred ground." Carroll then mentioned a sculpture and parts of the United Airlines airplane, which lay nearby: "Somber signs that the work here goes on." Allee: "I think of the people who died, and that's what we're trying to do. We're trying to connect with that."[87]

The police officer and Staten Islander Frank Marra referred to the operations at Fresh Kills as "not a traditional crime scene."[88] "This small working city," he said, "became a grave, a burial ground, for people who perished on that day."[89] He later observed:

> Even after over thirteen years, when I drive by the Landfill, I no longer see the "dump." I see a holy ground, a hallowed ground with bright lights that shine for miles, but all I can think about every time I pass is how many lives were affected there, mentally, physically, or both. I still think of it as a time that stood still. I think that I will begin to make the sign of the cross as I pass it, as I would if I were passing a Catholic church. I want to show my respect to those who gave so much of themselves on September 11, 2001, and to those whose final resting place is there, lost on "The Hill."[90]

For the families of victims whose remains were not recovered or identified, moving the WTC debris from Ground Zero to Fresh Kills seemed as if the city was throwing their loved ones onto an unmarked scrap heap at a historically odious site. Finding all the remains of everyone lost in the attack, of course, was impossible. But simply to comingle those remains with the rubble of the Twin Towers and then dispose of them at a landfill was offensive and sacrilegious. Families wanted closure as well as respect for the dead. They wanted a proper burial. Some charged that a quick cleanup had taken priority over a careful search for remains. And in subsequent years additional human remains were in fact discovered.[91]

Relatives of about 1,100 to 1,200 of the approximately three thousand people who died as a result of 9/11 had yet to receive any remains by 2004. Diane Horning's son Matthew died in the North Tower, where he worked for an insurance company. "We were promised the remains," she said, "and now we discover that my lost son is to spend eternity in a rubbish dump. This is morally reprehensible and emotionally unacceptable, and we are going to fight it all the way." Mrs. Horning and her husband, Kurt, received some small remains, but they wanted a more appropriate recognition of his death. They formed the World Trade Center Families for Proper Burial. The city offered to create two embankments of debris in the Fresh Kills area (not a formal cemetery) and construct a memorial where people could visit.[92]

For some, a memorial site was not enough. In August 2005 family members of 9/11 victims (including the Hornings) filed a lawsuit to restart the search for remains and to relocate those that were found. They were particularly distressed to hear from a recycling supervisor that debris was being used to pave roads and fill potholes.[93] The federal courts ultimately dismissed the case, and it was turned down on appeal to the U.S. Supreme Court. Sifting operations at the WTC site ceased in 2010.[94] The legal setbacks reinforced the families' frustrations. Oakwood resident Michael Mozzillio, who lost his firefighter son, stated, "I don't think [the materials] should have been in a dump in the first place. I'd like to think [Christopher's] not there. If he's been laying around there ten years, it's a terrible thought."[95]

This was an irresolvable standoff, and one that altered the perception of the Fresh Kills Landfill forever. In the fall of 2001, an exhibit entitled "Fresh Kills: Artists Respond to the Closure of the Staten Island Landfill" was to open at the Snug Harbor Cultural Center on Staten Island. It was nearly canceled after the 9/11 attacks. Instead, a billboard outside the center depicting Fresh Kills added the inscription "In Memoriam." Director of Visual Arts Olivia Georgia thought the exhibit should go forward: "We came to believe that it is crucial to continue." One artist, Mierle Laderman Ukeles, altered her multichannel video presentation, which included conversations with ecologists, engineers, art historians, sanitation workers, and island residents. Selected screens were left blank in tribute to the 9/11 casualties.[96]

In late October 2002, Congress began discussing the need for a national memorial at the Fresh Kills recovery site.[97] While many applauded the initiative,

FIGURE 18.4 *Postcards* is a 9/11 memorial on the St. George Esplanade on Staten Island, which includes two white-marble wing sculptures (each thirty feet) representing postcards to loved ones.

Source: Wikipedia.

others could not conceive of the juxtaposition of honored 9/11 dead and a landfill. One woman proclaimed, "I really do believe that for the sake of their souls and their families, to have the 'Dump' be a Memorial is a disgrace. Please don't even consider such an idea of this sort, the dump is the 'DUMP.'"[98]

FROM GIULIANI TO BLOOMBERG

Before Ground Zero was cleared and before the Fresh Kills recovery project was ended, New York City's mayoralty was handed from Rudy Giuliani to the billionaire Michael R. Bloomberg. Although both Republicans, Bloomberg's style and priorities were noticeably different. The primary campaign was postponed because of the attack, but eventually politicking resumed. Before the primaries, there were calls for a third consecutive term for the highly visible Giuliani, which would require overturning the term-limit laws. He was not totally opposed to it. Eventually, the need to complete the recovery and to deal with the economic slump that hit the city after 9/11 led Giuliani to propose staying on as mayor for an additional three months only. Bloomberg agreed, as did the Democratic candidate, Public Advocate Mark J. Green. Bronx Borough President Fernando Ferrer, running against Green in the primaries, used Green's stance on Giuliani against him and was able to shave off his opponent's support among black voters and white women.[99]

Green prevailed, nonetheless, as the Democrat's standard-bearer, aided by the backlash over Reverend Al Sharpton's (and others') racial and anti-Rudy politics in support of Ferrer. In his matchup against the Republican Bloomberg (who bested Herman Badillo in the primary) in the general election, he fell to Bloomberg. Giuliani's endorsement—and a campaign that spent more than $50 million—were central.[100] Bloomberg had made clear that he did not mind if Giuliani stayed in City Hall a little while more, as the sitting mayor wished, because Bloomberg desperately needed the current mayor's endorsement.[101]

Bloomberg, in his first venture in politics, eked out a razor-thin win, with just over 50 percent of the vote. He won only Queens and, overwhelmingly, Staten Island.[102] Writing some years later, the blogger John-Charles Hewitt wrote about the slim victory: "It's kinda funny that a large city like NYC can have its mayoral election decided by a paltry 1.4m people, and that an inconsequential suburb of like [sic] Staten Island can decide the fate of more populous and economically significant boroughs, but that's democracy for you."[103] It was not the first time. Ask Rudy Giuliani.

Bloomberg was a fiscal conservative and social liberal. A former Democrat, he was born into a family of Russian and Polish Jewish immigrants in 1942 in Medford, Massachusetts. His father was a bookkeeper for a local dairy. Mike Bloomberg attended Johns Hopkins University, where he studied electrical engineering, and obtained an MBA from Harvard Business School in 1966. He was hired by Salomon Brothers on Wall Street, becoming a partner in 1972. Ultimately, he supervised stock-trading sales and information systems. Dismissed in 1981 after the company was bought out, he started his own company, Bloomberg L.P., a financial-services software firm. His business blossomed when it entered the media world, launching *Bloomberg News* and then radio, television, internet, and publishing operations.[104] He presented himself as someone who got things done—not unlike his predecessor in that respect, but without Giuliani's tone. Often portrayed as bland, Bloomberg was nevertheless ambitious and pursued projects that some considered politically risky.[105] He would need all the skill he could muster to bring New York City back from the devastation of 9/11.

A NEW DISPOSAL PLAN: REOPEN FRESH KILLS?

The September 11 attack threw New York City's solid waste program into disarray. The fiscal crisis caused by the event (a $4 billion deficit) put many programs in jeopardy, including those at the DSNY.[106] The city still was working under Giuliani's "interim plan," initially dependent on truck and rail transportation to move as much refuse as possible over state lines. Plans to rely more fully on MTSs, barges, and rail transit hit some snags and kept countless trucks on the road. Fresh Kills' temporary reopening to accept debris from the WTC site added immensely to the DSNY's workload and taxed existing infrastructure.

The Linden facility, to be opened in New Jersey, was delayed or possibly faced termination. In September 2001 New Jersey's attorney general concluded the investigation of Linden Mayor Gregorio and his alleged crooked land deal.[107] In response, the state's Department of Environmental Protection refused to approve the project going forward and sent it back to local authorities for further review. The best-case scenario had been an opening of the facility in 2005; now the entire project was in jeopardy.[108]

Ben Miller again raised what Staten Islanders clearly viewed as unthinkable—reopening Fresh Kills.[109] In an October 6, 2001, talk, he stressed that Fresh Kills had been

> the fulcrum on which New York's waste-management "policy" has been precariously balanced for more than half a century. And now that we have completely lost our balance—and decided to close it precipitously with no replacement in place, making generations of New Yorkers yet-unborn dependent on the kindness of strangers, the ghost of the Twin Towers will not be the only parts of those four towering piles that will haunt us in the years to come.[110]

In an op-ed piece in the *New York Post* on November 30, Benjamin Smith of Smartertimes.com also suggested that one of six ways for New York to increase revenues or cut expenses in the aftermath of 9/11 was to reopen Fresh Kills. "Staten Island would holler," he said, "but budget watchers across the political spectrum . . . back the measure. New York now spends about $200 million a year exporting garbage, the Independent Budget Office reported."[111]

The immediate responses to reopening the landfill (especially as advocated in Smith's piece) were emphatically political and emotional. Senator Marchi: "An insult to all Staten Islanders." Borough President Guy Molinari: "The permanent closure of Fresh Kills landfill is non-negotiable." Representative Fossella: "I'll lead the charge for secession if they ever contemplate reopening the dump." Assembly Speaker Sheldon Silver: "The chances of overturning the law are nil." Assemblyman Eric Vitaliano: "The idea that this would save New York City money is probably the grossest insult of all." Councilman Oddo: "Benjamin Smith has told Staten Island to drop dead. Staten Islanders say to Mr. Smith: 'Go to hell.'" Mayor-elect Bloomberg stated, "I can never envision reopening the landfill under any circumstances. Mayor Giuliani and Borough President Molinari fought intensely to close the landfill and improve the quality of life in Staten Island and I promise to keep their work from becoming undone."[112]

The city's Independent Budget Office examined "nearly 60 options" of anticipated revenue savings for the city, of which reopening of Fresh Kills was one.[113] It also was looking for more than a temporary fix, and the necessary state legislation to reverse the closing would drag things on even more. One popular survey showed that only 26 percent of city residents favored the reopening.[114]

Going back on the city's word would only breed resentment on Staten Island.[115] Governor Pataki also made his strong feelings known on a trip to Staten Island to pay respects to outgoing borough president Molinari: "It's the wrong thing to do. Staten Islanders have put up with enough for too long."[116]

The Staten Island media, of course, made its feelings known, as well. An editorial in the December 6 issue of the *Staten Island Advance* quipped, "Mr. Miller is entitled to his views. So are people who say the earth is flat and that aliens run the government." Questioning why Miller had been included as part of an orientation panel for newly elected Staten Island Council representatives, the editorial went on to warn that "the sudden appearance of the notion of re-opening Fresh Kills as an economically sensible and political [sic] viable policy in several places just as the city confronts historic deficits is cause for grave concern, not complacency."[117]

A report prepared by Columbia University's Earth Institute outlined in detail New York City's *Life After Fresh Kills*. It stated that while it took no position on the city's current waste-export plan in the short term, it did conclude that "long-term this plan leaves the city vulnerable to a wide range of problems including dramatic cost increases and environmental degradation."[118] The report added that post–Fresh Kills contracts would impose "significant cost increases on City Taxpayers," with disposal charges rising from $42 per ton at Fresh Kills to $70 to $100 per ton through exporting. The city also would be more susceptible to changes in market control, such as further consolidation of the waste industry and landfill operations, rising tipping fees, and increased transportation costs. Political vulnerabilities included possible restrictions to flow control, renegotiating contracts, and unforeseen disruptions of the transportation infrastructure. Environmental concerns covered a spectrum of potential problems.[119] Looming over the report were dislocations already caused by the WTC disaster, everything from the possible need to push back contractual obligations to apprehension about unforeseen catastrophes.[120]

Faced with the formidable task of dealing with the city's refuse stream in the midst of a scorching controversy and post-9/11 recovery, the DSNY needed to revamp its ten-year waste management plan. Sanitation Commissioner Doherty did not believe that reopening Fresh Kills to save money was a viable option: "As the commissioner and a Staten Islander, I have no intention of reopening it. I don't even think about that, even in the worst conditions. I think the mayor will find ways of handling the budget that do not include reopening Fresh Kills. I mean, that is out of the question. That is not going to happen."[121]

The DSNY looked to short-term fixes. A proposal to divert some Queens refuse temporarily to the Bronx might reduce costs, but it violated the promise that boroughs would be self-sufficient in disposal matters. Department officials also began to explore alternatives to the Linden facility and other possible stations in New York City. The City Council's newly formed Sanitation and Solid

Waste Management Committee—chaired by Staten Island's Michael E. McMahon, a Democrat—sought alternatives to what the Giuliani administration had formulated. McMahon was aware that developing a new plan would help ease tensions in the wake of Fresh Kills' closing and the demands for reopening it.[122]

Before a new solid waste plan was in place, Mayor Bloomberg began to set a new direction. He decided to scale back the city's curbside recycling program as part of a series of cuts to reduce a projected $4.8 billion deficit in the preliminary budget of $41.4 billion. He wanted to suspend temporarily the collection of metal, glass, and plastic but to continue paper collection. "We must now meet the challenge of this new fiscal reality," he stated. "I am proposing a fiscal plan and preliminary budget that calls on all New Yorkers to make sacrifices, to cooperate in finding responsible and innovative ways to maintain the financial integrity of our city." The preliminary 2003 budget also called for a reduction in the DSNY's allocation from $1.1 billion in 2002 to $978 million. It proposed a reduction in the number of sanitation employees from a little over ten thousand to about 9,500.

Environmental groups obviously were not happy, especially after fighting the Giuliani administration on recycling for many years. In subsequent years, however, the potential success of maintaining the paper and glass program was questioned.[123] But at the time, as John Halenar of Region 2 (including New York City) of the New York State Association of Reduction Reuse and Recycling argued, "You can't turn a program on and off. It's just way too confusing for people. It's going to be deadly for the program to shut it off and try to turn it back on."[124]

On March 2 a *New York Times* editorial set off another firestorm. It began: "One of the most fundamental responsibilities of any city government is picking up the trash. Unfortunately, New York City's system of disposing of the 11,000 tons of garbage its residents generate every day is in a state of disarray." It charged that Giuliani had been "overly optimistic" about his disposal program and that "all in all, the current system is dirty, inefficient, costly and largely dependent on the good will of other states. It's hard to imagine how the city could do worse." Mayor Bloomberg was being "forced to face the true cost of life without Fresh Kills." While the editorial conceded that parts of the original Giuliani plan "could be revived and refined," reopening Fresh Kills "would help [Bloomberg] close the city's $4.7 billion budget gap."[125]

Bill Franz, in the *Staten Island Register*, which had been most dubious about the closure of landfill, piled on:

> There has never been a political performance like it in the history of Staten Island. Last year a self-congratulatory army of elected officials, accompanied by a vast entourage of patronage appointees, press secretaries, campaign operatives and rally-on-demand cheerleaders, ascended, Moses-like, a mountain

of fetid waste and, to a deafening chorus of orchestrated applause, grandly declared the Fresh Kills landfill closed forever.

"All the partying had been premature," he added. Commissioner Doherty himself revealed, "We are basically starting over again. Fresh Kills was really closed without an awful lot of thought, you know, if the story be told. We are still faced with a big problem."[126]

The call for reopening Fresh Kills and Doherty's frank reactions carried the disposal debate into the spring of 2002. Ben Miller claimed that "a consensus was emerging that the Fresh Kills landfill must be reopened and the city's trash plan has to start from scratch." While he did not have evidence of such a consensus, support from the *New York Times*, the *New York Daily News*, and others was putting on the pressure. Miller recommended that Mayor Bloomberg "squeeze every drop of waste out of our trash system." "A rationally designed collection and processing system," he added, could be competitive with waste exporting, ship waste in "a cost-effective and environmentally sound way," and help rethink site selection for transfer stations. These measures, he surmised (and had stated before), would take time to implement, "so reopening Fresh Kills immediately could give the city breathing room." It could also provide "bargaining leverage we desperately need to take control of the city's trash costs."[127]

Staten Island Borough President Molinaro defended the current system and reminded readers how Fresh Kills had precipitated the secession movement not so long ago.[128] Fossella questioned the environmental implications of reopening the landfill: "Environmental responsibility does not stop at the Verrazano-Narrows Bridge."[129] In a letter he planned to send to the *Times* but for some reason did not, Marchi intoned, "When I reflect on your editorial and its implied condescension toward the smallest borough in the city, I better understand why the secession movement carried a bill through the Senate years ago."[130] U.S. Senator Charles E. Schumer (a Democrat from New York) also stated emphatically that Fresh Kills could not reopen.[131]

In early April all five borough presidents agreed that Fresh Kills should not be reopened. Molinaro stated, "This is the first time in the history of Fresh Kills that all five borough presidents have joined together to announce their support for dump closure."[132]

How much of the debacle over Fresh Kills closing and reopening was attributable to politics? Politics was politics, and few were willing to poke at Staten Island in this way. An *Advance* story on March 10 asserted that "Even talk about a temporary re-opening would be a death wish here for Pataki as he revs up his re-election campaign." Bloomberg's election success on Staten Island was "a powerful incentive as well." Giuliani and Molinari also were mentioned.[133] Likewise, how much of the criticism of the Giuliani disposal plan arose because of the financial impact of the WTC disaster and its aftermath? This issue too, like politics, often masked the raw feelings about the recently shuttered landfill and the uncertainty (again) over what sort of disposal plan could realistically be employed for the city in 2002 and beyond. Any practical argument for reopening Fresh Kills had to overcome the political momentum of closure and Staten Island's history with "the dump."

With the reopening of Fresh Kills a moot point, the mayor and the city had to settle for other disposal solutions already mapped out—and some previously dismissed. The latest disposal debate even produced renewed speculation about incineration. In the March 5 *New York Daily News*, Ben Miller stated what was almost heresy for others: "We should stop pretending that burning garbage to generate energy is worse for the environment than landfills are."[134] Miller had raised the issue in various statements, and Commissioner Doherty brought it up before his City Council meeting. Bloomberg noted on his radio show, "We're going to have to take a look at all different possibilities. No matter what you hear about people yelling and screaming, an awful lot of this technology has gotten very much better." Environmentalists obviously were wary. James T. B. Tripp, the general counsel for Environmental Defense, argued, "If we're going to think about [incineration], we've got to be prepared to think about everything." That meant recycling, too.[135]

It did not take long for the mayor to back off from incineration. "The technology is there," Bloomberg said in his radio show in May, "but the politics are such that it would be phenomenally difficult to site incinerators in the New York City area, and other places don't want to have it either."[136]

A MODIFIED DISPOSAL PLAN?

In July Bloomberg revealed his proposed general approach to the city's garbage woes. About 11,000 to 14,000 tons per day of refuse would be transported to eight marine transfer stations in four boroughs, with a ninth station to be built on Staten Island (all equipped with giant compactors). From the MTSs, the material would be containerized and placed on barges that would bring the refuse to rail cars or oceangoing ships. The final destinations were not specified.[137]

A story by Kirk Johnson in the *New York Times* aptly discussed how the proposed plan "would reshuffle the winners and losers in New York's garbage game." "Garbage in New York City has always been about pain and gain," he stated. "Some residents and neighborhoods are plagued by stink and noise and the other irksome effluvia of trash collection. Others see only clean streets, with the clanking machinery of transport and disposal all but invisible." Several neighborhoods, such as Williamsburg, Brooklyn; the South Bronx; and Jamaica, Queens, would benefit from the reduction in truck traffic and diesel fumes by eliminating the need for land-based transfer stations. Staten Island "would be spared any hint of reopening the Fresh Kills landfill." The MTSs were to be city operated and thus "a victory for municipal workers." Unlike under the Giuliani plan, Manhattan would be a loser because it would have to become a site for trash compacting at two of its three transfer stations, which would have to expand into nearby property. The Linden plant was dead for the moment. What might happen to commercial waste and recycling was not so clear.[138]

Bloomberg's proposal received a good deal of praise for overcoming some of the shortcomings of the "interim plan." But like all ideas for dealing with New York City's disposal dilemma, the proof was in the implementation. The DSNY budget for 2003 was reduced by slightly less than preliminary requests, largely because the transfer of refuse from Queens to the Bronx was scrapped. Some savings were achieved by postponing capping work at Fresh Kills and reducing the frequency of garbage and recycling pickup outside Manhattan.[139] In 2002 the city reinstated the recycling of plastics and planned to reinstate recycling glass the following year.[140] But any major upgrade to the recycling program seemed more tenuous. It was unlikely to get much enthusiastic support from the mayor's office, especially given that Bloomberg had stated during the debates over the city budget that the recycling program "does not do anything for the environment" and reduces money for other services.[141]

If one listed on a sheet of paper the various disposal plans that New York City had attempted or proposed over the years, they would fill the page and then some. It would be difficult, however, to discern a clear pattern. Some ideas followed the trends of the time, some were the product of desperation, and others were simply politically motivated. At the moment, several questions remained unanswered under the Bloomberg plan. It dealt with "end-of-the-pipe" solutions, that is, it confronted the unwanted discards from items already produced. What about the *generation* of waste and an operational waste reduction goal? How did any of New York City's plans address the realities of the throwaway society?[142] Also, what responsibility did New York City bear for where its waste was dumped or incinerated—or contaminated? Did that responsibility end at the city limits?[143]

The closing of Fresh Kills was at the heart of the decisions about a new solid waste plan for New York City. In most respects, the Staten Island landfill was now cleaved from whatever future disposal ideas might be concocted. And Fresh Kills post-9/11 was not the same place it had been before the WTC disaster. The physical alterations were hardly as obvious as the salt marsh–turned-landfill in 1948. But the perception of the place had become muddled and complicated. Fresh Kills was now both cemetery and dumping ground—a strange spatial and emotional juxtaposition. Jim Johnson, in a September 2002 issue of *Waste News*, observed:

> A calm is returning to a Staten Island hillside that cradles the remains of one of America's darkest days.
>
> This is a nondescript place, part of Fresh Kills landfill on Staten Island. Just miles from the bustle of lower Manhattan, but somehow far away.
>
> Buried here, in a 40-acre graveyard, is Sept. 11.
>
> Buried here is what's left of the World Trade Center and the remains of those who never went home after terror struck that sunny New York City morning.
>
> Buried here is part of whatever innocence America had left before that morning.[144]

In the years that followed 9/11, Fresh Kills was reinvented—or at least acquired yet another persona—once again, with the dawn of Freshkills Park. In some respects, refuse disposal also was reinvented as the city turned from the single solution that was Fresh Kills to a borough-based and export-focused system—a system that created its own set of trials and encounters for Gotham and its neighbors.

CHAPTER 19

REGENERATION

The closing of Fresh Kills Landfill did little to erase "the dump" from Staten Island's memory. Closed or open, its post-1948 aspects were forever part of the island. Yet some stubbornly believed that transformation of that despoiled landscape was achievable, possibly as a regenerated ecosystem and recreational area. Two objectives emerged: consider ways to mend and invigorate Fresh Kills as a natural place and memorialize the people of the World Trade Center disaster. Clouded and often lost were the impulses to restore Fresh Kills to its pre-landfill state, commemorate its early human history, or effectively document its role as materially important to understanding New York City's modern lifestyle.

The artist Mierle Laderman Ukeles argued that a landfill is a "social sculpture" that shared its history with those who created it. The NYU anthropologist Robin Nagle considered the landfill an artificial geography, or "a commons," revealing "unexpected details about the society that creates it." In my own efforts to get the Fresno Sanitary Landfill named as a national historic landmark in 2000 and 2001, I stated that a landfill should be viewed "as an integral part of the process of living, and thus to view it as culturally and historically important." The University of Arizona (and later Stanford University) anthropologist and "garbologist" William L. Rathje reminded us that "our garbage heritage" is worth remembering. As part of the team that chose the finalists in the Fresh Kills park competition, he asserted, "I don't believe that we should try to sweep Fresh Kills under a 'natural' rug . . . it can never be returned to Mother Nature's bosom." What is, after all, the "natural" ecology of a closed landfill? He preferred to turn "the world's largest symbol of New York's and America's wastefulness into the world's largest symbol of New York's and America's new environmental ethic of reducing, reusing, and recycling waste."[1]

Such assessments, while valid, were made without fully taking into account the historical moment. The great relief among Staten Islanders at seeing the landfill closed in 2001, after decades of protest and threats of secession, made it politically impossible for the city to pursue any approach other than developing a recreational or open greenspace—or simply ignoring the area as closed, unreclaimable, and derelict land. In reality, the idea of developing a park or natural area at Fresh Kills predated the WTC disaster by many years and was integral to the closure discussion.

It was unlikely, as well, that Staten Islanders would take kindly to any monument to New York's consumer past on their soil. A 9/11 memorial was one thing, but a garbage museum was something else, even though many on Staten Island had made their livelihood at and devoted their working years to the refuse facility.

The mere presence of the landfill was a reminder of the scars it left on the island. To some, even a regenerated space did not erase the recollection of Fresh Kills. A park might promise some tangible future uses for the locals, maybe even increase local property values. But planning a park had pitfalls as well. Some Staten Islanders were skeptical of outsiders deciding on how to remake the space or dubious about the advantage of attracting hordes of people with no stake in the island's communities tramping around the made-over eyesore.[2] Jason Muss, of the Muss Development Corporation, was among those who saw economic opportunity in the closing of the landfill. But he also observed that any positive change of attitude concerning the closing might actually lag behind that of nonresidents.[3]

Some questioned how long it might take before the debased landscape could become useful as a recreational area; some estimated at least twenty years. Nicholas Dmytryszyn, an environmental engineer for Staten Island's Borough Hall, was more optimistic: "Areas like that could be used for passive recreation sometime in the next three to five years." Rosemarie Scura had recently moved with her family from Richmond to Arden Heights, into a new townhouse less than a half-mile from the landfill. She was happy to see the landfill closed and asserted, "I know this side of the island is going to be up and coming. I don't care what anybody says."[4] One thing was for certain: reopening the landfill was out of the question.[5]

Freshkills Park (modifying the name of the place to distinguish it from the landfill) would be the largest landfill-to-park transformation ever undertaken in the United States. But would the new use(s) for Fresh Kills be grounded in ecological restoration or in something else?

ECOLOGICAL RESTORATION

There is a long history of converting landfills into parkland, recreational areas, and other useable space. Conversion of landfills into parks goes back at least to 1916 in the United States, when Seattle built Rainier Playfield on Rainier Dump. While no one has counted all of the parks and recreational sites built on old landfills, the number could be anywhere from 250 to over a thousand.[6] In the twenty-first century the majority of people around the world live in cities, and usable open space in cities has become quite limited. Reuse of available urban land is now critical, and closed landfill sites offer real opportunities for exploitation.[7]

Fresh Kills Landfill, given its scale and location within a highly populated metropolis, is valuable real estate. The decision to convert it to parkland brought into question how that was to be accomplished and to what degree efforts would be made to restore the wastescape to its pre-landfill state. Fresh Kills was an important case study in the tension between creating usable land (and what that meant) for future generations and the desire for ecological restoration of an intensely human-modified space. This posed issues different from those Robert Moses had addressed in the previous century, when there was little debate about the value of converting worthless marshland into taxable property or parks.

The advantages of using old landfills for new development are linked to size, location, and cost. According to experts at the Center for City Park Excellence at the Trust for Public Lands, "In theory, turning a landfill into a park transforms a noxious liability into an attractive asset."[8] However, the process for making this change requires a great deal of time and planning. A variety of existing environmental and landfill regulations, as well as closure practices and requirements, complicate efforts at land development or any type of ecological restoration program.[9] Also crucial: What is to be accomplished in remaking the landfill space? How does ecological restoration play a part?

Despite the challenges, retired landfills seem especially suitable for some form of ecological restoration. While definitions vary, the Society for Ecological Restoration International Science & Policy Working Group (SER) stated: "Ecological restoration is an intentional activity that initiates or accelerates the recovery of an ecosystem with respect to its health, integrity and sustainability. Frequently, the ecosystem that requires restoration has been degraded, damaged, transformed or entirely destroyed as the direct or indirect result of human activities."[10]

The SER also emphasized that while "restoration endeavors to return an ecosystem to its historic trajectory"—that is, the one it had been following in the natural world—it will not necessarily be able to return the restored ecosystem to its former state. In the simplest cases, restoration consists of removing or modifying "a specific disturbance" to aid in recovery of the ecosystem. Under more difficult conditions, there may be a need to reintroduce native animal and plant species that have been lost and eliminate "harmful, invasive exotic species" that have taken their place. In all cases, ecological restoration is the process of "assisting the recovery of an ecosystem that has been degraded, damaged, or destroyed."[11]

Ecological restoration has been around for many years and practiced in many ways, such as erosion control, reforestation, and range improvement.[12] "The term *restoration* may be used broadly to refer to attempts to restore selected attributes of a landscape, such as its productivity, its beauty, its historic value, or its value as habitat for certain species of plants or animals."[13] Especially in early practices but also more recently, human interests have played a large role in such processes. Some environmentalists and environmental managers opposed restoration for fear that it would threaten efforts to preserve natural areas.[14]

The science of restoration ecology, however, is much newer, in terms of developing concepts, models, methodologies, and tools for reconciling theory with actual practices.[15] Richard Hobbs, of Australia's School of Environmental Sciences at Murdoch University, stated, "To restore an ecosystem, we need to understand how it worked before it was modified or degraded, and then use this understanding to reassemble it and reinstate essential processes."[16] A variety of difficulties and challenges emerge when attempting to apply ecological theory to the practice of restoration. Among them is determining when an ecosystem is so degraded or damaged that it needs restoration.[17] All ecosystems "run down" in some way and at some point, but how do we determine if and when human intervention—and by what means—should take place in the restoration process?

William Halvorson, of the School of Renewable Natural Resources at the University of Arizona, declared, "In dealing with societal expectations for ecosystem restoration, and therefore restoration ecology, I have observed that, for the most part, these are more about making systems productive for our use than about returning ecosystems to some theoretical, pre-human disturbance state."[18] But in recent years, many ecologists have disputed Halvorson's claim, promoting instead the idea of "ecocentric restoration." This suggests that humans should treat nature as intrinsically valuable and not necessarily linked to human needs and wants.[19] In moving toward the development of a park at Fresh Kills, accommodating the desire to meet peoples' needs and wants and at the same time protect the natural elements of the area would be seriously tested.

THE INTERNATIONAL DESIGN COMPETITION

The remaking of Fresh Kills paid some homage to ecological restoration, but it never was intended to embrace wholeheartedly as its primary goal either ecocentric restoration (few, if any, landfill conversions did) or memorializing or preserving the landfill as a human artifact. From 2001 to 2006 the City of New York's Department of City Planning (DCP) led a master-planning process to convert the landfill and environs into "a world class park." At the time, about 45 percent of the site was once landfill and landfill operations, and the remaining acreage at the site was wetlands, open waterways, and unfilled lowland.[20]

In the 1990s the ecologist Steven Handel, of Rutgers University, began considering whether a coastal scrub habitat typical of the region could be restored at Fresh Kills. Could the site become a woodland refuge? He observed that engineers "were just planting grass to prevent erosion" but that there was no biodiversity. After catching the attention of city officials at that time, Handel and his team planted seven hundred trees and shrubs of seven different species. Birds began to roost there, and seeds that the birds carried took root. Soon the site had more than twenty plant species.[21] In 2000 Handel became the first director of the Center of Urban Restoration Ecology (CURE), a joint venture between Rutgers and the Brooklyn Botanical Garden. Speaking about the proposed park at Fresh Kills a few years later, he remarked, "I feel blessed because I can see what this [Fresh Kills] will be like." "Someday," he added, "Fresh Kills is going to be three times bigger than Central Park. It's going to be a destination, not a place to avoid and make fun of."[22]

Brendan Sexton, the former Commissioner of the Department of Sanitation, was among the first to promote the idea for what became Freshkills Park. Sexton stated in an interview—a bit tongue in cheek—that it was "a lunatic idea" that would have to be encouraged quietly at first because of the skepticism for such a venture on Staten Island and in the rest of New York City. The Fresh Kills site, however, had great potential as a park. Only about half of it had been filled with refuse (or had been filled many years before), and it still contained—aside from landfill infrastructure and roads—intact wetlands, wildlife habitats, and open spaces. In addition, New York City had at least six other parks and several golf courses within the city limits that once were landfills, including the Fountain Avenue site in the National Gateway Recreation Area and the Marine Park Golf Course in Brooklyn.[23]

One story has it that in 2000 the landscape architect Lee Weintraub suggested to Sexton, then president of the Municipal Art Society (MAS), on the occasion of the closing of Fresh Kills and also the 150th anniversary of Central Park, that the landfill and environs might make a great park. Weintraub also had in mind a design competition like the one that had been held for Central Park in 1858. At the MAS, Director of Planning Issues Ellen Ryan joined the two to help move the idea forward.[24] They made calls to the Department of Planning, the Parks Department, the DSNY, and other entities. The Parks Department seemed an obvious place for the Fresh Kills park concept to grow, but Sexton's ties to both the Municipal Arts Society and the DSNY were also useful.

Sexton then worked in association with Commissioner Amanda M. Burden of the Department of City Planning, among others, to raise money for public outreach and to solicit proposals for the park. The proposed park idea and international competition was presented to Mayor Rudolph W. L. Giuliani, on the grounds that there never again would be this much land available for such a venture. Mayor Michael R. Bloomberg followed through when he entered office. Support also came from the New York State Department of State, the Division of Coastal Resources, and other entities.[25]

On September 5, 2001, the City of New York announced the start of an international design competition for a "phased end use" for Fresh Kills Landfill. (The subsequent process was delayed somewhat because of the World Trade Center disaster, but the idea to have a park was not challenged.) A request for proposals (RFP) was prepared to attract multidisciplinary teams composed of designers, engineers, ecologists, artists, and planners.[26]

Forty-eight teams (representing more than two hundred firms) responded to the RFP, and six finalists were selected to develop proposals for a draft master plan. Each proposal would speak to issues of land use, natural ecologies, and landfill operations and address the phasing of the project and its necessary flexibility. The task of the participating teams was complex, especially because the DCP brief proposed a natural study area encompassing a site much larger than the landfill, including parts of eastern New Jersey across the Arthur Kill. Natural systems would include geology, hydrology, soils, and plant and wildlife habitats, along with land uses such as transportation networks, residential neighborhoods, and other infrastructure. Historic uses would have to be taken into account, such as Native American and immigrant settlements, farming, and industrial development.[27]

All candidates were asked to "generate ideas and innovative designs to meet the needs of the city's communities and that respond to the natural and constructed nature of the site." The call for proposals did not specifically request that the space be transformed into a public park. The teams were challenged to "set new standards for landscape, architectural, and environmental design." They were to retain large areas of open land while improving routes of movement and providing active recreation; be aware of the site's size, topography, urban location, and potential for a variety of land uses; and recognize the site's ecological sustainability.[28] The promoters hoped that an open competition would generate public enthusiasm for the proposed makeover of Fresh Kills.[29]

The competition's design brief reflected the lofty expectations for the project:

> Fresh Kills Landfill received its last barge of garbage on March 22, 2001, marking the beginning of a new era for the landfill. Whether seen as a community blight, a majestic landscape or an engineering marvel, the landfill in all its varied aspects must be fully understood by the planners and designers contemplating

its future. All the individual components of the 2,200-acre site, and the context in which they fit, form an intricate mosaic of constraints, possibilities and intriguing opportunities. A site of this size, situated near the heart of a vibrant metropolis, has the potential for becoming one of the 21st century's masterpieces of design and planning responsive to the needs of its neighbors and environment.[30]

Additional background information provided for and used by the competition finalists emphasized that "the potential exists for these areas [in and around the highly engineered site at Fresh Kills], and eventually, the mounds themselves, to support broader and more active uses. With effective preparation now, the city can, over time, transform this controversial site into an important asset for Staten Island, the city and the region." It added, "Although Fresh Kills is not a wholly *natural* environment, the site has developed its own unique *ecology*." The potential value of the site also was increased because the Fresh Kills Estuary was located along the Atlantic Flyway—a route used by many bird species during the spring and fall as they migrated north and south.[31]

Given the multifaceted nature of the site, a literal plan for returning it to nature was neither possible nor practical. In this context, the DCP's objective to seek "a world-class recreational and scenic amenity" and a "model of land regeneration" was understandable. The plan also veered away from a strictly ecocentric approach because of where the site was located. While envisioned as a park to be added to the existing inventory of the City of New York, it also would be—at least in the minds of the developers and supporters of the plan in the forgotten borough—an important recreation and tourist destination specifically for Staten Island. However, it was uncertain as to whether a park located almost as far as possible from the other boroughs while still being within the city limits could attract a sufficient number of people to make it worthwhile.

Whatever the proposed uses, Staten Islanders would have to concur (at least in some general way), and this would become clearer as the project sought input from locals. On the one hand, as a source of recreation, tourism, and aesthetic beauty, the reformed space could make a tangible claim for turning an eyesore and a health risk into a variety of amenities. On the other hand, efforts to build in programs of ecological restoration and regeneration would convey a sense of recovery for the site. A memorial for the fallen on September 11 also would be included. But achieving all things for all people would be a formidable task.

Some observers saw deeper meaning in the justification for the park plan beyond simply erasing the dump from view—a kind of ecological atonement on Staten Island or a symbol of healing for years of enduring its blight.[32] Yet it is difficult to imagine that making amends to locals drove the decision for the new park. The opportunity to turn one kind of asset into another is more plausible. What was perceived as unproductive salt marshes was turned into a giant dumping ground for the city, and that degraded land was now to be replaced with a recreational and ecological wonderland. Rather than turning it back into what it had been before 1948, Fresh Kills Landfill would be a departure point for creating something grander.

THE FINALISTS

Each of the six finalists received $50,000 to pay for project expenses. On December 5, after touring the site and revising their documents, the six proposals were publicly displayed at the Staten Island Institute for Arts and Sciences in St. George, and the groups were scheduled to meet before an internationally prominent jury to make formal presentations. Frank E. Sanchis III, executive director of the Municipal Art Society, proclaimed, "What you see on the walls is New York City's chance to catch up with the rest of the world. One of the projects here is a future landmark."[33] (After January 18, 2002, the designs would move to the MAS's Urban Center Galleries in Manhattan.) Later in December, three finalists were picked to compete for selection as the planning consultant: Field Operations (Philadelphia and New York City), JMP Landscape and John McAslan + Partners (London), and RIOS Associates (Los Angeles). That put the three others—Hargreaves Associates (Cambridge, Massachusetts), Mathur/Da Cunha + Tom Leader Studio (Berkeley, California), and Sasaki Associates (Watertown, Massachusetts)—out of the running.[34]

The last time there had been a city-sponsored competition for a large-scale park was in 1858, when Frederick Law Olmsted (who had lived on Staten Island) and Calvert Vaux were selected for their design of Central Park.[35] The Fresh Kills project, however, was likely to be much larger. Some people attending the gala display at the Staten Island Institute were not as kind as Sanchis. Gersh Kuntzman from the *New York Post* asserted that he was overwhelmed by the "pretentiousness" (apparently for the design features of a recreational park and/or the proposal representation). "Often," he added, "a bad idea can look good 'on paper.' Well, in a design competition, even the best ideas can look lousy on paper." A critic for *Metropolis*, Philip Nobel, echoed Kuntzman's views: "It's typical pretentiousness. These designers are just building a wall of attitude between their actual ideas—whatever they are—and the public."[36]

Reviews were mixed, but now that the proposals were on display, the project felt more concrete and tangible. In the background was the shadow of 9/11. Holly Leicht, the MAS's director of design, planning, and advocacy stated, "Fresh Kills has been indelibly affected by the World Trade Center disaster and we are confident the teams will find a way to acknowledge Fresh Kills as a new place in New York City's history, if appropriate, within this opportunity for ecological rebirth."[37]

Jeff Sugarman, the competition's coordinator for the DCP, declared, "We're hoping that the [three] finalists' scope of services will be available by early spring [so that] a final decision can be made sometime during the summer." Aside from the delays caused by 9/11, the final results of the competition had been postponed by the change in mayoral administration. Given the city's uncertain economy after the WTC disaster (a $4 billion deficit), the question whether money would be available for the gargantuan project remained unclear.[38] There was lots of speculation about what elements such a park should include (beyond the suggestions of the finalists) and when its construction would begin. Homeowners on Staten Island, including those in new housing developments along the edges of the landfill, were impatient to see their property values rise. There was considerable anticipation about the announcement of the competition winner.[39]

THE SCALE OF THE PLACE

The key challenges for all of the finalists were to deal with the immense space designated for the new park, to decide on an ecological strategy, to determine what amenities should be included, and to resolve how to memorialize 9/11. With few exceptions, little or no attention was paid to preserving or addressing the long history of Fresh Kills Landfill. To what extent that history was to be embedded in the park plans going forward was hazy at best.

Some observers, including Ukeles, Nagle, and Rathje, were sensitive about the importance of the retention of the memory of what Fresh Kills was and its role in preserving the community's history. The architect Linda Pollak more pointedly was concerned about the omission of Fresh Kills' topography from the competition entries. She regarded the interpretation of the site as an opportunity to consider "the conceptual re-engineering of these mountains." For the most part, however, the proposals did not discuss the mounds of garbage *as* mounds of *garbage*. Echoing others, she believed that because the Fresh Kills landscape was "a consequence of our own material desires and consumption," it also reflected a willingness to ignore our mountains of waste and "to look in a different direction rather than risk being identified with them, to have them go away."

This "urge to purify" in the competition continued a tradition of "deploying landscape as camouflage or mask for abused land." The pretext that the site could be returned to a natural state "feeds the myth" that nature is separate from people, culture, technology, and history. Pollak even proposed that given "the ethical dilemmas" the site raised, it might be possible to reframe proposals around themes of environmental justice "to acknowledge the fact that the site's neighbors deserve some indemnity for having had to absorb the rest of the city's waste for fifty years." Here, again, was the atonement theme.[40]

City leaders and competition participants, however, were not looking to past failures and shortcomings to define the new park experience. It would have been surprising if they did, given the strong impulse to forget about the tons of buried refuse rather than reconsider what they represented. Instead they sought something uplifting that blurred more than retained the human past associated with "the dump."

Of the six finalists, Mathur/Da Cunha + Tom Leader Studio took on remembering the landfill *as landfill* most directly. Its proposal reflected on the prevailing desire "to quickly heal things throughout the city," which included rebuilding the WTC site. "We need to make space for reflection prior to action, for thoughtful investigation into the nature of the problems to be solved [at Fresh Kills] and the topographies to be altered."

The project focused on vertically layered terrains and depositions that exposed both the natural and human-made attributes of the space: on the surface "WTC Debris," a site for staging events; a layer featuring "City Garbage," depicting deposits of household garbage as part of political conflict, cultural commentary, and scientific study; "Marsh Detritus," featuring soil and detritus from the salt marshes that could be turned into benches on the mounds to intercept freshwater runoff and to act as nurseries for plants; "Glacial Till," showing glacial deposits from ten thousand years ago; and "Crushed Rock," connecting the crushed rock along a geologic fault line and its interface with current road traffic.

As a starting point, the park would focus on human engagement with the five material depositions, "continuing many years into the future to realize a new and evolving landscape growing out of tendencies latent in the Fresh Kills of today." "Garbage," it added, "is the reason this project came about. The purpose of this competition is to transform the world's biggest landfill into a different kind of civic resource."[41]

The two other proposals not selected among the top three, Sasaki Associates and Hargreaves Associates, more closely followed the historical path leading to a new park, that is, some connection between past and future. Sasaki would adopt "a simple strategy of repairing the ecosystem and layering a network of access points and corridors to reveal the site and integrate it into the metropolitan matrix." It would transform the site over time around its "two identities"—marine habitat and landfill—and eventually introduce recreational and other uses "to ensure the public's enjoyment of the site, the longevity of its productive use, and ultimately, its regional prominence."[42] Sasaki's XPark plan saw the site as evoking "nineteenth-century showcase parks" like Central Park in New York City, Prospect Park in Brooklyn, and Fairmont Park in Philadelphia. XPark would include venues for active and passive recreation, incorporate a civic center, and develop a circulation system for pedestrians, cars, bicycles, horses, and nonmotorized boats.[43]

The Hargreaves plan for a Fresh Kills Parklands would develop a variety of spaces or sites within the larger landscape. There would be the Domain (a communal space), the Meadows (a forest and grassland edge), Lake Island (a bird sanctuary), the Fields (a central sporting complex), the Hall (for exhibits and collaborative projects), and the Preserve ("a place for the global community to reflect, remember, and seek peace"). Hargreaves hoped that the parkland "can be realized soon with a relative economy of means."[44]

Two of the final three proposals (John McAslan + Partners and RIOS Associates) incorporated strong cores to be used for instructional purposes. McAslan (the only non-U.S. finalist) wanted to transform Fresh Kills into "our planet's greatest new self-sustaining landscape environment" and "a model for sustainable regeneration and nature-led redevelopment." Seeking "a phased program of renewed accessibility," the transformation would begin with a series of three perimeter parks. There also would be "the Hub"—the center of the park—which would include the Migration Center (wildlife migration), the Energy Center (energy resources and recycling), and the Earth Center (native plants). The three elements represented by the centers would be synthesized into "Eco-spheres" containing world-climate zones and serving a variety of educational objectives. The aspirations for the proposed park would be built around new ways for people to relate to their land: ecosystem restoration and regeneration, public use of the land, and researching facilities for new technologies for environmental and architectural design.[45] It envisioned Fresh Kills as a "vast working resource."[46] Possibly because the proposal came from outside the country, memorializing the WTC remains was not highlighted.

RIOS's re-Park was a bit of an outlier among the proposals: "The site at Fresh Kills tells an extraordinary story of evolution and change. Just as its history began long ago before its use as a landfill—now covered—its next incarnation will likely not be its last. In fact, at the moment, this ever-shifting landscape is measurable and even visible in the settlement of the mounds themselves." The renamed rePark intended "to make change itself the theme, experience, and lesson of the place." The first part of its organization was to establish eight unique ecologies—Walking Wetland, Roadside, Woodland, Tidal Wetland, Freshwater Wetland, Commercial Berm, Landfill Mounds, and World Trade Center Memorial Forest. In a second phase, the project would turn to developing a schedule of programs ("transects") for each ecology. The transects (temporary or longer term) could be anything from alternative energy "farms" to garbage tours or garbage exhibits. The rePark was not meant to develop toward "the maintenance of a single, static appearance." Its only real uniformity was the changing nature of ecosystems per se and the role of the park as a cultural institution, looped through the actual Fresh Kills landscape. Dealing with the history of the landfill itself, however, seemed beyond its purview.[47]

LIFESCAPE

Field Operations' *Lifescape* won the competition and in June 2003 was formally selected as the planning and design consulting firm for the Fresh Kills project. The next step would be to prepare a master plan to implement the proposal. This was an exciting and innovative submission, but it was neither regeneration in a traditional sense (not meant to restore what came before) nor a demonstration of a connection to the landfill's history. The design did appear to meet the criteria set out by the city to take Fresh Kills beyond its past. Planning Commission Chair Joseph B. Rose believed that the Field Operations proposal was "clearly the most responsive to the community input, sensitive to the ecology of Fresh Kills, and attuned to the site's potential for great beauty."[48]

Lifescape was "both a place and a process," composed of nature, recreation, and residential programs. As its project director, James Corner, suggested, Fresh Kills was to be "creatively transformed" into a 2,200-acre of public parkland, featuring "extensive and beautiful" tidal marshes and creeks; more than forty miles of trails and pathways; and "significant recreational, cultural and educational amenities," including a "dramatic hilltop earthwork monument" to honor the September 11 recovery effort carried out on the site.[49] He also acknowledged that there would be resistance to the project. "For skeptics, and there are plenty of them . . . this can be done."[50]

Treated as a process of "environmental reclamation and renewal on a vast scale," *Lifescape* placed its greatest emphasis on renewal, with a "long-term strategy based on natural processes, agricultural practice and plant life cycles" meant to rehabilitate the seriously degraded space over a thirty-year period. Aside for the 9/11 memorial, there was little in the plan meant explicitly to recall Fresh Kills' long and complex history. The focus of Field Operations' plan was decidedly more abstract and theoretical than the other proposals: "Design at Fresh Kills is as much about the design of a method and process of transformation as it is about the design of specific places."[51]

In several respects, by retaining few of its original landscape features or its more sweeping physical characteristics, the objective of *Lifescape* resembled Olmsted and Vaux's goal in remaking and reengineering what became Central Park. Corner stated:

> Rather than erasing the past (landfill) on the one hand, or recreating a long-lost environment (nature), on the other hand, Lifescape instead proposes a growth emergence from past and present conditions toward a new and unique future. The result will be a synthetic, integrative nature, simultaneously wild and cultivated, emergent and engineered. In this way, the plan seeks to change how we experience reclaimed landscapes in the city and demonstrate new potentials for closed landfills and other post-industrial sites around the world.[52]

The new expansive "nature sprawl" at Fresh Kills was meant to create "a new form of public-ecological landscape, an alternative paradigm of human creativity, biologically informed, guided more by time and process than by space and form." The team would create "a matrix of lines (threads), surfaces (mats), and clusters (islands)" to maximize "opportunities for movement" of seeds and biota, people and activities. Line threads would direct water, matter, and energy flows. Surface mats would create mostly porous surfaces for coverage, erosion control, and native habitat. Clusters of islands would provide protected habitats, seed sources, and program activities.

The conceptual framework that Field Operations provided was the preface to a more detailed examination of a variety of general requirements, design objectives, ecological processes, and practical considerations, such as constructing the new environmental setting amid ongoing landfill operations.[53] Placing itself squarely in the middle of questions regarding the relationship between nature and culture, the proposal confidently claimed: "Nature, traditionally conceived as separate from cultural endeavor, can now be fully integrated into the man-made landscape."[54]

PERCENT FOR ART

A set of projects was added to the park venture before a master plan was finalized. These projects (not included in the Field Operations proposal) were intended to connect the memory of Fresh Kills Landfill to the new Freshkills Park. In 1989 Mierle Laderman Ukeles officially became the DSNY's Percent for Art artist for Fresh Kills. Since 1982, the city, under the Percent for Art law, was required to spend 1 percent of its budget for eligible city-funded construction projects on public artwork.[55]

Mierle Laderman grew up in Denver, Colorado. Her father was an Orthodox rabbi, philosophy professor, and chair of the local Board of Health; her mother was an active civic and community builder. Mierle went on to receive a BA in history and international relations in 1961 from Barnard College. While she was an undergraduate she returned home to Colorado and attended art school at the University of Colorado, where her father taught. Art became her passion. In the early 1960s she attended Pratt Institute, but it was not a good fit. She received an art teacher's certificate at the University of Denver and then returned to New York in 1967, where she attended NYU and received her master's degree in InterRelated art in 1973. She married Jack Ukeles, an urbanist and city planner, in 1966.[56]

In 1968 Mierle Ukeles had her first of three children, Yael Sara. She then was wedged between motherhood and a budding career as an artist. When she became pregnant, her sculpture teacher told her, "Well, I guess now you can't be an artist." Even as "the favorite student of a famous sculptor," she ran right up against the gender prejudice that so markedly structured the working world. She reflected later:

> You know, things were different then. It threw me. I got amazingly angry and disappointed in him. So we had our baby, that we wanted so much. Then I entered this, this time of "Who am I? and "How am I going to do this?" I literally divided my life in half. Half of my time I was The Mother. I was afraid to go away from my baby. I was very nervous about leaving her with people that I didn't know. And half of the time, I was The Artist, because I was in a panic that if I stopped doing my work I would lose it. I just had this feeling. Maybe it was a lack of confidence, or maybe I had struggled for so many years to become an artist, I felt that if I stopped working, it—the magic "it"—would evaporate, because that happens to a lot of people.[57]

Out of her dual life and her artistic proclivities toward public and performance art, Ukeles embraced an aesthetic that was both profound and profoundly mundane in her theoretical *Manifesto for Maintenance Art* (1969). As the author and activist Don Krug explained, the manifesto "questioned binary systems of opposition that articulate differences between art/life, nature/culture, and public/private. The manifesto proposed undoing boundaries that separate the maintenance of everyday life from the role of an artist in society." Ukeles was enamored with the everyday routines of life expressed as art.[58] She raised hard questions about gender and class in the art world and also about the very foundations of the avant-garde.[59] "I am an artist," she wrote. "I am a woman. I am a wife. I am a mother. (Random order.) I do a hell of a lot of washing, cleaning, cooking, supporting, preserving, etc. Also (up to now separately) I do Art. Now, I will simply do these everyday things and flush them up to consciousness, exhibit them, as Art."[60] And so she did. In 1973 and 1974 Ukeles made several maintenance-art performance works. For *Washing* (1973), for example, she scrubbed the steps and entrance to the Wadsworth Athenaeum Museum of Art in Hartford, Connecticut.[61]

In the mid-1970s Ukeles was deeply involved in *Art<>World*, a series of exhibits and performances at New York City's Whitney Museum Downtown. In the fall of 1976, the city was in the throes of its financial crisis. (Among the various maintenance-arts performances and activities, she staged *Maintaining NYC in Crisis: What Keeps NYC Alive?*.) In a review of *Art<>World* in the *Village Voice*, David Bourdon stated that Ukeles's work gave the exhibition "real soul" and speculated (it is difficult to know how seriously) that "if the Department of Sanitation, for instance, could turn its regular work into a conceptual performance, the city might qualify for a grant from the National Endowment for the Arts."[62]

Ukeles almost immediately wrote to DSNY Commissioner Anthony Vaccarello, sent him Bourdon's review, and suggested that the department might like to have an artist-in-residence. She was installed as the first-ever artist-in-residence for the DSNY (nonpaid) within a short time (although the exact confirmation date is not clear) and was escorted around the various work sites so she could familiarize herself with all aspects of the department's work. Vaccarello was replaced in 1977 by Norman Steisel, who became a great advocate and facilitator of Ukeles's work at the DSNY. She quickly began writing grant proposals dealing with maintenance art (including some mentioning Fresh Kills). In 1978 she received two National Endowment for the Arts grants, one devoted to planning urban earthworks for the city's landfills and the other for the "Re-Raw Recovery" feasibility study for a design competition to open landfills to a variety of creators.[63] Originally a surprising pairing, Ukeles and the DSNY had become an obvious marriage.

From 1978 to 1980 Ukeles undertook her first major work at the DSNY: *Touch Sanitation Performance*. For its first part, *Handshake Ritual*, she visited all fifty-nine community districts in the five boroughs over an eleven-month period and shook the hands of 8,500 sanitation workers (averaging forty per day). "Thank you for keeping New York City alive," she told them. It was important not to overlook the work of the sanitation crew, she believed, and to recognize the great value in what they did for the city. *Follow in Your Footsteps* consisted of just that: replicating sanmen's actions as they carried out their jobs. In the fall of 1984 Ukeles presented *Touch Sanitation Show*, which was a large-scale program including an exhibition at the DSNY's Fifty-Ninth Street Station, an environmental exhibition at the Ronald Feldman Gallery, and a garbage-barge ballet on the Hudson River.[64] In a letter to sanitation workers explaining her *Touch Sanitation* project, she began, "Nobody understands enough about what you do, how tough it is to work day after day on a job like this, about how hard it can be in lousy weather."[65]

Art critics were impressed with Ukeles's creativity and the grounding of her work in the real world. And even those closest to projects like *Touch Sanitation* and the thirty-two-block "dance" of mechanical street sweepers in *The Ballet Mechanique* (1984) came around to admiring what she did. Ralph DeCicco, a thirty-three-year DSNY veteran, stated, "I must be honest, for the longest time, I thought she was nuts." But she won him over.[66] In 1993 Sanitation Commissioner Emily Lloyd said that when she heard that the department had hired an artist-in-residence "I thought it was odd and frivolous." She added, however, "Her philosophy is my own. She's saying, 'We have to understand that waste is an extension of ourselves and how we inhabit the planet.'"[67] Even Commissioner Steisel, who had supported Ukeles in her work, had been skeptical of projects like *Touch Sanitation*. He first thought it would be a "disaster," but the project even brought hardened sanmen to tears "because she opened up a hurt that each

FIGURE 19.1 Mierle Laderman Ukeles, *Touch Sanitation Performance*, 1979–1980 (distant shot).

Source: Courtesy the artist and Ronald Feldman Gallery, New York.

one was still carrying from when he felt humiliated." Steisel asserted, "Mierle really made me understand the power of art."[68]

The Social Mirror (1983) was another straightforward and powerful way Ukeles made visible the work of sanitation workers and the DSNY, and it also brought home the connection of people to their waste. A twelve-ton, twenty-eight-foot-long garbage truck was fitted with giant mirrors, which reflected the viewer's image back to them. As the truck traveled the streets of the city, onlookers were directly confronted with their relationship to solid waste.[69]

Ukeles's most ambitious project, one intended to bring the public most intimately to the disposal issue, was *Flow City* (1983–2001). It was never completed because of the closure of Fresh Kills and management issues at the DSNY. At its core was the redesign of the Fifty-Ninth Street marine transfer station. This was meant to be a point of public access to a facility that formed the intersection of the collection and disposal of the city's solid waste. She proposed that *Flow City* become an integral part of the MTS (which opened in 1991). The elements included the "Passage Ramp," a block-long covered passage that ran alongside the ramp used by garbage trucks; the "Glass Bridge," where people could watch garbage being dumped from trucks into barges; and a "Media Flow Wall," with cameras at the station and at the landfill to track the flow of the river and the flow of the refuse.[70] In a March 1996 interview, Ukeles stated,

FIGURE 19.2 Mierle Laderman Ukeles, *Touch Sanitation Performance*, 1979–1980 (close-up).

Source: Courtesy the artist and Ronald Feldman Gallery, New York.

I call it FLOW CITY because it embodies a multiplicity of flows: from the endless flow of waste material through the common and heroic work of transferring it from land to water and back to land, to the flow of the Hudson River, to the physical flow of the visitors themselves. . . . The word "flow" is also the connection between me—the artist—and the engineers, the quantifiers who designed this facility.[71]

Much of Ukeles's work with the DSNY led her to Fresh Kills—physically and artistically. Dean Patricia C. Phillips of Moore College, a writer and curator who focuses on public art, accurately stated that Ukeles had an "obsessive attraction" to Fresh Kills (and possibly to all landfills) because of what it suggested about the human condition. Phillips added, "Landfills are uncomfortably 'in our face,' bearing witness to the scale of our cities, consumption, and appetites for new products and things. . . . Ukeles is passionate about landfills because they are sites of dynamic transition and imminent transformations."[72]

Ukeles concluded, "We have created [landfills] in common, and it shows." "Landfills sit face-to-face with the public, every day, in continual disharmony, literally across the street from living and working areas."[73] An interviewer asked her, "Don't you think it's funny that most people don't know where their food comes from or where their garbage goes?" She answered, "They're cut off. And the more cut off, the more alienated you are."[74]

Ukeles worked directly with three different landfill sites: a former landfill at Danehy Park in Cambridge, Massachusetts; the Hiriya landfill near Tel Aviv; and Fresh Kills, which she first visited in 1977. Her work at Danehy Park began when she was commissioned by the Cambridge Arts Council in 1990 to create an art project for the landfill's central mound. She also proposed a nighttime light show and other features at Hiriya and developed a large art installation for the Tel Aviv Museum of Art in 1999–2000.[75]

At the DSNY Ukeles was involved in a variety of meetings, especially those dealing with end-use issues. She interacted with many staff members and leadership, including Phillip J. Gleason, who served as chief engineer at Fresh Kills and eventually was named assistant commissioner for waste management engineering. Gleason supervised improving environmental standards at the landfill and was responsible for the compliance with the consent orders and regulations governing closure and remediation. Relevant to park preparations, he oversaw the placement of a final clay barrier (or geothermal cover) to seal all the surfaces of the landfill—a process he called the "final burial of the garbage." The seal was then covered with a layer of soil and the overall site contoured for drainage.[76] Gleason was an important confederate of Ukeles at the DSNY; she admired him for his extensive expertise, and he was a welcome complement to her artistic interests.[77]

FIGURE 19.3 Mierle Laderman Ukeles, *Marine Transfer Station Transformation*, 1984.

Source: Courtesy the artist and Ronald Feldman Gallery, New York.

Since 1989 Ukeles's contracts with the DSNY had called for developing projects for Fresh Kills focusing on participating in the end-use and master plan for the site and the design of "constructible" permanent artworks. In the 1990s she became more deeply immersed in the end-use programs, busily researching the site and its connections to the past. "[The research] goes all the way back to sacred earth mounds, cultures transforming degraded places into sacred places. Dirty and clean in the earth are very close to each other." Here, her strongly held Jewish faith is evident even in the most common context: "There's a blessing that you say every day when you come out of the bathroom . . . Thank you, God, for keeping open all the channels of my body, because I know if they were blocked I wouldn't be able to be alive."[78]

Her connection with Fresh Kills raised her hopes that the infamous dump could one day serve a very different public purpose than its original role. When the closure of the landfill was announced in 1996 (with the terminal date of December 31, 2001), a change order was registered aligning Ukeles's projects with the closure decision and the announced competition for park proposals. Before the submission of the park finalists' proposals, Ukeles submitted several conceptual designs to agencies involved in the project. Between 1990 and 2002 she participated in three exhibitions on Fresh Kills. In the earliest (1990), the Municipal Art Society's *Garbage Out Front: A New Era of Public Design*, she was guest curator.[79]

In the spring of 2002, Ukeles declared, "I've been waiting to get to work [at Fresh Kills] for 24 years. Even though I've been absorbing the place all these years and creating various art works and texts about it, I've been waiting for a master design team to be organized to begin permanent work here. I never imagined I'd be working on a cemetery." She was upset that New York City had "mingle[d] human remains in a place where they put garbage; that would collapse a taboo in our whole culture. That crosses the line." But she came to realize that once that decision had been made, the "entire site cannot be turned into a memorial." "It is also important to remember that this garbage landfill has been a 50-year burden to the people of Staten Island. . . . Something was 'taken' from Staten Island and should be returned." It was necessary, she argued, to separate a memorial "that returns identity to, not strips identity from, each perished person" from the "social sculpture" that "we have all produced"—"the un-differentiated, un-named, no-value garbage, whose every iota of material identity has been banished."[80]

Ukeles had a consulting role in the international design competition but withdrew because of the possible conflict of interest if she eventually joined a design team. When Field Operations won the Fresh Kills contract, New York City Planning Commissioner Burden and Commissioner of New York City's Department of Cultural Affairs Kate Levin pressed to have Corner include Ukeles in developing the master plan. In 2003 Ukeles presented to Field Operations

"Phase II. Early Conceptual Design," ideas she had been working on since 1990 for some park installations. She would work side by side but not directly with the Field Operations team in developing her projects. "The [Fresh Kills Landfill]," she stated in the introduction, "becomes a world-class symbol of our power to create transformation in three stages." She viewed a "transformation process" as material, spiritual, and social.[81]

CORNER'S VISION

Field Operations' immediate task after winning the bid for the park design was to develop the master plan under James Corner's leadership. Corner came with impressive credentials. Born in the United Kingdom, he received an undergraduate degree from Manchester Metropolitan University and a master's degree in the influential landscape-architecture program at the University of Pennsylvania in 1986. At that time, the prominent landscape architect and writer Ian McHarg chaired the program. Corner was employed at the University of Pennsylvania beginning in 1988, teaching courses in media and theory and supervising design studios. He acquired additional academic experience in Denmark and Sweden.[82]

Corner combined his academic training, writing, and teaching with professional practice. His teaching and research during the 1990s focused on postindustrial landscapes, emphasizing infrastructure, ecology, and urbanism. Field Operations was established in collaboration with the architect Stan Allen in 1998. (The pair went their separate ways in 2005, and Corner retained the name of the firm.) Corner's central role in his profession was elevated not only by winning the Fresh Kills consultancy but also because of his firm's ability one year later to win the contract for High Line Park (designed with Diller Scofidio + Renfro and the planting designer Piet Oudolf). The task was to transform obsolete elevated-railway infrastructure (a railroad viaduct of about 1.5 miles in length) on the west side of Manhattan into a park utilizing native species. Field Operations was regarded as one of the world's most influential design firms.[83]

Corner liked the Fresh Kills project because it allowed for transforming an existing site environmentally and aesthetically. "First," he stated, "you have to start with the place itself, its genius loci.... It is huge and hikeable, a remarkable sequence of journeys and panoramic vistas—of estuaries, New York Harbor, Lower Manhattan, and New Jersey. What we had to come up with were strategies for ecological restoration that take into account the unique conditions of the site." Thinking about the place from a development perspective, he added, "You have to think of it as a kind of farming, something like strip cropping, growing plants in ridges and furrows according to the contour lines of the slopes. . . . An agricultural landscape like this has its own integrity as scenery."[84]

Because of the relative remoteness of the park from much of New York, transportation access became very important both in terms of getting to the park and negotiating it once there. "It's an uphill battle," Corner added, "and there is a long way to go, but you could put a positive spin on it and argue that the Fresh Kills landfill is the best thing that has happened to Staten Island. What you have here is four square miles of land preservation, which will never be subjected to overdevelopment like the rest of the island."[85]

Field Operations' proposal received a great deal of acclaim, but not everyone was convinced that it offered the best approach for the new park. With respect to the finalist proposals as a group (in the conceptual stage at least), questions remained about the degree to which they downplayed or failed to engage the landfill contents in some meaningful way. As Pollak stated, "Because the technology for addressing the mounds is based on covering them, they are closed forms, literally bounded objects."[86] There might be some type of presentation that did not totally mask the human history of the site.

More pointedly, Mira Engler, the landscape architect and author of *Designing America's Waste Landscapes* (2004), is quoted as saying that the jury chose the wrong winner. *Lifescape*, to her, is

> a landscape in complete denial of its own origin and identity. It is restricted to the site itself, does not refer to cultural issues beyond, and shies from confrontation. "Lifescape" is about reconciliation; it panders to nervous neighbors, long-waiting developers, nature-thirsty urbanities, and green consumers. . . . This is a timid proposal, which will erase a landscape full of contradictions, its source of power.[87]

Corner insisted, and the jury agreed, "There is no state where [*Lifescape*] reaches a kind of peak. It should continue to be responsive to people's needs. A good park ought to be changeable. It shouldn't be a frozen picture." The great contradiction in building the new park was to develop an ecologically sound space, provide recreational amenities, and honor the dead of 9/11, while also ignoring or masking the landfill history of the place. As a *Staten Island Advance* editorial stated in December 2001: "Fresh Kills was such an enormous symbol of the city's contempt for Staten Island and the environment, that it's hard for islanders who have hated it for so long to think of it in other terms."[88] *Lifescape* was not meant to challenge the memory of the landfill but to swap those memories for something newer and possibly more hopeful. In this sense, it also camouflaged Staten Island's biggest eyesore—but at what cost?

MOVING TOWARD THE DRAFT MASTER PLAN

In September 2003 Mayor Bloomberg announced the commencement of the master planning process at the College of Staten Island.[89] Allocating $3.4 million for the venture, he stated, "We are here to begin a rare and exciting journey that may not be finished in the lifetimes of many of us." Borough President James P. Molinaro, with a mixed metaphor, told a reporter, "What's happening today is the final nail in the heart of Dracula."[90]

The Field Operations team worked closely with DCP, a committee of city and state agencies, elected officials, and civic groups to develop a draft master plan. The attempt to convert concepts into a tangible project was trying and had to compete for time and money against demands from other city agencies. The master plan would be "an evolving, long-term planning document" that would establish the framework and key elements of a site reflecting "a clear vision created and adopted in an open process." The four basic components of the plan would include the organization, design principles, and ecological techniques shaping the park; funding sources and a management plan; a development schedule; and guidelines for designers.[91]

Between the fall of 2004 and spring of 2006, a community advisory group was formed. Several meetings and workshops were conducted to allow New Yorkers the opportunity to present their own vision of the park. Initial discussions, especially at private meetings with elected officials, focused on transportation (the viability of landfill roads for connection to public streets) much more than park design. Mayor Bloomberg, however, tried to reinforce the future orientation of the park project when he stated, "It's sort of like being an artist with a large canvas, but a blank one, that we can paint anything we want on it."[92]

Despite the potential that the proposed new park represented, reticence among the locals was palpable. A March 2004 editorial in the *Staten Island Advance* stated, "What if we told you that even now, if you can ignore the methane vents and the rusting Sanitation Department equipment and other traces of its former use, Fresh Kills is a beautiful place? . . . What if we told you that nature, kept at bay in such a horrifying way for five decades, has reasserted itself in miraculous and thrilling fashion throughout this enormous tract?" It added, "Jaded Staten Islanders won't believe it. The members of the Advance Editorial Board didn't when [officials] involved in Fresh Kills' transition process invited us for a tour." The editorial, however, turned from cynical to optimistic, stating that "haters of the Fresh Kills landfill" became "believers in the possibilities of Fresh Kills' post-landfill future." "[The landfill] was, we thought, doomed forever to be a malign, poisoned no-man's land, regardless of how much garbage was dumped there. We were wrong."[93]

In a late March 2004 speech, Bloomberg declared, "My administration intends to build one of the great public parks in the world, and we want Staten

Islanders to be the driving force in its design. It is a once-in-a-lifetime opportunity to create an invaluable legacy for the next generation." The mayor was aware that because of environmental and safety concerns, the transformation "will take considerable time."[94] He did not forget, however, to heed the demands of local officials for more and better roads to ease traffic congestion in close proximity to Fresh Kills. Virtually every Staten Island elected official and the three community boards endorsed connecting the West Shore Expressway to Richmond Avenue for easier cross-island travel.[95]

Bloomberg also knew that not all of the locals welcomed post-landfill unabated development, preferring Staten Island to retain its long-standing reputation as a somewhat conventional suburban community. He formed a task force in July 2003 to deal with growth-management issues and zoning reforms. It consisted of elected officials, city commissioners, and representatives from Staten Island civic and community organizations and professional groups; it would deliberate for 120 days. Since the early 1990s, the population on the island had grown by 17 percent (an increase of 65,000 people between 1990 and 2000), which made it the fastest-growing county in the state. During the 1990s, housing units had increased by 23,000, or 14 percent. Many of those units were townhouses, which older residents who occupied single-family homes believed changed the character of the island. "According to many Staten Islanders," the task force report stated, "the very qualities that make the Island's neighborhoods so attractive are being diminished by what they perceive as inappropriate and haphazard development." The intention of the members of the task force was to continue to meet regularly, if necessary, through 2004, until the various zoning and planning recommendations were completed.[96] The effects of rampant development in the wake of the closing of the landfill and the construction of the park were on everyone's minds.

Public input into the park development proposal was useful, but the suggestions for the space were scattershot. On March 24, 2004, for example, more than three hundred Staten Islanders turned out at Holy Trinity–St. Nicholas Greek Orthodox Church in Bulls Head to discuss Field Operations' plans. Borough President Molinaro declared, "A night that many people said would never happen is happening." Representatives of veterans' organizations called for the preservation of a section of the park as a cemetery for New York State veterans of war. Diana Horning, whose son had died on September 11, again called for the proper burial of victims' remains. Some advocated for wildlife refuges, others for a variety of amenities.

Other meetings focused on planning the "North Park" near Travis and the "South Park" near Arden Heights. One meeting strictly dealt with a September 11 memorial. The memorial would continue to be one of the most contentious issues facing the park developers. Other proposals for the use of the vast space included ball fields, picnic and fishing areas, shops and restaurants, a museum, a dog run, an equestrian center, an ice-skating rink, an area for kite flying, a hot-air-balloon launch, an observatory, bocce courts, swimming pools, a golf course, and a mountain biking and bicycle motorcross (BMX) venue for the 2012 Summer Olympics (if New York won the bid).[97] Henry Salmon, co-chair of the economic development committee of the Chamber of Commerce, stated at a 2005 meeting, "The Chamber is interested in all activities that would support Staten Island business."[98]

The park would increase the percentage of parkland on Staten Island from 21 percent to 27 percent.[99] While suggestions on how to use Fresh Kills were many and varied, a number of guiding principles ultimately emerged from the meetings:

- Offer good transportation connections (a primary objective for most, if not all, local officials)
- Keep the site passive and natural
- Retain large-scale open spaces
- Build paths and trails for humans and horses
- Provide water (coastal) access
- Limit commercial activity to the site's core
- Offer sports and recreational facilities
- Demonstrate renewable energy and ecological techniques of land reclamation[100]

The first phase of development in the landfill area was to begin in 2007 or 2008 on the flatter areas around the capped North Mound facing Travis and in an area at the convergence of Richmond Creek and Fresh Kills Creek.[101]

In late August 2005, Mayor Bloomberg gathered with several other local officials to unveil the site that would become the first element of the new park, a twenty-eight-acre section called Owl Hollow Fields. It was adjacent to Arden Heights Woods and located at the southern tip of what would become Freshkills Park. The site would include soccer fields, nature trails, and picnic areas, but construction was not slated to start until the following spring. The announcement was good public relations for Bloomberg. "Owl Hollow Fields," he declared, "marks the beginning of one of the greatest environmental reclamation projects ever undertaken."[102] Molinaro stated, "We all took a chance on him. We know what the mayor has accomplished for Staten Island; we know he's a friend of Staten Island."[103] The opening of Owl Hollow Fields, however, was delayed until 2013.[104]

In September 2005 an inaugural one-hour bus tour of the closed landfill brought onlookers to the site of the anticipated redevelopment.[105] Herb Goldberger of Great Kills (standing on the South Mound with about thirty others) remarked, "I always wanted to sneak in here and see what this was like. It's hard to believe that I'm standing in New York City." Joyce Rivera of Arden Heights observed, "I think once it will continue to replenish itself it will be beautiful . . . I hope they keep it passive like this."[106] Curiosity, anticipation, and some trepidation were among the emotions people felt as the grand project got underway. Concerns ranged from the routine (roads and parking) to the sublime. By late 2005 or early 2006 the draft master plan would be released and hopefully answer everyone's questions.[107]

BLOOMBERG TO IMPLEMENT HIS OWN REFUSE PLAN

With Fresh Kills out of the picture as New York City's primary disposal option, the Bloomberg administration faced the ongoing challenges of refuse collection and disposal. A writer in *New York Magazine* in March 2002 (echoing some of Ben Miller's concerns) wondered why the city was spending $200 million a year to send its garbage to Virginia and elsewhere when Fresh Kills still had more space at about two-thirds the cost. "And now Mayor Bloomberg is cutting another $350 million from the public schools, curtailing weekend library hours, closing a few senior centers, and ending a recycling program that, while it may be costing the city money if Bloomberg's calculations are correct, is nevertheless obviously the right thing to do and should be fixed, not defenestrated." He concluded that the city's waste management plan "is in crisis" and quoted the *New York Daily News* as suggesting that the plan "has essentially collapsed."[108]

By late 2002 the Bloomberg administration was moving beyond Mayor Giuliani's "interim plan" in hopes of shoring up the disposal process for the city. The two administrations, however, shared one thing: reopening Fresh Kills Landfill was politically untenable. Bloomberg's proposed plan, as stated earlier, called for about 11,000 to 14,000 tons per day of refuse to be taken to the eight available marine transfer stations (MTSs) in the city, with a ninth station to be constructed on Staten Island. The existing MTSs would be updated and modified as compaction and containerization facilities. From the MTSs, the material would be placed in twenty-foot shipping containers; loaded onto boats, barges, or trains; and disposed of at out-of-state locations. The changes from the Giuliani program—especially reducing or ending the use of land-based transfer stations—were meant to improve disposal efficiency and to quiet protesters (especially in Queens, Brooklyn, and the Bronx) by removing hundreds of polluting trucks from the streets.[109]

Bloomberg was unable to find an alternative to relying on outside jurisdictions like New Jersey and other states to be responsible for so much of New York City's waste handling. In August 2002 he stated, "I don't think it's prudent for New York City to put itself at the mercy of any outside force any more than we absolutely have to."[110] But he did not have a better idea. The mayor's brief flirtation with incineration proved wildly unpopular, and in May 2002 he affirmed incineration was a dead issue.[111] Exporting of refuse, therefore, would remain central to Bloomberg's disposal plan, as it had been for Giuliani. In addition, the financial pressure to implement the new system would be great because of the city's shaky budget in the wake of the 9/11 disaster.[112]

Bloomberg's lukewarm support for recycling (especially his cutbacks on collection of recyclables in the summer of 2002) emboldened critics of his overall sanitation plans. In a compromise reached between the mayor's office and City Council, the city would contract to resume its plastics recycling and glass recycling in July 2003 and July 2004, respectively. Comptroller William C. Thompson Jr. had come to the conclusion that the city was not saving much money by suspending recycling, and this may have been persuasive for the mayor. The new contract at least would show a modest profit.[113]

Citizen activists continued to push for the city to become more aggressive in recycling. In 2004 the New York City Zero Waste Campaign called for "Reaching for Zero: The Citizen Plan for Zero Waste in New York City." It sought to reduce the city's waste exports to near zero in twenty years. Details of the plan included a zero-waste goal for New York City in 2024; developing viable programs for reuse, recycling, and composting and building or contracting its infrastructure; steadily increasing diversion rates; and minimizing environmental impacts and equitably distributing benefits of the program. In addition, it wanted to "change our focus from export and disposal to encompass economic development: building industry and creating jobs with materials that are recovered from our waste stream."[114]

The Zero Waste movement was spreading throughout the country, focusing on eliminating wastefulness on the "front end" rather than simply disposing waste by whatever means on the "back end." Zero Waste was an outgrowth of grassroots activism and policy-level advocacy but also quite different from the recycling-reform efforts of the 1970s. A major goal was to encourage (or enforce through government) product modifications and retooling industry to produce more goods of reusable and biodegradable materials. Though ambitions were great and the reorientation of the waste issue compelling, achievements would be hard to come by, given the cost factors and entrenched interests within the private sector and government in the current consuming society.[115]

The DSNY had reservations about how quickly the Bloomberg disposal plan might be fully implemented; some said six years. One official was concerned that stopping truck hauling could significantly raise collection costs and thus affect the department's budget. In January 2003 the city estimated that the cost of retrofitting the MTSs would be at least $240 million, or $30 million per station. At the City Council, Assistant Sanitation Commissioner Harry Szarpanski stated, "It was hoped that the footprints and existing structures [of the

MTSs] could remain as they were, and that would have reduced the implementation timeline. But as it turned out, we were not able to accomplish what needs to be accomplished within the existing footprints."

Criticism immediately arose from councilmen who strongly favored getting rid of the street-congesting and fumes-belching trucks from their neighborhoods. In the background were the ever-present budget concerns, which meant that the transition proposals could drag on for years.[116] Speaking to a *Waste Age* reporter in 2004, Bloomberg confessed, "I think sanitation is one of those problems that—I am pleased that we are going in the right direction with most things—but the sanitation problem, every time you think you have a solution, there's another part of the problem. It does not sound like it should be this expensive or this controversial, but it certainly is."[117]

By the summer of 2002, after the end of the WTC operations, the DSNY shifted its attention back to closure activities and preparations for end-use development at Fresh Kills. In fiscal year 2002, it had exported 11,140 tons of residential and institutional refuse per day under contracts with seven vendors.[118] The City Planning Department's ten-year capital strategy for 2004–2013 allotted about 5 percent of the city's budget to sanitation, with particular attention to the long-term waste disposal strategy. Approximately 1 percent of the DSNY's capital strategy would be dedicated to waste disposal infrastructure unrelated to the long-term export plan. Funding also would be available for end-use development at Fresh Kills and acquisition of wetlands near the landfill.[119]

On October 7, 2004, Mayor Bloomberg and Sanitation Commissioner John J. Doherty rolled out the Comprehensive Solid Waste Management Plan at the Blue Room in City Hall, accompanied by several other city officials. It emphasized "a long-term structure" for transporting the city's waste (based on proposals made in 2002) plus greater emphasis on recycling. A waterborne network would replace truck hauling and dependence on land-based transfer stations, reducing private-hauler truck trips by about 200,000 per year and eliminating truck miles by nearly three million each year. Bloomberg stated, "We have crafted a Solid Waste Management Plan that addresses all aspects of refuse removal in our City."[120]

The plan was built around eight general principles: environmental protection, fairness, efficiency, realism, innovation, reliability, collaboration, and service. More specifically, it addressed recycling more emphatically by calling on the private sector to build a modern recycling facility where the city would deliver all metal, glass, and plastic and some mixed paper. The recycling program was meant to augment the disposal program, increase diversion rates, and lower costs. For the first time, the city was focusing on objectives to improve commercial waste transport, and it also outlined plans for long-term export facilities in each of the boroughs (including a new facility for Staten Island).

For the four watersheds served by city-owned MTSs, the city planned to enter into twenty-year service agreements with private management companies.[121]

The tasks were great. In the city, about fifty thousand tons of waste and recyclables were collected each day; 25 percent was generated by residents and institutions and directly managed by the DSNY. Through the curbside program, the plan was committed to achieving a 25 percent diversion rate for recyclables by 2007.[122] "It's going to be a long haul," Doherty concluded, "but I think we have a really, really good plan."[123]

Also in October, Comptroller Thompson released a report—*No Room to Move: New York City's Impending Solid Waste Crisis*—which addressed New York City's "impending solid waste crisis." It demonstrated that "unfavorable forces and changing governmental regulations may lead to unprecedented expenses for New York City to dispose of its waste." The report was particularly wary of "this reliance on outside sources to manage the City's waste stream," which "significantly compromises the City's long-term financial interests." In a tone much different from the Comprehensive Solid Waste Plan, it declared that the city "must make immediate efforts to exercise control over its waste disposal process." This was in reference to the fact that in 2003 Pennsylvania and Virginia had absorbed more than 70 percent of New York City's public and private refuse, that is, 4.1 million and 1.8 million tons of waste, respectively. Both states had adopted regulatory measures concerning large landfills "that call into question the continuing ability of New York City to rely on these two states to meet its disposal needs." To fulfill its obligations the city would need to access landfills elsewhere that did not currently accept large volumes of out-of-state waste.[124] Thus the dilemma.

Closing Fresh Kills meant that New York City more than doubled the volume of putrescible waste it was exporting. Also, DSNY waste disposal contracts were competing with those in the private sector in the face of limited disposal space in several places. Rising exporting costs and "the questionable reliability" of the current network of disposal facilities meant "significant risks to both the public and private sectors."[125] Echoes of Ben Miller.

The report concluded that given the potential for more out-of-state problems, the current lack of a "viable rail or barge-based waste export system," and the need to increase recycling rates, the city had to address these issues "before escalating costs significantly impact the City's fiscal strength." It also recommended, among other things, that the city examine the role of publicly controlled waste disposal capacity, consider developing in-state disposal capacity, evaluate the purchase or control of out-of-state landfills, and reduce amounts of exported waste.[126] All of these observations and suggestions added a pall to the more optimistic tone of the Comprehensive Solid Waste Management Plan and, ominously, suggested that serious thought might need to be given about reopening Fresh Kills.

Even the dire conclusions of *No Room to Move*, however, would not alter Fresh Kills' immediate course. The Fresh Kills Park Project website proclaimed the new park to be "one of the most ambitious public works projects in the world, combining state of the art ecological restoration techniques with extraordinary settings for recreation, public art, and facilities for many sports and programs that are unusual in the city."[127] "Landfill to landscape" seemed inevitable.

While the planning of Freshkills Park was underway, its completion was hardly assured, and certainly not within the framework of a projected 2030 deadline. Field Operations, City Planning, and other entities had yet to finalize a master plan that would turn a park concept into reality. The DSNY and its partners had much left to do to implement the landfill's environmental and other end-use requirements, even assuming funds were available. By the end of 2005, closure design and construction continued at Section 1/9 and Section 6/7. Grading and contouring of Section 6/7 had been completed the previous year. In Section 1/9 grading and contouring continued, as were post-closure care requirements (inspection, monitoring, operation, and maintenance of environmental-control systems for the whole facility).[128] The transformation of Fresh Kills from salt marsh to landfill to cemetery to park was not yet complete.

CHAPTER 20

CROSSROADS

The year 2006 can be seen as a crossroads for New York City's disposal program. In that year the Department of Sanitation (DSNY) released the 2006 Comprehensive Solid Waste Management Plan (SWMP), which both in tone and procedure was meant to break with past practices in important ways. Disposal costs had almost quadrupled since the 1990s, skyrocketing from $81 million in 1991 to $320 million in 2011.[1] About fifty thousand tons of refuse and recyclable material was being collected in New York City each day.[2] Residents and institutions of various types produced about 25 percent of the total, which was directly collected and disposed of by the DSNY. The remainder was generated by businesses or through construction projects and privately managed with city approval. Solid waste disposal in New York City, as stated in the executive summary of the 2006 plan, was "vast and complex." "For years," the report added, "this complex network converged at the Fresh Kills Landfill in Staten Island."[3] The years since Fresh Kills' closure had forced a rethinking of citywide collection, recycling, and disposal practices, which led to the new plan.

Seeking a new identity for its notorious landfill on Staten Island, city leaders announced a competition for the development of what was to become Freshkills Park. Field Operations had been formally selected in June 2003 as the planning and design consulting firm for the project, which it had named *Lifescape*. Fields Operations, in cooperation with city agencies, released a draft master plan for the park in April 2006, "for reclaiming the largest landfill in the country for public use."[4]

FIGURE 20.1 Depiction of proposed Freshkills Park.

Source: Courtesy of Freshkills Park and the City of New York.

Expectations for the DSNY's master plan focused on establishing ways to move beyond the dependence on a single-borough refuse disposal site to a multiborough approach that was equitable for the city's entire citizenry. The hope for Freshkills Park was that a landfill could be converted into a hospitable, natural, and recreational landscape. Could these aspirations be met, and how?

THE DSNY BEYOND FRESH KILLS

In its final form, the new Comprehensive Solid Waste Management Plan intended to move beyond Mayor Rudolph W. L. Giuliani's "interim plan," which was primarily a truck-based system relying on land-based transfer stations in various boroughs and then shipment of refuse to facilities in neighboring states. The 2006 report declared that the Giuliani approach was "unsustainable as a cornerstone of any long-term disposal plan." The extensive reliance on trucking, as many parties had argued, had negative environmental impacts especially on communities along the major routes leading to the transfer stations. The intention of the new plan was to reduce markedly truck trips and miles related to disposal and establish "a cost-effective, reliable, and environmentally sound system for managing the City's wastes over the next 20 years."[5] It had taken from

2001 until 2006 to devise a new direction for the DSNY after the long Fresh Kills era and to address the criticisms leveled against the Giuliani administration's hastily designed plan.

The 2006 plan called for development of "state-of-the-art" (converted) marine transfer stations (MTSs) at four existing sites, which would receive and containerize the city's residential refuse for transport by barge from the stations. In addition, it proposed that the Staten Island transfer station and up to five existing private solid waste transfer stations (under long-term contracts) would use rail or barges to move the refuse. The converted MTSs and other facilities would serve the communities in which they were located throughout the boroughs and "accommodate on-site truck queuing." The plan included initiatives to reduce truck traffic by commercial waste management firms and also called for a public-private partnership to operate a centralized processing facility for recyclables at the South Brooklyn marine terminal.[6]

The report also proclaimed that the new approach to disposal would include "meaningful and groundbreaking" changes to the city's recycling program and proposed "concrete steps" in addressing concerns related to commercial waste management. The framework and principles of the SWMP promised efficiency, reliability, and "sound business principles" but also struck a conciliatory note about fair treatment to all boroughs and building collaborative relationships with community groups, environmental advocates, elected officials, and the private sector. The 2006 plan may have been the first of its kind to include language supportive of social-justice concerns, especially neighborhood pollution around the transfer stations. Its recognition of the need to address environmental problems, however, narrowly focused on air and noise issues related to truck traffic but not necessarily on issues of contamination by the refuse at the site, for example.[7]

The report treated recycling as if it had been an effective and long-standing part of the DSNY's solid waste management practices. Recognizing the "unique challenges" of recycling in New York City (including the large numbers of high-rise buildings and multifamily residences), it asserted that the city was "a leader in recycling among other large American cities."[8] This may be true considering the sheer scale of the activity, but data from the inception of the recycling program did not bear out that New York City had ever met U.S. Environmental Protection Agency guidelines for diversion rates, which in 1988 had been set at 25 percent.[9] The SWMP hoped to "bolster that leadership position" through "aggressive but realistic" diversion goals, new education initiatives, and commitment to new in-city processing facilities.

A primary objective was to reduce the cost of curbside recycling (currently paper, metal, glass, and plastic) especially through a twenty-year contract with Sims Hugo Neu Corporation to process and market metal, glass, and plastic. More privatization of recycling was meant to shift market risk for commodity prices to the private sector, attract private investment in infrastructure, and give

FIGURE 20.2 DSNY Staten Island Transfer Station, October 14, 2017.

Source: Wikimedia Commons.

partners a commitment to help them better market the materials. Based on this new approach, the SWMP hoped to achieve a target of 25 percent diversion by 2007.[10] Ten years earlier, the DSNY had hoped to achieve that 25 percent rate by 2001, but it had yet to reach 20 percent.

Commercial waste remained a complex problem for the city. Dramatic increases in tipping rates at Fresh Kills for private haulers since the late 1980s had pushed this group to export much of its refuse to New Jersey and elsewhere. The private haulers also had built an extensive network of land-based transfer stations in the city to store and then move the waste. When Fresh Kills closed, the DSNY became even more reliant on private transfer stations and landfills and waste-to-energy facilities outside of the city limits. Thus truck traffic in the boroughs to and from private transfer stations grew precipitously, creating intense protests.

The SWMP suggested three categories of action related to private haulers: improving conditions around the transfer stations, facilitating a change in the private waste industry from relying on truck transport to barge and rail transport, and redistributing or limiting private transfer capacity from the current communities where the effects of the land-based transfer stations was most acute. Of all of the challenges confronted by the DSNY, a change in the culture and practices of commercial waste disposal might very well have been the most difficult.[11]

EARLY TRIALS FACING THE SWMP

A tangible concern during the transition to the new SWMP was cost. In June 2005 the New York City Independent Budget Office (IBO) estimated that implementing the plan would require a capital investment of over $400 million to rebuild and reopen the city-owned MTSs. In the short run, it estimated that the plan would add $100 million per year to the cost of disposing city-collected waste.[12] Between 2005 and 2008 the actual cost of disposal rose from $300 million to about $400 million primarily because of rising exporting contracts. (The total DSNY budget rose from $658 million in 2000 to about $1.3 billion in 2010.)[13] The IBO assumed, however, that if the plan met its objectives of keeping down the cost of exporting refuse and also reduced the number of collection trucks traveling to transfer stations, the city possibly could save hundreds of millions of dollars over the long run. Lower fees for processing recyclables could make recycling more cost-competitive.[14]

Recycling remained a contentious issue, however, despite the apparent willingness of the city to upgrade its program substantially. The city was making the first major expansion of recycling since it was introduced in 1989, most importantly because diversion rates had stagnated for several years.[15] Budgeting for recycling was somewhat unstable, as new programs replaced old ones and funding shifted. For example, the DSNY's Bureau of Waste Prevention, Reuse, and Recycling was replaced with the Recycling Outreach & Public Education Programs in 2010—with a budget cut in its programs.[16]

Various groups disagreed on the actual diversion rates achieved by the city, the degree to which commercial diversion rates were monitored, and whether the city was in compliance with state recycling policies.[17] Potential markets for recyclables were always changing, and they were influenced by everything from concerns about energy security to climate change. Recyclable-paper prices rose and fell as supplies mounted and declined. Items made from glass often were too expensive to reprocess and were not accepted by many vendors.

Recycling economics were basically local and thus sometimes difficult to project. For example, variations in hauling and tipping fees even within one region skewed projections for recycling costs. The National Solid Wastes Management Association stated in 2008 that such fees ranged from $24 per ton in the south-central and west-central regions of the United States to $70 in the Northeast, and thus the price that recyclables received (about $100 per ton) could be a savings or a cost depending on tipping fees. In 2008 the national average tipping fee was $34.29. Thus most curbside programs cost their municipalities money.[18]

Not only the shifting value of recyclables but also the composition of waste affected all aspects of refuse collection, disposal, and recycling. In 2006 the composition of municipal solid waste (MSW) collected by the DSNY was

47 percent organics, 23 percent paper, 14 percent plastics, and 7 percent construction debris, with the remainder glass, appliances and electronics, household hazardous waste, and metal. These various discards had to be handled differently in many cases, and some items, such as e-waste (electronics), had yet to establish stable markets. The rules and regulations about which plastics could or could not be recycled, for example, were byzantine.[19]

Implementing a revised waste export plan also was difficult. Particularly significant was a new U.S. Supreme Court ruling on flow control issued on April 30, 2007. The *Carbone* decision in 1994 had declared that solid waste was a commodity under the Commerce Clause of the U.S. Constitution, and therefore municipalities could not establish flow-control laws (which forced haulers to transport refuse to designated sites) to determine where refuse could be taken or disposed of. In New York as elsewhere, the *Carbone* decision was a victory for private haulers seeking to dispose of their waste both in state and, more importantly, out of state.

Under the 2007 case, *United Haulers Association, Inc. v. Oneida-Herkimer Solid Waste Management Authority*, local governments held the right to direct the flow of solid waste to publicly owned facilities without violating the Commerce Clause. This decision allowed government-owned waste facilities to charge private haulers high tipping fees. Private haulers also faced a more complicated and confusing waste-export market.[20] Barry Shanoff, outside legal counsel for the Solid Waste Association of North America, argued, "questions will come up, for example, when the facility is publicly owned and privately operated. As a result of this decision, we're going to see all sorts of flow-control scenarios unfold."[21] As of the end of 2014, *United Haulers* was upheld, and only in one case did the courts find a flow-control ordinance unconstitutional.[22]

Daily operations for the DSNY did not simply include collection, disposal, and recycling duties but, in addition, the ongoing management of Fresh Kills Landfill. Despite the initiatives that led to Freshkills Park, the DSNY's administrative and financial responsibilities for long-term closure and monitoring activity at the site would continue for thirty years after the facility closed. About ten years into the process, Commissioner John J. Doherty reflected, "It was always one of those things, 'Well, Fresh Kills will be around for a long time.' And then, all of a sudden, [the DSNY] got the notice that it was going to close down. And that required a major change in the department's operation. How are we going to get rid of garbage?"[23]

Some good progress, however, had been made by 2008 in dealing with Fresh Kills. In 2007 the DSNY assumed full responsibility for landfill gas control and the purification system. A new leachate contract had been struck in 2004, and a new drainage system was put in place in June 2005. Closure design and construction moved forward at Section 6/7 and Section 1/9 (the landfill's two largest areas), and post-closure operations continued at Section 2/8 and Section 3/4

FIGURE 20.3 Installing pipes at Fresh Kills, September 19, 2008.

Source: Courtesy of Michael Anton, NYC Department of Sanitation.

FIGURE 20.4 Installing a mound cover, Fresh Kills, September 19, 2008.

Source: Courtesy of Michael Anton, NYC Department of Sanitation.

(closed in the 1990s). Contouring the mounds also continued apace. Materials from the World Trade Center were buried at Section 1/9.[24] The DSNY also participated directly in the end-use plan to develop the park site.

In 2006 Doherty had estimated that the total cost of closing the landfill over thirty years could exceed $1.4 billion—substantially higher than earlier projections. By 2008 that estimate was raised to $1.8 billion, with over $420 million of the work already completed.[25] As of 2017, however, the City Council had not funded a comprehensive study of potential health risks caused by the landfill, which would be a significant cost and possibly open up messy legal and political issues.[26]

WASTE TO ENERGY—AGAIN

In 2006 city leaders and the DSNY were convinced that, in response to the closure of Fresh Kills, they had finally established a sound roadmap for refuse disposal for the next twenty years. There were some quibbles and concerns with the new SWMP but no criticisms as serious as those of the Citizens Budget Commission (CBC), beginning in 2012. Founded in 1932, the CBC was "a nonprofit, nonpartisan civic organization devoted to influencing constructive change in the finances and services of New York State and New York City governments." Two major themes emerged in its 2012 report, *Taxes In, Garbage Out*: the price tag of $2 billion in tax dollars for refuse collection and disposal was just too high, and the emphasis on exporting waste did "enormous environmental harm." About three-fourths of city refuse went to landfills at the time, 98 percent of which was shipped to Ohio, Pennsylvania, South Carolina, and Virginia. Shipping refuse by trucks and trains, the report argued, generated large amounts of greenhouse gases, and utilizing landfills produced additional emissions—if not now in New York City, then certainly elsewhere. (All of this was pollution redirected outside the city but pollution nonetheless.) In addition, the cost of disposal was not rising because of greater volumes of waste being produced (the trend was moving in the other direction) but because of the high transport costs of exporting refuse.[27]

The CBC's report reopened an old debate, that is, shifting a much larger part of the city's solid waste management system away from long-distance exporting to landfills and toward waste-to-energy technology (WTE). The 2012 report claimed that WTE technology was cheaper and environmentally safer than landfilling. New York City, it argued, sent only 9 percent of its municipally managed refuse to WTE facilities, far less than other big cities, especially those in Western Europe and Scandinavia. For example, about 20 percent of the energy for Sweden's district heating system came from refuse burned in incinerators.[28] The CBC recognized opposition to the disposal method but argued that it was "rooted in misunderstanding." Rather than replacing recycling, the CBC argued, WTE complemented it.[29]

To make its point stronger, the CBC report discussed the complications of the city's "two trash system"—the DSNY and the private carters. While the two handled waste in similar ways, they diverged appreciably in the extent of recycling. About 63 percent of commercial waste was recycled, compared with only 15 percent for residential and governmental waste. The major reason for the disparity likely was the more significant financial incentives that private businesses have to recycle. The report criticized the city's system for its confusing regulations but also acknowledged that the smaller residential spaces it needed to access for recyclables created problems. The CBC encouraged more recycling to preserve virgin materials, reduce using landfills, and avoid long-distance transport. But it also supported the more vigorous waste reduction programs (composting, saving virgin materials, and so forth) that were becoming more popular in the country, even at the DSNY. "The best way to reduce the fiscal and environmental impact of garbage," the report stated, "is simply to not create it."[30]

The CBC released additional reports in subsequent years strengthening their fiscal and environmental critique of the city's collection and disposal policies. In 2014 it released *12 Things New Yorkers Should Know About Their Garbage*, which included its concerns about the high collection costs of the DSNY-run refuse system, the inefficiency of the recycling system, and the overlapping collection routes of city trucks and private haulers.[31] In 2014 and 2015 the CBC published two reports focusing on increased DSNY budget efficiencies and better ways to pay for solid waste management.[32]

WTE was on the table again. The round robin in extolling the virtues of resource recovery, landfills, incinerators, and WTE seemed endless. The closure of Fresh Kills—despite the DSNY's optimism over the 2006 SWMP—prompted even more debate over the best disposal options for the city.

In 2007 Mayor Bloomberg released PlaNYC, a sustainability project for "a greener, greater New York" aimed at better preparing the city by 2030 for challenges such as future population growth and climate change. The plan included a section on solid waste management (in line with the SWMP) outlining a number of initiatives, including improving overall efficiency of the city's waste management system, promoting waste prevention, developing recycling incentives, recovering organic materials from waste, increasing diversion of refuse from landfills by 75 percent, and the possibility of WTE. By 2012, 32 percent of New York City's waste was exported by rail, 23 percent by DSNY trucks, and 45 percent by long-haul trucks. Implementation of the 2006 SWMP, however, had yet to change significantly these rail-use and barge-use statistics.[33]

In his 2012 State of the City address, Bloomberg made clear that he wanted to "explore the possibility of cleanly converting trash into renewable energy."

This was done in the name of seeking jobs, energy independence, reduction in greenhouse gases, and savings in solid waste management costs. On March 8, 2012, he announced a call for proposals to private companies to construct a pilot waste-to-energy facility near or within New York City. The municipality would not provide any capital funding for the proposed facility but would pay a fee to the plant operator.[34] Even though a portion of the city's waste was being disposed of by WTE incineration, there were no WTE facilities within the New York City limits at the time. Nationally eighty-seven WTE plants were in operation as of 2012 (ten in New York State).[35]

The mayor considered a portion of the closed Fresh Kills Landfill as a possible site for the pilot project, which prompted immediate protest from Staten Island. Councilman James S. Oddo, a Republican from Mid-Island, asserted, "It is clear that this administration has no understanding or appreciation of what Fresh Kills did to this community for 55 years." Councilwoman Debi Rose, a Democrat from the North Shore and the first African American elected to higher office from Staten Island, added, "New York City Government thinks there is a welcome mat at the landfill. We need to let them know that we refuse to be their doormat."[36] Bloomberg removed the Fresh Kills site as a possible location for the WTE plant in April.[37] Environmental-justice advocates in other parts of the city also began protesting against the RFP, including the New York Public Interest Research Group (NYPIRG). Eddie Bautista, executive director of the NYC Environmental Justice Alliance, declared that New York City's recycling diversion rate (as he understood it) of about 15.4 percent was extremely low and that something should be done to increase it. He worried that a WTE plant in the city would be a serious problem. "There are only so many neighborhoods zoned for this type of activity. They're typically located in low-income communities and they're already over-burdened with industrial polluters."[38]

A few companies, including Covanta Energy, which operated seven facilities in New York State, answered the RFP. Legal and financial concerns complicated the possibility of utilizing WTE technology. And it was unclear if there was sufficient political support for such facilities, especially given the impending end of Bloomberg's term as mayor in 2013.[39] Such a project for New York City still awaits development.

Solid waste problems were never easy to deal with in New York City. But rational planning most especially seemed to bump up against the trials of unanticipated and sometimes extreme events. Hurricane Sandy began as a tropical wave in the central area of the tropical North Atlantic on October 19, 2012. The gigantic storm—measuring more than nine hundred miles at its greatest extent—brought wind and flooding to Jamaica, Cuba, Haiti, the Dominican Republic, and mid-Atlantic and New England states. The total death rate was 147. On October 29 the storm moved west and made landfall near Atlantic City. When cold air mixed with the warm air of the hurricane, the weather event became Post–Tropical Cyclone Sandy, also known as Superstorm Sandy. The track of the hurricane/cyclone resulted in what the National Weather Service called "a worst case scenario for storm surge for coastal regions from New Jersey north to Connecticut including New York City and Long Island."[40] In New York City the storm surge measured about fourteen feet (some say less), causing the Hudson River, New York Harbor, and the East River to flood streets, tunnels, and subways in Lower Manhattan, Brooklyn, Queens, and Staten Island. Damage along the New Jersey coastline was extensive.[41]

It was Staten Island, at the back of New York Harbor, that suffered some of the worst damage. (Fresh Kills absorbed a significant amount of the storm surge.) Peak storm tides reached sixteen feet, with the area along the eastern coastline suffering most of the twenty-four casualties (forty-eight in all of New York).[42] The loss of lives and extensive destruction of property were tragedies, but then a further indignity was Fresh Kills Landfill reopening as the temporary staging site for debris from Sandy. This prompted memories of September 11. The decision to use Fresh Kills, given the nature of the emergency, was understandable (and opportune), but many Staten Islanders also found it a bit too convenient. A newspaper story at the time observed, "Amid the clanging of dump trucks, a crane with a clamshell scoop hoisted a pile of debris as big as a minivan and dropped into a waiting barge—striking evidence that New York City has revived a place it just cannot seem to do without."[43]

A NEW ADMINISTRATION, A NEW DAY?

Bloomberg's stance on waste-to-energy technology may have obscured his evolving views on waste disposal for New York City. Be it public pressure or concerns over climate change, the mayor no longer favored the status quo. While he may not have been the "passionate convert" to recycling that the editorial board of the *New York Times* proclaimed in April 2013, with nine months left in his term he made great strides. Bloomberg announced that rigid plastics (food containers, toys, etc.) would be recycled by the city for the first time.[44] He also sought to see recycling diversion rates increase (to 30 percent by 2017) partially through the city's new "Recycle Everything" public-information campaign. (Diversion rates for the city had been static at about 15 to 18 percent for many years, even after the end of Bloomberg's term.)[45] The DSNY's 2006 SWMP and attempts to implement it also demonstrated more attention to the environmental implications of collection and disposal, waste prevention, composting, and recycling.[46]

The impetus to change (or at least modify) waste disposal perspectives and practices evident in the late stages of the Bloomberg administration continued with even more intense rhetoric under the new mayor, Warren Wilhelm "Bill" de Blasio Jr. The liberal Democrat beat the Republican Joseph J. Lhota in the November 2013 mayoral election in a landslide and became the first Democratic mayor of New York City in twenty years. Born in New York City in 1961, Bill de Blasio was not a lawyer or businessman but an activist and career politician. He worked on David N. Dinkins's mayoral campaign, led William J. "Bill" Clinton's reelection campaign in New York, and managed Hillary Rodham Clinton's 2000 campaign for the U.S. Senate. In 2001 he joined the New York City Council (representing Brooklyn's Thirty-Ninth District) and became the city's public advocate in 2009. He supported progressive causes and vowed to tax the rich to pay for prekindergarten programs.[47]

Not unlike some of his predecessors, de Blasio made no immediate changes at the DSNY. He kept Doherty as commissioner until after the winter snow season.[48] Doherty resigned in March 2014. As the long-time commissioner stated at one point, "Sometimes we have been the heroes and darlings of the city. Sometimes we have been criticized." Doherty certainly had been both.[49]

The mayor named Kathryn Garcia as the new commissioner. The native New Yorker had extensive operations experience. She had served for ten years as vice president for Appleseed, a nonprofit consulting firm and advocacy group. Garcia then moved into the public sector with the New York Department of Environmental Protection. She was chief operating officer there when appointed by de Blasio to sanitation commissioner. The mayor and his commissioner shared similar objectives for the department, seeking to remake its image.[50]

The annual reports of the DSNY, beginning with the 2014–2015 report, reflected the focus and style of the new administration. The appearance of the report alone was less institutional and less detailed and was built more around broader issues than a typical office or bureau report. In so doing, however, the reports seemed to be driven more by public relations and less by data. The heart of the pages devoted to disposal emphasized "Getting to Zero," that is, achieving zero waste to landfills by 2030 by "exploring incentives to reduce waste and encouraging businesses to vastly reduce the garbage they generate." The DSNY now promoted "breaking the cycle" by "cutting the amount we consume, reusing what we already have, donating what we no longer need and recycling as much as possible [to] keep garbage from landfills—and help build a healthy, safe and clean future."[51] It was an optimistic message meant to inspire new thinking on the refuse issue; in this it was in line with emerging national trends.

The "Zero Waste" philosophy proclaimed an alternative way of thinking about waste: taking a "whole-system" approach to the flow of resources and waste. "Zero Waste maximizes recycling, minimizes waste, reduces consumption and ensures that products are made to be reused, repaired or recycled back into nature of the marketplace."[52] For decades public and private collection and disposal programs had focused on finding effective means of dealing with discards. This "end-of-the-pipe" solution paid little attention to waste generation on the front end. Zero Waste put substantial pressure on producers of goods, not just on consumers. The proactive objective of Zero Waste disrupted standard operating procedure, questioned how and why products were manufactured and marketed, and envisioned a much broader conception of reuse and recycling than before. Zero Waste was controversial in both the public and private sectors, but a consensus view on this "front-end" philosophy (or its implications) has yet to be reached.[53] Many cities over several years have advocated confronting the problem of waste generation (including New York City), but only a few have made major strides in justifying, promoting, or implementing Zero Waste programs in the early twenty-first century. Could New York City be a leader? That was the goal of Mayor de Blasio and Commissioner Garcia.

The DSNY's 2016 strategic plan was the linchpin for charting a new direction in collection and disposal; it was meant to provide a "blueprint" for the department over the following four years. Commissioner Garcia stated that the

FIGURE 20.5 Promoting Zero Waste.

Source: NYC Sanitation, *Annual Report, 2016*.

strategic plan was the first of its kind in the department's 135-year history, and it built upon the 2006 SWMP; the *One New York* initiative "for the city's equitable, sustainable, and resilient development"; and employee engagement in the department.[54]

Most of the assumptions about the future of collection and disposal in the strategic plan were drawn from the DSNY's assessment of the city's current residential waste stream, which in 2013 was composed of 31 percent curbside recyclables, 31 percent organics suitable for composting, 10 percent other divertible materials (such as electronic waste, plastic shopping bags, and various household products), and 26 percent other materials, such as construction and demolition waste, wood, firm plastics, and disposal diapers.[55] Such a composition of refuse suggested several opportunities for greater emphasis on recycling, reuse, and waste prevention as opposed to landfilling or burning waste.

The strategic plan listed twelve goals for the department and forty-six initiatives to be taken. The initiatives included launching new services to reach zero waste to landfills, such as expanding curbside organics collection;[56] partnerships to increase recycling participation; policies to encourage waste reduction; programs for electronics and textile recycling; opening marine transfer stations at Hamilton Avenue, East Ninety-First Street, and Southwest Brooklyn; and much more. One of the initiatives (well underway at this point) was transforming Fresh Kills Landfill into Freshkills Park—a recycling-and-reuse program all its own. Despite its broad scale, the strategic plan devoted little space to conventional disposal methods (necessary to some degree, no matter how much waste was recycled or reused), preferring to put faith in its zero-waste vision.[57]

Skepticism followed the strategic plan, as one would expect, given the history of New York City's long struggle with the waste issue. Some thought the plan to recycle everything was just too ambitious. Julius Brewster, who hauled commercial waste for Metropolitan Recycling in Brooklyn, asserted, "My father told me, 'You stay in the garbage business and you'll never get laid off.' There's always more garbage." Kendall Christiansen, manager of the city chapter of the National Waste & Recycling Association, argued that unlike other cities that attempted to build zero-waste measures into their goals and plans, "New York's plan has been pretty loose, without much public discussion, just rhetoric."[58]

There was a feeling among some that New Yorkers had few incentives to throw away less. Many cities, for example, treated refuse as a utility, charging customers for what they generated. Gotham used general tax revenue to pay for residential and public waste collection, so the cost to the consumer was hidden. Problems with collecting recyclables always were difficult given the dense, vertical nature of much of the city, the tight spaces, the plethora of apartment buildings in some of the boroughs, and the variations in land use in others.

While New York City boasted the largest curbside recycling program in North America, such a claim did not take into account the scale of the task it faced versus other metropolitan areas.[59]

Another key issue was the cost of waste export. Concern that they would continue to rise started as far back as the closure of Fresh Kills and Benjamin Miller's subsequent warnings. A 2008 study prepared for the Natural Resources Defense Council concluded that there would continue to be "upward pressure" on export costs of refuse.[60] DSNY statistics on tons of MSW exported per day between 2006 and 2013 varied slightly from a high of 11,783 in 2006 to a low of 10,827 tons per day in 2012, suggesting little change in exporting practices. (Such figures were not included in the reports beginning in 2014–2015.)[61] However, the cost of exporting refuse continued to rise over the years. Between 2010 and 2014 they averaged $300 million per year. In 2015 they rose to $316 million and then to $351 million in 2016. These costs were expected to increase to $420 million by 2021. A major objective of the "Getting to Zero" plan was to reduce export costs by severely reducing the waste headed for landfills and WTE facilities. The de Blasio administration hoped that the Zero Waste plan, plus a potential leveling off or reduction in marine transfer station costs, would weaken the dependence on waste exporting.[62]

The plan de Blasio and Garcia put forward not only emphasized severe reduction in waste and thus a much reduced disposal responsibility for the city. It was coupled with a stronger emphasis on correlating the sanitation system with issues of social justice. While the 2006 SWMP had gone a ways in setting that tone, the mayor signed the Waste Equity Bill in August 2018, which would limit the amount of refuse that could be handled in transfer stations located in overburdened neighborhoods and prohibit the construction of new transfer stations in neighborhoods that handled at least 10 percent of the city's waste, namely, northern Brooklyn, the South Bronx, and southeastern Queens. The mayor and City Council responded to pressure from a broad coalition of environmental-justice organizations, community groups, and labor activists to get the legislation passed—legislation directed primarily at the commercial waste handling operations in the city.[63] For its part, the commercial waste industry and its allies were deeply concerned that such a proposal threatened their businesses.[64]

Zero Waste, climate change, and environmental justice were terms rarely if ever associated with the collection and disposal of solid waste in New York City before the late stages of the Bloomberg years and into the de Blasio administration.[65] To what degree they were merely hollow rhetorical devices, as some declared, rather than part of a concrete plan for action, only time would tell. Between 2013 and 2017 the refuse disposed by the city hovered above 3,200 tons per year. What could change this course?[66]

SHAPING OF FRESHKILLS PARK: FROM PLAN TO POSSIBILITIES

The DSNY had arrived at the crossroads with the 2006 SWMP and had moved into new territory by 2017. Freshkills Park also faced its own crossroads as the site entered its implementation phase with the 2006 Draft Master Plan. The DSNY's 2016 strategic plan had highlighted transforming Fresh Kills Landfill into Freshkills Park as one of its key initiatives. The DSNY viewed its own role, along with others, as central: "The Department has been working section by section, adding soil and vegetation to create a landscape that complements the adjacent meadow, woods and wetland environment."[67]

On April 6, 2006, Mayor Bloomberg and City Planning Director Amanda M. Burden announced the release of the Draft Master Plan (DMP) for Freshkills Park. It was meant to utilize Field Operations' *Lifescape* concepts as a blueprint to remake the landfill site into "a tangible symbol of renewal and an expression of how our society can tap into natural processes and help to restore the proper functioning of our landscape."[68] The DMP was the "first major milestone" in the park project and was the result of a site study; numerous discussions; and feedback from the consultant team, local state and regional agencies, stakeholders, and the Staten Island community. The implementation of the project was to be accomplished in three ten-year phases and in roughly twenty-acre segments. By the winter of 2006, it was hoped that the environmental and land-use reviews and the final master plan would be completed and that the implementation would get underway by the summer of 2007. The first major project—the Owl Hollow soccer-field complex in South Park—was to open by 2008–2009.[69]

The park site was divided into five subareas, and with the exception of the space at the center of the park (the Confluence), constructed around the four mounds of the landfill:

- The Confluence (100 acres) was the center of the park, shaped by the Richmond, Main, and Fresh Kills creeks, with a focus on water. Activities would center around recreation, community gatherings, culture, and limited commercial uses.
- The North Park (233 acres) was primarily a natural area with neighborhood recreation uses adjacent to the Travis community.
- The South Park (425 acres) was mostly in a lowland area that had not been landfilled. It would include recreational, natural, and neighborhood park uses near the Arden Heights community. The South Mound, in the same area, would be used for sports requiring varied terrain, such as mountain biking.

- The East Park (482 acres) would have large-scale open spaces with a possible golf course and include large public art installations, such as those proposed by Mierle Laderman Ukeles. Bird watching and eco-education could take place at the seventy-acre marsh in the area.
- The West Park (545 acres) would provide access to trails for cross-country activities such as running or skiing. The plan also called for an earthwork monument to the 9/11 World Trade Center Recovery Effort (which took place on the West Mound).

Aside from the subareas, there would be an extensive system of park drives and bikeways to provide access into and through the park. Expectations also were high about the return of a variety of birds and other wildlife on the grounds and about park efforts to demonstrate the utility of alternative energy sources such as wind and solar power.[70]

As the park proposal was unveiled in 2006, Mierle Laderman Ukeles's Percent for Art Project at Fresh Kills began to take form. After several years working at the DSNY, her projects now had shifted from "work focusing on people" to "Land Art" at the landfill.[71] In the section of the DMP on "Art and Culture," Field Operations restated its intention to work with the Department of Cultural Affairs in creating opportunities for artists at the site. It also acknowledged a collaboration with Ukeles on possibly four projects.

FIGURE 20.6 South Mound, Fresh Kills.

Source: Courtesy of Freshkills Park and the City of New York.

FIGURE 20.7 Mierle Laderman Ukeles, *Cantilevered Overlook* (2014).

Source: Courtesy the artist and Ronald Feldman Gallery, New York.

In the 2006 draft master plan Ukeles had proposed a series of berm overlooks to provide vantage points for the transformation of the landfill facility into a park. In 2013 she revised the original design and installed the experimental art work *Landing: Cantilevered Overlook*, her first project completed for the park. It sticks out over wetlands between the Big South Mound and Little South Mound in what will become South Park.[72] Ukeles envisioned the overlook "almost like an arrow pointing to the East Mound, which is one of the largest mounds at Fresh Kills . . . that has gone through a total engineering, technological transformation. So that points to how people have dealt with this space that had a lot of garbage . . . and how it's being turned into a safe public place."[73] Reflecting on the site in 2016, Ukeles said, "It's a new kind of earth. It was degraded and would make people go 'ewww,' but now it's going to be a safe, healthy park."[74]

Ukeles's work was an important part of the park's development, but other art and education projects also took hold there or were promoted as part of an ongoing dimension of this massive converted space. For example, the 2012 Land Art Generator Initiative called for public artwork that combined aesthetics with the ability to harness energy from natural sources and convert it into electricity.[75]

The Department of Parks and Recreation (DPR) began the construction of the park with an initial allocation of $100 million in city capital funds ($196 million to develop the park over ten years).[76] Field Operations was expected to

FIGURE 20.8 Freshkills Park, September 29, 2013.

Source: Courtesy of Michael Anton, NYC Department of Sanitation.

remain involved through the environmental and land-use reviews and as a design consultant for improvements during the first phase.[77] It also would maintain an ongoing relationship with the DSNY, especially with respect to final closure issues, monitoring, and maintenance of the landfill infrastructure.[78]

In September 2006 the DPR chose the eminently qualified Eloise Hirsh as Freshkills Park administrator (although she had never been to Fresh Kills before she took the job).[79] Hirsh had spent twenty years in New York City government, including as director of the city's first Labor Management Productivity Committee and as first deputy commissioner of the DPR. In 1988 she moved to Pittsburgh, where she spent eighteen years as director of city planning, director of the mayor's commission on public education, and as firm principal for Iron Hill Associates. She also had been on the faculty of the Heinz School for Public Policy at Carnegie Mellon University, the Graduate School of Public and International Affairs at the University of Pittsburgh, and the Wagner School of Public Service at NYU.[80]

In testimony before the City Council's Committee on State and Federal Legislation on June 5, 2007, Hirsh discussed the process for developing the park. As her department began to implement the project, she made clear that there were "various regulatory issues and legal hurdles" to be dealt with, including the environmental impact statement (EIS) and the uniform land-use review (ULURP, dealing with mapping and boundary issues).[81] She hoped public approvals would be obtained by 2008. (Apparently at the time of her testimony they were still pending.) During the first phase, a major issue was the road system—"the bones of the park" she called it; she was uncertain as to where the roads would be located or how wide they would be.[82] Her task also called for working closely with several agencies and other departments. Especially challenging was working with the DSNY, which at first was wary of having the public on the site.[83]

According to Carrie Grassi, who moved from the Department of City Planning to DPR in 2006 and became Freshkills Park's land-use and outreach manager, "We kind of saw it as Sanitation being responsible for the underground and Parks being responsible for the aboveground, with some exceptions."[84] The DSNY's mission was to continue monitoring and maintaining Fresh Kills Landfill, complementing the design and shaping of the site, for which DPR was responsible. The intent was to marry the engineering and solid waste management culture of DSNY with the land-use and parkland culture of DPR.

SELLING THE PARK TO STATEN ISLANDERS

Hirsh's task was daunting, but she was enthusiastic: "[Freshkills Park] is such a fabulous project. I'm in love with the site. I just think it's so beautiful. I understand it's going to take a while [to complete the park], but I'm so excited to be in this from the beginning."[85] She understood that while many Staten Islanders welcomed the new park, skepticism and cynicism still lingered even as the DMP was announced. The conflation of the landfill (and all it represented) with a recreational and natural park spawned long-standing mixed feelings on the island. Most Staten Island political leaders had applauded the idea of a park in the press and at numerous public hearings while the development plans were underway. However, there were chronic concerns over the lack of road development and over limited input in the project from local leaders.[86]

After a presentation on the DMP in 2006, Councilman Oddo stated, "[The landfill] defined Staten Island and who we as Staten Islanders were known as off the Island. To be able to take long steps in pushing that farther back in the rearview mirror is really an amazing thing. We're finally obliterating Fresh Kills defining us as a community."[87] Local citizens also welcomed the new park—despite concerns over more traffic—making Staten Island "a place to go."[88] But at about the same time, postings in the *Staten Island Advance* reflected different sentiments: "Parkland? Not in My Lifetime. Yeah, parkland, golf, hiking trails. And the pictures look beautiful. Too bad reality doesn't fit the picture." "They're going to put a park on the landfill with all that methane under ground. How stupid can you be?"[89]

A plethora of meetings on the island concerning various aspects of the park continued into 2007 and beyond. Perhaps these were guided by project mandates, good intentions, or a means to get local concurrence through outreach, and possibly all three. The hearings and meetings saw dwindling attendance, which suggested that residents had tired of these events or believed that their opinions were being ignored. Be it exhaustion or impatience, many Staten Islanders were ready for some tangible results.[90]

Several social science studies between 2008 and 2015 sought to gauge the response of Staten Islanders to the development of the park or visiting the park and found opinions vacillating between the park as an asset and as a problem or concern.[91] As Hirsh stated in 2012, especially concerning criticism about non–Staten Islanders "calling the shots" on developing the park, "There's been some skepticism of how you can take a landfill and make it into a park, and the answer is 'very carefully.'"[92]

Staten Islanders eventually would grow accustomed not only to the idea of a park but to the actual park. While outsiders were hard-pressed to see beyond the lingering stereotypes of Staten Island, locals knew that it was a blend of the old and the new—for better or for worse, depending who you asked. Its identity as a suburban enclave of New York City endured. Promotional material and various reports were fond of pointing out that it was "New York City's greenest borough," offering the best of both worlds: "A park-filled rural atmosphere with the excitement of a big city"—and, now, the city's best-kept tourist secret.[93]

Post–Fresh Kills Staten Island, however, was marked by significant changes. A 2001 study declared, "in many ways, no other borough has changed as much as Staten Island."[94] Its makeup (if not its politics) was changing more rapidly than in the past, not least because of demographics. Staten Island was the fastest growing borough in the city, with its share of white non-Hispanic residents falling from 80 percent in 1990 to 68 percent in 2010. African American, Latino, and Asian residents were on the rise. The number of the island's foreign-born residents leaped from 12 to 20 percent between 1990 and 2010. Such growth and demographic changes led to new housing development, though the new construction was still not sufficient to meet local needs. Even with the population increases, private-sector businesses had been shrinking in size rather than growing, and unemployment was rising.

While Staten Island had not achieved the density and population of its sister boroughs, it still experienced many of the social ills afflicting larger communities, including poverty, drug abuse, and racial tension. Tourism, for all of its issues, offered a welcome expansion of the economy.[95] Whether Freshkills Park would be an imposition and a burden or an economic and recreational boom was still to be seen.

Staten Island had other chronic problems stemming from its toxic heritage. It continued to be plagued by an array of environmental threats, especially air and water pollution, along its industrial corridor. The Interstate Environmental Commission, a tristate water and air pollution control agency, stated in its 2005 annual report, "Staten Island remains as the source of more citizens' complaints than any other area in the Commission's jurisdiction."

These complaints mostly came from the western portion of the island, in the vicinity of the New York–New Jersey border, and from the neighborhoods closest to Fresh Kills (although complaints from that area had dropped that year). There also were concerns about the debris-control measures in place around the area where the transfer station was constructed.[96] Along the North Shore, which was home to the highest number of communities of color and low-income families in the borough, was a strip of about five miles that contained approximately twenty-one sites identified by the EPA or others as contaminated. All of them were within seventy feet of homes and apartments.[97]

FRESHKILLS PARK: ONE STEP FORWARD, ONE BACK

The first major section of the park within the landfill site—North Park (twenty-one acres)—was not underway until 2017.[98] The DPR and its associates had been forced to take incremental steps until then for logistical, legal, political, and economic reasons. The strategy had become to develop small, usable pieces of what would become Freshkills Park, provide access to the public as quickly as possible, and not open any area until it had been tested and found safe. There also was strong interest within the park-building community and without to use the space as what Professor Nevin Cohn at Eugene Lang College called "an incredible laboratory."[99] Hirsh referred to it as a "field station" for a variety of scientific and educational activities.[100]

One hurdle for the park was cleared with the release of the Final Generic Environmental Impact Statement in March 2009.[101] The first large public event (and advertisement for the Corner plan) was held at the park site—"Sneak Peak at Freshkills Park"—on October 3, 2010, and drew 1,800 visitors, which was higher than anticipated. This was the first time that the gates were opened to the public. (Later years would see other public openings, such as "Discovery Days.") Public and private tours had been run for some time, and they served as a way to raise the profile of the park. But nothing beat actually setting foot in the site itself.[102]

Sneak Peek events included family activities such as kite flying and kayaking and nature-themed activities such as bird watching and nature talks.[103] Elizabeth Barlow Rogers, president of the Foundation for Landscape Studies, attended Sneak Peak in 2013:

> It was a beautiful autumn day, and two of the temporarily decommissioned phragmites-devouring goats were on display. . . . Snowy Egrets were poking their long bills into the shallow water next to mudflats, and Red-tailed Hawks and ospreys were cruising above revived marsh creeks and the former garbage mounds where garbage-scavenging gulls used to be the only birds to be seen. . . . Elsewhere the place had something of the air of a small-scale county fair. There

were booths for food and kiosks for information and education, including the recapture process that brings methane gas to the on-site reprocessing plant.[104]

Such were the contradictions of this at once familiar and alien place.

From 2011 to 2015 a variety of events related to the park kept it in the news. In July 2011 the New York City Department of Design and Construction released an RFP for the first phase of the Freshkills Park road system. In February 2012 the Freshkills Park visitor center opened in a refurbished trailer on site. A renovated Schmul Park (previously a blacktop playground), the first complete park

FIGURE 20.9 Discovery Day, Freshkills Park, 2015.

Source: Courtesy of Freshkills Park and the City of New York.

FIGURE 20.10 Discovery Day, Freshkills Park, 2017.

Source: Courtesy of Freshkills Park and the City of New York.

FIGURE 20.11 Marshes at Freshkills Park, September 9, 2015.

Source: Courtesy of Michael Anton, NYC Department of Sanitation.

project, opened in the Travis neighborhood in October 2012. Also in October the Land Art Generator Initiative produced artistic concepts for incorporating renewable energy on site. Owl Hollow Fields opened in April 2013, with the first hiking program initiated at the park. In June 2013 the Main Creek Wetland Restoration pilot project (two acres within the park site) was completed, which stabilized the shoreline, created new salt-marsh habitat, and removed invasive species. In November came the announcement that Freshkills Park would host New York City's largest solar array—enough to power two thousand homes. And the New Springville Greenway (3.2 miles along the eastern edge of the park) was completed in 2015.[105]

In the transition from Bloomberg to de Blasio, some questioned whether the desire of the city leadership to complete Freshkills Park remained strong. The political subtext, especially for heavily Republican-dominated Staten Island, was whether a Democratic mayor would feel the same obligation to them as Giuliani and Bloomberg had exhibited. "I'm not about letting city government off the hook in terms of supplying capital money," stated Borough President Oddo in December 2014, "but I think you shouldn't expect the de Blasio administration to be funding these initiatives any time soon." By that point $950 million had been spent on site development and efforts to cap the four mounds at the landfill, but a great deal of planning was left. Of this amount, only $50

FIGURE 20.12 Sneak Peak Postcard.

Source: Courtesy of Freshkills Park and the City of New York.

million was used on the park portion, with about $45 million more in capital funds allocated for the first phases of the park.[106]

Freshkills Park was supposed to receive $11.7 million in capital funds annually from 2010 to 2015, when the city initially committed $100 million in 2005. Because of cuts, this did not happen. Instead, capital funds were allocated by project. Part of the confusion was from uncertainty surrounding specific cost estimates of specific projects, and in other cases rising prices affected the budget. "With a cloudy future as to what the budget will be in coming years," Oddo added, "how do you make this unique site reach its full potential under that scenario? I don't think you can." The original completion date of 2030 had been pushed back to 2036, and officials were unable or unwilling to say if the whole project would be completed by then.[107]

A big question was whether Freshkills Park was a high priority to the mayor. On August 18, 2016, de Blasio announced that the city would spend $150 million to upgrade five large New York parks, with $30 million going to each borough. In line with his general philosophy, he believed such an action addressed the problems of an unequal system where parks in wealthier parts of the city were financed while those in poorer sections were ignored. "This is about fairness," he stated. "We talk about fighting inequality. This is one of the most basic ways to do it." Freshkills Park would be one of the five parks to receive funds in the mayor's new Anchor Parks program. Despite the mayor's good intentions, the premier status of the Staten Island park in the citywide park system appeared

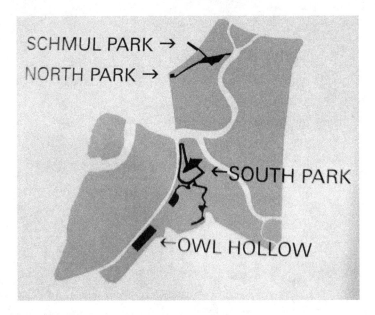

FIGURE 20.13 Park projects in Freshkills Park.

Source: Courtesy of Freshkills Park and the City of New York.

to be diminished. (In a strange way, this decision was like the move from a one-borough waste disposal program to a multiple-borough waste disposal program, but with a contrary aim.)[108] The biggest news for the park in 2017, however, was Lomma Construction Corporation being awarded the contract to carry out the first phase at the North Park, which would connect it to Schmul Park and the William T. Davis Wildlife Refuge.[109]

THE FUTURE OF FRESH KILLS LANDFILL AND FRESHKILLS PARK

Freshkills Park was unique, but in "growing" the park, the landfill remained—despite the camouflage.[110] In 2018 only small parcels of the ultimate plan for the park were complete. The overall site is vast and difficult to manage, and differences emerged over the years with respect to design issues and public expectations.[111] As Ryan Lavis of the *Staten Island Advance* stated, "while some guests [to the park] griped at the idea of building a park over a landfill, visitors . . . said that it was about time people accepted Fresh Kills as a park."[112]

It had been a "branding strategy" to change the name from "Fresh Kills" to "Freshkills." Raj Kottamasu, a community coordinator for the park, insisted

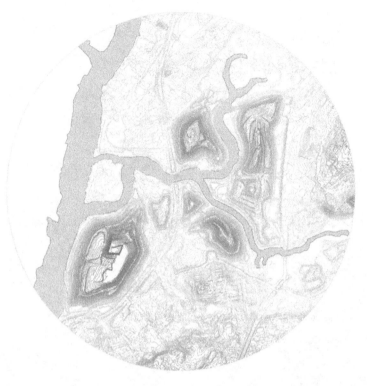

FIGURE 20.14 Fresh Kills topography, 2018.

Source: Courtesy of Jose Mario Lopez, Hines School of Architecture, University of Houston.

that the change in name was recommended to DPR by a consultant to "soften the word 'kills,'" because "in English 'fresh kills' is a harsh term."[113] Eloise Hirsh suggested that someone originally wanted to call it Phoenix Park, for its obvious rise from the symbolic ashes of the landfill.[114] Yet it seems most likely that the stigma of the landfill obliged some distance from the original name of the place. The new name, however, was more than rebranding; it was reinvention—a hybrid encompassing human intervention and natural processes.[115] Fresh Kills Landfill was a hybrid of human technology and natural marshland but also a space that had lost its dignity. As one observer noted, "Now Freshkills is on its way to becoming Staten Island's claim to fame rather than notoriety."[116] We shall see.

Freshkills Park seeks to replace Fresh Kills Landfill in the Staten Island landscape. But will the new refuse disposal plan for the City of New York also effectively replace the landfill's decades-long role as land sink for the metropolis? The crossroads from planning to implementation for both the new park and a new disposal system began but does not end with Fresh Kills/Freshkills.

CONCLUSION

New York City's geography, demography, and economic growth made its outer periphery—including Staten Island—subject and vulnerable to those institutions and practices not deemed suitable for the core. The placement of the Quarantine in Tompkinsville in the late nineteenth century, the erection of the Cobwell reduction plant on Lake's Island during World War I, and the establishment of a massive landfill at Fresh Kills after World War II proved that Staten Island was, in the minds of city officials, an appropriate location for unwanted people and refuse. All these decisions were made while the island's population was small, its land was plentiful compared to the core city (with abundant marshland considered worthless until developed), and politically anemic.

Given that its purpose was to hold the discards of the New York City metropolis, it is easy to dismiss Fresh Kills Landfill as nothing more than a mere dump. Carolyn Wheat, in her mystery novel *Fresh Kills*, called it "a great place to dump a body."[1] Bill Loehfelm, in a novel of the same name, called it, besides the Staten Island Mall, "Staten Island's other great contribution to modern American society."[2] And Christopher Hellstrom, an author and resident of Tottenville, wrote, "Queens has the World's Fair Unisphere. The Bronx has The Stadium. Manhattan is, well, Manhattan and Brooklyn has its Bridge. Our claim to fame is a giant pile of trash."[3]

John Byron Kuhner, a "sometime Latin and English teacher at the Staten Island Academy," took a wider perspective. He contrasted Staten Island's greenness with its industrial heritage: "For what is always urban about Staten Island is its air." He was referring to the toxins dumped into the atmosphere from nearby New Jersey factories (surrounding Staten Island on three sides), its heavy dependence on automobiles, and the gases escaping from Fresh Kills—"certainly Staten Island's most famous landmark." The equation of Staten Island with its landfill and its "unmistakable" smell was common even after its closure.[4] "It does not take a very intelligent person," he observed, "to realize that many of the items purchased at the mall return home and watch over future purchases from their vantage-point atop New York's mountains of trash."

What the future ecology of Fresh Kills might become was anyone's guess, but the possibility was always there for the landfill to reopen for reasons of proximity and economy. Kuhner looked beyond the immediate utility of Fresh Kills to the larger question of consumption and relentless waste. "Its closure only means that another Fresh Kills is being ruined on some other Staten Island somewhere else in the country," he wrote. "I have heard many islanders bitter at the rest of the city for dumping its garbage on them, but I have heard little guilt or sorrow that they are now dumping their garbage on someone else. The problem of our trash remains."[5]

Dietmar Offenhuber, of Northeastern University's departments of public policy and art and design, was correct when he claimed that waste is material information and that landfills are detailed records of consumption and behavior.[6] Such information provides insight into issues intrinsically connected to the wasting process but also to the society in which such actions are produced. The anthropologist/garbologist William Rathje referred to landfills as "mega-time-capsules" where insights about our lives were stored.[7]

Fresh Kills, therefore, is an iconic structure for New York City—maybe not in the same sense that the Statue of Liberty, Brooklyn Bridge, or Empire State Building are iconic, but it is nevertheless central to New York's modern history. Fresh Kills was at once a dramatic representation of the aftermath of consumption, a contested terrain between Staten Island and the remaining boroughs, a memorial to a tragic event, and a space of hopeful regeneration.

* * *

Verlyn Klinkenborg, writing in the *New York Times Magazine*, stated, "It's a revolting rhapsody, the saga of the city's garbage."[8] And so it was, in a way. In the case of refuse disposal, the city's strategy for discarding its waste had become less convenient since the end of ocean dumping. Experimenting with reduction and reuse, incineration, and landfilling each had its time—in all cases more than once. The revolving door of disposal methods was also subject to temporal, economic, environmental, and technical factors. Ocean dumping was expedient but proved to be a nuisance, environmentally risky, and ultimately illegal. Reduction provided resalable byproducts but was expensive and produced nauseating smells. (Later, various forms of recycling and reuse sought to take materials out of the waste stream, but their success was still uncertain.) Incineration went in and out of fashion depending on price and burning efficiency, but it eventually was seen as unsuitable for densely populated communities, often pitting borough against borough. Land dumping required a resource—open space—not readily available in much of New York as the years went by, triggered strong not-in-my-backyard responses, and eventually proved environmentally unsafe. A major flaw in the thinking of city leaders (not only in New York but elsewhere) was the mistaken belief that one best way for disposal was possible and that it was just a matter of discovering or fashioning it. Desperation sometimes led to impulsive decisions. And for the city's entire history, private and commercial haulers collected refuse too, and their relationship to the city complicated what disposal options might be most feasible and efficient.

At its opening, Fresh Kills Landfill had much to recommend it to the core city: a location distant enough as not to require the other boroughs to engage much longer in messy disposal practices (especially with the closing of other landfills, although transfer stations continued to plague the poorer neighborhoods outside Staten Island), close enough by barge to use the site on a daily basis, rich in valueless land to allow for a massive operation, and situated in a place not politically powerful enough to resist the decision.

From its opening in 1948, the landfill was an unwanted imposition on Staten Islanders, but it became increasingly essential to New York City's disposal plans. While city leaders outside of Staten Island considered the long-term use of Fresh Kills Landfill as a foregone conclusion, the temptation to flirt again and again with incineration—and sometimes reuse—produced a parallel possible disposal scenario for some.

The opening of the Verrazzano-Narrows Bridge in 1964 changed the relationship between Staten Island and the other boroughs. No longer physically isolated from the city core and growing in population, Staten Island began bringing its grievances directly to the city's notice. The environmental movement of the 1960s drew attention to the serious dangers of the landfill, such as methane and leachate. Fresh Kills was no longer just an eyesore or a blemish on the island's identity but a health risk and threat to the marshland ecology.

Despite the urgent environmental implications of the landfill, the city's attention was captured by the fiscal crisis of the mid-1970s. All the while, dependence on Fresh Kills mounted. By the 1980s the landfill was the city's only major depository for residential waste. Relentless demands for closure by Staten Islanders and a renewed interest in developing disposal alternatives (especially resource-recovery plants) clouded the future of the city's refuse disposal problem.

Contenders for the role of major waste sink for the city—whether waste-to-energy facilities or the rising tide of advocacy for recycling after the *Mobro* fiasco—waxed and waned in popularity, and all the while Fresh Kills inexorably accepted thousands of tons of refuse from all five boroughs each day. No other plan was able—or willing—to stop Fresh Kills' relentless expansion or the persistent trek of scows across New York Harbor. The landfill had become relatively inexpensive to operate compared to alternatives and thus precluded—for a time at least—other boroughs from bearing the brunt of the city's oppressively large waste load. One should not underestimate how important it was that Staten Island had assumed the role of sacrifice zone in this way.

In the end, politics eclipsed pragmatism in closing Fresh Kills. Although unsuccessful, the drive for Staten Island's secession in the 1990s was the prelude to the closure. The landfill was a cause célèbre for secession—a symbol of what separated the "forgotten borough" from the rest. The bitter taste of placing the landfill on Staten Island in 1948 persisted well into the twentieth century, not only because the islanders had to contend with the fetid, giant wound at the center of their island but also because it had been forced on them without their say. This was the rub—the helplessness, the enfeebling nature of the decision made for and without them.

Secession was proactive. Yet it took three Republican political leaders holding office simultaneously—not secession—to close the landfill. The decision was evidence of the newfound political power of Staten Island voters, who held the fate of Molinari, Giuliani, and Pataki largely in their hands. Staten Island got its wish at the precise moment that it had gained a political advantage over the other boroughs.

The political decision to close the landfill had consequences, however. Although Mayor Giuliani and his supporters claimed to have an alternative to Fresh Kills in mind (mainly by privatizing disposal), there was no easy transition between landfilling refuse and whatever would come next. An emphasis on exporting waste appeared reasonable in the abstract, but it demanded cooperation and agreement with political entities outside of the city and a reliance on a borough-wide approach to preparing refuse for transport inside the city. Staten Islanders were pleased not to have to bear the brunt of refuse disposal by themselves, but now the effects of disposal would fall most heavily on vulnerable neighborhoods throughout New York City and New Jersey. A multi-borough approach to refuse disposal in many ways was even more confounding than the longstanding single-borough solution Fresh Kills had made possible.

Before a new refuse disposal plan could be shaped and implemented—and soon after the last barge had made its way to Staten Island—the attack on the World Trade Center reshaped Fresh Kills' image. What was wastescape became a cemetery.

Fresh Kills the massive dump and Fresh Kills as hallowed ground existed in paradoxical relation. Nevertheless, this incongruity would permanently reshape its image. The decision to change Fresh Kills Landfill into Freshkills Park could not erase the past, but the commitment to the giant recreational space as a site of open nature was meant to look forward more than backward. Ecological

CONCLUSION ■ 551

restoration in any absolute sense was impossible. Field Operations' *Lifescape* instead favored environmental renewal without placing much attention on the site's history. Artists including Mierle Laderman Ukeles—independent of *Lifescape*'s overall plan—would try to find ways of connecting the site to some of its past. And 9/11 would need to be memorialized somehow. It was a tall order, one that continues today, for Freshkills Park to serve so many interests— remaking a natural place, remembering the dead, and preserving the human history (and abuse) of the site. All of this had to be done in a way that underplayed—or even eliminated—the stigma of the largest landfill in the world. While this formidable task was underway, New York City still needed to determine what path it would take in replacing that landfill with an effective waste disposal alternative.

The importance of the landfill to New York City's history required us to explore a broad sweep of time, from the 1850s to today. But this story has implications that go beyond this narrow timeline. It touches on the triad of issues—consumption, waste, and disposal/place—we first discussed in the introduction.

✳ ✳ ✳

A headline in the May 7, 2015, edition of the *Atlantic's* CityLab read, "New York Is the World's Most Wasteful Megacity." Tanvi Misra, the article's author, wrote that New York City "consumes more water and energy, and generates more waste, than any other huge metro. New York, many say, is the greatest city in the world. It also might be the most wasteful." Her data came from a National Academy of Science study of twenty-seven megacities (with ten million or more people) that stated that together these cities consumed 9.3 percent of the world's electricity and produced 12.6 percent of the world's waste, even though they had only 6.7 percent of the global population. New York led in all three categories: per capita energy use, water use, and solid waste disposal. Though she conceded that size, geography, demography, and economic activity played their parts[9] and that those variables made determining what is wasteful and how to measure wastefulness difficult, her criticism might still very well be true, as wastefulness does not automatically equate with consumption.

In his celebrated, complex, and much-discussed 1997 novel *Underworld*, Don DeLillo takes on the issues of consumption and waste, attending especially to Fresh Kills Landfill, "sort of the King Kong of American garbage mounds."[10] The protagonist, Nick Shay, was born in the Bronx, of Italian immigrants. He is a waste management executive for Waste Containment ("Whiz Co."), which designs and manages landfills. DeLillo comments on issues of consumption and waste—concretely and metaphorically—most often through this character. DeLillo unabashedly conflates waste with abundance (and affluence), which he views as fundamental to contemporary culture.[11]

The sociologist Martin O'Brien argues that there is "an important conceptual difference between talking about what people throw away and talking about what people waste." He questions the "throwaway society thesis," suggesting that it is "rooted in a family of ideas about the 'crisis of waste' that confuses a moral critique and a sociological analysis of consumerism."[12] The act of consumption produces residue because part of what is consumed is not usable (the packaging it comes in or the materials used to produce it, for example), is unwanted, or may indeed be wasteful. And consumer choices are often dependent on the form commodities take, as in the case of a manufacturer building in planned obsolescence or designing single-use consumable items.[13]

In *Fresh Kills: A History of Consuming and Discarding in New York City*, consumption plays two roles—one early in the book, when discussing the rising challenge of dealing with waste, and the other near the end of the book, with respect to plotting the implications and impact of a closed Fresh Kills Landfill. In the former role, consumption is directly linked to a quantitative assessment of the amounts of solid waste that the city must contend with. In the latter, the search for alternative methods in confronting residential refuse requires that more attention be paid to larger questions about how to deal with both the production of goods and potential limits on the waste. The rise of environmentalism, the environmental-justice movement, recycling, and Zero Waste opened up both practical and ethical questions about the preservation of virgin materials and responsible production practices and about the equitable distribution of health and other risk factors related to disposal. In both instances, the specific nature of the waste stream dictated concerns over methods of disposal. Waste as an end product of consumption spun off into consuming salt marshes and other land uses, exploiting waterways, and encroaching on neighborhoods and commercial properties. The dilemma of consuming confronted us well beyond consumer products.

* * *

Claims of a "throwaway society" might well be true, but finding a way to dispose of or reuse our waste requires asking some difficult questions: How much waste *should* we generate? What kinds of waste are acceptable? And how do we reduce the amount we generate, both before and after the product is consumed?

From a strictly statistical viewpoint, municipal refuse makes up only about 3 percent of all solid waste in the United States. Of course, the management of this miniscule portion of the waste stream is extremely important, especially in cities, where the functioning of daily life without collection and disposal of refuse can be calamitous. The sheer amount of municipal solid waste (MSW), despite its limited total share, is enormous—about 262.4 million tons nationally in 2015 (4.48 pounds per person per day) as compared to 88.1 million tons (or 2.68 pounds per person per day) in 1960. (For New York City alone in 2011,

14 million tons.) The total annual MSW generation in the United States since 1980 has increased by 73 percent.[14]

The waste stream's composition of organic material (food waste and yard trimmings), plastics, metals, wood, rubber, leather, textiles, and more has complicated implications for disposal. For each generation of sanitation workers, the composition posed unique problems, whether horse manure in the nineteenth century or obsolete electronics today. "Refuse is never a stable category," points out Professor Jani Scandura, cofounder of the Place & Space Research Collective.[15]

The simple disposal of whatever residents or businesses threw away has been challenged since at least the late nineteenth century. The straight line from garbage can to dump or incinerator was broken in earlier years by individuals sorting out recyclables such as clothing and a variety of other goods, which were then sold by junk dealers, secondhand stores, and scrap-metal concerns. Food waste became pig slop, and reduction plants produced saleable liquids such as fragrance base. In later years, incinerators doubled as waste-to-energy plants. Reuse and reduction never dominated or replaced other disposal options because of cost or environmental limits, but the idea of pulling items from the waste stream has been around as long as waste has been generated.

Recycling gained momentum in the 1960s as an alternative to disposal and as a virtuous way to demonstrate a commitment to environmentalism. Within ten years or so, people—including New Yorkers—were viewing recycling as a disposal option that could compete directly with landfills and incinerators. Nationally, recycling was progressing into the twenty-first century. In 2015, 66.6 percent of paper was recovered, as was 61.3 percent of yard waste, 33.3 percent of steel, 26.4 percent of glass, 18.5 percent of aluminum, 16.3 percent of wood, and 9.1 percent of plastics.[16] In 2015, the diversion rate for recyclables nationally was 34 percent, compared to less than 10 percent in 1980. In 2014 there were 633 material-recovery facilities in the United States accompanying the many curbside recycling programs.[17] For New York City, the diversion rate in recent years—about 17 percent—of total residential waste has remained substantially lower than the national average, though it has been improving in key areas, such as paper. (The rate in New York City also reflects a steady decline in curbside collections thanks to a decline in the total residential refuse generated.)[18]

When Fresh Kills closed in 2001, advocates of recycling hoped that it would be able to take up the position New York City's colossal landfill had held for many years and also challenge Giuliani's aspirations to privatize the export of waste. But recycling faced serious shortfalls in funding, access to potential markets for recyclables (such as China's decision not to accept imports of e-waste any longer), and difficulties in servicing a vast and diverse citizenry. Changing consumer habits had also altered the waste stream. For example, the amount of newspaper drastically declined as people turned to digital news sources. In

2005 New Yorkers recycled more than one hundred pounds of newspaper per capita; in 2017 that number had fallen to less than twenty pounds.[19]

A challenge to more conventional recycling thinking came from "Zero Waste"—a philosophy that incorporated recycling programs but was premised on a different view of bridging the gap between consumption and disposal. The environmental scientist and scholar Max Liboiron took a strong stand in questioning recycling as the answer to the waste problem. She saw a schism between the popular perception of recycling as a "good for the environment" and a process entrenched in industrialism and the drive to produce goods.[20] The focus, as Zero Waste purports, must move "upstream" into the realm of goods production.

In adopting the idea that "Zero Waste Starts Now," the Department of Sanitation (DSNY) and city officials in the de Blasio administration sought a new direction for their views of refuse collection and disposal. They proclaimed that New York City would be "rethinking garbage" by keeping more refuse out of garbage cans, composting organic waste, and recycling a wider range of materials.[21] While none of the DSNY's public statements advocated the most extreme positions on Zero Waste, they signaled a substantial modification in the department—and the city's—orientation.

As of the 2010s, however, the most that can be said about the new disposal approach was that it melded existing practices—exporting and recycling—with some efforts at composting and a reset of its education and public-relations campaigns to look beyond "end-of-the-pipe solutions." The degree to which any new practices would match the buoyant new language about connecting consumption with the act of discarding had yet to be determined. By several measures, the amount of material disposed of around the world was rising rather than declining, while recycling and reuse rates remained moderate or were not increasing significantly.[22] New York City hoped to buck the trend, though the task would be onerous.

* * *

At the start of this book, "place" and "disposal" were the central terms in Fresh Kills' backstory and story. As the book ends, both are taking on a new meaning for New York City.

"Disposal" has come to mean shipping refuse to someone else's backyard or finding ways to reuse and recycle what is reusable and recyclable. The World Trade Center attack and Hurricane Sandy demonstrated, however, that never saying never again is not a certainty. As a character in the noir story "Teenage Wasteland" comments, although Fresh Kills was only supposed to stay open for twenty years, "dumping is a hard habit to break."[23]

Fresh Kills has been a natural space, a private space, a public space, a desecrated space, and a sacred space. Through its incarnations—salt marsh, landfill,

CONCLUSION ■ 555

graveyard, and park—its meanings have changed. But as it changed, it accumulated memories and history. In its current state, it is convenient to refer to it as Freshkills Park, but it still contains original, undeveloped marshland; recreational facilities; four mounds and landfill infrastructure; new public art; and a September 11 memorial. Fresh Kills has been made and remade over and over again, and no single feature or historical memory—not even the giant landfill—easily defines it.

Were there a historic-preservation project to portray Fresh Kills without regard to time, what to preserve and what merited centrality would trigger intense debate. Instead, the exigencies of the moment dictate how New York City—more than just Staten Island—chooses to fix Fresh Kills' use and value. This currently means retaining what can be retained of the landscape's natural character, providing recreational amenities, and largely erasing the memory of the landfill. But in all of its eras, Fresh Kills is a massive human artifact, a collection of historical sites, and an archive.

The prism of Fresh Kills Landfill refracts and reflects much about the history of Staten Island and New York City. That history touches on the lives of key New York City figures from Andrew Haswell Green and Robert Moses to Barry Commoner and Mierle Laderman Ukeles. It also intertwines with historical moments—city consolidation, world wars, fiscal crises, and natural and human-made disasters.

* * *

Issues of consumption, waste, and place have been at the heart of the dilemmas faced by New York City for more than 150 years, and they echo the dilemmas found in other eras, in other cities, and in other cultures. The risks to health and the environment from the making and disposing of waste affect developed and developing societies alike, and these risks are never equitably apportioned. Human involvement in the production of garbage crises, even deepening the threat of climate change, is sometimes offset by human intervention in building parks or changing waste practices. Here too, the extremes are not limited to one place or time. In the end, there is nothing mundane or inherently exclusive about the history of Fresh Kills. It speaks to our daily lives, our ambitions, and our aspirations.

The sweeping story of Fresh Kills makes clear that we should not ignore the familiar, the disagreeable, or the unpleasant in making our past meaningful. The act of consuming in itself may be necessary and sometimes pleasurable, but it is not an isolated action without consequences. Fresh Kills is a monument to those consequences—materially, spatially, and culturally.

NOTES

PREFACE

1. "Ashes to Ashes," *Staten Island Advance*, March 31, 1991.
2. See "Fresno Sanitary Landfill," National Historic Landmark Nomination (NPS Form 10-900), U.S. Department of Interior, National Parks Service (August 2000); Martin V. Melosi, "The Fresno Sanitary Landfill in an American Cultural Context," *Public Historian* 24 (Summer 2002): 17–35.
3. Critical discard studies, an emerging interdisciplinary subfield, was defined by the founder of its blog, the anthropologist Robin Nagle, as taking "waste and wasting, broadly defined, as its topic of study. We use the 'discard studies' instead of 'waste studies' to ensure that the categories of what is systematically left out, devalued, left behind, and externalized are left open." Topics for inquiry include social customs, labor arrangements, resource stocks and flows, economic relationships, cultural norms, public health controversies, political histories, and geographies and circulation. This interdisciplinary subfield, therefore, deals with the materiality of waste as well as its cultural, social, economic, and political context.
4. Wallace Stegner, "The Town Dump," *Atlantic Monthly* (October 1959): 80.

INTRODUCTION: THE DILEMMA OF CONSUMING

1. In urban centers, especially, collection of solid wastes is integral to disposal from economic, environmental, political, logistic, and labor perspectives. Simply put, the act of discarding implies collection and disposal. In terms of dollars, collection has been and remains the most costly (and labor-intensive) part of the process, especially in developed countries. The focus of the book, however, relates more to the act of disposal, the ultimate stage of separating humans from their material castoffs.
2. Jeff Hearn and Sasha Roseneil, *Consuming Cultures: Power and Resistance* (New York: St. Martin's, 1999), 1.

3. Richard White, *The Republic for Which It Stands: The United States During Reconstruction and the Gilded Age, 1865–1896* (New York: Oxford University Press, 2017), 680.
4. Regina Lee Blaszczyk, *American Consumer Society, 1865–2005: From Hearth to HDTV* (Wheeling, IL: Harlan Davidson, 2009), 11–12, 16–19. See also Lizabeth Cohen, *A Consumers' Republic: The Politics of Mass Consumption in Postwar America* (New York: Knopf, 2003).
5. Blaszczyk, *American Consumer Society*, 21–27. See also Greg Kennedy, *An Ontology of Trash: The Disposable and Its Problematic Nature* (Albany: State University of New York Press, 2007), xv; Caroline Tauxe, "Consumerism," in *Encyclopedia of Consumption and Waste: The Social Science of Garbage*, vol. 1, ed. Carl A. Zimring and William L. Rathje (Los Angeles: Sage, 2012), 150–52.
6. Patrick Mullins et al., "Cities and Consumption Spaces," *Urban Affairs Review* 35 (September 1999): 44. See also Daniel Horowitz, *The Morality of Spending: Attitudes Toward the Consumer Society in America, 1875–1940* (Chicago: Ivan Dee, 1992), xvii–xxviii; Grant McCracken, *Culture and Consumption: New Approaches to the Symbolic Character of Consumer Goods and Activities* (Bloomington: Indiana University Press, 1988), 3, 22–23; Peter Dauvergne, *The Shadows of Consumption: Consequences for the Global Environment* (Cambridge, MA: The MIT Press, 2010); John C. Ryan and Alan Thein Durning, *Stuff: The Secret Lives of Everyday Things* (Seattle: Sightline Institute, 1997), 67.
7. Martin V. Melosi, *Garbage in the Cities: Refuse, Reform, and the Environment*, rev. ed. (Pittsburgh, PA: University of Pittsburgh Press, 2005), 12.
8. Blaszczyk, *American Consumer Society*, 54–55, 73–79, 84. See also Susan Strasser, *Satisfaction Guaranteed: The Making of the American Mass Market* (New York: Pantheon, 1989).
9. Strasser, *Satisfaction Guaranteed*, 6. See also White, *The Republic for Which It Stands*, 679.
10. See, for example, Lester R. Brown, "Throwaway Must Go Away," *USA Today* (November 2009); Gavin Lucas, "Disposability and Dispossession in the Twentieth Century," *Journal of Material Culture* 7 (2002): 5–22; Gillian Pye, ed., *Trash Culture: Objects and Obsolescence in Cultural Perspective* (Oxford: Peter Lang, 2010), 1, 77. For a critique of the concept of the throwaway society, see Nicky Gregson, Alan Metcalfe, and Louise Crewe, "Identity, Mobility, and the Throwaway Society," *Environment and Planning D: Society and Space* 25 (2007): 682–700.
11. Mullins et al., "Cities and Consumption Spaces," 45–49.
12. Blaszczyk, *American Consumer Society*, 99–101, 181–82.
13. See Edward Humes, *Garbology: Our Dirty Love Affair with Trash* (New York: Avery, 2012), 5–6; Heather Rogers, *Gone Tomorrow: The Hidden Life of Garbage* (New York: New Press, 2005), 2.
14. Stephanie Foote and Elizabeth Mazzolini, "Introduction: Histories of the Dustheap," in *Histories of the Dustheap: Waste, Material Cultures, Social Justice*, ed. Stephanie Foote and Elizabeth Mazzolini (Cambridge, MA: MIT Press, 2012), 2–3, 6–7; Foote and Mazzolini, "Conclusion: Object Lessons," 257, 259.
15. Kevin Lynch, *Wasting Away* (San Francisco: Sierra Club Books, 1990), 42, also 3, 28–29; Tim Dant, *Materiality and Society* (New York: Open University Press, 2005), ix–x, 3, 136–37; Ian Holder, *Entangled: An Archeology of the Relationships Between Humans and Things* (Chichester: Wiley, 2012), 17, 19; Gay Hawkins and Stephen Muecke, eds., *Culture and Waste: The Creation and Destruction of Value* (Lanham,

MD: Rowman and Littlefield, 2003), xiv; Kevin Hetherington, "Secondhandedness: Consumption, Disposal, and Absent Presence," *Environment and Planning D: Society and Space* 22 (2004): 158.

16. Rogers, *Gone Tomorrow*, 9.
17. Gay Hawkins, *The Ethics of Waste: How We Relate to Rubbish* (Lanham, MD: Rowman & Littlefield, 2006), 9, also vii, 8. This idea borrows from the French philosopher, anthropologist, and sociologist Bruno Latour. See also Susan Signe Morrison, *The Literature of Waste: Material Ecopoetics and Ethical Matter* (New York: Palgrave MacMillan, 2015), 1, 3, 9, 30, 159–60.
18. Michael Thompson, *Rubbish Theory: The Creation and Destruction of Value*, new ed. (London: Pluto, 2017), 10.
19. Thompson, *Rubbish Theory*, 27.
20. Thompson, *Rubbish Theory*, 29. See also Verena Winiwarter, "History of Waste," in *Waste in Ecological Economics*, ed. Katy Bisson and John Proops (Cheltenham: Edward Elgar, 2002), 2–5, 38–41; Sarah A. Moore, "Garbage Matters: Concepts in New Geographies of Waste," *Progress in Human Geography* 36 (2012): 780–93; Mikael Drackner, "What Is Waste? To Whom?—an Anthropological Perspective on Garbage," *Waste Management & Research* 23 (2005): 176, 179; Brian Neville and Johanne Villeneuve, eds., *Waste-Site Stories: The Recycling of Memory* (Albany: State University of New York Press, 2002), 1, 15, 34, 40–41; William A. Cohen and Ryan Johnson, eds., *Filth: Dirt, Disgust, and Modern Life* (Minneapolis: University of Minnesota Press, 2005), viii; Arjun Appadurai, *The Social Life of Things: Commodities in Cultural Perspective* (New York: Cambridge University Press, 1986), 3, 6; Maurizia Boscagli, *Stuff Theory: Everyday Objects, Radical Materialism* (New York: Bloomsbury, 2014), 52–56, 228, 231; Daniel Sosna and Lenka Brunclikova, eds., *Archaeologies of Waste: Encounters with the Unwanted* (Oxford: Oxbow, 2017), 17–20; Rogers, *Gone Tomorrow*, 3; John Scanlan, *On Garbage* (London: Reaktion, 2005), 41–42, 164.
21. Scanlan, *On Garbage*, 153, also 164. See also Derrick Jensen and Aric McBay, *What We Leave Behind* (New York: Seven Stories, 2009), 17, 25.
22. Kennedy, *An Ontology of Trash*, xi, xv. See also Jensen and McBay, *What We Leave Behind*, 25.
23. See not only Thompson, *Rubbish Theory*, but also Jensen and McBay, *What We Leave Behind*, 26. See also Martin O'Brien, *A Crisis of Waste? Understanding the Rubbish Society* (New York: Routledge, 2008), 10.
24. Hetherington, "Secondhandedness," 159. See also Eveline Durr and Rivke Jaffe, eds., *Urban Pollution: Cultural Meanings, Social Practices* (New York: Berghahn, 2010), 16.
25. Christopher Rootes and Liam Leonard, eds., *Environmental Movements and Waste Infrastructure* (London: Routledge, 2010), 1, 3–4, 6; Elizabeth V. Spelman, *Trash Talks: Revelations in the Rubbish* (New York: Oxford University Press, 2016), 2.
26. Greet DeBlock, "Ecological Infrastructure in a Critical-Historical Perspective: From Engineering 'Social' Territory to Encoding 'Natural' Topography," *Environment and Planning A* 48 (2016): 369. See also Sarah Hill, "Making Garbage, Making Land, Making Cities: A Global History of Waste in and out of Place," *Global Environment* 9, no. 1 (2016): 167.
27. Jani Scandura, *Down in the Dumps: Place, Modernity, American Depression* (Durham, NC: Duke University Press, 2008), 3–7.
28. Sarah S. Elkind, "View from the Dump: Stige O and the Question of Anti-Landscapes," in *The Anti-Landscape*, ed. David E. Nye and Sarah Elkind (Amsterdam: Rodopi, 2014), 203.

29. Anna Storm, "Landscapes of Waste: Malmberget and Ignatina as Cultural Tools in Heritage Process," in *The Anti-Landscape*, ed. Nye and Elkind, 161.
30. Rogers, *Gone Tomorrow*, 16.
31. Mira Engler, *Designing America's Waste Landscapes* (Baltimore, MD: Johns Hopkins University, 2004), 80–81, 100.
32. Engler, *Designing America's Waste Landscapes*, 94, 101; Joshua O. Reno, *Waste Away: Working and Living with a North American Landfill* (Berkeley: University of California Press, 2016), 13, 118.

1. ISLAND CITY

1. "Biography of Francis [sic] Trollope," *Essortment*, http://www.essortment.com/biography-francis-trollope-20549.html. Francois Weil, *A History of New York* (New York: Columbia University Press, 2004), 67–68. Weil also liked Trollope's comparison between New York and Venice.
2. Ted Steinberg, *Gotham Unbound: An Ecological History of Greater New York, 1609–2012* (New York: Simon & Schuster, 2014), 42. Thanks to Ted Steinberg for access to his book manuscript before publication. His insights about New York Harbor and environs were very useful for this chapter and elsewhere in my book.
3. Eric W. Sanderson, *Mannahatta: A Natural History of New York City* (New York: Abrams, 2009), 85, 87, 94–98.
4. City of New York, *Natural Hazard Mitigation Plan* (March 2009): 14, https://www1.nyc.gov/assets/em/downloads/pdf/hazard_mitigation/full_hmp_march_2009.pdf. See also Frederick R. Black, *Jamaica Bay: A History* (Prepared for the U.S. Department of Interior, National Park Service, August 1980).
5. Steinberg, *Gotham Unbound*, xvi, xvii.
6. David Leveson, "Describing the Topography of New York City," http://academic.brooklyn.cuny.edu/geology/leveson/core/linksa/nyc_describe/ny_top_describe.html; City of New York, *Natural Hazard Mitigation Plan*, 14; Christopher J. Schuberth, *The Geology of New York City and Environs* (Garden City, NY: Natural History Press, 1968), 8–20.
7. Anne-Marie Cantwell and Diana di Zerega Wall, *Unearthing Gotham: The Archaeology of New York City* (New Haven, CT: Yale University Press, 2001), 188.
8. Kenneth T. Jackson, Penelope Gelwicks, and Lisa Keller, "Manhattan," in *The Encyclopedia of New York City*, 2nd ed., ed. Kenneth T. Jackson (New Haven, CT: Yale University Press, 2010), 786.
9. City of New York Parks and Recreation, "Manhattan Schist in New York City Parks—J. Hood Wright Park," http://www.nycgovparks.org/parks/jhoodwrightpark/highlights/12369; "New York City Geographical Information," *ny.com*, http://www.ny.com/histfacts/geography.html; Sanderson, *Mannahatta*, 70–71; Elizabeth Barlow, *The Forests and Wetlands of New York City* (Boston: Little, Brown, 1971), 16, 18–19; Betsy McCully, *City at the Water's Edge: A Natural History of New York* (New Brunswick, NJ: Rivergate, 2007), 2, 6, 9–11.
10. City of New York, *Natural Hazard Mitigation Plan*, 12; Gary D. Hermalyn, "Bronx," in *The Encyclopedia of New York City*, ed. Jackson, 160.
11. City of New York, *Natural Hazard Mitigation Plan*, 12; "New York City Geographical Information"; Margaret Latimer, "Brooklyn," in *The Encyclopedia of New York City*, ed. Jackson, 168; Jeffrey A. Kroessler, Jon A. Peterson, and Vincent Seyfried, "Queens," in *The Encyclopedia of New York City*, ed. Jackson, 1063.

1. ISLAND CITY   ■   561

12. In 2018, the name was changed to Verrazzano-Narrows Bridge, with two z's, to spell correctly the name of the Italian explorer and navigator for whom the bridge was named. Throughout this book, in quoted material the older one-z spelling will be retained, but the newer spelling will be used otherwise. See https://en.wikipedia.org/wiki/Verrazzano-Narrows_Bridge#Official_name_with_one_%22z%22.
13. Charles L. Sachs, "Staten Island," in *The Encyclopedia of New York City*, ed. Jackson, 1231; Port Authority of New York and New Jersey, "Bayonne Bridge," http://www.panynj.gov/port-authority-ny-nj.html; City of New York, *Natural Hazard Mitigation Plan*, 12; "New York City Geographical Information."
14. Edward L. Glaeser, "Urban Colossus: Why Is New York America's Largest City?" Working Paper 11398 (Cambridge, MA: National Bureau of Economic Research, June 2005), 1. In a point-to-point transport system, for example, bales of tobacco could be picked up in Virginia and transported directly to England. But in this case, the exporting area did not import enough goods from England to fill their ships upon return to America. Triangular trade solved the problem. In the case of cotton, the commodity could be transported from Southern growers by coastal shipping to New York City, where it was loaded on transatlantic lines. Ships from Europe filled with imported goods could return to New York City, where the goods could be utilized locally or redistributed to several places. In this example, New York City became the hub connecting two spokes. Glaeser, "Urban Colossus," 12.
15. Glaeser, "Urban Colossus," 3. See also Edwin G. Burrows and Mike Wallace, *Gotham: A History of New York City to 1898* (New York: Oxford University Press, 1999), 335–36.
16. David C. Hammack, *Power and Society: Greater New York at the Turn of the Century* (New York: Columbia University Press, 1987), 51.
17. From *A Glimpse of High Society, from Society as I Have Found It* (1890), quoted in *Empire City: New York Through the Centuries*, ed. Kenneth T. Jackson and David S. Dunbar (New York: Columbia University Press, 20002), 358.
18. Poole won the first Pulitzer Prize for fiction with his next book, *His Family*, but the award was probably meant for the 1915 prolabor, prounion work. Ernest Poole, *The Harbor* (New York: The Macmillan Co., 1915), 9.
19. Glaeser, "Urban Colossus," 16–18.
20. Mona Domosh, *Invented Cities: The Creation of Landscape in Nineteenth-Century New York and Boston* (New Haven, CT: Yale University Press, 1996), 12–13. See also John R. Logan and Harvey L. Molotch, *Urban Fortunes: The Political Economy of Place* (Berkeley: University of California Press, 1987), 50–97.
21. Hammack, *Power and Society*, 37–38.
22. Selma Berrol, *The Empire City: New York and Its People, 1624–1996* (Westport, CT: Praeger, 2000), 75; "Port of New York," in *The Encyclopedia of New York City*, ed. Jackson, 1022–25.
23. Mary Beth Betts, "Masterplanning: Municipal Support of Maritime Transport and Commerce, 1870–1930s," in *The New York Waterfront: Evolution and Building Culture of the Port and Harbor*, rev. ed., ed. Kevin Bone (New York: Monacelli, 2004), 39–46.
24. Cited in William Grimes, "Down by the Riverside," *New York Times*, September 7, 1997.
25. Weil, *A History of New York*, 82.
26. Moses King, ed., *King's Handbook of New York City: A Outline History and Description of the American Metropolis* (Boston, MA: Moses King, 1893), 67.
27. Quoted in Ann L. Buttenwieser, *Manhattan Water-Bound: Manhattan's Waterfront from the Seventeenth Century to the Present*, 2nd ed. (Syracuse, NY: Syracuse University Press, 1999), 92.
28. Buttenwieser, *Manhattan Water-Bound*, 92.

29. Weil, *A History of New York*, 83, 86–87.
30. Berrol, *The Empire City*, 76; George J. Lankevich, *American Metropolis: A History of New York City* (New York: New York University Press, 1998), 126; Hammack, *Power and Society*, 41, 43–46, 49.
31. Weil, *A History of New York*, 88–89; Lankevich, *American Metropolis*, 126.
32. Adna Ferrin Weber, *The Growth of Cities in the Nineteenth Century: A Study in Statistics* (Ithaca, NY: Cornell University Press, 1967), 207. See also Burrows and Wallace, *Gotham*, 336, 873–75, 1153.
33. Matthew P. Drennan, "Economy," in *The Encyclopedia of New York City*, ed. Jackson, 395.
34. Joseph J. Salvo and Arun Peter Lobo, "Population," in *The Encyclopedia of New York City*, ed. Jackson, 1018–20.
35. Regina Lee Blaszezyk, *American Consumer Society, 1865–2005: From Hearth to HDTV* (Wheeling, IL: Harlan Davidson, 2009), 14–15.
36. Michael McGerr, *A Fierce Discontent: The Rise and Fall of the Progressive Movement in America* (New York: Oxford University Press, 2003), 7.
37. Hammack, *Power and Society*, 50–51.
38. Burrows and Wallace, *Gotham*, 722–23, 877–78.
39. Hammack, *Power and Society*, 86.
40. Hammack, *Power and Society*, 80–89.
41. Hammack, *Power and Society*, 89–91.
42. Glaeser, "Urban Colossus," 15, 17.
43. John F. Mariani, "Food," in *The Encyclopedia of New York City*, ed. Jackson, 464; Catherine McNeur, *Taming Manhattan: Environmental Battles in the Antebellum City* (Cambridge, MA: Harvard University Press, 2014), 26.
44. Martin V. Melosi, *Garbage in the Cities: Refuse, Reform, and the Environment*, rev. ed. (Pittsburgh, PA: University of Pittsburgh Press, 2005), 19–21.
45. Steinberg, *Gotham Unbound*, 183–84.
46. Edward K. Spann, "Grid Plan," in *The Encyclopedia of New York City*, ed. Jackson, 558; Steinberg, *Gotham Unbound*, 41–42, 56–63, 65. See also Gerard Koppel, *City on a Grid: How New York Became New York* (Boston: Da Capo, 2015); Jon Scott Logel, *Designing Gotham: West Point Engineers and the Rise of Modern New York, 1817–1898* (Baton Rouge: Louisiana State University, 2016).
47. Still, *Mirror for Gotham*, 212.
48. Weil, *A History of New York*, 98–99.
49. Burrows and Wallace, *Gotham*, 1041–49.
50. David M. Scobey, *Empire City: The Making and Meaning of the New York City Landscape* (Philadelphia: Temple University Press, 2002), 89, 91–93; Burrows and Wallace, *Gotham*, 1050.
51. Scobey, *Empire City*, 100–1.
52. "Population of the 20 Largest U.S. Cities, 1900–2010," Infoplease.com, http://www.infoplease.com/ipa/A0922422.html.
53. Weil, *A History of New York*, 95.
54. Edward F. Bergman and Thomas W. Pohl, *A Geography of the New York Metropolitan Region* (Dubuque, IA: Kendall/Hunt, 1975), 51.
55. Antonia Felix and the Editors of the *New York Post*, comps., *The Post's New York* (New York: HarperResource, 2001), 84.
56. Weil, *A History of New York*, 111–16; Ellis, *The Epic of New York City*, 416–22; Burrows and Wallace, *Gotham*, 1112.

57. "Quotes About Immigration," Goodreads, http://www.goodreads.com/quotes/tag/immigration.
58. Columbia University Digital Knowledge Ventures, "Dumbbell Tenements," *The Architecture and Development of New York City*, http://ci.columbia.edu/0240s/0243_2/0243_2_s1_2_text.html; Eric Homberger, *The Historical Atlas of New York City: A Visual Celebration of Four Hundred Years of New York City's History* (1994; New York: Henry Holt, 2005), 110.
59. Weil, *A History of New York*, 106; Homberger, *The Historical Atlas of New York City*, 110.
60. Wendell Cox Consultancy, "New York (Manhattan) Wards: Population & Density 1800–1910," *Demographia*, http://www.demographia.com/db-nyc-ward1800.htm; Weber, *The Growth of Cities in the Nineteenth Century*, 46–61.
61. Hammack, *Power and Society*, 36, 50.
62. Still, *Mirror for Gotham*, 211.
63. Domosh, *Invented Cities*, 14–15, 19; Still, *Mirror for Gotham*, 241.
64. Lankevich, *American Metropolis*, 119. See also Richard Florida, "Density and Class in Early Manhattan," *Citylab*, October 11, 2016, https://www.citylab.com/equity/2016/10/mapping-mid-nineteenth-century-manhattan/487937/.
65. Quoted in Jackson and Dunbar, eds., *Empire City*, 307.
66. Still, *Mirror for Gotham*, 206; Weil, *A History of New York*, 13–16; Domosh, *Invented Cities*, 16.
67. Weil, *A History of New York*, 96.
68. Domosh, *Invented Cities*, 7–8.
69. Weil, *A History of New York*, 90–92, 116; Domosh, *Invented Cities*, 36, 46. See also Blaszezyk, *American Consumer Society*, 75–77, 81–82, 103, 127, 208.
70. Robert Murray Haig, *Major Economic Factors in Metropolitan Growth and Arrangement: A Study of Trends and Tendencies in the Economic Activities with the Region of New York and Its Environs*, Regional Survey, vol. 1 (1927; New York: Arno, 1974), 54, 58, 64, 72, 74, 82, 84, 86, 88–90, 92, 96, 98–99.
71. Domosh, *Invented Cities*, 69.
72. Domosh, *Invented Cities*, 69.
73. Carol Willis, *Form Follows Finance: Skyscrapers and Skylines in New York and Chicago* (Princeton, NJ: Princeton Architectural Press, 1995), 36.
74. Willis, *Form Follows Finance*, 42, also 36–41, 43–48; Eugene P. Moehring, "Space, Economic Growth, and the Public Works Revolution in New York," *Essays in Public Works History* (December 1985): 51–52; Edward Robb Ellis, *The Epic of New York City* (1966; New York: Carroll & Graf, 2005), 405–6. Some have argued that skyscrapers were made possible, especially in parts of Manhattan, because of the solid rock formations underlying the island, but this may have been overstated.
75. Lloyd Morris, *Incredible New York: High Life and Low Life of the Last Hundred Years* (New York: Random House, 1951), 197–99; Domosh, *Invented Cities*, 72–76; Burrows and Wallace, *Gotham*, 1050–52.
76. Robert Andrews, *The Columbia Dictionary of Quotes* (New York: Columbia University Press, 1993), 159.
77. Quoted in Domosh, *Invented Cities*, 73.
78. Still, *Mirror for Gotham*, 206–7; Ellis, *The Epic of New York City*, 406.
79. Homberger, *The Historical Atlas of New York City*, 91.
80. City of New York, *Natural Hazard Mitigation Plan*, 14.
81. Sanderson, *Mannahatta*, 84–85.

1. ISLAND CITY

82. "New York: Geography and Climate," City-Data.com, http://www.city-data.com/us-cities/The-Northeast/New-York-Geography-and-Climate.html; City of New York, *Natural Hazard Mitigation Plan*, 12.
83. City of New York, *Natural Hazard Mitigation Plan*, 12.
84. Sharon Seitz and Stuart Miller, *The Other Islands of New York City: A History and Guide*, 3rd ed. (Woodstock, VT: Countryman, 2011), 12–14, 20–22, 34–41, 56–60; National Park Service, "Statue of Liberty," http://www.nps.gov/stli/index.htm; Statue of Liberty–Ellis Island Foundation, "Ellis Island-History," http://www.ellisisland.org/genealogy/ellis_island_history.asp.
85. Quoted in Seitz and Miller, *The Other Islands of New York City*, 157.
86. Seitz and Miller, *The Other Islands of New York City*, 154–59.
87. Seitz and Miller, *The Other Islands of New York City*, 138, 182, 201–2.
88. See Kara Murphy Schlichting, *New York Recentered: Building the Metropolis from the Shore* (Chicago: University of Chicago Press, 2019).
89. Manuel Lenio Jr., "Peter Stuyvesant and the History of Reclamation in the First World," *Revive Manila*, April 29, 2013, https://manilasolarcity.wordpress.com/2013/04/29/peter-stuyvesant-and-the-history-of-reclamation-in-the-first-world/; Ann L. Buttenwieser, "Landfill," in *The Encyclopedia of New York City*, ed. Jackson, 719.
90. Buttenwieser, "Landfill," 719.
91. Buttenwieser, *Manhattan Water-Bound*, 23.
92. Steinberg, *Gotham Unbound*, 101–8, 119–25.
93. Donald Squires and Kevin Bone, "The Beautiful Lake: The Promise of the Natural Systems," in *The New York Waterfront*, ed. Bone, 19–21.
94. Cantwell and di Zerega Wall, *Unearthing Gotham*, 225.
95. Quoted in Gina Pollara, "Transforming the Edge: Overview of Selected Plans and Projects," in *The New York Waterfront*, ed. Bone, 156.
96. Steinberg, *Gotham Unbound*, 22–38; Cantwell and di Zerega Wall, *Unearthing Gotham*, 225–26; Buttenwieser, *Manhattan Water-Bound*, 30–32, 37, 48–49; Pollara, "Transforming the Edge," 156–59.
97. Weil, *A History of New York*, 80.
98. "Hazardous Waste and Environmental Liability: A Historical Perspective," in *Effluent America: Cities, Industry, Energy, and the Environment*, ed. Martin V. Melosi (Pittsburgh, PA: University of Pittsburgh Press, 2001), 115.
99. Steinberg, *Gotham Unbound*, 34.
100. National Park Service, "Castle Clinton National Monument," http://www.nps.gov/cacl/index.htm; Seitz and Miller, *The Other Islands of New York City*, 80–82, 100.

2. WASTING AWAY

1. John Duffy, *The Sanitarians* (Urbana: University of Illinois Press, 1990), 33; Martin V. Melosi, *Garbage in the Cities: Refuse, Reform, and the Environment*, rev. ed. (Pittsburgh, PA: University of Pittsburgh Press, 2005), 13–15; Martin V. Melosi, *The Sanitary City: Environmental Services in Urban America from Colonial Times to the Present*, abridged ed. (Pittsburgh, PA: University of Pittsburgh Press, 2008), 12–13; Joel A. Tarr, "Urban Pollution: Many Long Years Ago," *American Heritage* 22 (October 1971): 64–69, 106; Garrick E. Louis, "A Historical Context of Municipal Solid Waste Management in the United States," *Waste Management & Research* 22 (2004): 306–7.
2. City of New York, Department of Sanitation, Office of Operations, *Outline History of the Department of Sanitation* (July 1954), 2.

3. *Our City—New York: A Textbook on City Government*, rev. Frank A. Rexford and Muriel Jean Drummond (Boston: Allyn and Bacon, 1937), 119.
4. Catherine McNeur, *Taming Manhattan: Environmental Battles in the Antebellum City* (Cambridge, MA: Harvard University Press, 2014), 97.
5. Graham Russell Hodges, *New York City Cartmen, 1667–1850* (New York: New York University Press, 1986).
6. McNeur, *Taming Manhattan*, 99–100.
7. Steven Hunt Corey, "King Garbage: A History of Solid Waste Management in New York City, 1881–1970," PhD diss., New York University, 1994, 12.
8. Corey, "King Garbage," 22–23; City of New York, Department of Sanitation, Office of Operations, *Outline History of the Department of Sanitation*, 2.
9. Corey, "King Garbage," 23.
10. Melosi, *Garbage in the Cities*, 25–29.
11. Ann L. Buttenwieser, *Manhattan Water-Bound: Manhattan's Waterfront from the Seventeenth Century to the Present*, 2nd ed. (Syracuse, NY: Syracuse University Press, 1999), 44–47; Corey, "King Garbage," 68–69. In 1903 the City of New York transferred the functions of the Department of Street Cleaning to the local government in Queens and Staten Island rather than take direct responsibility for their collection and street-cleaning services.
12. Dr. George A. Soper, "Report of the Disposal of the Refuse of New York City," *Public Health Papers and Reports* 28 (1902): 64.
13. Melosi, *Garbage in the Cities*, 29–31.
14. Melosi, *Garbage in the Cities*, 36–38; Corey, "King Garbage," 27–28.
15. Melosi, *Garbage in the Cities*, 18–20, 33, 36–38; Corey, "King Garbage," 27–28.
16. Melosi, *Garbage in the Cities*, 59. See also Carl Zimring, *Cash for Your Trash: Scrap Recycling in America* (New Brunswick, NJ: Rutgers University Press, 2005); Susan Strasser, *Waste and Want: A Social History of Trash* (New York: Metropolitan Books, 1999).
17. Corey, "King Garbage," xi.
18. Ted Steinberg, *Gotham Unbound: An Ecological History of Greater New York, 1609–2012* (New York: Simon & Schuster, 2014), 24–25; Anne-Marie Cantwell and Diana di Zerega Wall, *Unearthing Gotham: The Archaeology of New York City* (New Haven, CT: Yale University Press, 2001); Buttenwieser, *Manhattan Water-Bound*, 42–43.
19. A portion of the city waste that was used for fill was taken by scow to Blackwell's Island in the East River, which was regarded as the city's first significant organized landfill operation. Buttenwieser, *Manhattan Water-Bound*, 44–47; Corey, "King Garbage," 68–69.
20. Buttenwieser, *Manhattan Water-Bound*, 44.
21. Corey, "King Garbage," 72–79; Steinberg, *Gotham Unbound*, 121–22.
22. Cantwell and di Zerega Wall, *Unearthing Gotham*, 229–33, 236; Steinberg, *Gotham Unbound*, 24–34; Buttenwieser, *Manhattan Water-Bound*, 33; Donald Squires and Kevin Bone, "The Beautiful Lake: The Promise of the Natural Systems," in *The New York Waterfront: Evolution and Building Culture of the Port and Harbor*, rev. ed., ed. Kevin Bone (New York: Monacelli, 2004), 27.
23. Landfill would take on a different meaning, especially by the late 1930s, with the invention of the sanitary landfill. The sanitary landfill would not be a fill-dumping site but a systematic method of trenching accumulated solid wastes.
24. Ann L. Buttenwieser, "Landfill," in *The Encyclopedia of New York City*, 2nd ed., ed. Kenneth T. Jackson (New Haven, CT: Yale University Press, 2010), 719–20; Elizabeth

Barlow, *The Forests and Wetlands of New York City* (Boston, Little, Brown, 1971), 5–6; Cantwell and di Zerega Wall, *Unearthing Gotham*, 223–24.

Development of fill land continued into the twentieth century. What had been the picturesque Hunter and Twin Islands, where aristocrats built lavish country estates, were eventually connected to the Bronx by landfill. The islands went from private hands to the city in 1888, resulting in the building of Pelham Bay Park. The city also leased homes and mansions in the area to charity organizations until the 1930s. When the subway reached there in 1894, immigrant groups and others seeking recreation discovered the islands. The transformation continued into the 1930s when Robert Moses created Orchard Beach, which sported an extensive seashore. Fill was regularly used to reshape the site, and in 1947 Twin and Hunter Beach were linked, completing the refocus of the islands from forests and rocky terrain to a playground for sun worshippers.

Ellis Island was expanded by landfill for a new hospital and other buildings. Blizzard Island, at the mouth of the Hutchinson River, is now part of Pelham Bay Park because of landfill operations. North Brother Island was increased by four acres of fill at the turn of the twentieth century in order to erect several buildings for tuberculosis patients. In the 1920s Ruffle Bar near Brooklyn was transformed from a marshy landscape by the construction of about forty buildings after substantial filling. Other fill projects included such places as Codling Island, Wright's Island in the Bronx, East River islands like Mill Rock Island, and Barren Island in Jamaica Bay and the Rockaways. Jamaica Bay ceded a good deal of its area to fill. Even Coney Island was transformed by fill into an inlet off Gravesend Bay adjacent to Brooklyn. See Sharon Seitz and Stuart Miller, *The Other Islands of New York City: A History and Guide*, 3rd ed. (Woodstock, VT: Countryman, 2011), 60, 130–32, 146–49, 214, 230–35, 256–57, 271.

25. Corey, "King Garbage," 74, 79–82.
26. "Geography: The Islands: North & South Brother Island," *East River NYC*, 2008, http://www.eastrivernyc.org/content/geography/the-islands/north-south-brother.html.
27. "Barren Island," *Brooklyn Union*, September 4, 1876.
28. Benjamin Miller, *Fat of the Land: Garbage of New York the Last Two Hundred Years* (New York: Four Walls Eight Windows, 2000), 36–44.
29. Corey, "King Garbage," 82–86.
30. "Methods of Street Cleaning and Garbage Removal in New York," *Sanitary Engineer* 11 (1885): 541; New York City, Department of Street Cleaning, *Report for the Years 1887 and 1888* (1889), 63.
31. "Garbage Disposal for the City of New York," *Engineering News* 28 (October 27, 1892): 397.
32. New York City, Department of Street Cleaning, *Report for the Years 1887 and 1888*, 65. See also New York City, Department of Street Cleaning, *Report of the Operations of the Department of Street Cleaning for the Year 1886* (1886), 23; New York City, Department of Street Cleaning, *Report of Commissioner Andrews to His Honor the Mayor*, September 15, 1893 (1993).
33. "Methods of Street Cleaning and Garbage Removal in New York," 541.
34. New York City, Department of Street Cleaning, *Report for the Years 1887 and 1888*, 17, also 66.
35. New York City, Department of Street Cleaning, *Report for the Years 1887 and 1888*, 87–88.
36. Melosi, *Garbage in the Cities*, 30–32.

37. Melosi, *Garbage in the Cities*, 40.
38. Martin V. Melosi, "Technology Diffusion and Refuse Disposal," in *Technology and the Rise of the Networked City in Europe and America*, ed. Joel A. Tarr and Gabriel Dupuy (Philadelphia: Temple University Press, 1988), 207–22.
39. Corey, "King Garbage," 84.
40. New York City, Department of Street Cleaning, *Report for the Years 1887 and 1888*, 17–18.
41. "The New York City Deodorizing Plant of Riker's Island," *Engineering News* 32 (August 2, 1894): 90.
42. Corey, "King Garbage," 89–93.
43. Henry Smith Williams, "How New York Is Kept Partially Clean," *Harpers Weekly* 38 (October 13, 1894): 971. "Tammany Hall was a New York City political organization that endured for nearly two centuries. Formed in 1789 in opposition to the Federalist Party, [Tammany's] leadership often mirrored that of the local Democratic Party's executive committee. Although its popularity stemmed from a willingness to help the city's poor and immigrant populations, Tammany Hall became known for charges of corruption levied against leaders such as William M. 'Boss' Tweed. Its power waned during the tenure of New York City Mayor Fiorello La Guardia (1934–1945), and the organization was rendered extinct after John V. Lindsay took office in 1966." "Tammany Hall," *history.com*, http://www.history.com/topics/tammany-hall.
44. G. T. Ferris, "The Cleansing of Great Cities," *Harpers Weekly* 35 (January 10, 1891): 33.
45. Henry Smith Williams, "The Disposal of Garbage," *Harpers Weekly* 38 (September 1, 1894): 835.
46. Williams, "How New York Is Kept Partially Clean," 971.
47. Drs. Thomas H. Manley and Douglas H. Stewart, "The Economical and Efficient Disposal of the Household Garbage of New York," *The Sanitarian* 34, part 1 (January 1895): 4–5.
48. Drs. Thomas H. Manley and Douglas H. Stewart, "The Economical and Efficient Disposal of the Household Garbage of New York," *The Sanitarian* 34, part 2 (February 1895): 98.
49. New York City, Department of Street Cleaning, *Report of Commissioner Andrews to His Honor the Mayor*, September 15, 1893, 7.
50. Williams, "How New York Is Kept Partially Clean," 972.
51. Manley and Stewart, "The Economical and Efficient Disposal of the Household Garbage of New York," part 1, 7.
52. "Garbage Disposal for New York City," 397; Williams, "How New York Is Kept Partially Clean," 972.
53. Williams, "How New York Is Kept Partially Clean," 972.
54. Clay McShane and Joel Tarr, *The Horse in the City: Living Machines in the Nineteenth Century* (Baltimore, MD: Johns Hopkins University Press, 2007). See also McNeur, *Taming Manhattan*, 101–9.
55. Melosi, *Garbage in the Cities*, 20–21.
56. Steinberg, *Gotham Unbound*, 112–18; City of New York, Department of Sanitation, Office of Operations, *Outline History of the Department of Sanitation*, 2.
57. Susan Strasser, *Satisfaction Guaranteed: The Making of the American Mass Market* (Washington DC: Smithsonian Books, 1989), 6.
58. Gen. Emmons Clark, "Street-Cleaning in Large Cities," *Popular Science Monthly*, April 1891, 748.

59. Elizabeth Fee and Steven H. Corey, *Garbage! The History and Politics of Trash in New York City* (New York: New York Public Library, 1994), 13.
60. Strasser, *Waste and Want*, 12–15. For some innovative research about food consumption (especially meat) and food access in New York City in the eighteenth and nineteenth centuries, see Gergely Baics, *Feeding Gotham: The Political Economy and Geography of Food in New York, 1790–1860* (Princeton, NJ: Princeton University Press, 2016), 8, 13, 63, 199–200, 255.
61. Melosi, *Garbage in the Cities*, 33; Manley and Stewart, "The Economical and Efficient Disposal of the Household Garbage of New York," part 1, 4, 9–10.
62. Manley and Stewart, "The Economical and Efficient Disposal of the Household Garbage of New York," part 1, 6–7.
63. New York City, Department of Street Cleaning, *Report of Commissioner Andrews to His Honor the Mayor*, September 15, 1893, 6–7.
64. Soper, "Report of the Disposal of the Refuse of New York City," 64–65.
65. Corey, "King Garbage," 96–100.
66. David C. Hammack, *Power and Society: Greater New York at the Turn of the Century* (New York: Columbia University Press, 1987), 148–49; Joseph P. Viteritti, "Lexow Committee," in *The Encyclopedia of New York City*, ed. Jackson, 736; Lloyd Morris, *Incredible New York: High Life and Low Life of the Last Hundred Years* (New York: Random House, 1951), 232; "William L. Strong for Mayor," *New York Times*, October 6, 1894; "William L. Strong," *New York Times*, November 7, 1894; "Ex-Mayor Strong Has Passed Away," *New York Times*, November 3, 1900; "Mayors Talk on Reform," *New York Times*, December 21, 1894. See also Oliver E. Allen, *The Tiger: The Rise and Fall of Tammany Hall* (Reading, MA: Addison-Wesley, 1993).
67. Chicago Department of Public Works, *Annual Report* (1894); "Cost of Street Cleaning in Various Cities," *Engineering News* 33 (April 11, 1895): 247.
68. F. W. Hewes, "Street Cleaning," *Harper's Weekly*, March 9, 1895, 233–34. See also *Engineering News* 23 (January 4, 1890): 12–13; "Street Cleaning Statistics," *Municipality and County* 1 (April 1895): 153; Melosi, *Garbage in the Cities*, 38–39.
69. Melosi, *Garbage in the Cities*, 49; Morris, *Incredible New York*, 232.
70. Theodore Roosevelt, *Theodore Roosevelt: An Autobiography* (1913; New York: Charles Scribner's Sons, 1924), 168.
71. Melosi, *Garbage in the Cities*, 48–49.
72. Richard Skolnik, "George Edwin Waring, Jr.: A Model for Reformers," *New-York Historical Society Quarterly* 52 (October 1968): 357.
73. Melosi, *Garbage in the Cities*, 44–50.
74. Col. George E. Waring Jr., "The Necessity for Excluding Politics from Municipal Business," *Proceedings of the 4th National Conference for Good Government and of the 2nd Annual Meeting of the National Municipal League, Baltimore* (Philadelphia, PA: National Municipal League, June 6–8, 1896), 268.
75. George E. Waring Jr., "Government by Party," *North American Review* 163 (November 1896), 587.
76. George E. Waring Jr., "The Cleaning of the Streets of New York," *Harpers Weekly*, October 29, 1895, 1022. See also "Waring Holds the Broom," *New York Tribune*, January 16, 1895.
77. George E. Waring Jr., *Street Cleaning and the Disposal of a City's Wastes: Methods and Results and the Effect Upon Public Health, Public Morals, and Municipal Prosperity* (New York: Doubleday and McClure Co., 1898), 15–18.
78. Waring, *Street Cleaning and the Disposal of a City's Wastes*, 15–18.

79. Waring, "The Necessity for Excluding Politics from Municipal Business," 269.
80. "Col. Waring Declares Tammany to Be a Very Bad Lot," *New York Times*, May 28, 1898.
81. See *New York Times*, April 28, 1895; October 31, 1897; November 29, 1897; "Col. Waring on Street Cleaning," *City Government* 4 (June 1898): 223; "Hon. William S. Andrews and Col. George E. Waring, Jr., on the Cost of Street Cleaning in New York City," *Municipal Record and Advertiser*, June 26, 1897, 15; *Engineering News* 33 (January 3, 1895): 8; "Col. Waring Brings Suit: Charges Croker and 'The Morning Telegraph,'" *New York Tribune*, October 31, 1897.
82. "His Utterance About 'Pension Bummers' Arouses Indignation," *New York Times*, April 23, 1895.
83. Edwin G. Burrows and Mike Wallace, *Gotham: A History of New York City to 1898* (New York: Oxford University Press, 1999), 1194.
84. Melosi, *Garbage in the Cities*, 49–54.
85. Cited in Melosi, *Garbage in the Cities*, 54.
86. *New York Times*, October 11, 1896, Sunday Supplement.
87. David Willard, "The Juvenile Street-Cleaning Leagues," in Waring, *Street-Cleaning and the Disposal of a City's Wastes*, 177–86.
88. Quoted in Melosi, *Garbage in the Cities*, 53.
89. Melosi, *Garbage in the Cities*, 54–55.
90. Quoted in Daniel Eli Burnstein, *Next to Godliness: Confronting Dirt and Despair in Progressive Era New York City* (Urbana, IL: University of Illinois Press, 2006), 44.
91. Melosi, *Garbage in the Cities*, 54–55, 57–59; New York City, Department of Street Cleaning, *A Report on the Final Disposition of the Wastes of New York* (1896), 3–6.
92. George E. Waring Jr., "The Disposal of a City's Waste," *North American Review* 161 (July 1895): 56.
93. Waring, "The Disposal of a City's Waste," 51–56.
94. Melosi, *Garbage in the Cities*, 58–59.
95. "Mayor Seeks More Haste," *New York Times*, January 11, 1896; Miller, *Fat of the Land*, 75–76.
96. New York City, Department of Street Cleaning, *A Report on the Final Disposition of the Wastes of New York* (1896), 3.
97. Waring, "The Disposal of a City's Waste," 53.
98. New York City, Department of Street Cleaning, *Report on Final Disposition, 1895–1897* (December 16, 1897), 104–5.
99. New York City, Department of Street Cleaning, *Report on Final Disposition, 1895–1897*, 122–25.
100. George E. Waring Jr., "The Utilization of City Garbage," *Cosmopolitan Magazine* 24 (February 1898): 406.
101. Waring, "The Utilization of City Garbage," 408.
102. Burnstein, *Next to Godliness*, 41–42.
103. Waring, *Street Cleaning and the Disposal of a City's Wastes*, 68–73; Waring, "The Disposal of a City's Waste," 56; George E. Waring Jr., "The Cleaning of a Great City," *McClure's Magazine*, September 1897, 919–20.
104. Strasser, *Waste and Want*, 134.
105. William F. Morse, "Disposal of the City's Waste," *American City* 2 (May 1910): 271–72; Harry de Berkeley Parsons, *The Disposal of Municipal Refuse* (New York: John Wiley & Sons, 1906), 105.
106. "Disposal of Garbage," *City Government* 5 (August 1898): 67.

107. "Recent Refuse Disposal Practice," *Municipal Journal and Engineer* 37 (December 10, 1914): 848–49.
108. City of New York, Department of Sanitation, Office of Operations, *Outline History of the Department of Sanitation*, 3. Also on the island were a fish factory, a phosphate works, a rendering factory, a small population of workers and their families, four saloons, and one store. "The Utilization of New York City Garbage," *Scientific American* 77 (August 14, 1897): 102; Miller, *Fat of the Land*, 54, 87.
109. Morse, "Disposal of the City's Waste," 272. Companies representing Arnold, Holthaus, and Merz systems all tendered proposals, but the Arnold system won out with the lowest bid. See Waring, "The Utilization of City Garbage," 409–10; "May End Dumping at Sea," *New York Times*, June 3, 1896. See also Mike Wallace, *Greater Gotham: A History of New York City from 1898 to 1919* (New York: Oxford University Press, 2017), 216.
110. Waring, "The Utilization of City Garbage," 408.
111. Waring, "The Utilization of City Garbage," 408–10. The fumes from the other industrial activity on the island also contributed to the complaints, including those from the American Fisheries Company, E. F. Coe Company, White's rendering plant, and others. See New York City, Department of Health, *A Report as to the Existing Conditions on Barren Island* (April 1899), 5.
112. New York City, Department of Health, *A Report as to the Existing Conditions on Barren Island*, 5.
113. Soper, "Report of the Disposal of the Refuse of New York City," 66.
114. New York City, Department of Street Cleaning, *Report for the Year 1899* (April 1, 1900), 24. See also New York City, Department of Street Cleaning, *Report for the Year 1898* (January 2, 1899).
115. George J. Lankevich, *American Metropolis: A History of New York City* (New York: New York University Press, 1998), 131–32; Hammack, *Power and Society*, 151.
116. Morris, *Incredible New York*, 232.
117. Hammack, *Power and Society*, 178.
118. Antonia Felix and the Editors of the *New York Post*, comps., *The Post's New York* (New York: HarperResource, 2001), 81; Keith D. Revell, *Building Gotham: Civic Culture and Public Policy in New York City, 1898–1938* (Baltimore, MD: Johns Hopkins University Press, 2003), 1.
119. Andrew Sancton, *Merger Mania: The Assault on Local Government* (Montreal: McGill-Queen's University Press, 2000), 30.
120. "New York City: Organization and Governmental Structure," http://dls.virginia.gov/groups/consolidation/meetings/092204/NYCity.pdf.
121. Burrows and Wallace, *Gotham*, 1220.
122. Selma Berrol, *The Empire City: New York and Its People, 1624–1996* (Westport, CT: Praeger, 2000), 82.
123. Eric Homberger, *New York City: A Cultural History* (Northampton, MA: Interlink, 2008), 244.
124. Jim Yardley, "Not Just a New Year: A 100th Birthday for a Unified City," *New York Times*, January 1, 1998.
125. Anita Klutsch, "Andrew Haswell Green: The Father of Greater New York and His Dual Vision of a Cultivated and Consolidate Metropolis," PhD diss., Ludwig-Maximilians University, Munich, Germany, 2012, 45; Barry J. Kaplan, "Andrew H. Green and the Creation of a Planning Rationale: The Formation of Greater New York City, 1865–1890," *Urbanism Past & Present* 8 (Summer 1979): 32; Wallace, *Greater Gotham*, 51.

126. *Communication of Andrew H. Green to the Legislature of the State of New York*, March 4, 1890, 37–38.
127. Wallace, *Greater Gotham*, 49. Beyond his promotion of merger that culminated in the creation of Greater New York City in 1898, Green was a decisive figure in the development of Central Park, Riverside Park, Morningside Park, the New York Public Library, the Bronx Zoo, the American Museum of Natural History, the American Scenic and Historic Preservation Society, and the Metropolitan Museum of Art. See Clyde Haberman, "'Visionary' Was Forgotten, but Not by One Devotee," *New York Times*, March 15, 2010.
128. Kaplan, "Andrew H. Green and the Creation of a Planning Rationale," 39; Hammack, *Power and Society*, 195–96, 203–4; Rodgers and Rankin, *New York*, 91; "Consolidation and the Birth of Greater New York," *Inventing Gotham*, Department of History, Fieldston School, http://www.mapsites.net/gotham/Timelines/ConsolidationTimeline.htm.
129. Hammack, *Power and Society*, 206–8; Rodgers and Rankin, *New York*, 92; Burrows and Wallace, *Gotham*, 1230–31; Wallace, *Greater Gotham*, 49–60.
130. George A. Soper, "George Edwin Waring," *Dictionary of American Biography* (New York: Charles Scribner's Sons, 1936), 19:456.
131. *New York Times*, November 23, 1898.
132. Albert Shaw, *Life of Col. Geo. E. Waring, Jr.: The Greatest Apostle of Cleanliness* (New York: Patriotic League, 1899), 31.
133. "Col. George E. Waring, Jr.," *Engineering Record* 38 (November 5, 1898): 485.
134. John Brisben Walker, "The Street Cleaning Work of Colonel Waring in New York," *Cosmopolitan* 26 (December 1898): 235.
135. William Potts, "George Edwin Waring, Jr.," *Charities Review* 8 (1898): 462.
136. "Colonel Waring," *New York Times*, October 30, 1898.
137. Burnstein, *Next to Godliness*, 97; Soper, "Report of the Disposal of the Refuse of New York City," 67. See also W. F. Morse, "Street Cleaning Bureau of New York Under Tammany," *Municipal Journal and Engineer* 11 (November 1901): 222.
138. Strasser, *Waste and Want*, 125, 128–29.
139. Charles A. Meade, "City Cleansing in New York: Some Advances and Retreats," *Municipal Affairs* 4 (December 1900): 726.
140. "Barren Island Nuisances," *New York Times*, September 3, 1897; "To Purify Barren Island," *New York Times*, January 11, 1898; "The Barren Island Nuisance," *New York Times*, November 27, 1898.
141. "The Barren Island Nuisance."
142. Meade, "City Cleansing in New York: Some Advances and Retreats," 733–41; City of New York, Department of Sanitation, Office of Operations, *Outline History of the Department of Sanitation*, 4.
143. Miller, *Fat of the Land*, 87, 89.
144. Quoted in "Board of Health and Barren Island," *Brooklyn Eagle*, July 18, 1902, 4. See also "Dr. E. J. Lederle Dies in Sanitarium," *New York Times*, March 15, 1921; Charles F. Bolduan, *Over a Century of Health Administration in New York City*, Monograph Series 13 (City of New York, Department of Health of the City of New York, March 1916), 27.
145. "Woodbury Here Tomorrow," *Brooklyn Eagle*, January 20, 1902, 20.
146. "Barren Island Nuisance," *New York Times*, April 21, 1900; "Disposing of Garbage," *New York Times*, February 21, 1905; "Barren Island Reduction Plant Rebuilt on a Larger Scale," *New York Times*, September 13, 1906; "Building Cut in Two to Save It

From Sea," *New York Times*, November 28, 1905; "Garbage Problem and Barren Island," *New York Times*, November 30, 1905; "$1,500.00 Fire Loss on Barren Island," *New York Times*, May 21, 1906.
147. *New York Times*, May 17, 1907.
148. "Hunt Wild Hogs on Barren Island," *New York Times*, March 17, 1909.
149. New York City, Department of Street Cleaning, *Report for the Year 1898* (1899), 3.
150. *Engineering News* 46 (August 22, 1901): 120; John McGaw Woodbury, "The Wastes of a Great City," *Scribner's Magazine* 34 (October 1903): 396.
151. New York City, Department of Street Cleaning, *Report for the Year 1900* (1901), 22.
152. Corey, "King Garbage," 103; New York City, Department of Street Cleaning, *Report for the Year 1898*, 8.
153. New York City, Department of Street Cleaning, *Report of the Commission on Street Cleaning and Waste Disposal, The City of New York* (December 31, 1907); New York City, Department of Street Cleaning, *Report for the Years 1887 and 1888*, 65. See also New York City, Department of Street Cleaning, *Annual Report 1911* (December 31, 1911): 4; New York City, Department of Street Cleaning, *Annual Report 1913* (1913): 6–7.
154. Daniel C. Walsh, "Residential Refuse Composition and Generation Rates for the 20th Century," *Environmental Science & Technology* 36 (2002): 4938.

3. THE QUARANTINE

1. Raritan Bay to the south and east, Arthur Kill to the west, Kill Van Kull to the north, and the Narrows to the northeast. Ralph W. Tiner, *Wetlands of Staten Island, New York: Valuable Vanishing Urban Wetlands* (Washington, DC: National Wetlands Inventory Publication, January 2000), 8.
2. Industrial Bureau, Merchant's Association of New York, Staten Island, *New York City: Its Industrial Resources and Possibilities* (New York, 1922), 18–19, History Archives & Library, Staten Island Museum, Staten Island, New York.
3. Industrial Bureau, Merchant's Association of New York, Staten Island, *New York City*, 18–19.
4. Ted Steinberg, *Gotham Unbound: The Ecological History of Greater New York* (New York: Simon & Schuster, 2014), 226–27; Judith S. Weis and Carol A. Butler, *Salt Marshes: A Natural and Unnatural History* (New Brunswick, NJ: Rutgers University Press, 2009), 4; Tiner, *Wetlands of Staten Island*, 8.
5. Ted J. Williams, *The Villages of Staten Island: The Unofficial Guide to Township Boundaries* (May 15, 2007), 5–6. See also "Staten Island First Natives," *Staten Island Historian*, http://www.statenislandhistorian.com/staten-island-first-natives.html; "Happy 350th Birthday, Staten Island!" *Staten Island Advance*, August 22, 2011; Herbert C. Kraft, *The Lenape: Archeology, History, and Ethnography* (Newark, NJ: New Jersey Historical Society, 1986).
6. Jay Maeder, "Why Staten Island Joined NYC—Instead of New Jersey," *New York Daily News*, August 14, 2017.
7. Industrial Bureau, Merchant's Association of New York, Staten Island, *New York City*, 18–19, 34. See also *Illustrated Sketch Book, Staten Island, New York, Its Industries and Commerce* (New York: S. C. Judson, 1886), Pamphlets, History Archives & Library, Staten Island Museum; David J. Tysen, *Happenings Before and After Staten Island Became Part of Greater New York* (Staten Island: Staten Island Chamber of Commerce, 1924), History Archives & Library, Staten Island Museum.

3. THE QUARANTINE   ■   573

8. Charles L. Sachs, *Made on Staten Island: Agriculture, Industry, and Suburban Living in the City* (Richmondtown, Staten Island: Staten Island Historical Society, 1988), 15, 17–18, 20, 23.
9. Sachs, *Made on Staten Island*, 20.
10. See Clay McShane and Joel Tarr, *The Horse in the City: Living Machines in the Nineteenth Century* (Baltimore, MD: Johns Hopkins University Press, 2007), 130–44.
11. Sachs, *Made on Staten Island*, 35, 40–41, 52–65; Richard M. Bayles, ed., *History of Richmond County (Staten Island) New York, from Its Discovery to the Present Time* (New York: L. E. Preston & Co., 1887), 736.
12. Charles L. Sachs, "Business and the Economy," in *Discovering Staten Island: A 350th Anniversary Commemorative History*, ed. Kenneth M. Gold and Lori R. Weintrob (Charleston, SC: History Press, 2011), 48–49; Sachs, *Made on Staten Island*, 15, 18.
13. Sachs, *Made on Staten Island*, 45, 67.
14. Sachs, *Made on Staten Island*, 67, 71–72, 88.
15. Industrial Bureau, Merchants' Association of New York, *Staten Island*, 9. See also Vernon B. Hampton, *Staten Island's Claim to Fame: "The Garden Spot of New York Harbor"* (Staten Island: Richmond Borough Pub. & Printing Co., 1925), 9–11.
16. Industrial Bureau, Merchants' Association of New York, *Staten Island*, 10–14.
17. Industrial Bureau, Merchants' Association of New York, *Staten Island*, 10–14, 15.
18. Industrial Bureau, Merchants' Association of New York, *Staten Island*, 15.
19. Charles W. Leng and William T. Davis, *Staten Island and Its People: A History, 1609–1929* (New York: Lewis Historical Pub. Co., Inc., 1930), 1:4.
20. Quoted in Hampton, *Staten Island's Claim to Fame*, 15.
21. James Grant Wilson, ed., *The Memorial History of the City of New-York from Its First Settlement to the Year 1892* (New York: New-York History Company, 1893), 4:33, 35.
22. Quoted in Hampton, *Staten Island's Claim to Fame*, 9.
23. See Margaret Lundrigan and Tova Navarra, *Images of America: Staten Island in the Twentieth Century* (Charleston, SC: Arcadia, 1998).
24. "In Memoriam: Cornelius Geertruyus Kolff (September 2nd 1860–February 27th 1950)," *The Northeast Tolkien Society*, February 27, 2009, http://herenistarionnets.blogspot.com/2009/02/in-memorium-cornelius-geertruyu-kolff.html.
25. Cornelius G. Kolff, *Early History of Staten Island* (Rosebank, NY, May 1918). Under the same Library of Congress number is *A Short History of Staten Island* with a publication date of 1926. These are likely the same book.
26. Henry G. Steinmeyer, *Staten Island, 1524–1898*, rev. ed. (Richmondtown, Staten Island: Staten Island Historical Society, 1987), 89–95. See also Hon. Erastus Brooks, *Historical Records of Staten Island, Centennial and Bi-Centennial for Two Hundred Years and More* (November 1, 1883), 19.
27. Martin V. Melosi, *The Sanitary City: Urban Infrastructure in America from Colonial Times to the Present* (Baltimore, MD: Johns Hopkins University Press, 2000), 19, 60–61, 110–12; Philippe Martin Chatelain, "How a 19th Century Mob of Arsonists Burned Down Staten Island's Quarantine Hospital," *untapped cities*, December 3, 2013, http://untappedcities.com/2013/12/03/how-19th-century-mob-of-arsonists-burned-down-staten-islands-quarantine-hospital/.
28. "Meeting of Citizens of Staten Island, Committee Report," Executive Committee of Staten Island, *Facts and Documents Bearing Upon the Legal and Moral Questions Connected with the Recent Destruction of the Quarantine Buildings on Staten Island* (New York: W. M. C. Bryant & Co., 1858), 4, The Quarantine Collection, History Archives & Library, Staten Island Museum. See also Wilson, ed., *The Memorial*

History of the City of New York, 4:36. Wilson cites the move to Staten Island in 1801, but most sources accept 1799.

29. Cited in Kathryn Stephenson, "The Quarantine War: The Burning of the New York Marine Hospital in 1858," *Public Health Reports* 119 (January–February 2004): 82.
30. Gold and Weintrob, eds., *Discovering Staten Island*, 33.
31. See "The 19th Century Staten Island Quarantine" (Staten Island: Staten Island Institute of Archives & Sciences, n.d.). The primary source for the brochure was Leng and Davis, *Staten Island and Its People*.
32. The Seamen's Retreat was in nearby Stapleton, used for quarantining seamen and their families between 1831 and 1882.
33. Stephenson, "The Quarantine War," 79–83; Leng and Davis, *Staten Island and Its People*, 1:263; Chatelain, "How a 19th Century Mob of Arsonists Burned Down Staten Island's Quarantine Hospital." See also Steinmeyer, *Staten Island, 1524–1898*, 74.
34. The incident was sometimes referred to as the Sepoy Rebellion, named for a famous uprising by Indian troops against the British in Southeast Asia, which took place in 1857. The term was a derisive one, implicating the arsonists as barbarians and savages. Leng and Davis, *Staten Island and Its People*, 1:266–67; Marguerite Maria Rivas, "The Sepoy Rebellion: What Was It? An Historical Note," *Nearby Café*, http://www.nearbycafe.com/litandwriting/sepoy/what_was.html. See also Herbert G. Houze, "The Volcanic in Service During 'The Staten Island War' of 1858," *Man at Arms* (October 2005): 25, History Archives & Library, Staten Island Museum.
35. Stephenson, "The Quarantine War," 82–83. See also *An Act for Ascertaining and Collecting the Damages Caused by the Destruction of the Marine Hospitals and Other Building and Property at Quarantine*, April 17, 1860, 9–11; and *Argument of William Henry Anthon, Esq., in Behalf of the Defendants Messrs. Ray Tomkins, and John C. Thompson..., October 7, 1858* (New York: Wm. C. Bryant & Co., 1858), 8–11, The Quarantine Collection, History Archives & Library, Staten Island Museum.
36. Stephenson, "The Quarantine War," 82–83. The quote was from John Simonson, a real estate agent, speaking in 1849. See also Fielding H. Garrison, "The Destruction of the Quarantine Station on Station Island in 1858," *Bulletin of the New York Academy of Medicine* 2 (1926): 381.
37. "Meeting of Citizens of Staten Island, Committee Report," 5–6; Assembly Document no. 60 (1849), 21–46, The Quarantine Collection, History Archives & Library, Staten Island Museum. See also Stephenson, "The Quarantine War," 83–84.
38. "Meeting of Citizens of Staten Island, Committee Report," 6; Stephenson, "The Quarantine War," 83–84.
39. "Meeting of Citizens of Staten Island, Committee Report," 7.
40. "Meeting of Citizens of Staten Island, Committee Report," 8. See also State of New York, *Communication from the Governor Transmitting the Report of the Commissioners of Quarantine*, February 16, 1865, The Quarantine Collection, History Archives & Library, Staten Island Museum.
41. "Report of the Commissioners Relative to Removal of the Quarantine Station," March 10, 1858, The Quarantine Collection, History Archives & Library, Staten Island Museum.
42. Stephenson, "The Quarantine War," 84. See also "Report of the Commissioners Relative to Removal of the Quarantine Station," March 10, 1858; "The 19th Century Staten Island Quarantine"; Steinmeyer, *Staten Island, 1524–1898*, 74–75.
43. He was the grandson of Daniel D. Tomkins (New York's governor from 1807 to 1817 and vice president of the United States from 1817 to 1825), for whom Tomkinsville was

named. See David J. Krajicek, "How Arsonists Burned Down Staten Island's Hated Quarantine Hospital in the 19th Century with Little Resistance," *New York Daily News*, October 13, 2013.

44. Stephenson, "The Quarantine War," 85–87; Houze, "The Volcanic in Service During the 'Staten Island War' of 1858," 26. See also "The Staten Island Riot: The Quarantine Conflagration September 2, 1858," *History Box*, http://www.thehistorybox.com/ny_city/riots/sectionII/riots_article3b.htm. See also Steinmeyer, *Staten Island, 1524–1898*, 75.
45. "The 19th Century Staten Island Quarantine."
46. Quoted from Assembly Document no. 19, January 12, 1859, in Leng and Davis, *Staten Island and Its People*, 265.
47. New York, Commissioners of Emigration, "Marine Hospital and Quarantine Establishment at Staten Island," *Annual Report*, 1858, https://quod.lib.umich.edu/m/moa/AHN0806.0001.001.
48. "A Quarantine Victim," *New York Times*, October 9, 1858.
49. Yael Merkin, "Playing with Fire: The Staten Island Quarantine Riots of 1858" (2006–2007), The Quarantine Collection, History Archives & Library, Staten Island Museum.
50. Kevin McPartland, "The Forgotten Burial Ground," *New York Irish History* 18 (2004): 52. McPartland (49) claimed that beneath the eighteenth fairway of Staten Island's Silver Lake Golf Course was a burial ground with seven thousand Irish immigrants.
51. Merkin, "Playing with Fire: The Staten Island Quarantine Riots of 1858." McPartland, "The Forgotten Burial Ground," 51, however, claims that the majority of those in the Quarantine at the time of the fires were Irish.
52. Leng and Davis, *Staten Island and Its People*, 265.
53. *New York Herald*, September 3, 1858.
54. "The Quarantine War," *New York Times*, September 13, 1858.
55. Stephenson, "The Quarantine War," 88.
56. See *Argument of William Henry Anthon, Esq., in Behalf of the Defendants Messrs. Ray Tomkins, and John C. Thompson . . .* , October 7, 1858, Facts and Documents Bearing Upon the Legal and Moral Questions Connected with the Recent Destruction of the Quarantine Buildings on Staten Island, 26–34.
57. See "Meeting of Citizens of Staten Island, Committee Report," 10–16. See also Krajicek, "How Arsonists Burned Down Staten Island's Hated Quarantine Hospital in the 19th Century with Little Resistance."
58. Executive Committee of Staten Island, "Opinion of Judge Metcalfe," in *Facts and Documents Bearing Upon the Legal and Moral Questions Connected with the Recent Destruction of the Quarantine Buildings on Staten Island*, 48, The Quarantine Collection, History Archives & Library, Staten Island Museum.
59. Executive Committee of Staten Island, "Opinion of Judge Metcalfe," 70–71. See also *Testimony Taken Before Judge Metcalfe, in the Case of* The People Against Ray Tompkins and John C. Thompson, *Charged with the Destruction of the Quarantine Buildings*, 1859, The Quarantine Collection, History Archives & Library, Staten Island Museum; Leng and Davis, *Staten Island and Its People*, 266; William Franz, "Court Justified Act of Arsonists," *Staten Island Advance*, March 21, 1965; "The Legal Question of the Quarantine Controversy," *New York Evening Post*, September 16, 1858.
60. Stephenson, "The Quarantine War," 88. See also *Joel Wolfe, Respondent, Against the Supervisors of the County of Richmond, Appellants*, 1860; and *Proceedings Before the Commissioners, in the Matter of the Commission for Ascertaining and Collecting the Damages by the Destruction of the Marine Hospitals and Other Buildings and*

Property at the Quarantine, Staten Island, 1860, The Quarantine Collection, History Archives & Library, Staten Island Museum.
61. See Kolff, *Early History of Staten Island*; John Waldman, *Heartbeats in the Muck: The History, Sea Life, and Environment of New York Harbor*, rev. ed. (New York: Fordham University Press, 2013), 79; Steinmeyer, *Staten Island, 1524–1898*, 78.
62. Gold and Weintrob, eds., *Discovering Staten Island*, 126; Stephenson, "The Quarantine War," 88–89. See also *Description and Specifications for the Artificial Island Proposed for Quarantine Purposes, by the Commissioners of the State of New-York*, December 31, 1858, The Quarantine Collection, History Archives & Library, Staten Island Museum.

4. THE GARBAGE WAR

1. In 1788 the state legislature divided the island into four towns (Castleton, Northfield, Southfield, and Westfield), with a fifth (Middleton) added in 1860. There also were four independent villages (New Brighton; Edgewater; Port Richmond, incorporated in 1866; and Tottenville, incorporated in 1894). The population on the island grew slowly; its roads, sewerage system, and other infrastructure were among the poorest compared with neighboring counties. Schools were in short supply, and fire and police protection were at best mediocre. "History," *Staten Island History*, http://www.statenislandhistorian.com/history.html.
2. New York (NY) Board of Commissioners of the Central Park, *Twelfth Annual Report of the Board of Commissioners of the Central Park for the Year Ending December 31, 1868*, 163.
3. Andrew Haswell Green, "Communication of Andrew H. Green to the Legislature of the State of New York" (March 4, 1890), in *New York of the Future: Writings and Addresses by Andrew H. Green* (New York, 1893), 55.
4. "At a Meeting of the Commissioners, Held December 11, 1890, at the Office of the President," in Green, *New York of the Future*, 56–57.
5. Eric Homberger, *The Historical Atlas of New York City: A Visual Celebration of Four Hundred Years of New York City's History* (1994; New York: Henry Holt, 2005), 125.
6. "Bill Nye on Consolidation," *Staten Island Historian* 31 (April–June 1970): 13–14.
7. "Prohibition Park, Staten Island, New York," http://www.prohibitionists.org/Background/Party_Platform/Prohibition_Park.html; "Westerleigh Park," NYC Parks, http://www.nycgovparks.org/parks/westerleighpark/history; Kenneth M. Gold and Lori R. Weintrob, eds., *Discovering Staten Island: A 350th Anniversary Commemorative History* (Charleston, SC: History Press, 2011), 154–55.
8. See Charles W. Leng and William T. Davis, *Staten Island and Its People: A History, 1609–1929*, vol. 3 (New York: Lewis Historical Pub. Co., 1930), 40; and vol. 5 (1933), 143–44, 206–7, 217; "For Erastus Wiman, St. George Was a Golden Opportunity," March 27, 2011, *Staten Island Advance*, *SILive.com*, https://www.silive.com/specialreports/2011/03/for_erastus_wiman_st_george_wa.html; "Consolidation Discussed," *New-York Tribune*, April 3, 1896.
9. Quoted in "In 1898, Staten Island Embraced Joining 'the City,'" *Staten Island Advance*, March 27, 2011, *SILive.com*, http://www.silive.com/specialreports/index.ssf/2011/03/in_1898_staten_island_embraced.html. See also "Foundation Footnote: John De Morgan," *Victorian Footnotes*, May 8, 2011, http://victorianfootnotes.net/2011/05/08/foundation-footnote-john-de-morgan/; "Mr. Low on Staten Island," *New-York*

Tribune, October 29, 1897; "John de Morgan," *Encyclopedia of Science Fiction*, http://www.sf-encyclopedia.com/entry/de_morgan_john.
10. David C. Hammack, *Power and Society: Greater New York at the Turn of the Century* (New York: Columbia University Press, 1987), 206–8; Cleveland Rodgers and Rebecca B. Rankin, *New York: The World's Capital City* (New York: Harper & Brothers, 1948), 92; Edwin G. Burrows and Mike Wallace, *Gotham: A History of New York City to 1898* (New York: Oxford University Press, 1999), 1230–31.
11. "Blow at Consolidation," *New York Times*, January 9, 1896.
12. Rodgers and Rankin, *New York*, 92–93; Burrows and Wallace, *Gotham*, 1233–35.
13. Joseph P. Viteritti, "The Tradition of Municipal Reform: Charter Revision in Historical Context," *Proceedings of the Academy of Political Science* 37 (1989): 20–22. See also "Changes in the City Charter," in *The Encyclopedia of New York City*, 2nd ed., ed. Kenneth T. Jackson (New Haven, CT: Yale University Press, 2010), 232; Hammack, *Power and Society*, 224–25.
14. Irvin Leigh and Paul Matus, "Staten Island Rapid Transit: The Essential History," 1, *The Third Rail Online*, February 15, 2003, http://www.thethirdrail.net/0201/sirt1.html; Daniel C. Kramer and Richard M. Flanagan, *Staten Island: Conservative Bastion in a Liberal City* (Lanham, MD: University Press of America, 2012), 119.
15. "In 1898, Staten Island Embraced Joining 'the City.'"
16. The Republican trial attorney George M. Pinney Jr. was appointed by the governor to the Greater New York Commission in 1896 and served as secretary and editor of the charter. See Leng and Davis, *Staten Island and Its People*, 1:331, 5:84–85. See also Hammack, *Power and Society*, 357; "Index to Politicians," *PoliticalGraveyard.com*, http://politicalgraveyard.com/bio/greenaway-greenhut.html#903.37.03.
17. "First Borough President, George Cromwell, Is Elected by 6 Votes," *Staten Island Advance*, March 27, 2011, *SILive.com*, http://www.silive.com/specialreports/index.ssf/2011/03/first_borough_president_is_ele.html. See also Leng and Davis, *Staten Island and Its People*, 1:353–54.
18. Diane C. Lore, "Staten Island Goes from Bucolic Municipality to NYC Borough," *Staten Island Advance*, March 27, 2011, *SILive*, http://www.silive.com/specialreports/index.ssf/2011/03/staten_island_goes_from_bucoli.html; Edna Holden, "Chapter VIII: The Cosmopolitan Period: Since 1898," in *Holden's Staten Island: A History of Richmond County* (1964), Wiley Online Library, https://onlinelibrary.wiley.com/doi/pdf/10.1111/j.2050-411X.2003.tb00307.x.
19. Lore, "Staten Island Goes from Bucolic Municipality to NYC Borough"; Holden, "Chapter VIII"; Leng and Davis, *Staten Island and Its People*, 3:4–5; James Kaser, "Civic and Political Life," in *Discovering Staten Island*, ed. Gold and Weintrob, 31, 140, 152; Margaret Lundrigan and Tova Navarra, *Staten Island*, Images of America (Charleston, SC: Arcadia, 1997), 7; "Staten Island Borough Hall," *StatenIslandUSA.com*, https://www.statenislandusa.com/borough-hall.html; Dorothy Valentine Smith, *Staten Island: Gateway to New York* (Philadelphia, PA: Chilton Book Co., 1970), 188; Henry G. Steinmeyer, *Staten Island, 1524–1898* (Richmondtown, Staten Island: Staten Island Historical Society, 1987), 97–99.
20. Lore, "Staten Island Goes from Bucolic Municipality to NYC Borough"; Thomas W. Matteo, *Staten Island: I Didn't Know That!* (Virginia Beach: Donning, 2009), 16.
21. Kramer and Flanagan, *Staten Island*, 24–27. See also Raanan Geberer, "Opinion: Never Mind the Bikes: Subways to Staten Island!" *Brooklyn Daily Eagle*, April 8, 2013, http://www.brooklyneagle.com/articles/opinion-never-mind-bikes-subways-staten-island-2013-04-08-162000. Subways now reach all the boroughs save Richmond.

Instead, Staten Island Rapid Transit takes those arriving at the ferry terminal only to other parts of the island and back. In 1971, the local railway, running from St. George to Tottenville, technically joined the Metropolitan Transportation Authority (MTA), but it is not connected to the remainder of the MTA system. Leigh and Matus, "Staten Island Rapid Transit: The Essential History"; "Putting Trains Back on Track," ESRI, *ArcNews Online*, Summer 2006, http://www.esri.com/news/arcnews/summer06 articles/putting-trains-back.html.

22. Sayre and Kaufman, *Governing New York City*, 16, 626, 638; Jeffrey Underweiser, "The Legality of Staten Island's Attempt to Secede from New York City," *Fordham Urban Law Journal* 19 (1991): 150; Keith D. Revell, *Building Gotham: Civic Culture and Public Policy in New York City, 1898–1938* (Baltimore, MD: Johns Hopkins University Press, 2003), 246, 266. See also *The Charter of the City of New York: As Adopted by the Legislature of 1901, with Amendments* (June 1901).

23. Matteo, *Staten Island*, 16. See also "The Outlying City Districts Complain," *New York Times*, September 2, 1901; Leng and Davis, *Staten Island and Its People*, 1:331–32.

24. City of New York, Department of Street Cleaning, *Annual Report*, 1914 [draft/typed version], Office of the Mayor, John P. Mitchel Administration, Departmental Correspondence Received 1914–1917, Box 80, Folder 862, Department of Street Cleaning, October 1915–1917, New York City Municipal Archives and Library, New York City, NY.

25. Bill Thompson, "Garbage War of 1916–17: 'People Don't Get Out and Fight, Like We Did,'" *Staten Island Advance*, June 15, 1946, Archives & Special Collections, CSI Library, College of Staten Island, Staten Island, New York.

26. Thompson, "Garbage War of 1916–17."

27. Frederick L. Stearns, *The Work of the Department of Street Cleaning*, paper 83 (New York: Municipal Engineers of the City of New York, 1913), 212; "Reduction of New York's Garbage," *Municipal Journal* 39 (July 8, 1915): 35. While the plant was owned by the Sanitary Utilization Company, it was leased by New York Disposal Corporation in 1913 or 1914.

28. Benjamin Miller, *Fat of the Land: Garbage of New York, the Last Two Hundred Years* (New York: Four Walls Eight Windows, 2000), 89, 115–8.

29. The remainder of the waste for fill went to Newark Meadow, Arlington, Flushing Creek, Staten Island, and elsewhere. Hon. William H. Edwards, "The Work of the Street Cleaning Department of New York City," City Club of Philadelphia, *City Club Bulletin* (March 24, 1910): 116; New York City, Department of Docks and Ferries, *Report on the Disposal of City Wastes* (by Charles W. Staniford, February 1913), 7–8; Sharon Seitz and Stuart Miller, *The Other Islands of New York City: A History and Guide*, 3rd ed. (Woodstock, VT: Countryman, 2011), 261; Mike Wallace, *A History of New York City from 1898 to 1919* (New York: Oxford University Press, 2017), 215–16.

30. "Brooklyn Ash Removal Company" June 9, 2011, http://members.trainweb.com/bedt /indloco/barc.html.

31. Quoted in "The Corona Ash Dump: Brooklyn's Burden on Queens, a Vivid Literary Inspiration and Bleak, Rat-Filled Landscape," *Bowery Boys: New York City History*, May 9, 2013, http://theboweryboys.blogspot.com/2013/05/the-corona-ash-dumps -brooklyns-burden.html. See also Miller, *Fat of the Land*, 118, 120, about Brooklyn Ash's inflated costs of doing business and the debates over making waste disposal a public enterprise. And see Wallace, *A History of New York City from 1898 to 1919*, 215.

32. "Fatal Explosion on Barren Island," *New York Times*, May 1, 1910.

33. "Barren Island Is Out on a Strike," *New York Times*, April 12, 1913; "To Curb Garbage Odors," *New York Times*, July 22, 1913; "Will Continue War on Barren Island: Residents of Neponsit Plan Meeting to Further Campaign Against Disposal Plant," *New York Times*, November 6, 1915; "Hospital Menaced by Barren Island," *New York Times*, October 23, 1915; "Reduction of New York's Garbage," 37–39.
34. Seitz and Miller, *The Other Islands of New York City*, 263; Matthew Wills, "Outer Borough," July 6, 2007, *Brooklyn Rail*, http://www.brooklynrail.org/2007/07/local/outer-borough; Kirk Johnson, "Barren Island's Harsh Past Brought to Life," *Baltimore Sun*, December 3, 2000, http://articles.baltimoresun.com/2000-12-03/news/0012030148_1_barren-island-garbage-new-york; Miller, *Fat of the Land*, 204. Rikers Island inherited a good portion of the waste from Barren Island when its facilities finally were shut down in 1921. Such a shift only further degraded Rikers to such an extent that the rat-infested, pestilential site was proving to be an unattractive location even for the proposed prison. The new prison opened in 1935 nevertheless. The fill operations ended there in the 1940s, largely because of the 1939 World's Fair's proximate location, incessant complaints, and Robert Moses's urgings to seek alternative disposal methods. See Seitz and Miller, *The Other Islands of New York City*, 203; Miller, *Fat of the Land*, 134.
35. New York City, Street Cleaning Department, *Annual Report*, 1916, 28–29; Seitz and Miller, *The Other Islands of New York City*, 262.
36. New York City, Department of Docks and Ferries, *Report on the Disposal of City Wastes*, 1, 1–2.
37. New York City, Department of Docks and Ferries, *Report on the Disposal of City Wastes*, 13, 17.
38. "Excerpt from Affidavit of John Purroy Mitchel," January 26, 1916, *Staten Island Civic League Bulletin* 3 (July 1916): 4, Environmental Collection, Archives & Special Collections, CSI Library, College of Staten Island, CUNY.
39. Miller, *Fat of the Land*, 93–98, 130–31. See also Montrose Morris, "Walkabout: William H. Reynolds, Part 2," *Brownstoner: Brooklyn Inside and Out*, May 4, 2010, http://www.brownstoner.com/blog/2010/05/walkabout-willi/; New York City, Department of Street Cleaning, *Annual Report*, 1917 (December 31, 1917), 5.
40. Steven Hunt Corey, "King Garbage: A History of Solid Waste Management in New York City, 1881–1970," PhD diss., New York University, 1994, 122; Miller, *Fat of the Land*, 130–31.
41. Bill Franz, "Island 'Dump Wars' Began Almost 80 Years Ago!" *Staten Island Register*, April 20, 1993. See also Charles W. Leng and William T. Davis, *Staten Island and Its People: A History, 1609–1929* (New York: Lewis Historical Pub. Co., 1930), 1:368; "Chronology of the Garbage Fight," *Staten Island Civic League Bulletin* 3 (July 1916): 7. A plan to use Rikers Island for disposal passed the Board of Estimate but was defeated by the Board of Aldermen. See an example of the response by landowners in Queens County to the Rikers Island plan: Leach & Williams to John Purroy Mitchel, February 11, 1916, Office of the Mayor, John P. Mitchel Administration, Departmental Correspondence Received 1914–1917, Box 81, Folder 865, Department of Street Cleaning, January–February 1916, New York City Municipal Archives and Library.
42. Proponents claimed that in the new Cobwell reduction system garbage would lose 70 to 75 percent of its moisture in the first phase. Older reduction plants utilized direct tankage driers, which gave off odors far removed from the site. The Cobwell plant would not use such driers and thus were not likely to give off such intense odors. See

"Prof. Whipple Reports on New York's Proposed Garbage-Reduction Works," *Engineering Record* 75 (January 13, 1917): 50–51.
43. "Street Cleaning and Refuse Disposal: Location of New Garbage Plant," *Municipal Journal* 40 (April 27, 1916): 593–94; New York City, Street Cleaning Department, *Annual Report*, 1916, 29.
44. Quoted in Miller, *Fat of the Land*, 128, see also 127; "Chronology of the Garbage Fight," 7; George J. Lankevich, *American Metropolis: A History of New York City* (New York: New York University Press, 1998), 152–53.
45. Miller, *Fat of the Land*, 129–32.
46. "Protest of West New Brighton Board of Trade," March 9, 1916, *Staten Island Civic League Bulletin* 3 (July 1916): 11, History Archives & Library, Staten Island Museum, Staten Island, New York.
47. Leng and Davis, *Staten Island and Its People*, 1:368, 3:229; Franz, "Island 'Dump Wars' Began Almost 80 Years Ago!"; "Chronology of the Garbage Fight," 7. See also City of New York, Department of Sanitation, Office of Operations, *Outline History of the Department of Sanitation* (July 1954), 9.
48. "Garbage Problem," *Staten Islander*, March 22, 1916, Shirtboy Collection, Archives & Special Collections, CSI Library, College of Staten Island. After the meeting, Fetherston wrote to Mayor Mitchel that the people of Staten Island were "being aroused by agitation against the location of the plant in Richmond Borough." He was upset that an attempt had been made "to misconstrue my attitude, and some speakers ... declared that I would resign rather than sign a garbage contract for the plant on the island." This he denied. Fetherston to John Purroy Mitchel, March 23, 1916, Office of the Mayor, John P. Mitchel Administration, Departmental Correspondence Received 1914–1917, Box 81, Folder 866, Department of Street Cleaning, October 1915–1917, New York City Municipal Archives and Library.
49. William Wirt Mills to Calvin D. Van Name, March 23, 1916, John Mitchel Collection, Box 12, Folder 114. See also Charles B. Dullea to Calvin D. Van Name, Box 012, Folder 114, History Archives & Library, Staten Island Museum.
50. "Reply to the Mayor by Mr. Willcox," April 13, 1916, *Staten Island Civic League Bulletin* 3 (July 1916): 12.
51. "Fair Play for Staten Island," *Staten Island Civic League Bulletin* 3 (July 1916): 13.
52. Leng and Davis, *Staten Island and Its People*, 1:369; "Chronology of the Garbage Fight," 7.
53. "Mayor's Name Hissed," *New York Times*, April 15, 1916. See also "Chronology of the Garbage Fight," 8.
54. "Staten Islanders Talk Secession in Garbage Protest," *Evening World*, April 15, 1916.
55. Leng and Davis, *Staten Island and Its People*, 3:7.
56. "Warning to Mayor Mitchel," April 20, 1916, *Staten Island Civic League Bulletin* 3 (July 1916): 12.
57. "Mayor Would Build Island for Garbage," *New York Times*, May 4, 1916; Leng and Davis, *Staten Island and Its People*, 1:369, 3:5. The number of protesters had been estimated as 2,000 to 4,000 in various publications.
58. "Chronology of the Garbage Fight," 8; "Mayor Would Build Island for Garbage."
59. "Chronology of the Garbage Fight," 8.
60. For an example of new protests, see "President Van Name Did Protest," *Staten Islander*, May 27, 1916. All *Staten Islander* articles were obtained from the Staten Island Museum.
61. Miller, *Fat of the Land*, 131; Seitz and Miller, *The Other Islands of New York City*, 317.

4. THE GARBAGE WAR ■ 581

62. "Hon. E. P. Doyle," in A. Y. Hubbell, *Prominent Men of Staten Island, 1893* (New York: Hubbell, 1893), http://www.rootsweb.ancestry.com/~nyrichmo/prominent/doyle.pdf; Molly Greene, "A Mountain by the Sea: Waste-scapes, Life-scapes, and the Reinvention of Fresh Kills," https://hixon.yale.edu/sites/default/files/files/fellows/paper/greene_molly_2012_report.pdf.
63. Seitz and Miller, *The Other Islands of New York City*, 317.
64. Quoted in Thompson, "Garbage War of 1916–17." See also "Anti-Garbage Party Has Land After Raid," *New York Times*, May 19, 1916.
65. Quoted in Charles L. Sachs, "Staten Island Secessionism: Strained Threads of the Consolidated City Fabric," *culturefront* (Winter 1997/1998): 101.
66. "Anti-Garbage Party Has Land After Raid." See also "Chronology of the Garbage Fight," 8–9; "Garbage Difficulties," *Staten Islander*, May 24, 1916; "War on Garbage Has New Phases," *Staten Islander*, May 24, 1916; "Indicted in Garbage War," *New York Times*, June 14, 1916.
67. "War on Garbage Has New Phases"; "Armed Men Seize Island," *New York Times*, May 24, 1916; "Chronology of the Garbage Fight," 9; Seitz and Miller, *The Other Islands of New York City*, 318.
68. "Change of Site Agreed On," *Staten Islander*, May 27, 1916.
69. "Change of Site Agreed On."
70. "New Garbage Plant War," *New York Times*, May 28, 1916.
71. Interestingly, the seven men were soon discharged on a motion from the attorney for Gaffney, Gahegan & Van Etten. "Staten Island Backs Doyle," *New York Times*, June 2, 1916.
72. "New Garbage Plant War." See also "Warrant in Garbage Fight," *New York Times*, May 30, 1916; "Staten Island Backs Doyle"; Miller, *Fat of the Land*, 131; Thompson, "Garbage War of 1916–17."
73. "Doyle's Navy Gets Garbage Site Fort," *New York Times*, June 9, 1916. See also "With Derrick Doyle Rids Garbage Island of Foes," *New York Tribune*, June 9, 1916. The story in the *Tribune* dramatized in further detail the encounter at the cabin, relating that Doyle and six men rushed the men inside: "'This is my island and my cabin,' Doyle shouted to the six Drake guards. 'I warn you that I have come to remove the cabin, and give you this opportunity of getting out.' 'Go on, you piker,' came a voice from the inside. 'You're interrupting a gentlemen's game of poker.'" Doyle shouted a second warning to no avail, and his lawyer then "raised his voice" and read a copy of Doyle's deed to them. The cabin was soon after removed.
74. "Doyle's Navy Gets Garbage Site Fort"; "Chronology of the Garbage Fight," 9.
75. "With Derrick Doyle Rids Garbage Island of Foes."
76. "Chronology of the Garbage Fight," 9; Leng and Davis, *Staten Island and Its People*, 1:369.
77. Franz, "Island 'Dump Wars' Began Almost 80 Years Ago!"
78. "Lake's Island Realty Co., Inc.," *Staten Island Civic League Bulletin* 3 (July 1916): 2.
79. "Chronology of the Garbage Fight," 10. See also Leng and Davis, *Staten Island and Its People*, 1:369.
80. "Garbage Writs Refused," *New York Times*, July 6, 1916.
81. Thompson, "Garbage War of 1916–17"; "$300,000 Garbage War Suit," *New York Times*, June 18, 1916.
82. For several years, refuse on Staten Island was burned in two small incinerators, furnishing heat and light to run the machinery in the plant. See Charles Zueblin, *American Municipal Progress*, rev. ed. (New York: Macmillan, 1916), 83; William F.

Morse, "Progress in City Waste Disposal During 1910," *American City* 4 (January 1911): 33; George C. Whipple, "Disposal of Garbage of the City of New York at Staten Island," December 9, 1916, in State of New York, *Thirty-Seventh Annual Report of the State Department of Health*, February 28, 1917 [draft/typed version], in *Documents of the Assembly of the State of New York*, 140 Session, 1917, vol. 21, no. 57, part 2 (Albany, NY: J. B. Lyon Co., Printers, 1917), 755.
83. "Garbage and Infantile Paralysis," July 7, 1916, *Staten Island Civic League Bulletin* 3 (July 1916): 16.
84. "Whitman to Inquire Into Garbage Fight," *New York Times*, July 14, 1916.
85. "To the People of Staten Island," *Staten Island Civic League Bulletin* 3 (July 1916): 14.
86. "Whitman to Inquire Into Garbage Fight."
87. "Presentment to the Governor," *Staten Island Civic League Bulletin* 3 (July 1916): 4.
88. "Brief Submitted to the Governor," *Staten Island Civic League Bulletin* 3 (July 1916): 5.
89. "Charles S(eymour) Whitman," in *The Encyclopedia of New York State*, ed. Peter Eisenstadt (Syracuse: Syracuse University Press, 2005), 1699.
90. Charles S. Whitman to State Commissioner of Health, n.d., Hearing Files Relating to Proposed Garbage Reduction Plant to Be Erected on Staten Island, 1916, Records Series A0355, New York (State) Dept. of State, New York State Archives, Albany. Hereafter referred to as Hearing Files.
91. "Garbage Plans Approved," *New York Times*, July 14, 1916.
92. "Memorial Tribute to Linsly R. Williams, M.D., 1875–1934," *Bulletin of the New York Academy of Medicine* 61 (June 1985): 393–94; James Alexander Miller, "Memorial: Linsly R. William, M.D.," *Transactions of the American Clinical and Climatological Association* 50 (1934): lv–lvi.
93. Linsly R. Williams to Hermann M. Briggs, "Report on the Proposed Garbage Reduction Plant of the City of New York, Situated on Staten Island," December 27, 1916, Hearing Files.
94. "George C. Whipple (1866–1924)," *HSPH News*, Harvard School of Public Health, http://www.hsph.harvard.edu/news/centennial-george-c-whipple/.
95. "Garbage Firm Would Aid Biggs," *New York Times*, July 20, 1916.
96. Hearing Files, 2.
97. Whipple, "Disposal of Garbage of the City of New York at Staten Island," 753.
98. Martin V. Melosi, *Garbage in the Cities: Refuse, Reform, and the Environment*, rev. ed. (Pittsburgh, PA: University of Pittsburgh Press, 2005), 71–72, 125.
99. Whipple, "Disposal of Garbage of the City of New York at Staten Island," 754.
100. Hearing Files, 206–7.
101. Hearing Files, 150–51.
102. Hearing Files, 34–35.
103. Hearing Files, 50.
104. Hearing Files, 11.
105. Hearing Files, 102.
106. Hearing Files, 416.
107. Hearing Files, 457–58.
108. Hearing Files, 111. Hicks emphasized later that garbage only carried disease indirectly; it did not carry it itself. Thus garbage could be a disease culprit even if infantile paralysis had yet to be linked to it. Hearing Files, 11–12, 124.
109. Paul Hansen, Report to Albert C. Fach, November 18, 1916, Hearing Files, 1628–29.
110. Hearing Files, 1630–37.

111. Hearing Files, 1637.
112. Hearing Files, 1637–38.
113. Hearing Files, 1639. See also 1640–41.
114. Hearing Files, 432.
115. Hearing Files, 567.
116. Hearing Files, 602. See also 609.
117. Hearing Files, 611.
118. Hearing Files, 638, 640.
119. Hearing Files, 763–64.
120. Hearing Files, 984. Harry E. Bramley, a sanitary inspector, vouched for the Cobwell while never seeing it in operation. See 1305.
121. Hearing Files, 1444.
122. Hearing Files, 1446.
123. Hearing Files, 771–809.
124. Hearing Files, 1245.
125. Hearing Files, 1398.
126. Hearing Files, 1438.
127. Hearing Files, 595.
128. "New York Garbage-Reduction Works Controversy May Be Over," *Engineering News* 77 (January 18, 1917): 125.
129. Whipple, "Disposal of Garbage of the City of New York at Staten Island," 751–52, 754–66.
130. Whipple, "Disposal of Garbage of the City of New York at Staten Island," 752.
131. Whipple, "Disposal of Garbage of the City of New York at Staten Island," 752.
132. Linsly R. Williams, "Report on the Proposed Garbage Reduction Plant of the City of New York, Situated on Staten Island," December 27, 1916, State of New York, *Thirty-Seventh Annual Report of the State Department of Health*, February 28, 1917, in *Documents of the Assembly of the State of New York*, 140 Session, 1917, vol. 21, no. 57, part 2 (Albany, NY: J. B. Lyon Co., Printers, 1917), 749–50.
133. "Richmond Wins Garbage Fight," *New York Tribune*, December 29, 1916.
134. "Garbage Situation: Indication of a Clean-Cut Victory Says Dr. L. A. Dreyfus," *Staten Islander*, January 3, 1917.
135. State of New York, Office of Secretary of State, "Opinion of the Governor in the Matter of the Erection of a Garbage Plant on Staten Island," January 25, 1917, State of New York, Executive Chamber, Albany.
136. State of New York, Office of Secretary of State, "Opinion of the Governor." See also State of New York, Office of Secretary of State, Charles S. Whitman to Sheriff of the County of Richmond and District Attorney of the County of Richmond, January 25, 1917, State of New York, Executive Chamber, Albany.
137. "Sherriff of Richmond County May Act Against Garbage Plant," *Staten Islander*, January 13, 1917.
138. "Garbage Report: Governor Whitman Approves Health Commissioner's Findings," *Staten Islander*, January 27, 1917.
139. "Cromwell's Two Bills," *Staten Islander*, January 31, 1917; "To Support Cromwell," *Staten Islander*, February 17, 1917; "Bills Introduced by Senator Cromwell," *Staten Island Civic League Bulletin* 4 (January 1917): 15–16.
140. "Hearing on Garbage Bill," *Staten Islander*, February 14, 1917; "To Support Cromwell"; "To Albany To-Day," *Staten Islander*, February 21, 1917.
141. "Plea Against Garbage," *Staten Islander*, February 24, 1917.

142. "New York Is Powerless in Garbage Plant Fight: Cromwell Bill Hearing Reveals," *New York Tribune*, February 22, 1917.
143. "The City Wakes Up," *Staten Islander*, March 3, 1917.
144. "Garbage Fight at Albany," *Staten Islander*, April 7, 1917.
145. "Garbage Victory Sure," *Staten Islander*, April 14, 1917.
146. "Garbage Doomed!" *Staten Islander*, April 25, 1917.
147. "Anti-Garbage Defeated," *Staten Islander*, May 12, 1917; "Anti-Garbage Deferred," *Staten Islander*, May 12, 1917.
148. "The Fight Against the Garbage Plant," *Staten Island Civic League Bulletin* 4 (May 1917): 2; "Record of Work for Staten Island," *The Staten Island Civic League Bulletin* 4 (May 1917): 15.
149. The contractor received permission from the city to use the old Barren Island plant for three months from the original start date. At that point the Barren Island facility would be phased out.
150. New York City, Department of Street Cleaning, *Annual Report, 1917*, Office of the Mayor, John P. Mitchel Administration, Departmental Correspondence Received 1914–1917, Box 81, Folder 871, Department of Street Cleaning, July–December 1917; "Garbage Temporarily Disposed of at Barren Island," Affidavit of Commissioner MacStay, November 19, 1918, Office of the Mayor, John F. Hylan Administration, Departmental Correspondence received, 1918–1925, Box 154, Folder 1669, Department of Street Cleaning, February–March 1918, New York City Municipal Archives and Library. See also New York City, Department of Street Cleaning, *Annual Report*, 1916, 29.
151. "To Help Garbage," *Staten Islander*, March 3, 1917. The War Department was contacted because the cables could affect navigation in the area during the war.
152. "Walkout at Garbage Plant," *Staten Islander*, March 7, 1917.
153. "Claims for New Plant," *Staten Islander*, July 14, 1917; "Record of Work for Staten Island," *Staten Island Civic League Bulletin* 4 (May 1917): 4.
154. Leng and Davis, *Staten Island and Its People*, 1:370–71; Thompson, "Garbage War of 1916–17"; Franz, "Island 'Dump Wars' Began Almost 80 Years Ago!"; Whipple, "Disposal of Garbage of the City of New York at Staten Island," 759–61.
155. "Record of the Work for Staten Island," *Staten Island Civic League Bulletin* 4 (July 1917): 6.
156. "Garbage Situation," *Staten Islander*, May 26, 1917.
157. "Will Joffre Visit Garbageland?" *Staten Islander*, May 12, 1917.
158. "Separate from City," *Staten Islander*, May 12, 1917.
159. "Model Garbage Plant," *New York Times*, July 22, 1917.
160. "Visitors at Garbage Plant," *Staten Islander*, August 11, 1917.
161. "New Garbage Plant Emits Little Odor," *New York Tribune*, August 22, 1917. See also "Garbage Plant Inspection," *Staten Islander*, August 22, 1917.
162. Leng and Davis, *Staten Island and Its People*, 1:371.
163. Leng and Davis, *Staten Island and Its People*, 1:371.
164. "Fire Damages Staten Island Reduction Plant," *New York Evening Telegram*, January 2, 1918.
165. "Garbage Plant Inquiry," *New York Times*, June 11, 1918.
166. *Journal of Industrial and Engineering Chemistry* 10 (August 1, 1918): 661.
167. "New Demand to Close Garbage Plant Made," *New York Tribune*, July 17, 1918.
168. Seitz and Miller, *The Other Islands of New York City*, 262.
169. Lankevich, *American Metropolis*, 153–54; Miller, *Fat of the Land*, 132–33.
170. He apparently was called "Red Mike" because of his nasty temper. See Milton M. Klein, ed., *The Empire State: A History of New York* (Ithaca, NY: Cornell University

Press, 2001), 541. See also "Know Your Mayors: John F. Hylan," *The Bowery Boys: New York City History*, April 30, 2008, http://theboweryboys.blogspot.com/2008/04/know-your-mayors-john-f-hylan.html.
171. Miller, *Fat of the Land*, 133–34.
172. "The Women's Anti-Garbage League of Staten Island," *Staten Island Civic League Bulletin* 4 (July 1917): 12.
173. "Hylan in Richmond Hits 'Garbage Ring,' Promises 'Attack,'" *Brooklyn Daily Eagle*, October 28, 1917.
174. "Hylan Promises to Abate Garbage Plant 'Nuisance,'" *New York Tribune*, June 26, 1918.
175. Another view is that the Staten Island plant was briefly out of operation and that sea dumping was a short-term alternative. "Start Garbage Plant After Many Complain," *New York Times*, August 10, 1918.
176. Between 1890 and June 30, 1916 scows deposited 160,313,000 cubic yards of waste, about one-quarter of which was used as shore fill and the remainder dumped in deep water. See "Waste-Material Disposal at New York," *Engineering News* 77 (January 18, 1917): 115.
177. Seitz and Miller, *The Other Islands of New York City*, 262; Franz, "Island 'Dump Wars' Began Almost 80 Years Ago!"; Miller, *Fat of the Land*, 134; "Ask Government to Take Garbage Plant," *New York Times*, July 20, 1918; "Garbage Official Indicted as Federal Operation Is Urged," *New York Tribune*, July 20, 1918; "U.S. May Take Garbage Plant on Staten Island," *New York Tribune*, July 26, 1918; "Confer on Garbage Plant," *New York Times*, July 26, 1918; "May Take Garbage Plant," *New York Times*, July 27, 1918; "Negotiate for City Plant," *New York Times*, July 31, 1918. See also "Provisions of Contract Between the City and the Metropolitan By-Products Company, Inc., Repeatedly Violated by the Contractor," Affidavit of Commissioner MacStay, November 19, 1918, Office of the Mayor, John F. Hylan Administration, Departmental Correspondence Received, 1918–1925, Box 154, Folder 1669, Department of Street Cleaning, February–March 1918, New York City Municipal Archives and Library.
178. "Copeland Finds Garbage Plant Public Nuisance," *New York Tribune*, September 29, 1918.
179. "Garbage Plant on Staten Island Is Closed Officially," *New York Tribune*, October 9, 1918. In June 1920 the Health and Sanitation Committee of the Brooklyn Chamber of Commerce proclaimed displeasure with current disposal plans in their borough, and the president of the Bronx Taxpayers' Alliance vented similar complaints with contractors filling in swampland and "making a dumping ground out of the Bronx." The Brooklyn Chamber now called for a new reduction plant on Staten Island. "In the committee's opinion there is no better location for a reduction plant than on the shore of Staten Island, fronting the Kill Van Kull. This section has quiet water with no transportation difficulties and is farther removed from residential districts, and is already bordered with odor-producing plants." See "Waste and Garbage Disposal Charged," *New York Times*, June 27, 1920.

5. THE GO-AWAY SOCIETY

1. The idea of a waste sink must be credited to Joel A. Tarr, *The Search for the Ultimate Sink: Urban Pollution in Historical Perspective* (Akron, OH: University of Akron Press, 1996).

2. "Mayor Names MacStay," *New York Times*, January 31, 1918; "Admits Antiquity of City's Cleaning," *New York Times*, February 25, 1920.
3. Steven Hunt Corey, "King Garbage: A History of Solid Waste Management in New York City, 1881–1970," PhD diss., New York University, 1994, 123; "Opposes Mayor's Incinerator Plan," *New York Times*, March 30, 1919; *Staten Island Civic League Bulletin* 6 (November 1919): 10, Environmental Collection, Archives & Special Collections, CSI Library, College of Staten Island, Staten Island, New York.
4. "Waste in Garbage Disposal Charged," *New York Times*, June 27, 1920.
5. Brooklyn Chamber of Commerce, *The Need of a Comprehensive System for the Collection and Disposal of Municipal Waste in New York City* (Brooklyn, June 1920), 1–3.
6. Brooklyn Chamber of Commerce, *The Need of a Comprehensive System*, 1–3.
7. Brooklyn Chamber of Commerce, *The Need of a Comprehensive System*, 4, 8.
8. Brooklyn Chamber of Commerce, *The Need of a Comprehensive System*, 45.
9. Brooklyn Chamber of Commerce, *The Need of a Comprehensive System*, 43, 48.
10. Brooklyn Chamber of Commerce, *The Need of a Comprehensive System*, 30. See also 34.
11. Brooklyn Chamber of Commerce, *The Need of a Comprehensive System*, 35. See also "Waste in Garbage Disposal Charged."
12. Corey, "King Garbage," 125–27.
13. "Hearing on the Garbage Bills," *Staten Island Civic League Bulletin* 8 (May 1921): 10.
14. "'Tuning Up' Garbage Plant," *Staten Island Advance*, December 13, 1921.
15. *Seven Years of Progress: Important Public Improvements and Achievements by the Municipal and Borough Governments of the City of New York, 1918–1925* (Submitted to the Board of Aldermen, March 1925), 83, 85–86.
16. The city took over the unloading plant at Rikers Island in 1924. New York City, Department of Street Cleaning, *Annual Report, 1924*, 17.
17. In 1935, there were in effect forty-four permits issued for private dumps in New York City. William H. Best to F. H. La Guardia, August 13, 1935, Box 3157, Folder 1, Department of Sanitation, 1935, Fiorello H. La Guardia Collection, La Guardia and Wagner Archives, LaGuardia Community College, Long Island City, New York.
18. New York City, Department of Street Cleaning, *Annual Report, 1923*, Department of Records, Municipal Archives, New York City, 4–6, 19–20.
19. Brooklyn Chamber of Commerce, *The Need of a Comprehensive System*, 3.
20. Steven H. Corey, "Garbage in the Sea," *Seaport* 25 (Winter–Spring 1991): 23.
21. "Take Steps to End Beach Pollution," *New York Times*, July 25, 1923.
22. Corey, "King Garbage," 105–6; Corey, "Garbage in the Sea," 23. See also "Jersey in New War on Garbage at Sea," *New York Times*, September 5, 1926.
23. "Denies Garbage Is Dumped Inshore," *New York Times*, July 31, 1923.
24. Daniel C. Walsh, "Solid Wastes in New York City: A History," *Waste Age* 20 (April 1989): 118. The total of waste dumped at sea was a fraction of what it had been in the late nineteenth century. In the 1880s as much as 80 percent of Manhattan's refuse was dumped in the ocean. See Daniel C. Walsh, "The History of Waste Landfilling in New York City," *Ground Water* 29 (July–August 1991): 591.
25. "New York Must Solve Its Garbage Problem," *New York Times*, July 19, 1925.
26. "New York Must Solve Its Garbage Problem."
27. New York City, Department of Street Cleaning, *Annual Report, 1920*, 1; "Garbage Situation Better: Street Cleaning Tugs Operate Despite Harbor Strike," *New York Times*, January 28, 1920.
28. Corey, "King Garbage," 104.

29. Richard Fenton, "Current Trends in Municipal Solid Waste Disposal in New York City," *Resource Recovery and Conservation* 1 (1975): 170.
30. New York City, Department of Street Cleaning, *Annual Report, 1929*, 3; Charles A. Selden, "Looking Back on New York's Greatest Snowstorm," *New York Times*, March 7, 1920.
31. "Population of the 20 Largest U.S. Cities, 1900–2012," http://www.infoplease.com/ipa/A0922422.html.
32. Ira Rosenwaike, *Population History of New York City* (Syracuse, NY: Syracuse University Press, 1972), 133–34.
33. See Edward F. Bergman and Thomas W. Pohl, *A Geography of the New York Metropolitan Region* (Dubuque, IA: Kendall/Hunt, 1975), 55–66, 72; Rosenwaike, *Population History of New York City*, 132.
34. Rosenwaike, *Population History of New York City*, 132.
35. See Bergman and Pohl, *A Geography of the New York Metropolitan Region*, 77–86; Rosenwaike, *Population History of New York City*, 93, 132–33, 135; Francois Weil, *A History of New York* (New York: Columbia University Press, 2004), 201–19; Selam Berrol, *The Empire City: New York and Its People, 1624–1996* (Westport, CT: Praeger, 1997), 104–5.
36. Ann Folino White, "History of Consumption and Waste, U.S., 1900–1950," in *Encyclopedia of Consumption and Waste: The Social Science of Garbage*, ed. Carl A. Zimring and William L. Rathje (Los Angeles: Sage, 2012), 361.
37. White, "History of Consumption and Waste, U.S., 1900–1950," 361–62.
38. Martin V. Melosi, *Garbage in the Cities: Refuse, Reform, and the Environment*, rev. ed. (Pittsburgh, PA: University of Pittsburgh Press, 2005), 146–47, 176–77.
39. Walsh, "The History of Waste Landfilling in New York City," 591.
40. New York City, Department of Street Cleaning, *Annual Report, 1921*, 5; *Annual Report, 1922*, 5.
41. New York City, Department of Street Cleaning, *Annual Report, 1924*, 23.
42. "Resigns as D.S.C. Head," *New York Times*, January 4, 1921; "Hylan Reinstalled, Pledges Old Policy; Keeps His Old Staff," *New York Times*, January 3, 1921.
43. "John P. Leo Is Named to Clean the Streets," *New York Times*, January 6, 1921.
44. "Leo Quits Hylan: Attacks Inquiry," *New York Times*, November 19, 1921; "John P. Leo Dies from Heart Attack: Street Cleaning Commissioner Who Resigned After a Row with Hirshfield," *New York Times*, July 24, 1923; "Summon Hirshfield as News Falsifier," *New York Times*, November 24, 1912.
45. "Hirshfield Called City Whitewasher," *New York Times*, December 8, 1921.
46. "Hylan Inefficiency: Wallstein Amplifies His Charges of Waste and Corruption in the City Administration," *New York Times*, December 19, 1920.
47. "Curran Cites More Broken Promises," *New York Times*, October 19, 1921. See also "The Case of Hylan: Where His Administration Will Be Most Subject to Attack, as Critics See It," *New York Times*, May 1, 1921; "Waterman Assails City Garbage Plan," *New York Times*, September 5, 1925.
48. "Leo Quits Hylan: Attacks Inquiry"; "Alfred A. Taylor, Retired City Aide," *New York Times*, December 8, 1941; "Street Cleaning Job Goes to Taylor," *New York Times*, November 26, 1921.
49. "Higgins Suspends 19 City Employees," *New York Times*, April 22, 1928; "Walker and Higgins Push Graft Inquiry," *New York Times*, May 8, 1928; "Graft Inquiry Waits on Action in Bronx," *New York Times*, May 9, 1928; "Find New Evidence of Street Grafting," *New York Times*, May 15, 1928; "2 More Suspended in Street Inquiry," *New York*

Times, May 16, 1928; "More High Officers Face Graft Charges," *New York Times*, May 17, 1928; "Two More Arrested for Payroll Graft," *New York Times*, May 25, 1928; "High Bronx Officer Suspended in Graft," *New York Times*, May 27, 1928; "Gannon Forgetful at Hearing on Graft," *New York Times*, May 29, 1928; "Five in Manhattan Indicted for Graft," *New York Times*, June 19, 1928; "Street Cleaners Face New Inquiry," *New York Times*, August 30, 1928.

50. At the beginning of the twentieth century, the incinerators in the city were small-scale combustion types used to burn nonsalable rubbish at refuse-recycling and transfer stations. There were fourteen small private rubbish incinerators in Brooklyn that were used to reduce the volume of waste shipped by rail to landfills near Coney Island. Between 1895 and 1905, there were seven or more furnaces burning rubbish or generating steam in Manhattan. The Williamsburg Bridge Lighting Plant on the Lower East Side was purportedly the first major municipal energy-recovery incinerator in the United States. The first medical waste incinerator was built at a hospital in Manhattan in 1890, and a commercial onsite refuse incinerator in Manhattan—again the first in the country—was used by Macy's in 1902. See Daniel C. Walsh, "The Evolution of Refuse Incineration: What Led to the Rise and Fall of Incineration in New York City?" *Environmental Science & Technology* (August 1, 2002): 317A–318A.
51. Walsh, "The Evolution of Refuse Incineration," 318A; Martin V. Melosi, "The Viability of Incineration as a Disposal Option: The Evolution of a Niche Technology, 1885–1995," *Public Works Management & Policy* 1 (July 1996): 32–33.
52. Melosi, "The Viability of Incineration as a Disposal Option," 33–35; Walsh, "The Evolution of Refuse Incineration," 318A–319A.
53. Corey, "King Garbage," 138–44.
54. Corey, "King Garbage," 138–44.
55. Casimir A. Rogus, New York City, Department of Sanitation, Office of Engineering, "Refuse Disposal in New York City," paper presented before ASCE, Sanitary Division, Metropolitan Section, New York City, February 3, 1954, 5; Walsh, "The Evolution of Refuse Incineration," 318A–319A; Brooklyn Chamber of Commerce, *The Need of a Comprehensive System*, 30–31.
56. New York City, Department of Street Cleaning, *Annual Report, 1924*, 5.
57. New York City, Department of Street Cleaning, *Annual Report, 1924*, 4; *Annual Report, 1925*, 4, 17. See also "New Ways Are Sought to Keep City Clean," *New York Times*, February 1, 1925.
58. New York City, Department of Street Cleaning, *Annual Report, 1926*, 12.
59. New York City, Department of Street Cleaning, *Annual Report, 1927*, 3, 15.
60. New York City, Department of Street Cleaning, *Annual Report, 1928*, 3. See also New York City, Department of Street Cleaning, *Annual Report, 1927*, 3, 15.
61. New York City, Department of Street Cleaning, *Annual Report, 1928*, 8.
62. New York City, Department of Street Cleaning, *Annual Report, 1929*, 7–9.
63. "Incinerators Defended," *New York Times*, August 18, 1927.
64. Virginia Pope, "A New Sanitary System Awaits Voters' Decision," *New York Times*, October 20, 1929. See also Elmer Goodwin, "Keeping Streets Clean in Our Largest City," *Municipal Sanitation* 6 (June 1930): 317.
65. Corey, "King Garbage," 128, 132, 150. Sanitary engineers were the branch of the field trained in both engineering and public health, and they focused on such systems as water supplies, sewer systems, and solid waste collection and disposal. They almost universally supported municipal control of sanitation functions, which was their primary livelihood. Melosi, *Garbage in the Cities*, 78–84.

5. THE GO-AWAY SOCIETY ■ 589

66. Corey, "King Garbage," 144, 149.
67. "Schroeder Heads Sanitation Board," *New York Times*, December 1, 1929.
68. Corey, "King Garbage," 159–61.
69. "Shake Up Bureau of Street Cleaning," *New York Times*, June 13, 1930; "Schroeder Heads Sanitation Board"; "A. A. Taylor Gives Up $10,840 City Post," *New York Times*, August 3, 1933.
70. City of New York, Department of Sanitation, *First Annual Report* (1930), 34.
71. Derek Sylvan, "Municipal Solid Waste in New York City: An Economic and Environmental Analysis of Disposal Options," paper prepared for the New York League of Conservation Voters Education Fund, 2006, 1, https://nylcvef.org/wp-content/uploads/2017/08/Solid-Waste-Background-Paper.pdf.
72. Quoted in Corey, "King Garbage," 117.
73. *New York Times*, December 6, 1931.
74. Carl Wilhelm, "New Board of Sanitation Was Needed," *Brooklyn Daily Eagle*, December 22, 1929.
75. Corey, "King Garbage," 166–68.
76. City of New York, Department of Sanitation, *First Annual Report* (1930), 26.
77. Corey, "Garbage in the Sea," 23. Corey writes that the Supreme Court ordered the cessation of sea dumping beginning June 1, but several other sources cite July 1. See, for example, S. Tannenbaum, "A Brief History of Waste Disposal in New York City Since 1930," unpublished paper, New York City Municipal Archives, 1992, 1; Fenton, "Current Trends," 170. See also Corey, "King Garbage," 105.
78. Corey, "King Garbage," 166–68.
79. Incinerators began to gain a competitive advantage in several urban areas in the mid-1930s because open dumps had become increasingly unpopular in central cities. While incinerators held an important place among disposal options by the end of the 1930s, they were not dominant and were more typically located in smaller and medium-size cities. Melosi, "The Viability of Incineration as a Disposal Option," 36–37.
80. City of New York, Department of Sanitation, *First Annual Report* (1930), 26.
81. City of New York, Department of Sanitation, *First Annual Report* (1930), 26.
82. Kenneth Allen, "How New York Handles Her Garbage and Rubbish Problem," *Municipal Sanitation* 1 (January 1930): 16.
83. Tannenbaum, "A Brief History of Waste Disposal in New York City Since 1930," 1; Fenton, "Current Trends," 170.
84. Walsh, "The Evolution of Refuse Incineration," 319A.
85. George A. Soper, "Court's Decision Forces City to Solve Its Refuse Problems," *New York Times*, May 24, 1931.
86. Walter D. Binger, "Waste Disposal Progress in Nation's Metropolis," *Municipal Sanitation* 8 (October 1937): 506–7. See also Sharon Seitz and Stuart Miller, *The Other Islands of New York City: A History and Guide*, 3rd ed. (Woodstock, VT: Countryman, 2011), 202.
87. City of New York, Department of Sanitation, *Annual Report, 1934*.
88. William H. Best to F. H. La Guardia, August 2, 1935, and Best to La Guardia, August 13, 1935, Box 3157, Folder 1, Department of Sanitation, 1935, Fiorello H. La Guardia Collection, La Guardia and Wagner Archives.
89. Tannenbaum, "A Brief History of Waste Disposal in New York City Since 1930," 1; Ernest P. Goodrich, "The Opportunities for Refuse Salvage in New York City," *American City* 50 (January 1935): 58.

90. Frank Vos, "James J(ohn) Walker," in *The Encyclopedia of New York City*, 2nd ed., ed. Kenneth T. Jackson (New Haven, CT: Yale University Press, 2010), 1373; Berrol, *The Empire City*, 111; Weil, *A History of New York*, 223; Corey, "King Garbage," 171–75.
91. "Schroeder Heads Sanitation Board." See also "Dr. W. Schroeder, Led Sanitary Commission," *New York Times*, May 3, 1948; "To Fill Board of Sanitation Within Week," *Brooklyn Daily Eagle*, December 2, 1929; "Mayor Acts to Oust Schroeder at Once; McKay Also to Go," *New York Times*, March 21, 1933; Corey, "King Garbage," 169–70.
92. "Mayor Acts to Oust Schroeder at Once; McKay Also to Go."
93. City of New York, Department of Sanitation, *Second Annual Report* (1931), 71–72.
94. City of New York, Department of Sanitation, *Second Annual Report* (1931), 73.
95. Berrol, *The Empire City*, 111; Thomas Kessner, *Fiorello H. La Guardia and the Making of Modern New York* (New York: McGraw-Hill, 1989), 205; August Heckscher, *When LaGuardia Was Mayor: New York's Legendary Years* (New York: Norton, 1978), 60–66.
96. Thomas Kessner, "Fiorello La Guardia," in *The Encyclopedia of New York City*, ed. Jackson, 717–18; Charles Garrett, *The La Guardia Years: Machine and Reform Politics in New York City* (New Brunswick, NJ: Rutgers University Press, 1961), 225; Heckscher, *When LaGuardia Was Mayor*, 10, 15–18.
97. Garrett, *The La Guardia Years*, 178–80, 303. See also Arthur Mann, *LaGuardia Comes to Power, 1933* (New York: J. B. Lippincott, 1965), 22–23, 107, 156.
98. Kessner, *Fiorello H. La Guardia and the Making of Modern New York*, 451.
99. City of New York, Department of Sanitation, *Annual Report, 1935*, 1–3; Walsh, "The Evolution of Refuse Incineration," 319A; Garrett, *The La Guardia Years*, 140–41.
100. See Ernest P. Goodrich to Fiorello H. La Guardia, January 1, 1934, Box 3134, Folder 8, Department of Sanitation, 1934, La Guardia Collection, La Guardia and Wagner Archives.
101. "Ernest Goodrich, Engineer, 71, Dies," *New York Times*, October 9, 1955. See also "Goodrich Resigns Sanitation Post; Denounces Mayor," *New York Times*, March 8, 1934.
102. Corey, "King Garbage," 192.
103. "Goodrich Resigns Sanitation Post; Denounces Mayor."
104. "Col. Hammond Leaves Army for City Post," *New York Times*, February 7, 1934; Robert Moses, *La Guardia: A Salute and a Memoir* (New York: Simon and Schuster, 1957), 30.
105. "Hammond to Head Sanitation Bureau," *New York Times*, May 28, 1934. See also "Col. Hammond Sworn as Sanitation Head," *New York Times*, June 3, 1934; "City's Whitewings to Get New Garb," *New York Times*, April 17, 1935. Robert Moses claimed that La Guardia actually appointed Hammond as "strictly a matter of sentiment" for his friend. See Moses, *La Guardia*, 31.
106. "New Deal Pledged Sanitation Force," *New York Times*, March 13, 1934.
107. New York City, Department of Sanitation, *Annual Report, 1935*, 2.
108. "Hammond Ousts Tammany Deputy," *New York Times*, May 11, 1934; "Hammond Puts End to 300 Soft Jobs," *New York Times*, May 29, 1934.
109. New York City, Department of Sanitation, *Annual Report, 1935*, 6. See also Thomas W. Hammond to F. H. La Guardia, May 19, 1934; Commissioners Binger & Bromberger Presiding at a Meeting of Private Permit Men, April 10, 1934; Hammond to La Guardia, May 29, 1934; Hammond to La Guardia, June 27, 1934, Box 3134, Folder 8, Department of Sanitation, 1934; Hammond to La Guardia, October 8, 1934 and Address to be Delivered by Colonel Thomas W. Hammond, Commissioner of Sanitation, at

Luncheon of the City Club, on Wednesday, October 30, 1935, 10, Box 3157, Folder 1, La Guardia Collection, La Guardia and Wagner Archives. On contracts for scow trimmers and reclaiming waste materials, see Thomas W. Hammond to F. H. La Guardia, July 29, 1935; Hammond to La Guardia, August 8, 1935; Hammond to Waste Material Sorters, Trimmers and Handlers Union, April 30, 1935, Box 3157, Folder 1, Department of Sanitation, 1935, La Guardia and Wagner Archives.

110. "Sanitation Chief Starts Shake-Up," *New York Times*, March 22, 1934; "Cleanliness Drive by City Is Lauded," *New York Times*, May 10, 1934.
111. "City Drive to Seek Cleaner Streets," *New York Times*, March 7, 1935.
112. "Hammond Puts Big Force to Work as Answer to Critics When Temperature Rises," *New York Times*, January 27, 1936; "New Snow Fund Required by City," *New York Times*, February 4, 1936; "Litter in Streets Arouses Protest," *New York Times*, March 8, 1936.
113. Corey, "King Garbage," 195–97.
114. Rose C. Feld, "Disposal of Refuse Put on a New Basis," *New York Times*, July 1, 1934.
115. Feld, "Disposal of Refuse Put on a New Basis." See also "City Cleaning Cost $28,773,053 in 1934," *New York Times*, July 12, 1934; New York City, Department of Sanitation, *Annual Report, 1934*; *Annual Report, 1935*, 8; Corey, "King Garbage," 202–3.
116. Corey, "King Garbage," 204.
117. "Col. T. W. Hammond, City Official, Dies," *New York Times*, September 4, 1936. See also "W. F. Carey Named to Hammond Post," *New York Times*, May 17, 1936.
118. New York City, Department of Sanitation, *Annual Report, 1936*.
119. H. Allen Smith, "Angel of the White Wings," *Saturday Evening Post*, March 6, 1943, 24.
120. "William F. Carey 1878–1951," Hoosick Township Historical Society, online, http://www.hoosickhistory.com/biographies/carey.htm. See also "Carey in New Post," *New York Times*, May 27, 1936; Corey, "King Garbage," 210–11; Benjamin Miller, *Fat of the Land: Garbage of New York, the Last Two Hundred Years* (New York: Four Walls Eight Windows, 2000), 185.
121. "W. F. Carey Named to Hammond Post," *New York Times*, May 17, 1936.
122. Corey, "King Garbage," 210–11.
123. John W. Harrington, "City Sanitation Under New Chief," *New York Times*, May 24, 1936.

6. ONE BEST WAY

1. Daniel C. Walsh, "The Evolution of Refuse Incineration: What Led to the Rise and Fall of Incineration in New York City?" *Environmental Science & Technology* (August 1, 2002): 319A–320A.
2. Steven Hunt Corey, "King Garbage: A History of Solid Waste Management in New York City, 1881–1970," PhD diss., New York University, 1994, 204; C. A. Rogus, "New York City Turns to Incineration," *Civil Engineering* 21 (December 1952): 1030.
3. The Fifty-Sixth Street facility opened in 1924, closed in 1936, but reopened in 1937. The West Eighth Street and Georgia Avenue plants also opened in 1924 but closed in 1937. In addition, the Maspeth (1916), Averne (1918), and Jamaica (1921) plants all closed in 1937, and the Hammels (1925), Bergen Landing (1924), Betts Avenue (1926), and West New Brighton (1908) plants closed the following year, leaving only eleven in operation in 1939. See Daniel C. Walsh et al., "Refuse Incinerator Particulate Emissions and Combustion Residues for New York City During the 20th Century," *Environmental*

Science & Technology 35 (May 2001): 2443; City of New York, Department of Sanitation, *Annual Report, 1938*, 19. On closures and remaining operation of incinerators, also see William F. Carey to Henry H. Curran, Deputy Mayor, July 21, 1938; John B. Morton, Deputy Commissioner to Curran, July 26, 1938; Curran to Morton, July 29, 1938; Morton to Curran, August 9, 1938, Box 3219, Folder 4, Department of Sanitation, Fiorello H. La Guardia Collection, La Guardia and Wagner Archives, LaGuardia Community College, Long Island City, New York.

4. Martin V. Melosi, "The Viability of Incineration as a Disposal Option: The Evolution of a Niche Technology, 1885–1995," *Public Works Management & Policy* 1 (July 1996): 35–36.
5. Walsh, "Incineration," 319A–320A; Daniel C. Walsh, "Solid Wastes in New York City: A History," *Waste Age* 20 (April 1989): 118; S. Tannenbaum, "A Brief History of Waste Disposal in New York City Since 1930," unpublished paper, New York City Municipal Archives, 1992, 2.
6. In 1938 a substantial amount of material was removed from Rikers Island to be used in the expansion of the North Beach Airport. City of New York, Department of Sanitation, *Annual Report, 1938*, 20.
7. City of New York, Department of Sanitation, *Annual Report, 1937*, 2, 15, 23.
8. Corey, "King Garbage," 205–7, 211.
9. Daniel C. Walsh, "Solid Wastes in New York City: A History," *Waste Age* 20 (April 1989): 122; Richard Fenton, "Current Trends in Municipal Solid Waste Disposal in New York City," *Resource Recovery and Conservation* 1 (1975): 171; Corey, "King Garbage," 209–10.
10. City of New York, Department of Sanitation, *Annual Report, 1938*, 25.
11. E. J. Cleary, "Land Fills for Refuse Disposal," *Engineering News-Record* (September 1, 1938): 270.
12. Cleary, "Land Fills for Refuse Disposal," 273.
13. Walsh, "Solid Wastes in New York City," 122; Rolf Eliassen, "Sanitary Land Fills in New York City," *Civil Engineering* 12 (September 1942): 483.
14. Fenton, "Current Trends," 172.
15. Harrison P. Eddy Jr., "Cautions Regarding Land-Fill Disposal," *Engineering News-Record* (December 15, 1938): 766.
16. Eddy, "Cautions Regarding Land-Fill Disposal," 767.
17. This section was drawn in large measure from U.S. Department of the Interior, National Park Service, "National Historic Landfill Nomination: Fresno Sanitary Landfill," https://npgallery.nps.gov/GetAsset/d1ac905d-57dc-4c76-a019-9ee12d63c026.
18. See Harry R. Crohurst, "Municipal Wastes: Their Character, Collection, and Disposal," *Public Health Service Bulletin* 107 (October 1920): 43–45; H. deB. Parsons, *The Disposal of Municipal Refuse* (New York: John Wiley, 1906), 78–80; D. C. Faber, "Collection and Disposal of Refuse," *American Municipalities* 30 (February 1916): 185–86; Robert H. Wild, "Modern Methods of Municipal Refuse Disposal," *American City* 5 (October 1911): 207–8.
19. Martin V. Melosi, "Historic Development of Sanitary Landfills and Subtitle D," *Energy Laboratory Newsletter* 31 (1994): 20; Ellis L. Armstrong, Michael C. Robinson, and Suellen M. Hoy, eds., *History of Public Works in the United States, 1776–1976* (Chicago: American Public Works Association, 1976), 449–50; "An Interview with Jean Vincenz," *Public Works Historical Society Oral History Interview* 1 (Chicago: Public Works Historical Society, 1980), 9–10; John J. Casey, "Disposal of Mixed Refuse by Sanitary Fill Method at San Francisco," *Civil Engineering* 9 (October 1939): 590–92.

20. Rachel Maines and Joel Tarr, "Municipal Sanitation: Assessing Technological Cost, Risk, and Benefit," unpublished case study, Carnegie-Mellon University, December 1980, 16.
21. "An Interview with Jean Vincenz," 1.
22. "An Interview with Jean Vincenz," 1; Jean L. Vincenz, "Sanitary Fill at Fresno," *Engineering News-Record* 123 (October 26, 1939): 539–40; Vincenz, "The Sanitary Fill Method of Refuse Disposal," *Public Works Engineers' Yearbook* (1940): 187–201; Vincenz, "Refuse Disposal by the Sanitary Fill Method," *Public Works Engineers' Yearbook* (1944), 88–96; "The Sanitary Fill as Used in Fresno," *American City* 55 (February 1940): 42–43.
23. Vincenz, "The Sanitary Fill Method of Refuse Disposal," 199; Desmond P. Tynan, "Modern Garbage Disposal," *American City* 54 (June 1939): 100–1; Rolf Eliassen and Albert J. Lizee, "Sanitary Land Fills in New York City," *Civil Engineering* 12 (September 1942): 483–86.
24. Eliassen, "Sanitary Land Fills in New York City," 486.
25. Rice and Pincus, "Health Aspects of Land-Fills," 1396.
26. From 1924 to 1957, the total area in New York City devoted to landfills was 94.68 square kilometers (23,395 acres), and the next highest was 1954 to 1994, at 33.47 (8,270 acres) kilometers. In the latter period Richmond County had the highest acreage, at 12.32 square kilometers (3,044 acres). Daniel C. Walsh and Robert G. LaFleur, "Landfills in New York City: 1844–1994," *Ground Water* 33 (July–August 1995): 557. See also Daniel C. Walsh, "Urban Residential Refuse Composition and Generation Rates for the 20th Century," *Environmental Science & Technology* 36 (2002): 4938–41.
27. Corey, "King Garbage," 211–13.
28. Sharon Seitz and Stuart Miller, *The Other Islands of New York City: A History and Guide*, 3rd ed. (Woodstock, VT: Countryman, 2011), 202.
29. Ted Steinberg, *Gotham Unbound: An Ecological History of Greater New York, 1609–2012* (New York: Simon & Schuster, 2014), 212–14.
30. When Moses first got to know Carey, his reaction to how city officials responded to the larger-than-life Mayor La Guardia was: "Bill Carey, the big contractor and part-time Sanitation Commissioner, was one of the few unterrified department heads. He could always leave the job, and he regarded the Mayor's vaudeville stunts with a mixture of humor, friendly admiration and amazed incredulity." Robert Moses, *La Guardia: A Salute and a Memoir* (New York: Simon and Schuster, 1957), 30.
31. "Robert Moses," Biography.com, http://www.biography.com/people/robert-moses-9416268; Steinberg, *Gotham Unbound*, 189.
32. Beverly Smith, "Bigger Than Politics," *American Magazine* 118 (October 1, 1934): 55.
33. "Robert Moses," *New York Preservation Archive Project*, http://www.nypap.org/content/robert-moses; "The Man with Ten Heads," *Nation* 189 (October 31, 1959): 284–91.
34. Hilary Ballon and Kenneth T. Jackson, "Introduction," in *Robert Moses and the Modern City: The Transformation of New York*, ed. Hilary Ballon and Kenneth T. Jackson (New York: Norton, 2007), 65–66.
35. Paul Goldberger, "Robert Moses, Master Builder, Is Dead at 92," *New York Times*, July 30, 1981; Owen D. Gutfreund, "Rebuilding New York in the Auto Age: Robert Moses and His Highways," in *Robert Moses and the Modern City*, ed. Ballon and Jackson, 86.
36. The RPA published a ten-volume survey of the physical and social activities of the New York region along with a plan for the area's future development. In addition, it

undertook detailed studies of various parts of the city. Two additional plans in all were developed over the years. Ann L. Buttenwieser, *Manhattan Water-Bound: Manhattan's Waterfront from the Seventeenth Century to the Present*, 2nd ed. (Syracuse, NY: Syracuse University Press, 1999), 173–74; "The Regional Plans," Regional Planning Association, http://www.rpa.org/regional-plans.

37. David A. Johnson, "Regional Planning for the American Metropolis: New York Between the World Wars," in *Two Centuries of American Planning*, ed. Daniel Schaffer (Baltimore, MD: Johns Hopkins University Press, 1988), 186–92. See also Steinberg, *Gotham Unbound*, 187.

38. George J. Lankevich, *American Metropolis: A History of New York City* (New York: New York University Press, 1998), 173.

39. Teresa Carpenter, ed., *New York Diaries, 1609 to 2009* (New York: Modern Library, 2012), 380; Robert A. Caro, *The Power Broker: Robert Moses and the Fall of New York* (New York: Vintage, 1974), 426–42. La Guardia received an important endorsement from Nathan Strauss Jr., president of the Park Association of New York City, on the appointment of Moses as parks commissioner. Strauss, however, was wary of Moses holding too many state and city positions simultaneously, which would give him unfettered control of the park system. La Guardia favored unified control. See Nathan Strauss Jr. to F. La Guardia, December 14, 1933; "Statement by Nathan Straus, Jr., President of the Park Association of New York City, commenting on the appointment of Robert Moses as City Park Commissioner," press release, Office of the Mayor, January 6, 1934, Box 3129, Folder 1, Department of Parks, Fiorello H. La Guardia Collection, La Guardia and Wagner Archives.

40. Quoted in August Heckscher, *When LaGuardia Was Mayor: New York's Legendary Years* (New York: Norton, 1978), 74. See also Thomas Kessner, *Fiorello H. La Guardia and the Making of Modern New York* (New York: McGraw-Hill, 1989), 303–4; Ronald H. Bayor, *Fiorello La Guardia: Ethnicity and Reform* (Arlington Heights, IL: Harlan Davidson, 1993), 123–25; Caro, *The Power Broker*, 353–54, 444–53.

41. Robert Moses, *Public Works: A Dangerous Trade* (New York: McGraw-Hill, 1970), 851.

42. Joel Schwartz, "Robert Moses," in *The Encyclopedia of New York City*, 2nd ed., ed. Kenneth T. Jackson (New Haven, CT: Yale University Press, 2010), 856.

43. Martin Filler, "Moses and Megalopolis," *Art in America* 69 (September 1981): 126.

44. Quoted in Goldberger, "Robert Moses, Master Builder, Is Dead at 92." See also Jackson, "Robert Moses and the Rise of New York: The Power Broker in Perspective," 69–70; Hilary Ballon, "Robert Moses and Urban Renewal: The Title I Program," 94–115; Margaret Biondi, "Robert Moses, Race, and the Limits of An Activist State," 116; Joel Schwartz, "Robert Moses and City Planning," 130–33, all in *Robert Moses and the Modern City*, ed. Ballon and Jackson; Filler, "Moses and Megalopolis," 130.

45. Herbert Kaufman, "Robert Moses: Charismatic Bureaucrat," *Political Science Quarterly* 90 (Autumn 1975): 521, 530–31, 534.

46. S. J. Woolf, "How to Lose Friends and Influence People," *New York Times Magazine*, June 29, 1941, 13.

47. Joel Schwartz, *The New York Approach: Robert Moses, Urban Liberals, and Redevelopment of the Inner City* (Columbus: Ohio State University Press, 1993), xv–xix, 297–305; Ted Steinberg, *Down to Earth: Nature's Role in American History*, 2nd ed. (New York: Oxford University Press, 2009), 211–12; Roberta Brandes Gratz, *The Battle for Gotham: New York in the Shadow of Robert Moses and Jane Jacobs* (New York: Nation Books, 2010), xxi; Jameson W. Doig, "Regional Conflict in the New York Metropolis:

The Legend of Robert Moses and the Power of the Port Authority," *Urban Studies* 27 (April 1990); 201–2; Phillip Lopate, "Rethinking Robert Moses," *Metropolis*, December 1969, http://www.metropolismag.com/December-1969/Rethinking-Robert -Moses/.

48. Several years earlier, the suburban development of Corona had been marketed to residents of Little Italy and the Lower East Side of Manhattan as a possible residential community. See Benjamin Miller, "Fat of the Land: New York's Waste," *Social Research* 65 (Spring 1998): 92.
49. See Francis Morrone, "Robert Moses and 'The Great Gatsby,'" *New York Sun*, February 2, 2007, http://www.nysun.com/arts/robert-moses-and-the-great-gatsby /47902.
50. The Brooklyn Ash Removal Company originally owned and operated it.
51. See "Fishhooks McCarthy and the Importance of a Leader's Moral Philosophy," *Corridor Conversations*, May 14, 2014, http://stephenomeara.wordpress.com/2014/05 /14/fishhooks-mccarthy-and-the-importance-of-a-leaders-moral-philosophy/.
52. "Blanshard Favors Incinerator Deal," *New York Times*, May 11, 1934. Having the DSNY actually purchase the property kept Moses from paying for it out of his own budget. See Helen A. Harrison, "From Dump to Glory: Robert Moses and the Flushing Meadow Improvement," in *Robert Moses: Single-Minded Genius*, ed. Joann P. Krieg (Interlaken, NY: Heart of the Lakes, 1989), 91–92; Roger Starr, "The Valley of Ashes: F. Scott Fitzgerald and Robert Moses," *City Journal* (Autumn 1992), https://www.city -journal.org/html/valley-ashes-f-scott-fitzgerald-and-robert-moses-12680.html; Quennell Rothschild & Partners, *Flushing Meadows Corona Park: Strategic Framework Plan*, prepared for the New York City Department of Parks and Recreation, 52, http://www.nycgovparks.org/parks/flushing-meadows-corona-park; Grover A. Whalen, *Mr. New York: The Autobiography of Grover A. Whalen* (New York: G. P. Putnam's Sons, 1955), 199.
53. Mira Engler, *Designing America's Waste Landscapes* (Baltimore, MD: Johns Hopkins University Press, 2004), 93.
54. Martin Charles, "Robert Moses and the World's Fair," *Review of Reviews* (September 1936): 39; Robert Moses, "From Dump to Glory," *Saturday Evening Post* 210 (January 1938): 12.
55. "History of Mt. Corona," *Flushing Meadow Improvement* 1 (March 1937): 22.
56. Charles, "Robert Moses and the World's Fair," 41.
57. Robert Moses, "The Saga of Flushing Meadow," *nywf64.com*, April 11, 1966, http:// www.nywf64.com/saga02.shtml. See also "M'Aneny Accepts Sanitation Post," *New York Times*, April 19, 1933; Miller, *Fat of the Land*, 179–84; "M'Aneny Is Named to Controllership," *New York Times*, September 20, 1933; Caro, *The Power Broker*, 1082–85, 1092. See also Moses, "From Dump to Glory," 72; Caro, *The Power Broker*, 654; Harrison, "From Dump to Glory," 93.
58. "Robert Moses—Park Creator Extraordinary," *Recreation* 32 (August 1, 1938): 291; New York Public Library Manuscripts and Archives Division, "New York World's Fair 1939 and 1940 Incorporated Records, 1935–1945," vii, http://www.nypl.org/sites/default/files /archivalcollections/pdf/nywf39fa.pdf.
59. Quennell Rothschild & Partners, *Flushing Meadows*, 52; City of New York, Department of Sanitation, *Annual Report, 1938*, 26. See also Robert Moses to F. H. La Guardia, April 18, 1939, Box 3239, Folder 9, Department of Parks, 1939, La Guardia Collection, in which Moses is opposed to the city relinquishing any power over control of the World's Fair site.

60. Starr, "The Valley of Ashes"; James Miller, "Moses: Idealist in Action," *Current History* 51 (September 1, 1939): 54; Miller, *Fat of the Land*, 184.
61. Whalen, *Mr. New York*, 199. See also Steinberg, *Gotham Unbound*, 211–21.
62. "Sound View" is sometimes written as "Soundview."
63. In 1938 Works Progress Administration forces removed an estimated 14 million cubic yards of old fill for the extension of the airport. City of New York, Department of Sanitation, *Annual Report, 1938*, 20. For more on Moses's views on the cleanup of Rikers Island and transportation issues relative to the World's Fair, see Robert Moses to F. H. La Guardia, February 7, 1938; Grover A. Whalen to Moses, April 15, 1938; Joseph A. Palma to F. H. La Guardia, July 28, 1938, Box 3216, Folder 4, 1938, Department of Parks, La Guardia Collection.
64. Alexander P. Chopin, "Moses Attacks Rikers Island Garbage Dump Proposal," *Long Island Star-Journal*, June 15, 1939. See also "Queens Forces Set to Oppose Rikers Island Dump Plan," *Long Island Star-Journal*, September 5, 1939.
65. Seitz and Miller, *The Other Islands of New York City*, 203; Miller, *Fat of the Land*, 188–89; Corey, "King Garbage," 213. There was resistance to the nursery by the director of horticulture because he argued that the fermentation of the new fill made the ground impractical to grow trees. See Director of Horticulture David Schweizer to George E. Spargo, August 15, 1940, and Robert Moses to Spargo, August 16, 1940, Roll 6, Folder 14, NYC Department of Parks, Office of the Commissioner, 1940, New York City Municipal Archives and Library; C. C. Combs to George Spargo, May 28, 1943, Roll 17, Folder 44, Rikers Island, 1943, NYC Department of Parks, Office of the Commissioner, 1940, New York City Municipal Archives and Library.
66. Elizabeth Fee and Steven H. Corey, *Garbage! The History and Politics of Trash in New York City* (New York: New York Public Library, 1994), 51.
67. Mayank Teotia, *Managing New York City Municipal Solid Waste Using Anaerobic Digestion* (Brooklyn: Programs for Sustainable Planning and Development, Pratt Institute, School of Architecture, Spring 2013), 4.
68. Corey, "King Garbage," 214–16.
69. Seitz and Miller, *The Other Islands of New York City*, 242; Cleveland Rodgers, *Robert Moses: Builder for Democracy* (New York: Henry Holt, 1952), 124; Fee and Corey, *Garbage!*, 51; Miller, *Fat of the Land*, 189–92; Letter, Robert Moses to Fiorello La Guardia, July 18, 1938, in City of New York, Department of Parks, *The Future of Jamaica Bay* (July 18, 1938).
70. "Jamaica Bay Dump Fought by Moses," *New York Times*, April 20, 1938.
71. Letter, Robert Moses to Fiorello La Guardia, July 18, 1938, in City of New York, Department of Parks, *The Future of Jamaica Bay*.
72. Elizabeth Barlow, *The Forests and Wetlands of New York City* (Boston: Little, Brown, 1971), 112.
73. Letter, Robert Moses to Fiorello La Guardia, July 18, 1938, in City of New York, Department of Parks, *The Future of Jamaica Bay*.
74. Letter, Robert Moses to Fiorello La Guardia, July 18, 1938, in City of New York, Department of Parks, *The Future of Jamaica Bay*.
75. Corey, "King Garbage," 216–18.
76. "Carey Visits Site of Dumping Row," *New York Times*, July 14, 1938.
77. Corey, "King Garbage," 218–21.
78. Miller, *Fat of the Land*, 196, 199–203. See also "Bronx Area Benefited," *New York Times*, June 6, 1937; Robert Moses to F. H. La Guardia, May 22, 1939, Box 3239, Folder 9,

6. ONE BEST WAY   ■   597

Department of Parks, 1939, La Guardia Collection, concerning Moses's objections on how dredge material would be used for building a bulkhead at Sound View.

79. "Carey Bows to Protest on Jamaica Bay Dump," *New York Times*, September 23, 1938.
80. "Huge Waste Is Laid to Carey's Bureau," *New York Times*, October 17, 1938. On occasion, a positive story made the papers in these days. See, for example, L. H. Robbins, "Keeping House for the City's Millions," *New York Times*, October 23, 1938.
81. "Carey Says Moses Is a Propagandist," *New York Times*, October 11, 1938. See also "Legal Attack Aimed at Dump in Edgemere," *Brooklyn Eagle*, October 11, 1938.
82. Quoted in Corey, "King Garbage," 223–24.
83. See Miller, *Fat of the Land*, 189–203.
84. City of New York, Department of Sanitation, *Annual Report, 1939*, 51, 58. See also "The Department of Sanitation," an address by Honorable Mathew J. Diserio, assistant to the commissioner, New York University, December 21, 1939, Box 2622, Folder 8, Department of Sanitation, 1939, La Guardia Collection.
85. "Funds Are Voted to Start New Park," *New York Times*, March 3, 1939. See also "Extension of Park in Bronx Approved," *New York Times*, December 15, 1939.
86. "Carey Says Moses Is a Propagandist"; "Carey's Dump to Be Placed Near 3 Parks," *Staten Island Advance*, October 12, 1938.
87. Corey, "King Garbage," 224–25. Great Kills Park currently includes approximately 523 acres of salt marsh, beach, and woodlands in the vicinity of the Raritan and Lower Bays of Great Kills Harbor, on the South Shore of Staten Island. In 1929, the City of New York bought Crookes Point and adjacent lands for the purposes of building a public park. Between 1934 to 1951, the New York Department of Parks administered the Marine Park Project to develop the Great Kills Harbor as a shore-front recreation area. The city operated Great Kills as a city park until it was transferred to the National Park Service in 1972 and became part of the Staten Island Unit of Gateway National Recreation Area. See "Great Kills Park Environmental Cleanup Project," National Park Service, https://www.nps.gov/gate/learn/management/great killscleanup.htm.
88. Corey, "King Garbage," 224–25; Robin Nagle, "To Love a Landfill: The History and Future of Fresh Kills," in *Handbook of Regenerative Landscape Design*, ed. Robert L. France (Boca Raton, FL: CRC, 2008), 9.
89. Miller, *Fat of the Land*, 342.
90. "Carey's Dump to Be Placed Near 3 Parks."
91. "Carey Says Moses Is a Propagandist."
92. "Carey Is Indicted in Dumping Dispute," *New York Times*, January 11, 1939.
93. "Public Dump No. 1," *Staten Island Advance*, October 12, 1938.
94. Referring to the 1938 plans for the Willowbrook State School, a state school for children with intellectual disabilities, which was not opened until after World War II.
95. "Oakwood Raps Dump Proposal" and "Columbian Rotarians Assail Garbage Plan," *Staten Island Advance*, October 14, 1938.
96. City of New York, President of the Borough of Richmond, *Annual Report, 1938-1939*, 30, Staten Island Historical Museum, Staten Island, New York.
97. "Architects Protest Plan for Garbage Dump," *Staten Island Advance*, October 20, 1938.
98. "Garbage War Begun: 10,000 to Organize; Vigilantes Prepare," *Staten Island Advance*, October 20, 1938.
99. Staten Island Citizens Committee of 10,000, "Anti-Garbage Dumping Petition," October 1938, Staten Island Historical Museum.

100. "Complaints Listed in Dumping Row," *New York Times*, January 12, 1939.
101. The indictments were brought over presumed violations of state health statutes.
102. "Carey Is Indicted in Dumping Dispute." See also "Carey, Five Others Indicted in Dumping," *Staten Island Advance*, January 10, 1939; "LaGuardia, Irate, Blames Fetherston for Indictments in Dumping of Garbage," *Staten Island Advance*, January 11, 1939.
103. "Arrest Carey in Illegal S.I. Dumping," *Brooklyn Daily Eagle*, January 10, 1939. See also "Carey's Methods Defended by Aides," *New York Times*, February 19, 139; "Carey Is Indicted in Dumping Dispute."
104. "City Hall Gets a Whiff of Queens Garbage; Mayor Turns Cold Nose on Protest Group," *New York Times*, February 1, 1939.
105. "City Hall Gets a Whiff of Queens Garbage." See also "The Dumping Nuisance," *New York Times*, February 2, 1939; "Queens Area Planning Protest on Dumping," *Brooklyn Daily Eagle*, August 21, 1938.
106. "City Garbage Foes Ousted by Mayor," *New York Times*, February 17, 1939. See also "Dump Critics to Visit Rikers Fill Project," *Brooklyn Daily Eagle*, February 18, 1939.
107. "Anti-Dumping Suit Ordered to Trial," *New York Times*, March 8, 1939.
108. Corey, "King Garbage," 227.
109. "Anti-Dump Bill Given Council OK," *Staten Island Advance*, May 17, 1939. See also "Council Restricts Garbage Dumping," *New York Times*, March 15, 1939; Corey, "King Garbage," 230; Daniel C. Kramer and Richard M. Flanagan, *Staten Island: Conservative Bastion in a Liberal City* (Lanham, MD: University Press of America, 2012), 114–15.
110. "Johnson's Dump Bill Vetoed by Governor," *Staten Island Advance*, May 3, 1941.
111. They were Edgemere, South Ozone Park, Juniper Valley, and the Jamaica Bay section at Cornell Park and Bergen Landing.
112. "Carey and Dr. Rice Indicted in Queens," *New York Times*, March 29, 1939.
113. "Sanitation Job Is No Cinch," *Brooklyn Daily Eagle*, April 2, 1939. See also "The Dumping Controversy," *New York Times*, April 12, 1939.
114. "Carey and Rice Freed in Queens," *New York Times*, June 20, 1939. See also "Board Is Named on Queens Dumping," *New York Times*, July 3, 1939.
115. They included Dr. Eugene L. Bishop, director of health for the Tennessee Valley Authority; Dr. Milton J. Rosenau, director of the Division of Public Health at the University of North Carolina School of Medicine; Dr. Huntington Williams, health commissioner of Baltimore; Dr. Kenneth F. Maxcy, professor at Johns Hopkins University; and Ralph E. Tarbett, senior engineer of the U.S. Public Health Service.
116. Corey, "King Garbage," 229–31.
117. Richard Fenton, *Outline History of the Department of Sanitation* (New York: Department of Sanitation, 1954), 17; "Refuse Disposal by Sanitary Landfill Method Has Wartime Advantages," *Water and Sewer* 82 (August 1944): 18.
118. Quoted in John L. Rice and Sol Pincus, "Health Aspects of Land-Fills," *American Journal of Public Health and the Nation's Health* 30 (December 1940): 1397.
119. "Health Experts Endorse Landfills and Recommend Best Practice," *Engineering News-Record* 124 (March 28, 1940): 54–55 (vol. pp. 446–47). Best practices at the time, however, suggested a rather narrow definition of public health risks, which did not take into account leaching of toxic substances into nearby watercourses or the production of methane gas. See Martin V. Melosi, *Garbage in the Cities: Refuse, Reform, and the Environment*, rev. ed. (Pittsburgh, PA: University of Pittsburgh Press, 2005), 184–85; Martin Melosi, *The Sanitary City: Urban Infrastructure in America from*

Colonial Times to the Present (Baltimore, MD: Johns Hopkins University Press, 2000), 405.
120. See Corey, "King Garbage," 236.
121. Caro, *The Power Broker*, 336, 344, 363.
122. Caro, *The Power Broker*, 336, 344, 363. See also "Two Can Play at This Game," *Staten Island Advance*, May 20, 1940; *Report of Committee on the "Sanitation Fill Method" for Improving Marine Park, Great Kills, Staten Island*, proposed by Parks Commissioner Robert L. Moses to Mayor Fiorello H. La Guardia (New York: Richmond County Chapter, New York Society of Professional Engineers, Inc., 1940).
123. Robert Moses to Fiorello H. La Guardia, May 20, 1940, Box 3260, Folder 3, Department of Parks, Marine Park–Great Kills, 1940, La Guardia Collection.
124. C. L. Hall to Robert Moses, June 4, 1940, Box 3260, Folder 3, Department of Parks, Marine Park–Great Kills, 1940, La Guardia Collection.
125. Robert Moses to Fiorello H. La Guardia, June 4, 1940, Box 3260, Folder 3, Department of Parks, Marine Park–Great Kills, 1940, La Guardia Collection. See also C. L. Hall, District Engineer, U.S. Army Corps of Engineers, to Robert Moses, June 4, 1940, in the same folder; "Now or Never," *Staten Island Advance*, June 12, 1940.
126. Art O. Hedquist, Staten Island Chamber of Commerce, to Robert Moses, June 14, 1940; Robert Moses to F. H. La Guardia, June 15, 1940; Louis W. Kaufman, Staten Island Real Estate Board, to Robert Moses, June 19, 1940; Moses to La Guardia, July 11, 1940; Moses to La Guardia, September 10, 1940, Box 3260, Folder 3, Department of Parks, Marine Park–Great Kills, 1940, La Guardia Collection.
127. "Stamp of Approval," *Staten Island Advance*, June 7, 1940, Box 3260, Folder 3, Department of Parks, Marine Park–Great Kills, 1940, La Guardia Collection.
128. See "Two Can Play at This Game."
129. "Stamp of Approval."
130. Robert Moses to F. H. La Guardia, July 11, 1940; Robert Moses to Colonel C. L. Hall, September 3, 1940, Box 3260, Folder 3, Department of Parks, Marine Park–Great Kills, 1940, La Guardia Collection.
131. John C. Reidel, Chief Engineer to C. L. Hall, U.S. Army Corps of Engineers, July 2, 1940; L. S. Dillon, U.S. Army Corps of Engineers, to John C. Reidel, July 3, 1940; F. H. La Guardia to L. S. Dillon, July 11, 1940; Robert Moses to C. L. Hall, September 3, 1940; Robert Moses to F. H. La Guardia, September 10, 1940, Roll 3, Folder 14, NYC Department of Parks, Office of the Commissioner, 1940, New York City Municipal Archives and Library.
132. Editorial, *Staten Island Advance*, October 11, 1940, Box 3260, Folder 3, Department of Parks, Marine Park–Great Kills, 1940, La Guardia Collection. See also Robert Moses to F. H. La Guardia, September 10, 1940, Box 3260, Folder 3, Department of Parks, Marine Park–Great Kills, 1940, La Guardia Collection; Robert Moses to William F. Carey, August 13, 1943, Roll 17, Folder 51, NYC Department of Parks, Office of the Commissioner, Department of Sanitation, 1943, New York City Municipal Archives and Library.
133. Robert Moses to Fiorello H. La Guardia, October 16, 1940, Box 3260, Folder 3, Department of Parks, Marine Park–Great Kills, 1940, La Guardia Collection.
134. Frances Lehrich, Secretary, Board of Estimate, to Fiorello H. La Guardia, December 20, 1940, Box 3260, Folder 3, Department of Parks, Marine Park–Great Kills, 1940, La Guardia Collection.
135. Steinberg, *Gotham Unbound*, 244. See also Robert Moses to William F. Carey, January 9, 1942; William F. Carey to the Office of Production Management, January 1942;

and Robert Moses to William F. Carey, January 7, 1942, Roll 11, Folder 39, NYC Department of Parks, Office of the Commissioner, Department of Sanitation, 1942, New York City Municipal Archives and Library.
136. Robert Moses to Frederick J. Spender, June 15, 1943, Roll 14, Folder 8, NYC Department of Parks, Office of the Commissioner, Great Kills Park, 1943, New York City Municipal Archives and Library. Moses described the process as transporting waste material to Great Kills in enclosed barges, unloading it with mechanical equipment, and immediately covering it "with a foot of clean earth fill so that it is only exposed for a short period of time." The DSNY fill would only be placed in the interior sections of the park, with a strip of clean sand around Great Kills and the ocean front "to prevent any possible pollution of the waters of the bay or along the beach."
137. Robert Moses to Dan Harper, November 24, 1945, NYC Department of Parks, Office of the Commissioner, 1940–1956, Roll 28, Folder 17, 1945, City of New York, Department of Records and Information Services, Municipal Archives. See also Robert Moses to David Shepard, Great Kills Board of Commerce, September 13, 1945, NYC Department of Parks, Office of the Commissioner, Roll 23, Folder 14, Great Kills Board of Commerce, 1945, City of New York, Department of Records and Information Services, Municipal Archives; Maury Maverick to Robert Moses, March 24, 1943, Roll 17, Folder 51, NYC Department of Parks, Office of the Commissioner, Department of Sanitation, 1943, New York City Municipal Archives and Library.
138. William F. Carey to Robert Moses, November 21, 1945, NYC Department of Parks, Office of the Commissioner, 1940–1956, Roll 20, Folder 1, 1945, City of New York, Department of Records and Information Services, Municipal Archives. See also Robert Moses to Col. Clarence Renshaw, District Engineer, November 1, 1945, NYC Department of Parks, Office of the Commissioner, Roll 23, Folder 11, Great Kills Harbor, 1945, City of New York, Department of Records and Information Services, Municipal Archives.
139. "Lynch Calls Garbage Use in Landfill Job 'Disgrace,'" *Staten Island Advance*, July 24, 1945. See also "Mayor Gets Whiff of Land Fill Smell," *New York Times*, July 19, 1946. See also "From the Archives: Trail of Broken Promises Litters History of Former Fresh Kills Landfill," *SILive.com*, September 26, 2013, http://www.silive.com/news/2013/09/from_the_archives_trail_of_bro.html; "Great Kills Landfill Rats Attract Rare Snowy Owl," *Staten Island Advance*, January 5, 1946.
140. Robert Moses to Editor, *Staten Island Advance* (cover letter), June 22, 1946, NYC Department of Parks, Office of the Commissioner, 1940–1956, Box 107883, Folder 41, *Staten Island Advance*, 1946, City of New York, Department of Records and Information Services, Municipal Archives.
141. Daniel C. Walsh, "The History of Waste Landfilling in New York City," *Ground Water* 29 (July–August 1991): 591.
142. William O'Dwyer, *Beyond the Golden Door*, ed. Paul O'Dwyer (Jamaica, NY: St. John's University, 1987), 304.

7. FUTILE PROTESTS

1. "Connecting Staten Island with Manhattan," *Scientific American* 120 (May 24, 1919): 545, 558–59. See also "Reply to Radio Speech," *New York Times*, August 30, 1925.

7. FUTILE PROTESTS ■ 601

2. Calvin Van Name, *Staten Island: A Report by the President of the Borough of Richmond to the Mayor* (June 1921), 23–26, Staten Island Historical Museum, Staten Island, New York.
3. Daniel C. Kramer and Richard M. Flanagan, *Staten Island: Conservative Bastion in a Liberal City* (Lanham, MD: University Press of America, 2012), 25.
4. "The Verrazano Rail Tunnel (The Brooklyn-Richmond Freight & Passenger Tunnel)," Brooklyn Historic Railway Association, http://www.brooklynrail.net/verrazano_rail_tunnel.html; *Seven Years of Progress: Important Public Improvements and Achievements by the Municipal and Borough Governments of the City of New York, 1918–1925* (Submitted to the Board of Aldermen, March 1925), 176–77; Kramer and Flanagan, *Staten Island*, 25.
5. Charles W. Leng and William T. Davis, *Staten Island and Its People: A History, 1609–1929* (New York: Lewis Historical Pub. Co., 1930), 1:377–78.
6. Mayor John Hylan often was blamed for stopping the project, some believed because of his feud with Brooklyn-Manhattan Transit (BMT); others underscored competing business interests that preferred a freight route from Brooklyn to Bayonne, New Jersey. Alternatively, some argued that Governor Al Smith was responsible because of his interests in the Pennsylvania Railroad, a competitor of the B&O, which owned Staten Island Rapid Transit, or possibly because he did not want to undermine Port Authority planning. See Irvin Leigh and Paul Matus, "Staten Island Rapid Transit: The Essential History," December 23, 2001, *The Third Rail*, http://www.thethirdrail.net/0201/sirt9.html; Kramer and Flanagan, *Staten Island*, 25.
7. The Port Authority had opposed a Narrows tunnel as incompatible with its own transport vision. See Jameson W. Doig, *Empire on the Hudson: Entrepreneurial Vision and Political Power at the Port of New York Authority* (New York: Columbia University Press, 2001), 98–99, 113–14.
8. See Kramer and Flanagan, *Staten Island*, 26; Robert A. Caro, *The Power Broker: Robert Moses and the Fall of New York* (New York: Vintage, 1974), 341–44; Doig, *Empire on the Hudson*, 332–33; Kramer and Flanagan, *Staten Island*, 26–27.
9. J. Bernard Walker, "Placing Staten Island on the Map," *Scientific American* 137 (October 1, 1927): 306–8; Margaret Lundrigan and Tova Navarra, *Staten Island in the Twentieth Century* (Mount Pleasant, SC: Arcadia, 1998), 63. See also Betsy McCully, *City at the Water's Edge: A Natural History of New York* (New Brunswick, NJ: Rutgers University Press, 2007), 107.
10. Richard Dickenson, ed., *Holden's Staten Island: The History of Richmond County* (New York: Center for Migration Studies, January 2003), 157.
11. Committee on Regional Plan of New York and Its Environs, *Regional Plan of New York and Its Environs*, vol. 1, *The Graphic Regional Plan* (Philadelphia: Wm. F. Fell Printers, 1929), 320.
12. Leng and Davis, *Staten Island and Its People*, 2:638–41; Loring McMillen, *Staten Island: The Cosmopolitan Era from 1898* (Staten Island: Staten Island Historical Society, 1952), 12.
13. "Fresh Kills: Landfill to Landscape: International Design Competition: 2001," nyc.gov, https://www1.nyc.gov/assets/planning/download/pdf/plans/fkl/about_competition.pdf; Dickenson, ed., *Holden's Staten Island*, 200.
14. Dickenson, ed., *Holden's Staten Island*, 162–63, 171, 194–95; Charles L. Sachs, *Made on Staten Island: Agriculture, Industry, and Suburban Living in the City* (Richmondtown, Staten Island: Staten Island Historical Society, 1988), 101.

15. Quoted in Richard M. Bayles, ed., *History of Richmond County (Staten Island), New York, from Its Discovery to the Present Time* (New York: L. E. Preston & Co., 1887), 705.
16. Bayles, ed., *History of Richmond County*, 171–75.
17. "Staten Island's North Shore," "Staten Island's East Shore," "Staten Island's West Shore," "Staten Island's South Shore," *Old Staten Island*, http://www.secretstatenisland.com/.
18. "Fresh Kills: Landfill to Landscape."
19. "Fresh Kills: Landfill to Landscape"; "Travis, Staten Island," *Forgotten New York*, http://forgotten-ny.com/2006/03/travis-staten-island/; "Staten Island Neighborhood Profile: Travis," RealEstatesSINY.Com, http://www.realestatesiny.com/Staten-Island-Neighborhood-Profile-Travis.php.
20. Quoted in "New York's Worst Neighborhood Name Ever," January 28, 2010, *Ephemeral New York*, https://ephemeralnewyork.wordpress.com/2010/01/28/new-yorks-worst-neighborhood-name-ever/.
21. Ralph W. Tiner, *Wetlands of Staten Island, New York: Valuable Vanishing Urban Wetlands* (Hadley, MA: A Cooperative National Wetlands Inventory Publication, January 2000), 1.
22. Tiner, *Wetlands of Staten Island*, 10–13.
23. Ann Vileisis, *Discovering the Unknown Landscape: A History of America's Wetlands* (Washington, DC: Island, 1997), xi–xii, 1–17, 30–33, 47–50, 68–69, 94, 140–41, 151–52, 165–67. On the rollback of marshlands in New York and the role of Robert Moses, see Ted Steinberg, *Gotham Unbound: An Ecological History of Greater New York, 1609–2012* (New York: Simon & Schuster, 2014), 189–91, 199–210.
24. Judith S. Weis and Carol A. Butler, *Salt Marshes: A Natural and Unnatural History* (New Brunswick, NJ: Rutgers University Press, 2009), xiii, 3–4, 7, 18, 21–25; John Waldman, *Heartbeats in the Muck: The History, Sea Life, and Environment of New York Harbor*, rev. ed. (New York: Fordham University Press, 2013), 10; Elizabeth Barlow, *The Forests and Wetlands of New York City* (Boston: Little, Brown, 1971), 30–31, 36–37; Steinberg, *Gotham Unbound*, 184–85.
25. Tiner, *Wetlands of Staten Island*, 4–5.
26. Bayles, ed., *History of Richmond County (Staten Island)*, 17–18; Tiner, *Wetlands of Staten Island*, 8.
27. "Staten Island First Natives," *Staten Island History*, http://www.statenislandhistorian.com/staten-island-first-natives.html; Kenneth M. Gold and Lori R. Weintrob, eds., *Discovering Staten Island: A 350th Anniversary Commemorative History* (Charleston, SC: History Press, 2011), 80, 136.
28. Molly Greene, "A Mountain by the Sea: Waste-scapes, Life-scapes, and the Reinvention of Fresh Kills," 2, https://hixon.yale.edu/sites/default/files/files/fellows/paper/greene_molly_2012_report.pdf.
29. Waldman, *Heartbeats in the Muck*, 101; Weis and Butler, *Salt Marshes*, 22, 92–95. See also John Teal and Mildred Teal, *Life and Death of the Salt Marsh* (New York: Ballantine, 1969), 2, 25, 40–41; Steinberg, *Gotham Unbound*, 195, 198.
30. Clay McShane and Joel A. Tarr, *The Horse in the City: Living Machines in the Nineteenth Century* (Baltimore, MD: Johns Hopkins University Press, 2007), 130–36, 138–44.
31. "Marshes Yield Harvest," *Staten Island Advance*, January 30, 1932, William T. Davis Papers, Staten Island Historical Museum. The naturalist Davis commented on the number of people harvesting hay in the salt marshes.

7. FUTILE PROTESTS ■ 603

32. Vileisis, *Discovering the Unknown Landscape*, 111, 113, 205–6.
33. E. J. Cleary, "Land Fills for Refuse Disposal," *Engineering News-Record* 121 (September 1, 1938): 270.
34. Steven Hunt Corey, "King Garbage: A History of Solid Waste Management in New York City, 1881–1970," PhD diss., New York University, 1994, 257.
35. Corey, "King Garbage," 236–38. See also Suellen M. Hoy and Michael C. Robinson, *Recovering the Past: A Handbook of Community Recycling Programs* (Chicago: American Public Works Association, 1979).
36. Between 1924 and 1957, landfill acreage distributed by borough was Queens (11,040 acres), Brooklyn (5,640 acres), the Bronx (3,610 acres), Richmond (2,700 acres), and Manhattan (420 acres). See Daniel C. Walsh, "Reconnaissance Mapping of Landfills in New York City," 394–95, *Proceedings: FOCUS Conference on Eastern Regional Ground Water Issues*, National Ground Water Association, Portland, ME, October 29–31, 1991. See also Daniel C. Walsh, "The History of Waste Landfilling in New York City," *Ground Water* 29 (July–August 1991): 592; Richard Fenton, "An Analysis of the Problem of Sanitary Landfills in New York City," a report for the Bureau of Sanitary Engineering, New York City Health Department, August 1947, 6.
37. "Carey Aims to Quit for $4,000 City Job, But Mayor Says No," *New York Times*, August 3, 1940; "Carey Resigns $10,000 City Post, But It Is Only for a Day or Two," *New York Times*, October 17, 1944.
38. "Carey Admits He Violated Charter at Sanita, but for the City's Good," *New York Times*, October 19, 1943.
39. August Heckscher, *When LaGuardia Was Mayor: New York's Legendary Years* (New York: Norton, 1978), 355–56; "Carey Admits He Violated Charter at Sanita"; "Mayor and Carey Violated Charter, Council Body Finds," *New York Times*, December 16, 1943; Thomas Kessner, *Fiorello H. La Guardia and the Making of Modern New York* (New York: McGraw-Hill, 1989), 548–49. See also William B. Rhoads, "New York's White Wings and the Great Saga of Sanita," *New York History* 80 (April 1999): 153–84; H. Allen Smith, "Angel of the White Wings," *Saturday Evening Post*, March 6, 1943, 38, 40.
40. George J. Lankevich, *American Metropolis: A History of New York City* (New York: New York University Press, 1998), 182, 185–88; "William O'Dwyer," Arlington National Cemetery, http://www.arlingtoncemetery.net/williamo.htm; Michael J. O'Neill, "Surrounded by Rascals," *New York Times*, August 18, 1987; Chris McNickle, "William O'Dwyer," in *The Encyclopedia of New York City*, 2nd ed., ed. Kenneth T. Jackson (New Haven, CT: Yale University Press, 2010), 952–53; Francois Weil, *A History of New York* (New York: Columbia University Press, 2004), 282; Caro, *The Power Broker*, 699–700, 755–767.
41. Selma Berrol, *The Empire City: New York and Its People, 1624–1996* (Westport, CT: Praeger, 1997), 128–30; Lankevich, *American Metropolis*, 183–85; Joseph J. Salvo, "Population," in *The Encyclopedia of New York City*, ed. Jackson, 1019.
42. "Robert Moses, Master Builder, Is Dead at 92," On This Day, *New York Times*, July 30, 1981, http://www.nytimes.com/learning/general/onthisday/bday/1218.html; Charles Garrett, *The La Guardia Years: Machine and Reform Politics in New York City* (New Brunswick, NJ: Rutgers University Press, 1961), 309; Heckscher, *When LaGuardia Was Mayor*, 404–5; Caro, *The Power Broker*, 757–68.
43. "Powell Takes City Post," *New York Times*, August 28, 1936.
44. "William J. Powell Is Dead at 77: Ex-Commissioner of Sanitation," *New York Times*, May 10, 1966.

45. Janice Kabel, "The Fresh Kills Landfill: Thank Robert Moses for Idea of Transforming Marsh to Park," *Staten Island Advance*, October 2, 1978; Steinberg, *Gotham Unbound*, 244.
46. Quoted in Kabel, "The Fresh Kills Landfill," emphasis added.
47. His major complaint about Carey's landfill operations had to do with the careless management and operation of sanitary landfills.
48. Robert Moses to Editor, *Staten Island Advance* (cover letter), June 22, 1946, NYC Department of Parks, Office of the Commissioner, 1940–1956, Box 107883, Folder 41, *Staten Island Advance*, 1946, City of New York, Department of Records and Information Services, Municipal Archives.
49. See Harold H. Cassidy, "Moses Didn't Mention Garbage When He Asked for Marshland," *Staten Island Advance*, June 20, 1946.
50. Memorandum to Mr. Spargo from Commissioner Moses, January 7, 1946, NYC Department of Parks, Office of the Commissioner, 1940–1956, Box 107883, Folder 20, Department of Sanitation, 1946, City of New York, Department of Records and Information Services, Municipal Archives, New York City. Thanks to Ted Steinberg for providing me with some of the Moses correspondence.
51. William J. Powell to George E. Spargo, January 18, 1946, NYC Department of Parks, Office of the Commissioner, 1940–1956, Box 107883, Folder 20, Department of Sanitation, 1946, Municipal Archives.
52. See Robert Moses to William J. Powell, June 19, 1946, NYC Department of Parks, Office of the Commissioner, 1940–1956, Box 107883, Folder 20, Department of Sanitation, 1946, Municipal Archives.
53. Quoted in Kabel, "The Fresh Kills Landfill."
54. "Fresh Kills Dump Fund Again Put in Budget," *Staten Island Advance*, June 9, 1946.
55. "From the Archives: Trail of Broken Promises Litters History of Former Fresh Kills Landfill," *SILive.com*, September 26, 2013, http://www.silive.com/news/index.ssf/2013/09/from_the_archives_trail_of_bro.html.
56. "Fresh Kills Dump Fund Again Put in Budget."
57. "Moses and Hall," *Staten Island Advance*, June 5, 1946.
58. "Moses and Hall."
59. Kramer and Flanagan, *Staten Island*, 44–45.
60. "Moses and Hall." See also "Hall Backs Fresh Kills Dump," *Staten Island Advance*, June 8, 1946.
61. Kramer and Flanagan, *Staten Island*, 45–46.
62. "Mr. Morris and Mr. Moses," *Staten Island Advance*, June 15, 1946.
63. "Landfill . . . Again!" *Staten Island Advance*, June 9, 1946.
64. "Let's Fight!" *Staten Island Advance*, June 10, 1946.
65. See "Build Incinerators," *Staten Island Advance*, June 12, 1946. See also "Sewage Problem Is Cut," *New York Times*, June 17, 1946.
66. "Great Kills Landfill Rats Attract Rare Snowy Owl," *Staten Island Advance*, January 5, 1946.
67. "Great Kills–Bay Terrace Residents Score Landfill," *Staten Island Advance*, June 11, 1946; "Garbage Dumping Score by Engineers," *Staten Island Advance*, June 9, 1946.
68. "Watkins Post Joins Battle on Landfill," *Staten Island Advance*, June 12, 1946; "PTA Protests Landfill Plan," *Staten Island Advance*, June 14, 1946; "Landfill Plan Scored by VFW Post," *Staten Island Advance*, June 18, 1946; "Wanted: Leadership!" *Staten Island Advance*, June 19, 1946; "Schick Unit Joins Garbage Protest," *Staten Island Advance*, June 19, 1946; "Dump Project Condemned by County Legion," *Staten*

Island Advance, June 22, 1946; "Landfill Scored by Beach Group," *Staten Island Advance*, June 22, 1946.
69. S. S. McSheehy, "Palma Joins in Protest," *Staten Island Advance*, June 11, 1946. Even Newbold Morris, a member of the City Planning Commission and former president of the City Council, pledged support to Staten Island. See "Morris Raps Dump," *Staten Island Advance*, June 13, 1946.
70. "Mayor Silent on Fresh Kills Dump Program," *Staten Island Advance*, June 15, 1946.
71. "'Clean Fill' to Be Used, He Declares," *Staten Island Advance*, June 12, 1946.
72. "As Islanders Stormed City Hall to Protest Proposed Dump," *Staten Island Advance*, June 13, 1946; "City Planners Due to Vote OK Tomorrow," *Staten Island Advance*, June 20, 1946.
73. "Proposal Now Up to Board of Estimate," *Staten Island Advance*, June 22, 1946. See also "A Preview of Fresh Kills?" *Staten Island Advance*, June 22, 1946.
74. Robert Moses to Editor, *Staten Island Advance* (cover letter), June 22, 1946, NYC Department of Parks, Office of the Commissioner, 1940–1956, Box 107883, Folder 41, *Staten Island Advance*, 1946, City of New York, Department of Records and Information Services, Municipal Archives.
75. Robert Moses to Editor, *Staten Island Advance*. See also "'No Other Way of Meeting Problem' Says Moses, Defending Kills Dump," *Staten Island Advance*, June 25, 1946; Harvey Call, "Staten Island to Fight Plan to Set Up Dump at Fresh Kills," *New York Sun*, June 25, 1946.
76. "What We're Getting," *Staten Island Advance*, June 24, 1946.
77. "A Gift to Staten Island: Ten Years of This?" *Staten Island Advance*, June 24, 1946.
78. "To Mr. Moses," *Staten Island Advance*, June 26, 1946.
79. The following all appeared in the *Staten Island Advance*: "Estimate Board OK's Dump," June 28, 1946; "Republicans Rap Dump, Ignoring Hall and Ruppell," June 28, 1946; "Landfill Plan Is Scored by VFW Group," June 27, 1946; "Halls Highway: Richmond Dump!" June 27, 1946; "Moose Lodge Criticizes Fresh Kills Dump Plan," June 27, 1946; "Does This Need 'Improving'?" June 26, 1946; "ALP Scores Fresh Kills Dump Plan," June 26, 1946; "Do You Want It?" June 25, 1946; "Board Set for Dump Discussion Thursday," June 25, 1946; "Dump Foes Map Protest," June 25, 1946; "Editorial," June 28, 1946. See also Corey, "King Garbage," 259.
80. "The Inquiring Photographer," *Staten Island Advance*, June 26, 1946; "How Far Are You?" *Staten Island Advance*, June 26, 1946.
81. "Buck Asks U.S. Aid to Preserve Fresh Kills," *Staten Island Advance*, June 29, 1946.
82. "Will Limit Fresh Kills Dump, Says Mayor," *Staten Island Advance*, June 29, 1946. See also "'Dislike Dumping,' Says Hall," *Staten Island Advance*, July 5, 1946.
83. "Dump Inspection by Councilman Is Planned," *Staten Island Advance*, June 29, 1946. See also "Will Fight for Hearing on Garbage, Says Schick," *Staten Island Advance*, July 1, 1946; "'The Silent Knight,'" *Staten Island Advance*, July 2, 1946.
84. "Jury in Richmond to Act on Dumping," *New York Times*, July 2, 1946; "The First Dumping," *Staten Island Advance*, July 3, 1946.
85. "Anti-Garbage Rally Is Called to Plan Fight," *Staten Island Advance*, July 5, 1946; "Council to Hear Garbage Protest," *Staten Island Advance*, July 2, 1946; "Islandwide Dump Protest to Be Drawn Up Monday: United Group Joins Fight," *Staten Island Advance*, July 6, 1946.
86. "Block That Dump!" *Staten Island Advance*, July 8, 1946. See also "Hall's Choice," *Staten Island Advance*, July 6, 1946. See also "Staten Islanders' War on 'Landfill,'" *New York Times*, July 10, 1946.

7. FUTILE PROTESTS

87. "'Am Going to Fight,' Is Pledge by Schick," *Staten Island Advance*, July 9, 1946; "1938–39 Landfill Fight Is Recalled," *Staten Island Advance*, July 9, 1946; "Staten Islanders War on 'Landfill,'" *New York Times*, July 10, 1946; "'Clean Waters' to Be Shown to Kiwanians," *Staten Island Advance*, July 10, 1946.
88. "An Uphill Struggle," *Staten Island Advance*, July 10, 1946.
89. Jack Reycraft, "Other Boroughs Need Site Here, Keegan Claims," *Staten Island Advance*, July 12, 1946.
90. Rosemary Bower and Tom Stevens, "Must Use Island Site, He Insists," *Staten Island Advance*, July 12, 1946; "Dr. Weinstein, Too," *Staten Island Advance*, July 29, 1946; "Health Commissioner Approves Methods Used by City," *Staten Island Advance*, July 24, 1946. The new health commissioner, Dr. Israel Weinstein, also was circumspect about the health hazard that Fresh Kills might pose.
91. "Jury in Richmond to Act on Dumping."
92. "Dump Quiz Slated for Tomorrow," *Staten Island Advance*, July 11, 1946; "S.I. Grand Jury Inquires Into Garbage Dump Plan," *Staten Island Advance*, July 11, 1946; "Officials Appearing Today for Dump Quiz," *Staten Island Advance*, July 12, 1946; "Grand Jury Summons Hall as Dump Probe Nears End; Moses and Powell Quizzed," *Staten Island Advance*, July 13, 1946.
93. "Jury in Richmond Fights Dump Plan," July 17, 1946.
94. "Dewey Gets Copy of Grand Jury's Attack on Dump," *Staten Island Advance*, July 17, 1946.
95. Quoted in "Jury in Richmond Fights Dump Plan," *New York Times*, July 17, 1946.
96. "Hall Repeats Charge Palma Planned Dump," *Staten Island Advance*, July 20, 1946; "Hall Blames the *Advance* for Great Kills Dumping," *Staten Island Advance*, July 23, 1946. See also "Declares Proposal Originated in Palma Regime," *Staten Island Advance*, July 16, 1946; "How Big a Dump?" *Staten Island Advance*, July 17, 1946.
97. "Palma Denies Sponsorship of Garbage Dump," *Staten Island Advance*, July 18, 1946. See also "Borough President Replies to Hall; Hits Project," *Staten Island Advance*, July 18, 1946; "Palma and Hall," *Staten Island Advance*, July 19, 1946.
98. Kramer and Flanagan, *Staten Island*, 45.
99. Moses said as much in a letter to Mayor O'Dwyer. Robert Moses to William O'Dwyer, August 20, 1946, NYC Department of Parks, Office of the Commissioner, 1940–1956, Box 107883, Folder 28, Sewage Disposal, Port Richmond, 1946, City of New York, Department of Records and Information Services, Municipal Archives. See Benjamin Miller, *Fat of the Land: Garbage of New York, the Last Two Hundred Years* (New York: Four Walls Eight Windows, 2000), 205–6. See also *Life After Fresh Kills: Moving Beyond New York City's Current Waste Management Plan: Policy, Technical, and Environmental Considerations* (New York: Prepared by Earth Institute, Earth Engineering Center, and the Urban Habitat Project at the Center for Urban Research and Policy, School of International and Public Affairs, Columbia University, December 1, 2001), A-1.
100. "Dumping Project—The Truth Emerges!" *Staten Island Advance*, July 18, 1946. See also "Marshland for Garbage Dump, State Gift, Covers Wide Area," *Staten Island Advance*, July 19, 1946.
101. "Mayor Gets Whiff of Land Smell," *New York Times*, July 19, 1946. See also "O'Dwyer Sniffs S.I. Dump, and Backs Another," *New York Herald-Tribune*, July 19, 1946.
102. "Schick Speaks on Garbage Fight Plans," *Staten Island Advance*, July 24, 1946; "Dewey Counsel Acknowledges Dump Protest," *Staten Island Advance*, July 27, 1946. See also "Garbage Dump Offers a Threat to Entire Island," July 24, 1946; "Land Opposed,"

7. FUTILE PROTESTS ■ 607

Staten Island Advance, July 29, 1946; "'One of the Most Heinous Things Ever Perpetrated,'" *Staten Island Advance*, July 26, 1946; "3-Point Plan to Fight Dump Is Suggested," *Staten Island Advance*, August 5, 1946: "Fach Wins Approval of Appeal to Governor," *Staten Island Advance*, August 23, 1946.

103. "4 Attorneys Are Named in Garbage War," *Staten Island Advance*, July 30, 1946; McCallum Promises Continued Dump War, *Staten Island Advance*, August 1, 1946.

104. "Industry Build on Landfill?—Not for Many Years," *Staten Island Advance*, July 20, 1946; "Land Values Are Affected by Dumping," *Staten Island Advance*, July 22, 1946; "On Garbage Dumping: 'Amounts to Building a Roof on Rats,'" *Staten Island Advance*, August 15, 1946.

105. "Buck Seeks N.J. Aid in Dump War," *Staten Island Advance*, July 26, 1946.

106. "N.Y. Disposal Plan Feared by Mayor," *Carteret Press*, August 2, 1946.

107. "Carteret Gets Order to Halt Pollution," *Staten Island Advance*, July 29, 1946; "Asks State to Oppose Fresh Kills Dump Plan," *Staten Island Advance*, August 3, 1946; "City Will Map Fight," *Staten Island Advance*, August 6, 1946; "Councilman Receives Resolution," *Staten Island Advance*, August 8, 1946; "Four Jersey Towns Join War on Dump," *Staten Island Advance*, August 8, 1946; "N.Y. Garbage Dump Opposite Sewaren Stirs Big Battle," *Raritan Township Fords Beacon*, August 8, 1946, and *Woodbridge Independent-Leader*, August 8, 1946; "Serious Threat to Carteret Health Faced in N.Y. Garbage Dumping Plan," *Carteret Press*, August 9, 1946.

108. "New Jersey Protests," *Staten Island Advance*, August 17, 1946.

109. "Five Towns and Jersey County to Fight Plan," *Staten Island Advance*, August 14, 1946; "Elizabeth Business Group Hits Dump Plan," *Staten Island Advance*, August 16, 1946; "Court Fight by Jersey Assured," *Staten Island Advance*, August 20, 1946; "Jersey Town Joins in City Dump Fight," *New York Times*, August 21, 1946; "New Jersey in the Ring," *Staten Island Advance*, August 22, 1946; "Bankers Denounce Dumping Despite O'Connell Defense; N.J. Legal Fight Develops," *Staten Island Advance*, August 23, 1946; "More Help Is Pledged Jersey in Dump War," *Staten Island Advance*, August 26, 1946; "Map Fight on Garbage Dump Plan," *Woodbridge Independent-Leader*, August 15, 1946; "5 Communities Join in Fight on Garbage," *Carteret Press*, August 16, 1946; "Question of Pollution by New York Dump Still Skirted, Although Other Objections Told Amply," *Woodbridge Independent-Leader*, August 29, 1946; "Equal Vigor in Battling Fumes, Garbage Dump, Civic Head Plea," *Carteret Press*, September 6, 1946.

110. "Van Riper Is Ducking Dump Fight," *Carteret Press*, October 2, 1946; "Turk Hopes to Pursue Dump Fight," *Carteret Press*, October 11, 1946.

111. "Dumps Called Menace," *Staten Island Advance*, August 13, 1946; "Fresh Kills Plans Get Board's OK," *Staten Island Advance*, August 23, 1946.

112. City of New York, Department of Sanitation, *Annual Report*, 1946, 28. Another 180,244 cubic yards of steam ashes were sold to contractors.

113. "Waste Disposal in New York City," July 18, 1946, 2, NYC Department of Parks, Office of the Commissioner, 1940–1956, Box 107884, Folder 22, Waste Disposal in New York City, 1946, City of New York, Department of Records and Information Services, Municipal Archives.

114. "Waste Disposal in New York City," 5. See also 2–4. Emphasis added.

115. "Waste Disposal in New York City," 6. In a draft of a letter to New Jersey's Senator Albert W. Hawkes, Moses outlined his landfill objectives for Fresh Kills. He added, "It would be gratifying to know that the New Jersey communities contributing to the unsanitary condition of the Harbor are ready to embark on a program of sanitation

and sewage disposal which would ultimately result in a clearing up of the New York and New Jersey shore lines." See Robert Moses to Albert W. Hawkes, August 1946, Roll 33, Folder 11, NYC Department of Parks, Office of the Commissioner, Fresh Kills, 1946, City of New York, Department of Records and Information Services, Municipal Archives.
116. "Waste Disposal in New York City," 6.
117. "Waste Disposal in New York City," 8.
118. Robert Moses to William O'Dwyer, August 20, 1946, NYC Department of Parks, Office of the Commissioner, 1940–1956, Box 107883, Folder 28, Sewage Disposal, Port Richmond, 1946, City of New York, Department of Records and Information Services, Municipal Archives.
119. For an example of Moses treating Powell like a functionary, see Robert Moses to William J. Powell, December 14, 1946, NYC Department of Parks, Office of the Commissioner, 1940–1956, Box 107883, Folder 20, Sanitation Department, 1946, City of New York, Department of Records and Information Services, Municipal Archives.
120. "New York to Incinerate Refuse; Cease Dumping," *Sewage Works Engineering* 17 (October 1946): 532; Corey, "King Garbage," 254–56.
121. City of New York, Department of Sanitation, *Annual Report, 1946*, 30.
122. City of New York, Department of Sanitation, *Annual Report, 1946*, 31. Of the eight landfills in operation at the beginning of 1946—two were in the Bronx, two in Brooklyn, three in Queens, and Great Kills in Richmond. One in Queens was at capacity in late October, with another to close in January 1947; the two landfills serving Brooklyn were likely to be completed in 1947, with two more in the Bronx likely to close in 1948. Edgemere in Queens still had 148 acres to fill, and Great Kills had about ninety acres but was filling fast.
123. "Staten Islanders Favor Secession," *New York Times*, January 6, 1947. See also "Staten Islanders Talk of Secession," *New York Times*, January 7, 1947; "Officials Answer Ferry Complaints," *Staten Island Advance*, January 10, 1947; "Island Secession Backed by Many, Says Radigan," *Staten Island Advance*, January 8, 1947; "Will Back Secession from City If Islanders Want It, Says Radigan," *Staten Island Advance*, January 3, 1947; "Concord Group Backs Radigan Secession Plan," *Staten Island Advance*, February 19, 1947, "Radigan Says Fight for Plan Will Continue," *Staten Island Advance*, January 13, 1947.
124. "Will Back Secession from City If Islanders Want It, Says Radigan."
125. "Secession?" *New York Times*, January 12, 1947. See also Walter B. Hayward, "Father Knickerbocker's Other Island," *New York Times*, February 2, 1947.
126. "Johnson, Ruppell, Hall, and Schick Oppose Secession," *Staten Island Advance*, January 11, 1947.
127. "Radigan Scores Fellow Assemblyman for Attack on Plan to Quit City," *Staten Island Advance*, January 11, 1947.
128. "Johnson, Ruppell, Hall, and Schick Oppose Secession."
129. "New Appeal Sent to Governor on Anti-Dump Bills," *Staten Island Advance*, April 5, 1947.
130. "Queens Men at Albany Open Fight on Dumping," *Staten Island Advance*, January 20, 1947; "New Attack on Dumping Is Launched," *Staten Island Advance*, January 13, 1947.
131. "Radigan Readies Secession Bill on Wednesday," *Staten Island Advance*, February 21, 1947; "Manhattan Firm Given 'Go' Sign on Dump Plant," *Staten Island Advance*, February 21, 1947.

132. "Manhattan Firm Given 'Go' Sign on Dump Plant"; "City Calls for Speed on Fresh Kills Work," *Staten Island Advance*, March 14, 1947; Steinberg, *Gotham Unbound*, 247.
133. "Schick Hits Radigan's Secession Proposal," *Staten Island Advance*, February 24, 1947; Diane C. Lore, "Depression, War Give Way to Era of Growth for Staten Island," *SILive.com*, March 26, 2011, http://www.silive.com/specialreports/index.ssf/2011/03/depression_war_give_way_to_era.html.
134. Daniel C. Walsh, "The Evolution of Refuse Incineration: What Led to the Rise and Fall of Incineration in New York City?" *Environmental Science & Technology* (August 1, 2002): 320A; Corey, "King Garbage," 262; "Johnson, Berge Pledge Dumping Fight Renewal," *Staten Island Advance*, April 14, 1947.
135. Richard Fenton, *Analysis of the Problem of Sanitary Landfills in New York City: A Report for the Bureau of Sanitary Engineering, New York City Health Department* (August 1947), 63. See also 56–57.
136. "City Studies Dump Site in Another Borough; Mayor Renews Pledge of Fresh Kills Limit," *Staten Island Advance*, January 31, 1947; "Alternate Dumping Site Is Branded Unsuitable," *Staten Island Advance*, February 20, 1947.

8. THE BURNING QUESTION

1. "All Americans Get Bids to City's Fete," *New York Times*, January 27, 1948. See also "West New York Ends Its Golden Jubilee; 200,000 Crowd Helps Review Big Parade," *New York Times*, October 3, 1948.
2. "50,000 Will March in Jubilee Parade," *New York Times*, April 26, 1948. See also "Greater New York Marks 50th Year," *New York Times*, January 2, 1948.
3. "City's Exposition for Jubilee Opens in Blaze of Lights," *New York Times*, August 22, 1948. See also Robert W. Potter, "Three Big Shows for City's Jubilee," *New York Times*, June 6, 1948; Francois Weil, *A History of New York* (New York: Columbia University Press, 2004), 259–61; "Exhibit to Display City's Government," *New York Times*, March 22, 1948.
4. "New York City, 1898–1948," *New York Times*, June 12, 1948. See also Grover A. Whalen, *Mr. New York: The Autobiography of Grover A. Whalen* (New York: G. P. Putnam's Sons, 1955), 300; "Politics Ruled Out in City's Jubilee," *New York Times*, April 8, 1948.
5. Richard Dickenson, ed., *Holden's Staten Island: The History of Richmond County* (New York: Center for Migration Studies, January 2003), 157; Ira Rosenwaike, *Population History of New York City* (Syracuse, NY: Syracuse University Press, 1972), 133.
6. "Staten Island Problem," *New York Times*, July 23, 1946. See also Joel Cohen, "To Secede or Not Is a Question Bitterly Argued by Islanders," *Staten Island Advance*, August 17, 1951; Joel Cohen, "We Can Join New Jersey All Right but How About $1.50 a Ferry Ride?" *Staten Island Advance*, August 14, 1951; Joel Cohen, "Radigan, Secession's 'Father,' Still Favors 1947 Proposition," *Staten Island Advance*, August 16, 1951.
7. See City of New York, Department of Sanitation, *A Series of Three-Year Reports . . . No. 15*, October 2, 1949, 8; City of New York, Department of Sanitation, Office of the Director of Engineering, *Waste Disposal Facilities Program*, Report no. 13 (December 1950), v.
8. "Dump Getting 1st Scowload of Garbage," *Staten Island Advance*, April 17, 1948. See also Ted Steinberg, *Gotham Unbound: An Ecological History of Greater New York, 1609–2012* (New York: Simon & Schuster, 2014), 247.

9. Jeffrey C. Chen, "Closure, Remediation, and Maturation: The Case of Freshkills Landfill," submitted for the 2011 annual meeting of the European Association of Environmental and Resource Economists, June–July 2011, 4. See also Steven Hunt Corey, "King Garbage: A History of Solid Waste Management in New York City, 1881–1970," PhD diss., New York University, 1994, 263; Daniel C. Kramer and Richard M. Flanagan, *Staten Island: Conservative Bastion in a Liberal City* (Lanham, MD: University Press of America, 2012), 115; "Dump Getting 1st Scowload of Garbage."
10. See City of New York, Department of Sanitation, *Annual Report, 1946*, 30.
11. Robert Moses to William O'Dwyer, August 12, 1948, and attached Memorandum, "Disposal of Waste Material Problem," NYC Department of Parks, Office of the Commissioner, 1948, Roll 52, Folder 29, City of New York, Department of Records and Information Services, Municipal Archives; "City Backs Moses on Garbage Plan," *New York Times*, August 20, 1948; "$44,000,000 Sought for Disposal Plan," *New York Times*, August 16, 1948.
12. Robert Moses to William O'Dwyer, August 12, 1948, NYC Department of Parks, Office of the Commissioner, 1948, Roll 52, Folder 29, City of New York, Department of Records and Information Services, Municipal Archives. See also Robert Moses, *Working for the People: Promise and Performance in Public Service* (New York: Harper & Brothers, 1956), 161.
13. City of New York, Department of Sanitation, Office of Engineering, "Refuse Disposal in New York City," by Casimir A. Rogus, paper presented before ASCE Sanitary Division, Metropolitan Section, New York City, February 3, 1954, 5–11. See also "Asks Reorganizing of Sanitation Dept.," *New York Times*, July 16, 1948; "Sanitation Set-Up Being Reorganized," *New York Times*, September 25, 1948.
14. City of New York, Department of Sanitation, Office of Engineering, Casimir A. Rogus, "Refuse Disposal in New York City."
15. S. Tannenbaum, "A Brief History of Waste Disposal in New York City Since 1930," unpublished paper, 1992, New York City, Municipal Archives.
16. See Richard Fenton, *Outline History of the Department of Sanitation* (New York: Department of Sanitation, 1954), 18; C. A. Rogus, "New York City Turns to Incineration," *Civil Engineering* 26 (December 1952): 55.
17. *Remarks of Robert Moses at the Celebration of the Seventy-Fifth Anniversary of the Staten Island Institute of Arts and Sciences* (New York: Department of Parks, April 13, 1957): 1. See also City of New York, Department of Parks, *The Reclamation of Park Areas by Sanitary Fill and Synthetic Top Soil* (October 9, 1950).
18. "Contractor's Bids Sought on New Jobs," *Staten Island Advance*, March 14, 1947; "Manhattan Firm Given 'Go' Sign on Dump Plant," *Staten Island Advance*, February 21, 1947; Steinberg, *Gotham Unbound*, 247.
19. "Contractor's Bids Sought on New Jobs." See also "Dump Getting 1st Scowload of Garbage."
20. "Contractor's Bids Sought on New Jobs."
21. Steinberg, *Gotham Unbound*, 248–49.
22. "More Land to Be Taken for Dump at Fresh Kills," *Staten Island Advance*, March 4, 1948. See also Jim Hughes, "Landfill Volume to Double on Jan. 1," *Staten Island Advance*, July 21, 1985; Steinberg, *Gotham Unbound*, 247.
23. New York City, Department of Sanitation, *Duties of District Superintendent, Special Training Course*, 1949, 24. The document went on to state that the estimate was "based on the assumption that the incinerator program will be completed in five years and

that the present landfills will be limited to the city-owned, Spring Creek Park, Marine Park and Ferry Point Park, and private landfills, College Point and Edgmere."

24. "8 Scows a Day Dump Garbage at Fresh Kills," *Staten Island Advance*, March 3, 1949.
25. "8 Scows a Day Dump Garbage at Fresh Kills." See also City of New York, *President of the Borough of Richmond, Annual Report, 1951*, 3; "7,250,000 Tons Dumped," *Staten Island Advance*, September 17, 1952.
26. An Athey Wagon was a large metal cart used to transport refuse from the barge-unloading facility to an active dump site.
27. City of New York, Department of Sanitation, *Annual Report, 1949*, 22–23. See also City of New York, Department of Sanitation, Office of Engineering, *Sanitary Landfills in New York City*, May 12, 1954; Maurice Feldman, "Sanitary Landfills: Swamps =Refuse+Engineering+Muscle=Parks," *Sweep* (Winter 1959): 7, 21; "Completion of New Plant at Fresh Kills Is Delayed," *Staten Island Advance*, May 1, 1950; "New Fresh Kills Unloader to Start Operations Soon," *Staten Island Advance*, March 14, 1950.
28. City of New York, *President of the Borough of Richmond, Annual Report, 1951*, 3.
29. City of New York, *President of the Borough of Richmond, Annual Report, 1951*, 3–6. See also Press Release, Department of Parks, "Richmondtown," June 30, 1952.
30. Expansion of the landfill and the use of old barges always came with a concern about rat infestation. Hall had sought the help of the Board of Estimate to contract rat exterminators at Fresh Kills. While the board authorized several new barges for Fresh Kills, James Lyons, the Bronx's borough president, facetiously offered Hall two cats. See "Fresh Kills Rats: Pied Pipers Suggested," *Staten Island Advance*, October 10, 1952.
31. "Planners OK Extension of Fresh Kills Landfill," *Staten Island Advance*, November 20, 1952; "Chamber Calls for Curb on Fresh Kills Landfill," *Staten Island Advance*, October 17, 1952; "Fresh Kills Dumping Hit by Sportsmen," *Staten Island Advance*, October 10, 1952; "City Faces Battle on Fresh Kills Land Grab," *Staten Island Advance*, September 18, 1952; "Public's Views on Fresh Kills Plan Sought," *Staten Island Advance*, September 24, 1952; George O. Ready, "Fight Asked on Spread of Garbage as Landfill," *New York Daily News*, October 2, 1952; "GOP Candidate Assails Fresh Kills Proposal," *Staten Island Advance*, October 2, 1952.
32. "Acquisition of Area for Landfill OK'd," *Staten Island Advance*, March 13, 1953. See also "Change Approved to Restore Town," *Staten Island Advance*, May 20, 1954. In 1957 the city sought state title to lands under water near the Fresh Kills site to expand the landfill area further, and it continued to look for additional space. See "City to Acquire Underwater Lands at Fresh Kills," *Staten Island Advance*, August 23, 1957; "New Sector Earmarked for Landfill," *Staten Island Advance*, January 23, 1959.
33. On O'Dwyer's promise, see "Landfill Plan Seen Bringing Island Relief," *Staten Island Advance*, January 31, 1947.
34. The two smallest Kissena Park (59 acres) and Lefferts (51 acres) were fully reclaimed by December 31, 1948. Department of Sanitation, *Annual Report, 1948*, 26.
35. Daniel C. Walsh, "The History of Waste Landfilling in New York City," *Ground Water* 29 (July–August 1991): 592.
36. The measure of success with landfilling for sanitation administrators and political leaders alike was the development of taxable land. In 1940, the city reclaimed more than 200 acres of marshland, in 1947 it increased to 414 acres, and in 1949, 690 acres. See Daniel C. Walsh, "Reconnaissance Mapping of Landfills in New York City," in *Proceedings: FOCUS Conference on Eastern Regional Ground Water Issues, National Ground Water Association*, Portland, ME, October 29–31, 1991, 394, https://info.ngwa.org/GWOL/pdf/910155209.PDF.

37. "Top 100 Historical Events: Staten Island, Richmond County, NY," March 1, 2014, *Rootsweb*, http://www.rootsweb.ancestry.com/~nyrichmo/history.shtml; "From the Archives: Trail of Broken Promises Litters History of Former Fresh Kills Landfill," September 26, 2013, *SILive.com*, http://www.silive.com/news/index.ssf/2013/09/from_the_archives_trail_of_bro.html.
38. For example, "Landfill Trucks Deepen West Shore Dust Bowl," *Staten Island Advance*, September 9, 1955.
39. Arleen McNamara Brahn, "Simple Childhood Memories," *Staten Island Historian* 17, (Winter–Spring 2000), New York Public Library, http://legacy.www.nypl.org/branch/staten/history/sihistorianwinterspring2000.cfm.
40. "Hansine Bowe: Freshkills Backyard," *DSNY Oral History Archive*, Spring 2012, http://www.dsnyoralhistoryarchive.org/?p=37.
41. "Fumes Blanket Wide Area as Garbage Burns," *Staten Island Advance*, April 29, 1949. See also "Gusts Spread Dump Blaze to 10 Scows," *Staten Island Advance*, April 10, 1950; "4 Land Crews, Fireboat Fight Garbage Blaze," *Staten Island Advance*, March 7, 1950; "Garbage Aflame at Greenridge," *Staten Island Advance*, March 31, 1950.
42. "Garbage Scow Fire Clouds Bay," *Staten Island Advance*, May 16, 1949; "Garbage Dump Blaze Sends Odorous Smoke Over Travis Section," *Staten Island Advance*, June 13, 1949; "Fighting Garbage Fire," *Staten Island Advance*, June 14, 1949.
43. "Offensive Odor Brings Complaints in Rossville," *Staten Island Advance*, May 28, 1949. See also "Improvements at Fresh Kills Plant Asked," *Staten Island Advance*, September 13, 1949. Not just on Staten Island but elsewhere in the late 1940s, New Yorkers living near landfills (including Jamaica Bay) protested and sought direct political action to curb them. See Corey, "King Garbage," 252.
44. City of New York, Department of Sanitation, *Annual Report, 1948*, 6. See also City of New York, Division of Analysis, Bureau of the Budget, *Department of Sanitation: Study of the Organizational Structure*, May 21, 1968, 6.
45. City of New York, Department of Sanitation, Office of the Director of Engineering, *Waste Disposal Facilities Program*, Report no. 13, December 1950, 1.
46. City of New York, Department of Sanitation, Office of the Director of Engineering, *Waste Disposal Facilities Program*, 1. See also "Contracts Speed Garbage Program," *New York Times*, April 29, 1949.
47. For Manhattan this meant modernizing its four existing plants and adding the new Gansevoort facility. For Brooklyn it required refurbishing three existing plants and adding three more; in the Bronx the one existing burner would require modernizing. The Flushing incinerator in Queens had to be upgraded and the Betts Avenue incinerator completed. In Richmond, the existing plant was obsolete, and a new one of sufficient capacity was in order to account for the growth of Staten Island, especially after completion of the proposed Verrazzano-Narrows Bridge.
48. City of New York, Department of Sanitation, Office of the Director of Engineering, *Waste Disposal Facilities Program* (draft), Report no. 13, December 1950, ix, 3, 5–9.
49. The marine loading or transfer station is a building or processing site for the temporary holding of refuse after truck collection from the source. The refuse is held there until it is loaded onto barges for shipment to Fresh Kills.
50. City of New York, Department of Sanitation, Office of the Director of Engineering, *Waste Disposal Facilities Program* (draft), 17–20.
51. Paul Crowell, "City's Pier Program Is Key to Capital Budget Problems," *New York Times*, August 16, 1948; Tannenbaum, "A Brief History of Waste Disposal in New York City Since 1930," 2.

52. "Great Kills Seen as 'New Jones Beach,'" *Staten Island Advance*, July 2, 1949.
53. Martin V. Melosi, *Garbage in the Cities: Refuse, Reform, and the Environment*, rev. ed. (Pittsburgh, PA: University of Pittsburgh Press, 2005), 177.
54. The 750-ton capacity Betts Avenue incinerator in Queens was under construction. The Gansevoort Market destructor in Manhattan, the West Ninth Street plant in Brooklyn, and the South Shore facility in Queens all would have 750-ton capacities as well. The new incinerator planned for Richmond would be smaller, at five hundred tons. Eleven other incinerators, ranging in capacity from one hundred to 750 tons, already were in operation. In 1949 the city's incinerators were operating at near capacity. See Department of Sanitation, *Annual Report, 1948*, 21–22; *Annual Report, 1949*, 19. See also "Contracts Speed Garbage Program."
55. Richard Fenton, *Outline History of the Department of Sanitation* (New York: Department of Sanitation, 1954), 18. See also New York City Mayor, *Department of Sanitation, a Series of Three-Year Reports*, no. 15, October 2, 1949.
56. City of New York, President of the Borough of Richmond, *Annual Report*, 1948, 29.
57. Beginning in 1950, the DSNY listed refuse in tons rather than cubic yards.
58. The column for "landfills" represented material disposed of by truck in city landfills (truck-fills) save Fresh Kills; the column for "marine unloading"(waterfront disposal) represented waste shipped to Staten Island.
59. George J. Lankevich, *American Metropolis: A History of New York City* (New York: New York University Press, 1998), 185–88; "Vincent R(ichard) Impellitteri," in *The Encyclopedia of New York City*, 2nd ed., ed. Kenneth T. Jackson (New Haven, CT: Yale University Press, 2010), 644.
60. Melosi, *Garbage in the Cities*, 186–87.
61. "Mulrain Takes Over as Sanitation Chief," *New York Times*, December 22, 1949.
62. George Sessions Perry, "Can This Man Clean Up New York?" *Saturday Evening Post* 228 (November 12, 1955): 32.
63. City of New York, Department of Parks, *The Reclamation of Park Areas by Sanitation Fill and Synthetic Top Soil*, October 9, 1950, 1. See also City of New York, Department of Parks, *20 Years of Progress*, 1934.
64. City of New York, Department of Parks, *The Reclamation of Park Areas by Sanitation Fill and Synthetic Top Soil*, 2. See also City of New York, Department of Parks, *Reclamation*, October 9, 1950.
65. Department of Sanitation, *Annual Report, 1950*, 20. See also Fenton, *Outline History of the Department of Sanitation*, 19.
66. "New Incinerator Design Standards Set in World's Largest Plant," *American City* 65 (April 1, 1950): 108.
67. "Garbage Dumping Seen Ended by 1953," *New York Times*, February 16, 1950.
68. "Landfills—Present and Future," in City of New York, Department of Sanitation, Office of the Director of Engineering, *Waste Disposal Facilities Program* (draft), Report no. 13, December 1950.
69. At the time, the city disposed of approximately 5.43 million tons of refuse and incinerator residue. Incineration was responsible for 1.4 million tons of refuse. Fresh Kills accounted for 1.78 million tons of refuse and incinerator residue. Of the 2.24 million tons of material disposed at other landfills (refuse and incinerator residue), about 1.38 million tons were disposed of at the three city-owned sites (Ferry Point Park, the Bronx; Marine Park, Brooklyn; and Spring Creek Park, Queens.) The remainder went to authorized private facilities.
70. "Landfills—Present and Future."

71. "Moses Would Keep Garbage Fill Plan," *New York Times*, May 21, 1951. See also Corey, "King Garbage," 266.
72. Henry Liebman, "Sanitary Landfill Handles 9½ Million Yards Yearly," *Public Works* 82 (July 1951): 39.
73. See, for example, "How Did We Get in New York and How Do We Get out?" *Staten Island Advance*, August 15, 1951.
74. City of New York, Borough President of Richmond, Construction Co-ordinator, Department of Sanitation, and Department of Parks, *Fresh Kills Landfill* (November 1951), i. The city undertook similar operations in Brooklyn, Queens, and the Bronx.
75. City of New York, Borough President of Richmond, Construction Co-ordinator, Department of Sanitation, and Department of Parks, *Fresh Kills Landfill*, i, 1.
76. AKRF, Inc., *Phase 1A Archaeological Documentary Study: Fresh Kills Park, Richmond County, New York* (New York, March 2008), IV-6; Robert Moses, *Public Works: A Dangerous Trade* (New York: McGraw-Hill, 1970), 31–33.
77. Even before the brochure was issued, the Parks Department, the City Planning Commission, and Borough President Hall had talked about transforming landfill areas into extensive parkland. See City of New York, President of the Borough of Richmond, *Annual Report, 1951*, 4. See also "Plan, Park Boards OK Project at Fresh Kills," *Staten Island Advance*, June 27, 1951; "Fresh Kills Area to Be Surveyed," *Staten Island Advance*, April 19, 1951.
78. City of New York, Borough President of Richmond, Construction Co-ordinator, Department of Sanitation, and Department of Parks, *Fresh Kills Landfill*, 4, 8, 12. See also Department of Parks, "Richmondtown," press release, June 30, 1952, Staten Island Historical Museum, Staten Island, New York.
79. City of New York, Borough President of Richmond, Construction Co-ordinator, Department of Sanitation, and Department of Parks, *Fresh Kills Landfill*, 12. See also "Big Opportunity for City Planning Is Seen in Landfill on Staten Island," *New York Times*, November 26, 1951.
80. Eugene M. Itgen to Adolph Klein, December 4, 1951, John J. Marchi Papers, Box 7, Folder 9, Staten Island, Fresh Kills, Archives and Special Collections, College of Staten Island, CUNY. See also "Robert Moses on Fresh Kills," April 16, 2009, *Fresh Kills Park Blog*, https://freshkillspark.wordpress.com/2009/04/16/robert-moses-on-fresh-kills/.
81. Robert Moses to Cornelius A. Hall, June 14, 1950, in *Report on Preliminary Plan for the Future Use of the Fresh Kills Area, Richmond, New York*, December 4, 1951, Archives & Special Collections, CSI Library, College of Staten Island.
82. Cornelius A. Hall to John J. Bennett, March 7, 1951, in *Report on Preliminary Plan for the Future Use of the Fresh Kills Area, Richmond, New York*. For Moses's plans for the Richmond Parkway and his larger goal of a "Belt Parkway" system, see Molly Greene, "A Mountain by the Sea: Waste-scapes, Life-scapes, and the Reinvention of Fresh Kills," 16, https://hixon.yale.edu/sites/default/files/files/fellows/paper/greene_molly_2012_report.pdf.
83. Intradepartmental memorandum, Eugene M. Itjen, December 4, 1951, in *Report on Preliminary Plan for the Future Use of the Fresh Kills Area, Richmond, New York*.
84. Intradepartmental memorandum, Eugene M. Itjen, December 4, 1951, in *Report on Preliminary Plan for the Future Use of the Fresh Kills Area, Richmond, New York*. See also Itgen to Adolph Klein, December 4, 1951.
85. Intradepartmental memorandum, Eugene M. Itjen, December 4, 1951, in *Report on Preliminary Plan for the Future Use of the Fresh Kills Area, Richmond, New York*. See

8. THE BURNING QUESTION   ■   615

    also Housing and Land Use Section, Division of Planning, Department of City Planning and City Planning Commission, *Report on Preliminary Plan for the Future Use of the Fresh Kills Area, Richmond, New York*, July 3, 1951, Marchi Papers, Folder 9, Box 7, Staten Island, Fresh Kills, Archives and Special Collections, College of Staten Island, CUNY.
86. *Report on Preliminary Plan for the Future Use of the Fresh Kills Area, Richmond, New York*, 3–6.
87. *Report on Preliminary Plan for the Future Use of the Fresh Kills Area, Richmond, New York*, 2. During 1951, the DSNY also started a process of realigning collection districts. See City of New York, Department of Sanitation, Bureau of Street Cleaning & Waste Collection, *A Report of the Sanitation District Realignment*, July 1951.
88. "What a Lot of Garbage!" *Staten Island Advance*, April 17, 1988; "Geiger Counter Searches for Radium in Garbage," *Staten Island Advance*, March 15, 1949; "Fresh Kills Radium Hunt Is Given Up," *Staten Island Advance*, March 15, 1949.
89. New York City, Department of Sanitation, *Annual Report, 1951*; City of New York, President of the Borough of Richmond, *Annual Report, 1951*.
90. "Conservationists on Staten Island Score Plan to Fill in Wildlife Area," *New York Times*, November 14, 1951.
91. "Clay Pit Ponds State Park Preserve," New York State Department of Environmental Conservation, http://www.dec.ny.gov/outdoor/84326.html; Kenneth M. Gold and Lori R. Weintrob, eds., *Discovering Staten Island: A 350th Anniversary Commemorative History* (Charleston, SC: History Press), 131–32.
92. "200 Acres at Claypits to Be Made Into Park," *Staten Island Advance*, March 21, 1951.
93. The Federation of Sportsmen and Conservationists called for the city to discontinue dumping altogether at Fresh Kills.
94. Filling the ponds also provided better access to sources of topsoil for the landfill.
95. "Conservationists on Staten Island Score Plan to Fill in Wildlife Area." See also Steinberg, *Gotham Unbound*, 250.
96. "Feathered Folk to Face Eviction," *New York Times*, November 15, 1951. See also "Filling in Big Pond Held Necessary to Dump Operation, *Staten Island Advance*, November 28, 1951; "Moses Aide Defends Filling Claypit Ponds," *Staten Island Advance*, November 15, 1951; "Moses Asks Support of Fresh Kills Park Plans," *Staten Island Advance*, January 9, 1952; "Hall Asks Assurance City Won't Fill Ponds," *Staten Island Advance*, March 11, 1952; "Claypit Ponds Agreement on Fill Expected," *Staten Island Advance*, March 14, 1952; "Moses Pushes Fresh Kills Lagoon Job," *Staten Island Advance*, March 20, 1952.
97. "Moses Asks Support of Fresh Kills Park Plans," *Staten Island Advance*, January 9, 1952.
98. "Hall Asks Assurance City Won't Fill Ponds," *Staten Island Advance*, March 11, 1952.
99. "Hall Asks Assurance City Won't Fill Ponds," *Staten Island Advance*, March 11, 1952.
100. "City Calls Conference on Plan for Fresh Kills," *Staten Island Advance*, March 12, 1952.
101. "Danger Seen from Rats in Garbage Fill," *Staten Island Advance*, March 15, 1952.
102. "Claypit Ponds Agreement on Fill Expected," *Staten Island Advance*, March 14, 1952.
103. "Moses Pushes Fresh Kills Lagoon Job," *Staten Island Advance*, March 20, 1952.
104. New York City, Department of Sanitation, *Annual Report, 1952*, 5; New York City, Board of Management Improvement, *Report on a Survey of the Department of Sanitation to Honorable Vincent R. Impellitteri*, December 1952, 3–5.
105. New York City, Department of Sanitation, *Annual Report, 1952*, 41.
106. New York City, Department of Sanitation, *Annual Report, 1952*, 25. See also 22.

107. "What a Lot of Garbage!"; AKRF, Inc., *Phase 1A Archaeological Documentary Study*, IV-6.
108. "City Called Filthiest in the World by Edna Ferber on Arrival Here," *New York Times*, April 22, 1953.
109. See, for example, "To Clean Up New York," April 24, 1953; and "Neglect of City Discussed," April 27, 1953, in Letters to the *Times*, *New York Times*; "World's Dirtiest City!" *New York Times*, April 23, 1953; "Miss Ferber Is Winning," *New York Times*, April 29, 1953.
110. "City Called Filthiest in the World by Edna Ferber on Arrival Here."
111. "Sanitation Cleanup Drive," May 3, 1952, WNYC Collection, New York City Municipal Archives, http://www.wnyc.org/story/sanitation-cleanup-drive/.
112. "World's Dirtiest City!" See also Thomas Jefferson Miley, Executive Vice President of Commerce and Industry Association of New York, to Robert F. Wagner, December 22, 1954, Department of Sanitation, Box 124, Folder 1430, Robert F. Wagner Documents Collection, La Guardia and Wagner Archives, LaGuardia Community College, Long Island City, New York, which speaks to the seriousness with which Ferber's words were taken.
113. "Miss Ferber Shies at Clean-Up Role," *New York Times*, April 28, 1953. See also "Four Years Behind the Broom," Annual Report of the Citizens Committee to Keep New York City Clean, Inc., 1958–1959, Department of Sanitation, Citz Comm. Keep NYC Clean, 1959, Box 127, Folder 1462, Robert F. Wagner Documents Collection; Paul R. Screvane, "New York City Is Cleaner," *American City* 74 (April 1959): 96–98; Corey, "King Garbage," 277–84.
114. "'It's Not a Dirty City,'" *New York Times*, October 21, 1953.
115. City of New York, Department of Sanitation, *Annual Report*, 1953, 17. See also 18.
116. City of New York, Department of Sanitation, *Annual Report*, 42–43; City of New York, Department of Sanitation, Office of Engineering, *D.S. Barges, Existing and Proposed*, Report no. 39, September 18, 1953, v.
117. City of New York, Department of Sanitation, *Annual Report*, 1953, 43.
118. Table V, City of New York, Department of Sanitation, Office of Engineering, *Refuse Disposal in New York City*, February 3, 1954, 16. The South Shore incinerator, serving the southern portion of Queens and parts of Brooklyn, opened August 2, and added one thousand tons of capacity per twenty-four hours. On November 26, the North Shore marine transfer station also opened in Queens. See New York City, Department of Sanitation, *Annual Report*, 1954, 11, 23.
119. Harrison Salisbury, "City Wages Constant Battle to Keep Streets Litter-Free," *New York Times*, December 6, 1954.
120. See, for example, "City Garbage Men Threaten a Tie-Up," *New York Times*, June 8, 1954.
121. "Saga of Sanita," *Motorship* 39 (1954): 21.
122. Casimir A. Rogus, "Management Engineering in the New York City Department of Sanitation," *Municipal Engineers Journal* 41 (1955): 21.
123. Rogus, "Management Engineering in the New York City Department of Sanitation," 30–32, 35.
124. Robin Nagle, "To Love a Landfill: Dirt and the Environment," in *The Filthy Reality of Everyday Life*, ed. Nadine Monem (London: Profile, 2011), 194.
125. Text of Speech—"Sanitation Week" Inaugural, May 24, 1954, Speeches Series, Box 060070W, Folder 42, Robert F. Wagner Documents Collection.
126. "Robert F(erdinand) Wagner," in *The Encyclopedia of New York City*, ed. Jackson, 1371; Weil, *A History of New York*, 282–83; Lankevich, *American Metropolis*, 189–92.

127. Corey, "King Garbage," 270–73.
128. Corey, "King Garbage," 270–73.
129. Rogus, "New York City Makes Teammates of Sanitary Fills and Incinerators," 114.
130. Andrew W. Mulrain to Robert F. Wagner, February 23, 1955; and City of New York, Department of Sanitation, Office of Operations, *Trade Waste Service*, February 1955, Department of Sanitation, 1/1955–6/1955, Box 124, Folder 1432, Robert F. Wagner Documents Collection. See also several documents in Department of Sanitation, Private Carting, 1956, Box 127, Folder 1468, Robert F. Wagner Documents Collection.
131. "Site History," Freshkills Park: The Freshkills Park Alliance, http://freshkillspark.org/the-park/site-history; New York City, Department of Sanitation, *Annual Report, 1955*, 27.
132. New York City, Department of Sanitation, *Annual Report, 1955*, 62–63.
133. The Flushing incinerator, which had been closed down for rehabilitation, reopened in November, and the older Seventy-Third Street incinerator was finalizing its rehabilitation. The new Greenpoint incinerator was under construction, and the Southwest Brooklyn plant came off the drawing board. The Betts Avenue plant was set for major expansion. New York City, Department of Sanitation, *Annual Report, 1956*, 43.
134. New York City, Department of Sanitation, *Annual Report, 1956*, 43; also 19, 45.
135. Douglas Martin, "Paul R. Screvane Dies at 87; Held Many Political Offices," *New York Times*, November 7, 2001; Biography of Paul Screvane, Box 1, Folder 1, 1961–1964, Robert F. Wagner Documents Collection; City of New York, Department of Sanitation, Office of Engineering, *Refuse Disposal in New York City*, February 3, 1954, 12–13. Screvane went on to have a robust administrative and political career in the early 1960s as City Council president, a mayoral candidate, and campaign manager for Wagner and for Robert F. Kennedy's Senate campaign.
136. Corey "King Garbage," 283–84.
137. "All Dumping Permits Canceled by Mayor," *New York Times*, June 17, 1957.
138. Robert F. Wagner to Paul R. Screvane, June 27, 1957, Department of Sanitation, Box 124, Folder 1436, Department of Sanitation, Private Carting, 1/1957–6/1957, Robert F. Wagner Documents Collection. See also Press Release, February 18, 1957, Box 124, Folder 1436, Department of Sanitation, Private Carting, 1/1957–6/1957, November 15, 1957, Robert F. Wagner Documents Collection. See also Corey, "King Garbage," 285–86; New York City, Department of Sanitation, *Annual Report, 1957–1958*, 54–55.
139. New York City, Department of Sanitation, *Annual Report, 1957–1958*, 23, 27–28, 69–70.
140. "Biographical Sketch," Marchi Papers, Archives and Special Collections, College of Staten Island, CUNY; "John J. Marchi, Who Fought for Staten Island in Senate, Dies at 87," *New York Times*, April 27, 2009.
141. Marchi Papers, District Office Papers, Box 7 (Fresh Kills Landfill, 1958, 1996–2004), Archives and Special Collections, College of Staten Island, CUNY. See also "Anti-Dump Bill Gains at Albany," *Staten Island Advance*, March 12, 1958. Other bills to end dumping at Fresh Kills were authored before 1958. For example, see "Legislators Seek '55 Dumping Ban," *Staten Island Advance*, January 27, 1953; "Russo Claims Dewey Favors Landfill Bill," *Staten Island Advance*, February 19, 1954.
142. Marchi Papers, Folder 4, Box 7, District Office (Fresh Kills Landfill, 1958, 1996–2004), Archives and Special Collections, College of Staten Island, CUNY.
143. Quoted in "From the Archives: Trail of Broken Promises Litters History of Former Fresh Kills Landfill." See also "Harriman OKs Island 'Voice' Bill, Rebuffs Russo's

Anti-Dump Plan," *Staten Island Advance*, April 15, 1958; Veto, 1958, Marchi Papers, Folder 4, Box 9, District Office, Archives and Special Collections, College of Staten Island, CUNY.

144. Quoted in "From the Archives: Trail of Broken Promises Litters History of Former Fresh Kills Landfill."
145. City of New York, Department of Sanitation, *Annual Report, 1959–1960*, 28–34; Henry Liebman, "World's Greatest Housekeeping Job," *Sweep* (Spring 1959): 16–17, 20–21; Rogus, "Incineration in New York City," 5, 26.
146. In 1962 the department had thirty-six furnaces in eleven incinerator plants.
147. Daniel C. Walsh, "The Evolution of Refuse Incineration: What Led to the Rise and Fall of Incineration in New York City?" *Environmental Science & Technology* 36 (August 1, 2002): 320A, 322A; Daniel C. Walsh et al., "Refuse Incinerator Particulate Emissions and Combustion Residues for New York City During the 20th Century," *Environmental Science & Technology* 35 (June 15, 2001): 2443.
148. City of New York, Department of Sanitation, *Annual Report, 1961–1962*, 22. The new commissioner was Frank J. Lucia.
149. Mayank Teotia, *Managing New York City Municipal Solid Waste Using Anaerobic Digestion* (Brooklyn: Programs for Sustainable Planning and Development, Pratt Institute, School of Architecture, Spring 2013), 6; Walsh, "The History of Waste Landfilling in New York City," 592; Walsh, "The Evolution of Refuse Incineration," 321A; Tannenbaum, "A Brief History of Waste Disposal in New York City Since 1930," 2; Richard Fenton, "Current Trends in Municipal Solid Waste Disposal in New York City," *Resource Recovery and Conservation* 1 (1975): 172; Casimir A. Rogus, "New York City Makes Teammates of Sanitary Fills and Incinerators," *American City* 70 (March 1955): 115.
150. Walsh, "The Evolution of Refuse Incineration," 321A. See also Corey, "King Garbage," 269–70.
151. Martin V. Melosi, "The Viability of Incineration as a Disposal Option: The Evolution of a Niche Technology, 1885–1995," *Public Works Management & Policy* 1 (July 1996): 37.
152. See Julie Sze, *Noxious New York: The Racial Politics of Urban Health and Environmental Justice* (Cambridge, MA: MIT Press, 2007), 59.
153. It should be remembered that incineration did not eliminate all waste that was burned, since ashes and other incineration residue had to be landfilled.
154. Stanley Terkelsen, "Park May Not Rise on Fresh Kills Landfill Before 1980," *Staten Island Advance*, October 24, 1961.
155. *FreshKills: Artists Respond to the Closure of the Staten Island Landfill*, Snug Harbor Cultural Center, Newhouse Center for Contemporary Art, October 14, 2001–May 27, 2002, 27.
156. See City of New York, Department of Sanitation, Office of Engineering, *Sanitary Landfills in New York City*, May 12, 1954. See also Feldman, "Sanitary Landfills," 7, 21.
157. City of New York, Borough President of Richmond, Construction Co-ordinator, Department of Sanitation, and Department of Parks, *Fresh Kills Landfill*, 1. See also "Fresh Kills Dumping of City's Raw Garbage to Continue 8 Years," *Staten Island Advance*, September 23, 1952.
158. "15 Years to Go for Dumping at Fresh Kills," *Staten Island Advance*, February 15, 1958.

159. See Paul E. Zindel, "Island of the Future," *Staten Islander Magazine* (August 1958): 11–13; "Staten Island: Enigma in the City," *Telephone Review* (September 1954): n.p.
160. Eldon P. Koetter, "Keep That Sanitary Landfill Sanitary," *American City* 71 (November 1956): 130.

9. THE END OF ISOLATION

1. Triborough Bridge and Tunnel Authority, press release, November 19, 1964, Verrazano-Narrows Bridge Collection, Archives & Special Collections, CSI Library, College of Staten Island, CUNY.
2. *Spanning the Narrows* (New York: Triborough Bridge and Tunnel Authority, 1964), 12, St. George Library Center, New York Public Library, Staten Island.
3. See this volume, chapter 3; Edward M. Young, *The Great Bridge: The Verrazano-Narrows Bridge* (New York: Farrar, Straus and Giroux, 1965), 1–26; Fred E. Weiss, "Bridge Openings: A Look Back and a Look Forward," *Staten Island Advance*, November 14, 1963; *Spanning the Narrows*, 16–17.
4. Gay Talese, "Staten Island Link to Sister Boroughs Is Opening Today," *New York Times*, November 21, 1964; "Othmar Hermann Ammann," *Encyclopedia.com*, http://www.encyclopedia.com/topic/Othmar_Hermann_Ammann.aspx.
5. "Fact Sheet: The Verrazano-Narrows Bridge," Triborough Bridge and Tunnel Authority, November 1964, 1–4, Verrazano-Narrows Bridge Collection, Archives & Special Collections, CSI Library, College of Staten Island. See also Young, *The Great Bridge*, 46, 100–1; Robert A. Caro, *The Power Broker: Robert Moses and the Fall of New York* (New York: Vintage, 1974), 843–44, 925–26; Jerry Adler, "The History of the Verrazano-Narrows Bridge, 50 Years After Its Construction, *Smithsonian Magazine*, November 2014, *Smithsonian.com*, http://www.smithsonianmag.com/history/history-verrazano-narrows-bridge-50-years-after-its-construction-180953032/.
6. "Fact Sheet: The Verrazano-Narrows Bridge," 3.
7. *Spanning the Narrows*, 20.
8. Gay Talese, *The Bridge: The Building of the Verrazano-Narrows Bridge* (New York: Bloomsbury, 2014), 20–22.
9. Gay Talese, "Two Families, Displaced Five Years Ago by Narrows Bridge, Assess Their New Homes: Anger Lingering as Bridge Goes Up," *New York Times*, June 19, 1964. See also Talese, *The Bridge*, 22–29.
10. Paul Goldberger, "Robert Moses, Master Builder, Is Dead at 92," On This Day, *New York Times*, July 30, 1981, http://www.nytimes.com/learning/general/onthisday/bday/1218.html; Ted Steinberg, *Gotham Unbound: An Ecological History of Greater New York, 1609–2012* (New York: Simon & Schuster, 2014), 253–54; Vincent J. Cannato, *The Ungovernable City: John Lindsay and His Struggle to Save New York* (New York: Basic Books, 2001), 94.
11. *Verrazano-Narrows Bridge Dedication, November 21, 1964* (New York: Triborough Bridge and Tunnel Authority, 1964), Verrazano-Narrows Bridge Collection, Archives & Special Collections, CSI Library, College of Staten Island.
12. Triborough Bridge and Tunnel Authority, press release, November 19, 1964. See also Gay Talese, "Verrazano Bridge Opened to Traffic," *New York Times*, November 22, 1964.
13. *Verrazano-Narrows Bridge Dedication, November 21, 1964*.

14. "A Message from the President of the United States," November 20, 1964, Verrazano-Narrows Bridge Collection, Archives & Special Collections, CSI Library, College of Staten Island. See also Talese, "Verrazano Bridge Opened to Traffic"; Talese, *The Bridge*, 91–106.
15. Kenneth M. Gold and Lori R. Weintrob, eds., *Discovering Staten Island: A 350th Anniversary Commemorative History* (Charleston, SC: History Press), 19, 94. See also Talese, "Verrazano Bridge Opened to Traffic."
16. "Remarks by Honorable Albert V. Maniscalco, President, Borough of Richmond at the Opening Day Ceremonies of the Verrazano-Narrows Bridge on Staten Island, November 21st, 1965," in *Verrazano-Narrows Bridge Dedication, November 21, 1964*.
17. Talese, "Verrazano Bridge Opened to Traffic"; Diane C. Lore, "Verrazano-Narrows Bridge Opens, Forever Altering Staten Island Landscape," *SILive.com*, March 26, 2011, http://www.silive.com/specialreports/index.ssf/2011/03/v-n_opens_forever_altering_isl.html.
18. "Foreword," in Young, *The Great Bridge*.
19. "Land Rush on the City's Last Frontier," *Life*, October 4, 1963, NY4.
20. Nadia H. Youssef, *Population Dynamics on Staten Island: From Ethnic Homogeneity to Diversity, Occasional Papers and Documentation* (New York: Center for Migration Studies, 1991), 1, 4, 7–12, John J. Marchi Papers, Secession, Box 27, Folder 5, Archives & Special Collections, CSI Library, College of Staten Island.
21. "Remarks by Mayor Robert F. Wagner at Dedication of Verrazano-Narrows Bridge," November 21, 1964, Verrazano-Narrows Bridge Collection, Archives & Special Collections, CSI Library, College of Staten Island.
22. "A Message from the President of the United States," November 20, 1964, Verrazano-Narrows Bridge Collection, Archives & Special Collections, CSI Library, College of Staten Island.
23. "Remarks by Honorable Albert V. Maniscalco, President, Borough of Richmond at the Opening Day Ceremonies of the Verrazano-Narrows Bridge on Staten Island, November 21st, 1965," Verrazano-Narrows Bridge Collection, Archives & Special Collections, CSI Library, College of Staten Island.
24. "Fact Sheet: The Verrazano-Narrows Bridge," 4.
25. Jeffrey C. Chen, "Closure, Remediation, and Maturation: The Case of Freshkills Landfill," submitted for the 2011 annual meeting of the European Association of Environmental and Resource Economists, June–July 2011), 2–4.
26. Laurence G. O'Donnell, "Staten Island's Land Boom and Growing Pains Intensify as New York City Prepares to Open Verrazano Bridge," *Wall Street Journal*, November 18, 1964.
27. *Spanning the Narrows*, 14. See also Weiss, "Bridge Openings."
28. Ann Marie Barron, "Stunning Facts Tell the Tale of Explosive Growth, Vast Transformation After Verrazano Opens," *SILive.com*, November 2014, http://www.silive.com/news/index.ssf/2014/11/facts_tell_the_tale_of_islands.html.
29. Gold and Weintrob, eds., *Discovering Staten Island*, 55. The mall had a negative effect on neighborhood shopping districts across Staten Island, hurting particularly the mom-and-pop shops. Phillip Papas and Lori R. Weintrob, *Port Richmond* (Charleston, SC: Arcadia, 2009), 17. See also New York City, Economic Development Corporation and New York City, Department of City Planning, *Staten Island North Shore Land Use and Transportation Study: Existing Conditions Report*, December 2008, 2-2.

9. THE END OF ISOLATION   ■   621

30. City of New York, Department of City Planning, cover letter by John P. Riley of Lockwood, Kessler, in *Proposed Staten Island Industrial Park: Feasibility Study* (August 1, 1962), St. George Library Center.
31. Cover letter, in *Proposed Staten Island Industrial Park: Feasibility Study*, 2.
32. A 1955–1956 land-use inventory stated that 23.5 percent (9,086 acres) of the gross space on the island was devoted to mapped streets and highways. Land classified as "vacant" (15,606 acres) accounted for 40.4 percent of the gross acreage. This, however, included several hundred acres of farmland and parking lots. Some of the marshlands on the western shore that were being filled would turn vacant land to industrial use. No other borough had so much vacant land as Richmond. Community Council of Greater New York, Bureau of Community Statistical Services, Research Department, *Staten Island Communities: Population Characteristics and Neighborhood Social Resources* (May 1960), xi–xii.
33. City of New York, City Planning Commission, *Staten Island Development: Policies, Programs, and Priorities*, Comprehensive Planning Report (June 1966), 1, Archives & Special Collections, CSI Library, College of Staten Island.
34. City of New York, City Planning Commission, *Staten Island Development*, 4–21.
35. Thomas W. Matteo, *Staten Island: I Didn't Know That!* (Virginia Beach: Donning, 2009), 146–47.
36. Alina Durkovic, "From Beach Resort to Bedroom Community: Staten Island and the Impact of the Verrazano-Narrows Bridge," unpublished MA thesis, Department of Urban Studies, Fordham University, May 2014, 55.
37. AKRF, Inc., *Phase 1A Archaeological Documentary Study: Fresh Kills Park, Richmond County, New York* (New York, March 2008), iv-6, v-4. See also *Verrazano-Narrows Bridge: Lower Level Roadway* (New York: Triborough Bridge and Tunnel Authority, 1969), St. George Library Center.
38. "12-Area 'City' of 300,000 Proposed for Staten Island," *New York Times*, May 14, 1970.
39. Jeffrey A. Kroessler, "The Limits of Liberal Planning: The Lindsay Administration's Failed Plan to Control Development on Staten Island," *Journal of Planning History* (2016): 1–22; New York City Planning Commission, *Towns in South Richmond: A Proposal for Controlled Growth on Staten Island*, NYC DCP 75-06 (April 1975), St. George Library Center; Alfonso A. Narvaez, "S.I. Development Bill Is Offered to Meet Residents' Objections," *New York Times*, January 31, 1973; William Huus, "Group Sees Danger to Environment," *Staten Island Advance*, January 31, 1973.
40. City of New York, City Planning Commission, *Staten Island Development*, 22.
41. City of New York, City Planning Commission, *Staten Island Development*, 22. See also Port of New York Authority, Regional Studies Section, *Staten Island: A Study of Its Resources and Their Development* (Staten Island Chamber of Commerce, January 1971), 2, Archives & Special Collections, CSI Library, College of Staten Island.
42. See, for example, "Richmond," *Sweep* (Autumn 1961): 6–7.
43. *Fresh Kills: Landfill to Landscape: International Design Competition: 2001*, https://www1.nyc.gov/assets/planning/download/pdf/plans/fkl/about_fkl.pdf. See also William Rathje and Cullen Murphy, *Rubbish! The Archaeology of Garbage* (New York: Harper Collins, 1992), 119; "How Fresh Kills Became Notorious Staten Island Landfill," March 26, 2011, *SILive.com*, http://www.silive.com/specialreports/index.ssf/2011/03/how_fresh_kills_became_notorio.html.
44. O'Donnell, "Staten Island's Land Boom and Growing Pains Intensify."

45. Ian McHarg, "What Would You Do with, Say, Staten Island?" *Natural History* 78 (April 1969): 28, Staten Island Historical Museum, Staten Island, New York.
46. S. Sloan Colt and Robert Moses, *New York City: Verrazano-Narrows Bridge*, n.d., Archives & Special Collections, CSI Library, College of Staten Island. See also "Staten Island Can be Promised Land—Moses," *Ogdensburg Journal*, September 5, 1965.
47. William S. Paley et al., *The Threatened City: A Report on the Design of the City of New York by the Mayor's Task Force* (New York: Mayor's Task Force on Urban Design, 1967), section 1, 15.
48. Ann Marie Barron, "Verrazano-Narrows Bridge at 50: Before 1964, Staten Island was Rural Oasis, a World Away from 'The City,'" *SILive.com*, http://www.silive.com/news/index.ssf/2014/11/pre-bridge_island_was_an_oasis.html.
49. John E. Rossi, "Encounter with Doc," *Staten Island Historian* 17 (Winter–Spring 2000), http://archive.is/e6BPS.
50. Plant #1 served as the administrative headquarters for Fresh Kills and also as the main repair facility.
51. *Fresh Kills: Landfill to Landscape: International Design Competition: 2001*.
52. City of New York, Department of Sanitation, Bureau of Administrative Services, *Statistical Review and Progress Report, June 1964 to June 1965*, table 4i. See also *Audit Report on the Operating Practices of the Bureau of Plant Maintenance, New York City Department of Sanitation and EPA's Payroll and Personnel Operations (Section 4) as of June 30, 1973*, report no. NYC-15-74 (section 4), 13, Municipal Reference and Research Center, New York City.
53. *FreshKills: Artists Respond to the Closure of the Staten Island Landfill*, Snug Harbor Cultural Center, Newhouse Center for Contemporary Art, October 14, 2001–May 27, 2002, 27; Robin Nagle, "To Love a Landfill: The History and Future of Fresh Kills," in *Handbook of Regenerative Landscape Design*, ed. Robert L. France (Boca Raton, FL: CRC, 2008), 10. See also "From the Archives: Trail of Broken Promises Litters History of Former Fresh Kills Landfill," *SILive.com*, September 26, 2013, http://www.silive.com/news/index.ssf/2013/09/from_the_archives_trail_of_bro.html.
54. Tonnage figures throughout New York City's history really are rough estimates depending on the accuracy of the scales, diligence in tracking all disposals, what was counted, and so forth. The figure of 2.2 million includes incinerator residue from DSNY incinerators. City of New York, Department of Sanitation, Bureau of Administrative Services, *Statistical Review and Progress Report, June 1964 to June 1965*, table 1, Municipal Reference and Research Center. See also *Statistical Review and Progress Reports* for fiscal years 1960–1961, 1961–1962, 1962–1963, and 1963–1964; and City of New York, Department of Sanitation, *Annual Report*, 1961–1962, 22–23, 65. "Amateur" or private haulers sometimes did not drive onto scales before dumping their loads. See Alfred G. Haggerty, "They Should Not Steal, Not Even in City Dump," *Staten Island Advance*, October 11, 1963.
55. S. S. McSheehy, "Fresh Kills Landfill Job Has Years to Go," *Staten Island Advance*, July 25, 1965.
56. McSheehy, "Fresh Kills Landfill Job Has Years to Go."
57. See Martin V. Melosi, *Garbage in the Cities: Refuse, Reform, and the Environment*, rev. ed. (Pittsburgh, PA: University of Pittsburgh Press, 2005), 201–15.
58. "Use of Landfill Area Urged to Cut Illegal Dumping," *Staten Island Advance*, May 10, 1969. See also "Islanders Piling It On at the Dump," *Staten Island Advance*, May 19, 1969; Christopher M. Cook, "Oil Cleanup to Move Faster," *Staten Island Advance*, September 3, 1970; "Red Tape Cut at Dump," *Staten Island Advance*, April 24, 1963;

9. THE END OF ISOLATION ■ 623

        S. S. McSheehy, "Tons of Road Patch Discarded in City Dump," *Staten Island Advance*, February 2, 1965.

59. Some islanders did not recognize the changing nature of the landfill space. In 1966 the DSNY called for police protection for its workers from hunters armed with shotguns seeking pheasant on the landfill property near New Springville. See S. S. McSheey, "Bird Hunters Open 'Season' on Sanitation Workers," *Staten Island Advance*, October 18, 1966.

60. In 1968 a city official suggested that a two-thousand-foot ski slope be constructed at a completed location at Fresh Kills. See "'Mt. Landfill' Ski Slope Eyed for Island," *Staten Island Advance*, October 30, 1968.

61. "What a Lot of Garbage!" *Staten Island Advance*, April 17, 1988; "Landfill Tract Proposed for Small Planes," *Staten Island Advance*, April 9, 1964; Drew Fetherston, "City Reveals Plans for Site in Fresh Kills," *Staten Island Advance*, August 30, 1966; "City Airport Due on Staten Island," *Staten Island Advance*, August 31, 1966; "Red Tape Snarls Airport Proposal," *Staten Island Advance*, September 4, 1966; "Airport Plan Could Nix Fresh Kills Landfill Project," *Staten Island Advance*, September 18, 1966; "Fresh Kills Airport Not Doomed," *Staten Island Advance*, September 18, 1966; Brian Morris, "New Airport May Be Open in 2 Years," *Staten Island Advance*, July 30, 1967; "Budget Cut Grounds Fresh Kills Airport," *Staten Island Advance*, February 21, 1968; Janice Goldstein, "Outlook Fading for Airport at Garbage Site," *Staten Island Advance*, January 23, 1970; "FAA to Decide in 2 Weeks If Fresh Kills Gets Airport," *Staten Island Advance*, August 7, 1970; "Air Space for Airport Is Approved," *Staten Island Advance*, August 30, 1970; Christopher Cook, "Landfill Idea Might Speed Airport Work," *Staten Island Advance*, September 24, 1970.

62. "City Urged to Add Garbage Plants," *New York Times*, October 30, 1963.

63. McSheehy, "Fresh Kills Landfill Job Has Years to Go"; S. S. McSheehy, "3 Years May Hit 30 for Fresh Kills Landfill," *Staten Island Advance*, February 4, 1965. See also "City Urged to Add Garbage Plants," *New York Times*, October 30, 1963.

64. See Gael Hummel, "Which Direction Fresh Kills?" *Staten Island Advance*, December 9, 1969. See also City of New York, Department of Sanitation, *Annual Report, 1959–1960*, 28–34; Henry Liebman, "World's Greatest Housekeeping Job," *Sweep* (Spring 1959): 16–17, 20–21; Rogus, "Incineration in New York City," 5, 26.

65. See "Board Discusses 500 Ft. Garbage Pile," *Staten Island Advance*, November 24, 1982.

66. Hummel, "Which Direction Fresh Kills?"

67. "Fresh Kills Channel to Be Deepened," *Staten Island Advance*, December 30, 1964; "Landfill Project Bids Exceed City Estimate," *Staten Island Advance*, January 9, 1965; "Sanitation Dept. Seeks to Dredge New Channel," *Staten Island Advance*, May 21, 1969.

68. Rod Such, "Landfill Plan to Determine Airport $$," *Staten Island Advance*, November 23, 1969. See also "Speed Island Airport," *Staten Island Advance*, August 11, 1968; Rod Such, "City Nixes Airport for Island—Again," *Staten Island Advance*, January 19, 1969; "Airport Pushed for Fresh Kills Landfill Site," *Staten Island Advance*, August 21, 1969; Janice Goldstein, "Landfill Plans May Block Airport," *Staten Island Advance*, January 20, 1970.

69. Meg Morgenstern, "Garbage Exports Won't End Need for Fresh Kills," *Staten Island Advance*, December 30, 1967.

70. S. S. McSheehy, "Garbage Island a Long Way Off but Officials Are Wary," *Staten Island Advance*, May 21, 1966. See also "Sanitation Chief on Tour of Island," *Staten Island*

Advance, July 5, 1968; "City Proposes Off-Island Dump," *Staten Island Advance*, July 7, 1967.
71. See chapter 10 for more details.
72. "Proposed Dumping Area Would Take 3,000 Acres," *Staten Island Advance*, November 14, 1969. See also S. S. McSheehy, "Sanit Brass Still Trying to Extend Dump," *Staten Island Advance*, November 1, 1970.
73. "Proposed Dumping Area Would Take 3,000 Acres," *Staten Island Advance*, November 14, 1969. See also "Landfill Forever?" *Staten Island Advance*, November 14, 1969; Lawrence Auster, "City Seeks Underwater Land for Dump," *Staten Island Advance*, March 4, 1970.
74. See "Proposed Dumping Area Would Take in 3,000 Acres," *Staten Island Advance*, November 14, 1969. See also Gael Hummel, "New Landfill Plan: Alps of Garbage," *Staten Island Advance*, December 7, 1969. See Michael Azzara, "Landfill Along the Beachfront? *Staten Island Advance*, November 30, 1939, about a proposal to enlarge Staten Island through an offshore landfill project from South Beach to Tottenville.
75. Hummel, "New Landfill Plan." See also Michael Azzara, "More Dumping Space Asked," *Staten Island Advance*, October 26, 1969. In 1974, the State Assembly opened the possibility of establishing "Recap Island" (about 300 to 1,500 acres) about three miles off of Great Kills. The island would be developed by a private firm as a refuse-disposal site. The project never came to pass. See "Garbage Isle Vote Nearing in Assembly," *Staten Island Advance*, April 11, 1974; "Marchi Vows He'll Hold Up Garbage Isle," *Staten Island Advance*, April 12, 1974; "Trash Island Looms Again," *Staten Island Advance*, February 21, 1975.
76. City of New York, Department of Sanitation, Bureau of Administrative Services, *Statistical Review and Progress Report, June 1964 to June 1965*, i.
77. Robert D. McFadden, "John V. Lindsay, Mayor and Maverick, Dies at 79," *New York Times*, December 21, 2000.
78. McFadden, "John V. Lindsay, Mayor and Maverick, Dies at 79."
79. "John V(liet) Lindsay," in *The Encyclopedia of New York City*, 2nd ed., ed. Kenneth T. Jackson (New Haven, CT: Yale University Press, 2010), 750–51; McFadden, "John V. Lindsay, Mayor and Maverick, Dies at 79." See also George J. Lankevich, *American Metropolis: A History of New York City* (New York: New York University Press, 1998), 200–24.
80. Walter G. Farr Jr., Lance Liebman, and Jeffrey S. Wood, *Decentralizing City Government: A Practical Study of a Radical Proposal for New York City* (New York: Praeger, 1972), 199.
81. Themis Chronopoulos, "The Lindsay Administration and the Sanitation Crisis of New York City, 1966–1973," *Journal of Urban History* 40 (November 2014): 1138–39. See also Mariana Mogilevich, "Designing the Urban: Space and Politics in Lindsay's New York," PhD diss., Department of Architecture, Harvard University, April 2012, 203–4.
82. Chronopoulos, "The Lindsay Administration and the Sanitation Crisis of New York City, 1966–1973," 1140.
83. Chronopoulos, "The Lindsay Administration and the Sanitation Crisis of New York City," 1139–42.
84. Chronopoulos, "The Lindsay Administration and the Sanitation Crisis of New York City," 1143–44.
85. Steven Hunt Corey, "King Garbage: A History of Solid Waste Management in New York City, 1881–1970," PhD diss., New York University, 1994, 312–13.

9. THE END OF ISOLATION ■ 625

86. Roger Kahn, "On the Brink of Chaos," *Saturday Evening Post* 27 (July 27, 1968): 54.
87. Field Maloney, "The Collectors," *New York Times*, March 13, 2005. See also Charles R. Morris, *The Cost of Good Intentions: New York City and the Liberal Experiment, 1960–1975* (New York: Norton, 1980), 87.
88. Quoted in Robert Decherd, "Steering a Tight Ship in a Sinking City: John DeLury," *Harvard Crimson*, March 25, 1972, http://www.thecrimson.com/article/1972/3/25/steering-a-tight-ship-in-a/.
89. Corey, "King Garbage," 314–15.
90. Kahn, "On the Brink of Chaos," 54.
91. Kevin Rice, *Dignity and Respect: The History of Local 831* (New York: Uniformed Sanitationmen's Association, Local 831, 2009), 89–95; Chronopoulos, "The Lindsay Administration and the Sanitation Crisis of New York City," 1144.
92. Quoted in Rice, *Dignity and Respect*, 96–97. See also Corey, "King Garbage," 316; and Kahn, "On the Brink of Chaos," 54. For a remarkable look at the strike from the perspective of the USA, with plenty of pictures, transcripts, and clippings, see Uniformed Sanitationmen's Association, "Nine Days That Shook New York City," *USA Record*, n.d.
93. Quoted in Kahn, "On the Brink of Chaos," 55. See also "Shots Are Fired in Refuse Strike: Filth Litters City," *New York Times*, February 5, 1968.
94. "City Seeks Fining of Sanitationmen as Strike Goes On," *New York Times*, February 6, 1968.
95. "Union Head Gets 15 Days in Strike of Garbage Men," *New York Times*, February 7, 1968. See also Damon Stetson, "Sanitation Union Guilty of Contempt in Strike—DeLury Enters Jail," *New York Times*, February 8, 1968; "Judge Won't Free DeLury for Talks on Garbage Pact," *New York Times*, February 15, 1968.
96. The sanitation crisis and the enmity between Rockefeller and Lindsay stirred partisan politics beyond the state. Governor Ronald Reagan of California, a contender and rival for the Republican presidential nomination in 1968, charged that Rockefeller was "treading on pretty thin ice when he starts to try and run the major cities." The former vice president Richard M. Nixon, a declared candidate for president, stated that the strike "was patently illegal, and it was called simply as a tactic to force a fatter settlement by extortionate means." He added, "To acquiesce in this kind of blackmail is to invite repetition not only in New York but in every great city in America." Nixon believed that Lindsay had acted correctly and Rockefeller had not. See Peter Kihss, "Reagan Suggests Rockefeller Errs," *New York Times*, February 14, 1968; Robert B. Semple Jr., "Nixon Backs Lindsay on Strike; Implies Rockefeller Was Wrong," *New York Times*, February 15, 1968.
97. Rice, *Dignity and Respect*, 97–116.
98. Rice, *Dignity and Respect*, 119.
99. Kahn, "On the Brink of Chaos," 59.
100. Chronopoulos, "The Lindsay Administration and the Sanitation Crisis of New York City," 1144; Corey, "King Garbage," 316–19; Morris, *The Cost of Good Intentions*, 102–6; Cannato, *The Ungovernable City*, 196–204; "Sanitation Tieup Closes Travis Unit," *Staten Island Advance*, November 8, 1968; "Wage Rise Brings End to City Incinerator Strike," *New York Times*, November 15, 1968. See also "Today in NYC History: The Great Garbage Strike of 1968," *Untapped Cities*, February 11, 2015, http://untappedcities.com/2015/02/11/today-in-nyc-history-the-great-garbage-strike-of-1968/; "1968 New York and Memphis: Sanitation Workers on Strike," *Workers World*, January 8, 2011, http://www.workers.org/2011/us/1968_sanitation_workers_0113/; "Sanitation Talks

Making Progress," *New York Times*, February 17, 1968; Damon Stetson, "Governor Presses Talks to End Garbage Walkout and Resists Using Guard," *New York Times*, February 10, 1968; Damon Stetson, "Garbage Strike Is Ended on Rockefeller's Terms," *New York Times*, February 11, 1968; "Text of the Governor's Message to the Legislature on City's Sanitation Emergency," *New York Times*, February 13, 1968; "The City's Responsibility," *New York Times*, February 14, 1968.

101. Decherd, "Steering a Tight Ship in a Sinking City"; "Governor Backed by Sanitationmen," *New York Times*, September 25, 1970; Cannato, *The Ungovernable City*, 431. Days after the New York garbage strike ended, on February 12, 1968, almost one thousand sanitation workers in Memphis refused to report to work. They demanded higher wages, safer working conditions, and recognition of their union. The predominantly African American force had experienced relentless tensions with city officials that now erupted into a strike. Rev. Martin Luther King Jr. agreed to come to Memphis and to lead a nonviolent demonstration in support of the sanitation workers. Almost two months after the start of the strike, on April 3, 1968, he returned to Memphis and delivered his last public speech. The following night he was assassinated. "Memphis Sanitation Workers Strike," *Civil Rights Digital Library*, August 13, 2015, http://crdl.usg.edu/events/memphis_sanitation_strike/; "Teaching with Documents: Court Documents Related to Martin Luther King, Jr., and Memphis Sanitation Workers," National Archives, http://www.archives.gov/education/lessons/memphis-v-mlk.

102. Peter Lawrence Shaw, "Public Service Delivery System Overload and Clientele Impact: An Examination of New York City Department of Sanitation Refuse Collection, 1959–1970," PhD diss., New York University, 1971, 33, 152–55; Farr, Liebman, and Wood, *Decentralizing City Government*, 199. See also David T. Brown, "The Legacy of the Landfill: Perspectives on the Solid Waste Crisis," in *Plastics Waste Management: Disposal, Recycling, and Reuse*, ed. Nabil Mustafa (New York: Marcel Dekker, 1993), 22; Melosi, *Garbage in the Cities*, 206.

103. Martin Shefter, *Political Crisis/Fiscal Crisis of New York City: The Collapse and Revival of New York City* (New York: Basic Books, 1985), 114, 145.

10. AN ENVIRONMENTAL TURN

1. "Island Landfill Deadline Urged," *Staten Island Advance*, July 27, 1969.
2. Paul Wilkes, "The Garbage Apocalypse," *New York 2* (March 10, 1969): 23.
3. Martin V. Melosi, *Garbage in the Cities: Refuse, Reform, and the Environment*, rev. ed. (Pittsburgh, PA: University of Pittsburgh Press, 2005), 168, 190–91.
4. Melosi, *Garbage in the Cities*, 200–2. See also Samantha MacBride, *Recycling Reconsidered: The Present Failure and Future Promise of Environmental Action in the United States* (Cambridge, MA: MIT Press, 2012), 52.
5. Charles G. Bennett, "Top Aide Is Named Sanitation Chief: Selection of Lucia and New Post for Screvane Seen as Bid for Italian Support," *New York Times*, March 3, 1961.
6. "Frank J. Lucia is Dead at 58; Ex-Commissioner of Sanitation," *New York Times*, May 3, 1970.
7. Lindsay had failed to persuade Periconi not to resign but retained him as a consultant on legislative matters. See "Periconi to Be Appointed Sanitation Commissioner," *New York Times*, December 28, 1965. See also "Kearing Is Named Sanitation Chief," *New York Times*, November 16, 1966. See also Woody Klein, *Lindsay's Promise: The Dream That Failed* (New York: Macmillan, 1970), 264, 267.

8. James L. Marcus took over from Kearing, pending the appointment of a new commissioner. Fiorvante G. Perrotta replaced Marcus as seat-holder in December 1967, and then Maurice M. Feldman was appointed acting commissioner later that month. Griswold L. Moeller got the position in June 1968 and survived in the post until he resigned in the summer of 1970. He was the seventeenth high-ranking official in the Lindsay administration to resign at that point. See Peter Lawrence Shaw, "Public Service Delivery System Overload and Clientele Impact: An Examination of New York City Department of Sanitation Refuse Collection, 1959–1970," PhD diss., New York University, 1971, 121; "New Sanitation Chief; Griswold Lamour Moeller," *New York Times*, June 5, 1968; "Moeller, Sanitation Chief, Quits; 17th Aide of Lindsay to Resign," *New York Times*, July 15, 1970. See also Steven Hunt Corey, "King Garbage: A History of Solid Waste Management in New York City, 1881–1970," PhD diss., New York University, 1994, 305.
9. Themis Chronopoulos, "The Lindsay Administration and the Sanitation Crisis of New York City, 1966–1973," *Journal of Urban History* 40 (November 2014): 5–7.
10. Citizen Budget Commission, Inc., *Decentralization in the Department of Sanitation: A Case Study in Administrative Reorganization*, August 1972, 1, Department of Records and Information Services, Municipal Archives, New York City, New York.
11. Citizen Budget Commission, Inc., *Decentralization in the Department of Sanitation*, 2–3.
12. Martin V. Melosi, "The Viability of Incineration as a Disposal Option: The Evolution of a Niche Technology, 1885–1995," *Public Works Management & Policy* 1 (July 1996): 37–38.
13. Daniel C. Walsh, "The Evolution of Refuse Incineration: What Led to the Rise and Fall of Incineration in New York City?" *Environmental Science & Technology* (August 1, 2002): 321.
14. David Stradling, *The Nature of New York: An Environmental History of the Empire State* (Ithaca, NY: Cornell University Press, 2010), 206–8; Bruce L. R. Smith, "The Politics of Air Pollution in New York City," *Interplay* 3 (1970): 4. See also Mayank Teotia, *Managing New York City Municipal Solid Waste Using Anaerobic Digestion* (Brooklyn: Programs for Sustainable Planning and Development, Pratt Institute, School of Architecture, Spring 2013), 6–7. See also Klein, *Lindsay's Promise*, 135, 159; Harvey Lieber, "The Politics of Air and Water Pollution Control in the New York Metropolitan Area," PhD diss., Department of Political Science, Columbia University, 1968, 94.
15. Smith, "The Politics of Air Pollution in New York City," 6; Corey, "King Garbage," 305–7; Hannah Fons, "One Man's Trash . . . Is Still Trash: A Look at Sanitation in New York City," September 2003, *The Cooperator*, http://cooperator.com/article/one-mans-trashis-still-trash. On Con Edison, see Joseph A. Pratt, *A Managerial History of Consolidated Edison, 1936–1981* (New York: Consolidated Edison Company of New York, 1988), 270–88.
16. David Bird, "City Incinerator Is Closed," *New York Times*, May 21, 1969.
17. "Kearing Is Named Sanitation Chief," *New York Times*, November 16, 1966. See also Klein, *Lindsay's Promise*, 264, 267.
18. Henry Raymont, "Mayor, on Cleanup Tour, Is Irked by Con Ed Smoke," *New York Times*, December 5, 1966.
19. "Kearing Is Named Sanitation Chief."
20. Corey, "King Garbage," 309–10.
21. See David Bird, "Kearing Sees a Crisis in Garbage Disposal," *New York Times*, June 19, 1967; "Kearing Is Named Sanitation Chief." See also Wilkes, "The Garbage Apocalypse," 23.

22. Sandra Freed, "Islanders Seek a Pied Piper," *Staten Island Advance*, June 4, 1967; George Ledwith, "Are Woods, Landfill Going to the Dogs?" *Staten Island Advance*, July 9, 1967.
23. Interstate Sanitation Commission, *Odor Survey in the New Jersey–Staten Island Area and New Jersey–Upper Manhattan Area* (1967), 23, Interstate Environmental Commission, http://www.iec-nynjct.org/archive.htm.
24. "Garbage Scow Spillage Cited in ISC Report," *Staten Island Advance*, January 24, 1969.
25. "Landfill Methods Hit as Potential Hazard," *Staten Island Advance*, June 2, 1969.
26. John Scullin Jr., "Mounting Problems Threatening to Overwhelm Sanitation Dept.," *Staten Island Advance*, October 29, 1967.
27. S. S. McSheehy, "City Has Only 8 Years Left to Dump Garbage Here," *Staten Island Advance*, March 14, 1967.
28. Scullin Jr., "Mounting Problems Threatening to Overwhelm Sanitation Dept."
29. Samuel J. Kearing Jr., "The Politics of Garbage," *New York* (April 1970): 29.
30. Kearing Jr., "The Politics of Garbage," 29, 30.
31. Kearing Jr., "The Politics of Garbage," 31, 32.
32. City of New York, Office of the Mayor, Department of Administration, *Refuse Collection: Department of Sanitation vs. Private Carting*, November 1970.
33. "Landfill Forever?" *Staten Island Advance*, March 2, 1969.
34. Bird, "Kearing Sees a Crisis in Garbage Disposal."
35. Waste-to-energy plants burn municipal solid waste (MSW) to produce steam in a boiler that is then used to generate electricity.
36. Corey, "King Garbage," 307–12; Bird, "City Incinerator Is Closed."
37. At the time, the emissions from municipal incinerators in New York City exceeded sixfold the new state standard.
38. By 1970 about one-third of the apartment buildings turned to refuse compactors to handle their waste before disposal.
39. Richard Fenton, "Current Trends in Municipal Solid Waste Disposal in New York City," *Resource Recovery and Conservation* 1 (1975): 172; Teotia, *Managing New York City Municipal Solid Waste Using Anaerobic Digestion*, 7, 20; Daniel C. Walsh, "The History of Waste Landfilling in New York City," *Ground Water* 29 (July–August 1991): 592; Walsh, "The Evolution of Refuse Incineration," 321; Fons, "One Man's Trash . . . Is Still Trash." The SW Brooklyn plant closed in 1991, the Betts Avenue facility in 1993, and the Greenpoint plant in 1994.
40. Richard Fenton, "Report on Present Status of Municipal Refuse Incinerators with Particular Reference to Problems Related to Nonresidential Refuse Input," unpublished paper presented at the Solid Waste Processing Division Meeting, January 29, 1975, revised May 1975, 49–51, 55. See also Rodney R. Fleming, "Solid-Waste Disposal: Part II—Incineration and Composting," *American City* 81 (February 1966): 94–95.
41. See Lieber, "The Politics of Air and Water Pollution Control," 84, 177–78, 496–97.
42. George J. Kupchik, "Solid Waste Management: The Metropolitan View," *American Journal of Public Health* 61 (February 1971): 359.
43. Melosi, *Garbage in the Cities*, 168.
44. *Cities and the Nation's Disposal Crisis: A Report of the National League of Cities and the U.S. Conference of Mayors Solid Waste Management Task Force* (Washington, DC, March 1973), 1.
45. *Waste Management: Generation and Disposal of Solid, Liquid, and Gaseous Wastes in the New York Region: A Report of the Second Regional Plan* (New York: Regional Plan Association, March 1968), 8.

46. *Waste Management*, 9. See also Ted Steinberg, *Gotham Unbound: An Ecological History of Greater New York, 1609–2012* (New York: Simon & Schuster, 2014), 253.
47. Clifford B. Knight, *Basic Concepts of Ecology* (New York: Macmillan, 1965), 2.
48. Donald Worster, *Nature's Economy* (New York: Cambridge University Press, 1977), 339–40; Eugene Odum, "Ecology as a Science," in *The Encyclopedia of the Environment*, ed. Ruth A. Eblen and William R. Eblen (Boston: Houghton Mifflin, 1994), 171. See also Robert Gottlieb, *Forcing the Spring: The Transformation of the American Environmental Movement*, rev. ed. (Washington, DC: Island, 2005).
49. Victor B. Scheffer, *The Shaping of Environmentalism in America* (Seattle: University of Washington Press, 1991), 4.
50. Scheffer, *Shaping of Environmentalism in America*, 113; Martin V. Melosi, *Coping with Abundance: Energy and Environment in Industrial America* (New York: Knopf, 1984), 296–97.
51. Gottlieb, *Forcing the Spring*, 43ff.
52. Quoted in Melosi, *Coping with Abundance*, 297. See also Adam Rome, *The Genius of Earth Day: How a 1970 Teach-In Unexpectedly Made the First Green Generation* (New York: Hill and Wang, 2014).
53. Wallis E. McClain Jr., ed., *U.S. Environmental Laws: 1994 Edition* (Washington, DC: Bureau of National Affairs, 1994), 9–1; Melosi, *Coping with Abundance*, 297–98; Gottlieb, *Forcing the Spring*, 129–57.
54. Richard N. L. Andrews, "Environmental Protection Agency," in *Conservation and Environmentalism*, ed. Robert Paehlke (New York: Garland, 1995), 256; Gottlieb, *Forcing the Spring*, 134–35; Joseph Petulla, *Environmental Protection in the United States* (San Francisco: San Francisco Study Center, 1987), 48–49.
55. David Rogers, "Management Versus Bureaucracy," in *Summer in the City: John Lindsay, New York, and the American Dream*, ed. Joseph P. Viteritti (Baltimore, MD: Johns Hopkins University Press, 2014), 107.
56. Still the workforce was too small for the monumental tasks it faced. See Corey, "King Garbage," 319; "City Ends Freeze on Police, Fire, Sanitation Jobs," *New York Times*, November 1, 1972; "'Unbelievable,' New Sanitation Chief Says of Dirty Street," *New York Times*, June 11, 1968; "Pollution: New York Has It All," *Des Moines Register*, February 19, 1970.
57. Robert F. Keeler, "City Landfill Sites Facing a Crisis," *Staten Island Advance*, February 15, 1970.
58. NYEPA was established in January 1968 by Local Law no. 3.
59. A new section on noise abatement also was added.
60. Corey, "King Garbage," 320–21. See also Thelma E. Smith, ed., *Guide to the Municipal Government of the City of New York*, 10th ed. (New York: Meilen, 1973), 203, St. George Library Center; Joseph P. Viteritti, "Times a-Changin': A Mayor for the Great Society," in *Summer in the City*, ed. Viteritti, 16–17; Peter Lawrence Shaw, "Public Service Delivery System Overload and Clientele Impact: An Examination of New York City Department of Sanitation Refuse Collection, 1959–1970," PhD diss., New York University, 1971, 83, 85; Elizabeth Howe, "Where the Lindsay Reorganization Worked," *New York Affairs* 5 (Spring 1979): 47.
61. Wolfgang Saxon, "Merril Eisenbud, 82, Safety Expert Known for Work on Atomic Energy," *New York Times*, August 21, 1997.
62. Merril Eisenbud, "Environmental Pollution and Its Control," *Bulletin of the New York Academy of Medicine* 45 (May 1969): 447, 450.
63. Eisenbud, "Environmental Pollution and Its Control," 451.

64. David Bird, "Environment Superagency Asks City for Half Billion for Projects," *New York Times*, October 30, 1968.
65. Chronopoulos, "The Lindsay Administration and the Sanitation Crisis of New York City," 1146; Shaw, "Public Service Delivery System Overload and Clientele Impact," 267; Smith, ed., *Guide to the Municipal Government of the City of New York*, 212.
66. John V. Lindsay, *The City* (New York: Norton, 1969), 63.
67. Robert D. McFadden, "John V. Lindsay, Mayor and Maverick, Dies at 79," *New York Times*, December 21, 2000. See also Cannato, *The Ungovernable City*, 401–2, 408–9, 414, 416–18, 434, 437.
68. Lindsay, *The City*, 21.
69. "More Dumping Space Asked," *Staten Island Advance*, October 26, 1969.
70. S. S. McSheey, "BP Wants Cutoff Date for Dumping at Landfill," *Staten Island Advance*, October 28, 1969; "Landfill's Not Real Answer," *Staten Island Advance*, October 29, 1969.
71. "Proposed Dumping Area Would Take 3,000 Acres," *Staten Island Advance*, November 14, 1969. See also S. S. McSheehy, "Sanit Brass Still Trying to Extend Dump," *Staten Island Advance*, November 1, 1970; "Landfill Forever?" *Staten Island Advance*, November 14, 1969.
72. S. S. McSheehy, "Protests by 'Landfill Borough' Fall on Deaf Ears," *Staten Island Advance*, November 26, 1969. See also "Unity Could End Dumping," *Staten Island Advance*, November 26, 1969; "Landfill Plan Hit by Civic Group," *Staten Island Advance*, December 13, 1969.
73. Grace Lichtenstein, "Running for Mayor on a Garbage Truck," *New York Times*, April 27, 1971.
74. Kretchmer attended NYU and Columbia Law School and was the leader of a Democratic reform effort on the Upper West Side in 1960. In 1962 he began four terms in the New York State Assembly, promoting a variety of social programs. He gained the backing of Robert Kennedy and Lindsay, whose reelection he helped secure in 1969. Lichtenstein, "Running for Mayor on a Garbage Truck."
75. Lichtenstein, "Running for Mayor on a Garbage Truck."
76. Samantha MacBride, *Recycling Reconsidered: The Present Failure and Future Promise of Environmental Action in the United States* (Cambridge, MA: MIT Press, 2012), 54–55, 61–68; Morris, *The Cost of Good Intentions*, 160–62; Chronopoulos, "The Lindsay Administration and the Sanitation Crisis of New York City," 1146–48. See also Peter Kihss, "City to Test Private Pickups of Refuse," *New York Times*, February 7, 1972. On April 22, 1970, Earth Day legislation was signed creating the New York State Department of Environmental Conservation (DEC). It was one of the first governmental agencies on the state level formed to oversee all environmental issues through one body. Duties of the Conservation Department were melded with programs from the Department of Health and some state commissions. The DEC began operations on July 1, 1970. "History of DEC," New York State Department of Environmental Conservation, 2015, http://www.dec.ny.gov/about/9677.html.
77. "Garbage Collection Poses Mounting Political Problem for Lindsay," *New York Times*, July 26, 1970. See also "4 Community Groups Criticize Garbage Collections as 'Sporadic,'" *New York Times*, August 6, 1970.
78. Pyrolysis was a process of thermal degradation of refuse in the absence of air that produced recyclable products, including char, oil/wax, and combustible gases. See Dezhen Chen et al., "Pyrolysis Technologies for Municipal Solid Waste: A Review," *Waste Management* 34 (December 2014): 116. See also David O. Stewart, "Disposal Plant Studied for Pralls Island," *Staten Island Advance*, July 5, 1970.

10. AN ENVIRONMENTAL TURN ■ 631

79. Peter Kihss, "City Is Urged to Get Rid of Waste in New Ways," *New York Times*, November 30, 1970; Reginald Patrick, "Report Faults Incineration of Garbage," *Staten Island Advance*, November 30, 1970; S. S. McSheehy, "Mammoth Incinerator Is Planned for Island," *Staten Island Advance*, November 13, 1969; Kupchik, "Solid Waste Management," 359, 361.
80. Edward C. Burks, "Hill of Landfill Backed for S.I.," *New York Times*, December 1, 1970. See also "Rising Trash," *Syracuse Herald-Journal*, February 17, 1971.
81. The DSNY also regularly was seeking funds to modernize the facilities at Fresh Kills. See McSheehy, "Sanit Brass Still Trying to Extend Dump"; "City Again Seeking Bids for New Landfill Plant," *Staten Island Advance*, April 12, 1971.
82. S. S. McSheehy, "Planners Say Pile It Higher at Fresh Kills," *Staten Island Advance*, December 1, 1970; "BP Threatens Suit to Halt Dumping Plan," *Staten Island Advance*, December 3, 1970; "City Rejects Proposal on Waste Study," *Staten Island Advance*, December 29, 1970; S. S. McSheehy, "Garbage Study Plan Is Accepted by EPA," *Staten Island Advance*, March 1, 1971.
83. William Huus, "$82,000 Landfill Study Ordered," *Staten Island Advance*, August 18, 1971. See also "City to Open Fresh Kills Bids May 4," *Staten Island Advance*, April 21, 1971; "Fresh Kills Bids Put Off Again," *Staten Island Advance*, May 7, 1971; "L.I. Firm Gets Fresh Kills Job," *Staten Island Advance*, May 19, 1971.
84. One estimate surmised that over the years 160 sites, totaling more than 7,000 acres, had been filled by the DSNY. Kupchik, "Solid Waste Management," 361.
85. David Bird, "Landfill Crisis Nearing for City," *New York Times*, March 5, 1972. See also C. Gerald Fraser, "Kretchmer Shows Off City's Garbage," *New York Times*, March 26, 1972.
86. "Kretchmer Quits His Post to 'Explore' Mayoral Bid," *New York Times*, February 6, 1973. See also "Zealous Sanitation Commissioner: Herbert Elish," *New York Times*, April 27, 1971; "Garbage Crisis Feared in Bronx," *New York Times*, September 27, 1972; Rogers, "Management and Bureaucracy," 129–30.
87. Herbert Elish, "The Crisis in Solid Waste Disposal," *New York Affairs* 1 (1973): 95. See also Ned Houseman, "Ways Studied for Disposing of Garbage," *Staten Island Advance*, December 23, 1973; Polly Kummel, "Bound for the Old Waste-land," *Staten Island Advance*, February 14, 1973.
88. Elish, "The Crisis in Solid Waste Disposal," 96.
89. Elish, "The Crisis in Solid Waste Disposal," 96–104.
90. Marty Lipp, "What a Lot of Garbage! 40 Years Later, Landfill Keeps Growing," *Staten Island Advance*, April 17, 1988.
91. "City Has Only 8 Years Left to Dump Garbage Here," *Staten Island Advance*, March 14, 1967.
92. Hummel, "New Landfill Plan: Alps of Garbage." See also Bird, "City Incinerator is Closed"; Robert Sullivan, "Wall-E Park," November 23, 2008, *NYmag.com*, http://nymag.com/nymag/sponsored/green-design/52452/index2.html; Chen, "Closure, Remediation and Maturation: The Case of Freshkills Landfill," 4; Steinberg, *Gotham Unbound*, 254; "'Mt. Landfill' Ski Slope Eyed for Island," *Staten Island Advance*, October 30, 1968; "Forget Those Trash Hills," *Staten Island Advance*, July 29, 1969; Gael Hummel, "Which Direction Fresh Kills?" *Staten Island Advance*, December 9, 1969; "Landfill Plans May Block Airport," *Staten Island Advance*, January 20, 1970.
93. "Let's Keep It on the Level," *Staten Island Advance*, December 10, 1969.
94. "Mount Garbage," *Staten Island Advance*, January 25, 1970.
95. S. S. McSheehy, "Sanitation Boss Wants More Men for Island," *Staten Island Advance*, May 12, 1967; "$2.6 Million Asked for Landfill Work," *Staten Island Advance*,

October 19, 1967; "Landfill Project Funds Doubled," *Staten Island Advance*, November 10, 1967. See also E. Beltrami, N. Bhagat, and L. Bodin, *Refuse Disposal in New York City: An Analysis of Barge Dispatching*, USE Technical Report 71-10 (Stony Brook, NY: Urban & Policy Sciences Program, SUNY Stony Brook, July 1971).

96. Robin Nagle, "To Love a Landfill: The History and Future of Fresh Kills," in *Handbook of Regenerative Landscape Design*, ed. Robert L. France (Boca Raton, FL: CRC, 2008), 10; Matthew Gandy, *Recycling and the Politics of Urban Waste* (London: Earthscan, 1994), 75; *Fresh Kills: Landfill to Landscape: International Design Competition: 2001*, https://www1.nyc.gov/assets/planning/download/pdf/plans/fkl/about_fkl.pdf.
97. "From the Archives: Trail of Broken Promises Litters History of Former Fresh Kills Landfill," *SILive.com*, September 26, 2013, http://www.silive.com/news/index.ssf/2013/09/from_the_archives_trail_of_bro.html.
98. Lipp, "What a Lot of Garbage!"
99. See Bill Breen, "Landfills Are #1," *Garbage* 2 (September/October 1990): 45; William D. Ruckelshaus, "Solid Waste Management: An Overview," *Public Management* (October 1972): 1.
100. "Rising Costs Hit Landfill Construction," *Staten Island Advance*, June 22, 1970.
101. "Brass Section," *Sweep* (Fall 1971): 35; "Department of Sanitation, Bureau of Waste Disposal, Marine Division," *Tow Line* (Fall 1971): 7, both in the Cheryl Criaris-Bontales Collection, Holmdel, New Jersey; Cheryl Bontales to Martin Melosi, August 31, 2015; Martin Kruming, "Danger Is Their Daily Co-Worker," *Staten Island Advance*, May 2, 1969.
102. "Brass Section," 35.
103. "Department of Sanitation, Bureau of Waste Disposal, Marine Division," 7, 12–13; Lawrence De Maria, "Garbage Proves a Point," *Staten Island Advance*, November 3, 1972.
104. Kruming, "Danger Is Their Daily Co-Worker."
105. Inter-Departmental Correspondence, Department of Sanitation, William Criaris, Gen. Supt. MU Fresh Kills, Richmond to Arthur Price, Director, November 18, 1974, Cheryl Criaris-Bontales Collection.
106. Michael Belknap to William Criaris, June 11, 1971, Cheryl Criaris-Bontales Collection.
107. Linda Keller Brown, Rutgers University to William Criaris, April 11, 1974; Cynthia Jacobson, High Rock Park Conservation Center to William Criaris, May 1, 1971; Michael Bergman, Department of Sanitation to William Criaris, September 23, 1970, Cheryl Criaris-Bontales Collection.
108. Lieutenant General Edward J. O'Neill to Frank J. Lucia, Commissioner, Department of Sanitation, August 23, 1961; Robert E. Condon, Director of the Office of Civil Defense to Frank J. Lucia, Commissioner of Sanitation, October 26, 1964, Cheryl Criaris-Bontales Collection.
109. "Brass Section."
110. Sister Mary Agnes, Countess Moore High School to William Criaris, June 25, 1970; Rt. Rev. Msgr. George H. Guilfoyle, Director, Catholic Charities, to Frank J. Lucia, Commissioner, Department of Sanitation, November 19, 1962; Sr. Mary Genevieve, St. Michael's Home to William Criaris, November 3, 1962; Warren Williamsen, Willowbrook Development Center to Robert Groh, Commissioner, Department of Sanitation, July 23, 1975, Cheryl Criaris-Bontales Collection.
111. For other examples of workers' experiences at Fresh Kills, see Mark E. Hanley, "Sanitation Crane Operator Calls His Work 'Challenging,'" *Staten Island Advance*,

November 17, 1977; Janice Kabel, "Fresh Kills: For 300 Men, It's a Clean Living," *Staten Island Advance*, October 3, 1978.

112. SIGNAL, *Wetlands Aren't Wastelands: A Position Paper on the Present and Future Uses of Marshes* (1970), Staten Island Historical Museum. See also Steinberg, *Gotham Unbound*, 254, 256–57.
113. SIGNAL, *Wetlands Aren't Wastelands*.
114. SIGNAL, *Wetlands Aren't Wastelands*.
115. "Some Fear Race to Save Wetlands Is Lost," *Staten Island Advance*, July 18, 1967. See also "Tidal Wetlands," Department of Environmental Conservation, https://www.dec.ny.gov/lands/4940.html, especially concerning the 1974 Tidal Wetlands Inventory.
116. Michael Azzara, "Breakthrough Sought in Garbage Disposal Problem," *Staten Island Advance*, July 17, 1967.

11. FISCAL CRISIS AND DISPOSAL DILEMMA

1. Norman Steisel to Edward I. Koch, March 22, 1979, Departmental Correspondence Series, Edward I. Koch Collection, Box 252, Folder 1, April 1979–July 1979, La Guardia and Wagner Archives, LaGuardia Community College/CUNY, Long Island City, New York. See also oral interview, Martin Melosi with Norman Steisel, November 15, 2017.
2. James Trager, *The New York Chronology* (New York: HarperResource, 2003), 722.
3. Michael Stern, "City's Fiscal Crisis: The Guarantees Against Defaulting," *New York Times*, April 11, 1975. See also Roger E. Alcaly and David Mermelstein, eds., *The Fiscal Crisis of American Cities: Essays on the Political Economy of Urban America with Special Reference to New York* (New York: Vintage, 1977), 30–31; Sam Roberts, "When the City's Bankruptcy Was Just a Few Words Away," *New York Times*, December 31, 2006.
4. The material on the general urban crisis is adapted from Martin V. Melosi, *The Sanitary City: Urban Infrastructure in America from Colonial Times to the Present* (Baltimore, MD: Johns Hopkins University Press, 2000), 357. See also Howard Chernick and Andrew Reschovsky, "Urban Fiscal Problems," in *The Urban Crisis*, ed. Burton A. Weisbrod and James C. Worthy (Evanston, IL: Northwestern University Press, 1997), 132, 135–36; Jon C. Teaford, *The Metropolitan Revolution* (New York: Columbia University Press, 2006), 167–84.
5. Daniel T. Lichter and Martha L. Crowley, "Poverty Rates Vary Widely Across the United States," Population Reference Bureau, 2007, http://www.prb.org/Publications/Articles/2002/PovertyRatesVaryWidelyAcrosstheUnitedStates.aspx; Chernick and Reschovsky, "Urban Fiscal Problems," 138–41.
6. Jon C. Teaford, *The Twentieth-Century American City: Problem, Promise, and Reality* (Baltimore, MD: Johns Hopkins University Press, 1993), 142; Jon C. Teaford, *The Rough Road to Renaissance* (Baltimore, MD: Johns Hopkins University Press, 1990), 218, 225, 262, 265; Lawrence J. R. Herson and John M. Bolland, *The Urban Web* (Chicago: Nelson-Hall, 1990), 347.
7. Jonathan Soffer, *Ed Koch and the Rebuilding of New York City* (New York: Columbia University Press, 2010), 2–3.
8. Robert W. Bailey, *The Crisis Regime: The MAC, the EFCB, and the Political Impact of the New York City Financial Crisis* (Albany: State University of New York Press, 1984), 3–4. See also Steven R. Weisman, "Fiscal Crisis at a Glance," *New York Times*, May 28,

1975; Matthew Drennan, "The Decline and Rise of the New York Economy," in *Dual City: Restructuring New York*, ed. John Hull Mollenkopf and Manuel Castells (New York: Russell Sage Foundation, 1991), 26–28; Joseph J. Salvo, "Population," in *The Encyclopedia of New York City*, 2nd ed., ed. Kenneth T. Jackson (New Haven, CT: Yale University Press, 2010), 1019.

9. Ken Auletta, *The Streets Were Paved with Gold* (New York: Random House, 1979), 29, 31–32; Roger Dunstan, *Overview of New York City's Fiscal Crisis*, CRB note 3, March 1, 1995, https://www.library.ca.gov/crb/95/notes/v3n1.pdf.

10. The mandated costs in this period, however, proved to be only a small portion of total city outlays because of curbs on eligibility and other restrictions.

11. Charles Brecher et al., *Power Failure: New York City Politics and Policy Since 1960* (New York: Oxford University Press, 1993), 35–38.

12. Soffer, *Ed Koch and the Rebuilding of New York City*, 5; Felix G. Rohatyn, "Lessons of the '75 Fiscal Crisis," *New York Times*, June 10, 1985.

13. David R. Goldfield and Blaine A. Brownell, *Urban America*, 2nd ed. (Boston: Houghton Mifflin, 1990), 227–30; Teaford, *Rough Road to Renaissance*, 227–30; Teaford, *The Twentieth-Century American City*, 143–46; Carl Abbott, *Urban America in the Modern Age* (Arlington Heights, IL: Harlan Davidson, 1987), 130; Charles Brecher and Raymond D. Horton, eds., *Setting Municipal Priorities: American Cities and the New York Experience* (New York: New York University Press, 1984), 4; Brecher et al., *Power Failure*, 44. See also Donna E. Shalala and Carol Bellamy, "A State Saves a City: The New York Case," *Duke Law Journal* 6 (1976): 1119–23.

14. Correspondence, Jonathan Soffer to Martin Melosi, April 9, 2016.

15. "New York City's First Jewish Mayor," Richmond Hill Historical Society, http://www.richmondhillhistory.org/abeame.html; Joanne Reitano, *The Restless City: A Short History of New York from Colonial Times to the Present* (New York: Routledge, 2006), 188.

16. Quoted in Reitano, *The Restless City*, 188.

17. Chris McNickle, "Abraham D(avid) Beame," in *The Encyclopedia of New York City*, ed. Jackson, 106; "Mayoralty," in *The Encyclopedia of New York City*, ed. Jackson, 814.

18. Frank Lombard and Don Singleton, "Abe Beame, Mayor Who Faced '70s Crisis, Dies," *New York Daily News*, February 11, 2001. See also George J. Lankevich, *American Metropolis: A History of New York City* (New York: New York University Press, 1998), 213.

19. Quoted in Fuchs, *Mayors and Money*, 88. See John Darnton, "Bankers Warned Beame of a Crisis Early in 1974," *New York Times*, May 31, 1975; Ester R. Fuchs, *Mayors and Money: Fiscal Policy in New York and Chicago* (Chicago: University of Chicago Press, 1992), 87.

20. Robert D. McFadden, "Abraham Beame Is Dead at 94; Mayor During 70s Fiscal Crisis," *New York Times*, February 11, 2001. See also "How Large a Gap?" *New York Times*, May 27, 1975; Martin Mayer, "Plunging Into Bankruptcy: Or, How to Get New York Back Into the Swim," *New York Times*, May 19, 1975; Lombard and Singleton, "Abe Beame, Mayor Who Faced '70s Crisis, Dies"; Steven R. Weisman, "Beame, Accepting Risk of Failure, Now in Forefront of Fiscal Crisis," *New York Times*, March 4, 1977. For a good chronological overview of the early days of the crisis, see Ronald Smothers, "A Chronicle of the Crucial Dates in the City's Financial Crisis," *New York Times*, July 28, 1975.

21. Lankevich, *American Metropolis*, 214. See also "Beame Details New Plan to Cut 8,000 Employees," *New York Times*, November 12, 1975.

22. It should be remembered that the governor's thinking on the matter was in the context that New York State had absorbed less of the cost for social services than other states. See Reitano, *The Restless City*, 189.
23. Dunstan, "Overview of New York City's Fiscal Crisis"; Soffer, *Ed Koch and the Rebuilding of New York City*, 5. See also Chris Welles, "The Domino Scenario: The Day New York City Defaulted," *New York Magazine*, June 2, 1975, http://nymag.com/news/features/48290/; McFadden, "Abraham Beame Is Dead at 94"; Maurice Carroll, "Carey Aides Offer City Fiscal Plan," *New York Times*, May 29, 1975; Ronald Smothers, "Move by Albany Eases Cash-Flow Problem Through June," *New York Times*, May 30, 1975; Lankevich, *American Metropolis*, 217; Mollenkopf and Castells, ed., *Dual City*, 29. Over the years, after some reforms and eventual surpluses, the city regained most of its control over its finances.
24. Kim Phillips-Fein, "The Legacy of the 1970s Fiscal Crisis," *Nation*, April 16, 2013, http://www.thenation.com/article/legacy-1970s-fiscal-crisis/. See also Sam Roberts, "Infamous 'Drop Dead' Was Never Said by Ford," *New York Times*, December 28, 2006.
25. Gerald R. Ford to Abraham D. Beame, May 14, 1975, *American Presidency Project*, http://www.presidency.ucsb.edu/ws/?pid=4910.
26. Phillips-Fein, "The Legacy of the 1970s Fiscal Crisis"; Sam Roberts, "Infamous 'Drop Dead' Was Never Said by Ford"; Reitano, *The Restless City*, 189.
27. Roberts, "Infamous 'Drop Dead' Was Never Said by Ford." See also Lankevich, *American Metropolis*, 218–219.
28. Phillips-Fein, "The Legacy of the 1970s Fiscal Crisis." See also "Gerald Ford's Statement on New York City's $2.3 Billion Bailout Loan (11/26/1975)," July 1, 2013, *Distressed Volatility*, http://www.distressedvolatility.com/2013/07/gerald-fords-statement-nyc-bailout.html; Jeff Nussbaum, "The Night New York Saved Itself from Bankruptcy," October 16, 2015, *New Yorker*, http://www.newyorker.com/news/news-desk/the-night-new-york-saved-itself-from-bankruptcy; correspondence Soffer to Melosi, April 9, 2016.
29. See Leonard Silk, "Slump and Recovery," *New York Times*, December 31, 1975.
30. Dunstan, "Overview of New York City's Fiscal Crisis." See also Shalala and Bellamy, "A State Saves a City," 1123–32. See also E. J. McMahon and Fred Siegel, "Gotham's Fiscal Crisis: Lessons Unlearned," *Public Interest* 158 (Winter 2005): 96–110; Bailey, *The Crisis Regime*, 5–10; Brecher and Horton, eds., *Setting Municipal Priorities*, 2; Mollenkopf and Castells, ed., *Dual City*, 33–35.
31. Charlotte Curtis, "Abraham Beame Looks Back," *New York Times*, May 14, 1985. See also Phillips-Fein, "The Legacy of the 1970s Fiscal Crisis"; Comptroller General of the United States, *The Long-Term Fiscal Outlook for New York City* (1977).
32. McFadden, "Abraham Beame Is Dead at 94." See also Phillips-Fein, "The Legacy of the 1970s Fiscal Crisis."
33. At this point the mayor had ordered cuts of 11,985 city employees: 4,560 civil servants, 4,727 provisional appointees, and 2,700 employees over sixty-five years old. See Mary McCormick, "Labor Relations," in *Setting Municipal Priorities*, ed. Brecher and Horton, 318; Auletta, *The Streets Were Paved with Gold*, 150–51. See also Press Release, Office of the Mayor, June 27, 1975, and June 30, 1975, Press Releases, Abraham D. Beame Collection, Box 70012, Folder 9, La Guardia and Wagner Archives, LaGuardia Community College/CUNY; Damon Stetson, "City-Union Pacts Reached Quickly," *New York Times*, August 4, 1974; Fred Ferretti, "260 Police Ousted as PBA Rejects Beame Proposal," *New York Times*, January 31, 1975; "Wide

Impact Seen Should Arbitrator Rule for the PBA," *New York Times*, February 13, 1975.
34. Quoted in Auletta, *The Streets Were Paved with Gold*, 15.
35. James M. Hartman, "Sanitation," in *Setting Municipal Priorities*, ed. Brecher and Horton, 415–17.
36. William K. Tabb, *The Long Default: New York City and the Urban Fiscal Crisis* (New York: Monthly Review Press, 1982), 30, 42. Between January 1, 1975, and May 31, 1976, the city payroll was reduced by 15 percent, or 47,412 employees. Payrolls for parks and playgrounds were cut by 25 percent, schools by slightly more, and sanitation at about 25 percent. The cuts were greatest in the early years, since between 1978 and 1981, employment essentially remained stable. Raymond D. Horton and John Palmer Smith, "Expenditures and Services," in *Setting Municipal Priorities*, ed. Brecher and Horton, 358.
37. Tabb, *The Long Default*, 42. In constant dollars (using 1967 as a base year), overall expenditures for the DSNY decreased from $173.2 million in 1975 to $137.9 million in 1980. In those same years, full-time employees dropped from 14,384 to 10,855. Of that number, the steadiest decrease was among uniformed sanitationmen (from 12,231 in 1974 to 8,311 in 1981). See Hartman, "Sanitation," 419–21.
38. Auletta, *The Streets Were Paved with Gold*, 278; Hartman, "Sanitation," 358–59.
39. The city had exhausted its major landfills—save Fresh Kills—since 1960. Pelham Bay landfill and the Fountain Avenue site were getting close to maximum use. Pelham Bay would close in 1979. Hartman, "Sanitation," 417–21, 438–39; Colin Campbell, "World's Biggest Dump for Garbage Just a Monumental Problem on S.I.," *New York Times*, May 28, 1981.
40. NYDS figures of average daily tonnage of refuse disposed in 1981 consisted of 10,000 tons (40 percent) to Fresh Kills, 8,000 tons to truck-fed landfills (32 percent), 2,000 tons to incinerators (8 percent), 2,000 tons of construction waste (8 percent), and 3,000 tons privately carted to New Jersey (12 percent). Hartman, "Sanitation," 439. See also Michael Azzara, "Garbage Disposal Problems Grow as Landfills Fill Up," *Staten Island Advance*, April 1, 1979.
41. New York City, Department of Sanitation, *Executive Management Report*, September 6, 1977, 36, New York City, Municipal Archives; Hartman, "Sanitation," 438–43; Daniel C. Walsh and Robert G. LaFleur, "Landfills in New York City: 1844–1994," *Ground Water* 33 (July–August 1995): 556.
42. He was christened Gaetano Kenneth Molinari but changed his name during his early teens because "I didn't like the way it sounded." He took Victor as his middle name. "It was a lot easier to say Guy V. Molinari. Guy K. Molinari just didn't roll off the tongue easily." Guy V. Molinari, *A Life of Service* (New York: Page, 2016), 7.
43. "Guy V. Molinari (Retired)," Scamardella Gervasi Thomson & Kasegrande Law Firm, https://www.statenlaw.com/attorney-profiles/guy-victor-molinari/; Robert D. McFadden, "Guy V. Molinari, Staten Island Power Broker, Is Dead at 89," *New York Times*, July 25, 2018; "Guy Molinari, 89, a Giant Who Transformed Staten Island's Civic and Political Landscape," *Staten Island Advance*, July 25, 2018; Molinari, *A Life of Service*, 115–16, 118–20.
44. Peter Harrigan, "Garbage Plan Fails in Assembly," *Staten Island Advance*, July 9, 1975; Peter Harrigan, "Garbage Landfill Legislation Passes Second Time Around," *Staten Island Advance*, July 13, 1975.
45. "Garbage Bill Gets Green Light; Island Gets a Guarantee," *Staten Island Advance*, May 26, 1976. See also Peter Harrigan, "Bill for Trash Sites Slowed in Albany," *Staten*

11. FISCAL CRISIS AND DISPOSAL DILEMMA   ■   637

Island Advance, May 28, 1976; "Bill Would Allow Landfills Offshore, Opponents Say," *Staten Island Advance*, May 7, 1976; "Garbage Remedies a Prime Concern," *Staten Island Advance*, May 9, 1976; Peter Harrigan, "Island Lawmakers Oppose City's Garbage Bill Again," *Staten Island Advance*, May 9, 1976; Peter Harrigan, "Action Delayed on Garbage Disposal Bill," *Staten Island Advance*, May 25, 1976.

46. "Legislature OKs Bill Limiting Trash to Be Dumped Here," *Staten Island Advance*, June 30, 1976. See also "Brooklyn Garbage Coming?" *Staten Island Advance*, June 2, 1976; "Garbage Cure Must Be Local," *Staten Island Advance*, June 23, 1976.
47. NYEPA had received inquiries from owners of such mines in Ohio to ship New York waste to them by rail.
48. Edward Ranzal, "City Asked for $8 Million to Pay for Out-of-State Trash Disposal," *New York Times*, January 5, 1974.
49. The U.S. Supreme Court ordered that the ban should be reviewed by the state court in light of new federal regulations under the Resource Conservation and Recovery Act (1976). See Terri Schultz, "Garbage Is No Longer Treated Lightly," *New York Times*, March 6, 1977.
50. Private carters essentially collected commercial refuse. They were issued permits through the Department of Consumer Affairs but regulated by a variety of municipal agencies. See Clyde Haberman, "Rise Is Expected in Carting Rates After the Strike," *New York Times*, December 3, 1981; Paul L. Montgomery, "Angry Carting Concerns to Act on 'Unfair' Rates," *New York Times*, February 13, 1982.
51. David Bird, "City's Garbage Problem Eased by Dumping Ruling in Jersey," *New York Times*, March 31, 1974. See also Schultz, "Garbage Is No Longer Treated Lightly"; Peter Hellman, "In the Big Dump, a Sweet Smell of Success," *New York Times*, January 26, 1975; David Bird, "New York City Defends Way It Disposes of Waste," *New York Times*, January 25, 1976; Frank J. Prial, "The Basic Problem Is 14,000 Tons of Waste Each Day," December 11, 1976; "'Landfill' to Increase," *Staten Island Advance*, January 10, 1974.
52. Leonard Baldassano, "Firefighters Battle Dump Fires," *Staten Island Advance*, July 2, 1975.
53. Press Release, Office of the Mayor, July 1, July 2, and July 3, 1975, Press Releases, Beame Collection, Box 70012, Folder 10. See also Memorandum, James A. Cavanaugh to Abraham D. Beame, July 1, 1975, Beame Collection, Box 17, Folder 280.
54. Selwyn Raab, "McFeeley Scores Accord on Sanitation Rehiring," *New York Times*, July 4, 1975. See also Donald Kirk, "Beame's Gimmick Ends N.Y. Garbage Strike," *Chicago Tribune*, July 4, 1975; Fred Ferretti, "Beame Restores 2,600 More Jobs; Asks Pay Freeze," *New York Times*, July 8, 1975; Mary Breasted, "Health, Fire, and Sanitation Departments Worried by High Rate of Staff Attrition," *New York Times*, July 5, 1976.
55. "The Garbage Mess . . . ," *New York Times*, August 7, 1975. See also "Beame's Message to City Council and Estimate Board on 1978 Budget," *New York Times*, April 23, 1977; John E. Hurley, "Dump Fire Spreads Stench," *Staten Island Advance*, July 3, 1975.
56. Charles Kaiser, "Little Likelihood of Improved Sanitation Is Foreseen," *New York Times*, September 2, 1975.
57. Martin Lang to Abraham D. Beame, October 17, 1975, Beame Collection, Box 17, Folder 280.
58. Robert D. McFadden, "Groh Resigns Job, Citing Pressures on City Cleanup," *New York Times*, September 21, 1975. See also Press Release, Office of the Mayor, September 22, 1975, and December 12, 1975, Press Releases, Beame Collection, Box 70012, Folder 10.

59. Larry Celona, "Former New York City Mayor Ed Koch at Age 88," *New York Post*, February 1, 2013, http://www.usatoday.com/story/news/nation/2013/02/01/ed-koch-mayor-who-became-a-symbol-of-nyc-dies/1882459/; Robert D. McFadden, "Edward I. Koch, a Mayor as Brash, Shrewd and Colorful as the City He Led, Dies at 88," *New York Times*, February 1, 2013, http://www.nytimes.com/2013/02/02/nyregion/edward-i-koch-ex-mayor-of-new-york-dies.html; "Ed Koch, Mayor Who Became a Symbol of NYC, Dies at 88," *USA Today*, February 1, 2013, http://www.usatoday.com/story/news/nation/2013/02/01/ed-koch-mayor-who-became-a-symbol-of-nyc-dies/1882459/.
60. Edward I. Koch with Daniel Paisner, *Citizen Koch: An Autobiography* (New York: St. Martin's, 1992), 122, 126. See also Edward I. Koch, *Mayor: An Autobiography* (New York: Simon and Schuster, 1984), 29.
61. Martin Shefter, "Koch, Edward I(rving)," in *The Encyclopedia of New York City*, ed. Jackson, 704–5; McFadden, "Edward I. Koch"; Soffer, *Ed Koch and the Rebuilding of New York City*, 4.
62. Soffer, *Ed Koch and the Rebuilding of New York City*, 3–5. See also John Hull Mollenkopf, *A Phoenix in the Ashes: The Rise and Fall of the Koch Coalition in New York City Politics* (Princeton, NJ: Princeton University Press, 1992), 4. See also Reitano, *The Restless City*, 193–95.
63. Mollenkopf, *A Phoenix in the Ashes*, 4–9, 15, 21. See also Koch, *Mayor*, 97–110; Mollenkopf and Castells, ed., *Dual City*, 33, 35.
64. Mollenkopf, *A Phoenix in the Ashes*, 4.
65. Soffer, *Ed Koch and the Rebuilding of New York City*, 5–11.
66. See Reitano, *The Restless City*, 193.
67. McFadden, "Edward I. Koch." See also Soffer, *Ed Koch and the Rebuilding of New York City*, 10–11.
68. Richard D. Lyons, "Anthony T. Vaccarello, 67; Former Sanitation Commissioner," *New York Times*, November 10, 1993.
69. Anthony T. Vaccarello to Abraham D. Beame, December 6, 1977, Beame Collection, Box 10, Folder 297. See also Bob Monroe, "Bronx Needs a Garbage Dump; Vaccarello Eyes Staten Island," *Staten Island Advance*, February 24, 1977; Douglas Patrick, "Lamberti Stalls Bronx Plan to Ship Garbage to Island," *Staten Island Advance*, March 2, 1978.
70. Bob Monroe, "Bronx Needs a Garbage Dump; Vaccarello Eyes Staten Island," *Staten Island Advance*, February 24, 1977; "Don't Increase Dumping Here," *Staten Island Advance*, February 26, 1977; Terence J. Kivlan, "Bronx-to-Island Garbage Haul Seen Too Expensive," *Staten Island Advance*, March 10, 1977; David Berry, "City to Spend Million at Landfill," *Staten Island Advance*, May 22, 1977; Reginald Patrick, "Molinari Doubts Reported Cuts in Garbage Sent to Island," *Staten Island Advance*, May 22, 1977; Peter Harrigan, "Bill Slipped in to Dump More Garbage Here," *Staten Island Advance*, June 25, 1977; "Garbage Limit Must Continue," *Staten Island Advance*, June 28, 1977.
71. "Garbage Plant Bids Being Accepted," *Staten Island Advance*, September 15, 1978.
72. Auletta, *The Streets Were Paved with Gold*, 300–1. Under pressure from the sanitation union, Koch did not appoint his first choice, Nathan Leventhal, nor did he appoint Lang. Both Leventhal and Lang were viewed by the union as likely to bring too many changes to the department. He did appoint Frank Sisto, president of the USA, as first deputy commissioner.

11. FISCAL CRISIS AND DISPOSAL DILEMMA ■ 639

73. It was not unusual for new mayors initially to retain the sanitation commissioner through the winter. In case of a bad storm, he would have an excuse to let the commissioner go! Oral interview, Martin Melosi with Norman Steisel, November 15, 2017.
74. Soffer, *Ed Koch and the Rebuilding of New York City*, 162–63; oral interview, Martin Melosi with Norman Steisel, November 15, 2017.
75. Steisel taught courses in policy analysis and decision making at Yale's School of Organization and Management, Harvard's Kennedy School for Public Policy, Columbia University, NYU's School of Law, and Baruch College at CUNY.
76. "The Executive Career of Norman Steisel," http://www.norman-steisel.com/; Robert D. McFadden, "Man in the News; A Savvy and Efficient Manager," *New York Times*, December 9, 1989.
77. Deidre Carmody, "Steisel, Sanitation Chief, Resigning After 7 Years," *New York Times*, January 24, 1986. See also Benjamin Miller, *Fat of the Land: Garbage of New York, the Last Two Hundred Years* (New York: Four Walls Eight Windows, 2000), 235.
78. Norman Steisel, "Productivity in the New York City Department of Sanitation: The Role of the Public Sector Manager," *Public Productivity Review* 8 (Summer 1984): 103.
79. Steisel, "Productivity in the New York City Department of Sanitation," 104. However, between 1978 and 1982, the DSNY—along with police, fire, corrections, and education—avoided a hiring freeze placed on all other departments. See Soffer, *Ed Koch and the Rebuilding of New York City*, 155.
80. Oral interview, Martin Melosi with Norman Steisel, November 15, 2017.
81. Brecher et al., *Power Failure*, 229–30. See also Norman Steisel to Edward I. Koch, April 9, 1979, Departmental Correspondence Series, Koch Collection, Box 252, Folder 1, April 1979–July 1979, La Guardia and Wagner Archives.
82. Lucius J. Riccio, "Sweeping Change at NYC Sanitation," *Management Review* 76 (May 1978): 47–48. See also "The Executive Career of Norman Steisel"; Miller, *Fat of the Land*, 236; Soffer, *Ed Koch and the Rebuilding of New York City*, 7, 249–50.
83. McFadden, "Man in the News; A Savvy and Efficient Manager"; Carmody, "Steisel, Sanitation Chief, Resigning After 7 Years."
84. Steisel, "Productivity in the New York City Department of Sanitation," 106. See 106–125 for details of Steisel's departmental reforms.
85. Michael Sterne, "Garbage Problems Continue to Grow but Officials Try to Put the Lid On," *New York Times*, November 7, 1978.
86. Carmody, "Steisel, Sanitation Chief, Resigning After 7 Years." See also Soffer, *Ed Koch and the Rebuilding of New York City*, 188.
87. See J. J. Dunn Jr. and Penelope Hong, "Landfill Siting—An Old Skill in a New Setting," *APWA Reporter* 46 (June 1979): 12.
88. In 1975 the Borough of Richmond officially became the Borough of Staten Island.
89. DSNY efforts at landscaping along the fringes of the landfill (including the planting of trees and bushes) to shield private property from the dumping areas met with mixed reactions among locals. See Bruce Alpert, "Some of the Landfill's Problems May Soon Be Out of Sight," *Staten Island Advance*, December 12, 1979; Tom Robotham, "Sanit Brass Has a Hard Time Promising Landfill Neighbors," *Staten Island Advance*, May 8, 1979; Douglas Patrick, "Landscaping Project for Fresh Kills Landfill to Begin in March," *Staten Island Advance*, September 26, 1979. See also Janice Kabel, "The Fresh Kills Landfill: 2,200 Acres of Garbage: Is a Park Really in There?" *Staten Island Advance*, October 1, 1978; Douglas Patrick, "Landfill Opponents Unhappy with Improvements, Trial Likely," *Staten Island Advance*, November 4, 1978.

90. *Fresh Kills: Landfill to Landscape: International Design Competition: 2001*, https://www1.nyc.gov/assets/planning/download/pdf/plans/fkl/about_fkl.pdf; Daniel C. Walsh, "The History of Waste Landfilling in New York City," *Ground Water* 29 (July–August 1991): 593. See also "Booms to Trap Garbage Being Installed at Fresh Kills," *Staten Island Advance*, November 25, 1978.

91. Dena Kleiman, "Koch Decides Ferry Point Landfill Should Not Be Reopened After All," *New York Times*, May 20, 1978. See also Terence J. Kivlan, "Marchi Blocks Senate Bill to Keep Bronx Dump Closed," *Staten Island Advance*, March 10, 1978; Mary Beth Pfeiffer, "City to Dump Garbage from Bronx on Island," *Staten Island Advance*, May 20, 1978; "Molinari Blasts Koch's Decision to Ship Bronx Garbage to Island," *Staten Island Advance*, May 23, 1978; Michael Azzara, "Gaeta Gripe Stalls Board on Landfill $$," *Staten Island Advance*, May 26, 1978; Peter Harrigan, "We've Had Our Fill at Fresh Kills Landfill," *Staten Island Advance*, May 28, 1978; Bruce Alpert, "Council Vote Put Off on Funds to Expand Island Garbage Dump," *Staten Island Advance*, June 16, 1978.

92. "Bronx Dump 'Illegal'; Early Closing Ordered," *Staten Island Advance*, August 29, 1978. See also Janice Kabel, "Some Bronx Garbage Could Be Barged Here by Spring," *Staten Island Advance*, December 18, 1978.

93. Guy V. Molinari to Edward I. Koch, May 20, 1978, Departmental Correspondence Series, Koch Collection, Box 245, Folder 9, La Guardia and Wagner Archives. See also "Don't Expand the Landfill," *Staten Island Advance*, June 19, 1978; "Committee Again Tables Vote on Landfill Expansion," *Staten Island Advance*, June 21, 1978. Molinari noted that in addition to plans to expand Fresh Kills, the Power Authority of the State of New York (PASNY) was planning to build "a mammoth power plant" on Staten Island. "All this adds up to a very disturbing and distressing picture," he concluded. Despite all the talk on how to use garbage in the future, "Staten Island continues to be the loser."

94. Stephen B. Kaufman to Edward Koch, May 22, 1978, Departmental Correspondence Series, Koch Collection, Box 28, Folder 3.

95. Edward I. Koch to Stephen B. Kaufman, July 31, 1978, Departmental Correspondence Series, Koch Collection, Box 28, Folder 3. See also Koch to Kaufman, June 6, 1978; Kaufman to Koch, July 18, 1978.

96. 6 NYCRR Part 360, Solid Waste Management Facilities (Title 6 of the Official Compilation of Codes, Rules, and Regulation of the State of New York).

97. New York City, Department of Parks & Recreation, Freshkills Park Generic Environmental Impact Statement (March 13, 2009), v. 1, chap. 1, 1–8 to 1–9, http://www.nycgovparks.org/park-features/freshkills-park/public-review.

98. Martin V. Melosi, *Garbage in the Cities: Refuse, Reform, and the Environment*, rev. ed. (Pittsburgh, PA: University of Pittsburgh Press, 2005), 203. The "Criteria for Classification of Solid Waste Disposal Facilities and Practices" issue by the Environmental Protection Agency (Fall 1979) was meant to implement those portions of the 1977 law dealing with landfills, among other refuse facilities. Along with the EPA's "Guidelines for Development and Implementation of State Solid Waste Management Plans," published in July 1979, it represented the implementation of regulations related to solid waste–disposal facilities, especially landfills, outlined in Subtitle D of RCRA. The facilities that did not satisfy the criteria were defined as "open dumps," which would need to be upgraded or closed through state enforcement. The criteria also defined refuse-management practices that constituted "open dumping." Open dumping was prohibited by the act. The criteria had been published as proposed

regulations in the February 1978 *Federal Register*. John H. Skinner and Truett V. DeGeare, "The New Landfill Criteria: The Most Important Regulations to Ever Impact the U.S. Solid Waste Industry!" *Solid Waste Management* 22 (October 1979): 23–24.

99. "Connelly, Elizabeth A.," *Our Campaigns*, http://www.ourcampaigns.com/Candidate Detail.html?CandidateID=52115.
100. Elizabeth Connelly: Oral History Transcripts, February 9, 2004, 70, College of Staten Island Oral History Collection, v. 3, 2005, ed. Jeffery A. Kroessler, Archives & Special Collections, Department of the Library, College of Staten Island, CUNY.
101. Elizabeth Connelly: Oral History Transcripts, 71.
102. Mark E. Hanly, "Action Sought on Garbage Disposal," *Staten Island Advance*, January 15, 1978.
103. "Hearing to Investigate Garbage Problems," *Staten Island Advance*, September 27, 1978; "Garbage Plan Hearing Set by Marchi," *Staten Island Advance*, September 10, 1978.
104. New York State, Senate Committee on Finance, *Subcommittee Hearing in the Matter of Current and Projected Status of Solid Waste Disposal and Resource Recovery in New York City*, October 4, 1978.
105. Statement by Elizabeth A. Connelly, Public Hearing Before the Senate Standing Committee on Finance, October 4, 1978, Assemblywoman Elizabeth A. Connelly Papers, Series 1: Subject Files, 1978–1979, Box 20, Archives & Special Collections, Department of Library, College of Staten Island, CUNY. See also "News from Assemblywoman Elizabeth A. Connelly," September 26, 1978, Connelly Papers, Series 1: Subject Files, 1978–1979, Box 20. At about the same time, the *Staten Island Advance* reinforced Marchi's and Connelly's criticism of the city's disposal program with four successive pieces on Fresh Kills beginning on October 1. In the first installment, the reporter dredged up the origin story featuring Robert Moses, making clear that "the Islanders were powerless to stop Moses' plan for Fresh Kills." Janice Kabel, "The Fresh Kills Landfill: 2,200 Acres of Garbage: Is a Park Really in There?" *Staten Island Advance*, October 1, 1978. Kabel followed with a further elaboration of the Fresh Kills story, ending with: "Could it be that John Rubino, chief of the city's sanitary landfills, is right when he says: 'We will be finished at Fresh Kills when we finally cover it over and plant the grass.'" Kabel, "The Fresh Kills Landfill: Closing It Has Been Debated Since It Opened," *Staten Island Advance*, October 4, 1978.
106. "Assemblywoman Elizabeth A. Connelly: Reports to the People," March 1979, Connelly Papers, Series 1: Subject Files, 1978–1979, Box 20, Folder 1. See also "Dumped Garbage Will Be Covered, Landfill Boss Tells Travis," *Staten Island Advance*, August 31, 1976.
107. "Judge Refuses to Issue Temporary Ban on Fresh Kills Dumping," *Staten Island Advance*, February 17, 1979. See also "Piling Garbage on Fresh Kills," *Staten Island Advance*, February 22, 1979.
108. One consent order was negotiated by DEC and the other by Assemblywoman Connelly. The former was a binding two-year pact requiring the city to take temporary measures while a more detailed plan was developed. In his consent decree of October 24, 1978, relative to the Connelly lawsuit, State Supreme Court Justice Charles R. Rubin laid out a series of terms, including increasing the cover of refuse in nonactive dumping areas, monitoring methane-gas emissions, increasing the height of dirt berms along the West Shore Expressway, banning dumping in the Travis area, and prohibiting disposal within two hundred feet of any street or property line. See

Consent Order, *Elizabeth Connelly et al. v. The City of New York*, Index no. 314/78, October 24, 1980, Connelly Papers, Series 1: Subject Files, Lawsuits, 1978–1998, Box 20, Folder 7. See also State of New York, Department of Environmental Conservation, Order on Consent, DEC File no. 2-0527, October 24, 1980, Connelly Papers, Series 1: Subject Files, Lawsuits, 1978–1998, Box 20, Folder 7; Elizabeth Connelly: Oral History Transcripts, February 9, 2004, 71–72; Elizabeth A. Connelly to "Dear Friend," September 29, 1980, Connelly Papers, Series 1: Subject Files, Lawsuits, 1978–1998, Box 20, Folder 7; "Judge Puts on Pressure to Settle Landfill Suit," *Staten Island Advance*, October 31, 1979; "City Ordered to Clean Up Fresh Kills Landfill," *Staten Island Advance*, October 13, 1980; "Judge Eyes Trial for Permanent Ban on Fresh Kills Dumping," *Staten Island Advance*, October 28, 1978; "Landfill Opponents Unhappy with Improvements, Trial Likely," *Staten Island Advance*, November 4, 1978.

109. Leachate is liquid that drains or percolates from a landfill. It usually contains both dissolved and suspended material, some of which may be toxic.

110. Robin Nagle, "To Love a Landfill: The History and Future of Fresh Kills," in *Handbook of Regenerative Landscape Design*, ed. Robert L. France (Boca Raton, FL: CRC, 2008), 10. See also New York City, Department of Parks & Recreation, *Freshkills Park Generic Environmental Impact Statement* (March 13, 2009), v. 1, chap. 1, 1–9; Martin V. Melosi, "Fresno Sanitary Landfill," National Historic Landmark Nomination (NPS Form 10-900), U.S. Department of Interior, National Parks Service (August 2000).

111. "Judge Signs Decree for Changes at Dump," *Staten Island Advance*, October 24, 1980.

112. Connelly also had been incensed when she claimed in July that the DSNY was withholding a report on the effects of leakage from Fresh Kills Landfill, which she believed had damaging information that the department did not want to disclose. See "News from Assemblywoman Elizabeth A. Connelly," July 2, 1980, Connelly Papers, Series 1: Subject Files, Lawsuits, 1978–1998, Box 20, Folder 7.

113. "Island Officials Angry at Dump's Prolonged Life," *Staten Island Advance*, October 13, 1980. See also "News from Assemblywoman Elizabeth A. Connelly," October 12, 1980, Connelly Papers, Series 1: Subject Files, 1980–1989, Box 20, Folder 2.

114. Elizabeth A. Connelly to Norman Steisel, October 6, 1980, Connelly Papers, Series 1: Subject Files, 1980–1989, Box 20, Folder 2.

115. Alan Richman, "A 'Forgotten' Staten Island Concentrates on Its Future," *New York Times*, March 9, 1979.

116. David Bird, "City Collects Some Garbage," *New York Times*, December 7, 1978.

117. Normal Seisel to Edward I. Koch, September 10, 1979, Departmental Correspondence Series, Koch Collection, Box 63, Folder 31. See also Nicholas Gaitanis, Chief Engineer, to Norman Steisel, September 10, 1979, Box 63, Folder 31; Norman Steisel to Edward I. Koch, February 2, 1979, Departmental Correspondence Series, Koch Collection, Box 251, Folder 8; "Breakdown at Fresh Kills Clogging Landfill Operations, *Staten Island Advance*, February 19, 1978; "Repair of Fresh Kills Digger to Clear Trash Backlog Soon," *Staten Island Advance*, February 21, 1978; "Booms to Trap Garbage Being Installed at Fresh Kills," *Staten Island Advance*, November 25, 1978; Janice Kabel, "Temporary Pump to Keep Poisons from Seeping Into Landfill Streams," *Staten Island Advance*, November 20, 1978; "8 Bids Submitted for Construction of Bridge at Landfill," *Staten Island Advance*, January 27, 1977; "Fresh Kills Garbage Boom Repaired," *Staten Island Advance*, April 14, 1979; "Landfill Operations Cut by Equipment Problems," *Staten Island Advance*, August 19, 1979; "Crew Begins Cleaning Litter Near Landfill," *Staten Island Advance*, March 27, 1979. See also Sterne, "Garbage Problems Continue to Grow."

118. "A Health Emergency Declared in New York in 19-Day Tug Strike," *New York Times*, April 20, 1979. See also "Tug Strike Means Less Garbage for Fresh Kills Dump," *Staten Island Advance*, April 3, 1979.
119. Lee A. Daniels, "Tug Union Ordered to Ferry Garbage," *New York Times*, April 21, 1979. See also Lesley Oelsner, "Tug Union Balking at Justice's Order to Remove Garbage," *New York Times*, April 22, 1979; "Judge Orders Tug Crews Back to Work to Haul Garbage to Landfills," *Staten Island Advance*, April 21, 1979; "Tug Crews to Start Island Garbage Runs," *Staten Island Advance*, April 24, 1979.
120. "Aid by Coast Guard in Towing Garbage Ordered by Carter," *New York Times*, May 6, 1979; Damon Stetson, "City Ferryboat Crews Are Put Aboard Tugs to Transport Garbage," *New York Times*, May 23, 1979; "Chartered Tugs Move City Garbage," *New York Times*, May 24, 1979; Peter Kihss, "City Hopes Coast Guard Can Haul a Third of Garbage," *New York Times*, May 7, 1979; Damon Stetson, "Coast Guard Tugboats Start to Haul Garbage to Fresh Kills Dump," *New York Times*, May 8, 1979; "Coast Guard Tugs Get Police Escort to Landfill," *Staten Island Advance*, May 8, 1979; "Coast Guard Tugs Unload Trash Woes," *Staten Island Advance*, May 8, 1979.
121. Matthew Gandy, *Recycling and the Politics of Urban Waste* (London: Earthscan, 1994), 84. See also Damon Stetson, "88-Day Tug Strike Ends, Easing Shipments of Gas," *New York Times*, June 28, 1979; "88-Day Tugboat Strike Ends," *Syracuse Herald-Journal*, June 28, 1979.
122. Other proposals included extracting metal from the landfill to produce steel. See Mary Beth Pfeiffer, "Firm Signs Contract to Extract Metal from Landfill Garbage," *Staten Island Advance*, January 12, 1979.
123. "Gas Test on Garbage Approved," *Staten Island Advance*, August 15, 1975. The story reported that "methane gas is the natural product of the decaying garbage, and without any recovery system the gas seeps harmlessly into the air, occasionally causing a mild stench." Such a claim failed to account for the fact that methane is the second most prevalent greenhouse gas and thus a threat—if not monitored—to climate change. See "Overview of Greenhouse Gases—Methane," U.S. Environmental Protection Agency, http://www3.epa.gov/climatechange/ghgemissions/gases/ch4.html. See also Robert E. Huber, "California Firm Wins Bid to Hunt Landfill Methane," *Staten Island Advance*, December 6, 1975; Terence J. Kivlan, "Gas 'Mining' at Fresh Kills Set to Start," *Staten Island Advance*, December 14, 1976; "Landfill Gas Quest to Start," *Staten Island Advance*, December 16, 1976; Terence J. Kivlan, "Problems Postpone Garbage Dump Gas Mining for a Year," *Staten Island Advance*, February 16, 1977; "Landfill Gas 'Mining' Delayed," *Staten Island Advance*, February 19, 1977; Terence J. Kivlan, "Gas Co. Revives Landfill 'Mine' Plan," *Staten Island Advance*, June 10, 1977.
124. "Methane Plan at Landfill Gets City OK," *Staten Island Advance*, July 22, 1977; "Landfill Gas Quest On Again," *Staten Island Advance*, July 25, 1977; Terence J. Kivlan, "Second Firm to Test Gas at Landfill," *Staten Island Advance*, August 11, 1977; Terence J. Kivlan, "Judge Postpones Action in Suit Against Landfill Mining," *Staten Island Advance*, August 12, 1977; "Many Minds on Mining," *Staten Island Advance*, August 16, 1977; Terence J. Kivlan, "Firm Wins Second Landfill Drill Contract," *Staten Island Advance*, November 3, 1977.
125. Terence J. Kivlan, "Is There Good News in Garbage?" *Staten Island Advance*, September 21, 1975. See also Robert E. Huber, "Firm to Tap Fresh Kills to Meet Natural Gas Needs," *Staten Island Advance*, October 15, 1975.
126. "Better Use for Garbage," *Staten Island Advance*, August 2, 1975.

127. The agreement was with Getty Synthetic Fuels, Inc. (later GSF Energy, Inc.), of Long Beach, California (and its predecessor companies), and the Methane Development Corporation (a subsidiary of BUG). See also Frances Cerra, "Methane Sought By 3 L.I. Towns from Landfills," *New York Times*, May 31, 1982.

128. Joseph A. Vaszily, "Commercialization of Landfill Gas Recovery: Fresh Kills Case Study," in *Energy from Biomass and Waste VII* (symposium papers presented at Lake Buena Vista, Florida, January 24–28, 1983), 863–66. See also Peter Kihss, "City to Get Methane from Refuse," *New York Times*, November 27, 1979; New York State Energy Research and Development Authority, *Methane Production Rate Studies and Gas Flow Modeling for the Fresh Kills Landfill*, ERDA Report 80-21, November 1980; Institute of Gas Technology, *Evaluation of Fresh Kills Landfill Gas for Industrial Applications*, NTSERDA 80-1, March 1980, Senator John J. Marchi Papers, Series 4: Media Files, Subseries C: News Clippings, Box 105, Folder 6, Archives & Special Collections, Department of the Library, College of Staten Island, CUNY; Peter Kihss, "State Studying Use of 38 Landfills for Getting Power from Garbage," *New York Times*, April 27, 1981; Anthony J. Giuliani, "Largest Landfill Gas Plant Supplying State Island Users," *Pipeline & Gas Journal* 210 (March 1983): 17–19; "State Plans New Gas Effort at Landfill," *Staten Island Advance*, July 16, 1979.

129. James R. Pena, "Promises and Problems in Landfill Gas Recovery at Fresh Kills," paper no. 428, *Municipal Engineers Journal* 76 (1988): 29–33. See also "Mining Tests Find Lots of Methane at Landfill," *Staten Island Advance*, June 14, 1978.

130. Tom Robotham, "Methane Gas to Be Mined at Fresh Kills Landfill," *Staten Island Advance*, March 29, 1979.

131. Terence J. Kivlan, "Agency Wants to Boost Island Garbage Import," *Staten Island Advance*, March 1, 1978. See also Steve Cross, "Down in the Dumps? Sell Your Garbage," *Staten Island Advance*, December 2, 1979; "Deal Made to Mine Methane at Fresh Kills," *Staten Island Advance*, November 27, 1979; "Garbage 'Gold' That Doesn't Glitter," *Staten Island Advance*, November 29, 1979.

132. Richard Fenton, "Report on Present Status of Municipal Refuse Incinerators with Particular Reference to Problems Related to Nonresidential Refuse Input," paper presented at the Solid Waste Processing Division Meeting, ASME Research Committee, January 29, 1975, http://www.seas.columbia.edu/earth/wtert/sofos/nawtec/Incinerator-and-Solid-Waste-Technology/Incinerator-and-Solid-Waste-Technology-08.pdf.

133. The Organization of Oil Exporting Countries (OPEC) oil embargo in 1973 and the subsequent Iran crisis in 1979 raised serious questions about American vulnerabilities, especially in the acquisition and control of energy resources, setting off discussions about the production of alternatives to oil and natural gas. See Martin V. Melosi, *Coping with Abundance: Energy and Environment in Industrial America* (New York: Knopf, 1984), 277–94; Schultz, "Garbage Is No Longer Treated Lightly."

134. See Martin V. Melosi, "The Viability of Incineration as a Disposal Option: The Evolution of a Niche Technology, 1885–1995," *Public Works Management & Policy* 1 (July 1996): 38. See also Ronald Smothers, "Cleaning Up the Garbage Problem," *New York Times*, April 2, 1978.

135. Martin V. Melosi, *The Sanitary City: Environmental Services in Urban America from Colonial Times to the Present*, abridged ed. (Pittsburgh, PA: Pittsburgh University Press, 2008), 208.

136. Melosi, *The Sanitary City*, 246, 250–51. For example, in late 1978, the DSNY sunk a well into the garbage at Fresh Kills to stop leachate from reaching area waterways. A pump was installed that drew the liquid and sprayed it back over the landfill. More

11. FISCAL CRISIS AND DISPOSAL DILEMMA ■ 645

permanent solutions were being studied. "$108,000 Sunk Into Fresh Kills Landfill," *Staten Island Advance*, March 11, 1979.

137. It received the full-time services of Region II Solid Waste Engineer James Reid and a contract for consulting services with the firm of Leonard S. Wegman Co., Inc.

138. Leonard F. O'Reilly, of Leonard F. O'Reilly and Associates, Inc., was named director of the task force, which studied various financing and institutional schemes. See Leonard F. O'Reilly, "Planning for Resource Recovery in the 'Big Apple,'" *American Society of Mechanical Engineers, Proceedings of the 1978 National Waste Reprocessing Conference*, "Energy Conservation Through Waste Utilization," Chicago, May 7–10, 1978, 511–12.

139. Dennis Ischia et al., "Markets for Refuse-Derived Energy in New York City," *Resource Recovery & Energy Review* (July/August 1976): 8–11.

140. O'Reilly, "Planning for Resource Recovery in the 'Big Apple,'" 513–18. See also Terence Kivlan, "City Plans to Convert Trash Into Fuel for Power Plants," *Staten Island Advance*, December 4, 1977; "One Alternative to Dumping," *Staten Island Advance*, December 5, 1977.

141. Terence Kivlan, "PASNY Wants to Double Garbage Flow to Island," *Staten Island Advance*, December 8, 1977; Terence J. Kivlan, "Kill Plan to Dump More Garbage on Island," *Staten Island Advance*, April 13, 1978. See also Terence J. Kivlan, "P.A. Planning Network of Garbage Fuel Plants," *Staten Island Advance*, March 2, 1978; "Throwaway Plan for Trash," *Staten Island Advance*, March 4, 1978; Terence J. Kivlan, P.A. Garbage-Fuel Plan Lacks Island Site," *Staten Island Advance*, March 4, 1978.

142. Robert Hanley, "Panel Gets Long View of Fuel Sources," *New York Times*, October 23, 1979. See also "Wrong Borough for a 700-Megawatt Power Plant," *New York Times*, October 25, 1979; Edward C. Burks, "S.I. Site Picked for Transit Power Plant," *New York Times*, December 13, 1974; Elizabeth A. Connelly, Public Hearing on Resource Recovery, December 11, 1979, Albany, New York, Connelly Papers, Series 1: Subject Files, 1978–1979, Box 20, Folder 1.

143. Dena Kleiman, "Trash Is Boon in Plan Before Estimate Unit," *New York Times*, December 7, 1978. See also Olga A. Mendez to Edward I. Koch, July 5, 1978, and Edward I. Koch to Olga A. Mendez, July 31, 1978, Departmental Correspondence Series, Koch Collection, Box 245, Folder 1.

144. "Revised Timetable for Implementation of the Waste Resource Recovery Project in New York City," Departmental Correspondence Series, Koch Collection, Box 251, Folder 8.

145. Norman Steisel to Edward I. Koch, January 2, 1979, Departmental Correspondence Series, Koch Collection, Box 251, Folder 8.

146. Casowitz was a career civil servant who had joined the department in May 1978.

147. Oral interview, Martin Melosi and Paul Casowitz, November 16, 2017.

148. Memorandum, Norman Steisel, January 17, 1979, Departmental Correspondence Series, Koch Collection, Box 251, Folder 8.

149. Press Release, Department of Sanitation, January 10, 1979, Departmental Correspondence Series, Koch Collection, Box 251, Folder 8.

150. Norman Steisel to Edward I. Koch, March 1, 1979, Departmental Correspondence Series, Koch Collection, Box 251, Folder 8. See Press Release, January 16, 1974, Press Releases, Beame Collection Box 70009, Folder 19. See also Matthew Gandy, *Concrete and Clay: Reworking Nature in New York City* (Cambridge, MA: MIT Press, 2002), 200.

151. Paul Casowitz to Norman Steisel, February 15, 1979, Departmental Correspondence Series, Koch Collection, Box 251, Folder 8. See also oral interview, Martin Melosi with

Norman Steisel, November 15, 2017. More recently, Casowitz admitted that the Brooklyn Navy Yard project was flawed: "It wasn't flawed because of dioxin or because you couldn't burn waste cleanly. It was a barge-fed facility tied into a marine transfer station using open barges. And the longer I spent time with Fresh Kills and the barge movement, I began to appreciate how much damage those barges could do." Oral interview, Martin Melosi and Paul Casowitz, November 16, 2017.

152. Norman Steisel to Edward I. Koch, March 1, 1979, Departmental Correspondence Series, Koch Collection, Box 251, Folder 8; Gandy, *Concrete and Clay*, 203. See also Norman Steisel to Edward I. Koch, March 22, 1979, Departmental Correspondence Series, Koch Collection, Box 251, Folder 8.

153. City of New York, Department of Sanitation, *An Overview of Refuse Disposal and Resource Recovery in New York City: Issues and New Directions* (March 1979), i, also ii, v.

154. City of New York, Department of Sanitation, *An Overview of Refuse Disposal and Resource Recovery in New York City*, v, also iii–iv.

155. City of New York, Department of Sanitation, *An Overview of Refuse Disposal and Resource Recovery in New York City*, vii.

156. Sanitation, "Press Release," n.d., 2, Departmental Correspondence Series, Koch Collection, Box 251, Folder 8.

157. Any changes in department plans for disposal were made within the financial and operational realities of 1979. Only 12 percent of that year's budget (plus some indirect costs) could be spent on disposal; the remainder had to service the other department responsibilities. The workforce utilized by the Bureau of Waste Disposal in 1978 was 1,078 out of 12,030 employees, or about 9 per cent. See City of New York, Department of Sanitation, *An Overview of Refuse Disposal and Resource Recovery in New York City*, 14.

158. City of New York, Department of Sanitation, *An Overview of Refuse Disposal and Resource Recovery in New York City*, 15–20, 28–36.

159. City of New York, Department of Sanitation, *An Overview of Refuse Disposal and Resource Recovery in New York City*, 56, 63–64. See also John J. Marchi to Edward I. Koch, September 14, 1978, Departmental Correspondence Series, Koch Collection, Box 245, Folder 1.

160. "No Dumping Deadline Yet for Fresh Kills," *Staten Island Advance*, April 1, 1979; Bruce Alpert, "Fresh Kills Operations Come Under Scrutiny in 2 Assembly Bills," *Staten Island Advance*, June 4, 1979. See also "Island's Dirty Landfill Doesn't Deserve New Lease on Life," *Staten Island Advance*, December 17, 1979; Bruce Alpert, "Cover Garbage or Close Up: Ultimatum Due on Fresh Kills," *Staten Island Advance*, December 16, 1979.

161. "City May Double Garbage Flow Into Fresh Kills Landfill," *Staten Island Advance*, July 25, 1979. See also Bruce Alpert, "Fresh Kills Poker Game—Stakes as High as Garbage," *Staten Island Advance*, July 26, 1979; "A Loophole for Garbage," *Staten Island Advance*, November 14, 1979; "Sanit Official Warns: Garbage Fight Could Bring Heap of Trouble," *Staten Island Advance*, December 12, 1979.

12. FRESH KILLS AT MIDLIFE

1. Until the promulgation of strict federal laws in the early 1990s, more than 75 percent of municipal solid waste ended up in landfills. While the amount of refuse directed to landfills was high, the number of facilities declined rapidly by the 1980s. One study

identified 15,577 in 1980; another reported a drop to fewer than 8,000 by 1989. Between 1979 and 1986, about 3,500 landfills were closed. In 1985 and 1986, eighty-three landfills were closed in the state of New York alone, and only three were opened as replacements. Cities in the Northeast (including New York City) faced the greatest attrition in available landfill sites and thus incurred the highest tipping fees. These problems, complicated by environmental risks from leachate, methane discharge, and other health and environmental hazards, made understandable the dilemma local officials confronted. See Martin V. Melosi, *The Sanitary City: Environmental Services in Urban America from Colonial Times to the Present*, abridged ed. (Pittsburgh, PA: University of Pittsburgh Press, 2008), 247; Ford Fessenden, "Landfills Head for the Scrap Heap," in Newsday, *Rush to Burn: Solving America's Garbage Crisis?* (Washington, DC: Island, 1989), 44; Ted Steinberg, *Down to Earth: Nature's Role in American History* (New York: Oxford University Press, 2009), 233–34. See also Hans Tammemagi, *The Waste Crisis: Landfills, Incinerators, and the Search for a Sustainable Future* (New York: Oxford University Press, 1999), 28–31.

2. The context for the chronic disposal problems in American cities rested upon a sharp increase in the amounts of discards and the cost of handling refuse. Between 1970 and 1980, the generation of municipal solid waste in the United States increased from 3.3 to 3.6 pounds per capita per day. In 1986, refuse from residences and businesses averaged 4.3 pounds per capita per day. In 1980, solid waste management exceeded total expenditures of $ 4 billion annually, up from $1 billion (in real dollars) in 1960. Melosi, *The Sanitary City*, 247.

3. "Incinerator Closing in South Brooklyn," *New York Times*, January 3, 1981.

4. Colin Campbell, "World's Biggest Dump for Garbage Just a Monumental Problem on S.I.," *New York Times*, May 28, 1981.

5. Campbell, "World's Biggest Dump for Garbage"; City of New York, Department of Sanitation, "Facts About Fresh Kills," Fall 1984, Box 36, The Assemblyman Eric N. Vitaliano Papers, Archives & Special Collections, Department of the Library, College of Staten Island, CUNY; "Landfill Volume to Double on Jan. 1," *Staten Island Advance*, July 21, 1985.

6. Casowitz was sensitive to the complaints of Staten Islanders but believed—like Norman Steisel and others—that the landfill should remain open for its useful life. Oral interview, Martin Melosi and Paul Casowitz, November 16, 2017.

7. Campbell, "World's Biggest Dump for Garbage." See also Janet Marinelli and Gail Robinson, "We Produce Enough Solid Waste to Fill the Superdome from Floor to Ceiling Twice a Day. That's the Problem. Is There a Solution?" *Progressive* 45 (December 1981): 23–27. See also Clyde Haberman, "15% of Sanitation Force Is Termed Misused," *New York Times*, May 8, 1980.

8. Michael Azzara, "Work Begins on Expressway Fence," *Staten Island Advance*, December 3, 1980.

9. See, for example, City of New York, Department of Sanitation, *Solid Waste Disposal Operations Plan, Fresh Kills Landfill: Alternative Grading and End-Use Plans* (July 21, 1985). Steisel suggested that the DSNY was working on an end-use plan in 1980 or 1981. Oral interview, Martin Melosi with Norman Steisel, November 15, 2017.

10. Campbell, "World's Biggest Dump for Garbage."

11. Edward I. Koch to Bill Bradley, November 17, 1980, Departmental Correspondence Series, Edward I. Koch Collection, Box 240, Folder 9, La Guardia and Wagner Archives, LaGuardia Community College/CUNY, Long Island City, New York.

12. Remarks by John J. Marchi, "DEC Meeting on the Future of New York City's Landfills, Wagner College," December 15, 1980, Box 37, The Assemblyman Eric N. Vitaliano

Papers, Archives & Special Collections, Department of the Library, College of Staten Island, CUNY. See also Norman Steisel to John J. Marchi, December 29, 1983, The Senator John J. Marchi Papers, Series 1: Subject Files, Box 92, Folder 11, Archives & Special Collections, Department of the Library, College of Staten Island, CUNY. See also "The City Is Running Out of Time," *New York Times*, April 11, 1981.

13. "Garbage Disposal Costs to Soar as Era of Landfill Ends," *New York Times*, July 7, 1980. A number of possible technologies were floated in 1980 about addressing the landfill predicament. The New York State Power Authority proposed a seven-hundred-megawatt electricity-generating plant at Arthur Kill that would burn coal and refuse as part of its multimillion-dollar antipollution program. Other plans—such as barging compost from Wards Island to Fresh Kills—seemed to exacerbate nuisance problems. Assemblywoman Elizabeth Connelly strongly opposed this part of the city's Interim Sludge Management Plan. See Peter Kihss, "Power Authority Presents New Antipollution Plan," *New York Times*, August 24, 1980; Elizabeth A. Connelly to Charles S. Warren, Regional Administrator, EPA, September 16, 1980, The Assemblywoman Elizabeth A. Connelly Papers, Series 1: Subject Files, Box 20, 1981–1989, Archives & Special Collections, Department of Library, College of Staten Island, CUNY. See also Minutes, Fresh Kills Advisory Committee, February 4, 1981, Connelly Papers, Series 1: Subject Files, Box 20, 1981–1989, Folder 2; "News from Assemblywoman Elizabeth A. Connelly," December 9, 1981, Connelly Papers, Series 1: Subject Files, 1981–1989, Box 6, Folder 2.

14. "LOBR Review of Request for Proposals for the Brooklyn Navy Yard Resource Recovery Facility," April 2, 1980, New York City, Municipal Archives. See also James Barron, "Garbage Is Garbage, Burning It for Energy Is Difficult," *New York Times*, August 2, 1981; Colin Campbell, "City's Street-Cleaning Effort Is Getting Better, Koch Says," *New York Times*, July 28, 1981; "Waste Disposal in New York City: An Issue Without a Constituency," *Waste Age* 12 (December 1981): 44–46.

15. City of New York, Office of the Corporate Counsel, "Impact of Norman Steisel's Employment by Lazard Freres on the Brooklyn Navy Yard Resource Recovery Project," December 1, 1986, 5–7; Memorandum from Norman Steisel to Members of the Board of Estimate, "Waste Disposal Planning for New York City: No Time to Waste," September 25, 1984, 1–2, Vitaliano Papers, Series 1, Subject Files, Box 64.

16. See Clyde Haberman, "A Steam Plant Burning Refuse Due in Brooklyn," *New York Times*, December 11, 1981; Michelle Slatalla, "The 'Trash' of Incineration," in *Rush to Burn*, 112; Davis and Meier, "Will It Go Up in Smoke?" 145; Matthew Gandy, *Concrete and Clay: Reworking Nature in New York City* (Cambridge, MA: MIT Press, 2002), 208; Heather Rogers, *Gone Tomorrow: The Hidden Life of Garbage* (New York: The New Press, 2005), 161–63.

17. Dioxin, a lethal carcinogen, is a component of the defoliant Agent Orange and also produced in the incineration process. Furan, another carcinogen, is a highly flammable and volatile liquid.

18. Harrison J. Goldin to "Dear Friend," January 21, 1983, Vitaliano Papers, Series 1, Subject Files, Box 36, Folder 15.

19. James Barron, "The Heat's on the City to Burn Its Trash," *New York Times*, December 11, 1983.

20. Norman Steisel and Paul D. Casowitz, "Incinerate New York Garbage," *New York Times*, August 20, 1983.

21. Rupert Cornwell, "Barry Commoner: Scientist Who Forces Environmentalism Into the World's Consciousness," *Independent*, October 5, 2012, http://www.independent

.co.uk/news/obituaries/barry-commoner-scientist-who-forced-environmentalism-into-the-worlds-consciousness-8200315.html.

22. Michael Egan, *Barry Commoner and the Science of Survival: The Remaking of American Environmentalism* (Cambridge, MA: MIT Press, 2007), 19–22; "Barry Commoner Facts," *Encyclopedia of World Biography*, http://biography.yourdictionary.com/barry-commoner; Peter Dreier, "Remembering Barry Commoner," *Nation*, October 1, 2012, http://www.thenation.com/article/remembering-barry-commoner/.

23. "Barry Commoner Facts"; Dreier, "Remembering Barry Commoner"; Simon Butler, "Barry Commoner: Scientist, Activist, Radical Ecologist," *Climate and Capitalism*, October 5, 2012, http://climateandcapitalism.com/2012/10/05/barry-commoner-scientist-activist-radical-ecologist/; MacBride, *Recycling Reconsidered*, 75. See also Egan, *Barry Commoner and the Science of Survival*.

24. Dreier, "Remembering Barry Commoner." See also Elaine Woo, "Barry Commoner Dies at 95; Pillar of Environmental Movement," *Los Angeles Times*, October 1, 2012, http://articles.latimes.com/keyword/barry-commoner; Daniel Lewis, "Scientist, Candidate, and Planet Earth's Lifeguard," *New York Times*, October 1, 2012. See also Thomas Vinciguerra, "At 90, an Environmentalist from the '70s Still Has Hope," *New York Times*, June 19, 2007; Andrew C. Revkin, "Barry Commoner's Uncommon Life," *New York Times*, October 1, 2012.

25. "Brooklyn Incinerators: Pointless Worry Over Dioxin Emission," *New York Times*, October 29, 1983.

26. See "Meeting of the Fresh Kills Landfill Advisory Committee," March 1, 1983, Vitaliano Papers, Series 1, Subject Files, Box 36, Folder 15.

27. City of New York, Department of Sanitation, Resource Recovery and Waste Disposal Planning, *Solid Waste Disposal Operations Plan, Interim Report*, January 1983, 1, Vitaliano Papers, Series 1, Subject Files, Box 36.

28. City of New York, Department of Sanitation, Resource Recovery and Waste Disposal Planning, *Solid Waste Disposal Operations Plan, Interim Report*, 5.

29. City of New York, Department of Sanitation, *Solid Waste Disposal Operations Plan: Fresh Kills Landfill, Alternative Grading and End-Use Plans*, February 2, 1983, 1-1, Vitaliano Papers, Series 1, Subject Files, Box 36, Folder 15.

30. City of New York, Department of Sanitation, *Solid Waste Disposal Operations Plan: Fresh Kills Landfill, Alternative Grading and End-Use Plans*, 5–9. See also "Brooklyn to Add Garbage Facility," *New York Times*, February 8, 1982.

31. City of New York, Department of Sanitation, Resource Recovery and Waste Disposal Planning, *Solid Waste Disposal Operations Plan, Interim Report*, January 1983, 10.

32. City of New York, Department of Sanitation, Resource Recovery and Waste Disposal Planning, *Solid Waste Disposal Operations Plan, Interim Report*, January 1983, 35. See also Marla Ucelli, "Fresh Kills: Garbage Mounts Faster Than Opposition?" *Staten Island Advance*, June 9, 1983. See also Mark Labaton, "Support for Waste Incinerators Given by City and State Officials," *Staten Island Advance*, June 3, 1983.

33. See Martin V. Melosi, ed., *Effluent America: Cities, Industry, Energy, and the Environment* (Pittsburgh, PA: Pittsburgh University Press, 2001), 18. See also Craig E. Colten and Peter N. Skinner, *The Road to Love Canal: Managing Industrial Waste Before EPA* (Austin: University of Texas Press, 1996); Elizabeth Blum, *Love Canal Revisited: Race, Class, and Gender in Environmental Activism* (Lawrence: University Press of Kansas, 2008); Richard S. Newman, *Love Canal, a Toxic History from Colonial Times to the Present* (New York: Oxford University Press, 2016).

34. Tom Robotham and Bruce Alpert, "Connelly Threatens Sanit Department: 'Keep Out Toxics or We'll Shut Landfill,'" *Staten Island Advance*, January 4, 1982.
35. Renamed the Quanta Resources Corporation.
36. Leslie Bennetts, "Dumping of Lethal Wastes at 3 City Landfills Charged," *New York Times*, January 5, 1982.
37. Leslie Bennetts, "City Studying Ways to Test for Illegal Waste in Dumps," *New York Times*, January 6, 1982.
38. Cassiliano had submitted retirement papers soon after the January reports of the dumping, but his in-house trial was completed before the paperwork could be finalized. See Michael Goodwin, "Sanitation Aide Dismissed in Dumping of Toxic Waste," *New York Times*, February 12, 1982; "Toxic Dumper Convicted," *Staten Island Register*, November 18, 1982.
39. "City Plans Screening of Landfill Workers for Toxic Chemicals," *New York Times*, May 9, 1982. See also Peter Kihss, "After 7 Years, Plan for Power Plant on Arthur Kill's S.I. Shore Still Disputed," *New York Times*, April 10, 1982.
40. Ralph Blumenthal, "S.I. Dump Survey Finds No Serious Health Peril," *New York Times*, November 25, 1982. See also Michael Oreskes, "New York Studies S.I. Landfill Odors: Brookfield Dump Found to Be Source of Health Problems in the Vicinity in 1982," *Staten Island Advance*, November 24, 1985; "Brookfield Group Confronts Koch," *Staten Island Register*, July 15, 1982; "IRATE Tours Brookfield Landfill," *Staten Island Register*, February 17, 1983; Marilyn Wisniewski, "Great Kills Residents Picket City Hall," *Staten Island Register*, June 10, 1982; Roz Appel, "Broken Promises on Brookfield Dump," *Staten Island Register*, May 27, 1982; "Brookfield Meeting Angers IRATE," *Staten Island Register*, December 9, 1982.
41. "Tour of Environmental Hazards Set," *Staten Island Advance*, March 31, 1983.
42. Mary Engels, "Bill Would Close City Landfills," *Staten Island Advance*, March 8, 1983. See also Jim Buell, "Landfill Plan Hit by Board," *Staten Island Advance*, May 18, 1983; "Hearing Set on Fresh Kills," *Staten Island Advance*, June 2, 1983.
43. Elizabeth A. Connelly to Harrison J. Goldin, May 11, 1983, Connelly Papers, Series 1: Subject Files, Box 20, 1978–1998, Lawsuits, Folder 7. See also "News from Assemblywoman Elizabeth A. Connelly," May 17, 1983; "News from Assemblywoman Elizabeth A. Connelly," June 16, 1983, Connelly Papers, Series 1: Subject Files, Box 20, 1980–1989, Folder 2; "Connelly Asks Probe of Landfill Plans," *Staten Island Advance*, May 18, 1983; Mary Engels, "'Don't Dump on Us,' Connelly Says to City," *Staten Island Advance*, May 23, 1983.
44. Raymond Wittek, "Connelly in Court to Force Compliance with 1980 Decree," *Staten Island Advance*, September 29, 1983. See also Summons, Supreme Court of the State of New York, County of Richmond, *In the Matter of the Application of Elizabeth A. Connelly, et al. Against the City of New York, et al.*, September 27, 1983, Index no. 2026/83; and Order to Show Cause, *In the Matter of the Application of Elizabeth A. Connelly, et al. Against the City of New York, et al.*, September 28, 1983, Index no. 2026/83, Connelly Papers, Series 1: Subject Files, 1978–1998, Box 20, Folder 7; "Landfill Suit May Get City's Attention," *Staten Island Advance*, October 4, 1983.
45. Bill Farrell, "Down in Dumps on a Lovely Day," *Staten Island Advance*, May 19, 1983. See also Ralph Blumenthal, "Toxic Dumping in City Landfills Cited in a Study," *New York Times*, May 19, 1983.
46. Chris Olert, "Garbage 'Crisis' Near, Official Says," *Staten Island Advance*, May 28, 1983. See also "Garbage: How High Is High?" *Staten Island Advance*, June 14, 1983; Jim Buell, "Sanit Officials Hear Opposition to Fresh Kills Plan," *Staten Island Advance*,

June 14, 1983. See also "State Must Watch Landfill," *Staten Island Advance*, April 20, 1983.
47. Mark Labaton, "BP Asks Citywide Garbage Solution," *Staten Island Advance*, June 8, 1983. See also Ruth Shereff, "No Sanit Plans for Landfill," *Staten Island Register*, March 22, 1984.
48. Terence J. Kivlan, "Molinari Asks Feds to Probe Brookfield Dump," *Staten Island Advance*, May 24, 1983.
49. "Get to Bottom of Landfill," *Staten Island Advance*, May 28, 1983. See also Tom Robotham, "Molinari Tours Toxic Dump Sites; Promises 'Leads,'" *Staten Island Advance*, May 31, 1983; "Let's Unite in Landfill Fight," *Staten Island Advance*, May 31, 1983; "Molinari Skeds Landfill Meeting," *Staten Island Advance*, July 18, 1983. See also "Officials Seek Documents Showing Hazardous Dumps," *Journal* (Ogdensburg, NY), January 3, 1984.
50. Reginald Patrick, "CB 2 Supports Fossella Plan to Force Fresh Kills Closing," *Staten Island Advance*, June 19, 1983.
51. Elizabeth Connelly: Oral History Transcripts, February 9, 2004, 75–76, College of Staten Island Oral History Collection, v. 3, 2005, ed. Jeffery A. Kroessler, Archives & Special Collections, Department of the Library, College of Staten Island, CUNY.
52. Many registered Democrats there were "Reagan Democrats."
53. Biographical Sketch, Vitaliano Papers. See also "Statement on Resource Recovery Plan, New York City Board of Estimate," August 15, 1985, Vitaliano Papers, Series 1, Subject Files, Box 36.
54. Eric N. Vitaliano, "Statement on the Freshkill Landfill, Community Board #2 Hearing," June 13, 1983, Vitaliano Papers, Series 1, Subject Files, Box 36. See also Vitaliano, "Statement at the Department of Environmental Conservation's Public Meeting on the New York City Landfills, Wagner College, Staten Island," December 15, 1983, Vitaliano Papers, Series 1, Subject Files, Box 36. Even before 1982, Vitaliano was active in addressing the Fresh Kills issue. See "Statement of Eric N. Vitaliano, Re: Upgrading of the Fresh Kills Landfill, May 21, 1981, Public School 32, S.I.," Vitaliano Papers, Series 1, Subject Files, Box 40.
55. "State to Join in Garbage Dump Debate," *Staten Island Advance*, December 9, 1983. See also Wendy Diller, "Islandwide Movement to Oppose Landfill," *Staten Island Advance*, December 2, 1983; Ruth Shereff, "State Wants City to Clean Up Landfills," *Staten Island Advance*, December 7, 1983; "State to Join in Garbage Dump Debate," *Staten Island Advance*, December 9, 1983; "Whose Move on Garbage?" *Staten Island Advance*, December 22, 1983; "2 Contracts Affect Future of Fresh Kills Landfill," *Staten Island Advance*, December 14, 1983.
56. "Judge: I'll Close Fresh Kills If Garbage Isn't Contained," *Staten Island Advance*, June 2, 1983. See also "Wretched Refuse Suit Is Settled," *New York Times*, June 5, 1983.
57. "Judge Refuses to Shut Down Dump in New York–New Jersey Dispute," *New York Times*, April 22, 1984. See also Leo H. Carney, "The Environment," *New York Times*, May 20, 1984.
58. Charles Brecher et al., *Power Failure: New York City Politics and Policy Since 1960* (New York: Oxford University Press, 1993), 230–32.
59. Lucius J. Riccio, "Management Science in New York's Department of Sanitation," *Interfaces* 14 (March–April 1984): 11. See also Joseph J. Timpone and Paul E. Sussman, "Excellence Training for Productivity," *Public Productivity Review* 11 (Spring 1988): 105–14.
60. Riccio, "Management Science in New York's Department of Sanitation," 1.

61. Memorandum from Norman Steisel to Members of the Board of Estimate, "Waste Disposal Planning for New York City: No Time to Waste," September 25, 1984, 2, Vitaliano Papers, Series 1, Subject Files, Box 64. See also City of New York, Office of the Corporate Counsel, "Impact of Norman Steisel's Employment by Lazard Freres on the Brooklyn Navy Yard Resource Recovery Project," December 1, 1986, 5–7; David W. Dunlap, "City Plan Lists 8 Sites for Plants to Burn Trash and Make Power," *New York Times*, April 13, 1984. See Matthew L. Wald, "Is New York City's Incinerator Plan a Health Threat?" *New York Times*, November 26, 1984.
62. Memorandum from Norman Steisel to Members of the Board of Estimate, "Waste Disposal Planning for New York City: No Time to Waste," September 25, 1984, 4–5. See also Norman Steisel, Regina Morris, and Marjorie J. Clarke, "The Impact of the Dioxin Issue on Resource Recovery in the United States," *Waste Management & Research* 5 (1987): 381–94.
63. Steisel suggested that one of the advantages of the Brooklyn Navy Yard was that it was next to an existing Con Edison steam loop producing steam that went under the East River to Lower Manhattan and into a district heating system. Oral interview, Martin Melosi with Norman Steisel, November 15, 2017.
64. The primary distinction between mass-burn incineration and mass-burn water-wall incineration is that the former burns refuse directly to generate electricity; in the latter, refuse is burned in water-wall furnaces to generate steam.
65. Memorandum from Norman Steisel to Members of the Board of Estimate, "Waste Disposal Planning for New York City: No Time to Waste," September 25, 1984, 4; Matthew L. Wald, "City Studies See Little Dioxin from Garbage Fueled Plants," *New York Times*, September 26, 1984.
66. City of New York, Department of Sanitation, *The Waste Disposal Problem in New York City: A Proposal for Action* (April 1984), P-1-P-2, Science, Industry and Business Library, New York Public Library; Memorandum from Norman Steisel to Members of the Board of Estimate, "Waste Disposal Planning for New York City: No Time to Waste," 2.
67. City of New York, Department of Sanitation, *The Waste Disposal Problem in New York City: A Proposal for Action*, P-2.
68. City of New York, Department of Sanitation, *The Waste Disposal Problem in New York City: A Proposal for Action*, I-1–I-6.
69. City of New York, Department of Sanitation, *The Waste Disposal Problem in New York City: A Proposal for Action*, I-6–I-8.
70. City of New York, Department of Sanitation, *The Waste Disposal Problem in New York City: A Proposal for Action*, V-2, V-3. See also "Excerpts from Sanitation Dept. Report on Garbage Disposal," *New York Times*, April 13, 1984.
71. Memorandum from Norman Steisel to Members of the Board of Estimate, "Waste Disposal Planning for New York City: No Time to Waste," 11. See also Joan Gerstel, "NYPIRG: Recycling Garbage Best for City," *Staten Island Register*, December 20, 1984.
72. See, for example, Dan Janison, "Environmental Panel OKs 1990 Landfill Cutoff Date," *Staten Island Advance*, May 10, 1984; "NYPIRG: An Alternative to Koch's Incinerators," *Staten Island Register*, December 13, 1984; Jim Callaghan and Joan Gerstel, "Cancer Up If Garbage Plants Are Built," *Staten Island Register*, May 24, 1984.
73. Mary Engels, "Marchi Plan: Heap of Controversy," *Staten Island Advance*, January 16, 1984. See also Jim Buell, "Marchi Calls for 1st Trash Recovery Unit on Island," *Staten Island Advance*, December 16, 1984.

74. Norman Steisel to John J. Marchi, January 5, 1984, Marchi Papers, Series 1: Subject Files, Box 92, Folder 11.
75. "News from Senator John J. Marchi," January 10, 1984, Marchi Papers, Series 1: Subject Files, Box 92, Folder 11. See also John J. Marchi to Henry Williams, Department of Environmental Conservation, January 16, 1984, Marchi Papers, Series 1: Subject Files, Box 92, Folder 11.
76. "Remarks by Senator John J. Marchi, prepared for a public meeting on the Fresh Kills landfill, to be delivered at Susan Wagner High School, Thursday, February 9, 1984," Marchi Papers, Series 1: Subject Files, Box 92, Folder 11. See also Dan Janison, "State's $4M for Landfill Not a Waste," *Staten Island Advance*, April 10, 1984.
77. "News from Senator John J. Marchi," April 10, 1984, May 9, 1984; and May 16, 1984, Marchi Papers, Series 1: Subject Files, Box 92, Folder 11. See also John J. Marchi, "A New York Garbage Solution Isn't in the Bag," *New York Times*, June 16, 1984; Eric A. Goldstein to John J. Marchi, June 19, 1984, Marchi Papers, Series 1, Subject Files, Box 92, Folder 1.
78. "News from Assemblyman Eric N. Vitaliano," February 7, 1984, Vitaliano Papers, Series 1, Subject Files, Box 40. See also Dan Janison, "Vitaliano Pushes Conversion of Garbage Loading Stations," *Staten Island Advance*, February 8, 1984.
79. Eric Vitaliano, "Statement on Resource Recovery Plan, Public Hearings, City Hall," May 2, 1984, Vitaliano Papers, Series 1, Subject Files, Box 40.
80. Jim Buell, "Brookfield Area Residents to Ask for Public Hearing," *Staten Island Advance*, January 25, 1984. See also Jim Buell, "'Don't Dump on S.I.,' Official Rallying Call," *Staten Island Advance*, February 3, 1984; Sam Roberts, "City Public Works Plan Gets Mixed Report Card from Watchdog Group," *New York Times*, December 1, 1983; "City Garbage Plan a Necessity Now," *Staten Island Advance*, April 8, 1984.
81. Mike Hammer, "Options Given for Final Brookfield Closing," *Staten Island Advance*, February 17, 1984. See also Rick Korin to Eric N. Vitaliano, February 7, 1984; and Citizens Alliance, "Question for the Public Meeting," February 7, 1984, Vitaliano Papers, Series 1, Subject Files, 1984–1993, Box 38.
82. York Research Corporation, *Final Test Report for the First Air Quality Monitoring Program Conducted November 22/23, 1983, at the Brookfield Landfill*, March 19, 1984; and Orders on Consent, *In the Matter of the Development and Implementation of a Remedial Program for an Inactive Hazardous Waste Disposal Site*," Index #2-43-008, December 16, 1985, Vitaliano Papers, Series 1, Subject Files, 1984–1993, Box 38.
83. David Richardson, "Oil Dumps, New Tricks: Turning Landfills in Nature Preserves," *Grist.org*, January 24, 2012, https://grist.org/pollution/old-dumps-new-tricks-turning-landfills-into-nature-preserves/.
84. "The Creeping Landfill," *Staten Island Advance*, October 12, 1984. See also Elizabeth A. Connelly to Leonard A. Okoskin, Community Action Committee, November 5, 1985, Vitaliano Papers, Series 1, Subject Files, Box 36, Folder 15.
85. Eric N. Vitaliano to Henry G. Williams, October 5, 1984, Vitaliano Papers, Series 1, Subject Files, Box 37, Folder 1. See also "News from Assemblyman Eric N. Vitaliano," October 5, 1984, Vitaliano Papers, Series 1, Subject Files, Box 37, Folder 1. See also Bill Farrell, "Bigger Dump Rapped: Lawmaker Asks State Help at Fresh Kills," *Staten Island Advance*, October 7, 1984; "Vitaliano Asks State Official to Nix Fill Plan," *Staten Island Advance*, October 10, 1984.
86. Henry G. Williams to Assemblyman Vitaliano, December 13, 1984, Vitaliano Papers, Series 1, Subject Files, Box 37, Folder 1.
87. "City Must Utilize New Dumping Area," *Staten Island Advance*, November 8, 1984.

88. John E. Hurley, "Sanit Dept. Defends Use of New Landfill Area," *Staten Island Advance*, November 16, 1984.
89. Leonard A. Okoskin to Edward I. Koch, December 12, 1984, Vitaliano Papers, Series 1, Subject Files, Box 37, Folder 1. See also Nancy A. Tracey, Public Services Committee and Maxine J. Spierer, Chairman of the Borough of Staten Island Community Board 3 to Norman Steisel, December 17, 1984, Vitaliano Papers, Series 1, Subject Files, Box 37, Folder 1.
90. Mrs. Laura Mannino to Elected Official, January 17, 1985, Vitaliano Papers, Series 1, Subject Files, Box 37, Folder 1. See also Denise Reinaldo, "Only Steisel Can Halt Dump Plan," *Staten Island Advance*, January 16, 1985; Jack and Gloria Cohen to Eric N. Vitaliano, January 18, 1985, Vitaliano Papers, Series 1, Subject Files, Box 37, Folder 1; Richard and Paulette Young to Ralph J. Lamberti, January 29, 1985, Vitaliano Papers, Series 1, Subject Files, Box 36.
91. "Lamberti Asks Sanitation Boss to Alter Landfill Dumping Plan," *Staten Island Advance*, January 18, 1985. See also "Better Places to Dump," *Staten Island Advance*, January 20, 1985. Vitaliano also had kept up the pressure. See Phyllis Cirillo, *Report to Assemblyman Eric Vitaliano Regarding Sections 2 & 8 of the Fresh Kills Landfill*, January 27, 1985, Vitaliano Papers, Series 1, Subject Files, Box 37, Folder 1.
92. "Some Battles Are Worth Forgetting," *Staten Island Advance*, February 2, 1985. See also Michael Carpinello to Eric Vitaliano, February 1, 1985, Vitaliano Papers, Series 1, Subject Files, Box 36, Folder 15.
93. See also Jim Hughes, "Activist Group Calls for Landfill Protests," *Staten Island Advance*, April 24, 1985. See also Dennis S. McKeon, President of Citizens for a Cleaner Staten Island, Inc., to Eric Vitaliano, May 4, 1985, Vitaliano Papers, Series 1, Subject Files, Box 36, Folder 4.
94. Eric N. Vitaliano to Mr. and Mrs. Vincent Becastro, February 12, 1985, Vitaliano Papers, Series 1, Subject Files, Box 37, Folder 1.
95. Jim Hughes, "Expanded Landfill Dumping in Works," *Staten Island Advance*, November 24, 1985.
96. See "News from Assemblyman Eric N. Vitaliano," February 6, 1986; and Eric N. Vitaliano to Kathleen Dodd, Village Greens Residents Association, May 7, 1986, Vitaliano Papers, Series 1, Subject Files, Box 36, Folder 15; Norman Steisel to Eric Vitaliano, December 27, 1985, Vitaliano Papers, Series 1, Subject Files, Box 35, Folder 18. See also Eric N. Vitaliano to Kathleen Dodd, Village Greens Residents Association, May 7, 1986, Vitaliano Papers, Series 1, Subject Files, Box 35, Folder 18. See Dan Janison, "Assembly OKs Bill on Landfill," *Staten Island Advance*, March 14, 1985, concerning the Pennsylvania Avenue landfill with respect to toxic-waste cleanup. See "The Landfill Deadline," *Staten Island Advance*, March 17, 1985, concerning the Fountain Avenue landfill.
97. Josh Barbanel, "Officials Fault Plans to Build 8 Trash Plants," *New York Times*, December 7, 1984. See also Alexander Williams, "Trash Talk: Solid Waste Disposal in New York City," 2013, *Student Theses 2001–2013*, paper 18, http://fordham.bepress.com/environ_theses/18/.
98. "The City's Dilemma: 5 Boroughs, 1 Dump," *Staten Island Advance*, December 10, 1984. See also "Fresh Kills to Be Only City Dump," *Staten Island Advance*, December 7, 1984. See also Dan Janison, "Assembly OKs Bill on Landfill," *Staten Island Advance*, March 14, 1985; "The Landfill Deadline," *Staten Island Advance*, March 17, 1985.

99. Tevere MacFadyen, "Where Will All the Garbage Go?" *Atlantic* 255 (1985): 29, 32–33; Citizens Advisory Committee for Resource Recovery in Brooklyn, "Statement of Concerns and Recommendations Regarding the Proposed Brooklyn Navy Yard Resource Recovery Facility," April 4, 1985, 10, Vitaliano Papers, Series 1, Subject Files, Box 64, Folder 6 or 8(?). See also Alf Fischbein and Stephen Levin, "Comments on Health Risks Posed by Emissions from the Proposed Resource Recovery Facility for the Citizens' Advisory Committee on Resource Recovery in Brooklyn," March 20, 1985; Arthur J. Fossa to Councilman Abraham Gerges, April 3, 1985, Vitaliano Papers, Series 1, Subject Files, Box 64.
100. Citizens Advisory Committee for Resource Recovery in Brooklyn, "Statement of Concerns and Recommendations Regarding the Proposed Brooklyn Navy Yard Resource Recovery Facility," 1–11, Vitaliano Papers, Series 1, Subject Files, Box 64, Folder 6 or 8(?). See also Martin Gottlieb, "Public Projects: Are They Viable in the City Anymore?" *New York Times*, August 14, 1985.
101. Daniel Hays, "Incinerator Plan Is Fought on Environmental Grounds," *New York Daily News*, April 30, 1985. See also "What Are the Alternatives?" *New York Post*, May 1, 1985.
102. City Council President Carol Bellamy, Comptroller Goldin, and Brooklyn Borough President Golden voted against it. Manhattan Borough President Andrew J. Stein and Staten Island Borough President Lamberti voted for it.
103. Josh Barbanel, "Approval Is Given for Incinerator at Brooklyn Site," *New York Times*, August 16, 1985. See also Jeffrey Schmalz, "Sorting Out Solutions to City's Trash," *New York Times*, August 17, 1985; Eric N. Vitaliano to Barbara Ferrante, Treasurer, August 21, 1985, Vitaliano Papers, Series 1, Subject Files, Box 36, Folder 15; Eric Vitaliano, Statement on Resource Recovery Plan, New York City Board of Estimate, City Hall, August 15, 1985, Vitaliano Papers, Series 1, Subject Files, Box 36, Folder 15; Statement by Assemblywoman Elizabeth A. Connelly Before the New York City Board of Estimate as It Considers the Development of a Resource Recovery Plant at the Brooklyn Navy Yard, August 15, 1985, Connelly Papers, Series 1, Subject Files, Box 20, Folder 12; Josh Barbanel, "Disputed Incinerator Plan for Brooklyn Navy Yard Near Vote," *New York Times*, August 14, 1985; Williams, "Trash Talk."
104. MacBride, *Recycling Reconsidered*, 76–77. See also Chris Olert, "Trash Recycling Plan," *Staten Island Advance*, July 15, 1985.
105. Martin V. Melosi, *Garbage in the Cities: Refuse, Reform, and the Environment*, rev. ed. (Pittsburgh, PA: University of Pittsburgh Press, 2005), 220–22. The Solid Waste Management Act passed by the state legislature in 1988 called for an unrealistic statewide recycling target of 40 to 42 percent for municipal waste. See Matthew Gandy, "Political Conflict Over Waste-to-Energy Schemes: The Case of Incineration in New York," *Land Use Policy* 12 (1995): 32.
106. MacBride, *Recycling Reconsidered*, 76–77.
107. By August 1983 about one billion cubic feet of methane had been extracted from Fresh Kills, and Getty Synthetic Fuels, Inc., had drilled more than one hundred wells. However, the plant's operation sometimes led to gas releases that caused some short-term nuisances. See "Utility Cites Success in Methane Recovery," *New York Times*, August 5, 1983. Efforts in New Jersey lagged well behind. See Leo H. Carney, "Few Tap Methane Source," *New York Times*, January 20, 1985; James R. Pena, "Promises and Problems in Landfill Gas Recovery at Fresh Kills," *Municipal Engineers Journal* 76 (1988): 23–33.

108. Benjamin Miller, *Fat of the Land: Garbage of New York, the Last Two Hundred Years* (New York: Four Walls Eight Windows, 2000), 226–27, 235–41.
109. Barry Commoner, "Incinerators: The City's Half-Baked and Hazardous Solution to the Solid Waste Problem," *New York Affairs* 2 (1985): 30.
110. Steisel and Casowitz, "The Waste Disposal Crisis and Promise of Resource Recovery," *New York Affairs* 2 (1985): 133–39. For more on Steisel's impressions of Commoner and his role in the fight over the Brooklyn Navy Yards, see oral interview, Martin Melosi with Norman Steisel, November 15, 2017.
111. In a statement before the board at the August 15, 1985, meeting, Vitaliano said the following about the dioxin issue: "Dioxin was the next shibboleth to be hurled by the nay-sayers [concerning the safety of mass-burn resource-recovery facilities]. What about dioxin? Based on studies I have seen, fears of dioxin contamination from the type of process used at Peekskill and proposed for the Brooklyn Navy Yard are unfounded." See "Statement on Resource Recovery Plan, New York City Board of Estimate," August 15, 1985, Vitaliano Papers, Series 1, Subject Files, Box 36.
112. Miller, *Fat of the Land*, 242–51.
113. Dennis S. McKeon to Eric Vitaliano, May 4, 1985, Vitaliano Papers, Series 1, Subject Files, Box 36, Folder 4. See also Flyer, "Citizens for a Cleaner Staten Island, Inc., Enough is Enough!" June 1, 1985, Vitaliano Papers, Series 1, Subject Files, Box 36, Folder 4.
114. This was not the only time anxiety over dumping of medical waste had been voiced. See "Bellamy Criticizes Dumping of Waste," *Staten Island Advance*, June 3, 1985; Barbara Naness, "Medical Waste Prompts Landfill Demonstration," *Staten Island Advance*, June 13, 1985; Sara Rimer, "2 Hospitals Are Barred from Using City's Dumps," *New York Times*, June 29, 1985; Fresh Kills Landfill Citizens Advisory Committee, Minutes of Meeting, June 10, 1985, and July 8, 1985, Vitaliano Papers, Series 1, Subject Files, Box 36, Folder 4. See also "Report to Assemblyman Eric Vitaliano Regarding Sections 2 & 8 of the Fresh Kills Landfill," January 27, 1985, Vitaliano Papers, Series 1, Subject Files, Box 36. See also "Hospitals' Wastes Found in Landfill," *Newsday*, June 3, 1985; "3 Hospitals Dumping 'Infectious' Wastes," *New York Post*, June 3, 1985; Bill Farrell and Paul Meskil, "Hosp Dumping Probe Urged," *New York Daily News*, June 3, 1985; News Release, Office of the City Council President Carol Bellamy, June 2, 1985; and Carol Bellamy to Henry G. Williams, June 6, 1985, B2520-90A, Box 7, Folder: Fresh Kills, New York State Archives, Albany; "Some Hospitals Criticized for Waste Disposal Methods," *Times Record*, March 26, 1986.
115. Dennis S. McKeon to Eric Vitaliano, August 29, 1985, Vitaliano Papers, Series 1, Subject Files, Box 36, Folder 4. See also Mary Engels, "Straniere Seeks to Shrink Fresh Kills Landfill," *Staten Island Advance*, July 31, 1985; Jim Hughes, "Landfill Volume to Double on Jan. 1," *Staten Island Advance*, July 21, 1985; Kathy McCormick, "Travis Resident Want Buffer from Landfill," *Staten Island Advance*, July 24, 1985.
116. Dennis S. McKeon to Eric Vitaliano, October 16, 1985, Vitaliano Papers, Series 1, Subject Files, Box 36, Folder 4. See also Maxine J. Spierer to Harrison J. Goldin, April 9, 1984, Vitaliano Papers, Series 1, Subject Files, Box 36, Folder 15.
117. Jim Hughes, "Activist Group Calls for Landfill Protests," *Staten Island Advance*, April 24, 1985.
118. Vincent J. Moore to Leslie Fasciano, October 12, 1984, B2520-90A, Box 7, Folder: Fresh Kills, New York State Archives.
119. Edgemere would be permanently closed in 1991, but vast amounts of hazardous materials had been discovered there by the early 1980s, and it would become a Superfund

site. Allan R. Gold, "Dump's Close Ends Malodorous Era," *New York Times*, July 15, 1991.

120. Henry G. Williams to Norman Steisel, October 18, 1984, accompanied by the Orders on Consent, DEC File no. 2-0527 (Fresh Kills), DEC File nos. 2-0953 and 2-0953A (Pennsylvania Avenue), DEC File no. 2-0956 (Pelham Bay), DEC File no. 2-0955 (Edgemere), Dec File no. 2-0954 (Fountain Avenue), and DEC File no. 2-0952 (Brookfield), B2520-90A, Box 7, Folder: Fresh Kills, New York State Archives. Other versions of the Orders on Consent were in the files: DEC File no. 2-0957 (Fresh Kills), DEC File no. 2-0955 (Edgemere), DEC File no. 2-0954 (Fountain Avenue). See also Order on Consent, Fresh Kills LF Draft, April 27, 1983, Connelly Papers, Series 1: Subject Files, 1978–1998, Lawsuits, Box 20, Folder 7.

121. Norman Steisel to Henry G. Williams, November 1, 1984, B2520-90A, Box 7, Folder: Fresh Kills, New York State Archives.

122. Oral interview, Martin Melosi and Paul Casowitz, November 16, 2017.

123. News Release, New York State Department of Environmental Conservation, December 16, 1985, Vitaliano Papers, Series 1, Subject Files, Box 38, 1984–1993. See also Memorandum, Department of Sanitation, Public Policy Unit, Robin Geller via Phil Gleason to Martin Dembitz, June 25, 1986; Attachment, "Status of the Brookfield Landfill Closure as of 6/86"; and Martin A. Dembitz to Eric N. Vitaliano, June 30, 1986, Vitaliano Papers, Series 1, Subject Files, Box 36, Folder 15; Dan Janison, "Old Data Used to Declare Brookfield Dump 'Hazard,'" *Staten Island Advance*, July 10, 1986. New York City filed suit in Manhattan Federal Court in early 1985 to force fifteen companies to clean up illegally dumped toxic materials at Brookfield and four other dumpsites in Brooklyn, Queens, and the Bronx. See "City Sues Companies to Pay for Cleanup of Toxic Waste Sites," *Staten Island Advance*, March 14, 1985.

124. Eric N. Vitaliano and Elizabeth A. Connelly to Henry C. Williams, July 24, 1986, Vitaliano Papers, Series 1, Subject Files, Box 36, Folder 15. See also Eric N. Vitaliano to Brendan Sexton, August 21, 1986, Vitaliano Papers, Series 1, Subject Files, Box 36, Folder 15; Elizabeth A. Connelly to Harvey Schultz, DEC, August 21, 1986, Connelly Papers, Series 1, Subject Files, Box 20, 1978–1998, Lawsuits, Folder 7.

125. Robin Nagle, "To Love a Landfill: The History and Future of Fresh Kills," in *Handbook of Regenerative Landscape Design*, ed. Robert L. France (Boca Raton, FL: CRC, 2008), 10. See also Gandy, "Political Conflict Over Waste-to-Energy Schemes," 31; Sam Howe Verhovek, "Sailing 18,000 Tons of Trash a Day Across the Bay," *New York Times*, December 6, 1986.

126. On December 31, 1986, less than a month before Steisel announced his resignation and before he accepted employment with the investment-banking firm of Lazard Freres & Company, the DSNY executed an agreement with The Signal Companies, Inc., for construction and operation of the Brooklyn Navy Yard incinerator. Signal requested that Lazard be appointed as a lead underwriter for the financing of the plant. An investigation by the city's Office of the Corporate Counsel to determine how the underwriters were chosen, and to what degree Steisel had played a role in the process, submitted its findings on December 1, 1986. A major issue addressed was whether Steisel's job with Lazard was a payoff for selecting a Signal subsidiary as a vendor on the project. In its findings, the investigators concluded that "there is no evidence to support the suggestion that Steisel did anything corrupt, unethical, or improper." City of New York, Office of the Corporate Counsel, "Impact of Norman Steisel's Employment by Lazard Freres on the Brooklyn Navy Yard Resource Recovery Project," December 1, 1986, 3. For a not-so-flattering view, see "Fire Norman Steisel

Not Our Trash," *Panix.com*, April 4, 1999, http://www.panix.com/~danielc/nyc/steisel .htm. See also Norman Steisel to Edward I. Koch, December 13, 1985, B2520-90A, Box 7, Folder: Fresh Kills, New York State Archives. See also Norman Steisel to Fellow New Yorker, December 17, 1985, B2520-90A, Box 7, Folder: Fresh Kills, New York State Archives.

127. Oral interview, Martin Melosi with Brendan Sexton, November 17, 2018.
128. Elizabeth Kolbert, "New Sanitation Chief: Brendan John Sexton," *New York Times*, January 25, 1986. See also Nancy Murray-Bott, "An Analysis of Organizational Structure and Financial Management Decision-Making in New York City's Sanitation Department," PhD diss., New York University, 1986; Jonathan Soffer to Martin Melosi, correspondence, April 9, 2016.
129. Karen Phillips, "Landfill's Neighbors Get Dumping Reprieve," *Staten Island Advance*, May 2, 1986. See also oral interview, Martin Melosi with Brendan Sexton, November 17, 2018.
130. William Kleinknecht, "$3.2M Dredging Job Starts to Keep Fresh Kills Access," *Staten Island Advance*, February 5, 1986.
131. Brendan Sexton to Maxine J. Speirer and Nancy Tracey, June 23, 1986, Vitaliano Papers, Series 1, Subject Files, Box 37, Folder 1. See also oral interview, Martin Melosi with Brendan Sexton, November 17, 2018.
132. "News from Assemblyman Eric N. Vitaliano," May 7, 1986, Vitaliano Papers, Series 1, Subject Files, Box 36, Folder 15. See also "News from Assemblyman Eric N. Vitaliano," August 21, 1986, Vitaliano Papers, Series 1, Subject Files, Box 36, Folder 15; Maurice D. Hinchey, Legislative Commission on Solid Waste Management to Eric N. Vitaliano, May 7, 1986, Vitaliano Papers, Series 1, Subject Files, Box 36, File 15; "Statement by Assemblywoman Elizabeth A. Connelly Before the Legislative Commission on Solid Waste Management, September 24, 1986," Connelly Papers, Series 1, Subject Files, Box 20; Kathleen Dodd, Village Greens Residents Association, Inc., *Statement to the Legislative Commission on Solid Waste Management*, September 24, 1986, Vitaliano Papers, Series 1, Subject Files, Box 36.
133. *Legislative Commission on Solid Waste Management: Fresh Kills Landfill, Public Hearings*, September 24, 1986. See also Kathy McCormack, "Travis Residents Want Buffer from Landfill," *Staten Island Advance*, July 24, 1985.
134. Jim Hughes, "Fresh Kills Landfill 'Will Never Close,'" *Staten Island Advance*, May 13, 1986.
135. "Contract Awarded for Garbage Plant," *New York Times*, January 1, 1986. "A Solid Record on Solid Waste," *New York Times*, January 30, 1986; Herbert Mitgang, "New York's Water and Trash Problems," *New York Times*, January 30, 1986; Edward Hudson, "Garbage Crisis: Landfills Are Nearly Out of Space," *New York Times*, April 4, 1986.
136. "Civic Groups Vow to Fight Landfill," *Staten Island Advance*, January 10, 1986. See also Marsha Meyer, Village Greens Resident Association, to Brendan Sexton, June 22, 1986; and "Public Service Committee Meeting," June 23, 1986, Vitaliano Papers, Series 1, Subject Files, Box 36, Folder 15.
137. Eric N. Vitaliano to John N. Solazzo, Groups Against Garbage, February 7, 1986, and Maxine J. Spierer, Groups Against Garbage to Staten Islander, May 1, 1986, Vitaliano Papers, Series 1, Subject Files, Box 36; Karen Phillips, "Group Urges Islanders to Fight Landfill Plans," *Staten Island Advance*, May 13, 1986; Maxine J. Speirer to Members of the New York Assembly, September 24, 1986, Vitaliano Papers, Series 1, Subject Files, Box 36.

138. Hearing Testimony, February 1986, Vitaliano Papers, Series 1, Subject Files, Box 40, Folder 11–14(?).
139. See *The Economics of Recycling Municipal Waste: Background Analysis and Policy Approaches for The State and Local Governments*, Staff Report, New York State Legislative Commission on Solid Waste Management (Albany, 1986), 2.
140. *The Economics of Recycling Municipal Waste*, 114.

13. BARGE TO NOWHERE

1. "Ashes to Ashes," *Poughkeepsie Journal*, June 26, 1988.
2. City of New York, Department of Sanitation, *White Paper, New York City Recycling Strategy*, January 1988.
3. Harrelson argued that tightly compressed bales of garbage would be more efficient in producing methane gas than the loosely packed waste in landfills. Experts in the field were not so convinced. Philip S. Gutis, "Seeking Methane in Bales of Trash," *New York Times*, June 10, 1987.
4. Benjamin Miller, *Fat of the Land: Garbage of New York, the Last Two Hundred Years* (New York: Four Walls Eight Windows, 2000), 1–4; Shirley E. Perlman, "In the Barge's Wake," in Newsday, *Rush to Burn: Solving America's Garbage Crisis?* (Washington, DC: Island, 1989), 243–44.
5. Harrelson had financial backing ($300,000) from Tommy Gesuale—the only person in New York City who privately barged garbage—and supposedly from Salvatore Avellino, a reputed boss of the Lucchese family. Hroncich also brought in four carting companies as partners. See Alexander Williams, "Trash Talk: Solid Waste Disposal in New York City," DigitalResearch@Fordham, 2013, http://fordham.bepress.com/environ_theses/18/. See also Alex Pasternack, "The Most Watched Load of Garbage in the Memory of Man," *Motherboard*, May 13, 2013, http://motherboard.vice.com/blog/the-mobro-4000; Perlman, "In the Barge's Wake," 244–48.
6. The *Mobro* and *Break of Dawn* were leased from Harvey Gulf International Marine of Harvey, Louisiana. The partners initially envisioned a fleet of four barges hauling ten thousand tons a day. See Perlman, "In the Barge's Wake," 245.
7. Miller, *Fat of the Land*, 4–6. Williams, "Trash Talk."
8. Philip S. Gutis, "For Trash Barge Crew, Empty Days and Flies," *New York Times*, May 4, 1987.
9. Perlman, "In the Barge's Wake," 245–46.
10. Jonah Winter, *Here Comes the Garbage Barge!* (New York: Schwartz & Wade, 2010).
11. Eric Schmitt, "Path Cleared for the Return of L.I.'s Trash," *New York Times*, May 13, 1987.
12. Philip S. Gutis, "Wandering Trash Barge to Return to New York," *New York Times*, May 15, 1987.
13. Schmitt, "Path Cleared for the Return of L.I.'s Trash"; Williams, "Trash Talk."
14. Philip S. Gutis, "New York City Shuns Barge Bearing Well-Travelled Trash," *New York City*, May 16, 1987. See also "Trash Barge Enters New York Harbor, Heads for Brooklyn," *New York Times*, May 17, 1987.
15. Robert D. McFadden, "Garbage Barge Returns in Search of a Dump," *New York Times*, May 18, 1987. See also "Captain Just Wants to Dock Unwanted Garbage Barge," *Casa Grande, Arizona, Dispatch*, May 19, 1987.

16. Philip S. Gutis, "Judge Lifts Ban in Garbage Case but City Doesn't," *New York Times*, May 29, 1987; "Garbage Barge to Get Anchorage on Hudson," *New York Times*, June 2, 1987.
17. John J. Marchi to Edward I. Koch, June 2, 1987, The Senator John J. Marchi Papers, Series 1: Subject Files, Box 101, Folder 8, Archives & Special Collections, Department of the Library, College of Staten Island, CUNY. See also "It Won't Be Coming to Fresh Kills," *Staten Island Advance*, June 8, 1987.
18. "Municipal Waste," *Science* 236 (June 12, 1987): 1409. See also "The Garbage Barge: Symbol of a Problem," *New York Times*, June 29, 1987.
19. Philip S. Gutis, "Trash Barge to End Trip in Brooklyn," *New York Times*, July 11, 1987. See also Mary Connelly and Carlyle C. Douglas, "The Barge Stops Here," *New York Times*, July 12, 1987.
20. Philip S. Gutis, "The End Begins for Trash No One Wanted," *New York Times*, September 2, 1987. See also Miller, *Fat of the Land*, 6–11; "Garbage Barge's 155-Day Odyssey Comes to an End," *New York Times*, August 25, 1987; "Court Again Postpones a Plan on Garbage Barge," *New York Times*, July 18, 1987; Ponnampalam Veeravagu, "Long Island Opinion: To Waste Not, We Must Want Not Incineration, but Planning," *New York Times*, August 2, 1987; Eric Schmitt, "Brooklyn Judge Clears Burning of Barge Trash," *New York Times*, August 11, 1987; "Brooklyn Won't Fight Ruling on Trash Barge," *New York Times*, August 13, 1987.
21. Thomas C, Jorling Papers, Archives and Special Collections, Williams College, http://archives.williams.edu/manuscriptguides/jorling.php. Jorling resigned the commissionership in 1994 (much to the surprise of many) to take a position with International Paper. Some environmentalists questioned whether decisions late in his term favored the company and the paper industry in general, but no formal inquiry was held. James Dao, "A Resignation Sends Albany Into a Debate," *New York Times*, February 4, 1994.
22. William Bunch, "Where Will All the Garbage Go?" in Newsday, *Rush to Burn*, 76; Carl A. Zimring and William L. Rathje, eds., *Encyclopedia of Consumption and Waste: The Social Science of Garbage* (Los Angeles: Sage, 2012), 1:64–65.
23. Chaz Miller, "The Garbage Barge," *Waste 360*, February 2, 2007, http://waste360.com/author/chaz-miller; Norman H. Nosenchuck, "Key Events of the New York State Solid Waste Management Program: 1970–1995," *Albany Law Environmental Outlook* 35 (1995–1996): 37–38; "'Barge to Nowhere' Proves Timeless," *Lowell Sun*, October 22, 2012; John Mulliner, "NY Garbage Barge and PPEC History," *Canadian Packaging*, January 28, 2015, http://www.canadianpackaging.com/sustainability/ny-garbage-barge-ppec-history-139114/; Jane Katz, "What a Waste," Federal Reserve Bank of Boston, Q1, 2002, https://www.bostonfed.org/economic/nerr/rr2002/q1/waste.htm; "To the Editor: A Specter of Wandering Garbage Barges Is Haunting America," *New York Times*, February 7, 1988; Philip S. Gutis, "Closed Beaches and Wandering Barge: Two Chapters in the Same Story," *New York Times*, July 13, 1988; Vivian E. Thomson, *Garbage In, Garbage Out: Solving the Problems with Long-Distance Trash Transport* (Charlottesville: University of Virginia Press, 2009), 4–5; Newsday, *Rush to Burn*, ix–x; Ted Steinberg, *Down to Earth: Nature's Role in American History*, 2nd ed. (New York: Oxford University Press, 2009), 235–36.
24. Thomas C. Jorling to the *New Yorker*, January 11, 1991, David N. Dinkins Collection, Box 139, Folder 114, La Guardia and Wagner Archives, LaGuardia Community College/CUNY, Long Island City, New York.
25. New York State Legislative Commission on Solid Waste Management, *Where Will the Garbage Go? Update 1991*, March 1991, Series 1, Subject Files, Box 65, Folder 18, The

Assemblyman Eric N. Vitaliano Papers, Archives & Special Collections, Department of the Library, College of Staten Island, CUNY. See also Carl Campanile, "Landfill Fees Causing Garbage to Be Shipped to Other States," *Staten Island Advance*, April 5, 1991. See also Eric A. Goldstein and Mark A. Izeman, *The New York Environment Book* (Washington, DC: Island, 1990), 4.

26. U.S. Environmental Protection Agency, *Municipal Solid Waste Generation, Recycling, and Disposal in the United States: Facts and Figures for 2012*, 1–2, https://www.epa.gov/sites/production/files/2015-09/documents/2012_msw_fs.pdf; Heather Rogers, *Gone Tomorrow: The Hidden Life of Garbage* (New York: The New Press, 2005), 155–56.

27. Louis Blumberg and Robert Gottlieb, *War on Waste: Can American Win Its Battle with Garbage?* (Washington, DC: Island, 1989), 125–126. Waste-to-energy facilities were financially risky, requiring a high return on investment for private companies. Most companies, therefore, sought substantial public funding given the capital requirements and uncertain technical operation. There were questions about the ability to secure sufficient quantities of waste and stable markets for the sale of energy. See 126–27, 130–34, 139, 141–42, 144, 147.

28. Rogers, *Gone Tomorrow*, 155–56; U.S. Environmental Protection Agency, *The Solid Waste Dilemma: Agenda for Action, Background Document*, Draft Report, September 1988, 2.D-72.D-8. See also U.S. Environmental Protection Agency, Office of Resource Conservation and Recovery, *Economic Data and Indicators Scoping Analysis*, December 2013.

29. The other two were the Division of Hazardous Waste Remediation and the Division of Hazardous Substances Regulation.

30. During the John Lindsay administration measures dealing with source reduction got a hearing, if no action. Under Abraham Beame and Koch it was a nonissue. Koch opposed the bottle bill, for example. See Samantha MacBride, *Recycling Reconsidered: The Present Failure and Future Promise of Environmental Action in the United States* (Cambridge, MA: MIT Press, 2012), 72–78.

31. Carl Campanile, "'Garbage'? What Garbage?" *Staten Island Advance*, May 12, 1988.

32. There were only 253 active landfills in New York State in 1989.

33. New York State Legislative Commission on Solid Waste Management, *Where Will the Garbage Go? New York's Looming Crisis in Disposal Capacity, Update 1988*, Series 1, Subject Files, Box 65, Folder 19, Vitaliano Papers; New York State, Department of Environmental Conservation, Division of Solid Waste, *New York State Waste Management Plan, 1988/1989 Update*, March 1989, 25–33. See also New York State Legislative Commission on Solid Waste Management, *Where Will the Garbage Go? A Status Report on Solid Waste Management and Disposal in New York State* (January 1990), Vitaliano Papers, Series 1, Subject Files, Box 65, Folder 18; MacBride, *Recycling Reconsidered*, 72–78; "New York City on Verge of Recycling," *Syracuse Herald-Journal*, March 29, 1989.

34. Matthew Gandy, *Recycling and the Politics of Urban Waste* (London: Earthscan, 1994), 78–79; Nosenchuck, "Key Events of the New York State Solid Waste Management Program," 38–39. See also Elizabeth Kolbert, "New York Sets Stiff Rules on Waste," *New York Times*, September 1, 1988; "Compromising on Solid Waste," *New York Times*, January 23, 1989; New York State, Department of Environmental Conservation, *Beyond Waste: A Sustainable Materials Management Strategy for New York State*, December 27, 2010, 16–21.

35. Benjamin Miller to Martin Melosi, April 13, 2016. See also Miller, *Fat of the Land*, 268.

36. MacBride, *Recycling Reconsidered*, 77–78; Miller, *Fat of the Land*, 261.

37. Deidre Carmody, "Environmentalists' Report Faults City's Plans to Incinerate Garbage," *New York Times*, February 21, 1986. For a detailed defense of waste-to-energy technology in New York City, see Norman Steisel, "Waste-to-Energy: The Problem of the Solution, Lessons from New York City," presented to a Regional Workshop on Waste-to-Energy and the Management of Special Wastes, Denver, Colorado, December 13–14, 1988, Dinkins Collection, Box 71, Folder 577.
38. New York State Legislative Commission on Solid Waste Management, *Where Will the Garbage Go? Update 1991*; Campanile, "Landfill Fees Causing Garbage to Be Shipped to Other States"; Goldstein and Izeman, *The New York Environment Book*, 4. In 1988, about 30 percent of the solid waste that was disposed of came from businesses and commercial firms. This percentage dropped as local tipping fees rose. See Goldstein and Izeman, *The New York Environment Book*, 24–25.
39. Edgemere landfill took in 7.6 percent, but only for another year. Goldstein and Izeman, *The New York Environment Book*, 5. See also "Our World: How Bad Is New York's Environment?" *New York* (April 16, 1990): 30–31.
40. Carl Campanile, "Waste Produced by City Jumps 30% in 2-Year Period," *Staten Island Advance*, July 14, 1988. See also Charles Brecher et al., *Power Failure: New York City Politics and Policy Since 1960* (New York: Oxford University Press, 1993), 265.
41. Gandy, *Recycling and the Politics of Urban Waste*, 79; Brad Edmondson, *Environmental Affairs in New York State: An Historical Overview*, Publication no. 72 (New York State Archives, 2001), 45.
42. Curbside recycling systems were adopted a several cities in the 1980s. See Rogers, *Gone Tomorrow*, 167.
43. "Move to Require Recycling Gains in City Council," *New York Times*, January 11, 1988. See also "It's Time New York City Moved Ahead on Waste Recycling," *New York Times*, January 21, 1987; Elizabeth Neuffer, "Newspaper Recycling Catches On in West Village Experiment," *New York Times*, March 15, 1987; Richard Levine, "Koch Budget Provides More for Bridges," *New York Times*, April 29, 1988; David C. Anderson, "For Lack of Options, New York Gets Serious About Recycling," *New York Times*, May 15, 1988; "Recyclists Disband," *New York Times*, May 15, 1988; Newsday, *Rush to Burn*, 227; Miller, *Fat of the Land*, 252–53.
44. Marc Holzer, "Productivity In, Garbage Out: Sanitation Gains in New York," *Public Productivity Review* 11 (Spring 1988): 37–50. See also City of New York, Department of Sanitation, *White Paper, New York City Recycling Strategy*, January 1988.
45. Brecher et al., *Power Failure*, 265, also 269–70.
46. For disposal statistics from 1987 to 1989, see Brecher et al., *Power Failure*, 266–71.
47. "Recycling, Long Neglected, Is Now Seen as Vital to New York," *New York Times*, March 27, 1988. See also James Barron, "Now the Recyclable Trash Is Overwhelming New York," *New York Times*, December 10, 1989; Richard Severo, "New York City Confronts Prospect of Recycling Law," *New York Times*, February 23, 1989; Constance L. Hays, "Recycling, Reality, and the New York City Apartment," *New York Times*, March 19, 1989; "Recycling Law's Mechanics: Symphony of Sorting, Maybe," *New York Times*, March 27, 1989; William Kleinknecht, "Garbage May One Day Be a City Export," *Staten Island Advance*, March 28, 1987.
48. Goldstein and Izeman, *The New York Environment Book*, 22.
49. The City Council announced in 1989 that apartment-house incinerators would be banned in the city in four years. David W. Dunlap, "Panel Votes Bill to Ban Incinerators," *New York Times*, May 23, 1989.
50. Interstate Sanitation Commission, *Annual Report on the Water Pollution Control Activities and the Interstate Air Pollution Program, 1988*, 84–85. See also Kevin P.

Doering, EPA to Eric N. Vitaliano, April 22, 1988; U.S. EPA, Notice no. 88-27, May 5, 1988; William J. Muszynski to Roger Anderson, August 26, 1988; William J. Muszynski to Eric N. Vitaliano, August 26, 1988, Vitaliano Papers, Series 1, Subject Files, Box 64, Folder 8, about various pollution issues.
51. Press Release, Department of Sanitation, November 29, 1988, Vitaliano Papers, Series 1, Subject File, Box 64, Folder 10.
52. Carl Campanile, "City Stung by the DEC's Denial of Dump Exemption," *Staten Island Advance*, November 27, 1988. See also William J. Muszynski to Eric N. Vitaliano, September 26, 1989; William J. Muszynski to Eric N. Vitaliano, October 11, 1989; Constantine Sidamon-Eristoff to Eric N. Vitaliano, December 28, 1989, Vitaliano Papers, Series 1, Subject File, Box 64, Folder 6.
53. "Safety and Solid Waste," *New York Times*, November 25, 1989.
54. Williams, "Trash Talk."
55. See CAC, Minutes of Meeting, January 28, 1987, February 17, 1988, January 20, 1988, March 16, 1988, Vitaliano Papers, Series 1, Subject Files, Box 64, Folder 10; Environmental Risk Limited and RAM TRAC Corporation, *Technical Evaluation of EIS Chapter III, Section D, Titled Preliminary Draft Description of the Proposed Resource Recovery Facility, Arthur Kill, Staten Island*, May 9, 1988, New York City, Municipal Archives.
56. Michael McGarry, "Environmental Scientists Blast Incinerator Plans," *Staten Island Advance*, June 8, 1988.
57. "To the Editor: Eric N. Vitaliano to Jack Rosenthal," *New York Times*, February 27, 1989, Vitaliano Papers, Series 1, Subject Files, Box 36, Folder 15. Vitaliano insisted on responding to Commoner's January 29 article "Don't Let City Garbage Go Up in Smoke" in the *New York Times*, arguing that "an integrated solid waste management strategy which recognizes the perpetual need for carefully planned and sized, properly operated and maintained disposal capacity—beyond Fresh Kills—is the only prudent and responsible approach to tackling the City's trash burden."
58. William Kleinknecht, "Landfill Problems Refuse to Go Away," *Staten Island Advance*, January 18, 1987.
59. See Susan Molinari to Phyllis Cirillo, Huguenot Heights Civic Association, August 26, 1987, and Ralph J. Lamberti to Phyllis Cirillo, August 27, 1987, Vitaliano Papers, Series 1, Subject Files, Box 36, Folder 15; Martin Lipp, "City Weighs Plan to Handle Landfill Seepage," *Staten Island Advance*, May 4, 1987; Gordon Bishop, "Landfill's Leachate Out of Control?" *Staten Island Advance*, November 24, 1987; Joan Gerstel, "Leachate Is Sampled at Fresh Kills," *Staten Island Register*, January 28, 1988; "Sanitation Dept. Starts Leachate Control," *Staten Island Advance*, May 19, 1988; Marty Lipp, "3-Part Plan to Divert Leachate from Landfill," *Staten Island Advance*, June 8, 1988; William D. Marbach with Susan E. Katz, *Newsweek*, July 27, 1987, 51.
60. Kleinknecht, "Landfill Problems Refuse to Go Away."
61. Joan Gerstel, "Molinari Demands Answers About Garbage Mountain," *Staten Island Register*, January 14, 1988. See also Jim Meyer, DSNY, to Members of the Staten Island Citizens Advisory Committee on Resource Recovery, May 12, 1988, Vitaliano Papers, Series 1, Subject Files, Box 64, Folder 10; Wendy Greenfield, "Dump: How High Is Up?" *Staten Island Advance*, August 9, 1987; Karen Tumulty, "No Dumping (There's No More Dump)," *Los Angeles Times*, September 2, 1988.
62. "World's Largest Dump Is at Fresh Kills," *Staten Island Advance*, April 26, 1987; William Kleinknecht, "Landfill Facelift: Beauty Grows from the Garbage," *Staten Island Advance*, November 3, 1987.
63. Press Release, Department of Sanitation, "Fresh Kills to Get Breath of Fresh Air," July 11, 1989, Vitaliano Papers, Series 1, Subject Files, Box 36, Folder 15.

64. William Kleinknecht, "Reopening of Brooklyn Landfill Unlikely," *Staten Island Advance*, January 19, 1987.
65. William Kleinknecht, "Housing Sales Rise Despite Garbage Dump," *Staten Island Advance*, January 20, 1987. See also Beth J. Harpaz, "The World's Largest Dump Grows; Neighbors Say It's a Real Stinker," *Schenectady Gazette*, May 8, 1989.
66. "Landfill Odor Is Moving North," *Staten Island Register*, February 19, 1987.
67. Carl Campanile, "DEC Boss Cites Fresh Kills as One of Best Run in State," *Staten Island Advance*, May 19, 1988. See also Terence J. Kivlan, "Fresh Kills Nearly Within Proposed Rules," *Staten Island Advance*, August 26, 1988. For a detailed physical evaluation of Fresh Kills, see Joseph M. Suflita et al., "The World's Largest Landfill: A Multidisciplinary Investigation," *Municipal Engineers Journal* 76 (1988): 23–33.
68. "Tale of Two Landfills," *Staten Island Advance*, January 25, 1987.
69. "The Garbage Squeeze," *Staten Island Advance*, June 15, 1987. See also Francine Hirsch, "Island to Get More Garbage," *Staten Island Advance*, June 13, 1987.
70. Lori E. Miller, "Landfill in Queens to Close by 1991 Under Agreement," *Staten Island Advance*, July 9, 1987. See also Susan Molinari, "To the Editor: Don't Dump All of New York's Garbage Problem on Staten Island," *Staten Island Advance*, July 21, 1987; "Queens Dump, Divert Trash Here," *Staten Island Advance*, December 24, 1987; Joan Gerstel, "Landfill Lawsuit," *Staten Island Register*, March 17, 1988. New Jersey was having problems of its own with landfill space. See Bob Narus, "Landfill Shortage Putting Pressure on Budgets," *New York Times*, June 5, 1988. A story in the *New York Times* noted, "Two organized-crime families are so worried about running out of the landfill space for their waste-hauling companies in the New York area that they are seeking to gain control of valuable dump sites in Pennsylvania, law enforcement officials say." Selwyn Raab, "Mafia Reported to Be Seeking New Trash Sites," *New York Times*, November 11, 1989.
71. Miller, *Fat of the Land*, 256.
72. Carl Campanile, "Fresh Kills Landfill Could Become Site for Separate Dump with Toxic Ash Waste," *Staten Island Advance*, April 28, 1988.
73. See *Encyclopedia of Consumption and Waste*, ed. Zimring and Rathje, 1:485–87; Martin V. Melosi, *Garbage in the Cities: Refuse, Reform, and the Environment*, rev. ed. (Pittsburgh, PA: University of Pittsburgh Press, 2005), 203, 219.
74. Carl Campanile, "DEC Neglecting State's 2nd Worst Hot Spot?" *Staten Island Advance*, November 26, 1989.
75. Campanile, "Fresh Kills Landfill Could Become Site for Separate Dump with Toxic Ash Waste."
76. Matthew Gandy, *Concrete and Clay: Reworking Nature in New York City* (Cambridge, MA: MIT Press, 2002), 205.
77. "A Promise in Ashes," *Staten Island Advance*, May 1, 1988.
78. Carl Campanile, "Fresh Kills Ash Dump Has Powerful Foe," *Staten Island Advance*, May 3, 1988. See also CAC Minutes of May 4, 1988, Vitaliano Papers, Series 1, Subject Files, Box 64, Folder 10.
79. Carl Campanile, "Straniere: Ban Incinerator Ash from Fresh Kills," *Staten Island Advance*, May 4, 1988.
80. Vin Di Tizio Jr. to Assemblyman Vitaliano, May 5, 1988, Vitaliano Papers, Series 1, Subject Files, Box 36, Folder 15. See also Barbara Chinitz, Staten Island Citizens for Clean Air, to Eric Vitaliano, May 28, 1988, Vitaliano Papers, Series 1, Subject Files, Box 36, Folder 15.
81. See Eric N. Vitaliano to John Deane, June 14, 1988, Vitaliano Papers, Series 1, Subject Files, Box 36, Folder 15.

82. Carl Campanile, "Fresh Kills Is Lone Site Mentioned for Ash," *Staten Island Advance*, May 27, 1988.
83. Marty Lipp, "Plan to Dump Ash Draws Fire," *Staten Island Advance*, June 9, 1988. See also Joan Gerstel, "Molinari Blasts Plan for Dump," *Staten Island Register*, June 15, 1988; Terence J. Kivlan, "Molinari Pushing Ash Bill," *Staten Island Advance*, June 25, 1988; Elizabeth A. Connelly to Thomas C. Jorling, September 7, 1988, Vitaliano Papers, Series 1, Subject Files, Box 36, Folder 15; Iver Peterson, "Toxic Ash and Costs Worry the Neighbors of Incinerators," *New York Times*, November 15, 1987; Barbara Maness, "Merchants Oppose Ash Dumping," *Staten Island Advance*, November 17, 1988; Arthur Kell to Eric Vitaliano, December 1, 1988, Vitaliano Papers, Series 1, Subject Files, Box 36, Folder 15.
84. Michael McGarry, "Dumping Ash at Landfill Possible Despite Decision," *Staten Island Advance*, November 9, 1988. See also Carl Campanile, "Ash Ruling Smoky Victory," *Staten Island Advance*, November 30, 1988; "Lamberti Blasts Sanit Plan for an Ash Dump at Fresh Kills," *Staten Island Advance*, December 13, 1988; "Disposal of Ash Key to Final OK on Incinerator," *Staten Island Advance*, November 16, 1989; "State OKs Garbage Unit but Attaches Strings," *Engineering News-Record* 223 (November 30, 1989): 9.
85. Philip Shabecoff, "Ashes of Garbage Are Found Toxic," *New York Times*, November 26, 1989.
86. See Marty Lipp, "What a Lot of Garbage!" *Staten Island Advance*, April 17, 1988; Carl Campanile, "Island Environmentalists Attack Lawmakers," *Staten Island Advance*, May 17, 1988; Carl Campanile, "Landfill Might Never Close," *Staten Island Advance*, June 5, 1988.
87. "Judge: I'll Close Fresh Kills If Garbage Isn't Contained," *Staten Island Advance*, June 2, 1983. See also "Wretched Refuse Suit Is Settled," *New York Times*, June 5, 1983.
88. "Judge Refuses to Shut Down Dump in New York–New Jersey Dispute," *New York Times*, April 22, 1984. See also Leo H. Carney, "The Environment," *New York Times*, May 20, 1984.
89. Derrick Jensen and Aric McBay, *What We Leave Behind* (New York: Seven Stories, 2009), 130; Joseph F. Sullivan, "U.S. Links Slick in New Jersey to New York Waste," *New York Times*, September 16, 1987; "New Jersey Spoils for Scrap with Polluter," *New York Times*, September 20, 1987; Robert Hanley, "U.S. Subpoenas 7 Sanitation Workers," *New York Times*, September 12, 1987. See also Nancy Zeldis, "Burning of Hospital's Waste Irks Neighbors," *New York Times*, February 8, 1987, about concerns over the burning of infectious hospital waste on Long Island.
90. "Judge Finds New York in Contempt of Court," *Daily Intelligencer/Montgomery County Record*, October 27, 1987. See also Alfonso A. Narvaez, "New York City Loses a Ruling in Trash Battle," *New York Times*, October 27, 1987.
91. DSNY Commissioner Brendan Sexton, however, continued to deny that New York City was the source of the flotsam and jetsam washing ashore in New Jersey. Alfonso A. Narvaez, "New York City Reaches Accord with Jersey on Floating Garbage," *New York Times*, November 17, 1987. See also "New Jersey Fights the Trash Tide," *New York Times*, November 27, 1987; Bob Narus, "The Environment," *New York Times*, December 20, 1987.
92. "2nd Fresh Kills Suit Filed by New Jersey," *Staten Island Advance*, November 19, 1987. See also *Arthur Kill Oil Discharge Study: Volume 1, Final Report* (West Orange, NJ: Louis Berger and Associates, Inc., October 1991).
93. William Kleinknecht, "Trash Truce Ripped as Unfair to Island," *Staten Island Advance*, November 25, 1987; "Agreement with New York City on Beach Pollution

Assailed," *New York Times*, November 29, 1987; Press Release, Groups Against Garbage, December 7, 1987; and U.S. District Court for the District of New Jersey, *Town of Woodridge et al. and Groups Against Garbage v. City of New York*, Civil Action no. 79-1060, September 28, 1987, Vitaliano Papers, Series 1, Subject Files, Box 40, Folders 11–14; U.S. District Court for the District of New Jersey, *Town of Woodridge et al. v. City of New York*, Civil Action no. 79-1060, November 15, 1988.

94. Press Release, *Joint Statement by the New Jersey Attorney General, the Township of Woodridge, the Interstate Sanitation Commission, "Save Our Shores," "Groups Against Garbage," and the City of New York*, December 7, 1987, Vitaliano Papers, Series 1, Subject Files, Box 40, Folders 11–14. See also Jean Levine, "City Agrees to Clean Up Trash Handling," *Staten Island Advance*, December 8, 1987; Alfonso A. Narvaez, "New York City to Pay Jersey Town $1 Million Over Shore Pollution," *New York Times*, December 8, 1987; Joan Gerstel, "ISC Reverse Stand on Landfill," *Staten Island Register*, September 15, 1988. The pact did not include the building of an enclosed barge facility, although the issue continued to surface. Throughout 1988 Vitaliano fought to have barges covered to reduce odors and reduce spills (which was as much in the harbor's and New Jersey's interests as Staten Island's), but the Koch administration saw no reason to comply. A Vitaliano-Marchi bill became law in 1989, forcing the city's hand. See "Vitaliano Wants Barges Covered," *Staten Island Advance*, April 7, 1988; "Mayor Against Covering Barges," *Staten Island Advance*, April 26, 1988; Jane Gerstel, "Demand City Containerize Garbage," *Staten Island Register*, September 22, 1988; Carl Campanile, "Vitaliano-Marchi Bill Puts Lid on Barge Garbage," *Staten Island Advance*, January 11, 1989; Sam Howe Verhovek, "New York, Because of Spills, to Cover Its Garbage Barges," *New York Times*, January 14, 1989; Marty Lipp, "City Covering Garbage Barges," *Staten Island Advance*, January 19, 1989; Joseph F. Sullivan, "Less Debris on Beaches Is Predicted," *New York Times*, February 25, 1989.

95. Interstate Sanitation Commission, *Annual Report on the Water Pollution Control Activities and the Interstate Air Pollution Program, 1988*, 84. See also Daniel J. Wakin, "N.J. Judge Lauds Progress on Landfill Trash Control," *Staten Island Advance*, March 3, 1988; *Public Hearing Before the Senate Special Committee to Study Coastal and Ocean Pollution Concerning the Settlement of the Suit Against New York Regarding the Fresh Kills Landfill on Staten Island*, January 19, 1988; W. Cary Edwards, "Status and Chronology of Events," Consent Order Entered December 7, 1987, *Township of Woodbridge v. City of New York*.

96. "Not Everyone Is Happy with Garbage Accord," *New York Times*, January 24, 1988; "N.J. Edict: Show Us Fresh Kills," *Staten Island Advance*, May 25, 1988; Carl Campanile, "Some Improvements Seen in Containing Fresh Kills Garbage," *Staten Island Advance*, July 30, 1988; Mary Lipp, "Judge in N.J. Trash Suit Tours Fresh Kills Landfill," *Staten Island Advance*, June 30, 1988; Sarah Lyall, "Jersey's Beach Whodunit: The Slick of Summer '87," *New York Times*, July 8, 1988. Complaints about medical waste washing up on New Jersey beaches continued into the 1990s. See James E. McGreevey, Mayor of Woodbridge, New Jersey, to David N. Dinkins, January 7, 1992, Dinkins Collection, Box 262, Folder 2470.

97. Issues about refuse and medical waste washing ashore also dogged Long Island in the summer of 1988, fueled by the AIDS hysteria. Fresh Kills was thought to be one source of the refuse. See Philip S. Gutis, "Trash, Some of It Medical Waste, Closes Beaches on L.I. for 2nd Day," *New York Times*, July 8, 1988; "Mystery of Fouled Beaches: Clues, but No 'Smoking Gun,'" *New York Times*, July 25, 1988; "Beach Waste Has Multiple Sources, DEC Chief Says," *Post-Standard*, July 23, 1988; "DEC: Beach

Waste Had 5 Key Sources," *Post-Standard*, December 20, 1988. On issues related to the EPA, see Clifford D. May, "Jersey Demands, and Awaits, Action from EPA," *New York Times*, August 7, 1988; Bernard Weinraub, "Jackson, Campaigning in Jersey, Criticizes Reagan on Environment," *New York Times*, June 2, 1988. On sludge dumping, see Clifford D. May, "New York City Offers to End Dumping at Sea by 1998," *New York Times*, August 3, 1988; Clifford D. May, "Behind New York's Offer on Dumping," *New York Times*, August 4, 1988; "Feds Say No to Sludge at Landfill," *Staten Island Advance*, October 7, 1988; Edward McCann, "Terminating Ocean Dumping of Municipal Sewage Sludge: A Political Solution to an Environmental Problem," *Temple Environmental Law & Technology Journal* 69 (1990): 69–105; Terrence J. Kivlan, "No Sludge for Landfill," *Staten Island Advance*, July 19, 1991. Attention to the issue of medical waste led to more study and continued efforts to enforce medical waste laws. See Robert J. Owens, "Criminal Enforcement of New York's Medical Waste Laws: An Overview," *Hofstra Environmental Law Digest* 24 (1989): 24–31; "Congress Enacts Legislation to Curb Solid Waste and Medical Waste Dumping by Vessels," *Hofstra Environmental Law Digest* 19 (1989): 19–20; John A. Moore, "About Medical Waste," *EPA Journal*, March/April 1989, 48–49; Alan Burdick, "Hype Tide," *New Republic* 200 (June 12, 1989): 15.

98. Press Release, "State Wetlands Still Being Mapped Eleven Years After Passage of Freshwater Wetlands Act," January 19, 1987, Marchi Papers, Series 1, Box 101, Folder 8.
99. "News from Senator John J. Marchi," January 29, 1987, Marchi Papers, Series 1, Box 101, Folder 8. See also "News from Senator John J. Marchi," February 10, 1987, Marchi Papers, Series 1, Box 101, Folder 8.
100. "News from Senator John J. Marchi," March 3, 1987, Marchi Papers, Series 1, Box 101, Folder 8.
101. See Daniel L. Master to Susan Molinari, July 8, 1991, The Assemblywoman Elizabeth A. Connelly Papers, Series 1, Subject Files, Box 53, Folder 13, Archives & Special Collections, College of Staten Island, CUNY. Susan Molinari held a BA and an MA from SUNY, Albany, and worked around government and politics much of her adult life. She served on the City Council from 1986 to 1990 and then was elected as a Republican to the U.S. House of Representatives, filling the vacancy left by her father. She became a national figure in the GOP, deeply involved in Staten Island affairs (except secession), but her Republican orthodoxy stopped with the abortion issue, where she was pro-choice. Daniel C. Kramer and Richard M. Flanagan, *Staten Island: Conservative Bastion in a Liberal City* (Lanham, MD: University Press of America, 2012), 92, 99, 112, 138, 151. See also Susan Molinari, History, Art & Architecture, U.S. House of Representatives, http://history.house.gov/People/Detail/18389; Susan Molinari, *Representative Mom: Balancing Budgets, Bill, and Baby in the U.S. Congress* (New York: Doubleday, 1999); Oral History Interview, Honorable Susan Molinari, U.S. Representative of New York (1990–1997), final edited transcript, January 8, 2016, https://history.house.gov/Oral-History/Women/Women-Transcripts/molinari-transcript/.
102. David Bauder, "Court: DEC Can Post Wetlands," *Staten Island Advance*, March 23, 1988; "Power to Declare Wetlands Is Upheld," *Staten Island Advance*, March 24, 1988. Hardship cases sometimes favored landowners, however. See Anne Fanciullo, "Couple Wins Wetland Case," *Staten Island Advance*, July 1, 1989.
103. Dan Janison, "Wetlands Moratorium Proposed," *Staten Island Advance*, March 4, 1987.

104. Sam Howe Verhovek, "State Action on Wetlands Is Opposed by S.I. Residents," *Staten Island Advance*, March 30, 1987.
105. See "Not the Wetlands Solution at All," *Staten Island Advance*, February 2, 1987.
106. Carl Campanile, "Bill to Expand Wetland Rule OK'd," *Staten Island Advance*, March 10, 1989.
107. David Bauder, "Court: DEC Can Post Wetlands," *Staten Island Advance*, March 23, 1990. See also Barbara Shields to Mrs. Connelly, Memo from New York State Assembly, "Stable/Wetlands," May 14, 1990, Connelly Papers, Series 1, Subject Files, Box 20, Folder 3; "Statement by Assemblywoman Elizabeth A. Connelly Before the New York State Department of Environmental Conservation Regarding the Programmatic Impact Statement on Staten Island Freshwater Wetlands," June 20, 1990, Vitaliano Papers, Series 1, Subject Files, Box 41, Folder 7.
108. Jim Glenn and David Riggle, "The State of Garbage, Part I," *BioCycle* 32 (April 1911): 34.
109. Memo: Eric N. Vitaliano, May 27, 1988, Vitaliano Papers, Series 1, Subject Files, Box 36. See also Interstate Sanitation Commission, *Annual Report, 1990 on the Water Pollution Control Activities and the Interstate Air Pollution Program*, 63–65.
110. Elizabeth A. Connelly to Francis J. Murray Jr., September 6, 1989, Connelly Papers, Series 1, Subject Files, Box 20, Folder 8.
111. Richard Severo, "A Wonder of Waste Rises on S.I.," *New York Times*, April 13, 1989; Richard Severo, "Monument for Today: 'Alp' of Trash Rises on S.I.," *Staten Island Advance*, April 13, 1989; "Staten Island Dump Becoming Coast's Top Peak," *Syracuse Herald American*, May 7, 1989. Tension increased over the anticipated closure of the Edgemere landfill in 1991, the only remaining landfill in the city other than Fresh Kills. See "Queens Landfill Probably Will Close by Next Year," *Staten Island Advance*, January 14, 1989. See also Steven M. Polan to Robert P. Lamieux, Deputy Commissioner, DSNY, February 13, 1991, Dinkins Collection, Box 139, Folder 1149.
112. "The Endless Landfill," *Staten Island Advance*, June 4, 1989.
113. See Jean Gerstel, "Report Raps Pollution at Fresh Kills," *Staten Island Register*, August 24, 1989; Carl Campanile, "Agreement to Tighten Landfill Rules," *Staten Island Advance*, October 2, 1989; "New Landfill Bridge to Mean Less Pollution," *Staten Island Advance*, October 13, 1989.
114. Marty Lipp, "State Says Fresh Kills Has to Go," *Staten Island Advance*, November 9, 1989.
115. David W. Dunlap, "Irate Koch Jabs at State Threat to Shut Landfill," *New York Times*, November 10, 1989. See also "Koch Dumps on Landfill Threat," *Staten Island Advance*, November 10, 1989.
116. See Constantine Sidamon-Eristoff, EPA to Roger Anderson, SES Brooklyn Company, September 7, 1990, Vitaliano Papers, Series 1, Subject Files, Box 64, Folder 6; Ken Paulsen, "Army Delays Permit for Trash Incinerator," *Staten Island Advance*, December 1, 1990.
117. Throughout the early 1990s, however, the DSNY had engaged consultants for a variety of technical reports on the status of the landfill. See, for example, Woodward-Clyde Consultants, Inc.: City of New York, Department of Sanitation, Office of Waste Recovery and Disposal Planning, *Fresh Kills Landfill: Report on Preliminary Geotechnical Evaluation, Appendices*, January 1990; City of New York, Department of Sanitation, Office of Waste Recovery and Disposal Planning, *Fresh Kills Landfill: Report on Preliminary Geotechnical Evaluation*, March 1990; City of New York, Department of Sanitation, Office of Waste Recovery and Disposal Planning, *Fresh Kills Landfill:*

Report on Field Investigation; IT Corp., *Fresh Kills: Leachate Mitigation System Project, Interim Surface Water and Sediment Investigation Report*, vol. 1, Document no. 529363-00146, April 15, 1992; Marchi Papers, Series 1, Subject Files, Fresh Kills, 1996–2006, Box 8, Folder 8; City of New York, Department of Sanitation, Bureau of Waste Management and Facilities Development, *Fresh Kills Landfill, Addendum: Stability and Reliability Analyses Report*, January 1991; IT Analytical Services, *Analytical Quality Assurance Project Plan*, September 1990; BWD Engineering, *Final Contingency Plan for the Fresh Kills Landfill*, July 1992; City of New York, Department of Sanitation, *Final Contingency Plan for the Fresh Kills Landfill*, July 1992; City of New York, Department of Sanitation, *Results of Grain Size Analysis*, November 1990 and February 1991, and *Sediment Samples for the Fresh Kills Ash/Residue Landfill*, March 1991; City of New York, Department of Sanitation, et al., *Amendment no. 1 to the Final Landfill Gas Migration Investigation Report and October 31, 1991 Addendum*, November 27, 1991; New York City, Department of Sanitation and SCS Engineers, *Final Cover Design for All Sections of the Fresh Kills Landfill*, September 9, 1991; SCS Engineers, *Final Cover System Upgrade of Section 1/9, South of Arden Avenue*, May 6, 1991, Archives & Special Collections, Department of the Library, College of Staten Island; New York State, Department of Environmental Conservation, *Emissions Study for Fresh Kills Landfill*, Connelly Papers, Series 1, Subject Files, 1990–1995, Box 20, Folder 3; IT Corp., *Interim Leachate Mitigation Report, Executive Summary*, September 11, 1992, Marchi Papers, Series 1, Subject Files, Fresh Kills, 1958, 1996–2004, Box 7, Folder 9.

118. Marty Lipp, "City Says Landfill Good for 40 Years," *Staten Island Advance*, November 10, 1989. See also "Landfill Revelations," *Staten Island Advance*, November 13, 1989.
119. Sam Howe Verhovek, "Large Fines Are Sought Over Landfill," *New York Times*, November 9, 1989.
120. Marty Lipp, "City Tries to Imagine Life After the Landfill," *Staten Island Advance*, January 22, 1990.
121. Carolyn Rushefsky, "Hopes Pinned on New Landfill Studies," *Staten Island Advance*, April 15, 1990; "Bill Would Allow Citizens to Sue Polluters," *Staten Island Advance*, January 30, 1990. In one example, Joseph Woods, who lived one mile from Fresh Kills, sued the city primarily over leachate discharge. Staten Island Citizens for Clean Air (SICCA) was hoping to join the lawsuit. Since this was an enforcement action as opposed to a permit hearing, the DEC was opposed to SICCA as an intervenor. See Donna M. Cline to Mrs. Connelly, June 6, 1990, Connelly Papers. Series 1, Subject Files, 1990–1995, Box 20, Folder 3; Marty Lipp, "Future of Fresh Kills Might Be Set in Court," *Staten Island Advance*, April 14, 1990; Tom Wrobleski, "Clean Air Group Joins Suit Against Landfill," *Staten Island Advance*, May 22, 1990; U.S. District Court, Eastern District of New York, *Complaint of Plaintiffs-Intervenors*, Joseph P. Woods, Staten Island Citizens for Clean Air, Inc., and Anthony Cantalupo against City of New York, Edward I. Koch, and Brendan Sexton, May 8, 1990; Affidavit of Barbara Chinitz, May 8, 1990; and Affidavit of Anthony Cantalupo, Connelly Papers, Series 1, Subject Files, *Woods v. CNY*, Box 20, Folder 8; "Staff Report, New York City Financial Plan," First Quarter Review, November 21, 1989, Vitaliano Papers, Series 1, Subject Files, Box 36; Elizabeth A. Connelly to Matthew S. Pavis, June 28, 1990, Connelly Papers, Series 1, Subject Files, 1990–1995, Box 20, Folder 3.
122. Carl Campanile, "Marchi: Set Aside $$ to Close Landfill," *Staten Island Advance*, December 7, 1989; "Marchi Resubmits Measure to Make City Pay for Fresh Kills," *Staten Island Advance*, December 5, 1990. On the ashfill debate, see Barbara Chinitz

to Brendon Sexton, January 2, 1990; Brendan Sexton to Barbara Chinitz, January 18, 1990; Barbara Chinitz to Mario Cuomo, January 20, 1990; Eric N. Vitaliano to Mario M. Cuomo, February 5, 1990, Vitaliano Papers, Series 1, Subject Files, Box 36, Folder 9; Matthew S. Pavis to Thomas C. Jorling, December 3, 1990, and Matthew S. Pavis to Raymond J. Kordish, DEC, December 3, 1990, Dinkins Collection, Box 140, Folder 1150.

123. Patrick Joyce, "DEC Admits It Can't Order Landfill Shut," *Staten Island Advance*, February 15, 1990; "Landfill Pipedreams," *Staten Island Advance*, February 16, 1990.

124. "April 24, 1990, Fresh Kills Consent Order Summary," Connelly Papers, Series 1, Subject Files, 1990–1995, Box 20, Folder 3; *Order on Consent: Fresh Kills Landfill*, April 19, 1990, Connelly Papers, Series 1, Subject Files, Consent Order 1990, Box 20, Folder 9, and Vitaliano Papers, Series 1, Subject Files, Order on Consent, Box 38, Folder 11. See also Erica Heinz, Legislative Commission on Solid Waste Management to Eric N. Vitaliano, May 11, 1990, Connelly Papers, Series 1, Subject Files, 1990–1995, Box 20, Folder 3; "New York City Settles Waste Lawsuit," *Engineering-News Record* 224 (May 3, 1990): 13: "There's Gold in Those Hills," *Engineering News-Record* 224 (May 3, 1990): 182; "What a Dump: Fresh Start at Fresh Kills?" *Newsday*, May 4, 1990; Allan R. Gold, "A $600 Million Landfill Cleanup Is Set by New York City and State," *New York Times*, April 27, 1990; City of New York, Department of Sanitation, *Operations and Maintenance Plan for the Fresh Kills Landfill*, July 1990, Archives & Special Collections, Department of the Library, College of Staten Island, CUNY. Major site investigations were required under the consent order, including the following conducted by IT Corporation entitled *Fresh Kills Leachate Mitigation System Project, Draft Landfill Leachate Mitigation Investigation Workplan*, Document no. 529363-00094, September 14, 1990; and *Groundwater Monitoring Wall and Piezometer Survey Report*, Document no. 529363-00193, December 31, 1990, Archives & Special Collections, Department of the Library, College of Staten Island. See also Don Gross, "$41.7 Million for Landfill Leachate Fix," *Staten Island Advance*, July 14, 1990; City of New York, Department of Sanitation, Office of Waste Recovery and Disposal Planning, *Fresh Kills Landfill: Report on Preliminary Geotechnical Evaluation, Addendum*, August 1990; City of New York, Department of Sanitation, Office of Waste Recovery and Disposal Planning, *Fresh Kills Landfill: Report on Geotechnical Testing*, August 1990, Marchi Papers, Series 1, Subject Files, 1996–2006, Box 8, Folder 8; City of New York, Department of Sanitation, *Fresh Kills Landfill, Staten Island: Final Cover, As-Built Final Cover Performance Analysis, Sections 1/9, 2/8 and 3/4, 1986/1990 Construction Seasons*, August 1990, Archives & Special Collections, Department of the Library, College of Staten Island; City of New York, Department of Sanitation, *Draft Report, Waste Transport and Unloading Facilities at the Fresh Kills Landfill*, December 31, 1990, Marchi Papers, Series 1, Subject Files, 1996–2006, Box 7, Folder 9; Carl T. Ferrentino, Compliance Counsel for Environmental Quality, DEC to Steven M. Polan, Victor A. Kovner, and Jane Levine, September 21, 1990; and State of New York, Department of Environmental Conservation, Order on Consent, Modification #1, September 21, 1990, Vitaliano Papers, Series 1, Subject Files, Order on Consent, Box 38, Folder 11.

125. City of New York, Department of Sanitation, "Summary of April 24, 1990, Consent Order for Fresh Kills Landfill," Connelly Papers, Series 1, Subject Files, 1990–1995, Box 20, Folder 3.

126. Allan R. Gold, "Lower Fees Are Voted for Fresh Kills," *New York Times*, June 27, 1990. See also Don Gross, "Sanitation Boss: Expect More Trash," *Staten Island Advance*, June 12, 1990.

127. "News from Senator John J. Marchi," June 27, 1990, Marchi Papers, Series 1, Subject Files, Box 7. See also Reginald Patrick, "Board of Estimate Blasted on Landfill," *Staten Island Advance*, June 28, 1990; Sheldon S. Leffler, "Don't Litter This Dump," *New York Times*, June 29, 1990.
128. "Kill the Fresh Kills Plan," *New York Times*, August 16, 1990; Reginald Patrick, "Swing Vote Dooms City Plan to Cut Dump Fees," *Staten Island Advance*, August 9, 1990; Reginald Patrick, "Board Delays Landfill Fee Vote Until Today," *Staten Island Advance*, August 21, 1990.
129. "New York City Department of Sanitation: Fresh Kills Land Fill Consent Order," Statement of City Councilmember Jerome X. O'Donovan, *News from Council Member Jerome X. O'Donovan*, n.d.
130. Gold, "A $600 Million Landfill Cleanup." Molinari resigned his seat in Congress December 31, 1989, to become borough president. At his swearing-in ceremony on January 14, 1990, Rudy Giuliani and David Dinkins, newly elected mayor, were present. According to Guy Molinari in his memoirs: "While it was an unlikely cast of characters, it was a truly wonderful and inspiring day for my family and me." Guy V. Molinari, *A Life of Service* (New York: Page, 2016), 363.
131. Reginald Patrick and Carl Campanile, "City, State Deal Offers $500M to Clean Up Landfill," *Staten Island Advance*, May 27, 1990.
132. Eric N. Vitaliano, "Testimony for Public Hearing on the 1990 Fresh Kills Consent Order," n.d., Vitaliano Papers, Series 1, Subject Files, Box 40, Folder 11. See also Eric N. Vitaliano, "Annual First Friday Club and Masonic Order Joint Meeting, Luncheon Address," May 4, 1990, Vitaliano Papers, Series 1, Subject Files, Box 40, Folder 9; Joan Gerstel, "Islanders Blast Landfill Agreement," *Staten Island Register*, June 5, 1990; "News from Assemblyman Eric N. Vitaliano," June 12, 1990, Vitaliano Papers, Series 1, Subject Files, Box 37; Carl Campanile, "Bill Would End Garbage Dumping at Fresh Kills," *Staten Island Advance*, June 13, 1990; Mary Engels, "Vitaliano Moves for Shutdown of Landfill by 1998," *Staten Island Advance*, June 14, 1990.
133. Arthur S. Kell, *Setting the Record Straight: A Fiscal Analysis of the City of New York's Solid Waste Management Programs and the Proposed Brooklyn Navy Yard* (May 1992).

14. A NEW PLAN

1. For technical reports in 1991, see New York City, Department of Sanitation, *Fresh Kills Leachate Mitigation System Project*, Document no. 529363-00317, March 15, 1991; City of New York, Department of Sanitation, *Final Report: Waste Transport and Unloading Facilities at the Fresh Kills Landfill*, May 8, 1991, The Senator John J. Marchi Papers, Series 1, Subject Files, 1958, 1996–2004, Box 7, Folder 9, Archives & Special Collections, Department of the Library, College of Staten Island, CUNY; City of New York, Department of Sanitation, *Fresh Kills Leachate Mitigation System Project*, Hydrogeological Site Investigation Plan, Revision 1, Document no. 529363-00194, Revision 1, July 26, 1991; Louis Berger & Associates, Inc., *Arthur Kill Oil Discharge Study*, vol. 1, Final Report, October 1991; City of New York, Department of Sanitation, Office of Resource Recovery and Waste Disposal Planning, *Fresh Kills Landfill: Geotechnical Site Characterization Report*, May 1991; *Fresh Kills Landfill, Addendum, Stability and Reliability Analyses Report*, August 1991; City of New York, Department of Sanitation, Office of Resource Recovery and Waste Disposal Planning, *Fresh Kills Landfill, Draft, Monitoring System Design Report*, August 1991; City of New York, Department of Sanitation, *Fresh Kills: Leachate Mitigation System*

Project, Surface Water and Sediment Investigation Plan, Document no. 529363-00196, Revision 1, July 26, 1991; City of New York, Department of Sanitation, *Fresh Kills Landfill: Final Landfill Gas Migration Investigation Report*, Document no. 529363-00449, September 1, 1991, Archives & Special Collections, Department of the Library, College of Staten Island.

2. Jim Meyer, Department of Sanitation to Norman Nosenchuck, DEC, January 14, 1991, David N. Dinkins Collection, Box 139, Folder 1148, La Guardia and Wagner Archives, LaGuardia Community College/CUNY, Long Island City, New York. See also *Compliance Progress Report, Number 2, Fresh Kills Landfill*, Case no. D2-9001-89-03, February 1, 1991, The Assemblyman Eric N. Vitaliano Papers, Archives & Special Collections, Department of the Library, College of Staten Island, Series 1, Subject Files, Box 36, Folder 9.

3. For example, see Steven Polan to Normal Steisel, October 22, 1991, and Jane Levine, Deputy Commissioner for Legal Affairs to Carol Ash, DEC, October 22, 1991, Dinkins Collection, Box 141, Folder 1157; Sid Davidoff to Steven M. Polan, February 21, 1991, Dinkins Collection, Box 139, Folder 1149; Steven M. Polan to Thomas Jorling, October 22, 1991, Dinkins Collection, Box 141, Folder 1157.

4. Barbara Warren-Chinitz to Thomas Jorling, March 11, 1991, Vitaliano Papers, Series 1, Subject Files, Box 36, Folder 12; "Public to Hear Status of Landfill Compliance," *Staten Island Advance*, February 13, 1991; Patrick Joyce, "Engineer: Public Needs More Input on Landfill," *Staten Island Advance*. See also Steven Polan to Norman Steisel, February 15, 1991, Dinkins Collection, Box 139, Folder 1149.

5. Martin Shefter, "David N(orman) Dinkins," in *The Encyclopedia of New York City*, 2nd ed., ed. Kenneth T. Jackson (New Haven, CT: Yale University Press, 2010), 65, 68, 105–9, 131–35; George J. Lankevich, *American Metropolis: A History of New York City* (New York: New York University Press, 1998), 237, 239. See also John Hull Mollenkopf, *A Phoenix in the Ashes: The Rise and Fall of the Koch Coalition in New York City Politics* (Princeton, NJ: Princeton University Press, 1992), 198–99; Francois Weil, *A History of New York* (New York: Columbia University Press, 2004), 284–85; Chris McNickle, *The Power of the Mayor: David Dinkins, 1990–1993* (New Brunswick, NJ: Transaction, 2013), 302–3, 329–31; John Leland, "The Complex Legacy of David Dinkins, Now 90, Is Still a Matter of Debate," *New York Times*, November 12, 2017.

6. Shefter, "David N(orman) Dinkins," 367; "David Dinkins," *Encyclopedia Britannica*, http://www.britannica.com/biography/David-Dinkins; "David N. Dinkins (1927)," *Blackpast.org*, http://www.blackpast.org/aah/dinkins-david-n-1927.

7. R. W. Apple Jr., "The 1989 Elections: Black Success with Measured Approach," *New York Times*, November 8, 1989.

8. "The 1989 Elections; Excerpts from Speech by Dinkins: A New Link," *New York Times*, November 8, 1989.

9. Don Terry, "The 1989 Elections: Mayoral Election: Win by Dinkins Prompts Dancing and Skepticism," *New York Times*, November 8, 1989.

10. Michael Powell, "Another Look at the Dinkins Administration, and Not by Giuliani," *New York Times*, October 29, 2009; "David N. Dinkins (1927)," *Blackpast.org*; "David Dinkins," *Encyclopedia Britannica*; "David Dinkins," Columbia School of International and Public Affairs, https://sipa.columbia.edu/faculty/david-n-dinkins; Lankevich, *American Metropolis*, 238–45; Shefter, "David N(orman) Dinkins," 367.

11. Center for an Urban Future, *Staten Island: Then and Now*, May 2011, 8, 10, http://www.nycfuture.org.

12. Daniel C. Kramer and Richard M. Flanagan, *Staten Island: Conservative Bastion in a Liberal City* (Lanham, MD: University Press of America, 2012), 127–28.
13. Lankevich, *American Metropolis*, 241–43; Anthony Shorris, "What Kind of People Are We?" *New York Times*, October 15, 1990. In 1991, the Dinkins administration promoted 120 black and Hispanic sanitation workers to settle a lawsuit that charged that a promotions test offered by the Department of Sanitation was discriminatory. Felicia R. Lee, "New York to Promote Minority Sanitation Workers," *New York Times*, February 10, 1991.
14. Todd S. Purdum, "Dinkins Names Sanitation Head," *New York Times*, April 19, 1990.
15. Thomas C. Jorling to Assemblywoman Connelly, June 11, 1990, The Assemblywoman Elizabeth A. Connelly Papers, Series 1, Subject Files, Box 20, Folder 3, Archives & Special Collections, College of Staten Island.
16. "Ash Disposal Bill Passed," *Staten Island Advance*, July 3, 1990.
17. SCS Engineers, *Interim Final Cover Design Report for the Fresh Kills Landfill*, October 1990, 1–4, Marchi Papers, Series 1, Subject Files, 1996–2006, Box 8, Folder 8.
18. Patrick Joyce, "City Asks for Permit to Take Ash at Landfill," *Staten Island Advance*, October 25, 1990. See also Barbara Warren-Chinitz, "City's Incinerator Obsession Burns $$," *Staten Island Advance*, November 20, 1990; Tom Berman, "Speakers Blast City's Plan for Ash Dump at Fresh Kills," *Staten Island Advance*, December 4, 1990; Matthew S. Pavis to Mario Cuomo et al., December 31, 1990, Marchi Papers, Series 1, Subject Files, Box 7, Folder 6. On some of Kell's activities with respect to incinerators and protecting the Arthur Kill, see also Benjamin Miller, *Fat of the Land: Garbage of New York, the Last Two Hundred Years* (New York: Four Walls Eight Windows, 2000), 256–61.
19. Wehran EnviroTech, *Site Investigation Plan for the Fresh Kills Ash/Residue Disposal Site*, November 1990, Revision 1, Archives & Special Collections, Department of the Library, College of Staten Island.
20. Don Gross, "Plans for Ashfill Moving Ahead," *Staten Island Advance*, December 9, 1990; David N. Dinkins to Mark P. Jarret, Richmond County Medical Society, February 6, 1991, Dinkins Collection, Box 139, Folder 1149.
21. Allan R. Gold, "Dinkins Word on Garbage Is Questioned," *New York Times*, January 22, 1991. See also Don Gross, "Incineration Sparks Heated Debate," *Staten Island Advance*, February 27, 1991; Matthew Gandy, *Concrete and Clay: Reworking Nature in New York City* (Cambridge, MA: MIT Press, 2002), 204.
22. Don Gross, "Group Fights Ashfill at Fresh Kills Dump," *Staten Island Advance*, January 19, 1991. See also Jona Gerstel, "Ashfill: An Insidious Threat," *Staten Island Register*, February 5, 1991; "Fight Against Ashfill Gets Financial Boost," *Staten Island Advance*, February 15, 1991; Don Gross, "Chamber Airs Its Doubts About Ashes at Landfill," *Staten Island Advance*, February 16, 1991; "Anti-ash Group Plans Campaign," *Staten Island Advance*, March 16, 1991; Matthew S. Pavis to To Whom It May Concern, February 22, 1991, ND Richard P. Buegler and Ellen O'Flaherty Pratt, Protectors of Pine Oak Woods to To Whom It May Concern, March 3, 1991, Dinkins Collection, Box 140, Folder 1150; Guy V. Molinari to Ben Bennett, New York State Department of State, March 19, 1991; R. M. Danielson, Corps of Engineers, to Susan Molinari, March 21, 1991; R. M. Danielson to Steven M. Polan, March 21, 1991, Marchi Papers, Secession: Reports/Research Papers/Studies, Box 30, Folder 9; numerous protest letters over the ashfill issue in Vitaliano Papers, Series 1, Subject Files, Box 36, Folder 10; "News from Assemblyman Eric N. Vitaliano," February 5, 1991, Vitaliano Papers, Series 1, Subject Files, Box 37.

23. Stephen J. Solarz et al. to Mario Cuomo, March 25, 1991, Marchi Papers, Box 7, Folder 6.
24. Patrick Joyce, "Another Challenge to Ashes in Landfill," *Staten Island Advance*, February 6, 1991.
25. Barbara Naness, "Ash No Problem—DOS," *Staten Island Register*, February 19, 1991.
26. "Vitaliano Asks Cuomo to Kill Ashfill," *Staten Island Advance*, March 14, 1991. See also Reginald Patrick and Carl Campanile, "Officials Join Chorus Against Ashfill," *Staten Island Advance*, March 28, 1991; Eric N. Vitaliano to Mario M. Cuomo, March 21, 1991, Vitaliano Papers, Series 1, Subject Files, Box 37; Elizabeth A. Connelly to Thomas C. Jorling, March 20, 1991, Connelly Papers, Series 1, Subject Files, 1990–1995, Box 20, Folder 3; Alfred C. Cerullo and Jerome X. O'Donovan to Mario M. Cuomo, March 21, 1991, and Howard Golden, President of the Borough of Brooklyn to Mario Cuomo, March 25, 1991, Dinkins Collection, Box 140, Folder 1150; John J. Marchi to Mario Cuomo, March 22, 1991, and Guy V. Molinari to Mario M. Cuomo, March 27, 1991, Marchi Papers, Secession: Reports/Research Papers/Studies, Box 30, Folder 9.
27. Gerald Benjamin, "Mario M(atthew) Cuomo," in *The Encyclopedia of New York State*, ed. Peter Eisenstadt (Syracuse, NY: Syracuse University Press, 2005), 429–30. See also Leslie Gourse, "Mario M(atthew) Cuomo," in *The Encyclopedia of New York City*, ed. Jackson, 336; Kenneth Lovett and Larry McShane, "Mario Cuomo, Former New York Governor, Dead at 82," *New York Daily News*, January 5, 2015.
28. Rhea Kemble Dignam, Executive Deputy Comptroller, to Thomas C. Jorling, March 28, 1991, Vitaliano Papers, Series 1, Subject Files, Box 36, Folder 10.
29. Joan Gerstel, "DEC Accepts Ashfill Application," *Staten Island Register*, April 2, 1991. See also Terence J. Kivlan, "Army Corps Deals Ashfill Plans a Blow," *Staten Island Advance*, March 23, 1991.
30. Don Gross, "DEC OKs Submission on Ashfill," *Staten Island Advance*, March 30, 1991; Richard Eisen, "Rally Serves City Notice on Ashfill Plan," *Staten Island Advance*, May 6, 1991. See also Judy L. Randall, "Susan: I Want Ashfill Put Someplace Else," *Staten Island Advance*, March 31, 1991; Raymond A. Wittek, "Ashfill Proposal Violating Charter?" *Staten Island Advance*, April 4, 1991; "Anti-ashfill Rally Scheduled for May 5," *Staten Island Advance*, April 26, 1991; "Anti-ashfill Rally Set for Sunday," *Staten Island Advance*, May 3, 1991; Reginald Patrick, "Hearings Postponed on Ashfill Proposal," *Staten Island Advance*, May 4, 1991; Terence J. Kivlan, "Sue Proposed Hazardous-Ash Legislation," *Staten Island Advance*, May 10, 1991; Eileen A. J. Connelly, "Bill Could Block City on Incinerator, Ashfill Plans," *Staten Island Advance*, December 26, 1991; William F. Barton to Guy V. Molinari, April 30, 1991, Marchi Papers, Series 1, Secession: Reports/Research Papers/Studies, Box 30, Folder 9.
31. Don Gross, "Official: Ashfill in Doubt If Recycling on Hold," *Staten Island Advance*, May 11, 1991. See also "Recycling Champion," *Staten Island Advance*, May 14, 1991; Sharon Hoey, "Sanit Chief Polan Defends Ashfill Proposal," *Staten Island Advance*, May 26, 1991; Karen O'Shea, "Environmental Forum Tackles Ashfill Problem," *Staten Island Register*, October 1, 1991; Eileen A. J. Connelly, "Community Boards Unite Against Ashfill," *Staten Island Advance*, June 5, 1991; Joan Gerstel, "Another Ashfill Suit Is Filed," *Staten Island Register*, April 9, 1991.
32. Stephen Di Brienza to Norman Steisel, January 7, 1991, Dinkins Collection, Box 139, Folder 1148. See also Allan R. Gold, "Study Says Recycling Effort Could Fail in New York," *New York Times*, October 12, 1990; Allan R. Gold, "Cutbacks Delay Recycling in New York City," *New York Times*, November 11, 1990.

33. Stephen Di Brienza to Norman Steisel, January 7, 1991. See also Todd S. Purdum, "The Dinkins Budget," *New York Times*, January 17, 1991. Steisel had joined Dinkins's staff as a deputy mayor in 1990. The mayor came to know Steisel when he was Manhattan borough president and chair of the School Construction Authority. "It didn't hurt that we were both tennis players," Dinkins wrote in his memoirs. "I knew how he carried himself on the court, which I believe helps considerably in estimating a person's character." He added, "Norman is also a cigar chomper. He looks like a man who would be at home in the inner workings of city government, and he was." David N. Dinkins, *A Mayor's Life: Governing New York's Gorgeous Mosaic* (New York: Public Affairs, 2013), 174. See also McNickle, *The Power of the Mayor*, 56–57.
34. Allan R. Gold, "Recycling Program in New York Falls Behind Second-Year Goals," *New York Times*, March 9, 1991. See also Don Gross, "Officials Say Recycling Isn't Garbage Cure-All," *Staten Island Advance*, February 10, 1991; Don Gross, "Cost of Recycling Trash More Than City Expected," *Staten Island Advance*, February 24, 1991; "Sanitation: Recycling Slows the Mountain," *Staten Island Advance*, April 28, 1991; Reginald Patrick, "More Garbage and Less Light in Worst Budget Plan," *Staten Island Advance*, May 4, 1991.
35. Steven M. Polan to Norman Steisel, May 3, 1991, Dinkins Collection, Box 140, Folder 1152.
36. Reginald Patrick, "Junking Recycling Will Harm City, Polan Warns," *Staten Island Advance*, March 29, 1991; Don Gross, "Sanit Chief: Suspension of Recycling Possible," *Staten Island Advance*, June 7, 1991. See also Steven M. Polan to Harvey Robins, May 29, 1991, Dinkins Collection, Box 140, Folder 1152. Steven M. Polan to the *New Yorker*, *New Yorker*, February 1991.
37. "Sanitation Boss Denies Pushing Incinerators," *Staten Island Advance*, April 2, 1991.
38. See Gunner A. Sievert to David N. Dinkins, May 7, 1991, and Geri Moss and Kathy Ralph, West 102nd & 103rd Streets Block Association, Dinkins Collection, Box 140, Folder 1152; Reginald Patrick, "New Issue: Recycling or What?" *Staten Island Advance*, May 8, 1991; Patrick, "Junking Recycling Will Harm City, Polan Warns"; Jim Tripp and John Ruston, Environmental Defense Fund, June 11, 1911, Dinkins Collection, Box 140, Folder 1152; John A. Boyd, American Material Recycling, to David Dinkins, October 4, 1991, Dinkins Collection, Box 141, Folder 1158.
39. "Mayor Says Recycling Program to Continue," *New York Times*, October 29, 1991.
40. "Little Waste in the Dinkins Trash Plan," *New York Times*, September 9, 1991. See also Ken Paulsen, "Council Goes Against Incineration," *Staten Island Advance*, June 28, 1991.
41. Steven M. Polan to David N. Dinkins, March 21, 1991, Dinkins Collection, Box 140, Folder 1151.
42. Steven M. Polan to David N. Dinkins, March 21, 1991.
43. Steven M. Polan to David N. Dinkins, March 21, 1991.
44. Steven M. Polan to David N. Dinkins, March 21, 1991.
45. Terence J. Kivlan, "Fed Agency to Look Into Landfill Risk," *Staten Island Advance*, April 2, 1991; Barbara Naness, "Landfill Will Be Studied," *Staten Island Register*, April 9, 1991; Barbara Naness, "Landfill Will Be Studied by the Government," *Staten Island Register*, August 27, 1991; Terence J. Kivlan, "Feds Agree to Study Landfill Health Risk," *Staten Island Advance*, August 22, 1991. See also Allan R. Gold, "Sanitation Workers Begin Slowdown," *New York Times*, June 21, 1991; Carl Campanile, "Straniere Sick of Landfill," *Staten Island Advance*, June 26, 1991.

46. Carl Campanile, "Senate OKs Bill Preparing for Landfill Closing," *Staten Island Advance*, April 4, 1991. See also Carl Campanile, "Senate OKs Bill on Closing Landfill," *Staten Island Advance*, Match 4, 1992.
47. *A Comprehensive Solid Waste Management Plan for New York City and Generic Environmental Impact Statement: Scope and Process* (May 1991), Dinkins Collection, Box 140, Folder 1152.
48. *A Comprehensive Solid Waste Management Plan for New York City.* See also "Positive Declaration," Proposal no. DS 1-91, June 11, 1991; and "A Comprehensive Solid Waste Management Plan for the City of New York," Proposal no. DS 1-91, June 11, 1991, Dinkins Collection, Box 140, Folder 1152.
49. Steven M. Polan to David N. Dinkins, November 26, 1991, Dinkins Collection, Box 11, Folder 145. See also in same folder the mayor's response, David N. Dinkins to Steven N. Polan, November 27, 1991.
50. James C. McKinley Jr., "Mayor Praises Dinkins's Sanitation Chief, and Reappoints Her," *New York Times*, January 15, 1994; Calvin Sims, "New Sanitation Chief Hopes to Push Recycling Business," *New York Times*, January 16, 1992; Todd S. Purdum, "Dinkins Fills 2 Top Administration Posts," *New York Times*, January 15, 1992. See also Miller, *Fat of the Land*, 275–77.
51. Michael Specter, "Dinkins's Role in Sanitation Is Faulted," *New York Times*, January 18, 1992. See also "Topics of *The Times*; New York City's Loss," *New York Times*, December 1, 1991.
52. Emily Lloyd, DOS Commissioner, to David N. Dinkins, March 18, 1992, Dinkins Collection, Box 9, Folder 172.
53. Specter, "Dinkins's Role in Sanitation Is Faulted."
54. Elizabeth Holtzman, *Burn, Baby, Burn: How to Dispose of Garbage by Polluting Land, Sea, and Air at Enormous Cost* (City of New York, Office of the Comptroller, January 1992), 1, 4, Marchi Papers, Secession: Reports/Research Papers/Studies, Box 30, Folder 9.
55. Holtzman, *Burn, Baby, Burn*, 93. The Dinkins administration also faced resistance to the incineration plan from a variety of sources, including Ruth W. Messinger, borough president of Manhattan. See Ruth W. Messinger to Deputy Mayor Barbara Fife and Commissioner Stephen Polan, October 11, 1991, Dinkins Collection, Box 141, Folder 1157.
56. Elizabeth Holtzman, *Fire and Ice: How Garbage Incineration Contributes to Global Warning* (City of New York, Office of the Comptroller, March 1992); and *A Tale of Two Incinerators: How New York City Opposes Incineration in New Jersey While Supporting It at Home* (City of New York, Office of the Comptroller, May 1992), Marchi Papers, Secession: Reports/Research Papers/Studies, Box 30, Folder 9.
57. Joan Gerstel, "Waste Panel Objects to Higher Trash Mountain," *Staten Island Register*, February 11, 1992; Joan Gerstel, "Island Pols Blast Ash Plan," *Staten Island Register*, February 16, 1992; "Fed Probers Will Hear Testimony on High Rate of Cancer in Nearby Area," *Staten Island Advance*, February 17, 1992; Remarks of Assemblyman Eric N. Vitaliano, Revisions to the Fresh Kills Consent Order, College of Staten Island, February 26, 1992, Vitaliano, Series 1, Subject Files, Box 36, Folder 12; Karen O'Shea, "Dirty, Dangerous, Disgraceful Dump Decried—Won't Close by the Year 2000," *Staten Island Register*, March 3, 1992; Robert A. Straniere to Mario Cuomo, March 5, 1992, and Matthew S. Pavis to Carol Ash, March 16, 1992, Marchi Papers, Series 1, Subject Files, Box 7, Folder 9; "News from Assemblyman Robert A. Straniere," March 6, 1992, Connelly Papers, Series 1, Subject Files, 1990–1995, Box 20, Folder 3;

Terence J. Kivlan, "Feds Urged to Get Tough on Landfill Rules," *Staten Island Advance*, March 11, 1992; Carl Campanile, "Island Legislators Seeking 250Gs for Study of Landfill," *Staten Island Advance*, March 15, 1992; Vivian Schlesinger, "Legislators Asked to Find Funding to Shut Dump," *Staten Island Advance*, March 21, 1992; Don Gross, "Limit on Landfill's Height," *Staten Island Advance*, March 26, 1992; Carl Campanile, "Bid to Shift Incinerator $$ to Recycling, Health Studies," *Staten Island Advance*, March 28, 1992; Carl Campanile, "State to Fund $250,000 Study of Landfill Hazards," *Staten Island Advance*, March 31, 1992; "Studies in Inertia," *Staten Island Register*, April 14, 1992; Susan Molinari, "Is This the Death for Dumps?" *Staten Island Register*, April 21, 1992; State of New York, Department of Environmental Conservation, Order on Consent Modification #2, April 22, 1992, and News Release, "DEC Modifies Fresh Kills Consent Order," April 23, 1992, Connelly Papers, Series 1, Subject Files, 1990–1995, Box 20, Folder 3; Joan Gerstel, "S. Molinari Demands Analysis of Landfill," *Staten Island Register*, April 28, 1992; Barbara Naness, "Trash Mountain to Go to 170'," *Staten Island Register*, April 28, 1992; Carl Campanile, "Sanit Aide: Fed Study of Landfill 'Just Silly,'" *Staten Island Advance*, April 29, 1992; Mary B. W. Tabor, "Opposition Grows to Incinerator Planned for Brooklyn Navy Yard," *New York Times*, May 3, 1992; Don Gross, "All Eyes Shift to Landfill," *Staten Island Advance*, May 3, 1992; Don Gross, "City Counting on Landfill as Site for Ash Disposal," *Staten Island Advance*, May 11, 1992; Carl Campanile, "Why Is Landfill Kept Open?" *Staten Island Advance*, May 11, 1992.
58. Jeff O'Heir, "Angry Islanders Demand a Closing Date for Landfill," *Staten Island Advance*, February 27, 1992. See also "DEC: Landfill Won't Change Without Review," *Staten Island Advance*, March 24, 1992.
59. New York City, Department of Sanitation, *Final Closure Plan for Sections 2/8 and 3/4, Appendix A-3,* Order on Consent, Fresh Kills Landfill, May 18, 1992, Archives & Special Collections, Department of the Library, College of Staten Island. See also Matthew Pavis to Carol Ash, May 1, 1992, Marchi Papers, Series 1, Subject Files, Box 7, Folder 9.
60. Don Gross, "Landfill Ash Plan Blocked," *Staten Island Advance*, May 15, 1992.
61. *A Comprehensive Solid Waste Management Plan for New York City and Final Generic Environmental Impact Statement: Executive Summary*, August 1992, ES-1, Marchi Papers, Secession: Reports/Research Papers/Studies, Box 30, Folder 9.
62. *A Comprehensive Solid Waste Management Plan for New York City*, ES-1.
63. *A Comprehensive Solid Waste Management Plan for New York City*, ES-1-ES-2.
64. *A Comprehensive Solid Waste Management Plan for New York City*, ES-2.
65. *A Comprehensive Solid Waste Management Plan for New York City*, ES-2-ES-4.
66. *A Comprehensive Solid Waste Management Plan for New York City*, ES-12.
67. *A Comprehensive Solid Waste Management Plan for New York City*, ES-22-ES-26.
68. *A Comprehensive Solid Waste Management Plan for New York City*, ES-27-ES-34. See also *A Comprehensive Solid Waste Management Plan for New York City, Draft Executive Summary* [for Internal Review], December 18, 1991. This dissertation regarded the combination of recycling and waste-to-energy facilities to produce the best net benefits for the DSNY. John James Gerace, "The Net Benefits of the New York City Solid Waste Management Program," PhD diss., New School for Social Research, 1991.
69. *Proposed Final Citywide Solid Waste Management Plan, Chapter 19, The Implementation Process for the Plan*, June 26, 1992, Archives & Special Collections, Department of the Library, College of Staten Island.
70. "Trashing Us . . . Again," *Staten Island Register*, April 21, 1992.

678 ■ 14. A NEW PLAN

71. Assemblyman Eric N. Vitaliano, Remarks on the Comprehensive Solid Waste Management Plan and Draft Generic Environmental Impact Statement for the City of New York, May 18, 1992, Marchi Papers, Secession: Reports/Research Papers/Studies, Box 30, Folder 9. See also Eric N. Vitaliano to Duane C. Felton, April 1, 1992, Connelly Papers, Series 1, Subject Files, 1990–1995, Box 20, Folder 3.
72. Staten Island Coalition for Survival to Solid Waste Management Hearing, May 18, 1992, and Staten Island Environmental Caucus to Solid Waste Management Hearing, May 18, 1992, Marchi Papers, Secession: Reports/Research Papers/Studies, Box 30, Folder 9.
73. *Supplemental Testimony by Congresswoman Susan Molinari, Generic Environmental Impact Statement (GEIS), New York City Comprehensive Solid Waste Management Plan*, May 18, 1992, Marchi Papers, Secession: Reports/Research Papers/Studies, Box 30, Folder 9.
74. *Statement of the Honorable Alfred C. Cerullo, III, Council Member/Minority Leader, New York City Council, Before the Staten Island Borough Hearing on the New York City Solid Waste Management Plan*, May 18, 1992, Marchi Papers, Secession: Reports/Research Papers/Studies, Box 30, Folder 9.
75. Statement by Assemblywoman Elizabeth Connelly to the Joint Public Hearing Regarding the Solid Waste Management Plan for the City of New York, May 18, 1992, Marchi Papers, Secession: Reports/Research Papers/Studies, Box 30, Folder 9. See also Guy V. Molinari to Benjamin Miller, May 29, 1992, Marchi Papers, Secession: Reports/Research Papers/Studies, Box 30, Folder 9.
76. Don Gross, "Hundreds Turn Out to Blast Dump," *Staten Island Advance*, May 19, 1992. The City Council had to approve the plan by July 10, and the city had to submit it to the state by July 30. See Joan Gerstel, "Hundreds Declare: 'Dump the Dump,'" *Staten Island Register*, May 26, 1992.
77. Barbara Warren Chinitz to Benjamin Miller, May 24, 1992, Marchi Papers, Secession: Reports/Research Papers/Studies, Box 30, Folder 9. See also Barbara Warren Chinitz to Benjamin Miller, May 24, 1992; Alfred J. Brumme and Dennis J. Sarlo, Community Board 3, to Benjamin Miller, May 29, 1992; Lou Caravone and Tim Duffy, Community Board 2, to Benjamin Miller, May 29, 1992; Elizabeth A. Connelly to Benjamin Miller, June 1, 1992; Susan Molinari to Benjamin Miller, May 26, 1992, Marchi Papers, Secession: Reports/Research Papers/Studies, Box 30, Folder 9.
78. Barbara Warren Chinitz to Peter Vallone, May 26, 1992, Marchi Papers, Secession: Reports/Research Papers/Studies, Box 30, Folder 9.
79. Statement of Elizabeth Holtzman at Committee on Environmental Protection, the New York City Council, May 20, 1992, Marchi Papers, Secession: Reports/Research Papers/Studies, Box 30, Folder 9. See also Robert Harris and Suzanne Mattei to Elizabeth Holtzman, May 20, 1992, Marchi Papers, Secession: Reports/Research Papers/Studies, Box 30, Folder 9.
80. Staff of the Center for the Biology of Natural Systems and Staff of the Bronx Borough President, "Analysis of *A Comprehensive Solid Waste Management Plan for New York City and Draft Generic Environmental Impact Statement, March 1992*," May 28, 1992, 1–2, Marchi Papers, Secession: Reports/Research Papers/Studies, Box 30, Folder 9.
81. Staff of the Center for the Biology of Natural Systems and Staff of the Bronx Borough President, "Analysis of *A Comprehensive Solid Waste Management Plan for New York City*," 10. See also Don Gross, "Garbage Showdown," *Staten Island Advance*, May 31, 1992.

82. *Revisions to New York City's Comprehensive Solid Waste Management Plan*, August 26, 1992, Marchi Papers, Secession: Reports/Research Papers/Studies, Box 30, Folder 9. See also Robert P. Lemieux to Eric Vitaliano, October 29, 1992, Vitaliano Papers, Series 1, Subject Files, Box 36, Folder 10; Don Gross, "Ashfill for Island Looking Less Likely," *Staten Island Advance*, August 15, 1992; Don Gross, "State Might Allow Ash in Landfill Cover," *Staten Island Advance*, September 23, 1992.
83. See Ted Steinberg, *Down to Earth: Nature's Role in American History*, 2nd ed. (New York: Oxford University Press, 2009), 235.
84. Reginald Patrick, "City Likely to Scrap Ashfill, Incinerator Plan for Island," *Staten Island Advance*, August 11, 1992; Reginald Patrick, "Plenty of Other Places to Stash Incinerator Ash," *Staten Island Advance*, August 16, 1992. See also efforts concerning the revival of the project: Howard Golden, "Cutting Corners on Clean Air," *New York Newsday*, November 2, 1992; Howard Golden to Mario M. Cuomo, November 5, 1992, Vitaliano Papers, Series 1, Subject Files, Box 64, Folder 6. See also Miller, *Fat of the Land*, 277–79.
85. Dennis Hevesi, "Dinkins Delays Building Incinerator in Brooklyn," *New York Times*, March 25, 1992. See also Gandy, *Concrete and Clay*, 209.
86. See, for example, Jeff O'Heir, "City Council Committee OKs Sludge Facility," *Staten Island Advance*, June 26, 1992; Jeff O'Heir, "An Incinerator for Island?" *Staten Island Advance*, June 26, 1992; Don Gross, "A New Sludge Era Dawns," *Staten Island Advance*, June 30, 1992; Carolyn Rushefsky, "Landfill Passes Toxic Test," *Staten Island Advance*, July 30, 1992; "Claims About Landfill Must Be Based on Facts," *Staten Island Advance*, August 2, 1992; Carl Campanile, "Landfill Complaints Piling Up in Albany," *Staten Island Advance*, August 23, 1992; Carl Campanile, "Landfill Is Becoming a Political Liability for Cuomo," *Staten Island Advance*, August 23, 1992; Don Gross, "Toxic Landfill Study Widened," *Staten Island Advance*, September 15, 1992; Cyprian Cox to Eric N. Vitaliano, September 15, 1992, Vitaliano Papers, Series 1, Subject Files, Box 6, Folder 9; Don Gross, "Coordinated Health Study of Landfill Gets Rolling," *Staten Island Advance*, September 17, 1992; "Brookfield Added to Dump Study," *Staten Island Register*, September 22, 1992; Paul M. McPolin, "The Mountain of Trash Keeps Getting Higher," *Staten Island Advance*, November 15, 1992; "New York City: Garbage Ostrich," *New York Times*, April 26, 1992; David Riggle, "Implementing the World's Largest Recycling Program," *BioCycle* 33 (November 1992): 53–56.
87. "Transforming a Dump: Makeover at Fresh Kills," *Conservationist* 47 (November 1992) 54–55.
88. Carl Campanile, "Bill to Protect Landfill Buffer Areas Approved," *Staten Island Advance*, June 23, 1992; Carl Campanile, "Landfill 'Shrink' Bill OK'd," *Staten Island Advance*, June 27, 1992; Carl Campanile, "State, City Discussing Ways to 'Shrink' Landfill," *Staten Island Advance*, August 20, 1992.
89. Carl Campanile, "Landfill Is Becoming a Political Liability for Cuomo," *Staten Island Advance*, August 23, 1992.
90. Emily Lloyd to Eric N. Vitaliano, September 11, 1992, Vitaliano Papers, Series 1, Subject Files, Box 36. See also Carl Campanile, "Environment Officials Hope to Shrink Landfill," *Staten Island Advance*, September 18, 1992.
91. Carl Campanile, "City Agrees to Buffer at Landfill," *Staten Island Advance*, December 13, 1992.
92. Terence J. Kivlan, "Cuomo, Fearing Fed Control, Hits Landfill Bill," *Staten Island Advance*, July 3, 1992.

93. Richard Eisen, "2 Sections of Landfill to Close," *Staten Island Advance*, November 13, 1992; Don Gross, "2 Landfill Areas to Close in 1993," *Staten Island Advance*, December 11, 1992. For information on the *Final Closure Plan* for sections 1/9 and 6/7, see Phillip J. Gleason to Norman H. Nosenchuck, December 4, 1992, and SCS Engineers, *Final Closure Plan for Sections 1/9 and 6/7, Appendix A-3, Milestone 11, Order on Consent, Fresh Kills Landfill*, December 4, 1992, Vitaliano Papers, Subject Files, Order on Consent, Box 38, Folder 11.
94. James C. McKinley Jr., "Dinkins Tells Staten Island He Cares," *New York Times*, July 31, 1992.
95. Elizabeth A. Connelly to Jeffrey S. d'Auguste, September 16, 1991, Connelly Papers, Box 20, Folder 3.
96. John Holusha, "In Some Parts the Battle Cry Is 'Don't Dump on Me,'" *New York Times*, September 8, 1991.

15. SECESSION

1. Memorandum, Larry to Senator Marchi, September 30, 1993, and "Background for Staten Island Secession," The Senator John J. Marchi Papers, Secession, Subject File, 1983–1996, Box 33, Folder 4, Archives & Special Collections, Department of the Library, College of Staten Island, CUNY. Fresh Kills' connection to the secession issue was repeated by many people. See, for example, Abraham Lackman: Oral History Transcripts, November 11, 2004, 168, College of Staten Island Oral History Collection, v. 1, ed. Jeffery A. Kroessler, Archives & Special Collections, Department of the Library, College of Staten Island, CUNY. Lackman was an analyst for the State Senate Finance Committee from 1980 to 1993 and worked with Marchi.
2. Charles L. Sachs, "Staten Island Secessionism: Strained Threads of the Consolidated City Fabric," *Culturefront* (Winter 1997–1998): 101.
3. "Shall Staten Island Secede to the United States?" *New York Sun*, November 18, 1935.
4. Letters from Readers, *Staten Island Advance*, April 8, 1940.
5. "'Secession' Advocated by Lawyer," *Staten Island Advance*, April 3, 1940.
6. Editorial, *Staten Island Advance*, April 4, 1940. See also "Two Proposals," *Staten Island Advance*, April 6, 1940.
7. Aside from chapter 8, see Eileen P. Kavanaugh, "Staten Island: 'Taking a Bite Out of the Big Apple,'" BA thesis, Princeton University, 1991, 32–35; "Proposal Hit as 'Unsound,'" *Staten Island Advance*, April 16, 1947; Sachs, "Staten Island Secessionism," 101–2.
8. See Joel Cohen, "We Can Join New Jersey All Right but How About $1.50 a Ferry Ride?" *Staten Island Advance*, August 14, 1951; Joel Cohen, "How Did We Get in New York and How Do We Get Out?" *Staten Island Advance*, August 15, 1951; Joel Cohen, "Radigan, Secession's 'Father,' Still Favors 1947 Proposition," *Staten Island Advance*, August 16, 1951; Joel Cohen, "To Secede or Not Is a Question Bitterly Argued by Islanders," *Staten Island Advance*, August 17, 1951; Joel Cohen, "Robert E. Lee All Set to Lead Civil War Against New York," *Staten Island Advance*, August 18, 1951.
9. Norman Adler: Oral History Transcripts, January 19, 2005, 11, College of Staten Island Oral History Collection, v. 1.
10. Jeffrey Underweiser, "The Legality of Staten Island's Attempt to Secede from New York City," *Fordham Urban Law Journal* 19 (1991): 147–48, 150–51; George J. Lankevich, *American Metropolis: A History of New York City* (New York: New York University

Press, 1998), 237; Terry Golway, "The War of Secession," *Empire State Report*, May 1989, 39. See also Daniel C. Kramer and Richard M. Flanagan, *Staten Island: Conservative Bastion in a Liberal City* (Lanham, MD: University Press of America, 2012), 119–21. In 1958 a voting arrangement that favored the larger boroughs of Manhattan and Brooklyn was changed, resulting in one vote per borough regardless of size. See Joseph P. Viteritti, "Municipal Home Rule and the Conditions of Justifiable Secession," *Fordham Urban Law Journal* 23 (1995–1996): 53.

11. Richard Briffault, "Voting Rights, Home Rule, and Metropolitan Governance: The Secession of Staten Island as a Case Study in the Dilemma of Local Self-Determination," *Columbia Law Review* 92 (May 1991): 784. See also Kramer and Flanagan, *Staten Island: Conservative Bastion in a Liberal City*, 121–22.
12. Kramer and Flanagan, *Staten Island: Conservative Bastion in a Liberal City*, 122–23.
13. "One Man, One Vote—and One City," *New York Times*, May 20, 1983. See also Kavanaugh, "Staten Island: 'Taking a Bite Out of the Big Apple,'" 43–47.
14. John J. Marchi to Edward I. Koch, May 23, 1983, Marchi Papers, Secession, Home Rule, *Morris v. Board of Estimate*, 1983–1987, Box 2, Folder 16. See also "News from Senator John J. Marchi," May 24, 1983, Marchi Papers, Secession, Home Rule, *Morris v. Board of Estimate*, 1988–1989, Box 5, Folder 2.
15. See Briffault, "Voting Rights, Home Rule, and Metropolitan Governance," 788–91.
16. Kramer and Flanagan, *Staten Island: Conservative Bastion in a Liberal City*, 123. See also "Straniere: Put Secession to a Vote," *Staten Island Advance*, July 17, 1983; Raymond Fasano: Oral History Transcripts, March 22, 2002, 82–91, College of Staten Island Oral History Collection, v. 1.
17. John J. Marchi to Gene Tyksinski, June 1, 1983, Marchi Papers, Secession, Correspondence, 1983–2002, Box 39, Folder 2. See also Kavanaugh, "Staten Island: 'Taking a Bite Out of the Big Apple,'" 57–59.
18. Edward I. Koch to John J. Marchi, June 17, 1983, Marchi Papers, Secession, Home Rule, *Morris v. Board of Estimate*, 1983–1987, Box 2, Folder 16. See also John Waugaman, Editorial, "A Golden Opportunity," *WINS Radio 1010*, June 17, 1983, Marchi Papers, Secession, Home Rule, General Files, 1989–1995, Box 9, Folder 5.
19. "News from Senator John J. Marchi," June 20, 1983, and "News from Senator John J. Marchi," July 5, 1983, Marchi Papers, Secession, Home Rule, Board of Estimate, 1895–1991, Box 8, Folder 11; Duncan S. MacAffer to John J. Marchi, June 29, 1983, Marchi Papers, Secession, Correspondence, 1983–2002, Box 39, Folder 1. See also Kramer and Flanagan, *Staten Island: Conservative Bastion in a Liberal City*, 123.
20. Maurice Carroll, "Staten Island Wonders If Secession Can Succeed," *New York Times*, August 7, 1983.
21. "Staten Islanders Want to Secede from the Big Apple," *UPI*, July 5, 1983, http://www.upi.com/Archives/1983/07/05/StatenIslanders-want-to-secede-from-the-Big-Apple/4093426225600/; "'We Want Out!' Say Staten Islanders," *Christian Science Monitor*, July 5, 1983, http://www.csmonitor.com/1983/0705/070522.html.
22. *An Independent Staten Island: Fiscal Implications*, August 31, 1983, Marchi Papers, Secession, Correspondence, 1983–2002, Box 39, Folder 5.
23. Seth Mydans, "At Tercentennial Fete, Staten Islanders Brood," *New York Times*, September 12, 1983. See also Kavanaugh, "Staten Island: 'Taking a Bite Out of the Big Apple,'" 64–66.
24. Guy V. Molinari, *A Life of Service* (New York: Page, 2016), 400–1.
25. John J. Marchi to Mario M. Cuomo, July 17, 1983, Marchi Papers, Secession, Correspondence, 1983–2002, Box 39, Folder 5.

26. "News from Senator John J. Marchi," August 2, 1983, Marchi Papers, Secession, Home Rule, Board of Estimate, 1895–1991, Box 8, Folder 11.
27. "News from Senator John J. Marchi," July 25, 1983, Marchi Papers, Secession, Home Rule, Board of Estimate, 1895–1991, Box 8, Folder 11. See also Kavanaugh, "Staten Island: 'Taking a Bite Out of the Big Apple,'" 51–53.
28. "Break Up New York City?" *New York Times*, August 21, 1983. See also "A Staten Island Solution Fit for All of New York," *New York Times*, August 29, 1983.
29. "News from Senator John J. Marchi," November 17, 1983, and November 18, 1983, Marchi Papers, Secession, Home Rule, Board of Estimate, 1895–1991, Box 8, Folder 11.
30. "News from Assemblyman Eric N. Vitaliano," January 3, 1984, Series 1, Subject Files, Box 40, The Assemblyman Eric N. Vitaliano Papers, Archives & Special Collections, Department of the Library, College of Staten Island. Marchi also linked secession and the garbage crisis at times. He criticized Senator Carol Bellamy (a mayoral candidate) for trying to expand the role of Brooklyn and Queens on the Board of Estimate at the expense of Staten Island. That, coupled with her vote against the resource-recovery plan, "make it plain to me that her mayoral candidacy represents a serious threat to the stability of New York City." See "News from Senator John J. Marchi," February 11, 1985, Marchi Papers, Secession, Home Rule, Board of Estimate, 1895–1991, Box 8, Folder 11.
31. In general, home rule refers to the ability of a local government to prevent state government from intervening in its operations. The extent of its power, of course, is subject to limitations imposed by the state constitution and state statutes. The courts, in the case of the Board of Estimate, were grappling with the city's ability to apply new voting representation to this body on its own. Later, the effort at secession by Staten Island would also raise home-rule issues, that is, the ability of a new state law to allow Staten Island to become independent without New York City's approval.
32. Kramer and Flanagan, *Staten Island: Conservative Bastion in a Liberal City*, 123.
33. Alan Rothstein, Associate Director of the Citizens Union, to Edward R. Neaher, March 23, 1984, and Alan Rothstein to Edward R. Neaher, October 26, 1984, Marchi Papers, Secession, Home Rule, *Morris v. Board of Estimate*, 1983–1987, Box 2, Folder 16. For other examples of the debate over the Board of Estimate and the pending court case, see Loring McMillen, Staten Island Borough Historian, March 7, 1985, Marchi Papers, Secession, Correspondence, 1985–1998, Box 38, Folder 1; "A Voice Barely Heard," *Staten Island Advance*, April 17, 1985; Kathryn K. Rooney to Anthony I. Giacobbe, February 15, 1985; Richard Emery, NYCLU to Edward R. Neaher, May 17, 1985; John F. Banzhaf III to Edward Neaher, May 22, 1985; Steven J. Brams to Richard Emery, May 30, 1985; Richard Emery to Edward Neaher, May 31, 1985; Susan R. Rosenberg to Edward R. Neaher, June 11, 1985; John J. Marchi to Edward R. Neaher, June 11, 1985, Marchi Papers, Secession, Home Rule, *Morris v. Board of Estimate*, 1983–1987, Box 2, Folder 16; John C. Dearie to John J. Marchi, November 14, 1986, and John J. Marchi to John C. Dearie, November 25, 1986, Marchi Papers, Secession, Home Rule, Board of Estimate, 1895–1991, Box 8, Folder 11.
34. Kramer and Flanagan, *Staten Island: Conservative Bastion in a Liberal City*, 123.
35. Paul E. Proske to Ralph J. Lamberti, "The Mission of the Staten Island Committee for the Review of Secession," n.d., Vitaliano Papers, Series 1, Subject Files, Box 36, Folder 15.
36. *An Inquiry Into Self-Determination: A Report to Borough President Ralph J. Lamberti from the Staten Island Committee for the Review of Secession*, December 14, 1987, Staten Island/Richmond Borough, Box 5A, Folder 62, History Archives & Library, Staten Island Museum.

37. John J. Marchi to Editor, *Staten Island Advance*, January 5, 1987, Marchi Papers, Secession, Correspondence, 1983–2002, Box 39, Folder 2.
38. "News from Senator John J. Marchi," January 27, 1987, Marchi Papers, Secession, Home Rule, Board of Estimate, 1895–1991, Box 8, Folder 11.
39. "Important Dates in Secession," *Staten Island Advance*, March 1, 1994. See also Elizabeth Connelly: Oral History Transcripts, April 9, 2001, 80, College of Staten Island Oral History Collection, v. 3.
40. The appeal had been brought by the City of New York. Marchi thanked Mayor Koch for keeping his promise and also praised Richard Ravitch, chair of the New York City Charter Revision Commission. "News from Senator John J. Marchi," April 4, 1988, Marchi Papers, Secession, Subject Files, 1983–1996, Box 33, Folder 18.
41. James Barron, "Staten Island: In a Seceding State of Mind," *New York Times*, December 12, 1988.
42. Barron, "Staten Island: In a Seceding State of Mind."
43. State of New York, 1989–1990 Regular Sessions in Senate, February 21, 1989, Marchi Papers, Series 1, Subject Files, Box 47, Folder 5. State of New York, 1989–1990 Regular Sessions in Assembly, February 15, 1989, The Assemblywoman Elizabeth A. Connelly Papers, Series 1, Subject Files, Box 43, Folder 7, Archives & Special Collections, Department of the Library, College of Staten Island, CUNY.
44. "News from Senator John J. Marchi," March 22, 1989, Marchi Papers, Secession, Subject Files, Box 33, Folder 17. See also Martin Connor to his constituency, April 3, 1989, Box 8, Folder 11.
45. Some speculated that in the upper house, senators simply deferred to a respected senior leader or believed that the bill would not survive a vote in the Assembly. The bill did survive in the Assembly, despite the fact that Speaker Mel Miller opposed it. He was aware that Democrats from Staten Island (Vitaliano and Connelly) needed to support the bill for reelection purposes and thus wanted to shield them. Others simply may have assumed that Governor Mario M. Cuomo would veto the bill. Connelly worked very hard for the bill. While Vitaliano supported it, he did not devote major effort to its passage. See Kramer and Flanagan, *Staten Island: Conservative Bastion in a Liberal City*, 124.
46. Carl Campanile, "Secession Referendum OK'd," *Staten Island Advance*, July 1, 1989; Briffault, "Voting Rights, Home Rule, and Metropolitan Governance," 785; "Staten Island Secession Bill Passes in 1989," *New York Daily News*, June 20, 2015. See also "News from Senator John J. Marchi," April 17, 1989, June 1, 1989, June 27, 1989, Marchi Papers, Secession, Subject Files, Box 33, Folder 17.
47. "News from Senator John J. Marchi," July 1, 1989, Marchi Papers, Secession, Subject Files, Box 33, Folder 17. See also John J. Marchi to Mario Cuomo, July 10, 1989, in same box and folder.
48. "Staten Island Awaits Gov on Secession," *Staten Island Advance*, July 1, 1989. For negative reaction to the bill by Peter L. Zimroth, corporation counsel of the city of New York, see Peter L. Zimroth, "Reflections on My Years as Corporation Counsel," *New York Law School Law Review* 53 (2008/2009): 420–25.
49. James N. Baldwin to Evan A. Davis, Counsel to the Governor, July 5, 1989, Marchi Papers, Secession, Correspondence, Box 39, Folder 5.
50. Edward C. Farrell to Evan Davis, July 11, 1989, Marchi Papers, Secession, Correspondence, Box 39, Folder 5. See also in this box and folder Howard A. Jack, Energy Research and Development Authority, to Evan Davis, July 13, 1989.

51. Peter F. Vallone to Mario M. Cuomo, July 28, 1989, Marchi Papers, Series 1, Subject Files, Box 92, Folder 11. See also Reginald Patrick, "Vallone Urges Cuomo to Veto Bill on Secession Vote," *Staten Island Advance*, August 1, 1989.
52. See "News from Senator John J. Marchi," July 24, 1989, Marchi Papers, Secession, Subject Files, Box 33, Folder 17; Carl Campanile and Tom Cocola, "Molinari Urges Veto of Secession Bill," *Staten Island Advance*, November 11, 1989.
53. John J. Marchi to Mario M. Cuomo, December 7, 1989, Marchi Papers, Secession, Home Rule, 1989–1990, Box 6, Folder 7. See also "News from Senator John J. Marchi," December 7, 1989, Marchi Papers, Series 1, Subject Files, Box 92, Folder 11.
54. Eric N. Vitaliano to Mario M. Cuomo, July 11, 1989, Vitaliano Papers, Series 1, Subject Files, Box 36, Folder 15.
55. Elizabeth A. Connelly to Mario M. Cuomo, November 15, 1989, Vitaliano Papers, Series 1, Subject Files, Box 36, Folder 15. See also in the same box and folder, Peter L. Zimroth to Edward I. Koch, November 22, 1989.
56. David Goldfarb to Mario Cuomo, July 6, 1989, Marchi Papers, Secession, Home Rule, *Morris*, Box 2, Folder 16. See also "David Goldfarb," http://www.panix.com/~goldfarb/david.htm.
57. Street homelessness was a serious problem—and a more visible one—throughout New York City beginning in the 1980s. While Staten Island did not contain communities in the city that experienced the highest incidence, the problem existed there nonetheless. Unfortunately, homelessness surveys often undercounted people facing this desperate problem. Most available data, however, does not pick up until the early 2000s. For some insight on the problem, see "Staten Island: Street Homelessness in Staten Island," 2004, *NYC.org*, http://www.nyc.gov/html/endinghomelessness/downloads/pdf/statenisland.pdf; "Briefing Paper: Undercounting the Homeless," Coalition for the Homeless, May 24, 2004, http://www.coalitionforthehomeless.org/wp-content/uploads/2014/06/briefing-Undercountingthehomeless-2004.pdf; Roy Rowan, "Homeless Bound," *People.com*, March 5, 1990, https://people.com/archive/homeless-bound-vol-33-no-9/; Diane Jeantet, "A Brief History of Homelessness in New York," *citylimits.org*, March 11, 2013, https://citylimits.org/2013/03/11/a-brief-history-of-homelessness-in-new-york/.
58. David Goldfarb: Oral History Transcripts, December 1, 2004, College of Staten Island Oral History Collection, v. 3, 35, 42, also 24–34, 36–49. See also Victor A. Kovner, "Staten Island Secession," *New York Law Journal* (April 17, 1991).
59. Nadia H. Youssef, *Population Dynamics on Staten Island: From Ethnic Homogeneity to Diversity* (New York: Center for Migration Studies, 1991), 8, Marchi Papers, Secession, Home Rule, 1989–1995, Box 9, Folder 7.
60. See John Maggs, "Breakin' Up Is Hard to Do," *National Journal* (June 12, 1999): 1608.
61. Youssef, *Population Dynamics on Staten Island*, 9.
62. Youssef, *Population Dynamics on Staten Island*, 11–13, 1. See also Chip Brown, "Separatism Surging," *New York Times Magazine*, January 30, 1994, 30.
63. Russell W. Baker, "Staten Island Considers Secession," *Christian Science Monitor*, August 2, 1989.
64. Joseph P. Viteritti: Oral History Transcripts, April 9, 2001, 138, College of Staten Island Oral History Collection, v. 1.
65. Joseph P. Viteritti: Oral History Transcripts, April 9, 2001, 141, also 143. See also Senator John Marchi: Oral History Transcripts, February 1, 2002, 171, College of Staten Island Oral History Collection, v. 1.

66. Viteritti: Oral History Transcripts, April 9, 2001, 142. See also Kramer and Flanagan, *Staten Island: Conservative Bastion in a Liberal City*, 127.
67. Mary Engels, "Freedom Bus to Roll Into S.I.," *New York Daily News*, October 29, 1993.
68. Viteritti, "Municipal Home Rule and the Conditions of Justifiable Secession," 63–64. See also Goldfarb: Oral History Transcripts, December 1, 2004, 46.
69. Connelly: Oral History Transcripts, April 9, 2001, 83, also 85.
70. Oral interview, Martin Melosi with Norman Steisel, November 15, 2017; oral interview, Martin Melosi with Paul Casowitz, November 16, 2017.
71. The office estimated that such a plan could generate $421.3 million in revenue annually, with $166 million in expenses, thus a net gain of $255.3 million per year. Staten Island, however, also would have to buy water from the city. Hilary David Ring to John J. Marchi, November 9, 1989, Marchi Papers, Secession, Subject Files, Box 33, Folder 17. See also "News from Senator John J. Marchi," November 9, 1989, Box 33, Folder 17.
72. Eugene L. Salerni and Gordon M. Boyd to Abe Lackman, November 14, 1989, Marchi Papers, Series 1, Subject Files, Box 92, Folder 12.
73. Statement by Mayor Edward I. Koch, April 4, 1988, Marchi Papers, Secession, Subject Files, 1983–1996, Box 33, Folder 18.
74. Kramer and Flanagan, *Staten Island: Conservative Bastion in a Liberal City*, 126.
75. The decision was made special because Cuomo had called Marchi personally to relay the news. See "News from Senator John J. Marchi," December 15, 1989, Marchi Papers, Secession, Subject Files, Box 33, Folder 17.
76. Carl Campanile, "Cuomo Signs Secession Bill; 1st Island Vote Next Year; Called Democracy's Victory," *Staten Island Advance*, December 16, 1989.
77. Elizabeth Kolbert, "Cuomo Approves a Secession Vote for Staten Island," *New York Times*, December 16, 1989.
78. Viteritti, "Municipal Home Rule and the Conditions of Justifiable Secession," 57; Raphael J. Sonenshein and Tom Hogen-Esch, "Bringing the State (Government) Back In: Home Rule and the Politics of Secession in Los Angeles and New York City," *Urban Affairs Review* 41 (March 2006): 480. See also Molinari, *A Life of Service*, 402–3.
79. Mario M. Cuomo, "Memorandum Filed with Senate Bill Number 2655-A," December 15, 1989, Marchi Papers, Secession, Subject Files, Box 33, Folder 17.
80. Cuomo, "Memorandum Filed with Senate Bill Number 2655-A."
81. Kramer and Flanagan, *Staten Island: Conservative Bastion in a Liberal City*, 124, also 125. See also Anna Sanders, "Gov. Cuomo and the Staten Island Secession Movement," *Staten Island Advance*, January 2, 2015.
82. In identical letters to both Marchi and Connelly on December 20, 1989, Cuomo provided a justification for seeking the amendment: "After study, I concluded that the legal questions that concerned some legislators, can—and should—be dealt with quite simply by a law that gives Staten Island's decision to secede a review in the State Legislature, which represents not only the best interests of Staten Island but of the entire State." Cuomo promised to send a proposed law early in the next session. In September 1990 the State Court of Appeals upheld the constitutionality of what was called the Marchi-Connelly bill, asserting that this step toward independence did not require a home-rule declaration. See Mario M. Cuomo to Elizabeth Connelly, December 20, 1989; and Mario M. Cuomo to John Marchi, December 20, 1989, Marchi Papers, Secession, Correspondence, 1983–2002, Box 39, Folder 5.
83. Kramer and Flanagan, *Staten Island: Conservative Bastion in a Liberal City*, 125. See also Briffault, "Voting Rights, Home Rule, and Metropolitan Governance," 786;

Lester D. Steinman, "Staten Island Secession Proceeds," *Municipal Lawyer* 4 (September/October 1990): 1–2, 4; Cuomo, "Memorandum Filed with Senate Bill Number 2655-A"; Elizabeth Kolbert, "Cuomo Approves a Secession Vote for Staten Island," *New York Times*, December 16, 1989; "Assemblyman Robert A. Straniere Reports to the People," Fall 1989, Marchi Papers, Secession, Subject Files, 1983–1996, Box 33, Folder 18.

84. Reginald Patrick, "Cuomo Proud of His Role in Island Vote on Secession," *Staten Island Advance*, September 9, 1994.
85. "A Needless Cloud Over New York City," *New York Times*, December 21, 1989.
86. Briffault, "Voting Rights, Home Rule, and Metropolitan Governance," 786.
87. Kramer and Flanagan, *Staten Island: Conservative Bastion in a Liberal City*, 126; "News from Senator John J. Marchi," January 5, 1990, Marchi Papers, Secession, Subject Files, Press Releases, 1983–1996, Box 33, Folder 17.
88. Supreme Court of the State of New York, County of New York, Complaint for Declaratory Judgment, *The City of New York et al. v. The State of New York*, December 28, 1989, 1, 10, Marchi Papers, Secession, Home Rule, Morris, 1989–1990, Box 6, Folder 1.
89. "News from Senator John J. Marchi," December 29, 1984, Marchi Papers, Secession, Subject Files, Box 33, Folder 17.
90. State of New York, Court of Appeals, Opinion, *The City of New York et al. v. The State of New York*, September 18, 1990, Marchi Papers, Secession, Home Rule, 1989–90, Box 6, Folder 27. See also Briffault, "Voting Rights, Home Rule, and Metropolitan Governance," 787; "Secession Vote OK'd," *Staten Island Advance*, September 18, 1990; Underweiser, "The Legality of Staten Island's Attempt to Secede," 147.
91. Kramer and Flanagan, *Staten Island: Conservative Bastion in a Liberal City*, 126; Briffault, "Voting Rights, Home Rule, and Metropolitan Governance," 787; Viteritti, "Municipal Home Rule and the Conditions of Justifiable Secession," 58; Mireya Navarro, "The 1990 Elections: Secession; Staten Island Turns to Cost of Parting," *New York Times*, November 8, 1990; Sachs, "Staten Island Secessionism: Strained Threads of the Consolidated City Fabric," 103–4. See also Carl Campanile, "After the Vote: What's Next for Secession?" *Staten Island Advance*, November 7, 1990.
92. Alessandra Stanley, "Staten Island Votes a Resounding Yes on Taking Step Toward Separation," *New York Times*, November 7, 1990.
93. The original members included Senator Marchi (chair); Assemblywoman Connelly (vice chair); Assemblymen Straniere and Vitaliano; Senator Connor; former Staten Island Democratic Association (SIDA) president John W. Lavelle; Koch aide and former SIDA member Paul J. Henry; Cuomo administration official and former SIDA member Allen P. Cappelli; Marchi's counsel Kathryn R. Rooney; Francis H. Powers, board member of the MTA and Wall Street executive; Paul E. Proske, former chair of the Staten Island Chamber of Commerce; and Richard Thomas (a Republican and the only African American member of the commission); and Martin D. Lubin, associate director of District Council 37. After redistricting, Senator Christopher Mega replaced Connor on January 1, 1993. Mega was replaced in July 1993 by Senator Robert DiCarlo. Joseph P. Viteritti, "Municipal Home Rule and the Conditions of Justifiable Secession: Applications to Staten Island, New York," April 1995, Marchi Papers, Secession, Home Rule, 1989–1995, Box 9, Folder 2; Kramer and Flanagan, *Staten Island: Conservative Bastion in a Liberal City*, 128–29; Craig Schneider, "Secession Panel Begins Framing a Charter," *Staten Island Advance*, September 15, 1991; Patrick Joyce, "Secession Commission Is Sworn In," *Staten Island Advance*, December 28, 1991.

15. SECESSION ■ 687

94. *Viteritti: Oral History Transcripts*, April 9, 2001, 137–47; Carl Campanile, "Studying 'Divorce' from the City," *Staten Island Advance*, June 29, 1991. See also "Secession Panel Picks Director," *Staten Island Advance*, June 23, 1991.
95. *Public Hearings and Documents of the Staten Island Charter Commission* is available on the website of the Archives and Special Collections, Department of the Library, College of Staten Island, CUNY, http://163.238.8.180/archives/. Various minutes of the Staten Island Charter Commission are scattered through the Marchi secession files, but they tend to be largely procedural in emphasis.
96. Craig Schneider and Carl Campanile, "BP: Costs a Must on Secession," *Staten Island Advance*, November 9, 1992.
97. Joseph P. Viteritti to Members of the Commission, May 18, 1992, and "Why We Oppose Secession Now," Marchi Papers, Secession, Subject Files, 1983–1996, Box 33, Folder 4.
98. City of New York, Mayor's Task Force on Staten Island Secession, Office of Management and Budget, Department of Finance, *Staten Island Secession: The Price of Independence* (July 1992), 1, Archives and Special Collections, Department of the Library, College of Staten Island. See also Robert Berne et al., "Estimating the Fiscal Impact of Secession: The Case of Staten Island and New York City," *Public Budgeting & Financial Management* 7 (Summer 1995): 159.
99. City of New York, Mayor's Task Force on Staten Island Secession, Office of Management and Budget, Department of Finance, *Staten Island Secession: The Price of Independence*, 27.
100. City of New York, Mayor's Task Force on Staten Island Secession, Office of Management and Budget, Department of Finance, *Staten Island Secession: The Price of Independence*, 23.
101. Douglas Muzzio, "Staten Island on Secession," Department of Political Science, Baruch College, CUNY, March 1991, 1, Marchi Papers, Secession, Subject Files, 1983–1994, Box 34, Folder 3. See also Viteritti, "Municipal Home Rule and the Conditions of Justifiable Secession," 58–61.
102. Muzzio, "Staten Island on Secession," 2–21. See also Viteritti, "Municipal Home Rule and the Conditions of Justifiable Secession," 63–67.
103. Viteritti, "Municipal Home Rule and the Conditions of Justifiable Secession," 58–61; Mary Engels, "S.I. Secede? Tell 'Em More," *Daily News*, March 2, 1992. See also Joseph Viteritti, "Should Staten Island Leave the City?" *City Journal*, Autumn 1992, http://www.city-journal.org/html/should-staten-island-leave-city-12675.html.
104. Major areas of the study included governance, economics, and labor.
105. New York State, Charter Commission for Staten Island, *Report to the Governor and State Legislature/State of New York Charter Commission for Staten Island* (February 2, 1993), i–iii, Archives & Special Collections, City University of New York, CUNY Academic Works, http://academicworks.cuny.edu/si_arch_chart/. See also State of New York, Charter Commission for Staten Island, *Q&A*, June 23, 1993, revised, Marchi Papers, Secession, Subject Files, 1983–1994, Box 34, Folder 4.
106. David N. Dinkins to Mario M. Cuomo, February 11, 1993, Marchi Papers, Secession, Correspondence, 1983–2002, Box 39, Folder 5.
107. Carl Campanile and Craig Schneider, "Secession Foes Rip City Effort," *Staten Island Advance*, April 5, 1993.
108. See "Hey, Wait! Can't We Work It Out?" *Newsday*, October 28, 1993; "If at First You Do Secede," *Daily News*, November 1, 1993; James Dao, "The 1993 Elections: Staten Island; Secession Is Approved; Next Move Is Albany," *New York Times*, November 3, 1993.

109. "Staten Island: A Mini–New York City or a Downstate Urban County? A Report to the People from Borough President Guy Molinari," April 15, 1993, Marchi Papers, Secession, Subject Files, 1983–1996, Box 33, Folder 17.
110. "Staten Island Flashback: The Vote for Secession," *Staten Island Advance*, November 3, 2007, *SILive.com*, https://www.silive.com/news/2007/11/staten_island_flasback_the_vot.html. See also Bill Ferrell, Michael Finnegan, and Dick Sheridan, "From S.I., Parting Shots," *New York Daily News*, November 4, 1993; Bill Ferrell and Jere Hester, "Early S.I. Returns: Bye!" *New York Daily News*, November 3, 1993; "A Borough Gives Notice (or Hadn't You Noticed?)," *New York Times*, November 3, 1993; Gregg Birnbaum, "S.I.: Cut Us from Apple!" *New York Post*, November 3, 1993; Craig Schneider and Carl Campanile, "SECESSION: Vote Heard 'Round the World," *Staten Island Advance*, November 3, 1993; "Should Staten Island Secede from New York City?" *Newsday*, November 2, 1993.
111. Gregg Birnbaum, "Tough Road Ahead for S.I. Secessionists," *New York Post*, October 28, 1993. See also Al Efron: Oral History Transcripts, May 12, 2004, 44, College of Staten Island Oral History Collection, v. 7. District Leader Al Efron believed that Weprin favored secession, but he died in office.
112. Guy V. Molinari to Elizabeth Connelly, November 26, 1993, Marchi Papers, Secession, Home Rule, 1989–1995, Box 9, Folder 1. See also Goldfarb: Oral History Transcripts, December 1, 2004, 26; Molinari, *A Life of Service*, 401–3.
113. William Bunch, "Staten Island Wants to Go Its Own Way," *Newsday*, November 3, 1993. See also Molinari, *A Life of Service*, 404.
114. Vivienne Walt, "A Secession Vote for SI," *Newsday*, November 1, 1993.
115. "Staten Island Expected to Vote to Secede," *Los Angeles Times*, October 31, 1993.
116. Carl Campanile, "Experts Begin Working on Secession Bill," *Staten Island Advance*, December 16, 1993.
117. Viteritti, "Municipal Home Rule and the Conditions of Justifiable Secession," 2–3. See also "News from Senator John J. Marchi," March 24, 1994, May 9, 1994, Marchi Papers, Secession, Subject Files, 1983–1996, Box 33, Folder 17; Paul A. Crotty to Rudolph Giuliani, May 9, 1994; Elizabeth D. Moore to the Governor, March 10, 1994; Harold F. Kuffner Jr. to Sheldon Silver, September 27, 1994; Sheldon Silver to Harold F. Kuffner Jr., July 26, 1994, Marchi Papers, Secession, Home Rule, 1989–1995, Box 9, Folder 1; Paul A. Crotty to Ralph J. Marino, May 9, 1994, Marchi Papers, Secession, Home Rule, 1989–1995, Box 9, Folder 5; Carl Campanile, "Mayor's Counsel Spells Out Home Rule Defense," *Staten Island Advance*, May 10, 1994; Robert D. McFadden, "Staten Island Secession, the Debate That Wouldn't Die, Now Reaches Albany," *New York Times*, March 1, 1994; Maggs, "Breakin' Up Is Hard to Do," 1608–9; Viteritti: Oral History Transcripts, April 9, 2001, 148.
118. Craig Schneider, Carl Campanile, and Reginald Patrick, "Secret Memo Infuriates Lawmakers," *Staten Island Advance*, March 10, 1994.
119. Viteritti, "Municipal Home Rule and the Conditions of Justifiable Secession," 3.
120. John J. Marchi to Staten Island Charter Commission Members, January 19, 1995; court decision, January 17, 1993, Marchi Papers, Secession, Home Rule, 1994–1996, Box 7, Folder 5. See also Craig Schneider and Carl Campanile, "Island's Destiny on the Line," *Staten Island Advance*, January 18, 1995.
121. Carl Campanile and Craig Schneider, "Another Setback for Secession," *Staten Island Advance*, February 23, 1996. See also "Another Blow for S.I. Secession," *New York Post*, February 23, 1996.
122. Carl Campanile, "Last Pitch for Secession," *Staten Island Advance*, October 17, 1996; Kimberly Schaye, "Court Spoils Secession Try," *New York Daily News*, November 15,

1996. See also Kathryn K. Rooney to Senator Marchi, November 12, 1996, Marchi Papers, Secession, Subject Files, Box 32, Folder 12. A cache of court materials, law-journal essays, and some government documents dealing with Staten Island secession from 1989 to 1996 (some duplicated in the chapter which were located in other depositories) can be found at the Newman Library, Baruch College, CUNY. Also see Gary Spencer, "Staten Island Secession Ruled a Legislative Matter," *New York Law Journal*, February 23, 1996; Gary Spencer, "Staten Island Secessionists Suffer Defeat," *New York Law Journal*, November 15, 1996.

16. CLOSURE

1. "News from Senator John J. Marchi," May 9, 1994, The Senator John J. Marchi Papers, Secession, Subject Files, 1983–1996, Box 33, Folder 17, Archives & Special Collections, Department of the Library, College of Staten Island, CUNY.
2. See Elizabeth A. Connelly: Oral History Transcripts, October 10, 200, 84, College of Staten Island Oral History Collection, v. 3, ed. Jeffery A. Kroessler, Archives & Special Collections, Department of the Library, College of Staten Island, CUNY; David Goldfarb: Oral History Transcripts, December 1, 2004, 40, College of Staten Island Oral History Collection, v. 3; Joseph P. Viteritti: Oral History Transcripts, April 9, 2001, 150, College of Staten Island Oral History Collection, v. 1; Charles L. Sachs, "Staten Island Secessionism: Strained Threads of the Consolidated City Fabric," *culturefront*, Winter 1997–1998, 104; Rocco Parascandola, "S.I. Issue a Sticky One for Rudy, Gov," *New York Post*, November 4, 1993; "Secession in Jeopardy?" *New York Times*, November 4, 1993.
3. "Mayoralty," in *The Encyclopedia of New York City*, 2nd ed., ed. Kenneth T. Jackson (New Haven, CT: Yale University Press, 2010), 814–15.
4. Guy V. Molinari, *A Life of Service* (New York: Page, 2016), 403.
5. Interestingly, Koch received a higher percentage of Staten Island votes in both 1981 and 1985 than did Giuliani in 1989. John Hull Mollenkopf, *A Phoenix in the Ashes: The Rise and Fall of the Koch Coalition in New York City Politics* (Princeton, NJ: Princeton University Press, 1992), 168, 174–77, 180–85, 192–93.
6. "Mayoralty," in *The Encyclopedia of New York City*, ed. Jackson, 814–15.
7. David Seifman, "Staten Island Boosted Rudy Into City Hall," *New York Post*, November 4, 1993. See also Maurice Carroll and Clay Richards, "Big SI Vote Helped Push Rudy Over Top," *Newsday*, November 4, 1993; Sam Roberts, "The 1993 Elections: News Analysis; The Tide Turns on Voter Turnout," *New York Times*, November 4, 1993.
8. Jim Sleeper, "The End of the Rainbow," *New Republic*, October 31, 2013, https://newrepublic.com/article/115430/new-republic-mayoral-election-1993; Todd S. Purdum, "The 1993 Elections: Giuliani; Overcoming Early Negative Image to Pull Ahead in Voters' Perceptions," *New York Times*, November 4, 1993; William Bunch and Jessie Mangaliman, "Term Limits, SI Secession: Calls to Arms," *Newsday*, November 4, 1993; "Three Dubious Political Fixes," *New York Times*, November 4, 1993; James Dao, "The 1993 Elections: Staten Island; Secession Is Approved; Next Move Is Albany's," *New York Times*, November 3, 1993. See also George J. Lankevich, *American Metropolis: A History of New York City* (New York: New York University Press, 1998), 246–48.
9. Alison Mitchell, "The 1993 Elections: The Transition; Dinkins and Giuliani Join in a Call for Healing and Unity," *New York Times*, November 4, 1993. See also John J. Goldman, "New York Voters Choose Mayor in Watershed Race," *Los Angeles Times*,

November 3, 1993; Janet Cawley, "Giuliani Defeats Dinkins in Down-to-Wire New York Mayor's Race," *Chicago Tribune*, November 3, 1993.

10. Purdum, "The 1993 Elections"; "Three Dubious Political Fixes," *New York Times*, November 4, 1993; Bunch and Mangaliman, "Term Limits, SI Secession: Calls to Arms"; Sleeper, "The End of the Rainbow." See also Goldman, "New York Voters Choose Mayor in Watershed Race."
11. Fred Siegel, *The Prince of the City: Giuliani, New York, and the Genius of American Life* (San Francisco: Encounter, 2005), 77–82, also 25–33, 69–72; Wayne Barrett, *Rudy: An Investigative Biography of Rudolph Giuliani* (New York: Basic Books, 2000), 265–86. See also Wilbur C. Rich, *David Dinkins and New York City Politics: Race, Images, and the Media* (Albany: State University of New York Press, 2007), 44–49, 143, 163, 196–97; Lankevich, *American Metropolis*, 246–48.
12. Richard Riordan won his election in Los Angeles just five months earlier than Giuliani in June. Cawley, "Giuliani Defeats Dinkins."
13. "Rudolph Giuliani," Biography.com, http://www.biography.com/people/rudolph-giuliani-9312674.
14. During his six years as U.S. attorney, he had 4,152 convictions with only twenty-five reversals. "Rudolph Giuliani," Biography.com; "A Biography of Mayor Rudolph W. Giuliani," NYC.gov, http://www.nyc.gov/html/records/rwg/html/bio.html; "Biography: Rudolph Giuliani," Academy of Achievement, http://www.achievement.org/autodoc/page/giuopro-1.
15. Lynne A. Weikart, "The Giuliani Administration and the New Public Management in New York City," *Urban Affairs Review* 36 (January 2001): 359–61, 375–78. See also Lankevich, *American Metropolis*, 249–51; Robert Polner, ed., *America's Mayor: The Hidden History of Rudy Giuliani's New York* (Brooklyn: Soft Skull, 2005), xxi–xxvi; Barrett, *Rudy*, 9–11.
16. "Biography: Rudolph Giuliani," Academy of Achievement; "Rudy Giuliani," History.com, http://www.history.com/topics/rudy-giuliani.
17. "Rudolph W. Giuliani," *Encyclopedia Britannica*, https://www.britannica.com/biography/Rudolph-W-Giuliani; "Biography: Rudolph Giuliani," Academy of Achievement; "Rudolph Giuliani," Biography.com. See also Fred Siegel and Harry Siegel, "Rudolph (Rudy) W(illiam Lewis) Giuliani," in *The Encyclopedia of New York City*, ed. Jackson, 510–11; Siegel, *The Prince of the City*, 75–76.
18. See, for example, City of New York, Department of Sanitation, Bureau of Cleaning and Collection, Analysis of Bureau of Cleaning and Collection Costs, June 1995, Giuliani Papers, Deputy Mayors, Richard Schwartz, Subject Files, Sanitation Department—Refuse Collection Report, Box 02/07/009, Folder 351.
19. See International Technology Corporation, *Fresh Kills: Leachate Mitigation System: Final Leachate Mitigation Report*, v. 1, Project no. 529363, October 12, 1993; International Technology Corporation, *Draft Final Surface Water and Sediment Report: Fresh Kills Leachate Mitigation System Project*, April 15, 1993; BWD Engineering, *Report on the Split Bank Operations for the Fresh Kills Landfill*, March 1993; Woodward-Clyde Consultants, Inc., *Fresh Kills Landfill: Final Updated Geotechnical Site Characterization Report*, 92C4235, August 1993; International Technology Corporation, *Fresh Kills Leachate Mitigation System: Final Hydrological Report*, v. 1, November 26, 1993; International Technology Corporation, *Fresh Kills Leachate Mitigation System Project: Final Surface Water and Sediment Report*, v. 1, Doc. no. 529363-01836, December 23, 1993; International Technology Corporation, *Fresh Kills Leachate Mitigation System: Final Leachate Mitigation Report & Corrective Measures Assessment Report*,

Doc. no. 529363-01839, March 7, 1994; International Technology Corporation, *Fresh Kills Leachate Mitigation System: Leachate Mitigation System, NYSDEC Permit Application*, Doc. no. 529363-01924, March 31, 1994; International Technology Corporation, *Fresh Kills Leachate Mitigation System: Responses to General Comments Related to the Draft Leachate Mitigation Evaluation for Sections 2/8 & 3/4—and Related Reports*, May 1994; International Technology Corporation, *Fresh Kills Leachate Mitigation System: Responses to Comments on the Final Engineering Report, Final Plans & Specifications and Final QA/QC Report*, November 18, 1994; Woodward-Clyde Consultants, Inc., *Fresh Kills Landfill: Phase 2, Monitoring System Design Report*, 92C4235, April 1994; O'Brien & Gere Engineers, Inc., *Final Engineering Report: Fresh Kills Landfill, Section 1/9 Leachate Collection and Containment System*, August 8, 1994; O'Brien & Gere Engineers, Inc., *Final Construction Quality Assurance/Construction Quality Control (CQA/CQC) Plan: Fresh Kills Landfill, Sections 1/9 and 6/7 Leachate Collection and Containment System*, August 8, 1994; O'Brien & Gere Engineers, Inc., *Final Specifications: Fresh Kills Landfill, Sections 1/9 and 6/7 Leachate Collection and Containment System*, August 8, 1994, Archives & Special Collections, Department of the Library, College of Staten Island, CUNY. See also U.S. Environmental Protection Agency, Region 2, *Determination of Landfill Gas Composition and Pollutant Emission Rates at Fresh Kills Landfill*, vol. 1, Project Report, November 1995.

20. Carolyn Rushefsky, "Judge Orders Landfill Probe," *Staten Island Advance*, March 19, 1993. See also Don Gross, "Is City Doing Its Job at Landfill?" *Staten Island Advance*, February 8, 1993; Joan Gerstel, "Court Orders Expert to Assess Leachate Progress at Landfill," *Staten Island Register*, February 9, 1993; Carolyn Rushefsky, "Ground Broken for Leachate Treatment Plant," *Staten Island Advance*, February 18, 1993.

21. Reginald Patrick, "City Calls for Contracting Cleanup," *Staten Island Advance*, January 2, 1993.

22. Carolyn Rushefsky, "Sanitation to Field Leachate Questions," *Staten Island Advance*, March 22, 1994; Fresh Kills Landfill: Leachate Mitigation/Corrective Measures, Public Meeting, March 23, 1994, Archives & Special Collections, Department of the Library, College of Staten Island; "Fresh Kills to Get Fresher Leachate," *Civil Engineering* 64 (June 1994): 20; Carolyn Rushefsky, "Citizens' Group Raps Leachate Expansion," *Staten Island Advance*, August 14, 1994.

23. Don Gross, "Molinari Criticizes Landfill Air Study," *Staten Island Advance*, August 18, 1993.

24. Joan Gerstel, "BP Airs Complaints on DEC Study," *Staten Island Register*, August 24, 1993. See also Carolyn Rushefsky, "Is Fresh Kills a Dangerous Landfill?" *Staten Island Advance*, October 4, 1993.

25. See Carolyn Rushefsky, "Coming Clean: Landfill Working to Eliminate Odors," *Staten Island Advance*, January 17, 1993; Reginald Patrick, "Fresh Kills Gets Priority Funding," *Staten Island Advance*, May 17, 1993; Michael Azzara, "City Report: Landfill Mismanaged," *Staten Island Advance*, June 9, 1993; City of New York, Department of Sanitation, *Contingency Plan for the Fresh Kills Landfill*, July 1993, Archives & Special Collections, Department of the Library, College of Staten Island; Elizabeth A. Connelly to Joseph Ferrara, May 17, 1994, The Assemblywoman Elizabeth A. Connelly Papers, Series 1, Subject Files, Box 20, Folder 3, Archives & Special Collections, Department of the Library, College of Staten Island, CUNY. See also Citizens Budget Commission, *The Performance of the New York City Department of Sanitation: Recommendations for Improvement* (February 1993).

26. Emily Lloyd to Raymond Horton, January 12, 1993, David N. Dinkins Collection, Box 14, Folder 158, La Guardia and Wagner Archives, LaGuardia Community College/CUNY, Long Island City, New York. According to the Independent Budget Office in New York City, the diversion rate in 1992 was 6.7 percent, 9.2 percent in 1993, 14.7 percent in 1994 (well short of the 25 percent target), and 28.1 percent in 1995. See Independent Budget Office, *Overview of the Waste Stream Managed by the NYC Department of Sanitation*, Background Paper, February 2001, 5. See also Matthew Gandy, *Recycling and the Politics of Urban Waste* (London: Earthscan, 1994), 78–79; Megan Ryan, "Los Angeles 21, New York 5 . . . (recycling garbage)," *World Watch Magazine* 6 (March/April 1993), http://www.worldwatch.org/node/359.
27. Emily Lloyd to Raymond Horton, January 12, 1993.
28. Emily Lloyd to Raymond Horton, January 12, 1993.
29. See Carolyn Rushefsky, "'We Want the Landfill Closed,'" *Staten Island Advance*, February 23, 1993. See also Interstate Sanitation Commission, *1995 Annual Report of the Interstate Sanitation Commission on the Water Pollution Control Activities and the Interstate Air Pollution Program*, 72–73.
30. Joan Gerstel, "FK Neighbors Ready to File $2M Suit Against City," *Staten Island Register*, January 26, 1993. See also "Statement by Assemblywoman Elizabeth A. Connelly at the Fresh Kills Landfill Consent Hearing," February 22, 1993, Connelly Papers, Series 1, Subject Files, Box 20, Folder 3.
31. "Another Affront in Our Backyard," *Staten Island Register*, June 3, 1993.
32. Letter, Andrew Stein, January 28, 1993, Connelly Papers, Series 1, Subject Files, Box 20, Folder 3.
33. "Joint Statement of Assemblywoman Elizabeth A. Connelly and Assemblyman Eric N. Vitaliano at the Public Hearing on Revisions to 6 NYC RR Part 360—Department of Environmental Conservation," February 9, 1993; and Barbara Warren Chinitz to Eric N. Vitaliano, January 29, 1993, Connelly Papers, Series 1, Subject Files, Box 20, Folder 3; Carolyn Rushefsky, "Officials Denounce Proposal to Use Ash," *Staten Island Advance*, February 10, 1993; Joan Gerstel, "Island Pols Blast Ash Plan," *Staten Island Register*, February 16, 1993. See also Norman H. Nosenchuck, "Key Events of the New York State Solid Waste Management Program: 1970–1995," *Albany Law Environmental Outlook* 35 (1995–1996): 39.
34. Chris Franz, "'No Ash at Fresh Kills,' Promises DOS," *Staten Island Register*, March 2, 1993.
35. Terrence J. Kivlan, "Ash Could Still Go to Fresh Kills," *Staten Island Advance*, May 5, 1994. See also Nosenchuck, "Key Events of the New York State Solid Waste Management Program," 43.
36. Don Gross, "Brookfield's Hidden Secrets to Be Uncovered?" *Staten Island Advance*, May 7, 1993. See also Anthony J. Germano to Guy V. Molinari, May 23, 1996, Marchi Papers, Series 1, Subject Files, Box 47, Folder 5.
37. Gross, "Brookfield's Hidden Secrets to Be Uncovered?"; Carolyn Rushefsky, "We Reported Landfill Problems All Along," *Staten Island Advance*, May 19, 1993; Carolyn Rushefsky, "Health Tests Urged," *Staten Island Advance*, May 22, 1993; Carolyn Rushefsky, "Neighbors Demand Brookfield Tests Now," *Staten Island Advance*, May 28, 1993.
38. Carolyn Rushefsky and Terence J. Kivlan, "Feds to Probe Brookfield?" *Staten Island Advance*, June 5, 1993.
39. Don Gross, "Breathing a Little Easier," *Staten Island Advance*, July 1, 1993; Letter, Albert F. Appleton, NYC Department of Environmental Protection, August 5, 1993, The Assemblyman Eric N. Vitaliano Papers, Series 1, Subject Files, Box 6, Folder 8,

16. CLOSURE ■ 693

Archives & Special Collections, Department of the Library, College of Staten Island.
40. Carolyn Rushefsky, "'City Should Have Told Them,'" *Staten Island Advance*, August 10, 1993. See also Kristin Choo, "Dump Neighbors Suing: No Toxics Found So Far," *Staten Island Register*, August 17, 1993.
41. Terence J. Kivlan, "Brookfield Tests Find No Toxins," *Staten Island Advance*, September 2, 1993; "EPA Clears Soil and Air Near Brookfield Dump," *Staten Island Register*, September 7, 1993. See also "Brookfield Avenue Landfill: Public Meeting," September 30, 1993, Vitaliano Papers, Series 1, Subject Files, Box 6, Folder 8; Don Gross, "DEC Resumes Digging in Great Kills," *Staten Island Advance*, August 5, 1993; Terence J. Kivlan, "Landfill Waste Not Hazardous," *Staten Island Advance*, August 14, 1993.
42. Kristin Choo, "Brookfield Work Begins," *Staten Island Register*, December 7, 1993. See also "Brookfield Avenue Landfill RI/FS, CAC Meeting," January 26, 1994, and March 1, 1994; "Brookfield Landfill Health Study—Update," March 1, 1994; "Brookfield Avenue Landfill: Citizens Advisory Committee Meeting," June 21, 1994; Joel S. Hirschhorn to Brookfield Avenue Landfill CAC, June 23, 1994; Brookfield Ave. Landfill CAC to Eric Vitaliano, July 8, 1994; Marilyn Gelber to Elected Official, August 2, 1994; Marilyn Gelber to CAC Members, September 15, 1994, Vitaliano Papers, Series 1, Subject Files, Box 6, Folder 6; Joel S. Hirschhorn, *Remedial Investigation (RI) Issues Report*, May 1994; and Phyllis Y. Atwater, Langdon Marsh Commissioner to CAC Members, July 28, 1994, Vitaliano Papers, Series 1, Subject Files, Box 6, Folder 13.
43. "Double Jeopardy," *Staten Island Advance*, April 12, 1994. See also Phyllis Y. Atwater, DEC to Susan Molinari, June 8, 1995, Vitaliano Papers, Subject Files, Box 6, Folder 9; "Feds to Oversee Brookfield Cleanup," *Staten Island Register*, June 13, 1995.
44. See Martin V. Melosi, "Environmental Justice, Political Agenda Setting, and the Myths of History," in *Effluent America: Cities, Industry, Energy, and the Environment* (Pittsburgh, PA: University of Pittsburgh Press, 2001), 246–47.
45. Joan Gerstel, "State OKs Navy Yard Incinerator," *Staten Island Register*, April 21, 1993.
46. Terence Kivlan, "City Lobbyists Yawn at Anti-Incinerator Bill," *Staten Island Advance*, July 4, 1993. See also Brad Edmonson, *Environmental Affairs in New York State: A Historical Overview*, pub. no. 72 (Albany: New York State Archives, 2001), 46.
47. Dennis Hevesi, "State Approves the Brooklyn Navy Yard Incinerator," *New York Times*, September 12, 1993.
48. Howard Golden to David N. Dinkins, September 30, 1993, Vitaliano Papers, Series 1, Subject Files, Box 64, Folder 6.
49. Late in 1994, for example, asbestos in loads of cover and road materials going to Fresh Kills was discovered. John J. Doherty to Peter Powers, August 26, 1994; Jeanne M. Fox, DEC to Susan Molinari, September 2, 1994; Susan Molinari to Conrad Simon, EPA, August 23, 1994; Susan Molinari to Conrad Simon, August 24, 1994, Office of the Mayor, Rudolph W. Giuliani, Deputy Mayors, Peter Powers, Sanitation Department of Fresh Kills Landfill, Box 02/01/021, Folder 500, La Guardia and Wagner Archives, LaGuardia Community College. See also Office of Management and Budget, Audit Coordination and Review, January 24, 1993, concerning deficiencies in Fresh Kills' operations in same box and folder.
50. Emily Lloyd to Deputy Mayor Powers, February 15, 1994, Rudolph W. Giuliani Collection, Deputy Mayors, Peter Powers, Sanitation Department of Fresh Kills Landfill, Box 02/01/021, Folder 500, La Guardia and Wagner Archives.
51. City of New York, Department of Sanitation, *Executive Summary: Project Information and Site Assessment Document, Fresh Kills Landfill*, September 23, 1994, St. George Library Center, New York Public Library, Staten Island, 4, 6, 16.

52. City of New York, Department of Sanitation, *Executive Summary: Project Information and Site Assessment Document, Fresh Kills Landfill*, 1–2, 6, 10, 22. See also City of New York, Department of Sanitation, *Draft 6 NYCRR Part 360, Permit Application for the Fresh Kills Landfill*, September 15, 1994; and City of New York, Department of Sanitation, *Environmental Assessment Statement and Draft Scoping Document for the Draft Environmental Impact Statement, Fresh Kills Landfill*, September 23, 1994, St. George Library Center, New York Public Library.
53. Carolyn Rushefsky, "Landfill Starting to 'Shrink' as City Keeps Buffer Pledge," *Staten Island Advance*, January 15, 1993. See also Carl Campanile, "State, City Discussing Ways to 'Shrink' Landfill," *Staten Island Advance*, August 20, 1992; Eric N. Vitaliano to Emily Lloyd, July 2, 1993, Emily Lloyd to Eric N. Vitaliano, July 29, 1993, Emily Lloyd to Eric N. Vitaliano, July 29, 1993, Vitaliano Papers, Series 1, Subject Files, Box 36, Folder 16.
54. Paul M. McPolin, "City Funds Sought to Pay for Closure of Landfill," *Staten Island Advance*, February 24, 1993. See also City of New York, Department of Sanitation, *Final Scoping Document for the Draft Environmental Impact Statement, Fresh Kills Landfill*, February 10, 1995, Marchi Papers, Staten Island, Fresh Kills, 1958, 1996, 2004, Box 7, Folder 14.
55. Carolyn Rushefsky, "Landscape Architect Has a Vision for Landfill," *Staten Island Advance*, March 8, 1993; Kerry Diamond, "Envisioning a Second Life for the Fresh Kills Landfill," *Staten Island Advance*, March 12, 1993.
56. William Young, "A Tree Grows on Fresh Kills," *Garbage* 6 (Summer 1994): 60–61.
57. Lynn Cowan, "When Landfill Closes, Where Will Rats Go? *Staten Island Advance*, July 25, 1993. See also William Young, "A Dump No More," *American Forests* 101 (Autumn 1995): 58–59.
58. New York State Legislative Commission on Solid Waste Management, *Where Will the Garbage Go? 1995*, April 1995; New York State Legislative Commission on Solid Waste Management, *Where Will the Garbage Go? 1994*, April 1994, 1, 3, Vitaliano Papers, Series 1, Subject Files, Box 65, Folder 19.
59. Carolyn Rushefsky, "Public Airs Landfill Concerns," *Staten Island Advance*, February 16, 1995.
60. *Annual Meeting to Provide Consent Order Compliance and Milestone Updates for the Fresh Kills Landfill, Staten Island, New York City Department of Sanitation*, Public Hearing, February 15, 1995, 7, Marchi Papers, Staten Island, Fresh Kills, Box 7, Folder 14.
61. *Annual Meeting to Provide Consent Order Compliance and Milestone Updates for the Fresh Kills Landfill, Staten Island, New York City Department of Sanitation*, 45.
62. "State Cites Fresh Kills Landfill for Leachate Runoff," *Solid Waste Digest*, March 1995, 11. See also Reginald Patrick, "Leachate Control Program Demanded for Entire Landfill," *Staten Island Advance*, March 8, 1995; Terrence J. Kivlan, "EPA Wants to See Closing Date for Fresh Kills," *Staten Island Advance*, February 16, 1995. See also David H. Minott, "Air Pollution Emissions from MSW Landfills—The 'Sleeper' Issue for Landfill Design and Regulation," paper presented at the Tenth Conference on Solid Waste Management & Materials Policy, New York City, February 19–22, 1995.
63. Carolyn Rushevsky, "Landfill Studies Reviewed," *Staten Island Advance*, March 7, 1995. See Don Gross, "Federal Lawsuit Filed to Close Fresh Kills Landfill," *Staten Island Advance*, April 4, 1995.
64. Terrence J. Kivlan, "City Details Plans to Phase Out Fresh Kills," *Staten Island Advance*, February 17, 1995. See also Carl Campanile, "City May Get Year to Comply with Landfill Regulations," *Staten Island Advance*, April 20, 1995.

65. See Lynee A. Weikart, "The Giuliani Administration and the New Public Management in New York City," *Urban Affairs Review* 36 (January 2001): 360.
66. Steven Lee Myers, "Sanitation Dept. Gets up-from-Ranks Chief," *New York Times*, August 12, 1994.
67. "Waste in New York Municipal Government," *Press Digest*, *Daily News* Special Series, January 10–January 20, 1994, Office of the Mayor, Rudolph W. Giuliani, Deputy Mayors, Richard Schwartz, Subject Files, Waste Box 02/07/010, Folder 423, La Guardia and Wagner Archives.
68. Jo Thomas, "After Growing in Success, Recycling Faces Obstacles," *New York Times*, November 27, 1994. See also Alexander Williams, "Trash Talk: Solid Waste Disposal in New York City," Fordham University, DigitalResearch@Fordham, 2013, http://fordham.bepress.com/environ_theses/18/; Roger Starr, "Recycling: Myths and Realities," *Public Interest*, March 22, 1995.
69. Thomas, "After Growing in Success, Recycling Faces Obstacles." See also "Fed Up with Landfill? The News Is Not Good," *Staten Island Advance*, May 8, 1994.
70. Samantha McBride, *Recycling Reconsidered: The Present Failure and Future Promise of Environmental Action in the United States* (Cambridge, MA: MIT Press, 2012), 143–44.
71. Long Island and upstate New York were crucial to his victory, as was Staten Island.
72. James Dao, "The New Governor: Man in the News; A Pragmatic Conservative: George Elmer Pataki," *New York Times*, January 2, 1995. See also James Dao, "Conservatives Choose Pataki for Governor," *New York Times*, January 5, 1994; Peter Slocum, "George E(lmer) Pataki," in *The Encyclopedia of New York State*, ed. Peter Eisenstadt (Syracuse, NY: Syracuse University Press, 2005), 1184.
73. Joe Sexton, "New York Signals End to Incinerator Plan," *New York Times*, September 20, 1994. See also Keith Schneider, "Incinerator Operators Say Ruling Will Be Costly," *New York Times*, May 3, 1994; Mayank Teotia, "Managing New York City Municipal Solid Waste—Using Anaerobic Digestion," Program for Sustainable Planning and Development, School of Architecture, Pratt Institute, Spring 2013, 3; Arthur S. Kell, *Setting the Record Straight: A Fiscal Analysis of the City of New York's Solid Waste Management Programs and the Proposed Brooklyn Navy Yard Incinerator* (New York: NYPIRG, 1992), Marchi Papers, Secession, Reports/Research Papers/Studies, S–Z, 1990–1994, Box 30, Folder 9.
74. Williams, "Trash Talk." See also New York City, Department of Investigation, *Report of Investigation Regarding an Environmental Assessment of the Site for the Brooklyn Navy Yard Refuse-to-Energy Facility*, May 4, 1994; "Borough President Golden Commends the Mayor's Decision to Require a Supplemental Environmental Impact Statement (SEIS) for the Proposed Brooklyn Navy Yard Incinerator," February 15, 1996; and Howard Golden to John Doherty, February 28, 1996, Giuliani Collection, Deputy Mayors, Peter Powers, Sanitation Department of Fresh Kills Landfill, Box 02/01/048, Folder 1282.
75. City of New York, Independent Budget Office, *Overview of the Waste Stream Managed by the NYC Department of Sanitation*, Background Paper, February 2001, 2–6, 9, 12.
76. Reginald Patrick, "City to Export Trash to Relieve Burden on Landfill," *Staten Island Advance*, November 24, 1993.
77. Matthew L. Wald, "In Trash Crunch, Giuliani Looks out of State," *New York Times*, March 4, 1994. See also Terrence J. Kivlan, "House Backs Plan to Boost Landfill Tonnage," *Staten Island Advance*, September 29, 1994.

78. Edward Repa, "Interstate Movement of Municipal Solid Waste—1992 Update," *Waste Age* 25 (January 1994): 37–38.
79. For instance, see Paul Williamson, WasteMasters, Inc., to John J. Marchi, October 26, 1995, Marchi Papers, Subject Files, Box 7; Waste Management of New York City to Peter J. Powers, August 8, 1995, Giuliani Collection, Deputy Mayors, Richard Schwartz, Subject Files, Box 02/07/010, Folder 424.
80. Andrew G. Keller and Mitch Renkow, "Public v. Private Garbage Disposal: The Economics of Solid Waste Flow Controls," *Growth and Change* 30 (Summer 1999): 430–31, 433–34; William A. Campbell, "Flow-Control Ordinances Held Unconstitutional: C&A Carbone, Inc., v. Town of Clarkstown," *Local Government Law Bulletin* 59 (June 1994): 1; Bick Brickwedde, "Interstate Garbage: The Carbone Case and the Commerce Clause," *ABA State and Local Law News* 18 (1995): 1; Martin V. Melosi, *Garbage in the Cities: Refuse, Reform, and the Environment*, rev. ed. (Pittsburgh, PA: University of Pittsburgh Press, 2005), 215–16. See also U.S. Environmental Protection Agency, Solid Waste and Emergency Response, *Municipal Solid Waste Flow Control, Summary of Public Comments*, February 8, 1994; "Flow Control Issues in Solid Waste Management," NCRA Policy Committee White Paper, October 1994; Eric S. Peterson and David N. Abramowitz, "Municipal Solid Waste Flow Control Post-Carbone World," *Fordham Urban Law Journal* 22 (1994): 361–416; John Turner, "The Flow Control of Solid Waste and the Commerce Clause: Carbone and Its Progeny," *Villanova Environmental Law Journal* 7 (1996): 203–61; Michael D. Diederich Jr., "Does Garbage Have Standing? Democracy, Flow Control, and a Principled Constitutional Approach to Municipal Solid Waste," *Pace Environmental Law Review* 11 (Fall 1993): 157–264.
81. Vivian E. Thomson, *Garbage In, Garbage Out: Solving the Problems with Long-Distance Transport* (Charlottesville: University of Virginia Press, 2009), 71–72, 96–97. See also Amy Terdiman, "Talking Trash," *Empire State Report* (November 1995): 41–42, 44, 46, 48, 50.
82. The bill was cosponsored by the mid-Island Republican John A. Fusco and the North Shore Democrat Jerome X. O'Donovan, as well as others outside of Staten Island. "From the Archives: Trail of Broken Promises Litters History of Former Fresh Kills Landfill," *Staten Island Advance*, September 26, 2013; "Testimony of the Honorable Susan Molinari Before the New York City Council," May 29, 1996, Vitaliano Papers, Series 1, Subject Files, Box 39, Folder 4. For a variety of documents concerning a closure law before the City Council, see Council of the City of New York Collection, Series: Committee Files, Box 50079, Folders 4 and 5, La Guardia and Wagner Archives, LaGuardia Community College/CUNY.
83. "Nothing to Lose but a Landfill," *Staten Island Register*, February 16, 1996.
84. Reginald Patrick, "The Beginning of a Fight for Fairness," *Staten Island Advance*, February 16, 1996.
85. Reginald Patrick, "More Garbage Headed to Fresh Kills?" *Staten Island Advance*, February 14, 1996.
86. City of New York, Department of Sanitation, *Comprehensive Solid Waste Management Plan: Final Update and Plan Modification*, February 15, 1996, 1-1. Emphasis added.
87. City of New York, Department of Sanitation, *Comprehensive Solid Waste Management Plan: Final Update and Plan Modification*, 2-55.
88. Reginald Patrick, "Solid Waste Plan Passes City Council," *Staten Island Advance*, February 16, 1996. See also Vivian S. Toy, "Solid-Waste Management Plan Wins Council

Approval," *New York Times*, February 16, 1996; Peter J. Powers to Beth Petrone, May 9, 1996, Giuliani Collection, Fresh Kills Meeting, Box 30157, Folder 6267; Carolyn Rushefsky, "Who's Minding the Landfill?" *Staten Island Advance*, February 19, 1996; "Readings Confirm Landfill Spewing Toxic Gases," *Staten Island Advance*, February 23, 1996.

89. Carolyn Rushefsky, "'You're Killing Our Children,'" *Staten Island Advance*, February 29, 1996.
90. "Testimony of Staten Island Borough President Guy V. Molinari Before the Annual Fresh Kills Consent Hearing Regarding the Conditions at the Landfill," February 28, 1996, Archives & Special Collections, Department of the Library, College of Staten Island. See also Guy V. Molinari to John Doherty, October 4, 1995, and John Doherty to Guy V. Molinari, October 5, 1995, Giuliani Collection, Deputy Mayors, Peter Powers, Sanitation Department of Fresh Kills Landfill, Box 02/01/038, Folder 960.
91. Remarks by Senator John J. Marchi, Prepared for the Annual Public Meeting of the New York City Sanitation Department, February 28, 1996, Concerning the Operation of the Fresh Kills Landfill on Staten Island, Vitaliano Papers, Series 1, Subject Files, Box 37, Folder 4. See also Michael D. Zagata, DEC, to John J. Marchi, March 11, 1996, Marchi Papers, Series 1, Subject Files, Box 47, Folder 6, with respect to Marchi's concerns that the City of New York might attempt to modify the Fresh Kills consent order.
92. Statement of Assemblyman Eric N. Vitaliano, Department of Sanitation Public Meeting on the Fresh Kills Landfill, February 28, 1996, Archives & Special Collections, Department of the Library, College of Staten Island.
93. Testimony of Councilman Vito J. Fossella, Department of Sanitation, Annual Consent Order Public Meeting, Fresh Kills Landfill, February 28, 1996, Archives & Special Collections, Department of the Library, College of Staten Island.
94. Comments of Councilman John A. Fusco at the Department of Sanitation Fresh Kills Landfill Annual Public Meeting, February 28, 1996, Archives & Special Collections, Department of the Library, College of Staten Island. The blizzard of technical reports continued into 1996. For example, see Exeter Supply Company, Inc., *The Use of the Multi-Flow Drainage System at the Fresh Kills Landfill, Staten Island, New York*, March 1996; and Woodward-Clyde, *Fresh Kills Landfill: 1996 Updated Geotechnical Monitoring System Report*, April 1996, Archives & Special Collections, Department of the Library, College of Staten Island.
95. New York State, Department of Environmental Conservation, Division of Solid & Hazardous Materials, *New York State Solid Waste Management Plan, 1995/96 Update*, March 1996, vii, Marchi Papers, Series 1, Subject Files, Box 47, Folder 3.
96. New York State, Department of Environmental Conservation, Division of Solid & Hazardous Materials, *New York State Solid Waste Management Plan, 1995/96 Update*, 30. See also New York State, Department of Environmental Conservation, Division of Solid & Hazardous Materials, *Responsiveness Summary to Public Comments on the New York State Solid Waste Management Plan, 1995/96 Update*, March 1996, Marchi Papers, Series 1, Subject Files, Box 47, Folder 3.
97. Reginald Patrick, "City's Record on Landfill," *Staten Island Advance*, March 3, 1996. See also City of New York, Department of Sanitation, *Fresh Kills Landfill, Draft Environmental Impact Statement*, March 15, 1996, Marchi Papers, Series 1, Staten Island, Fresh Kills, 1996–2006, Box 8, Folder 8; City of New York, Department of Sanitation, *Notice of Completion of the Draft Environmental Impact Statement for Continued Landfilling at Sections 1/9 and 6/7 Pursuant to a 6 NYCRR Part 360*

Permit, March 15, 1996, Division of Environmental Permits, New York State Department of Environmental Conservation, Albany, New York.

98. John J. Doherty to DEC, March 25, 1996, Giuliani Collection, Deputy Mayors, Peter Powers, Sanitation Department of Fresh Kills Landfill, Box 02/01/048, Folder 1282. See also John J. Doherty to Peter J. Powers, August 4, 1995, in Box 02/01/038, Folder 960. Some concerns had arisen in 1994 about asbestos in cover and road material at Fresh Kills. See John J. Doherty to Peter Powers, August 26, 1994; Jeanne M. Fox Regional Administrator, EPA to Susan Molinari, September 2, 1994; Susan Molinari to Conrad Simon, EPA, August 23, 1994; Susan Molinari to Conrad Simon, August 24, 1994, Giuliani Collection, Deputy Mayors, Peter Powers, Sanitation Department of Fresh Kills Landfill, Box 02/01/021, Folder 500.

99. They had supported secession in 1993. See Carl Campanile, "Realtors to State Lawmakers: Help Us Close Landfill," *Staten Island Advance*, March 20, 1996. See also "Realtors to Press Officials on Fresh Kills Landfill Closure," *Staten Island Register*, May 7, 1996; Carolyn Rushefsky, "Realtors: Fresh Kills Is Bad for Business," *Staten Island Advance*, May 24, 1996.

100. Reginald Patrick, "Landfill Closure Bill Wins Stunning Support," *Staten Island Advance*, March 21, 1996.

101. "Studies Done on Fresh Kills Landfill," March 4, 1996; Reginald Patrick, "What's Making Us Sick?" March 3, 1996; Don Gross, "Questions Pile High as Landfill Mountains," March 24, 1996, all in *Staten Island Advance*. One report stated that the emission of toxic gases from the landfill did not pose a local health problem, but this was contested. See Don Gross, "Landfill Study Comes Up Empty," *Staten Island Advance*, March 29, 1996. See also John A. Fusco to Rudolph Giuliani, March 7, 1996, Giuliani Collection, Deputy Mayors, Peter Powers, Sanitation Department of Fresh Kills Landfill, Box 02/01/048, Folder 1282.

102. In this case "fair share" implies an equal distribution of refuse disposal among the boroughs.

103. Don Gross, "City to Be Sued Over Landfill," *Staten Island Advance*, March 25, 1996; Vivian S. Toy, "Staten Island Leadership Sues to Close Fresh Kills," *New York Times*, March 26, 1996. See also Don Gross, "Sanit: We Do Our Best to Make Landfill Work," *Staten Island Advance*, March 26, 1996; Ron Scherer, "Staten Island Has Had Its Fill of Nation's Largest Landfill," *Christian Science Monitor*, April 26, 1996.

104. Carl Campanile, "State Legislators Open Second Front in War on Landfill," *Staten Island Advance*, March 27, 1996; State of New York, Assembly, *An Act to Amend the General Municipal Law, in Relation to the Termination of Municipal Landfills in the City of New York . . .* , March 26, 1996; "News from Senator John J. Marchi," March 26, 1996, Marchi Papers, Series 1, Subject Files, Closure 1996–1999, Box 47, Folder 6. See also U.S. District Court, Eastern District of New York, Draft Complaint, Guy Molinari et al., March 21, 1996, Giuliani Collection, Deputy Mayors, Peter Powers, Sanitation Department of Fresh Kills Landfill, Box 02/01/048, Folder 1282.

105. "News from Assemblyman Eric N. Vitaliano," April 11, 1996, Marchi Papers, Series 1, Subject Files, Fresh Kills 2002–2002, Box 47, Folder 4. See also Eric N. Vitaliano to Editor, *Staten Island Advance*, April 17, 1996, Vitaliano Papers, Series 1, Subject Files, Box 37.

106. Carl Campanile, "Silver Speaks: Close the Dump," *Staten Island Advance*, April 12, 1996.

107. John Reel, "Landfill, Sewage Top CB 3 Meeting," *Staten Island Advance*, March 27, 1996. See also Reginald Patrick, "O'Donovan Bill Spells Out Strategy to End Dumping,"

Staten Island Advance, April 18, 1996; Carl Campanile, "No Place to Dump? That's Garbage," *Staten Island Advance*, April 18, 1996.

108. See Jerome X. Donovan to Rudolph W. Giuliani, February 20, 1996, Giuliani Collection, Deputy Mayors, Peter Powers, Sanitation Department of Fresh Kills Landfill, Box 02/01/048, Folder 1282.
109. Peter J. Powers to Jerome X. O'Donovan, April 29, 1996, Giuliani Collection, Deputy Mayors, Peter Powers, Sanitation Department of Fresh Kills Landfill, Box 02/01/048, Folder 1282. See also Jerome X. O'Donovan to Rudolph Giuliani, May 3, 1996, Giuliani Collection, Deputy Mayors, Peter Powers, Sanitation Department of Fresh Kills Landfill, Box 02/01/048, Folder 1282, dealing with funding and alternatives to Fresh Kills.
110. Carl Campanile, "Rudy Won't Give Up on Tax Cut," *Staten Island Advance*, May 21, 1996.
111. Benjamin Miller, *Fat of the Land: Garbage of New York, the Last Two Hundred Years* (New York: Four Walls Eight Windows, 2000), 283.
112. Carl Campanile, "Pataki: I Will Review Fresh Kills Closure Plan," *Staten Island Advance*, March 29, 1996.
113. Carl Campanile, "Stars in Political Galaxy Aligned Just Right on Fresh Kills Landfill," *Staten Island Advance*, March 31, 1996.
114. Campanile, "Stars in Political Galaxy Aligned Just Right on Fresh Kills Landfill." For a meeting over the closure bill see also Kate to Mrs. Connelly, April 17, 1996, Connelly Papers, Series 1, Subject Files, Box 20, Folder 3; Carl Campanile, "Timetable for Closing Landfill Added to Pending Legislation," *Staten Island Advance*, April 17, 1996; Elizabeth Connelly to Lauren Salerno, April 19, 1996, Connelly Papers, Series 1, Subject Files, Box 20, Folder 3.
115. See Reginald Patrick, "Permit Process Won't Slow Push to Close Dump," *Staten Island Advance*, April 4, 1996; Carl Campanile, "Pataki: Set a Date to Shut Dump," *Staten Island Advance*, April 9, 1996.
116. By this time, Giuliani already had stated that the incinerator project would be shelved for the time being.
117. Campanile, "Silver Speaks: Close the Dump." See also "Killing Fresh Kills," *Staten Island Advance*, April 15, 1996; Eric N. Vitaliano to Democratic Colleagues, April 16, 1996, Vitaliano Papers, Series 1, Subject Files, Box 38, Folder 8. See also Carl Campanile, "State Bill to Close Dump Clears Another Hurdle," *Staten Island Advance*, May 23, 1996. Modification of Vitaliano's bill continued into mid-April, especially regarding attempts to consider changes proposed by DEC. See Kate to Mrs. Connelly, April 17, 1996, Connelly Papers, Series 1, Subject Files, Box 20, Folder 4.
118. Reginald Patrick, "Getting Serious on the Landfill," *Staten Island Advance*, April 23, 1996.
119. See Alice Tetelman, NYC Washington Office, to Peter Powers, March 28, 1996, Giuliani Collection, Deputy Mayors, Peter Powers, Sanitation Department of Fresh Kills Landfill, Box 02/01/048, Folder 1282.
120. Carl Campanile, "Island Environmentalists Press for Closure of Landfill," *Staten Island Advance*, April 23, 1996. See also Carl Campanile, "Environmental Groups Endorse Closing Landfill," *Staten Island Advance*, April 26, 1996. Some protesters were not so optimistic, as suggested in the exchange between the AARP University South Staten Island chapter and Assemblywoman Connelly: James P. Falkenburg to Connelly, April 21, 1996, and Elizabeth A. Connelly to James P. Falkenburg, April 25, 1996, Connelly Papers, Series 1, Subject Files, Box 20, Folder 4.

121. Rich Cirillo, "Rudy on Landfill: 'Trust Me,'" *Staten Island Advance*, April 25, 1996.
122. Bruce Weber, "Broadway, Statue of Liberty, and Fresh Kills?" *New York Times*, March 27, 1996.
123. Bruce Weber, "Landfill Will Not Be a Tourist Attraction," *New York Times*, April 3, 1996. See also Charles Osgood, "Fresh Kills Landfill Will Not Become an Official Tourist Attraction Despite Drawing Visitors Who Want to View the Huge Landfill," *CBS News Transcripts*, April 3, 1996; Judy L. Randall, "Landfill Tours Scrapped," *Staten Island Advance*, April 2, 1996; "Let Staten Island's Dump Go on Tour," *New York Times*, March 30, 1996; "Travelogue," *Guardian*, March 9, 1996; Vito J. Fossella and John A. Fusco to Rudolph W. Giuliani, February 29, 1996, Giuliani Collection, Deputy Mayors, Peter Powers, Sanitation Department of Fresh Kills Landfill, Box 02/01/048, Folder 1282.
124. Reginald Patrick, "Read Our Lips: Close the Dump," *Staten Island Advance*, May 2, 1996.
125. Carl Campanile, "Governor: It Can Be Done, but City Must Be Protected," *Staten Island Advance*, May 2, 1996. See also Reginald Patrick, "City Funding Plan Could Help Close Fresh Kills Landfill," *Staten Island Advance*, May 7, 1996.
126. "The Cost of Closure," *Staten Island Advance*, May 8, 1996; Don Gross, "Sanit Boss: Closing Dump Will Cost Us Money," *Staten Island Advance*, May 17, 1996.
127. Fred Buffa to Editor, *Staten Island Advance*, April 18, 1996, Vitaliano Papers, Series 1, Subject Files, Press Clippings, Box 39, Folder 4.
128. Campanile, "Dump to Close." See also "Pataki, Giuliani Vow to Shut Down Fresh Kills Landfill for Good," *Solid Waste Digest* (Northeast Edition) 6 (June 1996): 9–10; Toy, "Accord Is Reached to Close Staten Island Landfill by 2001"; Williams, "Trash Talk."
129. See Marjorie J. Clarke, Adam D. Read, and Paul S. Phillips, "Integrated Waste Management Planning and Decision-Making in New York City," *Resources, Conservation, and Recycling* 26 (1999): 136.
130. City of New York, Office of the Mayor, Press Office, "Mayor Giuliani and Governor Pataki Announce Fresh Kills Landfill to Close in Five Years; Waste Disposal Task Force Established," Marchi Papers, Series 1, Subject Files, Box 47, Folder 4. See also Statement of NYPIRG Senior Attorney Larry Shapiro Regarding Close of the Fresh Kills Landfill, Vitaliano Papers, Series 1, Subject Files, Box 36; oral interview, Martin Melosi with George E. Pataki, November 15, 2017.
131. See Laura Bruno, "Too Little, Too Late for Some in Dump's Shadow," *Staten Island Advance*, May 30, 1996; "A Mistake on a Massive Scale," *Staten Island Advance*, May 30, 1996; Jayne Gastaldo, "Do You Really Believe the Landfill Will Close?" *Staten Island Register*, May 28, 1996.
132. "Executive Chamber, Governor, Mayor Reach Agreement to Close Fresh Kills," May 29, 1996, Giuliani Collection, Deputy Mayors, Peter Powers, Sanitation Department of Fresh Kills Landfill, Box 02/01/048, Folder 1283. See also Vivian S. Toy, "Accord Is Reached to Close Staten Island Landfill by 2001," *New York Times*, May 30, 1996. See also Testimony of Manhattan Borough President Ruth W. Massinger on Intro. no. 733 and Intro. no. 768 in Relation to the Closure of the Fresh Kills Landfill on Staten Island, May 1, 1996, in same folder, on support for closing Fresh Kills.
133. See Dena Diorio, City Legislative Affairs, to Peter J. Powers, April 18, 1996, Giuliani Collection, Deputy Mayors, Peter Powers, Sanitation Department of Fresh Kills Landfill, Box 02/01/048, Folder 1282.

134. Giuliani had supported Cuomo over Pataki for governor, but the party alliance overruled that political act.
135. Michael McMahon: Oral History Transcripts, July 26, 2004, 126, College of Staten Island Oral History Collection, v. 1.
136. See Carl Campanile, "Dump to Close," *Staten Island Advance*, May 29, 1996; Jerry McLaughlin: Oral History Transcripts, November 14, 2000, 100, College of Staten Island Oral History Collection, v. 1. See also Daniel C. Kramer and Richard M. Flanagan, *Staten Island: Conservative Bastion in a Liberal City* (Lanham, MD: University Press of America, 2012), 117.
137. Campanile, "Dump to Close."
138. For example, see Guy V. Molinari to John J. Doherty, January 31, 1996, Giuliani Collection, Deputy Mayors, Peter Powers, Sanitation Department of Fresh Kills Landfill, Box 02/01/048, Folder 1282, dealing with violations of state law in Section 6 of Fresh Kills. See also in same folder, John J. Doherty to Guy V. Molinari, February 16, 1996; "Island Takes on the Fight of Its Life," *Staten Island Advance*, September 22, 1996.
139. Kramer and Flanagan, *Staten Island: Conservative Bastion in a Liberal City*, 104–5.
140. Dan Master Jr.: Oral History Transcript, April 23, 2003, 83–88, 97, College of Staten Island Oral History Collection, v. 3. See also Miller, *Fat of the Land*, 282.
141. "Press Release, NY GOP Leaders Bill Powers and Guy Molinari Endorse Giuliani," American Presidency Project, February 28, 2007, http://www.presidency.ucsb.edu/ws/?pid=p4822.
142. Dan Master Jr.: Oral History Transcripts, April 23, 2003, 86, 88. See also "Island Takes on Fight of Its Life."
143. "Closing the Fresh Kills Landfill—at Last," *Borough Report*, February 1997, 1.
144. See Molinari, *A Life of Service*, 436–43.
145. Statement by Assemblywoman Elizabeth A. Connelly, 59th A.D., Public Hearing—Committee on Environmental Protection—NYC Council, May 1, 1996, Connelly Papers, Series 1, Subject Files, Box 20, Folder 4. See also Robert Straniere: Oral History Transcripts, January 10, 2002, 21–25, College of Staten Island Oral History Collection, v. 1; oral interview, Martin Melosi with Norman Steisel, November 15, 2017, New York City.

17. NOW WHAT?

1. By this time, sanitary landfills were in serious decline. A New York State report in 1997 showed that only 45.5 percent of New York's refuse was being deposited in municipal solid waste landfills, while 21.3 percent was being recycled, 17 percent burned in waste-to-energy facilities, and 16.2 percent exported. New York State Assembly, Legislative Commission on Solid Waste Management, *Where Will the Garbage Go? 1997* (July 1997), 2, 3, 6–11. On the national level, sanitary landfilling became the prime symbol of the garbage crisis. Between 1990 and 2000 the number of landfills dropped from approximately 7,300 to around two thousand. The sharp reduction in landfill sites was attributable to numerous factors, but especially significant was Subtitle D of the Resource Conservation and Recovery Act (1976). It provided new landfill standards for the development of environmentally sound disposal methods and the protection of groundwater. Revised Subtitle D regulations in 1991 set minimum national standards for landfills, effective on October 9, 1993. The cost of meeting Subtitle D

for existing and new landfills was extremely high, resulting in constriction in the landfill supply. Because many communities could not bear the high costs, at least 40 percent were held in private hands by 1998. Martin V. Melosi, *Garbage in the Cities: Refuse, Reform, and the Environment*, rev. ed. (Pittsburgh, PA: University of Pittsburgh Press, 2005), 209–14; Martin V. Melosi, "Historic Development of Sanitary Landfills and Subtitle D," *University of Houston Energy Laboratory Newsletter* 31 (Spring 1994): 20–24.

2. Douglas Martin and Andrew C. Revkin, "The Last Landfill: A Special Report; As Deadline Looms for Dump, Alternate Plan Proves Elusive," *New York Times*, August 30, 1999.
3. For a strong affirmation of this observation, see "Commentary," *New York Times*, January 6, 1997.
4. David Seifman, "City to Shut Down S.I. Dump—at Yearly Cost of $200M," *New York Post*, May 30, 1996. See also Benjamin Miller, "Fat of the Land: New York's Waste," *Social Research* 65 (Spring 1998): 75.
5. See several articles under the banner "This Time for Real," *Staten Island Advance*, March 30, 1996; Statement of NYPIRG Senior Attorney Larry Shapiro Regarding Passage of S. 6669-B/A 10418-A on the Fresh Kills Landfill and the Brooklyn Navy Yard Incinerator, May 30, 1996, The Senator John J. Marchi Papers, Series 1, Subject Files, Fresh Kills 2000–2002, Box 47, Folder 4, Archives & Special Collections, Department of the Library, College of Staten Island, CUNY. See also Craig Schneider, "The Legal Arena," *Staten Island Advance*, June 2, 1996; Craig Schneider, "The Landfill: Change Is in the Air," *Staten Island Advance*, June 2, 1996.
6. "A Fake Odyssey," *Staten Island Register*, June 4, 1996.
7. "Mayor Giuliani and Governor Pataki Announce Fresh Kills Landfill to Close in Five Years: Waste Disposal Task Force Established," press release, May 29, 1996, Marchi Papers, Series 1, Subject Files, Box 7. See also "Fresh Kills Exportation Announcement," July 1, 1997, Archives of Rudolph W. Giuliani, http://www.nyc.gov/html/records/rwg/html/97/freshkls.html.
8. Vivian S. Toy, "Despite Years of Broken Promises, Accord Vows to Close S.I. Landfill," *New York Times*, May 30, 1996.
9. Eric N. Vitaliano: Oral History Transcripts, January 25, 2001, 166–67, College of Staten Island Oral History Collection, v. 1, ed. Jeffery A. Kroessler, Archives & Special Collections, Department of the Library, College of Staten Island, CUNY.
10. Robert M. Harding to George E. Pataki, June 6, 1996, Rudolph W. Giuliani Collection, Scheduling Events, Bill Signing—Fresh Kills, Box 30158, Folder 6369, La Guardia and Wagner Archives, LaGuardia Community College/CUNY, Long Island City, New York. Recalling his 1958 bill to close Fresh Kills, which Governor Averell Harriman vetoed, Marchi stated, "The governor's veto message said that the bill was unnecessary because Fresh Kills was going to close in just a few years. Governor Harriman was a great man but his crystal ball wasn't working then and we have been dealing with this issue over too long a span." See "News from Senator John J. Marchi," May 29, 1996, Marchi Papers, Series 1, Subject Files, Closure 1996–1999, Box 47, Folder 6.
11. The legislation passed by a vote of 141 to 2. See Carl Campanile, "There Oughta Be a Law," *Staten Island Advance*, May 31, 1996. See also Eric N. Vitaliano to George E. Pataki, June 4, 1996, The Assemblyman Eric N. Vitaliano Papers, Series 1, Subject Files, Box 37, Folder 10, Archives & Special Collections, Department of the Library,

College of Staten Island, CUNY; John J. Marchi to George E. Pataki, June 4, 1996, Series 1, Subject Files, Box 8.

12. On the City Council, two Staten Island members, Vito J. Fossella Jr. and Jerome X. O'Donovan, sought to pass similar legislation for "insurance." See Judy L. Randall, "Closing the Dump: Where the Action Is," *Staten Island Advance*, June 1, 1996. See also Reginald Patrick, "Veto-Proof Landfill Bill Filed," *Staten Island Advance*, June 6, 1996; "News from Senator John J. Marchi," Vitaliano Papers, Series 1, Subject Files, Box 39, Folder 4.

13. Guy Molinari told the *New York Times* that he had decided to file the lawsuit two hours before the March 29 announcement because a federal court would ensure that no future mayor or City Council would rescind the closure agreement. Vivian S. Toy, "Despite Years of Broken Promises, Accord Vows to Close S.I. Landfill," *New York Times*, May 30, 1996. Also, Staten Island officials were uncertain about what funds might be available for the closure after the upcoming 1997 election without more assurances. See Dan Master Jr.: Oral History Transcripts, April 23, 2003, 86–89, College of Staten Island Oral History Collection, v. 3.

14. Toy, "Despite Years of Broken Promises, Accord Vows to Close S.I. Landfill."

15. Carolyn Rushefsky, "How to Close the Dump: Recycle and Export," *Staten Island Advance*, May 30, 1996. See also Marjorie J. Clarke et al., "Integrated Waste Management Planning and Decision-Making in New York City," *Resource Conservation and Recycling* 26 (1999): 137–38; "Backward Steps on Garbage Disposal," *New York Times*, June 20, 1996.

16. Allen J. Adams to Elizabeth A. Connelly, June 2, 1996, The Assemblywoman Elizabeth A. Connelly Papers, Series 1, Subject Files, 1996, Box 20, Folder 4, Archives & Special Collections, Department of the Library, College of Staten Island, CUNY.

17. Elizabeth Connelly to Allen J. Adams, June 27, 1996, Connelly Papers, Series 1, Subject Files, 1996, Box 20, Folder 4.

18. Among others, the Agency for Toxic Substances and Disease Registry (ATSDR) in Atlanta conducted a study in the mid-1990s concerning possible respiratory effects associated with living near landfills. See Suzanne Freeman, "Dump's Effects to Be Studied," *New York Times*, April 27, 1997. See other ATSDR reports as well at http://www.atsdr.cdc.gov. See also "Shutting the Fresh Kills Dump," *New York Times*, June 9, 1996; Bill Franz, "O'Donovan, Green Seek Study of Fresh Kills Dump," *Staten Island Register*, May 19, 1998; "Health Consultation: Fresh Kills Landfill, Staten Island, Richmond County, New York," ATSDR, May 7, 1998, Marchi Papers, Series 1, Subject Files, Box 46, Folder 8; Terence J. Kivlan, "Study: Landfill's Mercury Releases Not a Threat," *Staten Island Advance*, March 6, 1997; "The Landfill on Trial," *Staten Island Advance*, April 22, 1997; Robin Eisner, "New Springville Residents Aren't Breathing Any Easier," *Staten Island Advance*, June 15, 1997; Robin Eisner, "An Environmental Disaster for Years to Come," *Staten Island Advance*, June 17, 1997; Terence J. Kivlan, "Federal Officials Join Inquiry Into Landfill Incident," *Staten Island Advance*, May 15, 1998; Don Gross, "Landfill Odor May Be Eased by Summer," *Staten Island Advance*, November 4, 1997; Robin Eisner, "New Stench Belching from the Landfill," *Staten Island Advance*, November 23, 1997; Jennifer M. Nelson, "Officials Update Residents on Landfill Emissions Study," *Staten Island Advance*, November 6, 1997; Reginald Patrick, "Island Residents to Be Notified of Toxic Fumes," *Staten Island Advance*, April 30, 1996; Robin Eisner, "Landfill Emissions Report to Be Issued This Spring," *Staten Island Advance*, September 25, 1997.

19. Connelly to Adams, June 27, 1996, Connelly Papers, Series 1, Subject Files, 1996, Box 20, Folder 4. See also Terence J. Kivlan, "New Study: Landfill Emitting Small Traces of Mercury," *Staten Island Advance*, June 25, 1996; "Federal Agency Sees No Evil at S.I. Landfill," *Star Reporter Newspapers*, July 1996; "Court Awaits Monitors Report on S.I. Landfill," *Star Reporter Newspapers*, July 1996.
20. Elizabeth A. Connelly to Marilyn Haggerty-Blohm, July 3, 1996, Connelly Papers, Series 1, Subject Files, Box 20, Folder 4.
21. Marilyn Haggerty-Blohm to Elizabeth A. Connelly, July 10, 1996, Connelly Papers, Series 1, Subject Files, Box 20, Folder 4.
22. Marchi, Vitaliano, Silver, Connelly, Fusco, O'Donovan, Fossella, Assemblyman Robert Straniere (Republican, South Shore), Senator Robert DiCarlo (Republican, Brooklyn/East Shore), Guy Molinari, and Giuliani were all in attendance.
23. Carl Campanile, "'That Does It. It's Law,'" *Staten Island Advance*, June 10, 1996. The governor had the concurrence of Mayor Giuliani in supporting the bill. See Robert M. Harding to George E. Pataki, June 6, 1996, Marchi Papers, Series 1, Subject Files, Fresh Kills, 2000–2002, Box 47, Folder 4. See also Carl Campanile and Craig Schneider, "Pataki to Sign Dump Bill," *Staten Island Advance*, June 4, 1996; Terence Kivlan, "City May Regret Giving Up on Incinerators," *Staten Island Advance*, July 10, 1996; "New York Governor Signs Bill to Close Fresh Kills," *Environmental Laboratory Washington Report 7* (June 24, 1996).
24. Giuliani recently had hired Haggerty-Blohm as deputy director of the city's Office of Operations, where she had been working on plans to close the landfill.
25. Carl Campanile, "Former Aide to BP to Head Dump Panel," *Staten Island Advance*, June 14, 1996. See also Press Release, Office of the Mayor, June 20, 1996, and "Fresh Kills Task Force Briefing," August 23, 1996, Giuliani Collection, Deputy Mayors, Peter Powers, Sanitation—Fresh Kills, Box 02/01/048, Folder 1283, La Guardia and Wagner Archives.
26. Responding to one such request very narrowly and not very persuasively, Deputy Mayor Randy M. Mastro stated that given the short timeframe to evaluate waste management options, "the membership of the Task Force has been comprised of City and State Agency personnel expert in program development, fiscal analysis, and the management of solid waste." Randy M. Mastro to Samuel Bea Jr., September 12, 1996, Giuliani Collection, Deputy Mayors, Randy Mastro, Project Files, Sanitation—Fresh Kills, Box 02/04/006, Folder 193, La Guardia and Wagner Archives. See also Howard Golden to Rudolph W. Giuliani, July 22, 1996, Giuliani Collection, Deputy Mayors, Peter Powers, Sanitation—Fresh Kills, Box 02/01/048, Folder 1283, La Guardia and Wagner Archives; Howard Fernando Ferrer, Borough President, The Bronx, to Rudolph W. Giuliani, July 24, 1996; Assemblyman Samuel Bea Jr. to Rudolph W. Giuliani, August 28, 1996, Giuliani Collection, Deputy Mayors, Randy Mastro, Project Files, Sanitation—Fresh Kills, Box 02/04/006, Folder 193, La Guardia and Wagner Archives; Pedro G. Espada to Rudolph Giuliani, August 8, 1996, Giuliani Collection, Deputy Mayors, Peter Powers, Sanitation—Fresh Kills, Box 02/01/048, Folder 1283, La Guardia and Wagner Archives; Peter J. Powers to Pedro G. Espada, August 28, 1996, Giuliani Collection, Deputy Mayors, Peter Powers, Sanitation—Fresh Kills, Box 02/01/048, Folder 1283, La Guardia and Wagner Archives.
27. Carl Campanile, "Landfill Task Force to Operate in Secret," *Staten Island Advance*, June 25, 1996; "It's a Public Matter," *Staten Island Advance*, June 26, 1996. See Howard Golden to Robert Freeman, Committee on Open Government, July 22, 1996, and Howard Golden to Robert Freeman, August 13, 1996, Giuliani Collection, Deputy

Mayors, Peter Powers, Sanitation Department, Box 02/01/048, Folder 1283, La Guardia and Wagner Archives. See also in same folder George E. Pataki to Howard Golden, July 26, 1996.

28. Craig Schneider, "The Task Force's Task," *Staten Island Advance*, July 2, 1996. See also Carl Campanile, "Dumping Fresh Kills," *Empire State Report* (July 1996): 19–21.

29. Craig Schneider, "Time Slipping By and Landfill Panel Has Yet to Meet," *Staten Island Advance*, July 22, 1996.

30. See "Flow Control and Municipal Solid Waste," EPA, https://archive.epa.gov/epawaste/nonhaz/municipal/web/html/flowctrl.html.

31. "An Issue That Won't Go Away," *Staten Island Advance*, September 19, 1996; Carl Campanile, "Pataki: Bottling Up Interstate Trash Export Bill Helps N.Y.," *Staten Island Advance*, September 20, 1996. See also Terence J. Kivlan, "Lawmakers Backing Limits to Trash Export Concede Defeat," *Staten Island Advance*, October 1, 1996; Terence J. Kivlan, "Rep. Molinari Had Help in Defeating the Interstate Trash Restriction Bill," *Staten Island Advance*, October 6, 1996; "Prepared Statement of Senator Dan Coats Before the Senate Environment and Public Works Committee," Federal News Service, March 18, 1997; Rob Shapard, "Solid Waste: Trash, Cash, and Landfill Gas Remain Part of the Debate," *American City & Country*, May 1, 1997, http://www.americancityandcountry.com/mag/government_solid_waste_trash/index.html; Terence Kivlan, "Proponents of Trash Export Limits Dumped On," *Staten Island Advance*, February 22, 1998; Terence Kivlan, "Trash Foes Still Hoping to Derail Our Exports," *Staten Island Advance*, November 19, 1997.

32. "Testimony, March 18, 1997, Randy M. Mastro, Deputy Mayor for Operations, the City of New York, Senate Environment State and Local Control of Waste Transport, March 18, 1997," Federal News Service. See also "Testimony, March 18, 1997, John P. Cahill, Acting Commissioner, New York State Department of Environmental Conservation, Senate Environment, State and Local Control of Waste Transport, March 18, 1997," Federal Document Clearing House; Testimony of Randy Mastro Before the Committee on Environment and Public Works, March 18, 1997, Giuliani Collection, Deputy Mayors, Randy Mastro, Project Files, Sanitation—Fresh Kills, Box 02/04/019, Folder 466, La Guardia and Wagner Archives.

33. An interstate trash regulation bill was stalled in the House of Representatives for one year (largely because of opposition from the New York delegation), although it had support from twenty-four governors. See Terence J. Kivlan, "Pa.: We Don't Want City's Trash," *Staten Island Advance*, June 1, 1996. See also Andrew S. Voros to Robert Helbrock, Pennsylvania Department of Environmental Protection, June 10, 1996, sent to all Senate and House members; Carl Campanile, "Exporting Trash Won't Be a Piece of Cake," *Staten Island Advance*, June 5, 1996; Guy Molinari, "After Fresh Kills Closes: Other States Want Our Trash," *New York Daily News*, June 20, 1996; M. Joel Bolstein, Pennsylvania Department of Environmental Protection, to Eric N. Vitaliano, July 3, 1996, and Senator Raphael J. Musto, Pennsylvania, to Eric Vitaliano, July 3, 1996, Vitaliano Papers, Series 1, Subject Files Box 36, Folder 11; Craig Schneider, "Finding a Place for the Garbage," *Staten Island Advance*, June 23, 1996; Jeffrey R. Sipe, "Proposal to Export Garbage Has New Yorkers Talkin' Trash," *Insight* 12 (July 29, 1996): 39; Terence J. Kivlan, "Trash Bill Becomes Stand-Alone Measure," *Staten Island Advance*, February 5, 1997; Terence J. Kivlan, "Compromise Trash Export Bill Offered," *Staten Island Advance*, September 4, 1997; Vivian S. Toy, "After Fresh Kills, New York Will Need a Partner in Garbage," *New York Times*, June 21, 1996.

34. Edward W. Repa, "Interstate Movement: 1995 Update," *Waste Age* 28 (June 1997): 41. See also Legislative Commission on Solid Waste Management, New York State Assembly, *Where Will the Garbage Go? 1997* (July 1997), 13.
35. The total amount of solid waste imported in 1995, however, represented less than 9 percent of all waste.
36. Repa, "Interstate Movement: 1995 Update," 42–44, 48, 50, 52.
37. Testimony of Manhattan Borough President Ruth W. Messinger on Intro. no. 733 and Intro. no. 768 in Relation to the Closure of the Fresh Kills Landfill on Staten Island, May 1, 1996, Giuliani Collection, Deputy Mayors, Peter Powers, Sanitation—Fresh Kills, Box 02/01/048, Folder 1283, La Guardia and Wagner Archives.
38. She urged an appointment to the task force, but the request was ignored. See Ruth W. Messinger to Rudolph Giuliani, June 26, 1996, Giuliani Collection, Deputy Mayors, Peter Powers, Sanitation—Fresh Kills, Box 02/01/048, Folder 1283, La Guardia and Wagner Archives. Some waste prevention policies were being carried out. See also City of New York, Office of the Mayor, "Directive to All Heads of Agencies and Departments, Waste Prevention and Efficient Materials Management Policies," no. 96-2, September 27, 1996, Marchi Papers, Series 4, Media Files, Subseries C, News Clippings, Box 105, Folder 6; Amy Waldman, "Concern Grows on Where Trash Will Go After Fresh Kills," *New York Times*, October 16, 1997.
39. Melosi, *Garbage in the Cities*, 220–23.
40. Staten Island members placed on the table an unsuccessful $360 million service-restoration package, which included $22 million to go back to the DSNY for mixed-paper recycling, weekly recycling pickups, bulk recycling centers, and outreach and education programs. See Reginald Patrick, "Islanders Pressing Council to Restore Money for Recycling," *Staten Island Advance*, June 7, 1996; Vivian S. Toy, "Untraditionally, Council Approves Recycling Cut," *New York Times*, June 10, 1996.
41. He also called for task-force membership for city solid waste advisory boards and for opening the task force's meetings to the public. See Thomas Outerbridge to Mayor Giuliani, June 21, 1996, and Thomas Outerbridge to Mayor Giuliani, July 19, 1996, Giuliani Collection, Deputy Mayors, Peter Powers, Sanitation—Fresh Kills, Box 02/01/048, Folder 1283, La Guardia and Wagner Archives.
42. Peter J. Powers to Thomas Outerbridge, August 27, 1996, Giuliani Collection, Deputy Mayors, Peter Powers, Sanitation—Fresh Kills, Box 02/01/048, Folder 1283, La Guardia and Wagner Archives.
43. Ruth W. Messinger to Rudolph Giuliani, June 26, 1996, Giuliani Collection, Deputy Mayors, Peter Powers, Sanitation—Fresh Kills, Box 02/01/048, Folder 1283, La Guardia and Wagner Archives.
44. Arthur S. Kell, "Give Recycling a Chance," *New York Daily News*, June 20, 1996.
45. Carl Campanile, "Where Will the Garbage Go?" *Staten Island Advance*, June 10, 1996.
46. "Recycling: Mystery Card in Closure," *Staten Island Advance*, June 23, 1996; Joan Griffine McCabe, "Life After the Landfill Closes," *Park Slope Courier*, June 21–27, 1996; Lisa Ng, "Residential Recycling Rates: Influence of Policy, Power, and Priority," unpublished paper, Brooklyn College, December 21, 2015. A report from the Office of the State Comptroller stated that the DSNY was not meeting the tonnage requirements of Local Law 19 or "enforcing the law consistently throughout the five boroughs." See Rosemary Scanlon, Deputy for the City of New York, to John J. Doherty, September 13, 1996, State of New York, Office of the State Comptroller.
47. Reginald Patrick, "Budget Cuts Expected to Stall Recycling Rate," *Staten Island Advance*, September 17, 1996.

48. Alan Hevesi to Rudolph Giuliani, July 12, 1996, Giuliani Collection, Deputy Mayors, Peter Powers, Sanitation—Fresh Kills, Box 02/01/048, Folder 1283, La Guardia and Wagner Archives.
49. John Tierney, "Recycling Is Garbage," *New York Times Magazine*, June 30, 1996, 24, 26.
50. See Allen Herskowitz, "In Defense of Recycling," *Garbage* 65 (March 22, 1998): 141–218. See also David Rothbard and Craig Rucker, "Recycling Revisited: Ten Years After the *Mobro*," briefing paper #106, Committee for a Constructive Tomorrow, December 1997, 1–15.
51. Melosi, *Garbage in the Cities*, 223–25. See also Craig Schneider, "Fresh Kills Realities May Be Forcing a Fresh Look at Recycling," *Staten Island Advance*, October 1, 1996; "Recycling a Real Option," *Staten Island Advance*, October 3, 1996; Reginald D. Patrick, "Landfill Panel Reaching Out to Recyclers," *Staten Island Advance*, October 10, 1996. See also Matthew Gandy, *Concrete and Clay: Reworking Nature in New York City* (Cambridge, MA: MIT Press, 2002), 206.
52. Craig Schneider, "Can City Afford It?" *Staten Island Advance*, September 24, 1996.
53. Craig Schneider, "Put a Lid on It," *Staten Island Advance*, September 25, 1996.
54. An additional issue had to be dealt with. Phillip J. Gleason, the DSNY's director of landfill engineering, wanted the city to withdraw its application to the state that would allow for operating the landfill indefinitely, that is, end the review of the permit. This meant modifying the consent decree (1990) to include a plan to close Fresh Kills by 2002. Phillip J. Gleason to John J. Ferguson, September 27, 1996, Marchi Papers, Secession, Box 2, Folder 16. See also "City Wants to End Application to Rub Dump Indefinitely," *Staten Island Advance*, September 28, 1996.
55. Steven H. Corey, "Exporting (Not Just Exploring) Gotham's Garbage: Environmental Activism and the Privatization of New York City's Waste Stream," keynote address, Ford Scholars Symposium, Vassar College, February 13, 2002, 11.
56. James B. Jacobs, *Gotham Unbound: How New York City Was Liberated from the Grip of Organized Crime* (New York: New York University Press, 1999), 80–91.
57. Jacobs, *Gotham Unbound*, 195, also 191–94, 197–98.
58. Jacobs, *Gotham Unbound*, 196–99. See Allen R. Myerson, "The Garbage Wars: Cracking the Cartel," *New York Times*, July 30, 1995.
59. Quoted in Marissa L. Morelle, "'Something Smells Fishy': The Giuliani Administration's Effort to Rid the Commercial Trade Waste Collection Industry of Organized Crime," *New York Law School Law Review* 42 (1998): 1214.
60. See National Solid Waste Management Association and Deloitte & Touche, *NYC Commercial Sold Waste Cost Analysis and Request for Rate Deregulation*, n.d., Giuliani Collection, Deputy Mayors, Tony Coles, Trade Waste, Box 02/10/014, Folder 336; *Resolution of the New York City Trade Waste Commission . . .*, n.d., and Karen Koslowitz, City Council to Randy Mastro, July 15, 1996, Giuliani Collection, Deputy Mayors, Rudy Washington, Trade Waste, Box 02/08/005, Folder, 143; "Trade Waste Commission Arrests Unlicensed Carter for Illegal Dumping in Brooklyn," press release, May 14, 1997, Giuliani Collection, Deputy Mayors, Rudy Washington, Trade Waste, Box 02/08/012, Folder 336.
61. Jacobs, *Gotham Unbound*, 200–5. See also Corey, "Exporting (Not Just Exploring) Gotham's Garbage," 11–20; Richard Behar and Rao Rajiv, "Talk About TOUGH COMPETITION," *Fortune* 133 (1996): 90–98; Morelle, "'Something Smells Fishy,'" 1215–38.
62. Peter J. Powers to Beth Petrone, August 16, 1996, Giuliani Collection, Deputy Mayors, Randy Mastro, Project Files, Sanitation—Fresh Kills, Box 02/04/006, Folder 193, La Guardia and Wagner Archives.

63. Craig Schneider, "Landfill Panel Misses Report Deadline," *Staten Island Advance*, September 28, 1996. See also "Missing a Deadline," *Staten Island Advance*, September 30, 1996; "Tripp Named to Task Force on Closing Fresh Kills Landfill," *EDF Letter* 27 (November 1996): 2.
64. Craig Schneider, "They're Promising, and We're Watching," *Staten Island Advance*, September 29, 1996. Others outside the committee, such as Steve Englebright, chair of the Legislative Commission on Solid Waste Management, were convinced that "several non-incineration technologies" showed promise and should be examined as to avoid "exclusive reliance on exports for non-recyclable waste." See Steve Englebright to George Pataki and Rudolph Giuliani, September 17, 1997, Giuliani Collection, Deputy Mayors, Randy Mastro, Project Files, Sanitation—Fresh Kills, Box 02/04/019, Folder 466, La Guardia and Wagner Archives.
65. Marilyn Haggerty-Blohm to Randy M. Mastro, October 11, 1996, Giuliani Collection, Deputy Mayors, Randy Mastro, Project Files, Sanitation—Fresh Kills Box 02/04/006, Folder 193, La Guardia and Wagner Archives.
66. Sadat Associates, Inc., *Summary of the Solid Waste Management Alternatives Analysis for New York City Municipal Waste*, October 1996, 1–3, Marchi Papers, Series 1, Subject Files, Box 7.
67. Sadat Associates, *Summary of the Solid Waste Management Alternatives Analysis*, 4.
68. The Sadat Report (6, 17, 20–22) called for a three-phase approach: a short term (1997–2002) plan that phased out Fresh Kills and relied on some innovative methods of transportation and existing transfer stations, a medium-term (2002–2007) plan that implemented and refined transfer transportation methods and identified new disposal sites and new disposal options, and a longer-term (2007–2017) plan based on "achieving self-sufficiency in disposal capacity within New York State." It also supported existing source reduction and residential recycling during the phasing out of Fresh Kills.
69. City of New York and State of New York, *Report of the Fresh Kills Task Force: A Plan to Phase Out the Fresh Kills Landfill*, November 1996, 1, 8–10, 18–19, Marchi Papers, Series 1, Subject Files, Box 47.
70. City of New York and State of New York, *Report of the Fresh Kills Task Force*, 90–102, 105–24.
71. City of New York and State of New York, *Report of the Fresh Kills Task Force*, 2–5. Between 1990 and 1996, material diverted from landfills and incinerators nationwide increased by 67 percent. Trends also indicated more emphasis on reductions than recycling. Brad Edmondson, *Environmental Affairs in New York State: A Historical Overview*, Publication no. 72 (Albany: New York State Archives, 2001), 45.
72. City of New York and State of New York, *Report of the Fresh Kills Task Force*, 89.
73. In one striking example, the president of Consolidated American Industries, Inc., wrote to First Deputy Mayor Powers about a "potential solution to the Freshkills problem," which meant that "Rudy would not have to rely on anyone else, nor cajole and plead with state or federal officials." Aside from saving the city money, her proposal, "handled quickly and effectively—could almost virtually insure RWG's re-election." Lisa Duperier to Peter Powers, August 16, 1996, Giuliani Collection, Deputy Mayors, Randy Mastro, Project Files, Sanitation—Fresh Kills, Box 02/04/006, Folder 193, La Guardia and Wagner Archives. See also, for example, William Omiucke, President, All Products Marketing, Ltd., to Rudolph W. Giuliani, July 15, 1996; Peter J. Powers to William Omiucke, August 2, 1996; Harrichand Persaud to Rudolph W. Giuliani, August 8, 1996; Martha K. Hirst to Harrichand Persaud, August 16, 1996; *Fresh Kills*

Landfill: A Solution from Meglyn Enterprises, Inc., Draft Proposal for Discussion, n.d., Giuliani Collection, Deputy Mayors, Peter Powers, Sanitation—Fresh Kills, Box 02/01/048, Folder 1283, La Guardia and Wagner Archives; Jim Dancy to Rudolph Giuliani, August 27, 1996; Joseph A. Martorana to Randy M. Mastro, September 4, 1996; Joseph A. Martorana to Randy M. Mastro, September 6, 1996; Donna Lynne to Jim Dancy, October 11, 1996; Martha K. Hirst to Lisa Duperier, October 24, 1996; Helmut Konecsny to Eugene McGrath, October 25, 1996, Giuliani Collection, Deputy Mayors, Randy Mastro, Project Files, Sanitation—Fresh Kills, Box 02/04/006, Folder 193, La Guardia and Wagner Archives. Contracts for the gas-recovery concession, for example, had to be worked out. See John J. Doherty to Randy M. Mastro, October 8, 1996; David Gmach to Randy Mastro, October 18, 1996; Steven M. Polan to Paul Crotty, November 7, 1996; Sally Hernandez-Pinero to Randy Mastro, November 11, 1996; David Gmach to Randy Mastro, November 13, 1996, Giuliani Collection, Deputy Mayors, Randy Mastro, Project Files, Sanitation—Fresh Kills, Box 02/04/006, Folder 193, La Guardia and Wagner Archives; Kalkines, Arky, Zall & Bernstein LLP to Ethan Geto, November 21, 1996; John J. Doherty to Randy Mastro, December 26, 1996; NRG Energy, Inc., to Deputy Mayor Fran Reiter, November 21, 1996, Giuliani Collection, Deputy Mayors, David Klasfeld, Sanitation—Fresh Kills, Box 02/09/1/013, Folder 401; Robin Eisner, "Public Review in Works for Landfill Methane Plan," *Staten Island Advance*, October 16, 1997.

74. Mark Francis Cohen, "Where Will All the Garbage Go?" *New York Times*, September 15, 1995; Robin Eisner, "Going for the Gold in Garbage," *Staten Island Advance*, January 8, 1997.
75. Carl Campanile, "N.Y. Won't Force Trash on Anyone," *Staten Island Advance*, December 3, 1996.
76. Carl Campanile, "Landfill High on Pataki's Must-Do Agenda," *Staten Island Advance*, January 8, 1997. See also Carl Campanile, "Dump Closure Included in DEC Budget," *Staten Island Advance*, January 31, 1997; Carl Campanile, "State Money for Landfill Can't Be Spent Just Yet," *Staten Island Advance*, February 4, 1997; Carl Campanile, "Borrowing Authority Formed," *Staten Island Advance*, February 12, 1997; Reginald Patrick, "Pataki Vows to Shepherd Landfill $$," *Staten Island Advance*, November 8, 1996.
77. Robin Eisner, "5 Chosen for Landfill Closure Team," *Staten Island Advance*, January 6, 1997. The five-member group included legal counsel Daniel Master Jr. (chair); Deputy Borough President James Molinaro; the borough president's environmental engineer Nick Dmytryszyn; Molinari's environmental aide Meagan Devereaux; and Barbara Warren, the chair of the Solid Waste Advisory Board. Molinari asked Senator John Marchi to participate, and he accepted. See Guy V. Molinari to John J. Marchi, January 17, 1997, and John J. Marchi to Guy V. Molinari, January 24, 1997, Marchi Papers, Series 1, Subject Files, Box 7. See also Judy L. Randall, "Landfill Closure Task Force to Meet," *Staten Island Advance*, January 25, 1997; Robin Eisner, "Island Must Lead the Way," *Staten Island Advance*, January 29, 1997.
78. Robin Eisner, "P&G Site Eyed for Trash," *Staten Island Advance*, January 12, 1997; "Good Idea, Bad Location," *Staten Island Advance*, January 16, 1997. See also Robin Eisner, "Waste Site Plan Draws a Big YUCK!" *Staten Island Advance*, December 23, 1997.
79. Carl Campanile, "Bill Would Bar Trash Transfer Stations," *Staten Island Advance*, April 8, 1997. See also John J. Marchi to Mr. and Mrs. Amadeo Parisi, April 15, 1998, Marchi Papers, Series 1, Box 8. Jennifer M. Nelson, "Marchi to Pataki: Bill Could

Impede Landfill Closing," *Staten Island Advance*, April 14, 1998. See also "News from Senator John J. Marchi," April 13, 1998; Mark Green to John J. Marchi, April 14, 1998; John J. Marchi to George E. Pataki, April 8, 1998, Marchi Papers, Series 1, Box 8; Robin Eisner, "BP Reiterates His Opposition to Garbage Transfer Station," *Staten Island Advance*, March 3, 1998.

80. Timothy Forker to Marilyn Haggerty-Blohm, December 13, 1996, Marchi Papers, Series 1, Subject Files, Box 7. See Tim Forker to Manhattan Borough Solid Waste Working Group, memorandum, February 10, 1997, and Tim Forker to Marilyn Blohm and Ben Schmerler, February 25, 1996, Marchi Papers, Series 1, Subject Files, Box 37.

81. "Ominous Signs," *Staten Island Advance*, December 17, 1996; Claire Schulman to George E. Pataki, August 7, 1996, Giuliani Collection, Deputy Mayors, Peter Powers, Sanitation—Fresh Kills, Box 02/01/048, Folder 1283, La Guardia and Wagner Archives.

82. Robin Eisner, "Bronx Debates What to Do with Its Trash," *Staten Island Advance*, January 22, 1997; Robin Eisner, "Bronx Trash Won't Be Dumped Here," *Staten Island Advance*, June 8, 1997. See also Robin Eisner, "Bronx Trash Is Closer to Being Outta Here," *Staten Island Advance*, January 23, 1997; Judy L. Randall, "Ferrer Promises He'd Stick to Landfill Timetable," *Staten Island Advance*, January 31, 1997; Robin Eisner, "City Chided for Ignoring Trash Plan," *Staten Island Advance*, April 10, 1997.

83. *Fresh Kills Exportation Announcement*, Archives of Rudolph W. Giuliani, July 1, 1997, http://www.nyc.gov/html/records/rwg/html/97/freshkls.html. Other plans were underway to export refuse from Brooklyn and Queens in mid-1998. See Robin Eisner, "More Trash to Be Diverted from Island," *Staten Island Advance*, November 18, 1997.

84. They were somewhat surprised that the city received bids ranging from $46.64 to $66.55 per ton.

85. Vivian S. Toy, "Bids for Exporting Trash Are Lower Than Expected," *New York Times*, March 3, 1997.

86. Robin Eisner, "Solutions Must Start Here," *Staten Island Advance*, January 16, 1997. Private waste companies like Browning Ferris, Inc., also tried to cash in on recycling. Robin Eisner, "Venture Fund Proposed for Recycling Efforts," *Staten Island Advance*, March 31, 1997.

87. Testimony of Borough President Howard Golden Before the New York City Council's Environmental Committee, December 16, 1996, Marchi Papers, Series 1, Subject Files, Box 7. See also Howard Golden to Philip Rooney, August 14, 1996, Giuliani Collection, Deputy Mayors, Peter Powers, Sanitation—Fresh Kills, Box 02/01/048, Folder 1283, La Guardia and Wagner Archives; Howard Golden to Randy Mastro, December 9, 1996; Alan G. Hevesi to Rudolph Giuliani, December 10, 1996, Giuliani Collection, Deputy Mayors, Randy Mastro, Project Files, Sanitation—Fresh Kills, Box 02/04/006, Folder 193, La Guardia and Wagner Archives.

88. Robin Eisner, "Landfill Plan Dumped on at City Council Hearing," *Staten Island Advance*, December 17, 1996; "The Debate Begins," *Staten Island Advance*, December 18, 1996.

89. Guy V. Molinari, "Manhattan BP Not Interested in Closing Fresh Kills Landfill," *Staten Island Advance*, January 15, 1997.

90. *New York Times*, January 6, 1997.

91. Robin Eisner, "Mayor Budgets $321M to Close Dump," *Staten Island Advance*, January 31, 1997. The figures varied a little, depending on the report. For example, see Toy, "Bids for Exporting Trash Are Lower Than Expected." City Council budget

discussions for 1998 suggested slightly more funds for recycling. City of New York, Independent Budget Office, "Fresh Kills Closing: Where's the Garbage Going?" *Inside the Budget*, March 10, 1997, Vitaliano Papers, Series 1, Subject Files, Box 39, Folder 4.

92. Cohen, "Where Will All the Garbage Go?" See also Toy, "Bids for Exporting Trash Are Lower Than Expected."
93. Corey, "Exporting (Not Just Exploring) Gotham's Garbage," 7–9.
94. MTSs were two-level structures. Arriving trucks entered on the second level and dumped loads into hopper barges located in slips on the first level. Refuse was taken to disposal sites from there. City of New York, Department of Sanitation, *Feasibility Report on MTS Conversion*, April 1999, Vitaliano Papers, Series 1, Subject Files, Box 36, Folder 10.
95. Robin Eisner, "Here's How City Will Close Dump," *Staten Island Advance*, February 24, 1998. See also Robin Eisner, "Even After It's Closed, Landfill to Handle Our Trash," *Staten Island Advance*, February 25, 1998; Julie Sze, *Noxious New York: The Racial Politics of Urban Health and Environmental Justice* (Cambridge, MA: MIT Press, 2007), 113.
96. James Bradley, "Garbage Wars," *City Limits*, January 1, 1998, http://citylimits.org/1998/01/01/garbage-wars/. See also Nancy Walby, Brooklyn Solid Waste Advisory Board to Fellow Brooklynite, October 22, 1997, Vitaliano Papers, Series 1, Subject Files, Box 36, Folder 2; Nancy Walby to Randy Mastro, February 27, 1998, Giuliani Collection, Deputy Mayors Randy Mastro, Project Files, Sanitation—Fresh Kills, Box 02/04/027, Folder 677; Barbara Stewart, "Waste (Not), Want (Not)," *New York Times*, February 1, 1998; Terence Kivlan, "Fossella Says No to Brooklyn Trash Transfer Station," *Staten Island Advance*, December 6, 1997.
97. See Sze, *Noxious New York*, 11.
98. Bradley, "Garbage Wars." See also Sze, *Noxious New York*, 113–16.
99. Quoted in Corey, "Exporting (Not Just Exploring) Gotham's Garbage," 10. For an example of an OWN protest, see Robin Eisner, "Firms Want to Cash In on Trash Export," *Staten Island Advance*, April 16, 1997. See also Robin Eisner, "Ruling Could Keep Waste Stations off Island," *Staten Island Advance*, April 5, 1997; "Fresh Kills' Neighbors Protest Post-Landfill Closure Plan," *Waste Age* 31 (September 2000): 28–31, 34–36.
100. Commissioner Doherty echoed his remarks. Robin Eisner, "Bronx, Brooklyn: Don't Dump on Us," *Staten Island Advance*, March 4, 1998.
101. Robin Eisner, "Crucial Fight Taking Shape on Trash Transfer Stations," *Staten Island Advance*, February 5, 1998. See also John J. Doherty to Interested Party, January 28, 1998, Giuliani Collection, Deputy Mayors Joseph Lhota and Jake Menges, Sanitation—Waste Station Siting Rules, Box 02/05/1/003, Folder 154.
102. Douglas Martin, "Trash-Station Proposal Greeted by Protesters," *New York Times*, March 4, 1998.
103. The department continued to tinker with the distance into July and beyond. See Leslie Allan and Valerie Budzik to George Davis, April 29, 1998; and "Waste Transfer Station Siting Rule: Key Considerations and Background," July 29, 1998, Giuliani Collection, Deputy Mayors Joseph Lhota and Jake Menges, Sanitation—Waste Station Siting Rules, Box 02/05/1/003, Folder 154. See also Sze, *Noxious New York*, 115.
104. "No Time for Arrogance," *Staten Island Advance*, April 14, 1997. See also Robin Eisner, "Public Not Convinced That Landfill Will Close," *Staten Island Advance*, April 16, 1997.

105. Besides Manhattan, Queens, the Bronx, and Staten Island met the deadline. Brooklyn's plan came later. See Robin Eisner, "Brooklyn Misses Deadline for Garbage Proposal," *Staten Island Advance*, May 6, 1997.
106. City of New York, Office of the Manhattan Borough President, *Goodbye, Fresh Kills! Or "How the City Can Stop Worrying and Learn to Reduce, Reuse and Recycle,"* April 30, 1997, 1, Marchi Papers, Series 1, Box 7. The title was a play on words from the 1964 film *Dr. Strangelove, or How I Learned to Stop Worrying and Love the Bomb*.
107. City of New York, Office of the Manhattan Borough President, *Goodbye, Fresh Kills!*, 3, also 5–6; Bill Franz, "Messenger Blasts 'Secrecy' of Fresh Kills Plan," *Staten Island Advance*, May 6, 1997.
108. City of New York, Office of the Manhattan Borough President, *Goodbye, Fresh Kills!*, 1.
109. City of New York, Office of the Manhattan Borough President, *Goodbye, Fresh Kills!*, 2.
110. Stanley E. Michels, Statement on Release of the Council's Preliminary Response to the Fresh Kills Task Force Report, City Hall, October 21, 1997, Marchi Papers, Secession, Subject Files, Box 32, Folder 12. See also "The Demise of the Fresh Kills Landfill," *Searchlight on City Council*, January 1997; New York City Council Preliminary Response to the Fresh Kills Task Force Report, *Without Fresh Kills: A Blueprint for Solid Waste Management*, October 27, 1997, 1, Archives & Special Collections, Department of the Library, College of Staten Island, CUNY. For details of the oversight hearings, see Council of the City of New York Collection, Committee Files, Oversight: Fresh Kills Task Force Report, Box 50073, Folder 3, La Guardia and Wagner Archives.
111. Statement by City Council Speaker Peter F. Vallone on Release of Council Response to Report of Fresh Kills Task Force, October 21, 1997, Marchi Papers, Secession, Subject Files, Box 32, Folder 12. See also Statement By: Assemblywoman Elizabeth A. Connelly-59 A.D. Re: Fresh Kills Landfill Report, November 12, 1997, Marchi Papers, Secession, Subject Files, Box 32, Folder 12; Council of the City of New York, Office of Communications, "Council Speaker Vallone and Environmental Protection Chair Michels Call for Weekly Recycling Pickups Citywide," October 21, 1997, Marchi Papers, Series 1, Subject Files, Box 7.
112. New York City Council Preliminary Response to the Fresh Kills Task Force Report, *Without Fresh Kills*, 1, 2–3.
113. New York City Council Preliminary Response to the Fresh Kills Task Force Report, *Without Fresh Kills*, 4. See also "NYC Trash Transport," *Mobilizing the Region: A Weekly Bulletin from the Tri-State Transportation Campaign*, August 29, 1997; Reginald Patrick, "Who Will Take Out the Garbage?" *Staten Island Advance*, November 26, 1997; "Talking Trash," *Staten Island Advance*, November 28, 1997. Later efforts to pass such a bill drew the ire of Borough President Molinari, who believed that Democrats were using the tactic to stop the closure of Fresh Kills. See Robin Eisner, "BP: Dem Bosses Want Landfill Open," *Staten Island Advance*, November 28, 1997. See also Reginald Patrick, "GOP Presses O'Donovan on Landfill Bills," *Staten Island Advance*, December 1, 1997.
114. "Closing in on Closure," *Staten Island Advance*, December 21, 1997. See also Robin Eisner, "Dump Shutdown a Year Closer," *Staten Island Advance*, June 15, 1997; "The Politics of Garbage," *Staten Island Advance*, December 30, 1997. For details on the borough-based plans submitted to the City Council, see Council of the City of New York Collection, Committee Files, "The Borough-Based Solid Waste Management and Disposal Plans . . . ," 1997, Box 50072, Folder 11; and Committee Files, "Oversight:

Borough Based Solid Waste Planning as Proposed in the Fresh Kills Closure Report by the Fresh Kills Task Force," Box 50074, Folder 5, La Guardia and Wagner Archives.

115. Giuliani's base of support mirrored the 1993 win. The proportion of victory among whites remained about the same as in 1993, with a modest upturn among African Americans (20 percent) and a rise from 37 to 43 percent among Latinos. He increased his vote among women from 46 to 54 percent. The difference was the small turnout. Often abrasive and overzealous, Giuliani's upside was a "can-do" style that kept his supporters loyal (and had them largely ignoring the problems in his marriage). He ran an aggressive campaign with plenty of money, emphasizing his strong stance against crime and his promotion of economic recovery. Messinger's focus on race, job creation, and education could not compete with his message. See David Firestone, "The 1997 Elections: The Voters; Big Victory, but Gains for Mayor Are Modest," *New York Times*, November 6, 1997. See also "Mayoralty," in *The Encyclopedia of New York City*, 2nd ed., ed. Kenneth T. Jackson (New Haven, CT: Yale University Press, 2010), 815–16; Justin Oppmann, "Giuliani Wins with Ease," All Politics, *CNN/Time*, November 4, 1997, http://www.cnn.com/ALLPOLITICS/1997/11/04/mayor/.

116. David Firestone, "Mayor Opens New City Ad, Then Says It's the Last," *New York Times*, August 13, 1997.

117. Randy M. Mastro, "Don't Trash Plans for Fresh Kills," *New York Daily News*, January 21, 1998.

118. "Naysaying Nonsense," *Staten Island Advance*, February 8, 1998. See also "'Support' We Don't Need," *Staten Island Advance*, January 13, 1998.

119. Andrew G. Wright, "Big Apple Is Dumping Its Last Active Landfill," *Focus on Environment* 240 (February 16, 1998): 28.

120. City of New York, Department of Sanitation, *Final: Operations and Maintenance Plan for the Fresh Kills Landfill*, v. 1, January 31, 1997, Marchi Papers, Series 1, Misc. Collection, Subject Files, Box 47, Folder 7.

121. Robin Eisner, "A Tall Order," *Staten Island Advance*, July 27, 1997.

122. Wright, "Big Apple Is Dumping Its Last Active Landfill, 28. See also "Mountain of Work Lies Ahead to Close World's Largest Garbage Dump," *Fresno Bee*, December 21, 1997; Vivian S. Toy, "Sealing Mount Garbage; Closing Staten Island's Fresh Kills Dump Is an Operation of Staggering Complexity," *New York Times*, December 21, 1997. On gas migration monitoring and extraction issues, see Roy F. Weston of New York, Inc., *Fresh Kills Landfill: Summary Report of Field Work—Landfill Gas Migration Monitoring System*, March 1998; Roy F. Weston of New York, Inc., *Draft Final, Fresh Kills Landfill, Landfill Gas Migration Monitoring System Evaluation Summary Report*, February 25, 1997; Roy F. Weston of New York, Inc., *LFG Migration Monitoring Systems Evaluation Summary Report, Responses to NYSDEC Comments, Letter Dated June 12, 1997*, September 1997, Archives & Special Collections, Department of the Library, College of Staten Island; John J. Marchi to John Doherty, July 25, 1997, and John Doherty to John Marchi, August 13, 1997, Marchi Papers, Series 1, Subject Files, Box 7. In response to a forthcoming press release from Albany concerning the Fresh Kills gas-collection system, staff in the mayor's office wrote, "It appears that the State wants to emphasize the air monitoring program.... The odor control aspect, however, is less controversial and shows that the Mayor is dealing with this quality of life issue for Staten Islanders. Also, air monitoring suggests that something is wrong and that DEC is forcing us to confront it." See George Davis to Colleen Roche, October 27, 1997, Giuliani Collection, Deputy Mayors, Randy Mastro, Project Files,

Sanitation—Fresh Kills, Box 02/04/019, Folder 466, La Guardia and Wagner Archives; Eastern Research Group, Inc., *Fresh Kills Air Monitoring Program*, July 1998, Archives & Special Collections, Department of the Library, College of Staten Island.

123. Eric N. Vitaliano, Remarks at the Annual Department of Sanitation Meeting on the Fresh Kills Consent Order, February 26, 1997, Vitaliano Papers, Series 1, Subject Files, Box 37. See also City of New York, the Council, A Local Law: To amend the administrative code of the city of New York, in relation to the preparation of Fresh Kills Landfill Environmental Plan, Int. no. 1083, November 25, 1997; and Public Advocate for the City of New York, Memorandum in Support, n.d., Giuliani Collection, Series: Committee Files, 1997, Fresh Kills Landfill Environmental Plan, Box 50080, Folder 11. See also "News from Assemblyman Eric N. Vitaliano," February 24, 1998; and Statement of Assemblyman Eric N. Vitaliano, Department of Sanitation Public Meeting on the Fresh Kills Landfill, February 24, 1998, Vitaliano Papers, Series 1, Subject Files, Box 37. See also Jennifer M. Nelson, "DEC Head: Talks to Close Landfill Are Progressing," *Staten Island Advance*, February 24, 1998.

124. City of New York, Department of Sanitation, *Comprehensive Solid Waste Management Plan, Draft Modification*, April 3, 1998, 1-1-1-2, Marchi Papers, Secession, Box 9, Folder 5. There also was a newer version issued on May 28, 1999.

125. City of New York, Department of Sanitation, *Comprehensive Solid Waste Management Plan, Draft Modification*, 1-2-1-3.

126. In 1996 the DSNY managed 3.9 million tons of MSW through landfilling and 1.2 million tons through recycling. Legislative Commission on Solid Waste Management, News York State Assembly, *Where Will the Garbage Go? 1997* (July 1997), 16.

127. In 1997, 3,810 million tons were disposed of at Fresh Kills; in 1998, 3,240; in 1999, 2,807; in 2000, 1,921; and in 2001 it was projected as 1,008 and 2002 as 0. City of New York, Independent Budget Office, *Overview of the Waste Stream Managed by the NYC Department of Sanitation*, Background Paper, February 2001, 10.

128. City of New York, Department of Sanitation, *Comprehensive Solid Waste Management Plan, Draft Modification*, April 3, 1998, 1-5, 2-50–2-55. In January 1998, however, the DSNY officially opened bids from thirteen companies as part of a plan to export 2,400 tons of refuse from Brooklyn and Queens later in the year. See Reginald Patrick, "Several Companies Bid to Export City's Garbage," *Staten Island Advance*, January 14, 1998.

129. Determining diversion rates was no simple matter. In a city like New York there was great variation between neighborhoods as well as boroughs, and the rate depended on what one was measuring. For example, in November 1995 the DSNY began collecting mixed paper and bulk metal on Staten Island, beginning somewhat later for the other boroughs. These materials were added to curbside recyclables to determine diversion rates for the city. Also, different sources used different measures for aggregate statistics. The public-radio station WNYC placed the diversion rate for New York City in 1998 at 16.8 percent. The city's Independent Budget Office, using DSNY statistics, listed the diversion rate in 1998 at 22.8 percent, down from 24.2 percent in 1997. In this case, the diversion rate was based on the total managed waste (including disposed and recycled materials). Measures of diversion rates for curbside and containerized waste showed 13.4 percent in 1997 and 15.9 percent in 1998. See City of New York, Department of Sanitation, *Comprehensive Solid Waste Management Plan, Draft Modification*, April 3, 1998; WNYC, New York City Recycling Rate, https://project.wnyc.org/nyc-recycling; City of New York, Independent Budget Office, *Overview of the Waste Stream Managed by the NYC Department of Sanitation*, 2, 5. Statewide,

recycled materials in 1996 stood at 21.3 percent of the waste stream. See Legislative Commission on Solid Waste Management, News York State Assembly, *Where Will the Garbage Go? 1997* (July 1997), 1, 16. The EPA stated a national recovery rate for all categories of recyclables at 16.2 percent in 1990 and 29.0 percent in 2000. See Samantha McBride, *Recycling Reconsidered: The Present Failure and Future Promise of Environmental Action in the United States* (Cambridge, MA: MIT Press, 2012), 248.

130. City of New York, Independent Budget Office, *Overview of the Waste Stream Managed by the NYC Department of Sanitation*, 3. The DSNY also intended to upgrade the composting facility at Fresh Kills. See City of New York, Department of Sanitation, *Comprehensive Solid Waste Management Plan, Draft Modification*, April 3, 1998, 1-5-1-6, 2-14, 2-28.

131. Maureen Seaberg, "50th Anniversary of Landfill Marked with Closure in Mind," *Staten Island Advance*, April 16, 1998.

132. "After Fresh Kills, What?" *New York Times*, May 2, 1998.

133. Robin Eisner, "Close Dump on Time—or Else," *Staten Island Advance*, August 31, 1997; "Green Sounds Warning About Landfill Closing," *New York Times*, September 18, 1997; Robin Eisner, "Council Listening as Other Boroughs Balk at Trash Plans," *Staten Island Advance*, November 13, 1997; George Pataki to Senator Marchi, May 11, 1998, John J. Marchi Papers, Series 1, Subject Files, Box 8, Archives & Special Collections, Department of the Library, College of Staten Island, CUNY. See also Jennifer M. Nelson, "Gov: Dump Closure Safe," *Staten Island Advance*, September 18, 1998.

134. "What We Don't Know," *Staten Island Advance*, May 11, 1998.

135. Robin Eisner, "Group: Compensate Island for Enduring Landfill," *Staten Island Advance*, May 27, 1998.

136. Sydney Fisher, "The End Nears. The Fears Don't Fade," *New York Times*, April 19, 1998.

137. Brad Edmonson, *Environmental Affairs in New York State: A Historical Overview*, Publication no. 72 (Albany: New York State Archives, 2001), 45. See also "Solid Waste Transfer Stations by Community Board," October 31, 2000, Giuliani Collection, Deputy Mayors, Joseph Lhota, Sanitation—Solid Waste Management Plan, Box 02/05/003, Folder 87.

138. Robin Eisner, "'Racial Push' on Landfill Is Derided as Garbage," *Staten Island Advance*, November 18, 1998.

139. Both were associated with Staten Island Citizens for Clear Air (SICCA). Tripp also served as chair of the Staten Island Solid Waste Advisory Board.

140. James T. B. Tripp and Thomas Outerbridge to David Jaffe, May 13, 1998, Marchi Papers, Series 1, Subject Files, Box 46, Folder 8.

141. James T. B. Tripp and Thomas Outerbridge to David Jaffe, May 13, 1998.

142. James T. B. Tripp and Thomas Outerbridge to David Jaffe, May 13, 1998. See also Thomas Outerbridge, "The Crisis of Closing Fresh Kills," *BioCycle* 41 (April 200): 47-50.

143. "Fresh Kills Closure Opportunity to Reduce, Not Increase Truck Impacts," *Mobilizing the Region* 173 (May 15, 1998): 3. See also "City's Head Is in Sand on Fresh Kills Closing," *New York Times*, May 17, 1998; Jim Yardley, "Angry Reception Greets Trash Plan in Brooklyn," *New York Times*, December 3, 1998; Elizabeth A. Connelly to George E. Pataki, December 18, 1998, and State of New York, In Assembly, June 1, 1998, Marchi Papers, Secession, Subject Files, Box 32, Folder 12.

144. Robin Eisner, "Backers Say Transfer Station Will Ensure Closing of Fresh Kills," *Staten Island Advance*, May 27, 1998. The Port Ivory Recycling & Transfer Alliance

(PIRTA) had submitted a proposal for such a facility on the 123-acre site. It was expected to receive twelve barges each day to four slips. See Port Ivory Recycling & Transfer Alliance, *Overview of a Proposed Redevelopment Plan & Community & Environmental Benefits Program at Port Ivory, Staten Island*, November 1998, Marchi Papers, Secession, General Files, Box 9, Folder 1.

145. Robin Eisner, "Waste Transfer Stations Are Key to Landfill Closure," *Staten Island Advance*, October 31, 1998. In the summer of 1997, Browning-Ferris Industries had approached the New York City Economic Development Corporation (EDC) to discuss the possible siting of an enclosed unloading facility at the Sixty-Fifth Street Rail Yard in Brooklyn. The inquiry was in connection with the DSNY's Waste Export Request for Proposals. The EDC was reluctant to act without "the full understanding" of City Hall. The barge-to-rail approach was attractive because the project did not involve more truck traffic, although there were typical risks involved in developing additional disposal sites in Brooklyn. While there was strong interest in the idea, the complications of the bidding process stalled the project. See Charles Millard to Randy Levine and Randy Mastro, June 29, 1998, Giuliani Collection, Deputy Mayors, Randy L. Levine, Project/Subject Files, Waste Site, Box 02/03/017, Folder 732.

146. "The 'Right Way'?" *Staten Island Advance*, September 13, 1998.

147. Robin Eisner, "Brooklyn BP Fights Garbage Exportation," *Staten Island Advance*, October 6, 1998. See also "Fresh Kills and Fair Share," *Staten Island Advance*, November 11, 1998.

148. Robin Eisner, "Brooklyn Groups Weigh Suit to Stop City Garbage Plan," *Staten Island Advance*, October 28, 1998.

149. Rebecca Cavanaugh, "Brooklyn Thwarts Trash Plan," *Staten Island Advance*, October 31, 1998. See also Robin Eisner, "Brooklyn Trash Restraining Order Expected to Be Overturned Today," *Staten Island Advance*, November 2, 1998; Robin Eisner and Rebecca Cavanaugh, "No Quick Fix on B'klyn Trash," *Staten Island Advance*, November 3, 1998.

150. Paul H. B. Sin, "B'klyn Loses Trash Fight in a Fresh Kills Appeal," *New York Daily News*, November 3, 1998; Robin Eisner, "Brooklyn Wins Again in Landfill Transfer Battle," *Staten Island Advance*, November 13, 1998; "The Fresh Kills Perspective," *Staten Island Advance*, November 28, 1998; "Landfill Closure: A Big Win," *Staten Island Advance*, December 20, 1998. See also Douglas Martin, "Boroughs Battle Over Trash as Last Landfill Nears Close," *New York Times*, November 16, 1998; Robin Eisner, "The Landfill: Are We About to Get Dumped On?" *Staten Island Advance*, November 16, 1998; Robin Eisner, "Islanders to Brooklyn: Back Off!" *Staten Island Advance*, November 22, 1998; "Brooklyn Replies," *New York Times*, November 26, 1998; Robin Eisner, "BP: Reopen Landfill in Brooklyn," *Staten Island Advance*, December 1, 1998; Reginald Patrick, "Fossella Begins Effort to Reopen Brooklyn Dump," *Staten Island Advance*, December 8, 1998; Rebecca Cavanaugh, "Golden: Rethink Landfill Plan," *Staten Island Advance*, December 16, 1998; Reginald Patrick, "Molinari Trashes Golden Garbage Plan," *Staten Island Advance*, December 17, 1998; Robin Eisner, "Island Comes Out on Top in Trash Suit," *Staten Island Advance*, December 19, 1998; Reginald Patrick, "Possible Delay Seen in Closure of Landfill," *Staten Island Advance*, December 13, 1998.

151. Advance Information Sheet, December 2, 1998, Giuliani Collection, Scheduling Events, Press Avail—Long Term Waste Management, Box 030189, Folder 10076. See also Paul H. B. Shin, "Rudy Bares Trash Plan," *New York Daily News*, December 3, 1998. See also City of New York, City Council, Transcript of the Minutes of the

Committee on Environmental Protection, September 16, 1998, Marchi Papers, Series 1, Subject Files, Box 8.

152. Martin and Revkin, "The Last Landfill: A Special Report." See also "Battle Cry Rises Loud and Clear in Essex, Middlesex, and Union," *Newark Star-Ledger*, December 4, 1998; "New York Trash May Be Headed to N.J.," *Easton* (Pennsylvania) *Express-Times*, December 4, 1998; Tracey Porpora, "In Carteret, They Smell a Rat," *Staten Island Advance*, December 4, 1998. See also Robin Eisner, "City Eases Trash Plan for Queens, Manhattan," *Staten Island Advance*, December 31, 1998; Douglas Martin, "Trash Plan Is Praised, but Faces Opposition," *New York Times*, December 13, 1998; Douglas Martin and Dan Barry, "Giuliani Stirs up Border Tensions with Trash Plan," *New York Times*, December 3, 1998.

153. "Giuliani Wants to Ship Trash to N.J.; Whitman Vows to Block Barge Plan," *New Jersey Record*, December 3, 1998.

154. "Talkin' Trash and Pride," *New Jersey Record*, January 31, 1999.

155. "In Trash Wars, Economics May Conquer Politics," *Newark Star-Ledger*, December 4, 1998.

156. "NJ Gov to Rudy: Your Trash Plan Stinks," *Staten Island Advance*, December 3, 1998.

157. Robin Eisner, "Mayor: No Sweat on Our Trash," *Staten Island Advance*, December 4, 1998. See also "Whitman Dumps on NYC Barge Idea," *Trenton Times*, December 4, 1998; "Christie and Rudy Talk Trash," *Newark Star-Ledger*, December 6, 1998; "Carteret States Its Opposition to Trash Plan," *Newark Star-Ledger*, December 9, 1998. Despite the rhetoric, there remained interest by some communities in New Jersey (such as Bergen County) possibly to accept New York City refuse. See "Bergen Might Try to Get N.Y.C. Trash," *New Jersey Record*, January 22, 1999.

158. Robin Eisner, "Citywide Alliance: Existing Sites Can Handle Garbage Plan," *Staten Island Advance*, December 14, 1998.

159. Reginald Patrick, "Rudy: Help Me Close the Dump," *Staten Island Advance*, December 15, 1998. See also Robin Eisner and Jennifer Nelson, "Trash War: Island Fighting Back," *Staten Island Advance*, December 19, 1998.

160. Yardley, "Angry Reception Greets Trash Plan in Brooklyn." See also Jim Yardley, "Garbage In . . . and In . . . and In; Greenpoint Residents United to Fight Influx of Trash," *New York Times*, April 18, 1998; Robin Eisner, "Challenge Could Delay Landfill Closure Date," *New York Times*, September 6, 1998.

161. For example, see Eve Martinez, "Fresh Kills Closing Highlights Need for High-Rise Recycling," *Real Estate Weekly*, September 16, 1998.

162. Bill Franz, "Rudy Trashes Recycling—the Hoax on Us," *Staten Island Register*, November 17, 1998.

163. Reginald Patrick, "Mayor Signs Bill Phasing in Weekly Recycling Pickups," *Staten Island Advance*, December 23, 1998. See also "Post-Holiday Haul," *New York Times*, December 29, 1998. By 2000, forty-nine of the fifty-nine collection districts had weekly recycling collection service. New York City, Department of Sanitation, *The DOS Report: Closing the Fresh Kills Landfill*, February 2000, 2, Archives & Special Collections, Department of the Library, College of Staten Island, CUNY.

164. In Washington, DC, movement on flow-control legislation was stalled, further confusing the various strategies for shipping waste out of the city. Terence J. Kivlan, "Bill to Limit Trash Importing Expected in Senate," *Staten Island Advance*, May 11, 1998; Terence J. Kivlan, "Vito: Federal Trash Ban Bill 'Close to Dead,'" *Staten Island Advance*, June 22, 1998; Patrick Lally to Jake Menges and Debbie Clinton, December 10, 1998, Giuliani Collection, Deputy Mayors Joseph Lhota and Jake Menges,

Waste Export, Box 02/05/1/004, Folder 178. In the same folder, see Patrick Lally to Jake Menges, December 14, 1998, and "Hypothetical Scenarios for an Interstate Waste Bill—106th Congress," December 10, 1998.
165. New York City, Department of Sanitation, *Annual Report 1998*, iv; Edmonson, *Environmental Affairs in New York State*.
166. Doherty announced his retirement in August 1998. Early speculation was that Deputy Commissioner Michael Carpinello would replace him. Guy Molinari pushed Annadale's Robert C. Avaltroni Jr., deputy commissioner for the city's Department of Environmental Protection, for the job. The borough president regarded Doherty's choice of Carpinello as part of the "old guard." "He [Carpinello] was typical of the bureaucrats you see in city agencies," Molinari stated. "And it was particularly galling that he was a Staten Islander and he was less than forthcoming." Giuliani did make an appointment immediately, largely because of Molinari's intercession. Giuliani eventually chose Assistant Police Chief Farrell in March 1999. Farrell had served thirty-seven years with the NYPD, two of those as the "top cop" on Staten Island, and most recently as executive officer of the Detective Bureau. The forty-first DSNY commissioner was a lifelong Brooklyn resident. On accepting the appointment, Farrell laid out a three-point action plan: close Fresh Kills, recycling, and recognition to DSNY workers. See Anne Marie Calzolari, "Sanit Boss, an Islander, Will Retire," *Staten Island Advance*, August 19, 1998; Judy L. Randall, "Molinari Pushes DEP Official for Top Sanitation Job," *Staten Island Advance*, October 6, 1998; Robin Eisner, "A Top Cop City's New Sanit Boss," *Staten Island Advance*, March 16, 1999. See also Abby Goodnough, "Giuliani Names Career Police Administrator as Sanitation Chief," *New York Times*, March 16, 1999; "Mayor Giuliani Appoints Kevin Farrell as Sanitation Commissioner," Archives of the Mayor's Press Office, Release 091-99, March 15, 1999, http://www.nyc.gov/html/om/html/99a/pr091-99.html.
167. New York City, Department of Sanitation, *Annual Report 1998*, 13–17, 27–37. See also the City of New York, Department of Sanitation, *Fresh Kills Landfill: Contingency Plan*, July 15, 1998, Archives & Special Collections, Department of the Library, College of Staten Island, CUNY.
168. See "New Year, Same Battle," *Staten Island Advance*, January 6, 1999; "City to Hold Talks on Garbage Plan," *Staten Island Advance*, January 20, 1999; Jennifer M. Nelson, "State Budget Includes Less for Ferry, More $$ for Closing Fresh Kills," *Staten Island Advance*, January 28, 1999; Jennifer M. Nelson, "Mayor: City Moving Towards Fresh Kills Closure," *Staten Island Advance*, February 3, 1999; "One More Reason to Close It," *Staten Island Advance*, February 15, 1999; Robin Eisner, "Wanted: Fresh Plans to Shut Fresh Kills," *Staten Island Advance*, February 23, 1999; Terence J. Kivlan, "Mayor Insists Landfill Will Close on Schedule," *Staten Island Advance*, February 25, 1999.
169. See Robin Eisner, "Opening Countdown on Landfill Closing," *Staten Island Advance*, March 19, 1999.
170. Robin Eisner, "Sanit Boss: Dump's End a Certainty," *Staten Island Advance*, April 6, 1999. Guy Molinari kept the city's feet to the fire, and Marchi and Connelly were keeping the pressure on the state government. They introduced a bill to limit the amount of refuse handled at any waste management facility in New York City. This approach was meant to limit diverting waste from Fresh Kills to vacant industrial land and other space near the city. See "A Man on a Mission," *Staten Island Advance*, January 24, 1999. "Return to Sender," *Staten Island Register*, February 9, 1999; Robin Eisner, "City Late on Landfill Order," *Staten Island Advance*, February 26, 1999; "News

from Senator John J. Marchi," February 2, 1999, Marchi Papers, Series 1, Subject Files, Box 47.

171. Congressional Research Service, *Interstate Shipment of Municipal Solid Waste: 1998 Update*, August 6, 1998, Giuliani Papers, Deputy Mayors Joseph Lhota and Jake Menges, Waste Export, 02/05/1/004, Folder 178.

172. Fax, Patrick Lally to Jake Menges, January 14, 1999, Giuliani Papers, Deputy Mayors Joseph Lhota and Jake Menges, Waste Export, 02/05/1/004, Folder 178. See also Patrick Lally to Jake Menges and Betsy Collins, April 5, 1999, in same file, on a summary of pending interstate waste bills.

173. Terence J. Kivlan, "Fossella Renews Vow to Fight Limit on Trash Shipments," *Staten Island Advance*, January 5, 1999.

174. "Virginia Gov. Wants to Put a Lid on Garbage Shipments from N.Y.," *Staten Island Advance*, January 13, 1999. See also Terry Pristin, "New York Trash Imports Have Virginia in a Tizzy," *New York Times*, January 13, 1999; R. H. Melton, "Gilmore Seeks Limit on Trash," *Washington Post*, January 14, 1999; Stephen Dinan and Jeremy Redmon, "Gilmore Hits Trash Barges," *Washington Times*, January 14, 1999.

175. Terence J. Kivlan and Reginald Patrick, "N.Y. Officials Dismiss Va. Call for Limits on Trash Imports," *State Island Advance*, January 14, 1999.

176. "Reciprocal Garbage," *New York Post*, January 15, 1999. See also "Trash-Talking," *Staten Island Advance*, January 15, 1999; Terence Kivlan, "New Push in War to End Waste Imports," *Staten Island Advance*, January 16, 1999.

177. "Enough with the Garbage, Mayor, You're Embarrassing the Rest of Us," *Westchester Journal News*, February 1, 1999.

178. Gersh Kuntzman, "Rudy's Trash Talk May Yield Heap of Trouble," *New York Post*, January 18, 1999. See also "Trash Talk: NYC Mayor Antagonizes Virginians," *Watertown Daily Times*, January 19, 1999; Janet Ward, "A Trash Odyssey," *American City & County*, May 1, 1999, http://americancityandcounty.com/mag/government_trash_odyssey.

179. Reginald Patrick, "Mayor: There Are Va. Officials Who Want City Garbage," *Staten Island Advance*, January 19, 1999.

180. Douglas Martin, "Talks Are Set on Processing of Garbage," *New York Times*, January 20, 1999. See also "Staten Island Pols Flee City's Sinking Garbage Barge," *City Limits*, March 1, 1999.

181. Dan Barry, "Giuliani Says Trash Remarks Got Twisted," *New York Times*, January 20, 1999. See also Ray McAllister, "Forget Talking Trash: Let's Take Some," *Richmond Times Dispatch*, January 25, 1999; Andrew Petkofsky, "Second Firm Brings Trash to Virginia," *Richmond Times Dispatch*, January 25, 1999; "Virginia's Proposed Barge Ban Could Hurt Environment, Feds Say," *Staten Island Advance*, January 26, 1999; Posturing and Pandering," *Staten Island Advance*, January 27, 1999; Larry O'Dell, "Some Virginia Residents Like Trash, Ask Officials to Keep It Coming," *Staten Island Advance*, February 2, 1999.

182. Terence Kivlan, "Va. Boosts Backer of Restrictions on Garbage Exports," *Staten Island Advance*, January 23, 1999. See also Terence Kivlan, "Giuliani's Trash Talk Making Enemies Fast in Landfill Country," *Staten Island Advance*, January 24, 1999.

183. Douglas Martin, "New York Moves on Plan to Ship Trash out of City," *New York Times*, January 21, 1999.

184. Paul H. B. Shin, "Court Clamps the Lid on Trash Firm's 'Shed,'" *New York Daily News*, January 22, 1999.

185. Andy Newman, "5 States Team Up to Fight Giuliani's Trash Proposal," *New York Times*, February 8, 1999. See also Dan Robrish, "Traveling Trash Gives Way to

Multi-State Inspections," *Staten Island Advance*, February 9, 1999; "Good Trash, Bad Trash?" *Staten Island Advance*, March 22, 1999.

186. "In Pennsylvania, Giuliani Draws Protest on Trash," *New York Times*, February 6, 1999.

187. Dominic Perella and Robin Eisner, "Trash Plan Takes a Bloody Hit," *Staten Island Advance*, February 12, 1999. See also Robin Eisner, "State Down on the Dumps," *Staten Island Advance*, February 28, 1999; Angela Logomasini, "Trashing New Yorkers, Trashing Solutions," *Staten Island Advance*, February 28, 1999; Robin Eisner, "A Port Ready for Trash That May Never Arrive," *Staten Island Advance*, March 1, 1999.

188. Terence J. Kivlan, "Meanwhile, the States Are Negotiating," *Staten Island Advance*, March 1, 1999.

189. "Door Closes on Garbage," *Rome Daily Sentinel*, March 3, 1999.

190. Robin Eisner, "80s Rulings Set Stage for Landfill Wars," *Staten Island Advance*, March 2, 1999. See also Robin Eisner, "Court and Congress to Hash Out Trash War," *Staten Island Advance*, March 3, 1999; Terence J. Kivlan and Robin Eisner, "Barge Export of NYC Trash Gains Support from Virginia Lawmaker," *Staten Island Advance*, March 9, 1999; Terence Kivlan, "It's Not All Trash Other States Oppose; It's NY's Trash," *Staten Island Advance*, March 21, 1999.

191. See David W. Chen, "Luster of New York's Trash Dims in Virginia," *New York Times*, March 11, 1999. See also "Voices of Reason," *Staten Island Advance*, March 12, 1999; Tom Deignan, "More Trash Talk—Virginia Grumbles; Will Upstate Take Our Garbage?" *Staten Island Register*, March 16, 1999.

192. Jennifer M. Nelson, "Pataki Welcomes Garbage Deal with Harrisburg, Pa.," *Staten Island Advance*, March 17, 1999. See also Judy L. Randall, "Harrisburg: Give Us Trash," *Staten Island Advance*, March 12, 1999; Terence J. Kivlan, "Bill Would Permit Governors to Limit Garbage Imports," *Staten Island Advance*, March 19, 1999; Richard Sisk, "Va. Sniffs at N.Y. Garbage," *New York Daily News*, April 4, 1999; Terence Kivlan, "Ominous Signs for the Future of Trash Exporting," *Staten Island Advance*, May 7, 1999; Terence J. Kivlan, "Schumer Reaffirms Opposition to Limits on N.Y. Garbage Exports," *Staten Island Advance*, January 22, 1999; "Lawmakers Begin Fight to Ban NYC Garbage," *Schenectady Daily Gazette*, April 23, 1999; Terence J. Kivlan, "Bill Would Limit Trash Shipments," *Staten Island Advance*, April 23, 1999.

193. City of New York, Department of Sanitation, Feasibility Report on MTS Conversion, April 1999, Vitaliano Papers, Series 1, Subject Files, Box 36, Folder 10.

194. Douglas Martin, "Sanitation Chief Seeking to Keep Transfer Stations in Service," *New York Times*, April 30, 1999. See also Robin Eisner, "Sanitation Dept. Pulls Surprise," *Staten Island Advance*, April 30, 1999.

195. It was prepared for the mayor's 1998 solid waste plan (Draft Plan Modification, *2001 and Beyond*). Most significantly the Plan EIS evaluated the feasibility and impact of alternatives to facilities/sites listed in the 1998 plan. This would include the potential for containerizing refuse at MTSs and the use of other facilities in the Bronx, Brooklyn, Manhattan, Staten Island, and Queens that might provide export by barge or rail. Out-of-city sites were not subject to Plan EIS review. Alternative sites included the eight MTSs and previously proposed sites not included among those in the 1998 plan (including the Port Ivory site on the Arthur Kill). See City of New York, Department of Sanitation, *Final Scoping Document for the Comprehensive Solid Waste Management Plan, Draft Modification, Draft Environmental Impact Statement*, CEQR no. 99DOS002Y, May 28, 1999, Archives & Special Collections, Department of the Library, College of Staten Island. See also Aaron William Comrov, "Fresh Kills

Dumped: A Policy Assessment for the Management of New York City's Residential Solid Waste in the Twenty-First Century," unpublished MS thesis in environmental policy studies, New Jersey Institute of Technology, May 2003, 37–38.
196. Statement at Scoping Hearing of New York Department of Sanitation Concerning the Environmental Impact Statement for the Solid Waste Management Plan, April 13, 1999, Marchi Papers, Series 1, Subject Files, Box 8, Folder 8. See also Robin Eisner, "A New Attack on Dump Closure," *Staten Island Advance*, April 13, 1999; Robin Eisner, "Public Cries Foul Over Fresh Kills," *Staten Island Advance*, April 14, 1999.
197. Robin Eisner, "Mayor's Garbage Plan Trashed," *Staten Island Advance*, April 7, 1999. See also Reginald Patrick, "Brooklyn BP Blasts City Plan to Close Fresh Kills," *Staten Island Advance*, February 11, 1999; "NYC's Trash Squeeze: INFORM Pushes for Waste Prevention," *INFORM Reports* 19 (Spring 1999): 1, 3.
198. Reginald Patrick, "A Nice Place in the Country, for Garbage," *Staten Island Advance*, March 11, 1999.
199. Martin and Revkin, "The Last Landfill: A Special Report."
200. Patrick, "A Nice Place in the Country, for Garbage." See also Nelson, "Pataki Welcomes Garbage Deal with Harrisburg, Pa."
201. Carolyn Thompson, "Upstate Sites Being Eyed for New York City Trash," *Troy Record*, April 8, 1999.
202. David Kibbe, "Newburgh Eyed for NYC Garbage Transfer Spot," *Middleton Times Herald Record*, April 2, 1999. See also April Hunt and David Kibbe, "City Balks at Trash Pact," *Middletown Times Herald Record*, April 7, 1999; David Kibbe, "Men Who Would Move Trash," *Middletown Times Herald Dispatch*, April 10, 1999; Jennifer M. Nelson, "Upstate Newburgh Lining Up for City Trash?" *Staten Island Advance*, April 19, 1999; Robert Riemann, "The Lure of the Garbage Barge," *Empire State Report*, July 1999.
203. Lisa Rein, "City Trash Heads North," *New York Daily News*, April 6, 1999. See also Manley J. Anderson, "County Cool to Accepting Waste," *Jamestown Post Journal*, April 8, 1999; "NYC's Garbage Could Seep Upstate; Ava Site Not a Possibility," *Utica Observer Dispatch*, April 8, 1999; Keith Benman, "Report: Many Upstate Landfills Have Room," *Rome Daily Sentinel*, April 8, 1999; "Tell New York City We Don't Want Its Trash," *Troy Record*, April 10, 1999; Wayne A. Hall, "NYC Seeks Upstate Outlet for Garbage," *Middletown Times Herald Record*, April 10, 1999; Michael Hudson, "Critics: Trash Idea Stinks," *Niagara Gazette*, April 11, 1999; April Hunt, "Could NYC Trash Come Here?" *Middletown Times Herald Dispatch*, May 3, 1999; Diane O'Donnell, "Proposed Trash Transfer Station Raises Concerns," *Staten Island Advance*, August 12, 1999.
204. Ed Koch, "Political Stink from Rudy's Trash Policy," *New York Daily News*, June 25, 1999. See also "Interstate Transportation of Municipal Solid Waste," Hearings Before the Committee on Environment and Public Works, U.S. Senate, 106th Congress, 1st sess., June 17, 1999.
205. See Douglas Martin, "City's Last Waste Incinerator Is Torn Down," *New York Times*, May 6, 1999.
206. "Closure: Two Big Wins," *Staten Island Advance*, July 12, 1999.
207. Robin Eisner, "Landfill Closure Facing Crisis," *Staten Island Advance*, June 27, 1999.
208. "News from Senator John J. Marchi," August 30, 1999, Marchi Papers, Series 1, Subject Files, Box 8, Folder 7. See also Robin Eisner, "The Road to Closure Is Still Bumpy," *Staten Island Advance*, August 30, 1999; Robin Eisner, "Close Dump on Time—or Else," *Staten Island Advance*, August 31, 1999; Robin Eisner, "Threat Stirs a Chorus

of Promises on Closing Dump," *Staten Island Advance*, September 1, 1999; "Closure and the Law," *Staten Island Advance*, September 1, 1999; Robin Eisner, "Mark Green: Close Landfill on Time," *Staten Island Advance*, September 2, 1999; "Dump-Closing Plans Seem to Stink," *Staten Island Register*, September 7, 1999.

209. Kim O'Connell, "The End of an Era," *Waste Age* 30 (December 1999): 58.

210. Reginald Patrick and Brian Damiano, "Sanit Set to Expand Recycling," *Staten Island Advance*, June 15, 1999; Robin Eisner, "We Recycle More Weekly," *Staten Island Advance*, August 30, 1999. See also "Good-Sense Recycling," *Staten Island Advance*, August 31, 1999; Reginald Patrick, "Is Commercial Waste Being Recycled?" *Staten Island Advance*, October 3, 1999; William Rathje, "Stop the Garbage," *Los Angeles Times*, February 28, 1999.

211. Tom Deignan, "Is City Rethinking Its Plan for Trash?" *New York Times*, June 8, 1999; Reginald Patrick and Robin Eisner, "New Garbage Truck Could Eliminate Transfer Stations," *Staten Island Advance*, October 20, 1999; "An Idea Worth Study," *Staten Island Advance*, October 21, 1999; Open Meeting, Economic Development/Transportation Committee, Staten Island Chamber of Commerce, October 8, 1999, Marchi Papers, Series 1, Subject Files, Box 8; Robin Eisner, "Sanitation Open to Trash-Export Ideas," *Staten Island Advance*, May 25, 1999.

212. Robin Eisner and Terence J. Kivlan, "Virginia Must Let Trash In," *Staten Island Advance*, June 30, 1999; Justin Pope, "Bill to Force City Into Trash Suit Is Blocked," *Associated Press*, August 3, 1999; Diana Yates, "Judge KOs Law That Limited Flow of Garbage to Va.," *Staten Island Advance*, February 4, 2000.

213. Robin Eisner, "Brooklyn: You Can't Dump Here!" *Staten Island Advance*, July 9, 1999.

214. Robin Eisner, "N.J. Cool on Trash, but Not Bayonne," *Staten Island Advance*, July 31, 1999. See also Robin Eisner, "Island Trash Headed Out of Town," *Staten Island Advance*, September 27, 1999; Robin Eisner, "N.J. County Joins Attack on Trash Flow," *Staten Island Advance*, November 9, 1999.

215. "We Don't Want Your Garbage," *Staten Island Advance*, November 15, 1999.

216. Robin Eisner, "We'll Carry Trash War to Jersey Malls," *Staten Island Advance*, November 16, 1999.

217. Robin Eisner, "City of Elizabeth Takes City Trash Case to Court," *Staten Island Advance*, November 17, 1999; Robin Eisner, "Sanitation Can Continue to Take Trash to Jersey—for Now," *Staten Island Advance*, November 20, 1999; "Across State Lines," *Staten Island Advance*, November 17, 1999.

218. Glenn Collins, "In Trash Fight, Aiming Pebbles at Goliath," *New York Times*, November 23, 1999.

219. Terence J. Kivlan, "Mayor, Governor Rule Out Testifying on Trash Bills," *Staten Island Advance*, June 3, 1999; Terence J. Kivlan, "State Seeks Legislation to Limit Trash Imports," *Staten Island Advance*, June 18, 1999. See also Statement of Robert Eisenbud, Director of Legislative Affairs, Waste Management Before the Committee on Environment and Public Works, U.S. Senate, June 17, 1999, http://www.epw.senate.gov/107th/eis_6-17.htm; "Cooler Heads," *Staten Island Advance*, June 21, 1999.

220. Terence J. Kivlan, "Bill Threatens Landfill Closing Plan," *Staten Island Advance*, July 10, 1999.

221. "Landfill Update . . . Your Chamber Goes to Washington," *Staten Island Chamber of Commerce Newsletter*, September 1999.

222. Terence J. Kivlan, "Congressman Says Virginia's Anti-Trash Export Bill Is Dead," *Staten Island Advance*, August 2, 1999. See also Terence Kivlan, "Senate Unlikely to

17. NOW WHAT?   ■   723

Approve Limits on Trash Exports," *Staten Island Advance*, January 3, 2000; Terence Kivlan, "A New Assault on Trash-Export Plan," *Staten Island Advance*, February 27, 2000; Terence J. Kivlan, "Trash Battle Continues," *Staten Island Advance*, April 29, 2000; "Trash Negotiations Come to a Standstill," *Associated Press*, April 30, 2000; John T. Aquino, "Waste Rolls on Capitol Hill," *Waste Age*, December 1, 2000, http://www.waste360.com/mag/waste_waste_rolls_capitol.
223. Sze, *Noxious New York*, 113–16.
224. Anne Raver, "Up from the Ruins, Red Hook Faces a Dump," *New York Times*, October 1, 1998.
225. Paul Zielbauer, "Garbage Transfer Stations Face Civil Rights Inquiry," *New York Times*, March 7, 1999; Congressman Jose E. Serrano and Melissa Iachan, "Seeking Environmental Justice in the South Bronx," *New York Slant*, October 26, 2016, http://nyslant.com/article/opinion/seeking-environmental-justice-in-the-south-bronx.html. See also Robin Eisner, "Trash/Rights Probe on Tap for the Fall," *Staten Island Advance*, March 9, 1999; *Inner City Press Environmental Justice Reporter*, 1999 archive, [some of] March 15–December 31, 1999, http://www.innercitypress.org/ejrep99.html; Juliana Maantay, "Asthma and Air Pollution in the Bronx: Methodological and Data Considerations in Using GIS for Environmental Justice and Health Research," *Health & Place* 13 (2007): 32–56.
226. Zielbauer, "Garbage Transfer Stations Face Civil Rights Inquiry." See also Dolores Greenberg, "Reconstructing Race and Protest: Environmental Justice in New York City," *Environmental History* 5 (April 2000): 242.
227. Robin Eisner, "Transfer Stations Bring Feds to Town," *Staten Island Advance*, March 6, 1999.
228. Zielbauer, "Garbage Transfer Stations Face Civil Rights Inquiry." See also Paul H. B. Shin, "Waste Probe Launched Feds Eye Civil Rights," *New York Daily News*, March 9, 1999; Barbara Stewart, "Bronx Loudly Opposes Waste Station Plan," *New York Times*, March 9, 2000.
229. Robin Eisner, "High-Level Conference Looks to Clean Air," *Staten Island Advance*, March 7, 1999. See also "In Pursuit of Justice," *Staten Island Advance*, March 8, 1999.
230. Terrence J. Kivlan, "Violate Rights? Suit Asks: What About Island's?" *Staten Island Advance*, March 10, 1999.
231. Terence J. Kivlan, "Fosella: Where Has EPA Been for 50 Years?" *Staten Island Advance*, March 11, 1999. See also "Want to Talk Burdens?" *Staten Island Advance*, March 11, 1999.
232. Terence J. Kivlan, "Fossella to Gore: Come See the Island's Environmental Injustice," *Staten Island Advance*, March 12, 1999; Terence J. Kivlan, "White House OKs Dump Hearing," *Staten Island Advance*, March 16, 1999; Terence J. Kivlan, "Fossella: White House Backs Off Fresh Kills Probe," *Staten Island Advance*, June 11, 1999; Terence J. Kivlan, "White House Official Says Hearing on Fresh Kills Was Never Promised," *Staten Island Advance*, June 12, 1999; Terence J. Kivlan, "Feds Promise Not to Slow Landfill's End," *Staten Island Advance*, June 16, 1999. See also Rep. Vito J. Fossella, "S.I.'s Landfill Is the Real Injustice," *New York Daily News*, April 16, 1999; "And Justice for All," *Staten Island Advance*, June 20, 1999.
233. Terence J. Kivlan, "White House Heavy to Tour Dump," *Staten Island Advance*, June 25, 1999. See also "Seeing Is Believing," *Staten Island Advance*, June 28, 1999; Robin Eisner, "Landfill Ripe for Federal Sniffers on Guided Tour," *Staten Island Advance*, June 30, 1999.

234. "Williamsburg/Greenpoint (Brooklyn) Case Study," n.d., online, http://faculty.virginia.edu/ejus/Green.htm. See also Terence Kivlan, "Fossella Reapplies for Federal Probe of Fresh Kills," *Staten Island Advance*, January 12, 2000.
235. Terence J. Kivlan, "EPA to Fossella: Landfill Complaint May Be Dismissed," *Staten Island Advance*, January 6, 2000.
236. Blake Eskin, "The End of the Line for Fresh Kills," *Metropolis* 16 (September 2996): 20.
237. See "Environmental Justice Should Include Justice for All," *Congressional Record-House* 145 (June 15, 1999), part 9, 12912–13; "Environmental Injustice," *Congressional Record-House* 145 (1999), part 4, 4895.
238. "Message from the Mayor," in New York City, Department of Sanitation, *Annual Report 1999*.
239. New York City, Department of Sanitation, *The DOS Report: Closing the Fresh Kills Landfill*, February 2000, 1. See also Reginald Patrick, "Landfill Area Near Mall Is Scheduled to Close," *Staten Island Advance*, April 18, 1999.
240. In addition to the plan modification, a draft environmental-impact statement was included—requirement by city and state law. See Martha K. Hirst to Alessandra Sumowicz, Director, Mayor's Office of Environmental Coordination, October 26, 2000, Giuliani Collection, Deputy Mayors, Joseph Lhota, Sanitation—Solid Waste Management Plan, Box 02/05/003, Folder, 87.
241. Terence J. Kivlan, "City's New Plan to Close Dump Faces Obstacles on Several Fronts," *Staten Island Advance*, May 10, 2000. See also "Upstate Town Struggling with New Landfill Proposal," *Associated Press*, May 5, 2000; Terence J. Kivlan, "State's Trash Exports to Pile Up," *Staten Island Advance*, February 21, 2000; Adam Chamberlain, "Fresh Kills," *Solid Waste & Recycling* 5 (May 2000): 38; "New York State Sues Big Apple Over Trash Export," *Waste Age* 31 (April 2000): 4; "As Fresh Kills Closure Approaches, Ohio Could Receive More Trash," *Associated Press*, July 17, 2000.
242. Eric Lipton, "New Trash Plan Gains Backing, but Problems Remain," *New York Times*, May 7, 2000. See also Ann M. Gynn, "NYC Mayor Unveils Trash Disposal Plan," *Waste News* (May 8, 2000): 1; Susanna Duff, "NYC's Budget Grows with Criticism; Mayor Wants to Add $134 Million for Sanitation," *Waste News* (May 8, 2000): 4; Reginald Patrick, "Council Receptive to Mayor's Trash Plan," *Staten Island Advance*, May 2, 2000; Barbara Stewart, "Mayor Drops Unpopular Plans for Trash Hubs," *New York Times*, May 1, 2000; Diana Yates, "New Trash Plans Stinks to Some Waste Firms," *Staten Island Advance*, May 1, 2000; Diana Yates, "City Divulges Some of Its Trash Disposal Plan," *Staten Island Advance*, April 30, 2000; Ann M. Gynn, "36 Bids; NYC Agency Views Hauler's Proposals," *Waste News* (February 21, 2000): 1.
243. Diana Yates, "City Panel Takes Heat on Garbage Plan," *Staten Island Advance*, November 20, 2000; Diana Yates, "City Council Approves Plan for Replacing Fresh Kills," *Staten Island Advance*, November 30, 2000; "At the Far Turn," *Staten Island Advance*, December 5, 2000; Martin and Revkin, "The Last Landfill: A Special Report." See also Waste Management, Inc., *After Fresh Kills: Highlights of Waste Management Inc.'s Plan to Dispose of New York City's Residential Waste*, n.d., Giuliani Collection, Deputy Mayors, Joseph Lhota, Sanitation—Solid Waste Management Plan, Box 02/05/003, Folder 87.
244. "City Proposes Garbage Export Plan," City of New York, Independent Budget Office, *Inside the Budget* 65 (July 17, 2000): 1–2, Marchi Papers, Series 1, Subject Files, Box 8.
245. New York City, Department of Sanitation, *NYC Recycles*, 17, 20, 25, 29, 32, appendix 1. See also Larry Cipollina to Executive Committee, Levels I & II, Borough

Superintendents, September 22, 2000, Giuliani Collection, Deputy Mayors, Joseph Lhota, Sanitation—Solid Waste Management Plan, Box 02/05/003, Folder 87.

246. New York City, Department of Sanitation, *NYC Recycles*, appendix 1.

247. New York City, Department of Sanitation, *NYC Recycles: More Than a Decade of Outreach Activities by the NYC Department of Sanitation, FY 1986–1999* (Fall 1999), 3.

248. Controversy surrounded the decision to use Linden because of concerns over possible corruption. Mayor John T. Gregorio was the father-in-law of the owner of the Linden site. A local board of the state's Department of Consumer Affairs fined the mayor for using undue influence in securing the land. BFI would construct the EBUF, but not without experiencing legal disputes. See Sze, *Noxious New York*, 227; "Linden Trash Deal Faces Court Battle," *Newark Star-Ledger*, December 26, 2000; "Waste Firm in Squabble Over Linden Transfer Site," *Staten Island Advance*, December 27, 2000.

249. Lipton, "New Trash Plan Gains Backing, but Problems Remain." See also "NYC Approves Linden Garbage Route," *Newark Star-Ledger*, November 30, 2000; "County Agrees to Build Transfer Station," *Newark Star-Ledger*, December 1, 2000; Sze, *Noxious New York*, 132; Robin Eisner, "Deal with N.J. Could Speed Dump Closure," *Staten Island Advance*, January 30, 2000; Tracey Tully, "New Fresh Kills Plan Smells Better to Foes," *New York Daily News*, April 30, 2000.

250. Barbara Stewart, "Seeing Garbage, Smelling Money; but Plans for Transfer Station Annoy New Jersey Neighbors," *New York Times*, May 25, 2000.

251. See Maria Newman, "Like the Turnpike, Trash Deal Divides Linden Into Two Cities," *New York Times*, December 4, 2000; "The Battle of Tremley Point Neighbors of Linden Industry Fighting Disposal Plant Plans," *Newark Star-Ledger*, December 19, 2000.

252. "County Approves Garbage Station in Linden," *New York Times*, December 1, 2000. See also "Linden Remains Opposed to Waste Incinerator Idea," *Newark Star-Ledger*, December 1, 2000; "Local News Briefs," *Newark Star-Ledger*, December 5, 2000.

253. Stewart, "Seeing Garbage, Smelling Money." See also "Metuchen Sues Over Linden Trash Station," January 3, 2000; "Edison Joins Fight Over Trash Trains," *Newark Star-Ledger*, January 25, 2001; "NOT SO FAST," *Trentonian*, January 23, 2001. The uneasy relationship between New York and New Jersey over refuse exports did not fade easily. In an exchange between Whitman and U.S. Senator Frank Lautenberg in January 2000 over importation of New York City's refuse into his state, the New Jersey Democrat stated, "Our state has had a long and unfortunate history with New York City trash, and I do not want to see that history repeated." See Frank Lautenberg to Christine Todd Whitman, January 6, 2000, and Christine Todd Whitman to Frank Lautenberg, January 27, 2000, DEP Commissioner Robert C. Shinn Jr. Subject Files, Solid Waste Disposal: Misc. 2000, Box 195, New Jersey State Archives, Trenton, New Jersey. DEP Commissioner Shin also wrote to Commissioner Cahill about his concerns over air pollution from New York City garbage trucks traveling to New Jersey. See Robert C. Shinn Jr. to John P. Cahill, November 28, 2000, also in Box 195.

254. A community board in Staten Island approved the planning of a transfer station to be built in Travis to handle island garbage. "Panel OKs Plan to Handle Island Garbage," *Staten Island Advance*, December 6, 2000.

255. Diana Yates, "City's Garbage Plan Gets Airing at Council Hearing," *Staten Island Advance*, May 23, 2000; Peter L. Grogan, "Return to Sender," *BioCycle* 41 (January 2000): 75. See also Diana Yates, "Coalition Proposing Recycling Center at Fresh Kills," *Staten Island Advance*, June 1, 2000; "The Beginning of the End," *Staten Island Advance*, May 14, 2000.

256. Peter F. Vallone and Stanley E. Michaels to Council Members, October 24, 2000, Giuliani Collection, Deputy Mayors, Joseph Lhota, Sanitation—Solid Waste Management Plan, Box 02/05/003, Folder 87. The Giuliani administration continued to grapple with commercial waste and monitoring the trade-waste problems. See "Commercial Waste," November 20, 2000, and Confidential, Draft for Discussion Only, November 17, 2000, A Local Law to amend the administrative code of the city of New York, in relation to requiring a comprehensive study of the commercial waste management system within New York City, Giuliani Collection, Deputy Mayors, Joseph Lhota, Sanitation—Solid Waste Management Plan, Box 02/05/003, Folder 87. See also Frank DiTommaso to Edward T. Ferguson, Trade Waste Commission, Giuliani Collection, Deputy Mayors, Rudy Washington, Trade Waste, Box 02/08/021, Folder 645.
257. Martha K. Hirst to Josh Filler, October 25, 2000, Giuliani Collection, Deputy Mayors, Joseph Lhota, Sanitation—Solid Waste Management Plan, Box 02/05/003, Folder 87.
258. Program Planners, Inc., "Overview of Current Waste Disposal Plans of the City of New York, Department of Sanitation," n.d., Giuliani Collection, Deputy Mayors, Joseph Lhota, Sanitation—Solid Waste Management Plan, Box 02/05/003, Folder 87. See also a variety of attachments to the overview dealing with the position of Local 831 IBT, the private carting industry, monopoly in the solid waste disposal industry, and more, plus a group of news clips. See also Reginald Patrick, "City: Firm's Ouster Won't Have Effect on Landfill Closing," *Staten Island Advance*, September 6, 2000, concerning the alleged mob ties of a Staten Island construction company.
259. Diana Yates, "Critics Object to Cost of Landfill Closure Plan," *Staten Island Advance*, December 4, 2000.
260. Diana Yates, "City Solid Waste Plan Should Be Junked, Critics Say," *Staten Island Advance*, November 15, 2000.
261. Barbara Warren, *Taking Out the Trash: A New Direction for New York City's Waste* (Organization of Waterfront Neighborhoods and Consumer Policy Institute/Consumers Union, May 31, 2000), i.
262. "Dump Closing—Again," *Staten Island Advance*, May 23, 2000. See also Joe Truini, "Officials Promise Closure: Pataki, Giuliani Say Fresh Kills Landfill Will Close on Time," *Waste News* (May 22, 2000): 3; Robert Gavin, "Another Nail in Dump's Coffin," *Staten Island Advance*, May 16, 2000.
263. John J. Marchi to George E. Pataki, May 16, 2000, Marchi Papers, Series 1, Subject Files, Box 8.
264. Gavin, "Another Nail in Dump's Coffin." See also Eric Lipton, "Effort to Close Fresh Kills Are Taking Unforeseen Tolls," *New York Times*, February 21, 2000; Diana Yates, "Residents Express Concerns to Sanitation Dept. Chief," *Staten Island Advance*, February 9, 2000.
265. "Tonnage and Timetables," *Staten Island Advance*, March 16, 2000. See also Diana Yates, "Dump Closure Near? Could Be," *Staten Island Advance*, March 10, 2000.
266. Diana Yates, "Sanit Chief: Landfill Waste at 5,000 Tons Daily," *Staten Island Advance*, February 18, 2000.
267. "Manhattan: The New Garbage Barge," *New York Observer*, February 28, 2000. See also Brian Lewis, "Exporting Trash Is No Long-Term Solution, Commission Says," *Staten Island Advance*, August 25, 1998; "Is NYC's 'Every Borough for Itself' Post-Fresh Kills Waste Plan Feasible?" *Waste News*, August 14, 2000.
268. Gersh Kuntzman, "S.I.'s Perfectly Good Dump to Waste," *New York Post*, December 1, 1999.

18. 9/11

1. Jameson W. Doig, Kenneth T. Jackson, and Lisa Keller, "September 11," in *The Encyclopedia of New York City*, 2nd ed., ed. Kenneth T. Jackson (New Haven, CT: Yale University Press, 2010), 1168; National Commission on Terrorist Attacks Upon the United States, *The 9/11 Report* (New York: St. Martin's Press, 2004).
2. "Oral Testimony from Survivors of the World Trade Center," in *The 9/11 Encyclopedia*, ed. Stephen E. Atkins (Westport, CT: Praeger Security International, 2008), 2:367.
3. "9/11 by the Numbers," *New York Magazine*, http://nymag.com/news/articles/wtc/1year/numbers.htm; "Pentagon Attack," in *The 9/11 Encyclopedia*, ed. Atkins, 1:55–58, 233–35; Doig, Jackson, and Keller, "September 11," 1169.
4. "9/11 by the Numbers." See also Don DeLillo, *Falling Man* (New York: Scribner, 2007).
5. Heidi Singer, "A Borough's Rebirth in a Barge's Wake," *Staten Island Advance*, March 23, 2001.
6. Kirk Johnson, "After 53 Years, Fresh Kills Gets Its Final Load of Trash," *New York Times*, March 23, 2001. See also Diana Yates, "Winding Down to Closure," *Staten Island Advance*, March 23, 2001.
7. Singer, "A Borough's Rebirth in a Barge's Wake."
8. Diana Yates, "The Dump: RIP April 16, 1948–March 22, 2001," *Staten Island Advance*, March 22, 2001.
9. "Mayor Giuliani, Governor Pataki, and Borough President Molinari Commemorate Arrival of Last Garbage Barge at Fresh Kills Landfill, Staten Island," press release, March 22, 2001, Archives of the Mayor's Press Office, http://www.nyc.gov/html/om/html/2001a/pr091-01.html. See also Robert Gavin, "Pick to Head DEC Says Closing Landfill Tops Island Agenda," *Staten Island Advance*, March 6, 2001; Reginald Patrick, "Molinari's Crowning Moment," *Staten Island Advance*, March 25, 2001.
10. Kirk Johnson, "Dumping Ends at Fresh Kills, Symbol of Throw-Away Era," *New York Times*, March 18, 2001.
11. Johnson, "Dumping Ends at Fresh Kills." See also Stephanie Slepian, "Dead and Buried," *Staten Island Advance*, March 22, 2001; "Fresh Kills Finale," *Staten Island Advance*, March 22, 2001. See Roy F. Weston of New York, Inc., *Operations and Maintenance Plan: Fresh Kills Landfill*, vol. A, *Landfill Operations* (January 30, 2001); City of New York, Department of Sanitation, *2001 Operations and Maintenance Plan/Contingency Plan for the Fresh Kills Landfill* (January 30, 2001); Roy F. Weston of New York, Inc., *Contingency Plan* (January 30, 2001); City of New York, Department of Sanitation, *The Fresh Kills Landfill, Section 6/7, Final Cover Design Report*, vol. 1, *Report Text* (January 2001), Archives & Special Collections, Department of the Library, College of Staten Island, CUNY.
12. Lisa L. Colangelo, "Fresh Kills Is Closing Today: City's Trash Will Go to N.J., Pa., and Va.," *New York Daily News*, March 22, 2001. See also Mayor Rudy Giuliani, "New York City's Landfill Closes Ahead of Schedule," Archives of the Mayor's Weekly Column, Archives of Rudolph W. Giuliani, n.d., http://www1.nyc.gov/html/rwg/html/2001a/weekly/wkly0326.html. See also City of New York, Independent Budget Office, *Inside the Budget* 77 (February 5, 2001): 2; "Judge Refuses to Block Lease on Site for New York Trash," *New York Times*, January 13, 2001; Ann M. Gynn, "NYC Trash Bill to Top $1 Billion," *Waste News*, March 19, 2001, 3.
13. Benjamin Miller, "The Folly of Closing Fresh Kills," *New York Daily News*, January 5, 2001.

14. See "Fresh Kills Forever?" *Staten Island Advance*, January 9, 2001.
15. See Heidi Singer, "Noses Won't Notice Dump's Demise," *Staten Island Advance*, March 21, 2001; Stephanie Slepian, "Dump's Neighbors Breathing Easier," *Staten Island Advance*, March 21, 2001; Karen O'Shea and Jill Gardiner, "Downwind of Dump, Windfalls Emerge," *Staten Island Advance*, March 26, 2001; Reginald Patrick, "A Fresh Start," *Staten Island Advance*, March 25, 2001; Diana Yates, "Eerie Quiet Settles Over Fresh Kills," *Staten Island Advance*, April 4, 2001; Karen O'Shea and Jill Gardiner, "Downwind of the Dump, Windfalls Emerge—Closure Has Homebuyers Near the Landfill Smelling Potential Profits," *Staten Island Advance*, March 26, 2001.
16. "A New Staten Island," *Staten Island Advance*, March 26, 2001.
17. "Knocking Off Ahead of Time," *Waste News*, January 15, 2001, 8. See also Jim Johnson, "Fresh Kills Plans Early Closure," *Waste News*, January 15, 2001, 3; Robert Gavin, "Mayor: City Awaits $75M from State to Close Landfill," *Staten Island Advance*, January 23, 2001; Judy L. Randall, "Dump Is Down to Final Days," *Staten Island Advance*, February 4, 2001; Diana Yates, "Goodbye Dump," *Staten Island Advance*, March 14, 2001.
18. Eric Lipton, "City Begins Its Last Phase in Closing Out Fresh Kills Site," *New York Times*, February 5, 2001. See also Reginald Patrick, "Closing of Landfill May Mean Garbage Truck Parade in Travis," *Staten Island Advance*, March 30, 2001.
19. Lisa L. Colangelo, "Fresh Kills Is Closing Today." See also City of New York, Independent Budget Office, "Closing Fresh Kills Means Mounting Costs to Dispose of New York City's Garbage," *Inside the Budget* 77 (February 5, 2001): 1–3; Diana Yates, "Sanitation Budget Rises with Dump Closing," *Staten Island Advance*, February 7, 2001; Press Release, "Mayor Giuliani and Sanitation Commissioner Farrell Announce Expanded Recycling Program in City Schools," February 21, 2001, Archives of the Mayor's Press Office, http://home.nyc.gov/html/om/html/2001a/pr051-01.html; "Fresh Kills Landfill Closes," *BioCycle* 42 (April 2001): 22–23; "City Bids Fresh Kills Farewell," *Bronx Beat*, March 26–April 2, 2001; Jim Johnson, "New York City 'Nightmare' Ends," *Waste News*, March 26, 2001, 1, 35; New York City, Department of Sanitation, *New York City Recycling—In Context: A Comprehensive Analysis of Recycling in Major U.S. Cities* (August 2001). Ironically, Staten Islanders would be the closest New Yorkers to the city's waste when it was deposited at Linden. See Chris Franz, "Largest Transfer Station to Replace Largest Landfill," *Staten Island Register*, January 23, 2001.
20. Kim A. O'Connell, "The Manhattan Transfer," *Waste Age* 32 (March 1, 2001): 182.
21. "Fresh Thinking After Fresh Kills," *New York Daily News*, March 24, 2001; Eric Lipton, "Trash Transfer to New Jersey Is Postponed," *New York Times*, September 19, 2001; Farnaz Fassihi, "State Delays Linden's Garbage Facility Plans," *Newark Star-Ledger*, September 19, 2001. See also Tom Haydon, "Carteret's Trash-Transfer Records Subpoenaed—State Investigating Company's Current Plan to Build State for N.Y. Garbage in Linden," *Newark Star-Ledger*, March 3, 2001.
22. Terence Kivlan, "Yes, Virginia, There Will Be Trash Shipments," *Staten Island Advance*, February 11, 2001; Susanna Duff, "Pa. Bills Restrict Interstate Waste," *Waste News*, March 26, 2001, 1; "A Contested Inheritance," *Waste News*, March 26, 2001, 8; Eric Lipton, "City Trash Follows Long and Winding Road," *New York Times*, March 24, 2001; Terence J. Kivlan, "No Effect on Trash Exports Seen in Dems' Senate Control," *Staten Island Advance*, June 6, 2001; Terence J. Kivlan, "Neighboring States Are Fighting Mad Over Our Trash," *Staten Island Advance*, July 8, 2001; "Solid Waste Shipments," *Federal Document Clearing House Congressional Testimony*,

August 1, 2001; Terence Kivlan, "Fresh Kills' Garbage Pretty Sweet Deal for Other States," *Staten Island Advance*, August 19, 2001; Bill Franz, "Stage Set for Battle Over City's Trash in N.J.," *Staten Island Register*, August 21, 2001.

23. Susanna Duff, "Interstate Waste Keeps Crossing the Lines," *Waste News*, August 6, 2001, 4.
24. Carolyn Rushefsky, "Landscape Architect Has a Vision for Landfill," *Staten Island Advance*, March 8, 1993; Kerry Diamond, "Envisioning a Second Life for the Fresh Kills Landfill," *Staten Island Advance*, March 12, 1993.
25. Joseph B. Rose and Michael T. Carpinello to Joseph J. Lhota and Randy L. Levine, December 30, 1998, Rudolph W. Giuliani Collection, Deputy Mayors, Randy Mastro, Project Files, 1998, Sanitation—Fresh Kills, Box 02/04/027, Folder, 667, La Guardia and Wagner Archives, LaGuardia Community College/CUNY, Long Island City, New York.
26. Joseph B. Rose and Michael T. Carpinello to Joseph J. Lhota and Randy L. Levine, December 30, 1998.
27. Kim A. O'Connell, "A Landfill's Second Career Proves Successful," *Waste Age* 30 (December 1, 1999): 6, 8.
28. "Getting It Done," *State Island Advance*, March 26, 2000. See also "Infamous Fresh Kills Landfill Will Soon Be Nothing but History," *Staten Island Advance*, April 30, 2000.
29. Jennifer Steinhauer, "Fresh Kills, Chapter 2; Sure It's Just a Dump Now, but Think About the Possibilities," *New York Times*, August 13, 2000.
30. Jim Johnson, "NYC Develops Competition for Fresh Kills Site," *Waste News*, November 13, 2000, 11. See also New York City, Department of Sanitation, *DOS Report: Closing the Fresh Kills Landfill*, February 2000, 5.
31. Alysha Sideman, "From Landfill to Landscape," *Staten Island Advance*, March 9, 2001; Diana Yates, "City Picks Post-Dump Design Judge," *Staten Island Advance*, March 13, 2001. See also O'Shea and Gardiner, "Downwind of Dump, Windfalls Emerge"; "A Fresh Start at Fresh Kills: Landfill to Landscape," *Municipal Art Society Newsletter* (May/June 2001): 1.
32. Alysha Sideman, "Officials Ponder New Uses for an Old Landfill," *Staten Island Advance*, March 30, 2001; Alysha Sideman, "The Landfill: Where Should Future Begin?" *Staten Island Advance*, May 18, 2001.
33. Alysha Sideman, "Forum on Landfill Is Scheduled," *Staten Island Advance*, May 24, 2001. See also Alysha Sideman, "Past and Future of Fresh Kills Are Discussed," *Staten Island Advance*, May 31, 2001; Diana Yates, "Students Celebrate 'Regreening' of Fresh Kills," *Staten Island Advance*, June 13, 2001.
34. The Port Authority of New York and New Jersey (PANYNJ) constructed and also operated the complex.
35. "American Society for Civil Engineers Report," 1:18–19, and "World Trade Center," 1:303–4, in *The 9/11 Encyclopedia*, ed. Atkins. See also "Testimony of Dr. W. Gene Corley on Behalf of the American Society of Civil Engineers Before the Subcommittee on Environment, Technology and Standards and Subcommittee on Research of the U.S. House of Representatives Committee on Science (May 1, 2000)," in *The 9/11 Encyclopedia*, ed. Atkins, 2:428–34.
36. William Langewiesche, *American Ground: Unbuilding the World Trade Center* (New York: North Point, 2002), 4. See also "Context of May 30, 2002: Ground Zero Cleanup Operation Officially Ends," *History Commons*, http://www.historycommons.org/context.jsp?item=a053002cleanupends.

37. Kirk Johnson, "After the Attacks: The Cleanup; Challenges and Dangers in Disposing of Two Fallen Giants," *New York Times*, September 13, 2001.
38. C. J. Chivers, "12 Days at Ground Zero," *Esquire*, September 11, 2016, http://www.esquire.com/news-politics/a628/cj-chivers-ground-zero/.
39. Doig, Jackson, and Keller, "September 11," 1169; Langewiesche, *American Ground*, 7–9.
40. Teresa Carpenter, ed., *New York Diaries, 1609 to 2009* (New York: Modern Library, 2012), 296–97.
41. Jim Johnson, "Debris Gone; Memories Remain; Sept. 1 Buried Deep in Staten Island Landfill," *Waste News*, September 2, 2002.
42. "Cleanup Operations at Ground Zero," in *The 9/11 Encyclopedia*, ed. Atkins, 1:68–69; "Cleaning Debris at Ground Zero," *twinzero.net*, 2008, http://www.twinzero.net/overview-regarding-the-process-of-cleaning.html.
43. Kirk Johnson and Eric Lipton, "After the Attacks: The Canyons; First Inspections Show Most Buildings Are Structurally Sound," *New York Times*, September 14, 2001. See also Dennis Overbye, "A Nation Challenged; Engineers Tackle Havoc Underground," *New York Times*, September 18, 2001; Barton Gellman and Laura Blumenfeld, "Moving, Sifting 16 Acres of Rubble," *Washington Post*, September 13, 2011.
44. "Cleanup Operations at Ground Zero," 68–69; "Cleaning Debris at Ground Zero."
45. Susan Sachs, "After the Attacks: The Site; A Delicate Removal of Debris, with Monstrous Machines and Gloved Hands," *New York Times*, September 14, 2001. Among the eyewitness and journalist cleanup accounts, see David W. Ausmus, *In the Midst of Chaos: My Thirty Days at Ground Zero* (Victoria, BC: Trafford, 2004); Mitchell Fink and Lois Mathias, *Never Forget: An Oral History of September 11, 2001* (New York: HarperCollins, 2002); James Glanz and Eric Lipton, *City in the Sky: The Rise and Fall of the World Trade Center* (New York: Henry Holt, 2003).
46. Eric Lipton and James Glanz, "A Nation Challenged: The Site: Slowed by Site's Fragility, the Heavy Lifting Has Only Begun," *New York Times*, October 13, 2001.
47. "Cleanup Operations at Ground Zero," in *The 9/11 Encyclopedia*, ed. Atkins, 1:69. See also "Context of May 30, 2002: Ground Zero Cleanup Operation Officially Ends"; Dennis Smith, *Report from Ground Zero* (New York: Penguin, 2002), 194; Josh Reno "September 11 Attacks (Aftermath)," in *Encyclopedia of Consumption and Waste: The Social Science of Garbage*, ed. Carl A. Zimring and William L. Rathje (Los Angeles: Sage, 2012), 2:795–97; Susan Sachs, "A Nation Challenged: The Site; At the Site, Little Hope of Uncovering Survivors," *New York Times*, September 19, 2001; "World Trade Center Recovery Project," *New York Construction* 50 (December 2002): 21; "9/11 Cleanup Memories Linger," *C&D Recycler* 4 (September 2002).
48. Diana Yates, "FBI Search Rubble for Clues to Attack," *Staten Island Advance*, September 13, 2001.
49. Martin J. Bellew, "Clearing the Way for Recovery at Ground Zero: The 9-11 Role of the NYC Department of Sanitation," *APWA Reporter*, March 2004, http://www3.apwa.net/Resources/Reporter/Articles/2004/3/Clearing-the-way-for-recovery-at-Ground-Zero. See also Charlie Leduff, "After the Attacks: The Disposal; Hauling the Debris, and Darker Burdens," *New York Times*, September 17, 2001. Governor Thomas Ridge of Pennsylvania—temporarily halting his dispute with New York City—offered to allow landfills and incinerators in the state to take WTC debris. Because Fresh Kills was to be a crime scene along with a potential dumping ground, the offer was not accepted. See Terence J. Kivlan, "Pennsylvania Governor Says He'll Take WTC Debris—So Far, However, the City Has No Plans to Ship the Material Anywhere but the Fresh Kills Landfill," *Staten Island Advance*, September 27, 2001.

50. Interview with Martin Bellew, March 22, 2012, Oral History Projects, DSNY—Freshkills Park, http://www.dsnyoralhistoryarchive.org/. See also interview with Frank Zito, April 3, 2012.
51. Bellew, "Clearing the Way for Recovery at Ground Zero"; Howard L. Green, *Continuing the Mission: U.S. Army Corps of Engineers: A History of the New York District, 1975–2005* (Washington, DC: U.S. Army Corps of Engineers, New York District, 2009), 304.
52. Elias, "U.S. Army Corps of Engineers: New York District's Worst Day and Its Finest Hour: Responding to the Sept. 11 Attacks on the World Trade Center," *DefenseMediaNetwork*, https://www.defensemedianetwork.com/stories/u-s-army-corps-of-engineers-new-york-district%E2%80%99s-worst-day-and-its-finest-hour/. See also U.S. Army Corps of Engineers, New York District, *9/11, a Look Back, and a Look Forward . . . 10 Years Later* (2011).
53. Bellew, "Clearing the Way for Recovery at Ground Zero." See also Elias, "U.S. Army Corps of Engineers: New York District's Worst Day and Its Finest Hour"; "Ground Zero Remains Recovery, Cleanup, Crime Scene . . . ," https://sites.google.com/site/wtc7lies/groundzerocleanup,freshkillssortingopera; Green, *Continuing the Mission*, 315–16; "FBI New York History," FBI New York Field Office, https://www.fbi.gov/history/field-office-histories/newyork; "Remembering Sept. 11, 2001," U.S. Army Corps of Engineers, *Engineer Update* 35 (September 2011): 4; Yates, "FBI Search Rubble for Clues to Attack"; Stephen Hart, "Sportsmen Assist with Rescue Effort," *Staten Island Advance*, September 14, 2001; James Glanz, "Fresh Kills Journal; Mountains of Twisted Steel, Evoking the Dead," *New York Times*, October 1, 2001; Jim Johnson, "Trade Center Cleanup Moves at Snail's Pace," *Waste News*, October 1, 2001, 1; Dan Barry and Amy Waldman, "A Nation Challenged: The Landfill; At Landfill, Tons of Debris, Slivers of Solace," *New York Times*, October 21, 2001; Jim Johnson, "Down to a Science; N.Y. Cleanup Becomes More Systematic, Picks Up Pace," *Waste News*, October 29, 2001, 1; Richard T. Pienciak, "Anguished Search for Traces of the Missing a Grim Task at Fresh Kills," *New York Daily News*, January 9, 2002; Kathleen Millar, "'I Hope You Find a Lot of People,'" *U.S. Customs Today*, March 2002; Larry Fish, "Far from Ground Zero, Search Continues for Remains in Landfill," *Philadelphia Inquirer*, June 23, 2002; Julie Sceifo and Peg Tyre, "N.Y. Closure: The Piles of Debris at the Staten Island Landfill Continue to Reveal Grisly Artifacts," *Newsweek*, July 22, 2002; Joan Blaucki, "Searching for Answers," *Pollution Engineering* 30 (January 2003): 43–44.
54. Diana Yates, "Landfill a Key Site in War on Terror," *Staten Island Advance*, September 19, 2001. See also Tina Hesman, "Experts Keep Birds from Looting NY Crime Scene," *St. Louis Post-Dispatch*, February 11, 2002; "Michael Mucci, Chief of the New York Sanitation Department, Discussed the World Trade Center Debris That Is Hauled to the Fresh Kills Landfill on Staten Island," *Today Show*, NBC News Transcripts, October 22, 2001.
55. Langewiesche, *American Ground*, 9–11; "Department of Design and Construction," in *The 9/11 Encyclopedia*, ed. Atkins, 1:82–84; "Statement of Ken Holden to the National Commission on Terrorist Attacks Upon the United States," April 1, 2003, National Commission on Terrorist Attacks Upon the United States, http://govinfo.library.unt.edu/911/hearings/hearing1/witness_holden.htm. See also "Profile: New York City Department of Design and Construction," *History Commons*, http://www.historycommons.org/entity.jsp?entity=department_of_design_and_construction_1; William Langewiesche, "American Ground: Unbuilding the World Trade Center," *Atlantic Monthly*, July/August 2002, 47–48.

56. Christina DiMartino, "Waste Industry, Others Help with Cleanup at World Trade Center Site," *Waste Age*, November 1, 2001; Bellew, "Clearing the Way for Recovery at Ground Zero." See also City of New York, Department of Sanitation, *Annual Report, 2001*, 6–11.
57. The initial Army Corps assignment went to the New England District outside of Boston but ultimately involved many other units. USACE engaged in evacuation of stranded citizens in Manhattan, transported emergency personnel and supplies onto the island, and assisted in restoring electrical power where needed. Dredging operations were necessary in the Hudson River to deal with large barges carrying debris, and the Corps worked with other groups to develop debris-removal plans. Elias, "U.S. Army Corps of Engineers: New York District's Worst Day and Its Finest Hour"; Diana Yates, "Dredging Done, More Trade Center Debris Set for Landfill," *Staten Island Advance*, October 5, 2001.
58. Critical Incident Stress Management (CISM) teams were sent to help workers cope with what they were finding. There were so many CISM team members that they sometimes got in the way of the recovery/cleanup workers.
59. "Federal Emergency Management Agency (FEMA)," in *The 9/11 Encyclopedia*, ed. Atkins, 1:109–11. The Office of Emergency Management, New York City's disaster-management body, could not operate during the crisis because its headquarters had been located in the World Trade Center. "Office of Emergency Management (OEM)," in *The 9/11 Encyclopedia*, ed. Atkins, 1:220–21. See also Federal Emergency Management Agency, *A Nation Remembers, a Nation Recovers: Responding to September 11, 2011, One Year Later* (2011).
60. "Biography: Rudolph Giuliani," Academy of Achievement, http://www.achievement.org/autodoc/page/giuopro-1; Robert Polner, ed., *America's Mayor: The Hidden History of Rudy Giuliani's New York* (Brooklyn: Soft Skull, 2005), xxxiv. See also Dan Barry, "The Giuliani Years: The Overview; A Man Who Became More Than a Mayor," *New York Times*, December 31, 2001; Doig, Jackson, and Keller, "September 11," 1169.
61. Quoted in "Rudolph Giuliani," Biography.com, http://www.biography.com/people/rudolph-giuliani-9312674#early-life.
62. Jennifer Steinhauer, "A Nation Challenged: The Mayor; In Crisis, Giuliani's Popularity Overflows City," *New York Times*, September 20, 2001; "Rudy Giuliani—Facts and Summary," History.com, http://www.history.com/topics/rudy-giuliani.
63. Quoted in *After September 11: New York and the World* (New York: Pearson, 2003), 6.
64. See Kevin Ryan, "A Closer Look at 9/11 Legend Rudy Giuliani," *DigWithin*, November 26, 2016, https://digwithin.net/2016/11/26/giuliani/; "Context of '(September 11, 2001–May 2002): Little-Known City Agency Takes Charge of Ground Zero Cleanup Operation,'" History Commons, http://www.historycommons.org/context.jsp?item=a091101ddc; "Rudolph William Louis 'Rudy' Giuliani III (1944–)," in *The 9/11 Encyclopedia*, ed. Atkins, 1:135–36; Langewiesche, *American Ground*, 145–46, 161.
65. Diane Cardwell, "A Nation Challenged: The Site; Few Answers on Possibility of Private Overseer at Cleanup," *New York Times*, December 7, 2001.
66. Jim Johnson, "Faces of Teamwork: Many Help in NYC Cleanup," *Waste News*, December 10, 2001, 1. See also "Recovery Operation Continues at Ground Zero in New York City," *National Public Radio*, December 19, 2001; Jim Johnson, "Attacks Leave Debris in NYC, D.C.," *Waste News*, December 24, 2001, 10.
67. "Cleanup Operations at Ground Zero," in *The 9/11 Encyclopedia*, ed. Atkins, 1:70. See also Michael B. Gerrard, "World Trade Center Response, Recovery and Reconstruction," *New York Law Journal*, October 4, 2001, http://911research.wtc7.net/cache/wtc/groundzero/NYLJ_response.html.

68. Andrew C. Revkin, "After Attacks, Studies of Dust and Its Effects," *New York Times*, October 16, 2001. See also Andrew C. Revkin, "A Nation Challenged; Dust Is a Problem, but the Risk Seems Small," *New York Times*, September 18, 2001.
69. U.S. Department of Labor, Office of Public Affairs, *OSHA National News Release*, "EPA and OSHA Web Sites Provide Environmental Monitoring Data from World Trade Center and Surrounding Areas," October 3, 2001.
70. James Nash, "Cleaning up After 9/11: Respirators, Power, and Politics," *EHS Today*, May 29, 2002, http://ehstoday.com/ppe/respirators/ehs_imp_35479. See also Michael B. Gerrard, "Environmental Law Implications of the World Trade Center Disaster," New York Committee for Occupational Safety and Health, April 27, 2007, https://www.nycosh.org/environment_wtc/gerrardarticle.html; Bruce Geiselman, "Study Lifts 9-11 Haze," *Waste News*, September 15, 2003, 1, 33.
71. "Forgotten victims" may not have been coined by the *New York Daily News*, but it was a term used by them. See "The Making of a Health Disaster: Officials to Act on Ground Zero Perils," *New York Daily News*, July 25, 2006.
72. On December 18, 2015, the James L. Zadroga 9/11 Health & Compensation Act provided health and compensation benefits to victims of 9/11 suffering from chronic respiratory conditions, lung diseases, and cancers. See "9/11 Health and Compensation Act," nyc.gov, https://www1.nyc.gov/site/911health/enrollees/your-care.page.
73. "Cleanup Operations at Ground Zero," in *The 9/11 Encyclopedia*, ed. Atkins, 1:71; "Health Department Announces New Research on Health Effects of the World Trade Center Disaster," *NYCHealth*, September 9, 2016, https://www1.nyc.gov/site/doh/about/press/pr2016/pr072-16.page. See also "Occupational Safety and Health Agency (OSHA)," in *The 9/11 Encyclopedia*, ed. Atkins, 1:228–30. For a deeper discussion of health risks and toxins at the attack site, see Paul J. Lioy, *Dust: The Inside Story of Its Role in the September 11th Aftermath* (Lanham, MD: Rowan & Littlefield, 2010). See also Sindhu Sundar, "NY Federal Judge OKs Deals in Dozens of WTC Debris Suits," *Law360*, https://www.law360.com/articles/633608/ny-federal-judge-oks-deals-in-dozens-of-wtc-debris-suits; Mireya Navarro, "Sept. 11 Workers Agree to Settle Health Lawsuits," *New York Times*, November 19, 2010; *September 11, Worker Task Force, Annual Report*, June 1, 2008, https://labor.ny.gov/agencyinfo/PDFs/9-11_WPTF_Annual_Report_2009_0601.pdf; *World Trade Center Health Registry 2016 Annual Report*, NYCHealth, https://labor.ny.gov/agencyinfo/PDFs/9-11_WPTF_Annual_Report_2009_0601.pdf; Michele McPhee, "1,700 Sue Over 9-11 Sickness; Bravest, Finest City Worked at WTC and Fresh Kills," *New York Daily News*, May 24, 2004; "World Trade Center Health Registry," *Sanitation News* 2 (October 2003): 1; "Lawsuits by Victims of Ground Zero Exposures," 9-11research.com, http://911research.wtc7.net/wtc/groundzero/lawsuits.html; World Trade Center Health Panel, *Addressing the Health Impacts of 9-11: Report and Recommendations to Mayor Michael R. Bloomberg*, 2011, nyc.gov, https://www1.nyc.gov/site/911health/about/addressing-the-health-impacts-of-9-11.page.
74. Daniel C. Kramer and Richard M. Flanagan, *Staten Island: Conservative Bastion in a Liberal City* (Lanham, MD: University Press of America, 2012), 10.
75. Diana Yates, "Is Debris at Landfill Hazardous?" *Staten Island Advance*, January 14, 2002. See also Diana Yates and Heidi Singer, "Disaster Debris at Fresh Kills: 90,000 Tons and Counting," *Staten Island Advance*, September 24, 2001.
76. Michael B. Gerrard, "World Trade Center Response, Recovery and Reconstruction," *New York Law Journal*, October 4, 2001.
77. James E. Cone et al., "Asthma Among Staten Island Fresh Kills Landfill and Barge Workers Following the September 11, 2001, World Trade Center Terrorist Attacks,"

American Journal of Industrial Medicine 59 (2016): 795. See also Christine C. Ekenga et al., "9/11-Related Experiences and Tasks of Landfill and Barge Workers: Qualitative Analysis from the World Trade Center Health Registry," *Public Health* 11 (2011): 321–28; Linda DeNenno, "An Emergency with Precedent: Protecting the Safety and Health of Workers at the Fresh Kills Landfill," *Professional Safety* (July 2003): 27–29.

78. U.S. Environmental Protection Agency, "EPA Initiates Emergency Response Activities, Reassures Public About Environmental Hazards," Government Information Regarding September 11th, September 13, 2001, http://www.immuneweb.org/911/government/oct01.html. See also other documents related to health at the disaster site in this grouping.

79. U.S. Environmental Protection Agency, Region 2, "EPA Response to September 11," https://archive.epa.gov/wtc/web/html/. See also U.S. Environmental Protection Agency, Office of Inspector General, *Evaluation Report: EPA's Response to the World Trade Center Collapse: Challenges, Successes, and Areas of Improvement*, Report no. 2003-P-00012, August 21, 2003.

80. Megan D. Nordgren, Eric A. Goldstein, and Mark A. Izeman, *The Environmental Impacts of the World Trade Center Attacks: A Preliminary Assessment* (Natural Resources Defense Council, February 2002), iv.

81. Nordgren, Goldstein, and Izeman, *The Environmental Impacts of the World Trade Center Attacks*, iv. See also FDNY World Trade Center Health Program, *Health Impact on FDNY Rescue/Recovery Workers: 15 Years: 2001 to 2016*; U.S. House of Representatives, 108th Congress, 2nd sess., *Assessing September 11th Health Effects*, hearing before the Subcommittee on National Security, Emerging Threats, and International Relations of the Committee on Government Reform, September 8, 2005 (serial no. 108-283); U.S. Environmental Protection Agency, *Evaluation Report: EPA's Response to the World Trade Center Collapse: Challenges, Successes, and Areas for Improvement*, Report no. 2003-P-00012, August 21, 2003.

82. Mark Schoofs, "Fresh Kills, Repository of New York Tales, Adds Twin Towers' Story to Its Legend," *Wall Street Journal*, September 28, 2001.

83. Bob Herbert, "Ground Zero Diagnosis," *New York Times*, June 3, 2002; "Cleanup Operations at Ground Zero," in *The 9/11 Encyclopedia*, ed. Atkins, 1:70–71.

84. "Sifting Through the WTC Rubble," *NPR*, May 30, 2002, http://www.npr.org/templates/story/story.php?storyId=1144157.

85. "Fresh Kills Landfill," in *The 9/11 Encyclopedia*, ed. Atkins, 1:130–32; "September 11 Recovery Effort," Landfill-to-Park Timeline, Freshkills Park, http://timeline.freshkillspark.org/2002/07/september-11-recovery-effort/; Robin Nagle, "The History and Future of Fresh Kills," in *Dirt: The Filthy Reality of Everyday Life*, ed. Rosie Cox et al. (London: Profile, 2011), 197; Bellew, "Clearing the Way for Recovery at Ground Zero"; Jim Johnson, "Ceremony Closes WTC Search," *Waste News*, July 22, 2002, 6. See also Green, *Continuing the Mission*, 320; "Closing Ceremony to Be Held at Staten Island Landfill," *PR Newswire*, July 12, 2002; Dan Barry, "Sifting the Last Tons of Sept. 11 Debris," *New York Times*, May 14, 2002; Paul Zahn, "NYPD Ceremony Marks End of Search for Human Remains at WTC," *CNN Live Event*, July 15, 2002.

86. Johnson, "Debris Gone; Memories Remain."

87. "A Somber Place: WTC Debris Sorted at Fresh Kills Landfill," *CNN Live at Daybreak*, January 15, 2002, http://www.cnn.com/TRANSCRIPTS/0201/15/lad.12.html. See also "WTC Debris Searched at NYC Landfill," *AP Online*, January 15, 2000, http://www.highbeam.com/doc/1P1-49533660.html; Kit R. Roane and Andrew Lechtenstein, "Burial Ground," *U.S. News and World Report* 133 (September 11, 2002): 84; Jef Akst,

"Hallowed Landfill," *Scientist*, January 1, 2012, http://www.the-scientist.com/?articles.view/articleNo/31553/title/Hallowed-Landfill/.

88. Frank Marra and Maria Bellia Abbate, *From Landfill to Hallowed Ground: The Largest Crime Scene in America* (Dallas: Brown Books, 2015), 16.
89. Marra and Abbate, *From Landfill to Hallowed Ground*, 24.
90. Marra and Abbate, *From Landfill to Hallowed Ground*, 83–84. See also Gerome Truc, "Ground Zero: Between Construction Site and Mass Grave," *Raison Politiques* (2011), https://www.cairn-int.info/article-E_RAI_041_0033--ground-zero-between-construction-site.htm.
91. Reno, "September 11 Attacks (Aftermath)," 795. See also "Possible Human Remains Discovered at the World Trade Center Site as New York City Re-examines Site of 9/11 Terrorist Attack," *Daily Mail*, April 3, 2013; "Possible Human Remains Found in New 9/11 Debris Uncovered at World Trade Center Site 12 Years After Attack," *New York Daily News*, April 3, 2013.
92. Charles Laurence, "Remains of 9/11 Victims 'to Spend Eternity' in City Rubbish Dump," *Telegraph*, September 10, 2004. See also Ann Leslie, "The Final Indignity," *Daily Mail*, February 4, 2004; David V. Johnson, "Despite the Years, Stuck at Ground Zero," *Staten Island Advance*, September 10, 2006.
93. Anemona Hartcollis, "Landfill Has 9/11 Remains, Medical Examiner Wrote," *New York Times*, March 24, 2007; Thomas Zambito, "9/11 Remains Fill Potholes, Worker Claims," *New York Daily News*, March 24, 2007; Nikki Schwab, "For 9/11 Kin, Emotions Still Run High," *U.S. News and World Report*, August 8, 2007; Joe Pompeo, "The Remains of Fresh Kills," *StrausMedia*, May 16, 2007, http://www.nypress.com/the-remains-of-fresh-kills/.
94. Marra and Abbate, *From Landfill to Hallowed Ground*, 92–99. See also Beverly Thomson, "Group of WTC Victims' Families Demand Proper Burials," *CTV Television*, April 6, 2010; Anthony Gardner and Diane Horning, "9/11 Victims Should Not Be Left in the Fresh Kills Dump, Families Say," *New York Daily News*, February 24, 2008; Alan Feuer, "Judge Hints He May Reject 9/11 Families' Plea to Sift Landfill for Remains," *New York Times*, February 23, 2008; Frank Donnelly, "At Fresh Kills Landfill, a Heartbreaking Effort After World Trade Center Attacks," *SILive.com*, September 11, 2011, https://www.silive.com/september-11/index.ssf/2011/09/at_the_landfill_a_heartbreakin.html.
95. Donnelly, "At Fresh Kills Landfill, a Heartbreaking Effort." See also Steven Brill, *After: How America Confronted the September 12 Era* (New York: Simon & Schuster, 2003), 240, 251, 254–55, 267–68, 277–79, 384, 418.
96. Michael J. Fressola, "Artists Offer Perspectives on Fresh Kills," *Staten Island Advance*, October 9, 2001.
97. David Andreatta, "Monument Status Sought for Landfill Site," *Staten Island Advance*, October 22, 2002.
98. Quoted in Nagle, "The History and Future of Fresh Kills," 198.
99. Fred Siegel, *The Prince of the City: Giuliani, New York, and the Genius of American Life* (San Francisco: Encounter, 2005), 309–13; Jennifer Steinhauer, "A Nation Challenged: The Mayor; In Crisis, Giuliani's Popularity Overflows City," *New York Times*, September 20, 2001.
100. Siegel, *The Prince of the City*, 314–21; "The Race for Mayor," *Gotham Gazette's Searchlight on Campaign 2001*, http://www.gothamgazette.com/searchlight2001/mayor1.html. See also Adam Nagourney, "The 2001 Elections: Mayor; Bloomberg Edges Green in Race for Mayor," *New York Times*, November 7, 2001; Jessica Reaves,

"Election 2001: Finally, It's Bloomberg," *Time*, November 7, 2001, http://content.time.com/time/nation/article/0,8599,183271,00.html; Manuel Perez-Rivas, "Bloomberg Wins 'Tough' NYC Mayoral Race," *CNN.com*, November 7, 2001, http://www.cnn.com/2001/ALLPOLITICS/11/06/2001.election.newyork.

101. Chris Smith, "Election, Mayoral: How America's Mayor Created New York's Mayor," *NYMag.com*, August 27, 2011, http://nymag.com/news/9-11/10th-anniversary/mayoral-election/. See also Kramer and Flanagan, *Staten Island: Conservative Bastion in a Liberal City*, 150–51.
102. "Mayoralty," in *The Encyclopedia of New York City*, ed. Jackson, 816.
103. "What Factors Led to Michael Bloomberg (business person, politician)'s Election as Mayor of New York City?" *Quora*, October 30, 2012, https://www.quora.com/What-factors-led-to-Michael-Bloomberg-business-person-politician-s-election-as-mayor-of-New-York-City. See also Chris McNickle, *Bloomberg: A Billionaire's Ambition* (New York: Skyhorse, 2017), 16, 23, 42–43.
104. Andrew Stevens, "Michael Bloomberg: Mayor of New York," *City Mayors*, August 2012, http://www.citymayors.com/usa/nyc.html; Kenneth T. Jackson, "Michael Rubens Bloomberg," in *The Encyclopedia of New York City*, ed. Jackson, 134–35.
105. See Joyce Purnick, *Mike Bloomberg: Money, Power, Politics* (New York: Public Affairs, 2009).
106. See Legislative Commission on Solid Waste Management, *Where Will the Garbage Go? 2001*, 1, http://nyassembly.gov/comm/SolidWaste/20020513/#Waste.
107. Gregorio was fined for ethics violations by using his influence to advance the waterfront transfer station in Linden in July 2002. See Ronald Smothers, "Linden Mayor Is Fined for Ethics Violations," *New York Times*, July 19, 2002.
108. Andrew Friedman, "Reprieve for Fresh Kills," *Planning* 67 (November 2001): 4. See also Eric Lipton, "Trash Transfer to New Jersey Is Postponed," *New York Times*, September 19, 2001; Michael Burger and Christopher Stewart, "Garbage After Fresh Kills," *GothamGazette.com*, January 28, 2002, http://www.gothamgazette.com/iotw/garbage/.
109. Friedman, "Reprieve for Fresh Kills."
110. Benjamin Miller, "The Garbage Behind, the Garbage Ahead," October 6, 2001 (Session title: "Gotham's Garbage: The Free Market and the Hidden Costs of Turning Trash Into Cash"), 1.
111. Benjamin Smith, "Where to Cut?" *New York Post*, November 30, 2001. See also "Assault on Closure?" *Staten Island Advance*, December 2, 2001; Andrew Friedman, "Reprieve for Fresh Kills," *Planning* 67 (November 2001): 4–9.
112. Reginald Patrick, "Officials: Dump Closed Forever," *Staten Island Advance*, December 4, 2001. See also "Marchi: Landfill Closed for a Reason," *New York Post*, December 5, 2001.
113. New York City, Independent Budget Office, *Budget Options for New York City*, February 2003.
114. Aaron William Comrov, "Fresh Kills Dumped: A Policy Assessment for the Management of New York City's Residential Solid Waste in the Twenty-First Century," unpublished MS thesis in environmental policy science, New Jersey Institute of Technology, May 2003, 55–56, 115–16.
115. The Bloomberg administration would not consider revisiting the Fresh Kills debate, at the very least because the mayor's election had owed so much to the forgotten borough.
116. Wrobleski, "Pataki Says It Clearly: Landfill Must Stay Closed."

117. "Fresh Kills 'in Play'?" *Staten Island Advance*, December 6, 2001. See also Reginald Patrick, "Landfill Heresy Gets Lots of Ink," *Staten Island Advance*, December 9, 2001.
118. *Life After Fresh Kills: Moving Beyond New York City's Current Waste Management Plan: Policy, Technical, and Environmental Considerations*, a joint research project of Columbia University's Earth Institute, Earth Engineering Center, and the Urban Habitat Project at the Center for Urban Research and Policy of Columbia's School of International and Public Affairs, December 1, 2001, i.
119. *Life After Fresh Kills*, i–ii.
120. *Life After Fresh Kills*, A5, A7–8.
121. "Fresh Kills," *Living on Earth*, February 8, 2002, http://www.loe.org/shows/segments.html?programID=02-P13-00006&segmentID=5.
122. Heidi Singer, "New Council Panel to Keep Watch Over Fresh Kills Landfill," *Staten Island Advance*, February 8, 2002; Kirk Johnson, "To City's Burden, Add 11,000 Tons of Daily Trash," *New York Times*, February 28, 2002.
123. Nickolas J. Themelis and Claire E. Todd, "Recycling in a Megacity," *Journal of Air and Waste Management Association* (February 21, 2012): 389–95. The authors argue that Bloomberg's decision to temporarily suspend glass and plastics recycling "was considered by many to be anti-environmental, but the results of this study show that for lack of markets, even at zero or negative prices, nearly 90% of the plastic and glass set aside by thoughtful New Yorkers was transported to materials recovery facilities (MRFs) and from there to landfills" (389).
124. Joe Triuni, "New York May Cut Curbside Recycling," *Waste News* 7 (February 18, 2002): 1. See also Kirk Johnson, "Commissioner Defends Plan for Garbage and Recycling," *New York Times*, March 21, 2002.
125. "Rethinking Garbage," *New York Times*, March 2, 2002.
126. Bill Franz, "We Were Right All Along About Fresh Kills," *Staten Island Register*, March 5, 2002. See also Michael R. Blood, "City's Been Forced to Talk Trash Again," *New York Daily News*, March 6, 2002.
127. Benjamin Miller, "City Needs to Reopen Fresh Kills," *New York Daily News*, March 5, 2002. See also Michael Tomasky, "Talking Trash," *New York Magazine*, March 26, 2002, http://nymag.com/nymetro/news/politics/columns/citypolitic/5787/; Heidi Singer, "Movement or Media Creation?" *Staten Island Advance*, April 1, 2002; "Reopen Fresh Kills: Brilliant or Boneheaded?" *New York Times*, April 14, 2002; "Killing Fresh Kills," *Staten Island Advance*, April 28, 2002; Robert Brune, "Trash Heap? Not in My Backyard," *New York Times*, March 6, 2002; "New York City's Solid Waste Dilemma," *WasteWatch* 13 (Spring 2002): 2.
128. The Conservative Party's Molinaro replaced Guy Molinari as borough president on January 1, 2002. See James P. Molinaro, "Trash Heap? Not in My Backyard," *New York Times*, March 6, 2002. See also James P. Molinaro, "Dump Opening No Option—Period," *Staten Island Advance*, March 8, 2002.
129. Vito J. Fossella, "Trash Heap? Not in My Backyard," *New York Times*, March 6, 2002.
130. Robert Gavin, "Officials Vow Never to Pull Legal Stake from Landfill's Heart," *Staten Island Advance*, March 5, 2002. See also Stephanie Slepian, "Councilman Has Last Word in TV Landfill Debate: No!" *Staten Island Advance*, March 14, 2002; "Garbage on Garbage," *Staten Island Advance*, March 18, 2002.
131. Mary Engels, "Keep Landfill Closed, Schumer Says," *New York Daily News*, March 12, 2002; Tom Wrobleski, "Schumer Says 'Forget It,'" *Staten Island Advance*, March 9, 2002.
132. Diana Yates, "5 BPs Agree: Keep Fresh Kills Closed," *Staten Island Advance*, April 11, 2002.

133. "Landfill Politics in Play in Pataki Re-election Bid?" *Staten Island Advance*, March 10, 2002. See also Joyce Purnick, "Metro Matters; The Politics of Garbage, Forever Ripe, *New York Times*, March 14, 2002; "With Friends Like These . . ." *Staten Island Register*, April 2, 2002.
134. Miller, "City Needs to Reopen Fresh Kills."
135. Kirk Johnson, "As Options Shrink, New York Revisits Idea of Incineration," *New York Times*, March 23, 2002. On continued flow-control debates in Washington, DC, see Terence Kivlan, "The Politics of N.Y. Trash in D.C.," *Staten Island Advance*, March 31, 2002; John T. Aquino, "On Deck in 2002," *Waste Age*, April 1, 2002; Terence Kivlan, "Pressure Builds to Block Import of Trash to Virginia," *Staten Island Advance*, April 23, 2002.
136. Diane Cardwell, "Mayor Drops Incinerator Plan," *New York Times*, May 18, 2002.
137. "New York Mayor Proposes Trash Plan," *Waste Age*, September 2002, 6. See also Eric Lipton, "Great Idea for Trash, If It Works, Most Say," *New York Times*, August 1, 2002; "N.Y.C. Seeks Alternative to Trucking Garbage," *Engineering News-Record* 249 (August 12, 2002): 7.
138. Kirk Johnson, "Trash Plan Alters Mix of Winners and Losers," *New York Times*, August 2, 2002. See also "3 Sites—One in Linden—Being Looked at for NYC Garbage," *Jersey* (Jersey City, NJ) *Journal*, December 24, 2003.
139. Kirk Johnson, "The Mayor's Budget Plan: Spending; Sanitation and Environment," *New York Times*, November 15, 2002.
140. Legislative Commission on Solid Waste Management, *Where Will the Garbage Go? 2001*, 1–2.
141. Michael Cooper, "A Divisive Budget Question: Does It Really Pay to Recycle?" *New York Times*, June 14, 2002. See also John Tierney, "The Big City; Try Ending Free Pickup of Trash," *New York Times*, April 16, 2002.
142. See "Solving New York's Garbage Problems," *Futurist*, September–October 2002, 14; Chaz Miller, "Trash Diet," *Waste Age*, September 2002, 9.
143. See Keith Kloor, "Wretched Refuse: We Don't Want It. They Have to Take It. The Truth About Where Your Garbage Goes," *City Limits*, High Beam Research, November 1, 2002, https://www.highbeam.com/doc/1G1-93611105.html. See also City of New York, Department of Sanitation, *DSNY Annual Report, 2002–2003*; Nickolas J. Themelis, "Integrated Management of Solid Waste for New York City," paper read at the tenth North American Waste to Energy Conference, ASME 2000, https://pdfs.semanticscholar.org/93bb/dc63c45d212fd9b8846f75681a42700f69a0.pdf.
144. Jim Johnson, "Debris Gone; Memories Remain; Sept. 11 Buried Deep in Staten Island Landfill," *Waste News*, September 2, 2002, 1.

19. REGENERATION

1. Mierle Laderman Ukeles, "Leftovers: It's About Time for Fresh Kills," *Cabinet* 6 (Spring 2002), http://www.cabinetmagazine.org/issues/6/index.php; Robin Nagle, "To Love a Landfill: The History and Future of Fresh Kills," in *Handbook of Regenerative Landscape Design*, ed. Robert L. France (Boca Raton, FL: CRC, 2008), 3–4; Martin Melosi quoted in W. L. Rathje, "Let Landfills Be Landfills," June 2002, http://www.mswmanagement.com/may-june-2002/let-landfills-be-landfills.aspx. See also Hayden Thomas, "Fields of Dreams," *U.S. News & World Report* 132 (January 21, 2002): 62.

19. REGENERATION ■ 739

2. See, for example, Lucy R. Lippard, "New York Comes Clean: The Controversial Story of the Fresh Kills Dumpsite," *Guardian*, October 28, 2016.
3. "A New Staten Island," *Staten Island Advance*, March 26, 2001.
4. Karen O'Shea and Jill Gardiner, "Downwind of the Dump, Windfalls Emerge," *Staten Island Advance*, March 26, 2001.
5. Aaron William Comrov, "Fresh Kills Dumped: A Policy Assessment for the Management of New York City's Residential Solid Waste in the Twenty-First Century," MA thesis, New Jersey Institute of Technology, May 2003, 115.
6. Peter Harnik, Michael Taylor, and Ben Welle, "From Dumps to Destinations: The Conversion of Landfills to Parks," *Places* 18 (2006): 84. See also Julia Czerniak and George Hargreaves, eds., *Large Parks* (New York: Princeton Architectural Press, 2007).
7. Wolfram Hoefer et al., "Environmental Reviews and Case Studies: Unique Landfill Restoration Designs Increase Opportunities to Create Urban Open Space," *Environmental Practice* 18 (June 1916): 106–15; Harnik, Taylor, and Welle, "From Dumps to Destinations," 85.
8. Harnik, Taylor, and Welle, "From Dumps to Destinations," 85.
9. Hoefer et al., "Environmental Reviews and Case Studies," 106. An alternative to landfill reuse is landfill reclamation. According to the Environmental Protection Agency, "Landfill Reclamation is a relatively new approach used to expand municipal solid waste (MSW) landfill capacity and avoid the high cost of acquiring additional land." U.S. Environmental Protection Agency, Solid Waste and Emergency Response, *Landfill Reclamation*, EPA530-F-97-001, July 1997.
10. Society for Ecological Restoration International Science & Policy Working Group, *The SER International Primer on Ecological Restoration* (2004), 3, https://cdn.ymaws.com/www.ser.org/resource/resmgr/custompages/publications/SER_Primer/ser_primer.pdf.
11. Society for Ecological Restoration International Science & Policy Working Group, *The SER International Primer on Ecological Restoration*, 3. See also William R. Jordan III and Marcus Hall, "Ecological Restoration," in *Encyclopedia of World Environmental History*, ed. Shepard Krech III et al. (New York: Routledge, 2004), 1:371; William R. Jordan III and George M. Lubick, *Making Nature Whole: A History of Ecological Restoration* (Washington, DC: Island, 2011), 3.
12. T. P. Young, D. A. Petersen, and J. J. Clay, "The Ecology of Restoration: Historical Links Emerging Issues and Unexplored Realms," *Ecology Letters* 8 (2005): 662.
13. Terms of definition have included "restoration," "rehabilitation," "reclamation," "reforestation," "revegetation," or "healing."
14. Jordan III and Hall, "Ecological Restoration," 371, 374.
15. Young, Petersen, and Clay, "The Ecology of Restoration," 662. See also Liana Wortley, Jean-Marc Hero, and Michael Howes, "Evaluating Ecological Restoration Success: A Review of the Literature," *Restoration Ecology* 21 (September 2013): 537–43; Maria C. Ruiz-Jaen and T. Mitchell Aide, "Restoration Success: How Is It Being Measured?" *Restoration Ecology* 13 (September 2005): 569–77; D. J. Walker and N. C. Kenkel, "Landscape Complexity in Space and Time," *Community Ecology* 2 (2001): 109–19; T. Cohen, "Ecological Restoration," *Technology Review* 95 (February/March 1992): 20; Evelyn A. Howell, John A. Harrington, and Stephen B. Glass, *Introduction to Restoration Ecology* (Washington, DC: Island, 2011), 1.
16. Ecological Society of America, "Restoration Ecology: The Challenge of Social Values and Expectations," *Frontiers in Ecology and the Environment*, February 1, 2004.

17. Stuart K. Allison, "What *Do* We Mean When We Talk About Ecological Restoration?" *Ecological Restoration* 22 (December 2004): 281.
18. Ecological Society of America, "Restoration Ecology." See also Jordan III and Hall, "Ecological Restoration," 373–78.
19. See Jordan III and Lubick, *Making Nature Whole*, 4.
20. The DCP was supported by the Municipal Arts Society and the New York State Department of State, Division of Coastal Resources. See "The Park Plan," Freshkills Park Alliance, http://freshkillspark.org/the-park/the-park-plan. The Freshkills Park Alliance is the not-for-profit partner with the city with respect to the development of Freshkills Park. See also "Fresh Kills Park Project: Introduction," http://www1.nyc.gov/assets/planning/download/pdf/plans/fkl/fkl.pdf.
21. See John L. Eliot, "A Dump Reviled, Revered," *National Geographic*, January 2003, insert 8–9. See also Jim Carlton, "Where Trash Reigned, Trees Sprout," *Wall Street Journal*, January 23, 2002, B1. In 1994 Handel and some associates calculated that of an original 1,082 plant species on Staten Island, 443 native vascular plant species had been lost and one hundred non-native species had been added. See Betsy McCully, *City at the Water's Edge: A Natural History of New York* (New Brunswick, NJ: Rutgers University Press, 2007), 107–8.
22. Heather Millar, "Let a Billion Flowers Bloom," *Sierra* 90 (November/December 2005): 44.
23. "Fresh Kills," *Waste Management World*, January 12, 2007, https://waste-management-world.com/a/fresh-kills; "A Fresh Start at Fresh Kills: Landfill to Landscape," *Municipal Art Society Newsletter*, May/June 2001, 1.
24. Some stories place the origins of the Freshkill Park idea at MAS earlier than 2000.
25. Robin Lynnard and Francis Morrone, *Guide to New York City Urban Landscapes* (New York: Norton, 2013), 133–35; oral interview, Martin Melosi with Brendan Sexton, November 17, 2017, New York City. See also oral interview, Martin Melosi with Mierle Laderman Ukeles, August 17, 2017, New York City; oral interview, Martin Melosi with Eloise Hirsch and Mariel Villere, November 17, 2017, New York City; interview with Raj Kottamasu, April 3, 2012, Oral History Projects, DSNY–Freshkills Park, http://www.dsnyoralhistoryarchive.org. An idea for a memorial park in honor of 9/11 victims was floated but never became an alternative to Freshkills Park. Diana Yates, "Landfill Might Become a Monument?" *Staten Island Advance*, September 27, 2001.
26. Ashley Schafer, Amanda Reese, and Megan Miller, "Fresh Kills Landfill to Landscape," *Praxis* 4 (October 1, 2002): 18.
27. Schafer, Reese, and Miller, "Fresh Kills Landfill to Landscape," 18.
28. Rebecca Krinke, "Fresh Ideas?" *Landscape Architecture* (June 2002): 76, 78–79.
29. Rosten Woo, "Fresh Kills? Can New York Afford to Turn Its Largest Landfill Into a Park?" *Metropolis* 21 (2002): 36.
30. *Fresh Kills: Landfill to Landscape*, Competition Design Brief, vol. 2: *Technical Planning & Design Report*, August 15, 2001, 5.
31. "About Fresh Kills," Fresh Kills: Landfill to Landscape, Fresh Kills Park Project, International Design Competition, http://www1.nyc.gov/assets/planning/download/pdf/plans/fkl/fkl.pdf.
32. See, for example, Marissa Reilly, "Ecological Atonement in Fresh Kills: From Landfill to Landscape," Senior Capstone Project, Vassar College, 2013, 5–6, http://digitalwindow.vassar.edu/cgi/viewcontent.cgi?article=1186&context=senior_capstone.

33. "At Fresh Kills Landfill, Garbage Out, Grand Plans," *New York Times*, December 9, 2001.
34. The jury included professionals from the fields of architecture, landscape architecture, and environmental science, plus senior public officials with knowledge of the site. See "Fresh Kills: Landfill to Landscape, International Design Competition: 2001," http://www1.nyc.gov/assets/planning/download/pdf/plans/fkl/about_competition.pdf. See also Diana Yates, "Panel Picks Top Design for Landfill," *Staten Island Advance*, December 19, 2001; Kara Fulfer, "Fresh Kills Redevelopment Offers Huge Ecological Promise," *Land Restoration*, November/December 2002, 21.
35. "At Fresh Kills Landfill, Garbage Out, Grand Plans." For information on the building of Central Park and common ground with more recent New York parks, see Roy Rosenzweig and Elizabeth Blackmar, *The Park and the People: A History of Central Park* (New York: Cornell University Press, 1992); Travis Beck, *Principles of Ecological Landscape Design* (Washington, DC: Island, 2013), 1.
36. Gersh Kuntzman, "Artistes Provide Landfill Overkill," *New York Post*, December 10, 2001. See also "Six Design Teams Named Finalists in Fresh Kills Landfill Plan," *Real Estate Weekly*, November 7, 2001.
37. "MAS Announces Design Teams for Fresh Kills," June 12, 2002, Municipal Arts Society of New York, https://www.mas.org/mas-announces-design-teams-for-fresh-kills/. See also Yates, "Landfill Might Become a Monument?"
38. "Fresh Kills Redesign Finalists Announced," *Waste Age* 31 (April 2002): 12–13. See also Jim Johnson, "Attack Resurrects NYC's Fresh Kills," *Waste News*, November 12, 2001, 13; Michael Burger and Christopher Stewart, "Garbage After Fresh Kills," January 28, 2002, *Gotham Gazette.com*, https://www.gothamgazette.com/iotw/garbage/.
39. See Barbara Stewart, "Landfill to Park? Give It Time; The Transformation of Fresh Kills Will Take Decades," *New York Times*, November 28, 2002; "What Should Be the Future Use of the Fresh Kills Landfill?" *Staten Island Advance*, October 29, 2002.
40. Linda Pollak, "Sublime Matters: Fresh Kills," *Praxis* 4 (2002): 58–60.
41. "Narrative: Mathur/Da Cunha + Tom Leader Studio," http://www1.nyc.gov/assets/planning/download/pdf/plans/fkl/lean1.pdf. See also "Mathur/Da Cunha + Tom Leader Studio," *Praxis* 4 (2002): 40. In another document, the firm labeled its five areas differently: Event Surface, Experimental Field, Material Datum, Depositional Edge, and Tectonic Zone. It stated, "The shifting nature of Fresh Kills confounds interpretations, predictions and conviction necessary for end or phased scenarios. It rather calls for seeds nurtured with agility, immediacy, and necessity. We recommend five seeds, each with a tendency. . . . Their trajectories will intersect . . . but each has the potential to negotiate its way through these intersections." "Narrative: Mathur/Da Cunha + Tom Leader Studio," https://www1.nyc.gov/assets/planning/download/pdf/plans/fkl/lean1.pdf.
42. Sasaki Team, "XPark: Planning Approach Narrative Summary," http://www1.nyc.gov/assets/planning/download/pdf/plans/fkl/sasn1.pdf.
43. "XPark, Sasaki Associates," *Praxis* 4 (2002): 55.
44. "Fresh Kills Parklands," http://www1.nyc.gov/assets/planning/download/pdf/plans/fkl/harn1.pdf. See also "Hargreaves Associates," *Praxis* 4 (2002): 28–33.
45. "Executive Summary," http://www1.nyc.gov/assets/planning/download/pdf/plans/fkl/jmpn1.pdf.
46. "John McAslan + Partners," *Praxis* 4 (2002): 35.
47. "rePark: recycle recollect recreate," http://www1.nyc.gov/assets/planning/download/pdf/plans/fkl/rion1.pdf. See also "RIOS," *Praxis* 4 (2002): 49–50; Caroline Fraser,

Rewilding the World: Dispatches from the Conservation Revolution (New York: Henry Holt, 2009), 284–86.

48. "Department of City Planning Announces the Winner of the International Competition for a Conceptual Design and Master Plan for the Future of Fresh Kills," press release, December 18, 2001, https://www1.nyc.gov/assets/planning/download/pdf/about/press-releases/pr121801.pdf. See also "Fresh Kills Park Project: Project History," http://www1.nyc.gov/assets/planning/download/pdf/plans/fkl/fkl.pdf; Frederick Steiner, "Frontiers in Urban Ecological Design and Planning Research," *Landscape and Urban Planning* 125 (2004): 308–9; Laura Bliss, "The Wild Comeback of New York's Legendary Landfill," CityLab, February 17, 2017, https://www.citylab.com/cityfixer/2017/02/the-wild-comeback-of-new-yorks-legendary-landfill/516822/.
49. James Corner, *Lifescape–Fresh Kills Parkland*, http://www.nextroom.at/data/media/med_binary/original/1121022434.pdf.
50. Glenn Nyback, "Dump Dreaming-Visionaries Look to Transform the Dump Into City's Biggest Park—and They're Open to Your Ideas," *Staten Island Advance*, March 14, 2004.
51. Corner, *Lifescape–Fresh Kills Parkland*.
52. Corner, *Lifescape–Fresh Kills Parkland*.
53. Field Operations, "Lifescape: Fresh Kills Reserve, Staten Island, New York," http://www1.nyc.gov/assets/planning/download/pdf/plans/fkl/fien1.pdf.
54. "Lifescape, Field Operations," *Praxis* 4 (2002): 18. See also Christoph Lindner, "New York Undead: Globalization, Landscape Urbanism, and the Afterlife of the Twin Towers," *Journal of American Culture* 31 (September 2008): 302–4, 306–9.
55. The NYC Percent for Art website states, "Managed by the City's Department of Cultural Affairs, the Percent for Art program has commissioned hundreds of site-specific projects in variety of media—painting, new technologies, lighting, mosaic, glass, textiles, sculpture, and works that are integrated into infrastructure and architecture—by artists whose sensibilities reflect the diversity of New York City.... The Percent for Art Program offers City agencies the opportunity to acquire, commission or restore works of art specifically for City-owned buildings throughout the five boroughs. By bringing artists into the design process, the City's civic and community buildings are enriched." "The Department of Cultural Affairs Percent for Art," NYC Percent for Art, http://www1.nyc.gov/site/dclapercentforart/index.page.
56. Robin Cembalest, "Talking Trash with Mierle Laderman Ukeles," *Forward*, November 25, 1994, 9; Patricia C. Phillips, "Making Necessity Art: Collisions of Maintenance and Freedom," in *Mierle Laderman Ukeles: Maintenance Art*, ed. Patricia C. Phillips et al. (Munich: Del Monico, 2016), 28–30, 37; Tom Finkelpearl, *Dialogues in Public Art* (Cambridge, MA: MIT Press, 2000), 299–300; Jeffrey Kastner, "The Department of Sanitation's Artist in Residence," *New York Times*, May 19, 2002.
57. Finkelpearl, *Dialogues in Public Art*, 302. See also Kastner, "The Department of Sanitation's Artist in Residence."
58. Don Krug, "Ecological Restoration: Mierle Ukeles, Flow City," 2006, *greenmuseum.org*, http://www.greenmuseum.org/c/aen/Issues/ukeles.php. See also Bartholomew Ryan, "Manifesto for Maintenance: A Conversation with Mierle Laderman Ukeles," *Art in America*, March 20, 2009, http://www.artinamericamagazine.com/news-features/interviews/draft-mierle-interview/; Mark B. Feldman, "Inside the Sanitation System: Mierle Ukeles, Urban Ecology, and the Social Circulation of Garbage," *Iowa Journal of Cultural Studies* 10/11 (Spring/Fall 2009): 47–48; Holland Cotter, "Maintenance Required," *New York Times*, June 20, 2013; Phillips, "Making Necessity Art," 37–41.

59. Randy Kennedy, "An Artist Who Calls the Sanitation Department Home," *New York Times*, September 21, 2016.
60. Quoted in Kastner, "The Department of Sanitation's Artist in Residence." See also Lucy R. Lippard, "Never Done: Women's Work by Mierle Laderman Ukeles," in *Mierle Laderman Ukeles*, ed. Phillips et al., 15–17, 20, 24.
61. Feldman, "Inside the Sanitation System," 48–49; Phillips, "Making Necessity Art," 65, 70.
62. Quoted in Phillips, "Making Necessity Art," 87. See also 80–83.
63. Phillips, "Making Necessity Art," 89–91, 95. See also Krug, "Ecological Restoration"; Ukeles, "Leftovers"; Finkelpearl, *Dialogues in Public Art*, 295, 311–12.
64. Finkelpearl, *Dialogues in Public Art*, 296. See also Carly Berwick, "What a Dump!" *ARTnews* 103 (2004): 89; Andy Battaglia, "'Maintenance Art' Puts Trash in Full View," *Wall Street Journal*, September 22, 2016; Ukeles, "Leftovers"; Krug, "Ecological Restoration"; Feldman, "Inside the Sanitation System," 50–51; Phillips, "Making Necessity Art," 97–99, 122–23, 126, 138.
65. Mierle Laderman Ukeles to Sanman, 1979, in Phillips, "Making Necessity Art," 101. Recently, the sociologist Robin Nagle has taken up the cause of New York City sanitation workers in her book *Picking Up: On the Streets and Behind the Trucks with the Sanitation Workers of New York City* (New York: Farrar, Straus and Giroux, 2013).
66. William E. Geist, "About New York; Down at 'The Dump,' the Healing Power of Art," *New York Times*, June 2, 1984.
67. James Barron, "Art Work Is (Yes, Really) Garbage," *New York Times*, June 10, 1993. See also "Why Sanitation Can Be Used as a Model for Public Art," statement by Mierle Laderman Ukeles, artist-in-residence, NYC Sanitation, May 8, 1984.
68. Oral interview, Martin Melosi with Norman Steisel, November 15, 2017, New York City.
69. Battaglia, "'Maintenance Art' Puts Trash in Full View"; Feldman, "Inside the Sanitation System," 52; Amy Zimmer, "One City's Trash Truck," *Metro NY*, February 14, 2007, 5; Finkelpearl, *Dialogues in Public Art*, 297–98.
70. Krug, "Ecological Restoration"; Mierle Laderman Ukeles, "Flow City," *Grand Street* 59 (Summer 1996): 209; Feldman, "Inside the Sanitation System," 53; Phillips, "Making Necessity Art," 139, 146, 150, 152.
71. Ukeles, "Flow City," 209.
72. Ukeles, "Flow City," 27, 170.
73. Mierle Laderman Ukeles, "A Journey: Earth/City/Flow," *Art Journal* (Summer 1992): 12.
74. Natalie Stanchfield, "Interview with Mierle Laderman Ukeles," *ArtSlant*, May 2008, https://www.artslant.com/ny/articles/show/1592-interview-with-mierle-laderman-ukeles.
75. Phillips, "Making Necessity Art," 169–70, 176–79.
76. Quoted in Phillips, "Making Necessity Art," 178. See also oral interview, Martin Melosi with Phillip Gleason and Robin Geller, August 14, 2017, New York City.
77. Phillips, "Making Necessity Art," 177–78.
78. Quoted in Cembalest, "Talking Trash with Mierle Laderman Ukeles." See also Phillips, "Making Necessity Art," 178; oral interview, Martin Melosi with Mierle Laderman Ukeles, August 17, 2017. Ukeles stated in the interview that in about 1989 there was an end-use design team in the DSNY composed of representatives from the Sasaki, Walker and Associates landscape firm, Andropogon Associates ecological landscape planners, and SCS engineering company, and an ornithologist. She participated in their deliberations, but it only lasted a few years into the 1990s without producing a report.

79. Ukeles, "Leftovers"; Phillips, "Making Necessity Art," 178–79. See also Kastner, "The Department of Sanitation's Artist in Residence."
80. Ukeles, "Leftovers." See also Emily Gertz, "Fresh Kills: An Unnatural Context," *WorldChanging*, April 2, 2004; Daniel Belasco, "Learning from Landfill: Artist Mierle Ukeles Finds Great Potential, and New Found Sadness, in Fresh Kills," *New York Jewish Week*, October 26, 2001.
81. Phillips, "Making Necessity Art," 183. See also oral interview, Martin Melosi with Mierle Laderman Ukeles, August 17, 2017.
82. "Studio Visit: James Corner Field Operations," World-Architects, Profiles of Selected Architects, November 2014, https://www.world-architects.com/en/architecture-news/insight/studio-visit-james-corner-field-operations.
83. "Studio Visit: James Corner Field Operations."
84. Quoted in Elizabeth Barlow Rogers, *Green Metropolis: The Extraordinary Landscapes of New York City as Nature, History, and Design* (New York: Knopf, 2016), 172, 174.
85. Rogers, *Green Metropolis*, 179.
86. Pollak, "Sublime Matters: Fresh Kills," 61.
87. Quoted in Krinke, "Fresh Ideas?" 84–85.
88. "Landfill to 'Lifescape,'" *Staten Island Advance*, December 20, 2001.
89. "Fresh Kills Park Project: Project History." See also Thomas P. Hayden, "A New Life for Old Trash," *U.S. News & World Report*, May 1, 2004.
90. "Green Dream: Mike's Park Plan for S.I. Dump," *New York Post*, September 30, 2003.
91. "New York's New Parkland, Fresh Kills," fact sheet, http://www1.nyc.gov/assets/planning/download/pdf/plans/fkl/factsheet2.pdf.
92. David Andreatta, "Landfill Priority: Roads—Transportation Will Be Addressed First by Panel of Officials, Civic Leaders," *Staten Island Advance*, September 20, 2003. See also Jim Johnson, "Killer Blueprint; NYC Invests $3 Million in Landfill Redevelopment," *Waste News* (October 13, 2003): 1; "New York Launches First Phase of Fresh Kills Reuse Plan," *Waste Age* 34 (November 2003): 10.
93. "Landfill to 'Lifescape'—Our Opinion—The Issue: Planners Sketching Fresh Kills' Future—Where We Stand Now, Instead of What Was, What Could Be Is What Matters," *Staten Island Advance*, March 14, 2004. See also Nyback, "Dump Dreaming-Visionaries."
94. "The Mayor's Message on Fresh Kills—Urges Islanders to Get Involved—Beginning Wednesday—In the Creation of the City's Biggest Park," *Staten Island Advance*, March 21, 2004.
95. "The Mayor's Message on Fresh Kills"; Andreatta, "Landfill Priority."
96. Staten Island Growth Management Task Force, *Final Report: Recommendations to Mayor Michael R. Bloomberg*, December 2, 2003, 3, 32.
97. Glenn Nyback and Seth Solomonow, "What's a Dump When Not a Dump?—Residents Supply Planners with a Myriad of Suggestions on How Best to Utilize Former Landfill Acreage," *Staten Island Advance*, March 25, 2004. See also Diane Horning, "Park, Landfill, and 9/11," *New York Times*, October 6, 2003; Glenn Nyback, "Planners Seek Islanders' Input on Fresh Kills Park," *Staten Island Advance*, May 9, 2004; Glenn Nyback, "Developers Want Your Opinion on Arden Heights Fresh Kills Park Plan," *Staten Island Advance*, June 7, 2004; Anthony Depalma, "Landfill, Park . . . Final Resting Place?" *New York Times*, June 14, 2004; Glenn Nyback, "Workshop on Fresh Kills 9/11 Memorial," *Staten Island Advance*, August 9, 2004; Glenn Nyback, "Today's Trash Mounds Could Be Tomorrow's Fairways," *Staten Island Advance*, August 22,

2004; Glenn Nyback, "Fresh Kills Landfill: New Olympics Spot," *Staten Island Advance*, November 9, 2004; Mike McIntire, "Gazing Upon a Landfill, the Mayor Sees a Park of Olympic Dreams," *New York Times*, November 10, 2004.

98. "Landfill Parkland a Potential Magnet for Business Patrons," *Staten Island Advance*, March 22, 2005.
99. Jim Johnson, "Fresh Kills' Next Life Begins to Take Shape," *Waste News*, November 10, 2003, 15.
100. "Public Review," Freshkills Park Alliance, http://freshkillspark.org/the-park/public-review.
101. Glenn Nyback, "Design Team to Present Draft of Landfill Park," *Staten Island Advance*, June 10, 2005.
102. "Mayor Bloomberg Announces First Park Project of the Fresh Kills Redevelopment Master Plan," press release, August 22, 2005, http://www1.nyc.gov/office-of-the-mayor/news/324-05/mayor-bloomberg-first-park-project-the-fresh-kills-redevelopment-master-plan. See also "Plans Progress for Fresh Kills," *New York Construction* 53 (November 1, 2005): 17.
103. Jim Rutenberg, "Bloomberg Announces Plan for Fresh Kills Park on Staten Island," *New York Times*, August 23, 2005. See also Julia Levy, "Mayor Says Park Plan for Fresh Kills Would Redress 'a Historic Wrong,'" *New York Sun*, September 30, 2003.
104. "Owl Hollow Fields—Completed 2013," freshkillspark.org, https://freshkillspark.org/design-construction/owl-hollow-fields.
105. Private bus tours continue to this day. They are offered on selected weekdays between April and November. "Private Bus Tours," freshkillspark.org, https://freshkillspark.org/tour-information.
106. Yoav Gonen, "Public Gets a Chance to Tour Infamous Landfill," *Staten Island Advance*, September 25, 2005.
107. Glenn Nyback, "The Greening of the Garbage—Fresh Kills: An Extreme Makeover," *Staten Island Advance*, October 30, 2005; "Fresh Kills Park Timeline," *Staten Island Advance*, October 30, 2005; Rob Hart, "CB3 Airs Concerns on Fresh Kills Park Plans," *Staten Island Advance*, November 16, 2005.
108. Michael Tomasky, "Talking Trash," *New York Magazine*, March 18, 2002. See also "Latest 'Where Will the Garbage Go' Report Released by the NYS Legislative Commission on Solid Waste Management," press release, December 11, 2003; Joyce Purnick, "The Politics of Garbage, Forever Ripe," *New York Times*, March 14, 2002; "Rethinking Garbage," *New York Times*, March 2, 2002; Kirk Johnson, "To City's Burden, Add 11,000 Tons of Daily Trash," *New York Times*, February 28, 2002.
109. Comrov, "Fresh Kills Dumped," 16, 41. Using transfer stations in Manhattan to move refuse to out-of-state disposal sites remained very unpopular. The group Environmental Defense, however, presented a plan to make it more palatable. See Ramon J. Cruz, Thomas Outerbridge, and James T. B. Tripp, *Trash and the City: Toward a Cleaner, More Equitable Waste Transfer System in Manhattan* (New York: Environmental Defense, 2004). Of the nineteen stations that take household refuse, six were in Queens, eight in Brooklyn, and five in the Bronx. None were in Manhattan in 2005. Some local leaders tried to keep it that way. See Charisse Jones, "A Whiff of Politics in Gotham's Garbage," *USA Today*, July 19, 2005. See also Ian Urbina, "City Trash Plan Forgoes Trucks, Favoring Barges," *New York Times*, October 7, 2004.
110. Michael Cooper, "A Plan to Ship Garbage, but No Destination," *New York Times*, August 1, 2002.
111. Comrov, "Fresh Kills Dumped," 15, 40–41.

112. "New York Mayor Proposes Trash Plan," *Waste Age*, September 2002, 6. See also Eric Lipton, "Great Idea for Trash, If It Works, Most Say," *New York Times*, August 1, 2002; "NYC Seeks Alternative to Trucking Garbage," *Engineering News-Record* 249 (August 12, 2002): 7; Comrov, "Fresh Kills Dumped," 15; Kirk Johnson, "As Options Shrink, New York Revisits Idea of Incineration," *New York Times*, March 23, 2002; Diane Cardwell, "Mayor Drops Incinerator Plan," *New York Times*, May 18, 2002.

113. Kirk Johnson, "Gold in Them Thar Tin Cans? Recycling Sees Money to Be Made from City's Containers," *New York Times*, January 11, 2003; Michael Cooper, "City to Resume Recycling of Plastics," *New York Times*, January 14, 2003. See also William C. Thompson Jr. to A. Gifford Miller (Speaker, New York City Council), May 19, 2003; and William Thompson Jr., "Projected Cost of Collection and Managing Residential Waste in Fiscal Year 2004," submitted to the Sanitation Committee of the New York City Council, City of New York, Office of the Comptroller. See also Joe Truini, "Study Highlights High Cost of NYC Recycling Program," *Waste News*, February 16, 2004, 1, 27. See also Joe Truini, "NYC Pact's Effect Expected to Ripple Out," *Waste Age*, September 27, 2004, 1, 25.

114. Resa Dimino and Barbara Warren, *Reaching for Zero: The Citizen Plan for Zero Waste in New York City* (New York City Zero Waste Campaign and Consumer Policy Institute/Consumers Union, June 2004), https://advocacy.consumerreports.org/research/reaching-for-zero-the-citizens-plan-for-zero-waste-in-new-york-city/. The DSNY did have a composting program in place since 1990, which it continued to pursue into the 2000s. It also conducted several waste prevention pilot programs for residential and commercial waste, with varying degrees of success. See Comrov, "Fresh Kills Dumped," 90–97; Rich Flammer, "Composting Comes Back to the Big Apple," *BioCycle* 46 (September 2005): 29.

115. Heather Rogers, *Gone Tomorrow: The Hidden Life of Garbage* (New York: The New Press, 2005), 224–27.

116. Nichole M. Christian, "Officials Say Mayor's Plan for Garbage Could Take 6 Years," *New York Times*, January 29, 2003. See also Don Costello, "Transfer Stations and Material Recovery Facilities: The Solution to Shrinking Solid Waste Disposal Options," *HDR Innovations* 11 (Summer 2003): 1–3; "Fine-Tuning NYC's Future," *Waste Age*, August 2004. Jim Johnson, "NYC Mayor Wrestles with Solid Waste Plans," *Waste Age*, August 2, 2004, 4.

117. Johnson, "NYC Mayor Wrestles with Solid Waste Plans."

118. New York City Department of Sanitation, *Annual Report, 2002–2003*, 15. See also "New York City Department of Sanitation, Solid Waste Disposal Contracts, 2003-N-9," in *A Report by the New York State Office of the State Comptroller*, 4.

119. City of New York, Department of City Planning, *Ten-Year Capital Strategy, Fiscal Years 2004–2013*, April 15, 2003, i, 49.

120. "Mayor Michael R. Bloomberg and Sanitation Commissioner John J. Doherty Present Solid Waste Management Plan," press release, October 7, 2004.

121. "Mayor Michael R. Bloomberg and Sanitation Commissioner John J. Doherty Present Solid Waste Management Plan."

122. City of New York, Department of Sanitation, *Draft Comprehensive Solid Waste Management Plan, Executive Summary*, October 2004, ES-1, ES-4. See also New York City Department of Sanitation, *Annual Report, 2004*, 20–22.

123. Jim Johnson, "Its Barges, Rail for NYC's Trash; Mayor's New Plan Sets Target for 2007," *Waste News*, October 11, 2004, 1. See also Stephen Ursery, "By Train or Barge,"

Waste Age 35 (November 2004): 6; "New York Garbage Adapts to Life After Fresh Kills," *New York Times*, August 12, 2005.
124. City of New York, Office of the Comptroller, Office of Policy Management, *No Room to Move: New York City's Impending Solid Waste Crisis*, October 2004, 1.
125. City of New York, Office of the Comptroller, Office of Policy Management, *No Room to Move*, 1.
126. City of New York, Office of the Comptroller, Office of Policy Management, *No Room to Move*, 3, 4–7. See also "Under New Plan Cost of Disposing Curbside Waste Grows, for Now," *Inside the Budget* 138 (June 2, 2005); City of New York, Office of Management and Budget and Department of City Planning, *Preliminary Ten-Year Capital Strategy, Fiscal Years 2006–2015*, i, 50, 123.
127. "Fresh Kills Park Project: Introduction," http://www1.nyc.gov/assets/planning/download/pdf/plans/fkl/fkl.pdf.
128. New York City Department of Sanitation, *Annual Report*, 2005, 20–22.

20. CROSSROADS

1. Mayank Teotia, "Managing New York City Municipal Solid Waste—Using Anaerobic Digestion," Program for Sustainable Planning and Development, School of Architecture, Pratt Institute, Spring 2013, 15–16.
2. In most cases, the greatest amounts came from Brooklyn, followed by Queens, Manhattan, the Bronx, and then Staten Island.
3. City of New York, Department of Sanitation, *Comprehensive Solid Waste Management Plan*, September 2006, ES-1. See also *DSNY Annual Report, 2011*, 21.
4. "Fresh Kills Park Transformation Takes Major Step Forward with Release of Draft Master Plan Illustrating Planned Features of the Park," press release, April 6, 2006. Brookfield Landfill also was turned into a park. Mark D. Stein, "Staten Island's Brookfield Landfill on Track to Be a City Park in 2017," *Staten Island Advance*, June 20, 2012; "Brookfield Park," NYC Parks, https://www.nycgovparks.org/parks/brookfield-park.
5. City of New York, Department of Sanitation, *Comprehensive Solid Waste Management Plan*, ES-1.
6. John J. Doherty, memo, February 13, 2006, New York City, Department of Sanitation, *Lead Agency Findings Statement for the New York City Comprehensive Solid Waste Management Plan*, February 2006, http://www.nyc.gov/sanitation.
7. City of New York, Department of Sanitation, *Comprehensive Solid Waste Management Plan*, ES-2. See also City of New York, Department of Sanitation, *Environmental Justice Informational Meeting*, Flushing, New York, April 17, 2007. See also Talia Groom, "Grassroots Voices in the Second Garbage War: Social Equity and NYC's Latest Solid Waste Management Plan," *Cities in the 21st Century* 3 (Fall 2012): 1–35.
8. City of New York, Department of Sanitation, *Comprehensive Solid Waste Management Plan*, ES-3–ES-4.
9. Martin V. Melosi, *Garbage in the Cities: Refuse, Reform, and the Environment*, rev. ed. (Pittsburgh, PA: University of Pittsburgh Press, 2005), 222.
10. City of New York, Department of Sanitation, *Comprehensive Solid Waste Management Plan*, ES-3–ES-4.
11. City of New York, Department of Sanitation, *Comprehensive Solid Waste Management Plan*, ES-8–ES-11. See also New York City, Department of Sanitation, *Lead*

Agency Findings Statement for the New York City Comprehensive Solid Waste Management Plan, February 2006, CEQR no. 03-DOS-004Y, 1, 8, 9–15, 45, 75–76, 88–89.

12. The city currently had seventeen contracts with private firms to dispose of city waste collected by the DSNY.
13. Steven Cohen, "Wasted: New York City's Giant Garbage Problem," *Observer*, April 3, 2008. See also Committee on Sanitation & Solid Waste Management, *Fiscal 2009: Preliminary Budget Hearings*, March 2008; Committee on Finance Jointly with the Committee on Sanitation & Solid Waste Management, *Fiscal 2009: Executive Budget Hearings*, May 2008; New York City Council, Finance Division, *Hearing on the Mayor's Fiscal Year 2011 Executive Budget, Department of Sanitation*, June 1, 2010, 1; City of New York, Office of Management and Budget, *February 2011 Financial Plan, Fiscal Years 2011–2015*, February 17, 2011, E-16.
14. "Under New Plan Cost of Disposing Curbside Waste Grows, for Now," *Inside the Budget* 138, June 2, 2005, New York City Independent Budget Office, http://www.ibo.nyc.ny.us/newsfax/insidethebudget138.pdf. See also Derek Sylvan, "Municipal Solid Waste in New York City: An Economic and Environmental Analysis of Disposal Options," 2017, 7, http://nylcvef.org/wp-content/uploads/2017/08/Solid-Waste-Background-Paper.pdf. The *Ten-Year Capital Strategy for Fiscal Years 2004–2013* showed sanitation received 5 percent of the total $49.3 billion in all funds (or $2.3 billion) for departmental management, equipment acquisition, garages and facilities, and disposal infrastructure not related to the long-term waste export plan. In FY 2005 $493.2 million was provided for construction of the marine transfer stations. City of New York, Office of Management and Budget, Department of City Planning, *Ten-Year Capital Strategy, Fiscal Years 2004–2013*, i, 49.
15. Amanda Smith-Teutsch, "NYC Looks to Expand Recycling," *Waste & Recycling*, April 26, 2010, 1, 39.
16. Sylvan, "Municipal Solid Waste in New York City," 6.
17. See, for example, Concerned Citizens of Cattaraugus County, Inc., "New York City's Garage Crisis," July 25, 2006, http://concernedcitizens.homestead.com/fkfacts.html.
18. Alex Hutchinson, "Is Recycling Worth It?" *Popular Mechanics*, November 12, 2008, https://www.popularmechanics.com/science/environment/a3752/4291566/. See also Norman Steisel and Benjamin Miller, "Power for Trash," *New York Times*, April 27, 2010; Sylvan, "Municipal Solid Waste in New York City," 8.
19. Hutchinson, "Is Recycling Worth It?"; Sylvan, "Municipal Solid Waste in New York City," 4.
20. *United Haulers Association, Inc. v. Oneida-Herkimer Solid Waste Management Authority*, https://caselaw.findlaw.com/us-2nd-circuit/1302178.html; Steve Barlas, "Legality of Flow-Control Ordinances Upheld," *Public Works*, July 2, 2007, https://www.pwmag.com/water-sewer/stormwater/legality-of-flow-control-ordinances-upheld_o.
21. Barlas, "Legality of Flow-Control Ordinances Upheld."
22. Tommy Lavender, "An Update of Flow Control Jurisprudence Since United Haulers," *Nexsen/Pruet*, August 29, 2014, https://www.nexsenpruet.com/insights/an-update-of-flow-control-jurisprudence-since-united-haulers. In the Fiscal 2009 Adopted Budget, the DSNY's waste export program had a budget of $315.8 million; an increase was forecast for 2010. See New York City Council, *Budget Report, Analysis of the Fiscal 2010 Preliminary Budget and Fiscal 2009 Preliminary Mayor's Management Report for the Department of Sanitation*, March 19, 2009, 7.
23. Amy Eddings, "Ten Years After Closure, Fresh Kills Is Still a Landfill in Transition," *WNYC News*, March 21, 2011, https://www.wnyc.org/story/118960-blog-fresh-kills/.

24. City of New York, Department of Sanitation, *Comprehensive Solid Waste Management Plan*, attachment X, 1–13. See also *DSNY Annual Report, 2006*, 20.
25. City of New York, Department of Sanitation, *Fresh Kills: From Landfill to Parkland, Closing the Fresh Kills Landfill*, Spring 2008; Sewell Chan, "Cost to Close S.I. Landfill Could Exceed $1.4 Billion," *New York Times*, May 18, 2006. See also NYC Sanitation, *Solid Waste Management Plan: Post Fresh Kills Landfill*, June 10, 2015.
26. Anna Sanders, "No Money for Fresh Kills Dump Health Risks Study Yet," *Staten Island Advance*, June 5, 2017; "Landfill Health Study: Did the Dump Make Us Sick?" *Staten Island Advance*, December 15, 2017. See also New York City, Department of Sanitation, *DSNY Annual Report 2012*, 25.
27. Citizens Budget Commission, *Taxes In, Garbage Out: The Need for Better Solid Waste Disposal Policies in New York City*, May 2012, foreword, i, 10–12, https://cbcny.org/sites/default/files/REPORT_SolidWaste_053312012.pdf.
28. Vilhelm Carlstron, "Foreign Media Reports Sweden Has Run Out of Garbage and Is Forced to Import—Here's What Is Really Going On," *Business Insider, Nordic*, December 14, 2016.
29. Citizens Budget Commission, *Taxes In, Garbage Out*, i–ii, 15–21.
30. Citizens Budget Commission, *Taxes In, Garbage Out*, 2–4, 26.
31. Citizens Budget Commission, *12 Things New Yorkers Should Know About Their Garbage*, May 21, 2014, https://cbcny.org/research/12-things-new-yorkers-should-know-about-their-garbage.
32. Citizens Budget Commission, *Getting the Fiscal Waste Out of Solid Waste Collection in New York City*, September 23, 2014, https://cbcny.org/sites/default/files/media/files/REPORT_SolidWaste_09232014.pdf; Citizens Budget Commission, *A Better Way to Pay for Solid Waste Management*, February 5, 2015, https://cbcny.org/research/better-way-pay-solid-waste-management.
33. Yana Manevich, "A TimeLine of Solid Waste Management in New York City," *Shaping the Future of New York City*, March 17, 2013, https://macaulay.cuny.edu/eportfolios/macbride13/research/a-timeline-of-solid-waste-management-in-new-york-city/. See also City of New York, *plaNYC, Update April 2011*, April 2011.
34. Christopher Rizzo and Michael K. Plumb, "Waste-to-Energy Facilities in New York City: Challenges and Opportunities," *New York Law Journal*, March 8, 2012, Carter Ledyard & Milburn LLP, https://www.clm.com/publication.cfm?ID=370. See also Teotia, "Managing New York City Municipal Solid Waste," 10.
35. Sylvan, "Municipal Solid Waste in New York City," 10; Bill Hughes, "Fiscal Woes, Long-Held Fears Spur Waste-to-Energy Debate," *citylimits.org*, October 12, 2012, https://citylimits.org/2012/10/10/fiscal-woes-long-held-fears-spur-waste-to-energy-debate/.
36. "Protests Target Staten Island Waste-to-Energy Plant: If You Care, Be There," *Staten Island Advance*, April 1, 2012.
37. Jon Lentz, "Bloomberg Pulls Fresh Kills from Waste-to-Energy RFP," *Strausmedia*, April 16, 2012, http://www.nypress.com/bloomberg-pulls-fresh-kills-site-from-waste-to-energy-rfp/.
38. Hughes, "Fiscal Woes, Long-Held Fears Spur Waste-to-Energy Debate." See also Mireya Navarro, "New York City Gets High Marks for Environmental Awareness, but Containers Add to a Mountain of Waste," *New York Times*, October 23, 2001.
39. Andrew J. Hawkins, "NYC Wants to Burn Its Trash Problem," *Waste & Recycling News* 18 (September 17, 2012): 3.
40. "Hurricane Sandy—October 29, 2012," National Weather Service, https://www.weather.gov/okx/HurricaneSandy.

41. John P. Rafferty, "Superstorm Sandy," *Britannica.com*, October 12, 2018, https://www.britannica.com/event/Superstorm-Sandy.
42. Crystal Gammon, "Why Hurricane Sandy Hit Staten Island So Hard," *Live Science*, November 7, 2012, https://www.livescience.com/24616-hurricane-sandy-staten-island-effects.html. See also Michael Kimmelman, "Former Landfill, a Park to Be, Proves a Savior in the Hurricane," *New York Times*, December 17, 2012.
43. Eric Lipton and Kirk Semple, "NYC Revives Fresh Kills Landfill for Storm Debris," *Star Tribune*, November 17, 2012, http://www.startribune.com/nyc-revives-fresh-kills-landfill-for-storm-debris/179820051/. See also Sgt. Ferdinand Detres, "Corps of Engineers Works with N.Y. Dept. of Sanitation in Hurricane Sandy Debris Removal," November 14, 2012, https://www.army.mil/article/91150/corps_of_engineers_works_with_ny_dept_of_sanitation_in_hurricane_sandy_debris_removal; New York City, Department of Sanitation, *DSNY 2013*, 42; Irina Vinnitskaya, "Landfill Reclamation: Fresh Kills Park Develops as a Natural Coastal Buffer and Parkland for Staten Island," *ArchDaily*, March 3, 2013, https://www.archdaily.com/339133/landfill-reclamation-fresh-kills-park-develops-as-a-natural-coastal-buffer-and-parkland-for-staten-island.
44. "The Mayor Rethinks Recycling," *New York Times*, April 29, 2013. See also Andy Newman, "City Expands Recycling Program to Include Hard Plastics," *New York Times*, April 24, 2013.
45. Citizens Budget Commission, "Testimony on the Department of Sanitation's Waste Characterization Study," April 24, 2018, https://cbcny.org/sites/default/files/media/files/TESTIMONY_04232018.pdf.
46. "Mayor Bloomberg Launches New 'Recycle Everything' Ads and Announces Expansion of the Organic Food Waste Recycling Program," nyc.gov, July 29, 2013, https://www1.nyc.gov/office-of-the-mayor/news/260-13/mayor-bloomberg-launches-new-recycle-everything-ads-announces-expansion-the-organic-food.
47. "Bill de Blasio," Biography.com, https://www.biography.com/people/bill-de-blasio-21388699.
48. Jillian Jorgensen, "Sanitation Commissioner John Doherty Stepping Down This Month," *Staten Island Advance*, March 14, 2014; Yoav Gonen, "NYC Sanitation Boss John Doherty Announces Retirement," *New York Post*, March 14, 2014.
49. Michael M. Grynbaum, "New York Sanitation Chief Closes a 54-Year Career," *New York Times*, March 28, 2014.
50. "Mayor de Blasio Appoints Kathryn Garcia as Department of Sanitation Commissioner," press release, March 15, 2014, https://www1.nyc.gov/office-of-the-mayor/news/091-14/mayor-de-blasio-appoints-kathryn-garcia-department-sanitation-commissioner#/0.
51. NYC Sanitation, 2014–2015 *Biennial Report*, 9.
52. "What Is Zero Waste?" *GrassRoots Recycling Network*, http://www.grrn.org/page/what-zero-waste.
53. "What Is Zero Waste?" Waste Management Media Room, January 9, 2018, http://mediaroom.wm.com/what-is-zero-waste/; Gary Liss, "What Is Zero Waste?" National Recycling Coalition, August 8, 2016, https://nrcrecycles.org/blog/2016/08/08/what-is-zero-waste/; EPA, "How Communities Have Defined Zero Waste," https://www.epa.gov/transforming-waste-tool/how-communities-have-defined-zero-waste.
54. NYC Sanitation, *2016 Strategic Plan*, 4. Garcia stated that the plan was the result of dozens of town-hall meetings, retreats, focus groups, presentations, and input from employees (now 2,700 strong).

55. In 2017, 34 percent curbside recyclables, 34 percent organics, 9 percent other divertible material, and 23 percent other materials. City of New York, Department of Sanitation, *2013 NYC Curbside Waste Characterization Study*, 8, https://dsny.cityofnewyork.us/wp-content/uploads/2017/12/2013-Waste-Characterization-Study.pdf. See also NYC Sanitation, *2017 NYC Residential, School, and NYCHA Waste Characterization Study*, 11, https://dsny.cityofnewyork.us/wp-content/uploads/2018/04/2017-Waste-Characterization-Study.pdf; Mariela Quintana, "The Future of NYC's Waste: Getting to Zero," *Street Easy*, March 1, 2017, https://streeteasy.com/blog/where-nyc-garbage-goes/.
56. In May 2013 the DSNY began a voluntary residential organics diversion pilot program with 3,250 households in Staten Island. Three major expansions followed, which included more than 186,600 households not only in Staten Island but also in the Bronx, Brooklyn, and Queens. The focus on organics was in major part environmental. A study showed in 2013 and in subsequent years that DSNY-landfilled refuse generated 1 million tons of greenhouse-gas emissions, largely because of organic waste. Citizens Budget Commission, *Can We Have Our Cake and Compost It Too?* February 2, 2016, https://cbcny.org/research/can-we-have-our-cake-and-compost-it-too. See also Benjamin Miller and Juliette Spertus, "Encouraging the Development of Processing Infrastructure for New York City's Organics Waste Stream," *Environmental Law in New York* 26 (May 2015): 75–85.
57. NYC Sanitation, *2016 Strategic Plan*, 5ff.
58. Emily Bobrow, "No Time to Waste," *Crain's New York Business* 32 (July 11, 2016): 13.
59. Bobrow, "No Time to Waste."
60. DSM Environmental, *Analysis of New York City Department of Sanitation Curbside Recycling and Refuse Costs, Final Report*, May 2008, ii.
61. City of New York, Department of Sanitation, *Annual Reports, 2006–2013*. DSNY figures on recycling per week during this same period showed a decline from 12,200 tons in 2006 to 10,114 tons in 2013. The de Blasio administration sought to reverse that trend.
62. "Exporting City's Trash Is Getting More Expensive Years After Staten Island's Dump," *Staten Island Advance*, March 24, 2017.
63. "Mayor de Blasio and Speaker Johnson Celebrate Signing of Waste Equity Legislation," press release, August 16, 2018, https://www1.nyc.gov/office-of-the-mayor/news/417-18/mayor-de-blasio-speaker-johnson-celebrate-signing-waste-equity-legislation; Eric A. Goldstein, "NYC Takes a Step Toward Environmental Justice," NRDC, July 18, 2018, https://www.nrdc.org/experts/eric-goldstein/nyc-takes-step-toward-environmental-justice.
64. Kendall Christiensen, "Don't Let Election-Year Politics Lay Waste to a Crucial Industry," *Crain's New York Business*, September 13, 2017, https://www.crainsnewyork.com/article/20170913/OPINION/170919963/op-ed-don-t-let-election-year-politics-lay-waste-to-a-crucial-industry.
65. See Benjamin Miller, "Managing New York's Municipal Solid Waste to Support the City's Goal of Reducing Greenhouse Gases by 80% by 2050," *Environmental Law in New York* 29 (February 2018): 19–30.
66. NYC Sanitation, *2017 Annual Report*, 7.
67. NYC Sanitation, *2016 Strategic Plan*, 24.
68. At this juncture, the name of the site was "Fresh Kills Park." It was later changed to "Freshkills Park." See *Fresh Kills Park: Lifescape, Staten Island, New York, Draft Master Plan*, March 2006, 2.

69. See *Fresh Kills Park: Lifescape, Staten Island, New York, Draft Master Plan*, March 2006, 4, 8.
70. See Laura Bliss, "The Wild Comeback of New York's Legendary Landfill," CityLab, https://www.citylab.com/solutions/2017/02/the-wild-comeback-of-new-yorks-legendary-landfill/516822/; "A Birder's Heaven: Just Follow the Stench to the Landfill," *New York Times*, January 28, 2017; Joe Trezza, "Where Coyotes, Foxes and Bobolinks Find a New Home: Freshkills Park," *New York Times*, June 9, 2016; Nick Kimbrell, "Can a Landfill Site Ever Return to Nature?" *Ecologist* 40 (August 2010): 4–6; "Turning Trash Piles Into a Bird-Watcher's Paradise," *New York Times*, January 26, 2010; Sam Williams, "Wind Power in NYC," *Gotham Gazette*, March 8, 2006, http://www.gothamgazette.com/index.php/environment/3181-wind-power-in-nyc; "A Wind-Win Situation," *Staten Island Advance*, August 26, 2007; "German Co. Looks to Build Solar on Fresh Kills," *Power Finance & Risk*, September 29, 2008; "Staten Island Wind Project Waits on Parks Dept. Approval," *Power Finance & Risk*, September 1, 2008; Erin Einhorn, "Windmills in Fresh Kills Hardly Hot Air," *New York Daily News*, May 20, 2011.
71. "Interview with Mierle Laderman Ukeles," *Public Art Dialogue* 9 (Winter 2017), https://publicartdialogue.org/newsletter/winter-2017/interview-mierle-laderman-ukeles.
72. *Landing* has three components: the cantilevered *Overlook* and the two earthworks, *Earth Bench* and *Earth Triangle*. As stated on the Freshkills Park Alliance website: "These three components will be connected by the newly graded and ADA-accessible landbridge which—at nearly 500 feet long and some 32 feet above the tidal inlet below—connects the two mounds that form South Park. Each component invites visitors to 'land' and to experience this new kind of public place in three different ways: perched and floating (*Overlook*), standing solidly on the land itself (*Earth Bench*) and being within the land which becomes a sheltering refuge (*Earth Triangle*).The *Overlook* is composed of two paths that form an intersection of the human-made and the ancient natural environments. Path One points to a clear unobstructed two-mile long view of the evolving technologically designed urban park landscape: what the former landfill areas will become. Path Two reveals the 'Ecological Theater' of wetlands and tidal inlet directly below: what this place was originally. Both exist in harmony to form the new park." See "The Landing," Freshkills Park Alliance, https://freshkillspark.org/os-art/landing. See also Patricia C. Phillips, "Making Necessity Art: Collisions of Maintenance and Freedom," in *Mierle Laderman Ukeles: Maintenance Art*, ed. Patricia C. Phillips et al. (Munich: Del Monico, 2016), 184–86; Mierle Laderman Ukeles, *Percent for Art Project, Fresh Kills Park: Alternative Conceptual Design of the Overlook, March 12, 2012*. Ukeles sensed some reluctance on the part of Field Operations to give up "big idea" planning at the site to an artist, but the relationship between the parties was compatible. See oral interview, Martin Melosi with Mierle Laderman Ukeles, August 17, 2017, New York City.
73. Oral interview, Martin Melosi with Mierle Laderman Ukeles, August 17, 2017, New York City.
74. Andy Battaglia, "'Maintenance Art' Puts Trash in Full View," *Wall Street Journal*, September 22, 2016.
75. Press Release, "LAGI Partnership with NYC Department of Parks & Recreation Announces Opening of Design Competition," Society for Cultural Exchange, Inc., January 3, 2012; Society for Cultural Exchange, *2012 Annual Report*, 1–5. See also Shannon Lee, "From 'Ew' to a View: How a Former Staten Island Landfill Could

Represent the Future of Art in Parks," *Artspace*, April 5, 2018, https://www.artspace.com/magazine/interviews_features/watch_this_space/how-a-former-staten-island-landfill-could-represent-the-future-of-art-in-parks-55364; Benjamin A. Lawson, "Garbage Mountains: The Use, Redevelopment, and Artistic Representations of New York City's Fresh Kills, Greater Toronto's Keele Valley, and Tel Aviv's Hiriya Landfills," PhD diss., University of Iowa, Fall 2015; oral interview, Martin Melosi with Eloise Hirsh and Mariel Villere, November 11, 2017, New York City.

76. Other agency partners included the DSNY, the Department of City Planning, the Department of Transportation, the New York State Department of Environmental Conservation, the New York State Department of Transportation, and the U.S. Department of State.

77. "Fresh Kills Park Transformation Takes Major Step Forward with Release of Draft Master Plan Illustrating Planned Features of the Park," press release, April 6, 2006, https://www1.nyc.gov/assets/planning/download/pdf/about/press-releases/pr040606.pdf. See also *Fresh Kills Park: Lifescape, Staten Island, New York, Draft Master Plan*, 9–10, 14–57.

78. *Fresh Kills Park: Lifescape, Staten Island, New York, Draft Master Plan*, 8.

79. Hirsh also took charge of Freshkills Park Alliance, a not-for-profit partner with the city to develop the park, when it was established in 2010.

80. "Eloise Hirsh," Freshkills Park Alliance, https://freshkillspark.org/people/eloise-hirsh. See also oral interview, Martin Melosi with Eloise Hirsh and Mariel Villere, November 11, 2017, New York City.

81. See New York City, Department of Parks & Recreation, *Statement of Findings: New York City Department of Parks & Recreation, Fresh Kills Park Project*, October 30, 2009; New York City, Department of Parks and Recreation, *Fresh Kills Park: Final Generic Environmental Impact Statement*, March 13, 2009; "Public Review," FreshkillsPark, https://freshkillspark.org/the-park/public-review.

82. New York City, Department of Parks & Recreation, Hearing Before the City Council, Committee on State and Federal Legislation, Fresh Kills Park, Testimony by Eloise Hirsh, Administrator of Fresh Kills Park, June 5, 2007.

83. Oral interview, Martin Melosi with Eloise Hirsh and Mariel Villere, November 11, 2017, New York City.

84. Oral interview, Martin Melosi with Carrie Grassi, September 11, 2018, New York City. See also Melissa Zavala, "Wild NYC: Building Biodiversity in Fresh Kills and City Parks," PhD diss., Anthropology, City University of New York, 2014, 174–75.

85. Glenn Nyback, "City Planner to Lead Transformation of Dump Into Oasis," *Staten Island Advance*, September 28, 2006.

86. See Liz Robbins, "Putrid Past, and a Still-Distant Future as Field of Play," *New York Times*, March 22, 2011. In 2009 Borough President James P. Molinaro criticized the Parks & Recreation Department over its roads policy, calling it "a farce." See "Fresh Kills Putting Its Bad, Old Days Behind It," *Staten Island Advance*, April 26, 2009. See also Glenn Nyback, "City Planners Studying Feasibility of Expanding Roads in Landfill," *Staten Island Advance*, April 23, 2006; "The Future Fresh Kills," *Staten Island Advance*, May 30, 2006.

87. Glenn Nyback, "Final Review of Landfill Park Plan Begins Wednesday—Parks Department Seeks Public Opinion on Its Draft Master Plan for the Massive Park in Fresh Kills," *Staten Island Advance*, May 21, 2006.

88. Glenn Nyback, "Change of Scenery Coming," *Staten Island Advance*, February 1, 2006.

89. Glenn Nyback, "Islanders Aren't Shy in Expressing Their Opinions on Former Landfill—Skeptics, Advocates Don't Hold Back in Discussing Former Dump's Transformation Into Parkland," *Staten Island Advance*, May 21, 2006. See also Glenn Nyack, "Park Plans for Landfill Heard by Civil Crowd," *Staten Island Advance*, May 25, 2006. Similar bitterness was directed at the construction of the new transfer station on Staten Island (constructed in 2005)—another intrusion like the reduction plant or the landfill. See Jeff Vandam, "The Dump Was Closed, but the Rancor Never Ends," *New York Times*, February 5, 2006; "Turning Fresh Kills Into Something We Can Admire," *Staten Island Advance*, April 29, 2007.
90. Zavala, "Wild NYC," 189, 194–95.
91. See Staten Island West Shore, Strategic Vision, *Summary of Public Comments*, March 13, 2008; *2011 Staten Island/Freshkills Parks Resident Survey: Data Report*, May 2012, https://www.nrs.fs.fed.us/nyc/local-resources/downloads/FK_DataReport_FINAL_5June2012.pdf; Christine A. Vogt et al., "Resident Support for a Landfill-to-Park Transformation," *Journal of Park and Recreation Administration* 33 (Winter 2015): 32–50. For the Parks & Recreation Department's assessment of public review, see "Freshkills Park," NYCParks, https://www.nycgovparks.org/park-features/freshkills-park.
92. Hilary Potkewitz, "Talking Trash: Huge Park Is Surprisingly Contentious," *Crain's New York Business*, August 27, 2012, 3.
93. *Staten Island Attractions*, 2007, 4. See also Parsons Brinkerhoff, *Staten Island North Shore, Land Use and Transportation Study, Existing Conditions Report*, December 2008; NYC Economic Development Corporation, NYC Department of City Planning, and Urbitran Team—Consultants, *Staten Island West Shore, Strategic Vision, Advisory Committee Update*, May 15, 2008.
94. Center for an Urban Future, *Staten Island: Then and Now*, May 2011, 3, http://www.nycfuture.org.
95. Center for an Urban Future, *Staten Island: Then and Now*, 4–5.
96. Interstate Environmental Commission, *2005 Annual Report of the Interstate Environmental Commission*, 45, 97, 102–3.
97. Melissa Check, "Staten Island's Toxic Stew," *Gotham Gazette*, May 26, 2009, http://www.gothamgazette.com/index.php/city/227-staten-islands-toxic-stew.
98. Tanay Warerkar, "Staten Island's Fresh Kills Landfill Is Officially on Its Way to Becoming a Public Park," *Curbed*, November 10, 2017, https://ny.curbed.com/2017/11/10/16633948/fresh-kills-north-park-groundbreaking.
99. "News: Garbage Greening in Fresh Kills," *Inprint: Eugene Lang College & New School University Newspaper*, May 9, 2007. Mark Hauber, who was studying various issues related to birds, was the first researcher Hirsh invited to the park site. See Elizabeth Royte, "New York's Fresh Kills Gets an Epic Facelift," *Audubon*, July/August 2015, https://www.audubon.org/magazine/july-august-2015/new-yorks-fresh-kills-landfill-gets-epic. Steven Handel, a restoration ecologist at Rutgers University, spent years at the site. Edward Toth, director of the New York City Parks and Recreation Department's Greenbelt Native Plant Center, called Handel "one of the real pioneers of exploring the urban environment." See Elisabeth Ginsburg, "When Nature Gets a Second Chance," *Christian Science Monitor*, June 18, 2009.
100. Oral interview, Martin Melosi with Eloise Hirsh and Mariel Villere, November 11, 2017, New York City.
101. According to Carrie Grassi, it was decided to develop a generic EIS for the master plan and then focus on specific projects for more detailed review. Remapping the site

20. CROSSROADS   ■   755

to be more in line with the vision of the park under the ULURP has been stalled. Oral interview, Martin Melosi with Carrie Grassi, September 11, 2018, New York City.

102. "Landfill to Park Timeline," Freshkills Park Alliance, https://freshkillspark.org/the-park/chronology-of-the-freshkills-park-site; http://timeline.freshkillspark.org/. See also Peter N. Spencer, "Meeting to Offer Sneak Peak of Future of Fresh Kills," *Staten Island Advance*, April 5, 2010.

103. See "From Behind the Mounds: First Days at Freshkills Park," Freshkills Park Alliance, https://freshkillspark.org/blog/first-days-freshkills.

104. Elizabeth Barlow Rogers, *Green Metropolis: The Extraordinary Landscapes of New York City as Nature, History, and Design* (New York: Knopf, 2016), 178.

105. "Landfill to Park Timeline"; "Landfill-to-Park Times." See also *DSNY Annual Report 2010*, 25–26; "Dynamic South Park Design Now Nearing Completion," *Fresh Perspectives: FreshkillsPark Newsletter*, Spring/Summer 2010, 3; Winter/Spring 2011, 1; "Design + Construction Updates," Freshkills Park Alliance, https://freshkillspark.org/design-construction; Peter N. Spencer, "Bloomberg Presides at Schmul Park Groundbreaking in Travis," *Staten Island Advance*, October 28, 2010; Jamie Lee, "Parks Department in Need of Some Fresh Ideas," *Staten Island Advance*, July 23, 2009; Ula Ilnytzky, "NYC Garbage Dump World's Largest Landfill-to-Park Project," *AP*, September 3, 2016; Ryan Lavis, "Taking a 'Sneak Peak' at Staten Island's Freshkills Park," *Staten Island Advance*, October 8, 2011.

106. For 2014 the city budgeted $4.8 million to the park, plus $3.6 million in noncity money, and committed $31.3 more through 2017.

107. Anna Sanders, "Uncertain Future," *Staten Island Advance*, December 28, 2014. See also FreshkillsPark and Freshkills Park Alliance, *2015 Annual Report*. See also "New Yorker Spotlight: Eloise Hirsh on Turning the Freshkills Landfill Into a Thriving Park," *6sqft.com*, September 26, 2014, https://www.6sqft.com/new-yorker-spotlight-eloise-hirsh-on-reclaiming-freshkills-and-developing-a-park/. Some questions were raised about whether the Freshkills Park project was overshadowed by Field Operations' High Line project, which received considerable public attention. See Stephen Kleege, "Parks Translate to Profits," *Crain's New York Business* 28, no. 43 (October 22, 2012): 15.

108. "De Blasio, Citing Fairness, Puts $150 Million Toward Improving 5 Neglected Parks," *New York Times*, August 19, 2016. See also Josh Dawsey, "New York City to Invest $150 in Five Parks," *Wall Street Journal*, August 17, 2016; FreshkillsPark and the Freshkills Park Alliance, *2016 Annual Report*; "NYC Parks Meets with Staten Island Community to Report Back on Anchor Park Scoping Ideas," *States News Service*, February 3, 2017.

109. Jason Sayer, "James Corner Field Operations' Freshkills Park Moves Closer to Realization," *Architects Newspaper*, August 17, 2017, https://archpaper.com/2017/08/james-corner-field-operations-freshkills-park/.

110. Robert Sullivan, "Wall-E Park," *New York*, December 1, 2008, 30–35, 106–7.

111. Zavala, "Wild NYC," 187, 190, 196, 198.

112. Lavis, "Taking a 'Sneak Peak' at Staten Island's Freshkills Park."

113. Raj Kottamasu, Oral History Projects, DSNY–Freshkills Park, April 3, 2012, http://www.dsnyoralhistoryarchive.org/.

114. Oral interview, Martin Melosi with Eloise Hirsh and Mariel Villere, November 11, 2017, New York City.

115. See Peter Gisolfi, *Reclaiming Spoiled Landscapes*, Planning 77 (February 2011): 25.

116. Ben Adler, "Once a Landfill, Now a Lush Landscape," *Architectural Record* 201 (August 2013): 1.

CONCLUSION

1. Carolyn Wheat, *Fresh Kills* (New York: Berkeley Prime Crime, 1995), 2.
2. Bill Loehfelm, *Fresh Kills* (New York: G. P. Putnam's Sons, 2008), 73.
3. Christopher Hellstrom, *Fresh Kills*, 2nd ed. (New York: Third Eve, 2009), 23.
4. John Byron Kuhner, *Staten Island, or Life in the Boroughs* (New York: Crumpled, 2010), 48–49.
5. Kuhner, *Staten Island, or Life in the Boroughs*, 51.
6. See Dietmar Offenhuber, *Waste Is Information: Infrastructure Legibility and Governance* (Cambridge, MA: MIT Press, 2017).
7. William Rathje, "Time Capsules: The Future's Lost and Found," *Scientific American*, November/December, 86–88. See also Robin Nagle, "The History and Future of Fresh Kills," in *Dirt: The Filthy Reality of Everyday Life*, ed. Rosie Cox et al. (London: Profile, 2011), 192–93.
8. Verlyn Klinhenborg, "Elegy to a Dumpscape," *New York Times Magazine*, October 10, 1999.
9. Tanvi Misra, "New York Is the World's Most Wasteful Megacity, in 3 Charts," *CityLab*, May 7, 2015, https://www.citylab.com/environment/2015/05/new-york-is-the-worlds-most-wasteful-megacity-in-3-charts/392636/. See also "A Crisis of Consumption," *Poughkeepsie Journal*, June 26, 1988.
10. Dom DeLillo, *Underworld* (New York: Simon and Schuster, 1997), 163. See also Di Cinzia Scarpino, "Ground Zero/Fresh Kills: Cataloguing Ruins, Garbage and Memory," *Altre Modernita*, November 2011, 243–45.
11. See Peter Boxall, "'There's No Lack of Void': Waste and Abundance in Beckett and DeLillo," *SubStance* 37, no. 2 (2008): 56–57. See also Margaret Robson, "Rubbish: Don DeLillo's Wastelands," *Irish Journal of American Studies*, http://ijas.iaas.ie/article-rubbish-don-delillos-wastelands/; David H. Evans, "Taking Out the Trash: Don DeLillo's *Underworld*, Liquid Modernity, and the End of Garbage," *Cambridge Quarterly* 35 (2006): 103–32; Alan J. Gravano, "New York City in Don DeLillo's Novels," *Italian Americana* 29 (Summer 2011): 181–89; Maurizia Boscagli, *Stuff Theory: Everyday Objects, Radical Materialism* (New York: Bloomsbury, 2014), 52–56; Rachele Dini, *Consumerism, Waste, and Reuse in Twentieth-Century Fiction* (New York: Palgrave, 2016), 143–58; Alan Joseph Gravano, "The Transfiguration of the Discarded: Consumption and Waste in Wallace Stevens, A. R. Ammons, Don DeLillo, and Martin Scorsese," PhD diss., University of Miami, December 2006, 107–16, 130–38.
12. Martin O'Brien, *A Crisis of Waste? Understanding the Rubbish Society* (New York: Routledge, 2008), 84.
13. See Max Liboiron, "Modern Waste as Strategy," *Lo Squaderno: Explorations in Space and Society* 29 (2013): 9.
14. Center for Sustainable Systems, University of Michigan, *Municipal Solid Waste Factsheet*, http://css.umich.edu/factsheets/municipal-solid-waste-factsheet. See also U.S. Environmental Protection Agency, *Municipal Solid Waste Generation, Recycling, and Disposal in the United States: Facts and Figures for 2012*, https://www.epa.gov/sites/production/files/2015-09/documents/2012_msw_fs.pdf; plaNYC, "Solid Waste," 2011, 137, http://s-media.nyc.gov/agencies/planyc2030/pdf/planyc_2011_solid_waste.pdf.
15. Jani Scandura, *Down in the Dumps: Place, Modernity, American Depression* (Durham, NC: Duke University Press, 2008), 12.
16. Center for Sustainable Systems, University of Michigan, *Municipal Solid Waste Factsheet*, http://css.umich.edu/factsheets/municipal-solid-waste-factsheet.

17. ASCE, *Infrastructure Report Card, 2017*, https://www.infrastructurereportcard.org/.
18. Grow NYC, "Recycling Facts," https://www.grownyc.org/recycling/facts; NYC Sanitation, *2017 NYC Residential, School, and NYCHA Waste Characterization Study*, 11, https://dsny.cityofnewyork.us/wp-content/uploads/2018/04/2017-Waste-Characterization-Study.pdf.
19. NYC Sanitation, *2017 NYC Residential, School, and NYCHA Waste Characterization Study*, 19.
20. See Max Liboiron, "Recycling as a Crisis of Meaning," September 20, 2012, https://maxliboiron.files.wordpress.com/2013/08/liboiron-recycling-etopia.pdf.
21. See NYC Sanitation, *2016 Annual Report*, 12–13.
22. Christof Mauch, ed., *A Future Without Waste? Zero Waste in Theory and Practice* (Munich: RCC Perspectives, 2016), 6. See also Robert Krausz, "Zero Waste," in *Encyclopedia of Consumption and Waste: The Social Science of Garbage*, ed. Carl A. Zimring and William L. Rathje (Los Angeles: Sage, 2012), 2:1015–16.
23. See Ashley Dawson, "Teenage Wasteland," in Patricia Smith, *Staten Island Noir* (New York: Akashic, 2012), 218.

INDEX

Italicized page numbers refer to illustrations

Abrams, Robert, 346
Adams, Allen J., 424
advertising, 5, 30
African Americans: Dinkins and, 351, 396; and environmentalism, 402; Giuliani and, 713n115; Koch and, 285; and Memphis sanitation strike, 626n101; in New York City, 27, 130; in Staten Island, 237, 381, 383, 529, 540; and transfer stations, 453. *See also* race
Agency for Toxic Substances and Disease Registry (ATSDR), 703n18
AIDS, 287, 323, 352, 428, 666n97
air pollution: action about, 253, 255–56, 259–62, 264–65, 308, 414, 424; and exporting waste, 459, 725n253; and Fresh Kills Landfill, 399, 425, 430; and incineration, 146, 226–27, 255–56, 260–61, 268; and industrialization, 172, 226–27, 239, 255–56, 261–63, 540; and resource recovery, 305; and September 11 cleanup, 478
Allee, William, 479–80
Allen, Kenneth, 138
Allen, Stan, 510
Allied Waste Industries, 457

Amec Construction Management, 472
American Legion, 183
Ammann, Othmar H., 234–35
Anchor Parks program, 544
Angelone, Linda, 441
Anti–Barren Island League, 64
Anti-Garbage League for Women, 101
Arthur Kill, 19, *171*, 183, *254*, *445*
asbestos, 267, 319, 471, 476–78, 693n50, 698n98
ash: from incineration, 245, 331, 338–42, 347, 353–54, 401; use as landfill, 92, 139, 153–54, 158, 353–55, 361, 365–67, 400–401; and waste stream, 25, 42–43, 49–51, 59, 66, 92, 130
Asians, 130, 540. *See also* race
Athey wagons, *149*, *341*, 611n26
Audubon Society, 218, 263
Avellino, Salvatore, 659n5

Bailey, Frank, 97
Bainbridge, Robert, 178
Baldwin, James N., 380
Ballet Mechanique, The (Ukeles), 505
Baltimore and Ohio Railroad (B&O), 171–72

barges, *155, 179, 283, 466, 480*
Barren Island, 44, 60–61, 64–66, 91–97, *93–94*, 108, 114–19, 132, 565–66n24
Barren Island Airport, 95
Barron, James, 379
Bautista, Eddie, 529
Baxter, Brady, Lent & Company, 44
Bayonne, 190
Bayonne Bridge, 20, 75, 170, 233
Beame, Abraham D. (mayor), 277–81, 284–87, 300
Bechtel Group, 476
Bedloe's Island, 32–33, 78
Beirut, 35
Belknap, Michael, 272–73
Bellamy, Carol, 682n30
Bellanca Aeroplane Corporation, 172
Bellew, Martin J., 473
Bellow, Bonnie, 454
Benish, Lee, 473
benzene, 401
Berge, Arthur T., 193–94
Berry, Charles W., 137
Betts Avenue incinerator, 211–12, *260–61*
Biggane, James L., 284
Biggs, Hermann M., 112–13
Birnbaum, Gregg, 391–92
Blackwell's Island, 32–33, 86, 565n19
Blizzard Island, 565–66n24
Bloomberg, Michael R. (mayor), 482–84, 486, 488–89, 496, 512, 515–19, 535
blue laws, 61
Board of Estimate (Board of Estimate and Apportionment), 89–90, 375–84
Board of Pilot Commissioners, 43
Board of Underwriters of New York, 81
Bollwage, J. Christian, 452
borough-specific waste disposal, 432–35, 442–44
Boston, 23, 35
Bovis Company, 472
Bowe, Hansine, 207
Boyd, Gordon, 326
Brahn, Arleen McNamara, 207
Break of Dawn (tug), 329–32
Brewster, Julius, 533
Brezenoff, Stanley, 330
Broadway, 29
Bronx, 19, 27

Brookfield landfill, 242–43, 256, 301, 304, 310–12, 317–19, 324, 401–2
Brooklyn, 19, 26–27, 62
Brooklyn Ash Removal Company, 43, 92, 127, 139
Brooklyn Chamber of Commerce, 124–26, 585n179
Brooklyn-Manhattan Transit (BMT), 90
Brooklyn Navy Yard, 259, 299–300, *307*
Brooklyn Navy Yard incinerator: ash disposal from, 339–40, 345–46, 353–54, 357, 361, 365, 367–68; delays and cancellation, 402–4, 410, 412, 414, 416, 423, 467, 645–46n151; original plan and controversy, 259, 306–10, 314–15, 317, 320–23, 326, 334–36, 400. *See also* incineration
Brooklyn Union Gas Company (BUG), 296
brownfields, 432
Browning Ferris Industries (BFI), 431, 468, 716n145
Bryan, Frederick P., 162
Buck, Ellsworth B., 187
Buffalo, 53
Burden, Amanda M., 496, 509, 535
Burn, Baby, Burn: How to Dispose of Garbage by Polluting Land, Sea, and Air at Enormous Cost (Holtzman), 360
Burton, Michael J., 474

C&A Carbone, Inc. v. Town of Clarkstown (1994), 411, 525
Cahill, John P., 459
Cambridge Arts Council, 508
Campanile, Carl, 336, 415–16
Campbell, Bradley A., 454
Campbell, Edward, 138
cancer, 401, 407, 477
Cantalupo, Anthony, 400
Canty, Anne, 405, 411
Carey, Hugh L. (governor), 278–79, 298
Carey, William F., 143–44, 146–50, 154–62, 164–65, 176, 593n30
Carpinello, Michael, 718n166
Carroll, Jason, 479–80
Carson, Rachel, 262
Carter, Jimmy, 295
Casowitz, Paul: background and appointment, 299; and Fresh Kills

Landfill, 304, 319, 324; and incinerators, 300, 307, 322, 338–39; on recycling, 334; and resource-recovery plants, 301–2; and toxic material dumping, 311
Cassilano, John (Johnny Cash), 311, 401
Castle Clinton, 32, 36–37
Cavanaugh, Thomas F., 124
Centennial Exhibition (Philadelphia, 1876), 24
Center for the Biology of Natural Systems (CBNS), 308, 367
Center of Urban Restoration Ecology (CURE), 495
Central Park, 53, 239, 498
Cerria, Philip, 313
Cerullo, Alfred, 353, 366, 410
Chalfen, Lucian F., 417, 440
Chambers, Wilber W., 105
Charity Hospital (City Hospital), 33
Charles, Martin, 153–54
Charter for the City of Staten Island, 386. *See also* secession
Cherry Lane factory, 74
Chew, Susan, 337
Chicago, 6, 22–23, 27, 35
Chinitz, Barbara Warren, 366, 369
Chittenden, R. P., 106
cholera, 44, 80–81, 84
Christiansen, Kendall, 533
Church of St. Andrew, 79
Cincinnati, 6
Citizens Advisory Committee for Resource Recovery (CAC), 320, 336
Citizens Budget Commission (CBC), 268, 400, 527–28
Citizens for a Cleaner Staten Island, Inc., 323
Citizens Union, 61, 378
Civic League, 116
Claro, Cesar, 467
Clay Pit Ponds, 218–20
Clean Air Act, 253, 260, 414, 424. *See also* air pollution
clean fill, 4, 50, 146, 149, 158–61, 165, 179–80, 183, 245
Clearwaters, Inc., 243
Cleary, E. J., 175
Cleaves, Howard, 182–83
Clinton, Dewitt (mayor and governor), 36
Clinton, William Jefferson (Bill), 403

C. O. Bartlett and Snow Company, 97
cobb wharf, 43
Cobwell reduction system, 107–12, 579n42. *See also* reduction plants
Codling Island, 565–66n24
Cohen, Neil, 445
Cohn, Nevin, 541
Coleman, J. S., 46–48
Collins, John F., 106–7
Colosi, Natale, 183–84
Colt, S. Sloan, 241
Columbia College, 29
Columbian Rotarians, 159
commercial waste: disposal regulations, 223, 534, 637n50, 726n256; and export outside New York City, 361, 400, 411, 426–27, 489, 517, 523; organized crime and, 329, 430–31; and recycling, 528
Committee for Nuclear Information, 308
Commoner, Barry, x, 307–8, 322, 336, 409
Community Councils of New York, 156, 161
compaction, 148, 244, 269, 515
composting: as alternative to incineration, 268–69; as alternative to landfills, 201, 432; and Bloomberg administration, 516, 528, 530, 533; and de Blasio administration, 554; and Dinkins administration, 357, 363, 746n114; and Giuliani administration, 432, 446, 455; and Resource Recovery Act (1970), 298
Comprehensive Solid Waste Management Plan (New York City, 1992), 361–67
Comprehensive Solid Waste Management Plan (New York City, 1996), 412–13, 440–41
Comprehensive Solid Waste Management Plan (New York City, 2004), 517–18
Comprehensive Solid Waste Management Plan (New York City, 2006), 520, 522–25, 527–28, 530, 533–39
Concerned Staten Island Realtors, 414
Coney Island, 32, 565–66n24
Connelly, Elizabeth A. (Betty): background, 291–92; and Fresh Kills Landfill closure, 405, 421, 424–25, 434, 718n170; and opposition to Fresh Kills Landfill, 291–94, 310, 312–13, 324, 345, 370, 648n13; and resource recovery, 316, 321, 338, 366, 414; and Staten Island secession, 377, 379, 383

Connor, Robert T., 243, 245, 257, 268, 270
Conservation Foundation, 263
Conservative Party, 240, 410, 420
Constable, Stuart, 218
consumerism, 5–7, 130–31, 209, 252, 262–63, 335, 489, 516, 548, 552
consumption-waste-disposal (place) system, 9, 548, 551–55
controlled tipping, 148. *See also* landfills
Copeland, Royal S., 118, 126
Cornell, Alonzo B. (governor), 46
Corner, James, 502–3, 509–11
Corona and Corona Ash Dump, 92, 153–54, 595n48
Cortelyou, Charles and Burton, 175
Council on Environmental Quality (CEQ), 264, 453–54
Covanta Energy, 529
Crash Clean-Up Campaign (1966), 247
Criaris, William (Bill), 271–73, 272
"Criteria for Classification of Solid Waste Disposal Facilities and Practices" (Environmental Protection Agency), 640n98
critical discard studies, xi
Croker, Richard, 22, 56, 61
Cromwell, George, 89, 99, 113–14
Cropsey, James C., 104
Crotty, Erin M., 465
Cunningham, Edward, 84
Cuomo, Mario M. (governor), 286, 354–55, 368, 384–88
Curran, Henry H., 132

Davis, Dwight F., 127–28
Davis, William T., 75–76, 84
Dawson, Ashley, 554
DDT, 308
de Blasio, Warren Wilhelm (Bill), Jr. (mayor), 531–32, 543–45, 554
Defusing the Garbage Time Bomb conference (1978), 292
Delancey Street incinerator, 92
DeLillo, Don, 551
DeLury, John J., 248–50, 267, 284–85
DeMaria, Lawrence, 467
Democratic Club of Oakwood Heights, 159
De Morgan, John, 88

Department of Consumer Affairs (DCA), 430–31, 637n50
Department of Corrections, 64, 474, 639n79
Department of Design and Construction (DDC), 474–76, 542
Department of Docks, 22
Department of Environmental Protection (New Jersey), 484
Department of Marine and Aviation, 243
Department of Parks and Recreation (DPR), 537–39
Department of Public Charities and Correction, 33, 48
Department of Public Works, 140, 203
Department of Sanitation (DSNY): creation and reorganizations, 136–44, 207–8, 265; and fiscal crisis (1970s), 275, 280–81, 284–85, 288–89; and gender and race, 140, 289, 673n13
Department of Street Cleaning (NYDSC), 40–41, 49–50, 52–53, 57, 64, 126–36
Department of the Army, 234
Department of the Interior (U.S.), 320
DeSapio, Carmine, 223
DeStefano, Louis, 249
Dewey, Thomas E. (governor), 178–79, 192, 194
Di Brienza, Stephen, 355–56
Dickens, Charles, 26, 33
Diggins, Dennis, 479
Diller Scofidio + Renfro, 510
Dimino, Resa, 516
Dinkins, David Norman (mayor), 351–52, 354, 357–58, 367–70, 385, 391, 410, 675n33
dioxin, 307–8, 314, 322, 471, 476, 645–46n151, 656n111
disposal sink, 123, 265, 299, 350, 361, 546, 549
Division of Coastal Resources, 496
Dix Island, 86. *See also* Swinburne Island
Dmytryszyn, Nicholas, 473, 492
"dock mudd," 42
Doherty, John J.: and 2004 DSNY plan, 517–18; background and career, 408, 531, 718n166; and Fresh Kills Landfill, 406, 413–14, 416–17, 465, 485, 487, 525, 527; and incineration, 488; on September 11 cleanup, 472

Domestic Manners of the Americans (Trollope), 15
Dongan, Thomas, 35
Dongan Charter, 35
Doyle, Edward P., 100–104
Doyle, Roger, 100–102
Drake, J. Sterling, 100–101, 103
Dreyfus, Louis A., 99, 112–13, 116
DSNY map of waste management facilities in 1992, 363
Dublin, 35
Duff, Susanna, 468
dumping boards, 42, 126

Early History of Staten Island (Kolff), 76, 573n25
Earth Day (1970), 263
Earth Institute (Columbia University), 485
East 91st Street marine transfer station, 533
Eastern Transfer, Inc., 448
East River, 17
ecology, 262, 406, 493–95, 497, 501–2, 510, 548–49
Eddy, Harrison P., Jr., 147
Ederle, Ernst J., 64
Edgemere landfill, 206, 256, 265, 301; closure of, 326, 338, 608n122, 656n119, 668n111; opposition to, 156–57, 324; and toxic waste, 656n119
Edwards, Joan, 335
Eisenbud, Merril, 265
Eliassen, Rolf, 149
Elish, Herbert, 268–69, 284
Elizabeth (New Jersey), 190
Elizabeth II (Queen), 475
Elliott, Donald H., 268
Ellis Island, 27, 32–33, 43, 86, 565–66n24
Emergency Financial Control Board (EFCB), 278
Emerson, Haven, 111
Emerson, Ralph Waldo, 76
Emerson, William, 76
Engler, Mira, 10, 511
Enviro-Chem Systems Inc., 267
Environmental Action, 263
Environmental Action Coalition, 292
Environmental Defense Fund (EDF), 263, 321, 745n109

Environmental Equity: Reducing Risk for All Communities (EPA), 403
environmental-impact statement (EIS), 263
environmentalism, 252, 261–64, 402, 552–53
environmental justice, 402–4, 552; and Fresh Kills Landfill closure, 454–55, 499, 529; and waste-to-energy plants, 529; and waste transfer stations, 445, 534
Environmental Justice Act, 403–4
environmental movement, 9, 252, 262–64, 307, 402, 549
Environmental Protection Agency (EPA): and Brookfield landfill, 401–2; creation of, 297; and environmental racism and justice, 403, 437, 453–54, 541; establishment and early functioning, 264; and Fresh Kills Landfill, 368, 407; and *Mobro 4000*, 329, 332; and recycling, 322, 428, 714–15n129; and September 11 cleanup, 474, 477–78; and toxic waste, 310, 338, 340, 541
environmental racism, 402–3, 437, 442, 453–55
Erie Canal, 21
Eustis, John R., 111
Executive Committee of Staten Island, 81
exporting waste: and Bloomberg administration, 487–89, 515–18, 521–23; and de Blasio administration, 534; and Dinkins administration, 404, 406, 410; flow-control laws, 411, 426–27, 447, 449, 452, 525, 717n164; and Giuliani administration, 410–11, 423, 426–27, 432–34, 439, 447–52, 456–59, 521; medical waste, 449

Fach, Albert C., 102, 106–8, 111, 113, 160
Failace, James, 444
Farmer, James and William, 101
Farrell, Edward C., 380, 447
Farrell, Kevin P., 446
"Federal Actions to Address Environmental Justice in Minority Populations and Low-Income Populations" (Executive Order 12898), 403
Federal Bureau of Investigation (FBI), 473–74

Federal Emergency Management Agency (FEMA), 475
Federated Sportmen's Clubs of Staten Island, 274
Feeney, John, 88–89
Feldman, Maurice M., 250, 257
Fenways and Brooksides Committee, 218
Ferber, Edna, 221
Ferrer, Fernando J., 434, 482–83
Ferry Point landfill, 209, 211, 290
Fetherston, John T., 97, 580n48
Fetherston, William T., 160
Field Operations, 498, 502–3, 509–12, 519–20, 536–38, 752n72
Fife, Barbara J., 359
Fifth Avenue (New York City), 29
Fifty-Ninth Street marine transfer station, 506
filth theory of disease, 54, 63
Fire and Ice: How Garbage Incineration Contributes to Global Warning (Holtzman), 361
fiscal crisis (1970s), 275–81
Fish and Game Protective Association, 218
Fitzgerald, F. Scott, 94, 153
Fitzgerald, Maurice A., 190
Five Points, 43
Florence Nightingale (hospital ship), 86
Flow City (Ukeles), 506–7
flow-control laws, 411, 426–27, 447, 449, 452, 525, 717n164. *See also* exporting waste
Floyd Bennett Field, 95
Flushing Meadows, 92, 94, 153–54, 156
Ford, Gerald R., 279, *280*
Forker, Timothy, 427, 434
Fossella, Frank, 312
Fossella, Vito J., Jr., 412, 441, 447, 454, 460, 484, 487
Foster D. Snell, Inc., 245
Fountain Avenue landfill: closure of, 304, 312, 320, 323–24, 326, 337–38; operation of, 256, 290, 301, 315, 317, 319, 636n39; and park construction, 495
Franz, Bill, 97, 486–87
Fred C. Hart Associates, 314
Fremont, John C., 128
Fresh Kills, 19, 70, 158–68, *195*, 219, 254, 407, *546*

Fresh Kills (Loehfelm), 547, 756n2
Fresh Kills (Wheat), 547, 756n1
"Fresh Kills: Artists Respond to the Closure of the Staten Island Landfill" (2001), 481
Fresh Kills Landfill: after closure, 2, 423–60, 468–70, 483–89, 491–519, 525–27, *526*, 545–46; and ashfill, 353–55, 361, 365–67, 400–401; closure of, 405, 412–26, 431–52, 455, 459, 464–70, 550; Consent Decrees (1980), 291–94, 304–5, 323–24, 459, 641n108; Consent Order (1985), 318, 324, 326, 336, 342; Consent Order (1990), 328, 344–48, 350, 353, 398, 404–6, 412–13, 424, 440, 508; design and construction, ix, 1–2, 16, 167–68, 178–96, 200–207, 268–70, 311; economic importance, 271–73, 424–25; and Hurricane Sandy, 530; and land reclamation, 200–203, 214–17; and leachate, 270–71, 293–94, 304–5, 336, 348, 373, 398–99, 407, 467, 644n136, 646–47n1, 669n121; maps and photos, *204–5*, *283*, *341–42*, *369*, *399*, *420*; and methane, 296–98, 301, 348, 373, 468, 539, 641n108, 655n107; reliance on in 1960s and 1970s, 227–31, 242–45, 266–70, 287, 294, 301; reliance on in 1980s and 1990s, 304–5, 361–62, 365–70, 549–50; and September 11 cleanup, 473–75, 477–82, 480, 489–90, 498–99; and Staten Island secession, 193–94, 383–84, 389–90, 394; and toxic material, 274, 310–12, 339, 366, 442; and wildlife preservation, 217–20
Fresh Kills Landfill (City of New York et al.), 214–17, *216*, 230, 614n74
Fresh Kills Landfill Advisory Committee, 293, 323
"Fresh Kills Landfill End Use Conceptual Master Plan Design Competition," 470
"Fresh Kills Landfill Study" (DSNY and DCP), 469
Freshkills Park: design and construction of, 492, 496–511, 535–39, 541–45; and environmental restoration, 2, 551; future of, 545–46, 550–51; photos and maps, *521*, *536*, *538*, *542–45*; public response to, 539–41; September 11 memorial, 513, 536, 551
Freshkills Park Alliance, 753n79

Freshwater Wetlands Act (New York State, 1975), 342–44. *See also* wetlands
Fresno Sanitary Landfill, x, 148, 491
Frisch, Max, 27
furans, 314
Future of Jamaica Bay, The (City of New York, Department of Parks), 156

Gaeta, Anthony R., 312
Gaffney, Gahagan & Van Etten, 97, 101, 104
Gaffney, James, 97, 101
Gansevoort incinerator, 222
Garbage in the Cities: Refuse, Reform, and the Environment, 1880–1920 (Melosi), ix
Garbage Out Front: A New Era of Public Design (Municipal Art Society, 1990), 509
Garbage Wars (1876–1880), 45–46, 92–119
Garcia, Kathryn, 531
Garcia, Yolanda, 437
garment industry, 6, 21, 23, 25, 29, 51
Garrett, Tom, 374
gender, 289, 403, 504
Genovesi, Anthony J., 379
George, Henry, 29
Georgia, Olivia, 481
Gerbounker, Richard, 457
germ theory of disease, 54, 78
Gerrard, Michael B., 401, 451, 477
Gerstel, Joan, 399
Gesuale, Thomas, 329, 659n5
Getting to Zero campaign, 531
Getty Synthetic Fuels, Inc., 655n107
Giardino-Lenart Corporation, 194
Gill, Robert, 244
Gilmore, James S., Jr., 447
Gilroy, Thomas (mayor), 48
Giuliani, Rudolph W. L. (Rudy) (mayor): background and election of, 395–98, 439–40; and exporting waste, 410–11, 423, 447–52, 521; and Fresh Kills Landfill closure, 398, 415–21, 424–25, 434–38, 444–49, 452, 455, 459, 465, 467; and Freshkills Park proposal, 496; and incineration, 410; and organized crime, 399, 430–31; and privatization, 408–15, 423; and race, 396; and recycling, 408–9, 428–29, 440, 446; and September 11 recovery and cleanup, 475–76, 482; and Staten Island secession, 395
Glaeser, Edward L., 20
Gleason, Phillip J., 508, 707n54
glycerin, 118
Goethals Bridge, 75, 170, *171*
Goldberger, Herb, 514
Golden, Howard, 404, 435, 443
Golden Anniversary Exposition (New York City, 1948), 199–200
Goldfarb, David, 380–81, 391
Goldin, Harrison J., 306–7, 335
Goldstein, Eric A., 347, 408, 422
Goode, Anne, 455
Goodrich, Ernest P., 142
Governors Island, 18, 32–33, 48, 78, 80, 133, 138, 142
Grace Church, 29
Grand Central Parkway, 150, 153–54
Grant, Hugh J. (mayor), 52
Grassi, Carrie, 539, 754n101
Gravesend Bay incinerator, 410
Great Depression (1930s), 130, 134, 140–41, 146, 172
Great Fresh Kill, 205, *254*
Great Gatsby, The (Fitzgerald), 94, 153
Great Kills landfill, 180, 182–83, 185, 187–90, 200, 206
Great Kills Park, 43, 146, 164–67, 178, 209, 214, 240, 597n87
Great Migration, 130
Green, Andrew Haswell, 62, 87–88, 153
Green, Mark J., 430, 441, 482–83
Greenbelt (Staten Island), 274
greenhouse gases, 527, 529, 643n123, 751n56
"Greening of Fresh Kills Landfill: Landscaping and End Use" (Young), 405
Greenwich Street, 43
Gregorio, John T., 457, 468, 484, 725n248
Greve, William M., 97, 113–15
Grey, Theresa, 379
Groh, Robert T., 285
Gross, Don, 401
Groups Against Garbage (GAG), 326, 437
Guidry, Robert, 330

Haggerty-Blohm, Marilyn, 425, 431, 434
Halenar, John, 486

Hall, Cornelius A. (Neil), 181, 189, 193, 202, 209, 214–20, 611n30
Halvorson, William, 494
Hamilton Avenue incinerator, 226, 260, 304, 533
Hamilton Avenue marine transfer station, 533
Hammack, David, 24
Hammond, Thomas W., 142–43
Hand, Charles S., 136
Handel, Steven, 495, 754n99
Hang, Walter, 336
Hansen, August E., 109–10
Hansen, Paul, 108–9
Harbor, The (Poole), 21, 561n18
Hargreaves Associates, 498, 500–501
Harlem River, 17
Harnedy, Tom, 474
Harrelson, Lowell, 329–32
Harriman, Averill (governor), 225
Harrington, F. C., 141
Harris, Elisha, 80
Hart Island, 32–33
Hauber, Mark, 754n99
Hawkins, Gay, 8
Health Department of New York State, 477
Hearst, William Randolph, 98
Heller, Austin N., 259
Hellstrom, Christopher, 547
Helsinki, 35
Here Comes the Garbage Barge! (Winter), 330, 659n10
Hering, Rudolph, 106, 110–11
Hetherington, Kevin, 9
Hevesi, Alan G., 429
Hewes, F. W., 53
Hewitt, John-Charles, 483
Hicks, John R., 108
Higgins, James A., 132
High Line Park, 510
High Rock Park Conservation Center, 273
Hillquit, Morris, 117
Hinchey, Maurice, 339
Hirsh, Eloise, 538–40, 546
Hirshfield, David, 131
Hobbs, Richard, 494
Hoboken, 26
Hoffman Island, 20, 86
Holden, Kenneth, 474

Holtzman, Elizabeth, 355, 360–61, 367
Holusha, John, 370
Holzmacher, McLendon and Murrel, 309
homelessness, 381
home rule, 392–93, 682n31. *See also* secession
Hooker Chemical, 310
Horning, Diane, 481, 513
Horning, Kurt, 481
Horning, Matthew, 481, 513
horses: and recreation, 469, 500, 514; and transportation, 39, 73, 126, 175; and waste, 7, 25, 51.44, 130, 553
Hroncich, Thomas, 329, 659n5
Hudson, Henry, 71
Hudson Oil Refining Company, 310–11
Hudson River, 17
Hughes, Charles Evans (governor), 117
Hughes, Richard J., 235
Hunter Island, 32, 565–66n24
Hurricane Donna, 226
Hurricane Sandy, 529–30
Hylan, John Francis (Red Mike) (mayor), 90, 117–18, 123, 126, 131–32, 134, 601n6

Illinois (hospital ship), 86
immigration, 24, 27, 81, 83–84, 130, 140
Immigration Act of 1924, 130
Immigration and Nationality Act of 1965, 237
Impellitteri, Vincent (Impy) (mayor), 210, 221, 223
incineration: in late 1800s and early 1900s, 47–48, 59, 65, 96, 133; in 1920s and 1930s, 123–24, 132–35, 137–40, 145–47; in late 1940s and 1950s, 207–13, 221–22; in late 1950s and 1960s, 200, 224–31, 244, 250, 258–61, 267–69; after 1960s, x, 296, 333–34, 410, 488; and air pollution, 226–27, 255–56; ash from, 245, 331, 338–42, 347, 353–54, 401; and Fresh Kills Landfill, 187–88, 192, 194, 196; of medical waste, 558n50; and recycling, 359–61, 403; and reduction plants, 65; technology of, 10, 47–48, 133, 211, 652n64. *See also* Brooklyn Navy Yard incinerator; waste-to-energy
"Incinerators: The City's Half-Baked and Hazardous Solution to the Solid Waste Problem" (Commoner), 322, 656n109

INDEX

Independent Budget Office, 484
industrialization, 5–7, 19, 49–52, 69, 172, 175, 204
infantile paralysis (poliomyelitis), 104, 108
Innes, Frank H., 160
Interstate Sanitation Commission (ISC), 190, 257, 293, 335
IRATE (Islanders Against a Toxic Environment), 311
Island (Isle) of Meadows, 20, *91*, 92, 103
Italian Historical Society, 235
Itjen, Eugene M., 216

Jamaica Bay: landfill proposals, 95–97, 245, 258; landfill *vs.* recreation controversy (1930s), 155–59, 161–62, 183, 194, 208, 213, 565–66n24; as major wetlands, 70, 96, 174; preservation plans (1960s), 258
James River Environmental Group, 449
Jeffries, Frederick, 118
Jersey City, 26, 59
Jesup, Morris K., 52
Jesup Committee, 52
Jewett White Lead Company, 74
JMP Landscape, 498
John McAslan + Partners, 498, 501
Johnson, Kirk, 465, 488
Johnson, Lyndon B., 235, 253
Johnson, Robert E., 193–94
Jorling, Thomas C., 331, 336–40, 345–48, 353, 355, 403–4, 659n10
Joyce, Jerry, 464
Juvenile Street Cleaning League, 56

Kabel, Janice, 641n105
Kahn, Roger, 250
Kane, Farrell M., 187–88
Kaufman, Stephen B., 290
Kausel, Eduardo, 472
Kazmiroff, Theodore, 259
Kearing, Samuel J., 254, 256–60
Keating, Kenneth B., 235
Kelby, Charles H., 117
Kell, Arthur S., 349, 355, 413, 428
Kennedy, Robert F., 235
Kenny, Andrew J., 157
Khian Sea (cargo ship), 331
Kiley, Robert R., 353
killer fog of London (1952), 255

Kill Van Kull, 19
King, Evelyn, 383
King, John A. (governor), 83
King, Martin Luther, Jr., 626n101
King, Moses, 22
Kissena Park landfill, 206, 611n34
Kleinknecht, William, 336–38
Klinkenborg, Verlyn, 548
Koch, Edward Irving (mayor): background and election of, 285–87; and Board of Estimate structure dispute, 376, 384–85; and Brooklyn Navy Yard incinerator, 307; and DSNY, 288; and fiscal crisis (1970s), 286; on Giuliani, 451, 475; and landfills, 290–91, 305, 345–46, 451; and *Mobro 4000*, 330; and recycling, 321–22, 661n30; and resource recovery, 299; and tugboat workers strike (1978), 295
Koetter, Eldon P., 231
Kohnke, Quitman, 60
Kolff, Cornelius Geertruyus, 76–77
Kottamasu, Raj, 545
Kretchmer, Jerome, 267–68
Kuhner, John Byron, 547–48
Kuntzman, Gersh, 498

Ladies' Health Protective Association, 40–41
LaFrance truck, *137*
La Guardia, Fiorello (mayor), 140, *141*, 151–52, 161–62, 170, 567n43
LaGuardia Airport, 95, 154, 166, 183
Lake's Island (Simonson Island), 91–92, 101–5, 108–9, 115–18, 125–26, 374, 547
Lake's Island Realty Company, 103
Lally, Patrick, 447
Lamberti, Ralph J., 290, 312, 319–20
Land Art Generator Initiative, 537, 543
landfills: in 1800s, 41–43, 48–49; in 1920s and early 1930s, 127–28, 143, 162; in late 1930s, 1940s, and 1950s, 144, 146–50, 209–17; in 1960s and 1970s, 256–58, 282; after 1970s, 332–33, 406, 531, 533–34, 646n1, 701n1; federal and state regulation of, 270; and land reclamation, 4, 33–37, 41–49, 611n36; and leachate, 298, 333, 549; park conversions, 469–70, 493–94; technology of, 10, 202, 243, 294, 565n23; and toxic waste, 310–11

Landing: Cantilevered Overlook (Ukeles), 537, 537
Landreth, Olin H., 108–9
Lang, Martin, 285, 287
Langewiesche, William, 471
Latinos, 286, 351, 383, 402, 453, 540, 713n115. *See also* race
Lautenberg, Frank, 725n253
Lavis, Ryan, 545
lazaretto, 78. *See also* quarantine station
leachate, 642n109; and Fresh Kills Landfill, 304–5, 336, 348, 373, 407, 467, 646–47n1, 669n121; mitigation and recovery of, 40, 293–94, 325, 344–45, 398–99, 430, 440, 455, 468, 525, 644n136; in sanitary landfills, 298, 333, 549
lead and lead poisoning, 306, 339, 476
Lefferts Boulevard landfill, 157, 206, 611n34
Lehman, Herbert H. (governor), 161
Leicht, Holly, 498
Leman, Francis F., 98–99
Lemieux, Robert P., 361
Lenape nation, 2, 71, 159, 175
Leng, Charles W., 75–76, 84
Leo, John P., 131
Leventhal, Nathan, 288
Levin, Kate, 509
Lewis, John R., 403
Lewis, Liston L., 106–7
Lexow Committee, 53
Lhota, Joseph, 444, 450, 456, 460, 465
Liberty Island, 32–33, 43, 78
Liboiron, Max, 554
Liebman, Henry, 208, 213–14
Life After Fresh Kills: Moving Beyond New York City's Current Waste Management Plan: Policy, Technical, and Environmental Considerations (Earth Institution et al.), 485, 737n118
Lifescape (Field Operations), 502–3, 511, 520, 551
Linden, 190; and Fresh Kills Landfill, 190; photo, 445; transfer stations, 456–58, 484–85, 489
Lindsay, John V. (mayor), 245–50, 254, 259, 265–66, 567n43
liquefied natural gas (LNG), 296
Liskamm, William H., 470
Little Fresh Kill, 205, 244, 254

Lloyd, Emily, 359, 367–68, 400–401, 405, 505
Local Law 3 (1968), 629n58
Local Law 14 (1966), 255–56
Local Law 19 (1989), 335, 356, 358, 429
Local Law 40 (1990), 453–55
Lockwood Kessler & Bartlett, Inc, 238–39
Loehfelm, Bill, 547
Loizeaux, Mark, 473
Lomma Construction Corporation, 545
Long Island Sound, 17
Lorca, Federico Garcia, 30–31
Love Canal, 310, 338–39
Low, Seth (mayor), 64
Lucia, Frank J., 244, 253
Lupis, Giuseppe, 235
Lynch, John A., 166
Lynch, Kevin, 7
Lynn, Andrew S., 470
Lyons, James, 611n30

MacStay, Arnold B., 123–24, 126, 131
Maintaining NYC in Crisis: What Keeps NYC Alive (Ukeles), 504
Manderville, Richard, 296
Manes, Donald R., 320
Manhattan, 16, 18–19, 25–27, 34
Manhattan Refuse Cremating Company, 48
Manifesto for Maintenance Art (Ukeles), 504
Manila, 35
Maniscalco, Albert V., 235, 237
Manley, Thomas H., 49–50
Mapes, James J., 53
Marchi, John J.: background and election, 225; and Fresh Kills Landfill closure, 434, 451, 459, 484, 487; and *Mobro 4000*, 331; opposition to Fresh Kills Landfill, 225–26, 292, 305, 346, 368, 405, 413, 416; and resource-recovery plants, 305, 316–17, 336; and Staten Island development, 240; and Staten Island secession, 375–80, 382, 385, 387–88, 393; and tipping fees, 347; and wetlands designations, 343
Marine Park (Brooklyn), 43, 95, 211, 495, 613n69

Marine Park (Staten Island), 146, 163–67. *See also* Great Kills Park
marine transfer stations (MTS), 612n48, 711n94. *See also* transfer stations
Marine Transfer Station Transformation (Ukeles), 508
Marino, Ralph, J., 311
Marra, Frank, 480
Mary Michael (Sister), 442
Masters, Deborah, 436
Mastro, Randy M., 426, 438, 440
material-recovery facilities, 432, 553, 737n123. *See also* recycling
Mathur/Da Cunha + Tom Leader Studio, 498, 500
McAllister, Ward, 21
McCallum, Joseph W., 187, 193
McCartney, James, 63
McClellan, George B., 22
McCracken, Lynn W., 218
McDonald, Joseph A., 160
McDonough, Peter, 444
McFeeley, Ken, 284
McGerr, Michael, 24
McGivney, Richard, 297
McHarg, Ian, 241
McKeon, Dennis S., 323
McKinley, William, 63
McMahon, Michael E., 419, 486
McNeur, Catherine, 39–40
McPartland, Kevin, 575nn50–51
McSheehy, S. S., 267
Meadowlands, 70, 174
Medicaid, 277
medical waste, 323, 329, 341–42, 362, 449, 558n50
Merchants' Association of New York, 74, 77, 94
Messinger, Ruth W., 427–28, 434–35, 676n55
Metcalfe, H. B., 85
methane: and Fresh Kills Landfill, 296–98, 301, 348, 373, 468, 539, 641n108, 655n107; and landfills, 149, 243, 270–71, 294, 549, 598n119, 646–47n1; recovery of, 296–97, 322, 329, 344–45, 440, 542
Metropolitan Board of Health, 40
Metropolitan By-Products Company, 97–98, 105–19
Metropolitan Health Act (1866), 40

Mexico City, 35
miasma theory of disease, 54, 78
Michaels, Stanley E., 409, 438–39, 458
Middlesex County, 190
Midland Beach Democratic Association, 183
Midler, Bette, 470
Miller, Benjamin, 358–59, 458–59, 466–67, 484–88
Miller, Mel, 683n45
Mill Rock Island, 565–66n24
Mills, William Wirt, 98, 103, 107, 374
Minneapolis, 6
Misra, Tanvi, 551
Mitchel, John Purroy (mayor), 95–98, 100, 113, 115–17
Mobro 4000 (garbage barge), 328–33
Moeller, Griswold L., 264–65
Molinari, Guy V.: and Brooklyn Navy Yard incinerator, 340; and exporting waste, 450; and Ferry Point landfill, 290; and Fountain Avenue landfill, 337; and Fresh Kills Landfill closure, 418–21, 425–26, 431–34, 450, 452, 465, 470; on island landfill construction, 282; on landfill technology, 336; opposition to Fresh Kills Landfill, 294, 312, 348, 399, 406–7, 412–14, 424, 454, 718n170; and recycling and resource recovery, 301, 311, 316, 409; and Staten Island secession, 376, 380, 387–88, 391–93
Molinari, Susan, 343, 354–55, 366, 368, 401, 414, 424, 667n101
Molinaro, James P., 465, 487, 512–14
Montevideo, 35
Montgomerie, John, 36
Montgomerie Charter, 36
Montgomery Ward, 6, 51
Montreal, 35
Moore, Vincent J., 324
Morris, Newbold, 185, 605n69
Morrison, Henry P., 98, 107
Morris v. Board of Estimate, 378–79
Morse, William F., 106, 109
Morton, Levi P. (governor), 89
Moses, Robert: background and career, 151–52, 177, 223, 235; and Fresh Kills Landfill, 158–59, 178–92, 202–3, 206, 214–20, 225, 230; and incineration,

Moses, Robert (*continued*)
201–2, 209, 211, 213, 230; and Jamaica Bay, 155–57; and land reclamation, 146, 158, 201–3; and Marine Park (Brooklyn), 95; and Marine Park (Staten Island), 163; and New York City infrastructure, 150–53, 158; and Orchard Beach, 565–66n24; photos, *151*, *184*; and race and social class, 152; and Rikers Island, 150, 579n34; and Staten Island rail connection, 170; and Verrazzano-Narrows Bridge, 234–37, 241; on William Carey, 593n30; and World's Fair (1939), 153–55
mosquitoes, 147, 162, 174–75, 178–79, 304
Mozzillio, Christopher, 481
Mozzillio, Michael, 481
Mulrain, Andrew W., 211, 214, 221, 223
Mumbai, 35
Municipal Art Society (MAS), 469, 495–96
Municipal Assistance Corporation (MAC), 278
Murphy, Charles F., 114
Muss, Jason, 492

Nagle, Percival E., 63–64, 92
Nagle, Robin, 491, 499
Nagyvathy, Rudolf, 242
naphtha, 60, 115
Napoleone, Tommy, 465
National Association for the Advancement of Colored People (NAACP), 383
National Board of Health, 54
National Commission on Materials Policy, 298
National Endowment for the Arts, 504–5
National Environmental Policy Act (1969, NEPA), 263–64
National League of Cities, 262
National Resources Defense Council (NRDC), 321
National Science Foundation, 292
National Solid Wastes Management Association, 475, 524
National Survey of Community Solid Waste Practices, 253
Native Americans, 403. *See also* race
nativism, 84. *See also* race

Natural Resources Defense Council (NRDC), 478
Neaher, Edward R., 378
Nelson, Harold, 189
Netherlands, 35
New Jersey: and Fresh Kills Landfill, 341–42; and New York City waste disposal, 138, 143, 284, 338; and Quarantine War (1850s), 80–81; and Staten Island, 72–75, 169–72, 200, 375
New Orleans, 6, 35
New York Bay, 16–20
New York Chamber of Commerce, 81
New York City Environmental Justice Alliance, 437
New York City geography and history, 15, 17–37, *18*, 61–63, 128–31, 199–200
New York Civil Liberties Union (NYCLU), 375
New York Dyeing and Printing, 74
New York Environmental Protection Administration (NYEPA), 265–68, 282, 284, 298
"New York Is the World's Most Wasteful Megacity" (Misra), 551, 756n9
New York Marine Hospital, 78–86, *79*, *83*
New York Public Interest Research Group (NYPIRG): and Fresh Kills Landfill, 355, 415, 529; and incinerators, 322, 333, 336, 410; and recycling, 349, 413, 428; and toxic waste, 311, 317, 336
New-York Sanitary Utilization Company (NYSUC), 60–61, 64–66, 91–97, *93–94*, 108, 114–19, 132
New York State Commission of Correction, 150
New York State Department of Environmental Conservation (DEC), 287, 291, 323–24, 330, 332–33, 353, 474, 630n76
New York State Department of State, 469, 496
New York State Legislative Commission on Solid Waste Management, 326
New York State Resource Recovery Policy Act (1977), 291
New York State Society of Professional Engineers, 183
New York State Solid Waste Management Plan (1987), 332–33

New York State Solid Waste Management Plan (1995/1996 Update), 413
New York Stock Exchange and Board, 23
New York University, 133
night soil, 51
Nixon, Richard M., 263, 277, 625n96
Nobel, Philip, 498
No Room to Move: New York City's Impending Solid Waste Crisis (Thompson), 518–19
North Beach Airport, 95, 143, 154, 158, 592n6
North Brother Island, 32–33, 44, 565–66n24
North Shore marine transfer station, 222, 616n118
Norton, Alfred V., 160
NRG New Fuel Company, 296
nuclear power and radioactive waste, 263, 265, 400, 436
Nutten Island, 78
Nye, Bill, 88
NYSUC (New-York Sanitary Utilization Company), 60–61, 93, 114–15, 118

O'Brien, John P., 139–40
O'Brien, Martin, 552
O'Brien Brothers, Inc., 131
Occupational Safety and Health Administration (OSHA), 476–77
ocean dumping: in late 1800s, 45, 45–47, 59, 61, 586n24; in early 1900s, 65, 91, 96, 118–19, 123, 127; ending of, 138, 362
O'Connell, Kim, 451, 469–70
Oddo, James S., 470, 484, 529, 539, 543–44
O'Donnell, Laurence, 241
O'Donovan, Jerome X., 347–48, 393, 414–15
O'Dwyer, Paul, 292
O'Dwyer, William (mayor), 168, 177–78, 183, 187–90, 194–95, 199, 202, 206, 209–10
Offenhuber, Dietmar, 548
Office of Emergency Management, 732n59
Office of Master Planning (OMP), 215–17
Office of Resource Recovery and Waste Disposal Planning, 309
O'Leary, Al, 336
Olmsted, Frederick Law, 53, 239, 498
One New York initiative, 533
open dumps, 133, 148, 161, 231, 291, 589n79, 640n98

Orchard Beach, 166, 565–66n24
Organization of Waterfront Neighborhoods (OWN), 437, 453, 459
organized crime: Giuliani and, 397, 399, 430, 443; and Impellitteri, 210; and waste disposal, 201, 258, 329, 399, 430–31, 443, 659n5, 664n70; and William O'Dwyer, 177
Oudolf, Piet, 510
Outerbridge, Thomas, 428, 442
Outerbridge Crossing, 75, 170, 171
Overview of Refuse Disposal and Resource Recovery in New York City: Issues and New Directions (City of New York, Department of Sanitation), 300–301
Owl Hollow Fields, 514, 543
Oyster Island, 43–44. *See also* Ellis Island; Liberty Island
oysters, 16, 36, 38, 46, 73, 172–73, 175

Pacific Islanders, 403. *See also* race
Palma, Joseph A., 159–60, 183, 189
Park Row Business Building, 30
Parren, Thomas, Jr., 162
Parren Commission, 162
Pataki, George E. (governor): background and election, 409; and Fresh Kills Landfill closure, 415–21, 424–25, 433, 452, 459, 465, 485, 487; and recycling, 409; and Staten Island secession, 393
Patrolmen's Benevolent Association (PBA), 248, 280–81
Pavis, Matthew S., 354–55, 398
Pearl Street, 29
Pelham Bay landfill, 209, 256, 265, 290–92, 304, 310, 324, 636n39
Pelham Bay Park, 259, 565–66n24
Penitentiary Hospital (later Island Hospital), 33
Pennsylvania Avenue landfill, 301, 310, 319, 324
pentachlorophenol (PCP), 317
Percent for Art law, 503
Periconi, Joseph F., 254
Perth Amboy, 190
Pfeiffer, Joseph, 463
Philadelphia, 6, 22–23, 331
Phillips, Patricia C., 507
Phillips & Jordan, Inc., 474

Pinney, George M., Jr., 577n16
planned obsolescence, 6, 252, 552. *See also* consumerism
PlaNYC, 528
Polan, Steven M., 352–53, 356–60
police: and corruption, 53, 105, 177, 210; and street cleaning, 40, 46
Pollak, Linda, 499, 511
polyvinyl chloride (PVC), 308
Poole, Ernest, 21
Poole, Gordon C., 142
Port Authority of New York and New Jersey (PANYNJ), 170–72, 234, 241
Port of New York, 21–22, 36
Port of New York and New Jersey, 172, 204
Port of New York Authority, 234, 241
Postcards 9/11 memorial, 482
post-traumatic stress disorder (PTSD), 477
Powell, William J., 177–78, 180, 192, 202, 208
Power Authority of the State of New York (PASNY), 298–99, 375, 640n93, 648n13
Powers, Peter J., 409, 415–16, 428
Prall's Island, 20, 100–103, 245, 267–68, 270, 445
Pratt Industries, Inc., 409
private carters: and commercial waste, 528, 637n50; and corruption, 142, 223–25, 247, 256, 258, 399; and exporting waste, 361, 400, 411; in nineteenth century, 39–40; and recycling, 347, 409, 528; role and regulation of, 637n50; and tipping fees, 347, 453
private dumps, 139, 201, 224–25, 586n17
privatization: of Brooklyn Navy Yard incinerator, 400; of city agencies, 397, 408, 410; exporting waste and, 423, 426, 430, 445, 458, 550, 553; recycling and, 356, 522
Procaccino, Mario A., 266
Procter & Gamble Company, 74, 172
Program Planners, Inc, 458
Progress and Poverty (George), 29
PTA (Parent Teacher Association), 183
Public Utility Regulatory Policies Act (PURPA, 1979), 297
Public Works Administration (PWA), 141, 146
pyrolysis, 267–69, 297, 630n78

quarantine station, 2, 11, 66, 78–86
Quarantine War (1850s), 78–86, 83
Queens, geography and history of, 19, 27
Queens-shore Board of Health, 44

race: Dinkins and, 351, 396; and environmentalism, 402–3, 453–55; and Fresh Kills Landfill, 442; Giuliani and, 396, 713n115; and immigration law, 130; Koch and, 285–86; Memphis sanitation strike, 626n101; and New York City, 27, 130; Robert Moses and, 152; social unrest of 1960s, 246; and Staten Island, 237, 380–83, 529, 540; and waste transfer stations, 437, 442, 453
Radigan, Edward P., 193–94, 374–75
radioactive waste and nuclear power, 263, 265, 400, 436
Rahway, 190
Rainier Playfield, 493
Randalls Island, 18, 32–33
Randol, Robert, 322
Rathje, William L., x, 491–92, 499, 548
Reaching for Zero: The Citizen Plan for Zero Waste in New York City (Dimino and Warren), 516, 746
Reagan, Ronald, 625n96
Real Estate Exchange and Auction Room, 26
Recap Island, 624n75
recycling: and Bloomberg administration, 486, 489, 516–17, 522–25, 530, 737n123; and de Blasio administration, 533–34; and Dinkins administration, 355–61; and early waste disposal, 41, 127–28; and Giuliani administration, 400, 403, 406, 408–9, 423, 427–30, 433, 438–39, 441, 446, 456; growing interest in after 1960s, 298, 321–22, 326, 332–35, 552–54; and incineration, 359–61, 403; and Koch administration, 321, 661n30; material-recovery facilities, 432, 553, 737n123; Returnable Containers Act (bottle law), 316, 661n30; and September 11 cleanup, 473–74, 476, 481; and waste-to-energy technology, 528; and Zero Waste, 554
Red Cross, 474
Red Nose Studio, 330
reduction plants, 10, 60–61, 64–65, 91–97, 93, 107–12, 114, 118–19, 579n42

Regional Plan Association (New York), 262
Regional Plan of New York and Its Environs (1929), 151
Reigada, Flora, 242
"Re-Raw Recovery" (Ukeles), 505
Reserve Synthetic Fuels, Inc., 296–97
Resource Conservation and Recovery Act (RCRA, 1976), 291, 333, 401, 637n49, 701n1
resource recovery, 268, 290–92, 295–301, 303–10, 314–18, 320–23, 326–27
Resource Recovery Act (1970), 297–98
Resources for the Future, 263
Retail Dealers Protective Association, 25
Returnable Container Act (RCA) (bottle law), 316, 661n30
Review Avenue Enterprises, 329
Reynolds, William H. (Billy), 97
Riccio, Lucius J., 314
Rice, Fred, 297
Rice, John L., 161–62
Richmond Airways, Inc., 172
Richmond Light & Railroad Company, 115
Richmondtown restoration project, 206
Rickard, Tex, 143
Ridge, Thomas, 426–27, 730n49
Rikers Island: early history of, 32–33; landfill, 48, 59–60, 64, 92, *110*, 150, 154–55, 579n34, 579n41; prison, 150, 579n34; reduction plant, 111
Ring, Hilary David, 383–84
Rio de Janeiro, 35
Riordan, Richard, 690n12
RIOS Associates, 498, 501
Rivera, Joyce, 515
Rivers and Harbors Act (Refuse Act, 1899), 36
Riverside Park, 43
Robb, Charles S., 452
Robin, Edward, 245
Robitaille, Rene, 161
Rockefeller, Nelson A. (governor), 235, 249–50, 276
Roe, Robert A., 299
Rogers, Elizabeth Barlow, 541–42
Rogers, Heather, 7, 10
Rogus, Casimir, 212–13, 222
Romer Shoal, 100

Roosevelt, Theodore (governor and president), 53, 64, 117
Roosevelt Island, 18, 33
Rose, Debi, 529
Rose, Joseph B., 502
Rossi, John E., 242
Rourke, L. D., 100
Rouse Company, 240
Rubbish Theory (Thompson), 8–9, 559n18
Rubin, Charles R., 293–94
Rubino, John, 641n105
Ruckelshaus, William, 264
Ruffle Bar, 565–66n24
Ruppell, Edward A., 183, 193
Ryan, Ellen, 495

sacrifice zone, 2, 4, 11, 33, 66, 435–36, 550
Sadat Associates, Inc., 432
Sadat Report, 432
Saint Paul's Church, 30
Saint Petersburg, 35
Salisbury, Harrison, 222
Salmon, Henry, 514
salt hay, 2, 73, 101, 175
salt marshes: and Corona Ash Dump, 92; as environmental resources, 173–75, 259, 273–74, 343, 543; and Fresh Kills Landfill, 1–4; origin and exploitation of, 35, 174–76; on Staten Island, 70, *73*, 293; as waste land, 168, 178, 180, 208, 259, 497, 552
Sanchis, Frank E., III, 498
San Francisco, 6, 35
Sanita Hills resort, 176
Sanitary City: Urban Infrastructure in America from Colonial Times to the Present (Melosi), x
Sanitary Commission, 136, 139
sanitary engineers, 588n65
sanitary landfills. *See* landfills
Sanitary Security Company, 55
sanitation workers: and Beame administration, 278, 281, 284; and Fresh Kills Landfill, 346, 424–25, 442, 623n59; and Giuliani administration, 408; and Lindsay administration, 246–50, 256, 264, 266, 281; in nineteenth century, 39–42; photos, *45*, *58*, *93–94*, *110*, *272*, *506–7*; and race and gender, 289, 673n13;

sanitation workers (*continued*)
  scow trimmers, 41, 58, 127; and
    September 11 cleanup, 475, 481; strikes,
    246–50, 284, 295, 626n101; and Ukeles's
    art, x, 289, 481, 505–6, *506–7*
Sasaki Associates, 498, 500
Savino, Barbara, 337
Say No to Ash coalition, 354–55
Scalero, Frank, 242
Scandura, Jani, 553
Scanlan, John, 8
Scher, Irving, 320
Schick, Fred, 183, 187–88, 193
Schillinger, Salerni and Boyd, Inc., 384
Schmidt, Jeff, 449
Schmittberger, Max F., 101
Schmittberger Detective Agency, 101
Schoofs, Mark, 479
Schroeder, William, Jr., 136, 138–39
Schulman, Claire, 434
Schumer, Charles E., 487
Schweizer, David, 596n65
Scofield, George, 104
scow trimmers, 41, 58, 127.
  See also sanitation workers
Screvane, Paul R., 224–25, 617n135
Scura, Rosemarie, 492
Seaberg, Maureen, 441
Sears, 6, 51
secession, 99–100, 116, 193–94, 373–74, 682n31
Seesselberg, Henry A., 114
Segwick, William T., 105
September 11: events of, 463–64, 471–72; and Freshkills Park, 513, 536; recovery and cleanup, 464, 472–90, 480
Serrano, Jose E., 453
Sexton, Brendan J., 325, 335, 340, 346, 495–96
Shanoff, Barry, 525
Shapiro, Karen, 307
Shapiro, Larry, 410, 415–16
Sharpton, Al, 483
Shaw, Albert, 63
Shears, Joseph A., 111
Shenzhen, 35
Shooters Island, 20
Sierra Club, 263, 449

Signal Companies, Inc, 306, 314, 657n126
Signal Environmental Systems, 326
Silent Spring (Carson), 262
Silver, Sheldon, 392–93, 416, 424, 484
Silzer, George S., 127–28
Simonson, Charles E., 103
Simonson, May (Mrs. Charles E.), 92, 101, 113
Sims Hugo Neu Corporation, 522
Singapore, 35
Singer, Heidi, 465
Sisto, Frank, 638n72
skyscrapers, 30–31
"Small Is Beautiful," 6
smallpox, 33, 44, 79, 81, 84, 86
Smith, Alfred E. (governor), 150, 601n6
Smith, Benjamin, 484
Smith, Beverly, 150
Smith, Leonard C. L., 136
Smith, Wessel S., 80
Snapp, L. Fletcher, 106
"Sneak Peak at Freshkills Park" (2010), 541–42, *544*
Snell, Foster D., 267
social class: and environmentalism, 402–3; Moses and, 152; and New York City history, 24–25, 27–29; and Staten Island, 237, 382–83; and waste disposal, 50–52, 57, 59
Social Mirror, The (Ukeles), 506
Society for Ecological Restoration International Science & Policy Working Group (SER), 493–94
Solid Waste Disposal Act (1965), 253
Solid Waste Management Act (New York State, 1988), 334, 655n105
Solid Waste Masterplan (1978), 298
Soper, George A., 40, 52, 61, 139
Sound View Park, 43, 154, 157–58, 161, 163–66, 183, 192
South Amboy, 190
South Avenue landfill, 301
South Bronx marine transfer station, 226, 410
South Brother Island, 44
South Shore incinerator, 222, 616n118
Southwest Brooklyn marine transfer station, 341, 533
Southwest incinerator (Brooklyn), 331

INDEX

Special Natural Area District (SNAD), 290
Spellman, Francis, 235
Springborn, W. J., 110
Spring Cleanup Campaign (1953), 221
Staniford, Charles W., 95–96
Staten Island: alienation from other boroughs, 2–3, 119; geography of, 19–20, 69–70, 70–73, 75, 77, 91, 170; history of, 19–20, 27, 70–78, 87–92, 172–73; isolation of, 75–78, 90, 169–72, 220, 232–33, 273–74; and New Jersey, 72–75, 169–72, 200, 375; social and demographic changes in, 540; as target for waste disposal, 66, 78, 95, 170, 547–48
Staten Island Airport, 244
Staten Island Anti-Garbage Dump Association, 186, 193
Staten Island Chamber of Commerce, 164, 235, 447, 452, 514
Staten Island Charter Commission, 388, 390
Staten Island Citizens Committee, 160
Staten Island Citizens for Clean Air (SICCA), 358, 366, 398–99
Staten Island Coalition for Survival, 365
Staten Island Environmental Coalition, 365–66
Staten Islanders for a Unified New York (SIUNY), 380, 388, 391
Staten Island Greenbelt–Natural Areas League (SIGNAL), 274
Staten Island Growers' Association, 172
Staten Island Historical Society, 206
Staten Island Industrial Park, 238
Staten Island Mall, 238, 241
Staten Island Railway, 90
Staten Island Rapid Transit, 577–78n21
Staten Island Real Estate Board, 164
Staten Island Rebellion, 80.
    See also Quarantine War (1850s)
Staten Island Savings Bank, 402
Staten Island Society of Architects, 160
Staten Island transfer station, 522, 523
Staten Island Vigilantes, 160
Statue of Liberty, 33, 37
Stegner, Wallace, xi
Stein, Andrew, 401
Steinberg, Ted, 16–17, 560n2, 604n50

Steisel, Norman: background and career, 288, 323, 657n126, 675n33; and DSNY, 288–89; and Fresh Kills Landfill, 305, 318; and resource recovery, 299–301, 314–16, 321–23; and tugboat workers strike (1978), 295; and Ukeles, 505–6
Stern, Herbert J., 313
Stevens, John Paul, 379
Stevenson, Adlai E., 235
Stewart, Douglas H., 49–50
St. Louis, 22
Storm, Anna, 10
Straniere, Robert A.: and Fresh Kills Landfill, 311, 323, 339, 414, 416, 460; and resource recovery, 316; and secession, 377, 379, 393
Strauss, Nathan, Jr., 594n39
Street-Cleaning Society, 53
Streit, Saul S., 249
Strong, William L. (mayor), 53, 55, 59, 61
strontium-90, 308
Sugarman, Jeff, 499
Sullivan, Daniel, 458
Sullivan, Jeremiah A., 183
Summer Olympics (2012), 514
Sweet, Thaddeus C., 114
Swinburne Island, 20, 86, 100
Symes, Lancaster, 100
Szarpanski, Harry, 516–17

Taft, William Howard, 117
Tale of Two Incinerators: How New York City Opposes Incineration in New Jersey While Supporting It at Home (Holtzman), 361
Talese, Gay, 234
Tammany Hall, 22, 49, 53, 55, 61, 63, 253–54, 567n43
Tarr, Joel A., x, 585n1
Task Force on Staten Island Secession, 388–89
Taxes In, Garbage Out: The Need for Better Solid Waste Disposal Policies in New York City (Citizens Budget Commission), 527, 749n27
Taxter, E. Steward, 117
Taylor, Alfred A., 126–28, 132, 134–36
Taylor, Henry W., 221

Taylor Law (Public Employees Fair Employment Act, 1967), 249
Teamsters Local 813 (private sanitation workers), 430–31
Teamsters Local 831 (DSNY workers), 356–57. *See also* Uniformed Sanitationmen's Association (USA)
"Teenage Wasteland" (Dawson), 554, 757n23
tenements, 26, 28, 51–52
terraforming, 37, 168. *See also* landfills
Thompson, John, 82
Thompson, Michael, 8–9
Thompson, William C., Jr., 516, 518
Thoreau, Henry David, 76
throwaway society, 6, 209, 262, 335, 489, 552. *See also* consumerism
Tian, Song, xi
Tierney, Brian, 479
Tierney, John, 429
tipping fees, 333–34, 347, 384, 453, 485, 524–25, 647n1
Tomkins, Calvin, 95
Tomkins, Ray, 82
Tompkins, Daniel D., 574n43
Tompkins Square Community Center, 247
Toronto, 35
Toth, Edward, 754n99
Touch Sanitation Performance (Ukeles), 505–6, 506–7
"Town Dump, The" (Stegner), xi
Towns, Edolphus, 403
toxic material: and Brookfield landfill, 310–11, 319, 401, 657n123; and Fresh Kills Landfill, 274, 310–12, 339, 366, 442; leachate, 642n109; and September 11 cleanup, 476
Trade Waste Commission (TWC), 431
transfer stations, 612n49, 711n94; and Bloomberg administration, 488, 515, 517, 521–24; and de Blasio administration, 533–34, 541; and environmental justice, 435, 437–38, 453–54, 534, 541, 549; and fiscal crisis (1970s), 281–88; and Giuliani administration, 435–58, 466–68, 487–89, 708n68; maps and photos, 340, 363–64, 523; private, 364, 427, 435–36, 439, 453, 523
Transport Workers Union, 248

Trautmann, Les, 385
Travis, Jacob, 172
Triborough Bridge and Tunnel Authority (TBTA), 232–42
Tribune Building, 30–31
Tribus, Louis L., 109
trichloroethane, 401
Trinity Church, 30
Tripp, James T. B., 431, 442–43, 488
Trollope, Frances (Fanny), 15
tugboats, 41, 127, 179, 295
tugboat workers strike (1979), 264, 295, 315
Tully Construction, 472
Turner Construction, 472
Tweed, William M. (Boss), 567n43
12 Things New Yorkers Should Know About Their Garbage (Citizens Budget Commission), 528, 749n31
Twin Island, 565–66n24
Typhoid Mary (Mary Mallon), 44

Udall, Stewart L., 259
Ukeles, Jack, 503
Ukeles, Mierle Laderman: background, x, 503–4; and DSNY, 289, 504–10; and Fresh Kills Landfill, 3, 551; and Freshkills Park, 491, 499, 536–37, 752n72; photos, 506–7; and September 11 commemoration, 481
Underworld (DeLillo), 551, 756n10
Uniformed Firefighters Association (UFA), 280
Uniformed Sanitationmen's Association (USA), 248–50, 280, 356–57, 638n72
Union Club, 29
United Haulers Association, Inc. v. Oneida-Herkimer Solid Waste Management Authority (2007), 525
United Oil Products, Inc. (UOP), 306
Urban Development Corporation, 276
Urban League, 383
U.S. Army Corps of Engineers (USACE), 474–75
U.S. Conference of Mayors, 262
U Thant Island, 18

Vaccarello, Anthony T., 287–88, 505
Vallone, Peter, 380, 438–39, 458
Vanderbilt, Cornelius, 85

Van Etten, Charles R., 107-8, 116
Van Name, Calvin D., 97-98, 102, 171
Van Wyck, Robert A. (mayor), 61
Velasquez, Nydia M., 437
Verrazzano, Giovanni da, 71
Verrazzano-Narrows Bridge, 20, 75, 233, 233-34, 236, 237-42, 273-74, 549
Veterans of Foreign Wars, 183
Vigilance Committee, 99-100, 102-5, 107, 115
Village Greens Residents Association, 319
Vincenz, Jean, 148
vinyl chloride, 401
Vitaliano, Eric N.: background and election, 312-13; on dioxin, 656n111; and Fresh Kills Landfill closure, 325-26, 368, 413-14, 416, 424, 440, 484; opposition to Fresh Kills Landfill, 291, 312-13, 318-19, 324, 348, 354, 365, 405; and resource recovery plants, 316-17, 321, 336, 340, 353; and Staten Island secession, 377-80, 386, 393; on the throwaway society, 335; and wetlands, 313, 343
Viteritti, Joseph P., 382, 388
Vonnegut, Kurt, 18

Wagner, Robert F., Jr. (mayor), 222-25, 235, 237, 253-56, 277
Wagner, Robert F., Sr., 223
Walker, James J. (Jimmy) (mayor), 136-37, 139
Wallstein, Leonard, 132
Wall Street, 23, 29-30, 35
Ward, Mitchell, 217
Wards Island, 18, 32-33, 86
Waring, Chapman, and Farquar, 55
Waring, George E., Jr., 38, 53-61, 54, 63, 78
Warren, Barbara, 431, 459, 516
Washing (Ukeles), 504
Washington Square, 43
Wasservogel, Isidor, 162
waste: in late 1800s and early 1900s, 50-52, 65-66; in 1950s and 1960s, 227-29, 230, 332; in 1970s and 1980s, 332-34, 647n2; and consumerism, 7-9; as material information, 8-9, 548; as pollution, 261-69
Waste Alternatives, Inc., 329

Waste Disposal Problem in New York City: A Proposal for Action (City of New York, Department of Sanitation), 315
Waste Disposal Task Force (New York City and New York State), 423-35
Waste Equity Bill (2018), 534
Waste Management, Inc, 431, 434, 446
waste-to-energy, 259, 296-98, 406, 527-30, 661n27
Waterfront Watch, 240
water pollution, 239, 264, 274, 308, 425, 478, 540
Water Street, 43
Wehran Engineering, 309
Wehran EnviroTech, 354
Weinstein, Israel, 190, 606n90
Weintraub, Lee, 495
Wells, Raymond, 111
Weprin, Saul, 392-93
Western Union Telegraph Building, 30
West Shore Expressway, 181, 215-20, 238-39, 513
wetlands, 70, 73, 96; preservation of, 173-76, 259, 312-13, 342-44, 377, 405, 459, 495, 517, 543; on Staten Island, 1-4; as waste land, 25, 35, 92, 147, 149, 168, 176, 178, 180, 208, 259, 292, 293, 497, 552
Whalen, Grover A., 154
Wheat, Carolyn, 547
Whipple, George C., 105-6, 108, 111-13
White, Alfred W., 44
Whitman, Charles S. (governor), 104-5, 113, 478
Whitman, Christine Todd, 444, 452
"Why We Oppose Secession Now" (SIUNY), 388
Wilkes, Paul, 252
Willcox, William G., 98-99, 103, 113
Williams, Henry G., 324
Williams, Henry Smith, 49-50
Williams, Linsly R., 105, 107
Williams, Robert, 393
Williamsburg Bridge Lighting Plant, 588n50
Willis, Carol, 30
Willowbrook State School, 597n94
Wilson, Woodrow, 117
Winter, Jonah, 330

Without Fresh Kills: A Blueprint for Solid Waste Management (New York City Council), 438–39, 712n110
Women's Anti-Garbage League, 92, 117
Woodbridge (New Jersey), 190, 270, 305, 313, 341–42, 344, 458
Woodbury, John McGaw, 64–65, 92
Woolf Electric Disinfecting Company, 48
Works Progress Administration (WPA), 141, 158, 596n63
World's Fair (1939), 150, 153–55, 158, 167, 579n34, 595n59
World Trade Center Families for Proper Burial, 481
World War II, 130, 146
Wright's Island, 565–66n24

yellow fever, 11, 63, 69, 78–86
Young, William, 405–6, 468–69

Zadroga, James, 477
Zero Waste, 3, 516, 531–32, 532, 552, 554
Zimroth, Peter L., 387
Zurmuhlen, Frederick H., 212